SPIN EXCITATIONS IN NUCLEI

SPIN EXCITATIONS IN NUCLEI

Edited by

F. Petrovich
Florida State University
Tallahassee, Florida

G. E. Brown
State University of New York at Stony Brook
Stony Brook, New York

G. T. Garvey
Argonne National Laboratory
Argonne, Illinois

C. D. Goodman
Indiana University
Bloomington, Indiana

R. A. Lindgren
University of Massachusetts
Amherst, Massachusetts

W. G. Love
University of Georgia
Athens, Georgia

PLENUM PRESS • NEW YORK AND LONDON

Library of Congress Cataloging in Publication Data

International Conference on Spin Excitations in Nuclei (1982: Telluride, Colo.)
 Spin excitations in nuclei.

 "Proceedings of the International Conference on Spin Excitations in Nuclei, held
March 25–27, 1982, in Telluride, Colorado"—T.p. verso.
 Includes bibliographical references and index.
 1.Spin excitations—Congresses. 2. Nuclear reactions—Congresses. I. Petrovich, F.
(Fred) II. Title.
QC794.6.E9I58 1982   539.7   83-24677

ISBN-13: 978-1-4684-4708-8   e-ISBN-13: 978-1-4684-4706-4
DOI: 10.1007/978-1-4684-4706-4

This material was prepared with the support of the U.S. Department of Energy under
Grant Number DE-AS05-82ER40056. Any opinions, findings, conclusions, or recom-
mendations are those of the authors and do not necessarily reflect the views of DOE.

Proceedings of the International Conference on Spin Excitations in Nuclei, held
March 25–27, 1982, in Telluride, Colorado

A Division of Plenum Publishing Corporation
233 Spring Street, New York, N.Y. 10013

PREFACE

This volume contains the proceedings of the "International Conference on Spin Excitations in Nuclei" held in Telluride, Colorado, March 25-27, 1982.

The motivation for the conference was, in a large part due to the recent development of new variable energy accelerators which produce high quality beams of electrons, protons, and pions that are providing the first precise information on spin excitations in nuclei over a large range of spin and mass. In the past such data had been restricted primarily to light nuclei and were generally resolution limited. Perhaps, the most exciting new result has been the clear observation of the elusive spin-dipole strength (Gamow-Teller and M1) in medium and heavy mass nuclei through the use of the (p,n) and (p,p') reactions at or near zero degrees with 100-200 MeV incident protons. Energy dependence in the isovector parts of the nucleon-nucleon interaction make the 100-200 MeV energy region particularly appropriate for such studies. The clean data from (e,e'), (π,π'), (p,p'), and (p,n) on high spin "stretched" states which have particularly simple structure has also been quite important.

The recent results contain important new information on the nature of the spin dependent forces in nuclei. These in turn are inherently related to the properties of the nuclear mesonic field and the underlying quantum chromodynamics. As a result of these features, the problem of spin excitations in nuclei has provided a common link between an unusually wide spectrum of nuclear physicists, e.g. experimentalists, reaction theorists, structure theorists, and field theorists. The conference was born with the purpose of bringing together as many of these investigators as possible to review and discuss the current status of the work in this exciting new area of research. Telluride seemed the natural place to hold this conference, since it had served as the site of the 1979 "Conference on the (p,n) Reaction and the Nucleon-Nucleon Force" where some of the earliest findings were presented.

The following people served as the conference organizing committee:

S. D. Bloom, Lawrence Livermore Laboratory
G. E. Brown, SUNY at Stony Brook
M. Ericson, Institut de Physique Nucléaire and CERN
C. Gaarde, Niels Bohr Institute
G. T. Garvey, Argonne National Laboratory
C. D. Goodman, Indiana University Cyclotron Facility
M. L. Johnson, Los Alamos National Laboratory
W. G. Love, University of Georgia
R. A. Lindgren, University of Massachusetts
H. McManus, Michigan State University
E. J. Moniz, Massachusetts Institute of Technology
F. Petrovich, Florida State University
  (Conference Chairman)
J. Speth, KFA Jülich
E. Vogt, TRIUMF

The conference enjoyed the sponsorship of the following
organizations:

Florida State University
Indiana University Cyclotron Facility
International Union of Pure and Applied Physics

It also was supported by grants from:

The National Science Foundation
United States Department of Energy

Of course, the major credit for the success of the conference
must go to the speakers who diligently prepared the stimulating
talks that are reproduced in this volume and the conference participants whose spirited interaction made the conference a truly
memorable event.

Credit for the nearly flawless conference arrangements is due
to many people. Joan Rissinger of the Florida State University
Center for Professional Development and Nancy Conley, Lois Crew,
Susan Harper, Art Carpenter, and Barbara Beckford of the Department
of Physics at Florida State University performed the bulk of the
work. Joan, as conference coordinator, was responsible for conference registration, travel, lodging, site reservations, and preparation and distribution of the conference brochure. Nancy administer-
ed the conference grants. Lois handled the conference typing and
conference communication which included the early mailings, proposal
preparation, collection of manuscripts, and, with the help of Susan,
the correction of manuscripts and the final preparation of this
volume. Art and Barbara, a graduate and undergraduate assistant,
assisted by preparing a computerized mailing list.

Alice Horpedahl of Los Alamos National Laboratory served as conference hostess and along with Joan Rissinger and Lois Crew operated the on-site conference office. Special mention goes to Lois who was the one person involved in all phases of the conference from beginning to end.

Thanks go out to several people in Telluride who assisted the conference in a variety of ways. These are (in no particular order and with apoligies to anyone who has inadvertently been forgotten) Bill and Stella Pence, Jim Bedford, Malcom Goldy, Sue Gray, Lou and Wayne Gustafson, Leslie Sherlock, Lucia Midbey, Tom Sharp, and Andrea Benda. Having such pleasant and cooperative people to deal with contributed significantly to the pleasure of holding the conference in Telluride.

Thanks are also due to the ORTEC and Lecroy corporations who contributed towards the expense of the fine pastries served at the coffee breaks.

The scheduling of the technical sessions was modeled to some degree after the Gordon Conferences with morning and evening sessions and free afternoons. Sessions were held in the Sheridan Opera House, with the participants housed in the adjoining Sheridan Hotel or nearby inns. The compact layout and beautiful surroundings of the historic town of Telluride did much to create the proper atmosphere for effective interaction among the conference participants.

Final thanks goes to that very important force of nature, the weather, which saw fit to provide us all with the most spectacular spring skiing conditions that had been seen in a number of years.

Fred Petrovich

CONTENTS

(The name in capitals indicates the author who presented the talk)

ix

OPENING REMARKS: BASIC QUARK CONSIDERATIONS

Edward Teller

Lawrence Livermore Laboratory
University of California
Livermore, CA   94550

INTRODUCTION[†]

All of us, in the days when we were studying nuclear physics,
heard about Gamow-Teller transitions. Recently with the experiments
using (p,n) reaction and other reactions, there has been a revived
interest in Gamow-Teller strength and, in fact, there is a big
question of where is the missing Gamow-Teller strength or why is it
missing? Our first speaker may not be able to tell us where it has
gone, but he can certainly tell us where the idea came from. It is
my pleasure to introduce Dr. Edward Teller.

QUARK PICTURE OF NUCLEONS AND DELTAS

I am afraid that I will disappoint you, but I hope to do it
fairly briefly. To me it seems that the connection between a little
paper on beta decay that has been written many years ago and the
important things that you are talking about now is a little tenuous.
If one concentrates on the magnitude of $\sigma\tau$ then, of course, the connection is obvious. At any rate it seems that you have introduced
the names of two mythological figures to designate your subject, and
I am in a way happy about it because one of them was by far the
nicest Russian I ever knew and in addition, as you all know, a very
good physicist. Now, I thought I would make a few introductory
remarks and tell you a few things that have been done. Most of these

---

[†]Delivered by D. Kurath

you know better than I, but perhaps I will talk about them in a
slightly different manner.

To begin with, if a proton transfers its charge in a near
miss to a somewhat (but not very) distant nucleus, the best way to
do it is by a single π-meson. Since the π-meson has negative parity,
the only way this can be done is to emit the π-meson in a p-state
and carry away an angular momentum of unit value which also has
negative parity. Then you have a spin-flip to conserve angular
momentum. On arrival in the nucleus the process is reversed and
in this way you have not only carried a unit of charge, but also a
unit of angular momentum which can be expressed by the operator στ.
This is the model that we are talking about. Taking scattering in
the forward direction and also using high energies (because at high
energies a given excitation can be accomplished by a relatively
small momentum change) the momentum change is relatively small as
is the case in the beta decays. Therefore, the use of the στ operator becomes a good approximation in describing the transfer of
charge from the incident particle.

Studying nuclear forces you should be able to predict the
total strength, which of course should be proportional to the excess
of neutrons which can make this change without violating the Pauli
principle; however, the total measured strength does not quite come
up to expectations. The question is: where has the rest of the
strength gone?

Hints of several people and a paper by Bohr and Mottleson have
suggested that the additional strength goes into the transformation
of a nucleon into a delta particle. I think that this is quite
plausible. While in a very few minutes it is not possible and hardly
worthwhile to calculate precisely how big that effect should be, I
can give you a very simple model, a quark model, which is the basis
of this whole consideration. This will at least qualitatively show
you how this process happens, how to do it, and also how not to do
it. (When I tried to work it out for myself I, of course, at first
did it the way how not to do it.)

I start with the simplest thing, which is the spin three-halves
state of the negative charge state of the delta particle. This state
consists of three down quarks. That a quark is "down" is shown by
writing a single stem ↑ indicating the charge -1/3. An "up" quark
will be shown by an arrow with two stems ⇑ showing the charge 2/3.
(In both cases, spins pointing upward are shown.) The delta state
of interest has a charge -1 and spin three halves. It is therefore

                              ↑↑↑.

All this is already symmetric, as it should be.

How to construct from this other states is a standard procedure. For instance, you can construct a state (and I will leave out normalization factors throughout) that is the same except that the component of the spin in the z direction is now no longer three halves, but one half and it is symmetric as it should be:

$$(\uparrow\uparrow\downarrow + \uparrow\downarrow\uparrow + \downarrow\uparrow\uparrow)$$

You might try to make out of that a particle with spin 1/2 (not only spin component one-half, but total spin one-half). The standard way of doing that is that you construct an orthogonal state which will be:

$$\uparrow\uparrow\downarrow - \tfrac{1}{2} (\uparrow\downarrow\uparrow + \downarrow\uparrow\uparrow)$$

Now all this is fine except that we know that we have to symmetrize it. If you symmetrize it, you get zero. So, a one-half spin state with a minus three-half value for τ_z does not exist.

Indeed, you can count very easily the number of symmetric states. There are 20 of them, which is 16 plus 4, as it should be corresponding to the 16 states of delta and the 4 to the proton-neutron isospin doublet.

Returning to the original state, I can go one step further and flip one of the τ values transforming one of the down quarks into an up quark. I will choose to do it here in order not to write too much, by making always the first quark an up quark, where up means just two strokes: $\Uparrow$ $\uparrow\uparrow$. Obviously, this is not yet symmetric. But even more obviously, I can symmetrize it by adding another two terms. Now I have z components of $\sigma_z = 3/2$ and $\tau_z = -1/2$. Proceeding further I can construct a delta state of $\sigma_z = 1/2$ and $\tau_z = -1/2$. I can make an orthogonal state to that which will actually be a neutron with an up-spin:

$$(\tfrac{1}{2} \Uparrow\uparrow\downarrow + \tfrac{1}{2} \Uparrow\downarrow\uparrow - \Downarrow\uparrow\uparrow) +\ldots +\ldots$$

It has a spin 1/2, and a $2/3 - 1/3 - 1/3 = 0$ charge.

The one way I want to use this is to show how to make out of it a delta state. To do this, I use a τ_+ operator, like $\tau_x + i\tau_y$, which gives a flip of a down isospin to an up isospin and zero if the isospin already was up. At the same time, I multiply by $\sigma_x + i\sigma_y$ because I need to transfer an angular momentum and σ_+ represents the angular momentum to transfer. But what I do is not applied to the nucleon as a whole state, but rather one by one to the quarks. This I do to each quark and symmetrize only after τ_+ and σ_+ have been applied.

You notice what is going to happen. In the last term of the

expression which I have written, by operating on the first factor
I get zero because τ_z points upward and by operating on the second
and third factor I get zero because σ_z points upward. If I, however,
operate on the first terms, then there is a down spin both in σ and
τ that I can flip up and I get:

$$\Uparrow \uparrow \Uparrow \; + \; \Uparrow \Uparrow \uparrow$$

Now if I symmetrize this, I clearly get a pure delta state and a
delta state with a charge one. Had I not used this special combination $(\tau_x + i\tau_y)(\sigma_x + i\sigma_y)$, then I would have gotten much more
complicated expressions. I have to average over the whole thing,
and only this way do I get what Bohr and Mottleson and others calculated and considered. At any rate, that there should be strength
in the delta state is completely obvious.

I would like to mention another point which is closely connected with all of this. And that is that if a single $\tau\sigma$ emission
or absorption can cause a transformation of a nucleon into a delta
state, then with a reasonably small probability amplitude, the
delta states should be present in common nuclei and should contribute to the nuclear forces.

Now I want to do something slightly funny which does not seem
to belong to this discussion but which I think might be worth considering. If one arbitrarily assigns to the masses of the quarks
1/3 the mass of the nucleons, one obtains for the magnetic moments
of the up and down quarks, 2 and -1 nuclear Bohr magnetons, respectively. Using wave functions such as have been displayed, one obtains for the proton and neutron nuclear magnetic moments of 3 and
-2, instead of the observed values 2.79 and -1.91. That is, one
approximates the real values which in a way is a stronger statement
than the usual one that the ratio of the two magnetic moments is 3
to -2.

The further question that arises is whether the delta states
may make a contribution to the magnetic moments of the proton and
the neutron. The pure delta states which we have been dealing with
should have the nuclear Bohr magneton values 6, 3, 0 and -3 corresponding to the states with charge 2, 1, 0 and -1. It would be of
further interest to extend these considerations to the t and ^{3}He
with nuclear Bohr magneton values of 2.97 and -2.12. Of course,
very many more magnetic moments are known, but in those cases there
can be no doubt that configuration mixing is important, which is less
important in the case of the t and ^{3}He.

The point is then that admixtures of states caused by the application of the $\sigma\tau$ operator will change the magnetic moments by
resulting in the presence of pions (with a small probability) and
corresponding changed states of neutron, proton, and Δ. Thus, the

admixture of delta states might be a factor of importance not only
in connection with the topic of this conference, but also in wider
applications. In any case the explanation of μ of -2.12 for ^{3}He
will remain difficult. I am afraid I have now badly overtalked,
so apologies for having said too little and talked too long.

ACKNOWLEDGEMENT

I wish to thank the Conference Organizing Committee for
making it possible for me to attend this meeting and Ms. Lois Crew
for her help in preparing this manuscript.

SHELL MODEL PICTURE OF NUCLEAR MAGNETIC MOMENTS

Akito Arima

Department of Physics
Faculty of Science
University of Tokyo
Hongo 7-3-1, Bunkyo-ku
Tokyo, Japan

1. INTRODUCTION

The problem of nuclear magnetic moments is old yet unsolved and interesting.[1] In this talk I would like to show the following points.

(1)   Deviations of observed magnetic dipole moments from the Schmidt values do not show any strong mass number dependence.

(2)   Shell model calculations in the $0\hbar\omega$ major shell give good agreement with observed values of magnetic dipole moments and transitions when the renormalization of the magnetic dipole operator is taken into account. The renormalization can be estimated by using the magnetic moments and transitions of nuclei with an LS closed shell $\pm$ one nucleon such as ^{17}O and ^{17}F.

(3)   Evidence for core polarization or so called Arima-Horie effect[2] is shown in the N = 28 isotones.

(4)   Evidence for the core polarization is shown in the vicinity of ^{208}Pb by using the N = 126 isotones.

(5)   The importance of tensor correlations is pointed out to explain the quenching of spin magnetic moments and Gamow-Teller transitions. The Δ-h contribution is shown to be small at least in nuclei lighter than 41.

(6)   The value of the Landau-Migdal parameter g' plays a very crucial role in determining how much the Δ-h mixing quenches the spin magnetic moments and the Gamow-Teller transitions. It is pointed out that the g' for $NN^{-1} \rightarrow \Delta N^{-1}$ is as small as 0.4 when the $\pi + \rho$ exchange model is adopted. With this value 0.4 for the g', one can show that the Δ-h mixing quenches the spin magnetic moments at most by 10 %.

2. A-DEPENDENCE OF MAGNETIC DIPOLE-MOMENTS OF $S_{1/2}$ NUCLEI

There has been a belief[3] that the quenching of the spin magnetic moment depends rather strongly on A. Some Calculations[3] indicated that while the quenching is very small in light nuclei, it increases and saturates after A = 40.

Table 1. Magnetic moments of $s_{1/2}$ nuclei.

	μ_s	μ^{obs}	$\gamma = \mu^{obs}/\mu_s$
^{29}P	2.79	1.23	0.44
^{197}Tℓ	2.79	1.58	0.57
^{199}Tℓ	2.79	1.60	0.57
^{201}Tℓ	2.79	1.61	0.58
^{203}Tℓ	2.79	1.62	0.58
^{205}Tℓ	2.79	1.64	0.59

	μ_s	μ^{obs}	$\gamma = \mu^{obs}/\mu_s$
^{29}Si	−1.91	−0.56	0.29
^{115}Sn	−1.91	−0.92	0.48
^{117}Sn	−1.91	−1.00	0.52
^{119}Sn	−1.91	−1.05	0.55
^{123}Te	−1.91	−0.74	0.39
^{125}Te	−1.91	−0.89	0.47
^{129}Xe	−1.91	−0.78	0.41
^{133}Ba	−1.91	−0.77	0.40

Table 1 shows the ratio γ of observed magnetic dipole moments to their corresponding Schmidt values. Here only odd-A nuclei with $s_{1/2}$ nucleon are chosen in order to avoid complication caused by orbital angular momenta. Contrary to the common belief, the γs for 1s-0d shell nuclei are even smaller than those for heavy nuclei.[3]

3. SHELL MODEL CALCULATIONS WITHIN $0\hbar\omega$ MAJOR SHELL

The first order core polarization (Arima-Horie effect) vanishes in nuclei with an LS closed shell ± one nucleon.[2] In reality, their magnetic moments and Gamow-Teller transitions deviate slightly from their single particle values.[1] For example the Gamow-Teller transition of ^{41}Sc to ^{41}Ca is 24% (in amplitude) smaller than its single particle value;

$$G^{eff}/G = 1 - 0.24.$$

Here G is the coupling constant of the Gamow-Teller transition and "effect" indicates that the operator is renormalized. Similarly the magnetic dipole moments of ^{41}Ca and ^{41}Sc are slightly different from their Schmidt values;

$$\mu(^{41}Ca) = -1.59 \text{ n.m.}$$

and

$$\mu(^{41}Sc) = 5.43 \text{ n.m.} \quad .$$

Their corresponding Schmidt values are

$$\mu_{Sch}(^{41}Ca) = -1.91 \text{ n.m.}$$

and

$$\mu_{Sch}(^{41}Sc) = 5.79 \text{ n.m.}$$

The lowest 7^+ state of ^{42}Sc decays into the lowest 6^+ state of ^{42}Ca. This transition is not quenched at all. This is puzzling in the first glance.[4] However, if one takes into account configuration mixing, the puzzle is solved.

The wave functions of those states in the 0f-1p shell model are

$$|T = 0, 7^+\rangle\rangle = |f_{7/2}^2 \, 0, 7^+\rangle \text{ for } ^{42}Sc \tag{1}$$

and

$$|T = 1, 6^+\rangle\rangle = |f_{7/2}^2 \, 1, 6^+\rangle + \alpha|f_{7/2}f_{5/2} \, 1, 6^+\rangle \text{ for } ^{42}Ca. \tag{2}$$

Here α is assumed to be very small.

Then the Gamow-Teller transition is given as

$$\frac{G^{eff} <<0, 7^+ \| \tau\sigma \| 1, 6^+>>}{G <0, 7^+ \| \tau\sigma \| 1, 6^+>}$$

$$= \frac{G^{eff}}{G} + \alpha \frac{G^{eff}}{G} \frac{<f_{7/2}^2 \; 7^+ \| \tau\sigma \| f_{7/2} \; f_{5/2} \; 6^+>}{<f_{7/2}^2 \; 7^+ \| \tau\sigma \| f_{7/2}^2 \; 6^+>} \approx 1. \qquad (3)$$

From this equation, one can estimate α quite easily and finds the following value

$$\alpha \approx -0.1. \qquad (4)$$

Using the Kuo-Brown interaction, one gets

$$\alpha_{cal} = -\frac{1}{\Delta E} <f_{7/2}^2 \; 6^+ \; |v| \; f_{7/2} \; f_{5/2} \; 6^+> \cong -0.08.$$

The same wave function (2) can be used to predict the magnetic moment of the 6^+ state;

$$\mu(6^+) = <f_{7/2}^2 \; 6^+ |\mu^{eff}| f_{7/2}^2 \; 6^+>$$

$$+ 2\alpha <f_{7/2}^2 \; 6^+ |\mu^{eff}| f_{7/2} \; f_{5/2} \; 6^+>.$$

We now introduced μ^{eff}. The diagonal matrix elements of μ^{eff} are assumed to be given by the observed values;

$$< nf_{7/2} \; m = 7/2 \; |\mu^{eff}| n \; f_{7/2} \; m = 7/2> = -1.59 \qquad (5)$$

and

$$<p \; f_{7/2} \; m = 7/2 \; |\mu^{eff}| p \; f_{7/2} \; m = 7/2> = 5.43. \qquad (6)$$

Since there is no observation of M1 transitions from $f_{5/2}$ to $f_{7/2}$, we assume that

$$<f_{7/2} \| \mu^{eff} \| f_{5/2}> = \frac{G^{eff}}{G} <f_{7/2} \| \mu \| f_{5/2}>. \qquad (7)$$

Then

$$\mu(6^+) = -2.73 + 0.27 = -2.46$$

which is in very good agreement with its observed value[5] $\mu(6^+)_{obs} = -2.49 \pm 0.09$. It is easy to show that the magnetic moment of the lowest $19/2^-$ state in ^{43}Sc can be written in terms of the same α as

$$\mu(19/2^-) = \mu(^{41}Sc) + <f_{7/2}^2 \, 6^+|\mu^{eff}|f_{7/2}^2 \, 6^+>$$

$$+ \, 3\alpha <f_{7/2}^2 \, 6^+|\mu^{eff}|f_{7/2} \, f_{5/1} \, 6^+>$$

Using eq. (4), one obtains

$$\mu(19/2^-) = 3.11$$

which is again in goood agreement with its observed value[6]

$$\mu(19/2^-)_{obs} = 3.15 \pm 0.02.$$

Instead of eq. (7), McGrory and Wildenthal[7] assumed

$$\frac{<f_{7/2}\|\mu^{eff}\|f_{5/2}>}{<f_{7/2}\|\mu\|f_{5/2}>} = \frac{<f_{7/2}\|\mu^{eff}\|f_{7/2}>}{<f_{7/2}\|\mu\|f_{7/2}>} = 0.83 \tag{8}$$

Using the effective magnetic dipole moment, they calculated magnetic dipole transitions of the Ca isotopes. Their results are given in Table 2.

Table 2. $\sum B(M1)_{cal}/\sum B(M1)_{S.P.}$

	Cal	Exp
^{42}Ca	0.43	$\approx$ 0.40
^{44}Ca	0.47	$\approx$ 0.33
^{46}Ca	0.48	
^{48}Ca	0.49	$\approx$ 0.43

The ground state correlation reduces roughly by 25% the B(M1) of ^{48}Ca. The total theoretical reduction is then

$$0.75 \times (0.83)^2 = 0.75 \times 0.59$$

$$= 0.51$$

which corresponds to 0.49 in Table 2.

One sees from this table that some major shell calculations are able to reproduce very well the B(M1) except for ^{44}Ca when μeff is used. A very similar conclusion has already been drawn in the 1s-od shell by Wildenthal and Chung.[8]

4. EVIDENCE FOR THE CORE POLARIZATION

The magnetic moments of the isotones with N = 28 have a
conspicious Z dependence. This can be easily understood from the
core polarization point of view (the Arima-Horie effect). Assuming
the following wave function for these isotones

$$|Z=20+z> = |f_{7/2}^z> + \alpha|f_{7/2}^{z-1} f_{5/2}>,$$

one can easily prove that

$$\delta\mu \propto z - 1.$$

This is the enhancement of the Arima-Horie effect depending on the
number of nucleons in $j = \ell + 1/2$ shell.

Yokoyama and Horie[9] calculated those magnetic moments using
μ instead of μ^{eff} and assuming the following configurations;

$$f_{7/2}^z$$

$$f_{7/2}^{z-1}(f_{5/2}, \text{ or } p_{3/2} \text{ or } p_{1/2})^1 .$$

Their result is shown in Fig. 1.

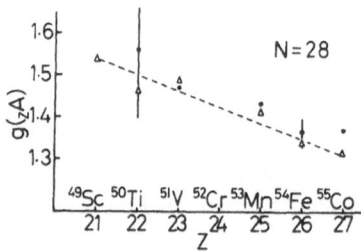

Fig. 1. The gyromagnetic ratios of the N = 28 Isotopes
         Δ stands for calculated values and • for observed values.

One sees that this calculation gives a good explanation to the
observed A-dependence of those magnetic moments. Namely it is clear
that the Arima-Horie effect is responsible for the z-dependence.
When μ^{eff} in eq. (8) is used, the A-dependence becomes slightly
weaker than that shown in Fig. 2. If one tries to explain the

quenching by the Δ-h state, one must expect an almost constant δg for all isotopes.

5. MAGNETIC DIPOLE MOMENTS IN THE VICINITY OF ^{208}Pb

When one calculates the effect of core polarization (See Fig. 2) one assumes that there are 1^+ excited states of a core nucleus which decay strongly into the ground state of the core nucleus. There has been doubt expressed as to the role of the core polarization in the vicinity of ^{208}Pb, because there has been no observation of 1^+ states by the electron scattering.[10]

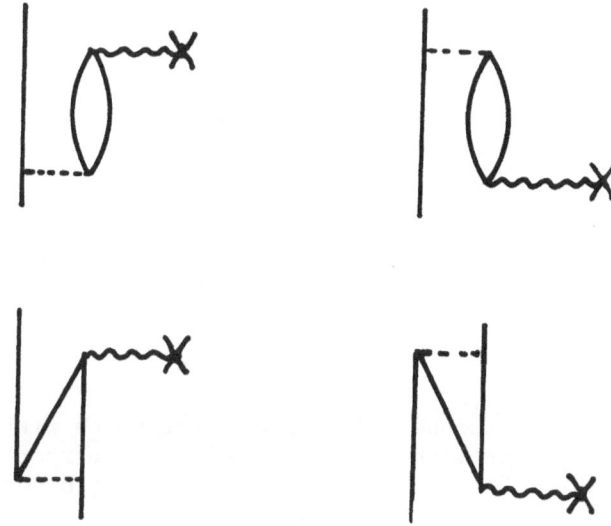

Fig. 2. Core polarization.

If this is the case, one expects that the gyromagnetic ratio g of the N = 126 isotones should be almost constant. The observation, however, shows clearly a Z dependence as seen in Fig. 3. If only the first order core polarization is assumed,

$$|\pi h_{9/2}{}^z> + \alpha|\pi h_{11/2}{}^{-1} h_{9/2}(1^+)\pi h_{9/2}{}^z>$$

the deviation δg must be proportional to z - 1 (z = Z - 82);

$$\delta g = \frac{(9-z)}{8} \delta g_{core}(\pi h).$$

Figure 3 shows that δg has a quadratic term z^2. This can be explained by the mixture of configurations;

$$h_{11/2}{}^{-2} h_{9/2}{}^{z+2} \tag{9}$$

which gives approximately

$$\frac{(9-z)(8-z)}{56} \, \delta g_{core}^{2nd}(\pi h) ;$$

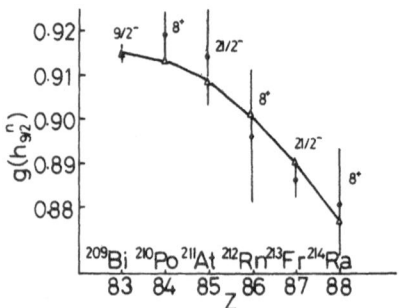

Fig. 3.   The gyromagnetic ratios of the N = 126 isotopes.

This is a part of the second order corrections which will be
discussed in the next section. This second order correction caused
by the configuration (9) partially cancels the first order correction as was pointed out by Mavromatis and Zamick.[11]

There are a few other configurations which contribute to the
magnetic moments;

$$\nu(i_{13/2}^{-1} \, i_{11/2})(1^+)\pi h_{9/2}^z$$

which gives rise to a constant change $\delta g_{core}(\nu i)$, and

$$\pi h_{9/2}^{z-1} \, j \, (j = f_{7/2}, \, f_{5/2} \cdots)$$

which gives rise to a second order correction

$$(z-1)\delta g^{cm}$$

Those corrections were studied by Arita[12] and the Chalk River
group.[13]   Finally meson exchange currents modify mainly the
gyromagnetic ratio g_ℓ of orbital angular momentium.[1],[14],[15]
The modification δg_ℓ is roughly

$$\delta g_{\ell} \gtrsim 0.1 \; . \tag{10}$$

Summing up all contributions, one finds

$$g(h_{9/2}{}^{z}) = g_{single}(h_{9/2}) + \frac{12}{11} \delta g_{\ell} + \delta g_{core}(\nu i)$$

$$+ \frac{9-z}{8} \delta g_{core}(\pi h) + (z-1)\delta g^{cm}$$

$$+ \frac{(9-z)(8-z)}{56} \delta g_{core}^{2nd}(\pi h) . \tag{11}$$

we first estimate $\delta g_{core}(\nu i)$. For this purpose one can use the observed magnetic moments of ^{202}Po and ^{210}Po. The observed gyromagnetic ratio of ^{202}Po is 0.93, while that of ^{210}Po is 0.91. The difference comes from the core polarization such as $\nu f_{7/2}{}^{-1} f_{5/2}$ and $\nu p_{3/2}{}^{-1} p_{1/2}$. Assuming that the core polarization due to $\nu_{13/2}{}^{-1} i_{11/2}$ gives rise to the same amount of δg, one obtains the following estimate

$$\delta g_{core}(\nu i) \stackrel{\sim}{\sim} 0.02. \tag{12}$$

The observed gyromagnetic ratios of the $N = 128$ isotopes are used to find the values of $\delta g_{core}(\pi h) + 8\delta g^{cm}$ and $\delta g_{core}(\pi h)$. Using the χ^2-fitting method, one finds the following values;

$$\delta g_{core}(\pi h) + 8 \; \delta g^{cm} = 0.199$$
$$\delta g_{core}^{2nd}(\pi h) = -0.082 \tag{13}$$

The δg^{cm} was estimated by Arita[12] and Towner et al,[13]

$$\delta g^{cm} \sim 0.0026.$$

Here this will be ignored. One thus finds that the modification of the gyromagnetic ratio in ^{209}Bi caused by the (1st order and 2nd order) core polarization δg_{core} is

$$\delta g_{core} = \delta g_{core}(\pi h) + \delta g_{core}^{2nd}(\pi h) + \delta g_{core}(\nu i)$$

$$= 0.137. \tag{14}$$

From this one derives the contribution of the core polarization to the magnetic moment of ^{209}Bi which is

$$\delta \mu_{core}(h_{9/2}) = \frac{9}{2} \delta g_{core} \tag{15}$$

$$= 0.62$$

Though this value is slightly smaller than a value predicted by an old calculation[1],[16] (∼0.8), we certainly cannot ignore the core polarization (or the so-called Arima-Horie effect). Without this mechanism, we cannot explain the observed z dependence at all. The gyromagnetic ratio of ^{209}Bi should be written as

$$g(^{209}\text{Bi}) = \delta g_{core} + \frac{12}{11} \delta g_\ell + g_{sing}(h_{9/2})$$

$$= 0.137 + 0.11 + g_{sing}(h_{9/2}) \qquad (16)$$

where equations (10), (11) and (14) are used. From eq. (16), one obtains $g_{sing}(h_{9/2})$ as

$$g_{sing}(h_{9/2}) = 0.67$$

which apparently differs from that for a free proton $g_{free}(h_{9/2})$ which is 0.583. The difference is as follows;

$$\Delta g_{sing}(h_{9/2}) = 0.67 - 0.58$$

$$= 0.09.$$

This Δg should reflect the renormalization caused by the tensor correlations, the Δ-h effects etc. Assuming the Δg comes from the quenching of the spin gyromagnetic ratio g_s, which we denote as δg_s, one finds the following relation;

$$\Delta g_{sing}(h_{9/2}) = -\delta g_s/11,$$

which gives

$$\delta g_s = -0.94.$$

The ratio of this δg_s to g_s for a free nucleon is

$$\delta g_s/g_s = -0.94/5.585 = -0.17, \qquad (17)$$

i.e. 17% reduction. It is very interesting to point out that this reduction is almost the same as that for ^{41}Ca.

Some magnetic dipole transitions in the vicinity of ^{208}Pb were observed by the Chalk River group. Their results are shown in Table 3. They found that all transitions are about a quarter of the pure single particle values. Thus the quenching of M1 transitions have been known for quite some time. They were nicely explained by Lie-Jeng Huang and Arima[16] and independently by the Chalk River group.[17] According to those calculations the core polarization reduces the transition matrix element by roughly fifty per cent; $\delta M_{CP} \overset{\sim}{=} -0.5\ M_{S.P.}$ In these calculations, however, they assumed the magnetic dipole operator μ for a free nucleon.

Table 3. The B(M1) values (in unites of $(n.m.)^2$).
(Take from ref. 16)

Nucleus	$j_i \rightarrow j_f$	$B(M1)_{SP}$	$B_1(M1)$	$B_2(M1)$	$B(M1)_{exp}$
^{207}Pb	$p_{3/2}^{-1} \rightarrow p_{1/2}^{-1}$	1.167	0.447	0.498	0.41 ± 0.07
	$f_{7/2}^{-1} \rightarrow f_{5/2}^{-1}$	1.500	0.420	0.452	0.49 ± 0.16
	$p_{3/2}^{-1} \rightarrow f_{5/2}^{-1}$	0	5.3×10^{-3}	0.03×10^{-3}	—
^{209}Pb	$i_{11/2} \rightarrow g_{9/2}$	0	0.21×10^{-3}	1.4×10^{-3}	<0.04
^{209}Bi	$f_{5/2} \rightarrow f_{7/2}$	2.862	0.755	0.826	0.76 ± 0.15
	$f_{7/2} \rightarrow h_{9/2}$	0	1.55×10^{-2}	3.6×10^{-3}	$(4.3\pm0.7)\times10^{-3}$
^{207}Tl	$d_{3/2}^{-1} \rightarrow s_{1/2}^{-1}$	0	1.68×10^{-2}	5.6×10^{-3}	—

$B(M1)_{SP}=M_{SP}^2/(2j_i+1)$, $B(M1)_{1,2}=M_{1,2}^2/(2j_i+1)$ The core polarization only is taken into account in M_1, and the mesonic effect is added in M_2.
$B(M1)_{exp}$ are taken from the measurements by O. Häusser, F.C.Khanna and D. Ward.[17]

As we found in the last section, the effective operator μ^{eff} must be used. In the previous section the matrix element of the effective operatot in the $0h_{9/2}$ proton state was estimated roughly as 80% of that of the operator for a free nucleon (See eq. (17)). Further we found that the effect of the core polarization (first order plus second order) should be slightly smaller than that predicted by old first order calculations.[16] Namely the former is about 0.63 while the latter is about 0.8. Thus we must reduce the effect of the core polarization found by the previous calculations by a factor 0.63/0.8. Now the dipole transition matrix element should be roughly written as

$$0.8\{M_{s.p}+\delta M_{cp}\times\frac{0.63}{0.8}\}$$

$$= 0.8\ M_{s.p}\{1 - 0.5\times\frac{0.63}{0.8}\} \overset{\sim}{=} 0.48\ M_{s.p}$$

where the first factor comes from the renormalization. The observed values are roughly a half of M_{sp}. Although more precise calculations must be carried out, it is already clear that core polarization including the renormalization operator will provide a good explanation for the observed M1 transitions.

The ground ·state correlation which corresponds to the first order core polarization should have smaller effects on magnetic dipole transitions in ^{208}Pb than those M1 transitions discussed above. A preliminary calculation gives rise to about 30% reduction of the M1 transitions in ^{208}Pb from their shell model values. We thus predict the following M1 transition strength in ^{208}Pb;

$$|M_{free}(0^+ \to 1^+)|^2 (0.8)^2 \times 0.70$$

$$= 0.48 |M_{free}(0^+ \to 1)|^2. \tag{18}$$

where M_{free} means that the operator for a free nucleon is used.

It should be pointed out here that the renormalization is not necessary common for non-diagonal matrix elements between $j = \ell + 1/2$ to $j' = \ell - 1/2$ and diagonal matrix elements. As will be shown in the next section, the Δ-h state has a larger reduction for the non-diagonal matrix elements than that for the diagonal elements. Assuming that this is the case and that the renormalization factor for $j \to j'$ is 0.7 (instead of 0.8), one finds instead of eq. (18)

$$|M_{free}(0^+ \to 1^+)| (0.7)^2 (0.70)$$

$$= 0.34 |M_{free}(0^+ \to 1^+)|^2.$$

Several 1^+ states were observed around 7^{MeV} by the (γ, n) reaction. The sum of those M1 strengths are about 20% $\sim$ 25% of the shell model values.[18] We must still find 14% $\sim$ 9 % strength somewhere.

A 1^+ state was found at 5.8 MeV in ^{208}Pb by the Osaka group who used the (p, p') reaction,[27] while Vergados predicted such a state[28] at 5.4 MeV. It is now very urgent to observe the M1 strength carried by this state.

6. RENORMALIZATION DUE TO HIGH ENERGY STATE

We have shown in the previous sections that shell model calculations can provide a good explanation of nuclear magnetic dipole moments and transitions. Here it is assumed that the renormalized operator is used and either $0\hbar\omega$-major shell calculations are carried out or the core-polarization (the Arima-Horie) effect is taken into account. In these calculations, one truncates the shell model space into a $0\hbar\omega$-major shell and omits higher shell model states and states involving Δ. In order to take into account those highly excited states, one has to renormalize operators.

Now one must ask what causes the renormalization of the magnetic dipole operator and the Gamow-Teller ($\tau\sigma$) operator. Ten years ago, Shimizu and his collaborators[19] pointed out the importance of the tensor correlation (second order), which was found to reduce Gamow-Teller transitions by 15% in amplitudes (by 25% in squared amplitudes). They could explain at least qualitatively the observed Gamow-Teller transition probabilities in light nuclei with an LS-closed shell ± one nucleon.

Recently, however, the effect of the Δ-hole state has been actively discussed by many authors.[3],[20],[21] While some results amounted to as much as a 30% reduction in amplitudes, the effects of tensor correlation seems to be omitted entirely.

I would like to first point out that one needs the tensor correlation to explain the quenching of isoscalar magnetic moments. Because their deviations from shell model values are very small, these moments have been usually ignored. This situation is, however, very misleading. These deviations are small, simply because the isoscalar magnetic dipole operator has a small factor;

$$\mu^S = 0.88\ S + L/2 = 0.38\ S + J/2.$$

The deviations of $\langle S_0 \rangle$ from their simple shell values are indeed very large as shown in Table 4. The observed matrix elements of

Table 4. Deviations of matrix elements of (S_0) from the single particle values.

A		$\langle S \rangle_s$	$\langle S \rangle_{obs}$	$\delta\langle S \rangle^{cal}/\langle S \rangle_s$			$\delta\langle S \rangle^{obs}/\langle S \rangle_s$
				Conf.mix.	$\Delta+\rho\pi\gamma$	sum	
15	$p_{1/2}$	−0.167	−0.085	−0.63	0.10	−0.53	−0.49
17	$d_{5/2}$	0.5	0.433	−0.18	0.03	−0.15	−0.13
39	$d_{3/2}$	−0.3	−0.118	−0.55	0.06	−0.49	−0.62
41	$f_{7/2}$	0.5	0.442	−0.19	0.03	−0.16	−0.12

S deviate from their single particle values in a very systematic
ways. They are always quenched.

The Sachs moment and other mesonic effects do not contribute
to the isoscalar magnetic moments, except for $\rho\pi\gamma$ and resonance
admixture (Δ-h). Their contributions have the sign opposite to the
observed values and are only one fifth (in magnitude) of those from
the tensor correlations.

The quenching of the isoscalar magnetic moments or spins (S)
are explained very well by the tensor correlations as shown in Table
4. The best evidence for the tensor correlation is thus provided
by the quenching of the isoscalar magnetic moments.

The next subject is the quenching of the Gamow-Teller moments.[22]
A very thorough calculation was acomplished by Towner and Khanna.
Their results are shown in Table 5. The nuclear second order correction (NSC) is mainly caused by the tensor correlation. This is
slightly cancelled by $\rho\pi\gamma$ and the nuclear second order correction
involving Δ state (NSCΔ). However it is clear from examination
of Table 5 that the nuclear second order correction dominates,
while, the effect of Δ-hole states which is estimated by using the
Random Phase Approximation does not seem very large according to
Towner and Khanna.

Isovector magnetic moments which are shown in Table 6 are more
complicated because the effect of the meson exchange current is
greatly enhanced by the tensor correlation (C in table). This
effect cancels partly the effect of the nuclear second order correction (the tensor correlation A in Table 6). Even if one sums
both effects, however, the tensor correlation still seems to play
an important role. In this table B stands for the contribution
of the meson exchange currents which includes Δ state too. The
Δ-hole quenches the matrix element of $\tau\sigma$ by only 5% according to
Towner and Khanna.[22] This quenching corresponds to about 0.1 n.m.
for $j = \ell + 1/2$. This seems to be less than a quarter of the contribution from the tensor correlation.

In this conference, Towner reported his very thorough calculation of the isovector magnetic moments.[23] From his result,
one confirms the statement obtained above. See Table 7.

I would like to point out that the form factor of the electron
scattering is sensitive to many effects. One should not be naive
enough to believe that this is modified only by the Δ-hole contribution. The ordinary core-polarization (particle-hole correlation)
also contributes. The importance of this[24] has been pointed out
for example in $^{12}C(e, e')^{12}C*(15$ MeV, $1^+)$.

Table 5. Various corrections of the Gamow-Teller
matrix elements calculated by Towner & Khanna.[22]

	A=15	17	39	41
NSC	-0.19	-0.14	-0.25	-0.17
NSCΔ	0.05	0.02	0.04	0.02
RPAΔ	0.04	-0.03	-0.03	-0.04
ρπγ	0.05	0.03	0.04	0.04
Relativistic	-0.04	-0.03	-0.04	-0.03
Sum	-0.09	-0.15	-0.26	-0.25
Exp	-0.12	-0.12	-0.32	-0.25

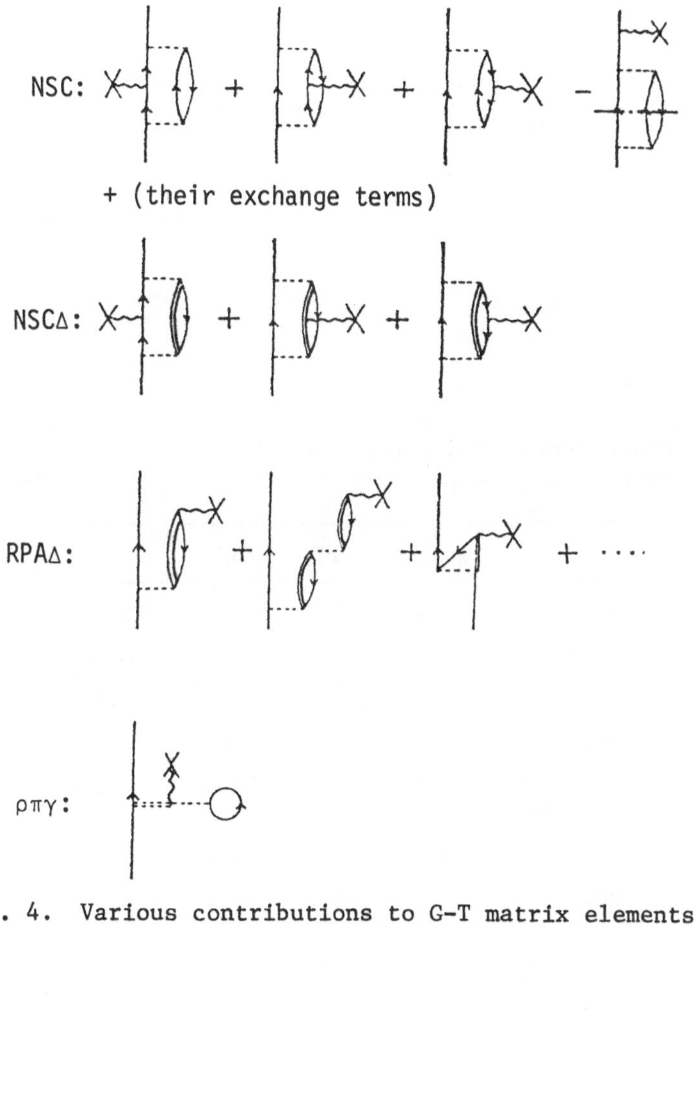

Fig. 4. Various contributions to G-T matrix elements.

Table 6. Contributions to the isovector magnetic moment from the various processes, $\mu_s^{(1)}$ stand for the single particle values.

A	$\mu_s^{(1)}$		$\delta\mu^{(1)}$	A	$\mu_s^{(1)}$		$\delta\mu^{(1)}$
15	0.451	A	0.032	39	0.512	A	0.227
		B	0.014			B	−0.162
		C	−0.007			C	−0.116
		sum	−0.02			sum	−0.06
		obs	0.05 .			obs	−0.20
17	−3.353	A	0.483	41	−3.853	A	0.762
		B	−0.258			B	−0.360
		C	−0.397			C	−0.352
		sum	−0.17			sum	0.05
		obs	0.05			obs	0.34

Table 7. Calculated correction to the isovector magnetic moment calculated by Towner and Khanna (invited talk in this conference) NSC corresponds to A, MEC to B and MEC with core pol. to C in Table 4. Here $\delta\mu/\mu$ are shown.

	A=15	17	39	41
NSC	9.2	−14.1	37.5	−19.5
NSCΔ	5.0	0.6	7.6	0.8
RPAΔ	11.3	−0.9	−2.5	−1.7
MEC	−17.6	5.1	−41.5	7.1
MEC with core pol.	1.6	6.7	−10.0	9.7
Sum	9.5	−2.6	−8.9	−3.6
Expt.	11.1	−1.3	−38.4	−8.9

In addition to the Chalk River calculation, older calculations found small corrections due to the Δ-hole state which was usually taken into account as a part of the meson exchange currents.[1] On the other hand, some calculations[3,20,21] found very large values for them. One[22] is 5% and the other[3,20,21] amounts to even 30%. Why is there such a large difference ?

This puzzles me very much. Large values have been obtained by calculations in which the Landau-Migdal method is used and a large value is assumed for $g'_{NN^{-1}-\Delta N^{-1}}$. This $g'_{NN^{-1}-\Delta N^{-1}}$ is very crucial. Shimizu and his collaborators[25] estimated a value for $g'_{NN^{-1}-\Delta N^{-1}}$ by taking the $\pi + \rho$ exchange model. They found a small value (≈ 0.4). This value gives rise to very similar corrections found by Towner and Khanna.[22] A similar small value of $g'_{NN^{-1}-\Delta N^{-1}}$ was reported by the Tübingen group.[26] There may be other mechanisms which increase the value of $g'_{NN^{-1}-\Delta N^{-1}}$. It is an interesting open problem. I would like to stress that one must settle this problem before one is convinced by the effect of the Δ-hole state.

Shimizu and his collaborators performed similar calculations as the Chalk River group taking the $\pi + \rho$ exchange model. Their results are shown in Table 8.

Table 8. Contribution to Gamow-Teller matrix elements from Δ-h excitations.
In this caclulation, $f^*/f=2$, $\Delta M=300$ MeV $f^2_\rho=4.86$ are used. As the short range correlation, we used a cut-off radius $c=0.7$ fm.

A		Reduction %	Towner-Khanna from Table 5
12	$\langle P_{1/2}\, P_{3/2}^{-1}\, 11\|\|\|\tau\sigma\|\|\|0\rangle$	-6.5	
15	$\langle p_{1/2}^{-1}\|\|\|\tau\sigma\|\|\|p_{1/2}^{-1}\rangle$	10.3	4
17	$\langle d_{5/2}\|\|\|\tau\sigma\|\|\|d_{5/2}\rangle$	-2.6	-3
39	$\langle d_{3/2}^{-1}\|\|\|\tau\sigma\|\|\|d_{3/2}^{-1}\rangle$	0.4	-3
41	$\langle f_{7/2}\|\|\|\tau\sigma\|\|\|f_{7/2}\rangle$	-3.7	-4
564	$\langle f_{5/2}\, f_{7/2}^{-1}\, 11\|\|\|\tau\sigma\|\|\|0\rangle$	-9.9	

This table shows an interesting fact. The correction due to the
Δ-hole state is small for the diagonal matrix elements (j↔j) as
found by the Chalk River group but is about two times larger for
the non-diagnonal matrix elements (j = $\ell + \frac{1}{2}$ ↔ j' = $\ell - \frac{1}{2}$ and
$0^+ \rightarrow 1^+$). In the precious calculations, we have been interested
in only the diagonal elements which are magnetic moments.
Therefore, we believed that the Δ-hole effect is very small and
amounts to only about 0.1 n.m. The effect can be larger in transitions such as the $0^+ \rightarrow 1^+$ excitation. Why does this situation
happen ? The reason seems to be very simple.

The effective magnetic dipole operator can be written as

$$\mu^{eff} = g_s^{eff} s + g_\ell^{eff} \ell + g_p [s \times Y^{(2)}]^{(1)}.$$

The matrix element of μ^{eff} is as follows;

$$<j\ m = j|\mu^{eff}|j\ m = j> = \pm \frac{1}{2} g_s^{eff} + \ell g_\ell^{eff} \pm \frac{1}{4\sqrt{2\pi}} g_p$$

where j = $\ell \pm \frac{1}{2}$, and ℓ is assumed to be very large

and

$$<j = \ell + \frac{1}{2} \| \mu^{eff} \| j' = \ell - \frac{1}{2}> = - \sqrt{\ell}(g_s^{eff} - g_\ell^{eff} - \frac{1}{4\sqrt{2\pi}} g_p).$$

The effect of the Δ-hole state gives rise for protons

$$g_s^\Delta < 0, \ g_\rho^\Delta > 0.$$

Thus they cancel each other in the diagonal matrix elements, while
they contribute coherently in the non-diagonal (transition) matrix
elements. Thus the Δ-hole state has a larger effect in the transitions. Yet this reduction is about 10% in amplitudes if one
assumes the π + ρ exchange model.

There is an argument that the contribution from the Δ-hole
excitation must be decomposed into a direct part and an exchange
part. Furthermore according to this argument, only the direct part
should be taken into account, because the exchange part should
be cancelled out by other complicated corrections such as the
tensor correlation. This argument, however, misses a very important
point. The advertised role of the Δ-h excitation is that the Δ-h
state absorbs a certain amount of M1 strengths, because the Δ-h
(1^+) state admixes into normal particle-hole 1^+ states. In order
to calculate the mixing amplitude of the Δ-h state, one must taken
into account both the direct and exchange matrix elements.

It does not make any sense to separate the mixing caused by the
direct interaction from that by the exchange interaction.

One more point which I would like to make here is how to
interpret the Landau-Migdal theory. According to a usual interpretation, g' is introduced to simulate the effect of the short-
range correlation and the exchange interaction. Thus I believe
that one must sum both the direct and exchange contributions
calculated by Towner and Khanna (Table 5). Then the sum of both
the direct and exchange contribution (not one of them) must be
compared with the results of calculations in which the Landau-
Migdal method is used. This comparison suggests that $g'_{NN^{-1}-\Delta N^{-1}}$
should be small as mentioned above.

ACKNOWLEDGEMENT

I would like to thank K. Shimizu for his cooperation and
enlighting discussions. My thanks are also due to K. Yazaki,
H. Hyuga, T. Suzuki and T. Cheon for their stimulating and
instructive discussions.

REFERENCES

1.   For example, A. Arima and H. Hyuga, Mesons in Nuclei,
      Eds. M. Rho and D. H. Wilkinson (North-Holland Publishing
      Company, 1979) p.718.
2.   A. Arima and H. Horie, Prog. Theor. Phys. 11 (1954) 509 and
      12 (1954) 623.
3.   For example, H. Toki and W. Weise, Phys. Lett. 97B (1980) 12.
4.   A. Arima, The structure of $1f_{7/2}$ nuclei, Ed. R. A. Ricci,
      Editorice Compositori, (Bologna 1971) p. 385.
5.   S. K. Bhattacherjee, R. Brenn, D. H. Fossan, G. D. Sprouse
      and L. E. Young, Phys. Rev. Lett. 35 (1975) 497.
6.   K. Nakai, B. Skaali, N. J. S. Hansen, B. Herskind and
      Z. Sawa, Phys. Rev. Lett. 27 (1971) 155.
7.   J. B. McGrory and B. H. Wildenthal, Phys. Lett. 103B (1981) 173.
8.   B. H. Wildenthal and W. Chung, Mesons in Nuclei, Eds. M. Rho
      and D. H. Wilkinson (North-Holland Publishing Company, 1979)
      p. 721.
9.   A. Yokoyama and H. Horie, private communication
10.   A. Richter, Nucl. Phys. A374 (1982) 177.
11.   H. A. Mavromatis and L. Zamick, Nucl. Phys. A104 (1967) 17.
12.   K. Arita, Proceedings of the International Conference on
      Nuclear Physics, Munich (1973), Eds. J. de Boer and H. J. Mang.
13.   I. S. Towner, F. C. Khanna and O. Häusser, Nucl. Phys.
      A277 (1977) 285.

14.   H. Miyazawa, Prog. Theor. Phys. 6 (1951) 801.
15.   T. Yamazaki in Mesons in Nuclei, Eds. M. Rho and D. H.
       Wilkinson (North-Holland Publishing Company, 1979) p. 651.
16.   A. Arima and L. J. Huang-Lin, Phys. Lett. 41B (1972) 429, 439.
17.   O. Häusser, F. C. Khanna and D. Ward, Nucl. Phys. A194 (1972)
       113.
18.   R. J. Holt, H. E. Jackson, R. M. Laszewski and J. R. Specht,
       Phys. Rev. C20 (1979) 93.
19.   K. Shimizu, M. Ichimura and A. Arima, Nucl. Phys. A226 (1974)
       282.
20.   E. Oset and M. Rho, Phys. Rev. Lett. 42 (1979) 47.
       95B (1980) 349.
21.   For example, W. Knüpfer, M. Dillig and A. Richter,
       Phys. Lett. 95B (1980) 349.
22.   I. S. Towner and F. C. Khanna, Phys. Rev. Lett. 42 (1979) 51
23.   I. S. Towner and F. C. Khanna, these proceedings.
24.   H. Sagawa, T. Suzuki, H. Hyuga and A. Arima, Nucl. Phys.
       A322 (1979) 361.
25.   K. Shimizu, private communication.
26.   W. H. Dickhoff, A. Faessler, J. Meyer-ter-Vehn and H. Müther
       Phys. Rev. C23 (1981) 1154.
27.   S-I. Hayakawa, private communication.
28.   J. D. Vergados, Phys. Lett. 36B (1971) 12.

SPIN-ISOSPIN RESPONSE FUNCTIONS IN NUCLEI

M. Ericson

Institut de Physique Nucléaire, Lyon
and
CERN, Geneva, Switzerland

INTRODUCTION

In this talk I will give a coherent survey of the question of the spin-isospin excitations in nuclei. It is a proper time for such a review since a unity to this field has emerged, which was not originally apparent.

My presentation will be oversimplified. While other speakers will give the detailed and quantitative treatments, it nevertheless contains the main physical features.

Let us consider as a starting point the response function of the nucleus to a spin-isospin perturbation. A concrete example is, for instance, the axial current of the weak interactions

$$\vec{A}^{\pm}(\vec{x}) = g_A \sum_i \vec{\sigma}_i \, \tau_i^{\pm} \, \delta(\vec{x} - \vec{x}_i)$$

or the isovector magnetic transition operator

$$\vec{J}_s(\vec{x}) = g_M \sum_i (\vec{\sigma}_i \times \vec{\nabla}) \, \tau_i^3 \, \delta(\vec{x} - \vec{x}_i) \ ,$$

in general any perturbation which couples to the spin isospin density fluctuations.

When such a perturbation is applied to the nuclear medium, this medium responds in such a way as to adjust to the presence of the perturbing field. If all interactions between the nucleons are switched off, they respond individually to the external field. The response will then occur in two ways. Firstly, the nucleons can flip

27

their spin and isospin. This process represents an adjustment of the
nucleonic orbits, in a way which is of course restricted by the Pauli
principle. Secondly, the field can flip the spin and isospin of an
internal constituent of the nucleons, of a quark, transforming the
nucleon into a Δ resonance for instance.

Let us consider the case of infinite nuclear matter. The total
response $\pi^0(q, \omega)$, which is a function of the momentum and frequency
of the exciting field, is the sum of these two components

$$\pi^0(q,\omega) = \pi^N(q,\omega) + \pi^\Delta(q,\omega) \ . \tag{1}$$

What is then the behaviour of π^0? Its imaginary part is fixed
by the excitations of the system. Restricting to single particle
excitations these are of two kinds. Firstly, we have the one nucleon
excitations at low energies. Secondly, we have the states with a
real Δ excited which are located at high energies, around $\omega_\Delta \approx$
≈ 300 MeV. Between these two types of excitations there is a no
man's land where Im π vanishes (see Fig. 1). The real part of π^0,
on the contrary, extends everywhere and has the usual dispersive
behaviour, so that Re π^N changes sign at low energy and Re π^Δ at
high energy (see Fig. 2).

When the interaction between the nucleons is switched on, the
force acting on one nucleon is transmitted to its neighbours. The
response then becomes collective. The random phase approximation
(RPA) takes into account the transmission of the force by summing
the chain of the ring diagrams shown in Fig. 3, where the heavy line
represents the particle-hole (p.h.) interaction. As you see, the

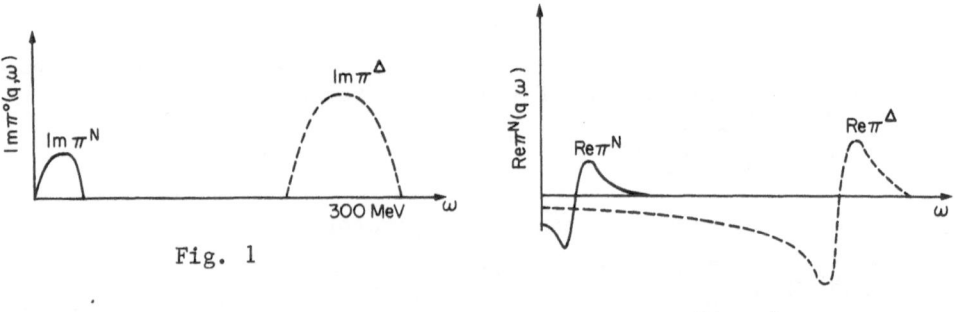

Fig. 1

Fig. 2

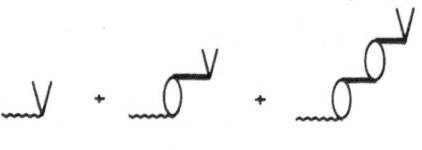

Fig. 3

field can act directly on one nucleon or on a neighbouring one which transmits the force and so on. By summing this geometrical series one obtains the RPA expression

$$\pi(q,\omega) = \frac{\pi^0(q,\omega)}{1 - V(q,\omega)\pi^0(q,\omega)} \, , \tag{2}$$

where $V(q,\omega)$ is the p.h. interaction. This expression is oversimplified in many respects. Firstly, it is a nuclear-matter expression, while in practice one has to deal with finite systems. Secondly, it is based on the RPA which does not represent all the physics. The effect of two particle-two-hole states (2p.2h.) for instance, is ignored. Finally, even in RPA the ring expression is not exact because it omits the antisymmetrization. In the following we omit in addition the Hartree-Fock effects and use the free nucleon mass. In spite of these restrictions, this expression is useful since it is so simple and compact. Keeping in mind its quantitative limitations we will use it to predict or interpret many physical phenomena.

A very obvious remark: if the RPA denominator is close to unity, the response keeps its single-particle character, while if it deviates appreciably from unity, the response becomes collective. In particular, it may happen that the denominator vanishes. If this is the case, there will be a finite response, even for an infinitesimal perturbation with all nucleons participating. This is the condition for a collective state which is written:

$$1 - V(q,\omega)\pi^0(q,\omega) = 0 \, . \tag{3}$$

PARTICLE-HOLE INTERACTION

The quantity V depends on the p.h. quantum numbers. It is not sufficient to fix the total spin and isospin of the p.h. pair, which in our case gives $S = T = 1$. In addition, one must also specify the magnetic quantum number M_S. The longitudinal channel corresponds to $M_S = 0$. It is explored with a $\vec{\sigma} \cdot \vec{q}$ coupling, which probes the component of the magnetization along the momentum $\vec{q}$. The transverse channel explores the transverse magnetization $(\vec{\sigma} \times \vec{q}$ coupling) and corresponds to $M_S = \pm 1$. In nuclear matter these two channels are completely decoupled and do not communicate.

In the longitudinal case the pion-exchange force is effective, providing the following force which is attractive at small energies:

$$- \frac{f^2}{m_\pi^2} \frac{q^2}{q^2 + m_\pi^2 - \omega^2}$$

where f is the πN coupling constant, $f^2/4\pi = 0.08$.

Since the pion mass is so small the attraction which vanishes at q = 0 builds up very quickly with increasing q. Pion exchange is not the only possible one. There are also heavy exchanges (equivalently short-range effects). These are simulated in the Landau-Migdal approach by a repulsive contact force characterized by a parameter g'. In this phenomenological approach one does not enquire about the origin of the force. It can be due to multimeson exchange as well as to the overlap of two quark bags: g' is taken as phenomenological and should be determined from experiments. The overall force is then

$$V(q,\omega) = \frac{f^2}{m_\pi^2} \left[g' - \frac{q^2}{q^2 + m_\pi^2 - \omega^2} \right] .$$ (4)

In the transverse channel there is no light meson exchange to provide an attraction so readily (the lightest exchange is the ρ meson) and the force remains repulsive over a larger range of momenta.

The determination of the g' has been the subject of very careful and thorough studies[1,2]. From the position of the nuclear states, in particular those of unnatural parity, which are sensitive to these forces, it was concluded that the repulsion is large with $g' \approx 0.6$ to 0.7 (in the same units, the maximum attraction from the pion is −1). Such large values have been accounted for theoretically[3,4]. The forces in the two channels are featured in Fig. 4, which displays a remarkable contrast between the two at moderate momenta, $1 < q < 2$ fm^{-1}. The longitudinal force is mildly attractive, while the transverse one stays repulsive.

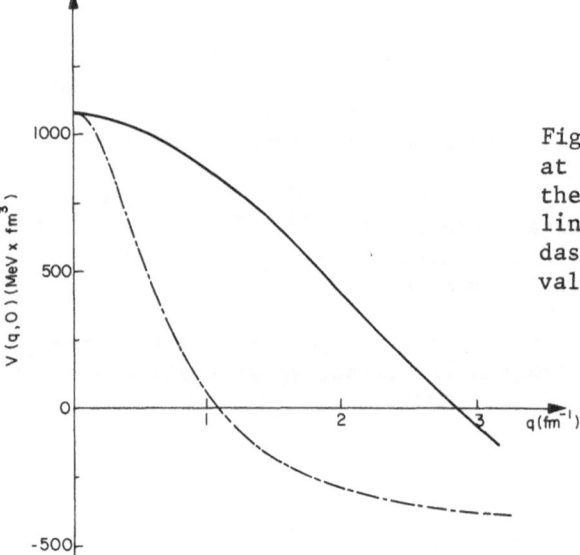

Fig. 4. The p.h. interaction at $\omega = 0$ as a function of q in the transverse (continuous line) and longitudinal (dot-dashed line) channel. The value of g' is fixed at 0.7.

We can exploit this contrast. Suppose we suspect that an effect in the spin-isospin response originates from the p.h. force.
We propose as a clean experimental signature for this the contrast
between the responses in the two channels.

Let us now explore how the responses are affected by the forces,
following the evolution with increasing momentum.

LOW MOMENTUM REGION (q ≤ 0.3 fm^{-1})

The forces being identical in the two channels there is no need
to distinguish between the two responses. The strong repulsion
creates a collective state, the zero sound, predicted by Ikeda et
al.[5]. Since V is real at low energies, the condition (3) for a collective state requires Im π^0 = 0. It is fulfilled outside the p.h.
continuum, i.e. in the no man's land region of Fig. 1, where Im π^0 = 0.
The collective state is pushed above the p.h. continuum by the repulsive interaction. The presence of the collective state produces
a complete reshaping of the response as shown in Fig. 5. The single-
particle states are completely depopulated in favour of the collective one.

The existence of this collective state has been beautifully displayed in the (p,n)[6,7] and (pp')[8] reactions. The (p,n) experiments
have in addition revealed a very interesting feature, which concerns
the total Gamow-Teller (GT) strength. Since all final states are
energetically accessible in the (p,n) experiments, contrary to β
decay, it is possible to, extract the total GT strength.

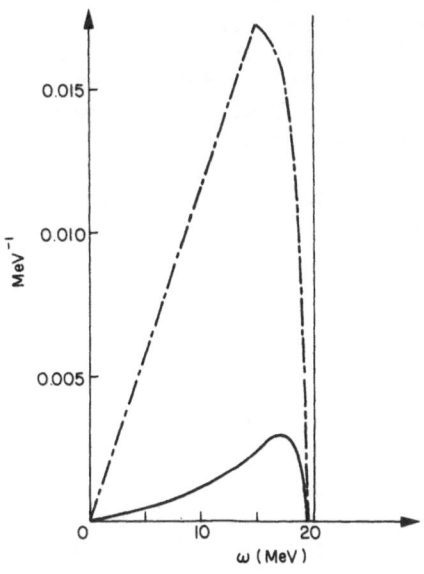

Fig. 5. The imaginary part of the
response function of infinite nuclear matter at q = 0.3 fm^{-1} as a
function of ω for a free Fermi gas
(dot-dashed line) and in the ring
approximation (continuous line).
From Ref. 29.

The result could be compared with an RPA calculation, but an even simpler comparison has been performed[9]. Take the difference of the total GT strengths for the $\beta^{\mp}$ branches, associated with the spin isospin operators $\sigma^3 \tau^{\pm}$.

$$S = S^- - S^+$$
$$= g_A^2 \sum_I \left\{ \left| \left\langle I \right| \sum_i \sigma_i^3 \tau_i^+ \left| 0 \right\rangle \right|^2 - \left| \left\langle I \right| \sum_i \sigma_i^3 \tau_i^- \left| 0 \right\rangle \right|^2 \right\} . \qquad (5)$$

Performing closure over the intermediate states I, one is left with the commutator of $(\tau^+, \tau^-) = \tau^3$ and

$$S = g_A^2 (N - Z) . \qquad (6)$$

The sum S thus depends only on the neutron excess of the nucleus. It is totally independent of the detailed structure of the initial state and its correlations.

The β^+ branch which would be obtained from an (n,p) reaction is not accessible at present. In nuclei with a neutron excess, S is small due to the Pauli blocking, and $S^- \approx S = g_A^2 (N - Z)$. Even if S^+ cannot be neglected, the sum rule provides, in the case of a neutron excess, a lower limit for S^- (if Z > N, it gives a limit for S^+ instead).

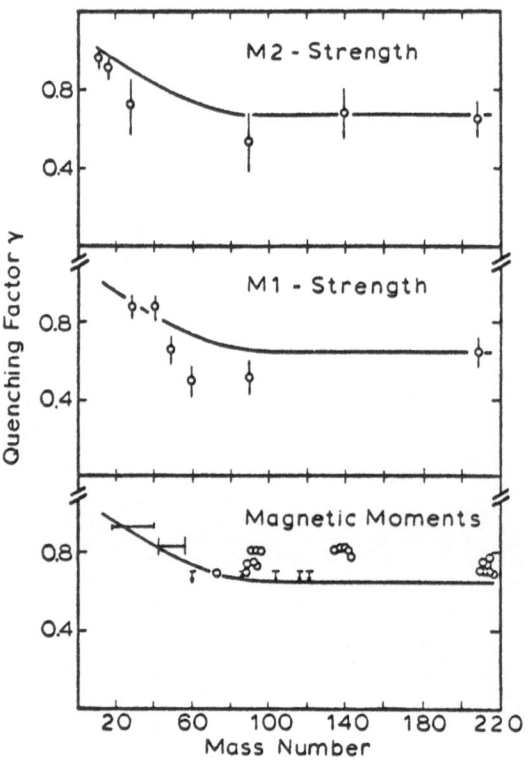

This comparison has been performed. The remarkable result is that the experimental values which should be above the (N − Z) line, lie in fact a factor $\approx$ 2 below[9]. What are the reasons for this discrepancy? One could try to blame the (p,n) reaction itself. After all it is not straightforward to extract the GT strength from a (p,n) reaction which involves a strongly interacting probe. However, similar quenching had already

Fig. 6. Quenching fator γ of the g_s factor as a function of the mass number A from the M_2 and M_1 strengths. From Ref. 10.

been observed in the M1 and M2 strengths[10], as shown in Fig. 6 (the quenching factor shown applies to the amplitude and should be squared for the strength). Here there is no similar sum rule and the quenching refers to a RPA evaluation. This comparison shows in addition that the quenching is A-dependent, disappearing in light nuclei. It is therefore natural to seek for a universal mechanism which applies to all spin transitions.

A POSSIBLE INTERPRETATION: THE LORENTZ-LORENZ (LL) EFFECT[11,12]

What could have gone wrong in our sum rule? We have assumed that the axial current couples to the nucleons with their free coupling constant, while the nucleons are not free in the nucleus. They are imbedded in the nuclear medium, which is not inert and becomes polarized under the influence of the weak field. Can this affect the sum rule? Suppose for a moment that there is no Δ excitation. The polarization represents the virtual excitations of the system, i.e. the mixing of nuclear configurations. But we have seen that the sum rule is independent of this mixing. Therefore as long as it respects the integrity of the nucleon the nuclear polarization does not affect the sum rule. This is no longer the case when the nucleonic excitations are introduced. Their effect can be pictured in the following way: under the influence of the weak field there appear polarization charges (axial charges) in the medium. In particular, there are surface charges at the surface of the correlation hole which surrounds a nucleon. This has the consequence that the field which couples to the nucleonic spins is not the average field in the medium but it is an effective one, as occurs for the e.m. field in a dielectric: $\vec{E}_{eff} = \vec{E}/(1 + \alpha/3)$ where α is the polarizability of the medium. This entails a renormalization of the axial coupling constant in the nucleus and therefore a modification of the sum rule*. Although the polarizability π in the expression of E_{eff} in principle involves all types of excitations, in the sum rule there survives only the effect of the Δ, the purely nuclear excitations cancelling their effect on the average. The renormalization factor of the sum is $1/(1 + f^2\pi^\Delta/3m_\pi^2)$. This renormalization appears clearly in the RPA expression, which allows the generalization to a phenomenological p.h. force[17,21]. The classical LL effect corresponds

---

*One may enquire how this geometrical description translates into a delta-hole (ΔN^{-1}) model. It is well known that the usual LL effect of pion scattering has an exact equivalence in that formalism[13]. The same occurs for the axial current. Indeed the geometrical picture originally proposed (in the first-order approximation) by Ericson et al.[11] was translated into the ΔN^{-1} formalism independently by Rho[14], and by Ohta and Wakamatsu[15]. Restricting to infinite nuclear matter they generalized our results to all orders. In the geometrical picture the A-dependence immediately follows from the existence of axial charges at the nuclear surface which suppress the LL effect in a small nucleus, as was shown in Ref. 11. In the ΔN^{-1} language this suppression requires the full RPA formalism.

to the value $g' = 1/3$ of the Migdal parameter. We want to separate
the roles of π^N and π^Δ in the RPA expression (2). It is rather natural to perform this separation since they behave quite differently:
at low momentum and small energy π^Δ is purely real and practically
constant with energy $\pi^\Delta(\omega) \approx \pi^\Delta$, while π^N, on the contrary, has both
a real and an imaginary part which vary rapidly with energy. We con-
sider for simplicity the very small momentum region where V reduced
to $V \approx G' = (f^2/m_\pi^2)g'$.

The imaginary part of the response function, proportional to the
inelastic cross-section is written:

$$\text{Im } \pi(q,\omega) = \frac{\text{Im } \pi^0(q,\omega)}{\left[1 - G' \text{ Re } \pi^0(q,\omega)\right]^2 + \left[G' \text{ Im } \pi^0(q,\omega)\right]^2}$$

$$\text{(7)}$$

$$= \frac{1}{(1 - G'\pi^\Delta)^2} \frac{\text{Im } \pi^N(q,\omega)}{\left[1 - \dfrac{G'}{1 - G'\pi^\Delta} \text{ Re } \pi^N(q,\omega)\right]^2 + \left[\dfrac{G'}{1 - G'\pi^\Delta} \text{ Im } \pi^N(q,\omega)\right]^2}$$

In this form the RPA denominator has been split in two pieces.
The second part depends on the energy through $\pi^N(\omega)$. It mixes the
different single-particle configurations with a force $G'/(1 - G' \pi^\Delta)$
renormalized by the presence of the Δ. But we have seen that the GT
sum rule is totally independent of this mixing and of the force. If
it were only for this term the sum rule would be conserved. But there
is the first constant factor $1/(1 - G' \pi^\Delta)^2$ which produces a renormalization of the GT strength (LL effect). Since π^Δ is negative it represents a quenching of the strength.

The renormalization factor depends strongly on the force, which
mixes NN^{-1} and ΔN^{-1} states (which for simplicity has been taken here
to be the same as the NN^{-1} force). For $g' \approx 0.6$ and $\pi^\Delta = -0.8$ the
quenching of the strength is strong: a factor ≈ 2. The LL effect
offers a likely interpretation for at least part of the missing
strength. If correct, it is also another indication for a strong
collectivity, involving this time Δ resonances. One cannot flip the
spin of one nucleon without flipping those of the virtual Δ cloud
which surrounds it. However, the present experimental situation cannot conclude decisively in favour of the LL interpretation. Other
mechanisms can also be partly responsible for the quenching[22].

Summarizing the previous results, there exists a nice sum rule
which is independent of the mixing force as long as the space is
restricted to nucleonic degrees of freedom. When the Δ degree of
freedom is introduced, the sum rule is violated and the result becomes dependent on the force.

It is then natural to ask the question: Is it possible to find another sum rule which is independent of this mixing as well. For this one has to enlarge the traditional perspective of nuclear physics so as to incorporate the Δ degree of freedom in the basic description. The natural framework to perform this is the quark model, where the N and Δ are different configurations of the same three quarks.

GAMOW TELLER NUCLEAR QUARK SUM RULES[23]

We take the simplest quark model, the constituent quark one where the axial current couples non-relativistically to the spin and isospin of the constituent quarks with a unit coupling constant

$$(A^3)^{\pm}(\vec{x}) = \sum_{\lambda} \sigma_{\lambda}^3 \, \tau_{\lambda}^{\pm} \, \delta(\vec{x} - \vec{x}_{\lambda}) \; . \tag{8}$$

This operator acts on the initial nucleus, which consists of 3A quarks. We do not assume that these quarks are free, but that they are instead strongly correlated due to the organization of the nucleus in nucleons and Δ. This idea is to find a sum rule which is independent of this structure.

We take as before the difference Σ of the total GT strengths for the β^- and β^+ branches, but this time for the operator (8). Performing closure we obtain

$$\Sigma = \Sigma^- - \Sigma^+ = N - Z \; . \tag{9}$$

As before the sum depends only on the neutron excess of the nucleus, but it is reduced as compared to the classical sum rule by the factor g_A^2, which should be taken in the quark model ($g_A = 5/3$), $g_A^2 = 25/9$. What is the origin of this difference? It is clear; we are not talking of the same sum as before, $\Sigma \neq S$. When we apply closure the intermediate states automatically incorporate all possible excitations. Since the quark operator can transform a N into a Δ, the intermediate states include the excitations of a real Δ, the high-energy part of the GT strength which was not incorporated in our previous quantity S.

The different contributions to the sum are illustrated by the example of a neutron as initial state. With the $\sigma\tau^+$ operator one can reach either a proton or a Δ^+. With $\sigma\tau^-$ one can reach a Δ^{--}. The contribution to Σ^- from the proton intermediate state is clearly $g_A^2 = (25/9)$, since the matrix corresponding element is that of the axial current between neutron and proton. The various contributions are given below:

$$\Sigma^- = \frac{25}{9} \text{ (p)} + \frac{8}{9} (\Delta^+) = \frac{33}{9}$$

$$\Sigma^+ = \frac{8}{3} (\Delta^{--})$$

$$\Sigma = \Sigma^- - \Sigma^+ = 1 .$$

The two branches are clearly necessary to saturate the sum. In a nucleus, if there is no interaction between nucleons and Δ, and with total Pauli blocking, the respective contributions are the previous ones multiplied by N - Z. When the interaction between N and Δ is switched on the respective contributions are modified. The nucleon sector is quenched by the LL effect, but this is compensated by a corresponding modification of the high-energy sector in such a way as to respect the sum rule.

With our quark sum rule we have thus linked the low-energy and high-energy part of the GT strength. This result applies only to the difference of the β^- and β^+ branches. A measurement of the GT strength at high energy would allow a check of this sum rule. This is not feasible at present, but we can make the following observation: the new sum rule is reduced by a factor of 25/9 as compared to the classical one. This happens to be close to the quenching factor of the classical sum observed in the (p,n) reactions. It appears then that the low-energy nucleon sector by itself nearly saturates the quark sum rule. This suggests a blocking of the Δ sector. The mixing forces seem to be such that the high-energy parts of the $\beta^{\mp}$ branches acquire more or less equal strength so as to approximately cancel in the quark sum rule. At present we have no possibility to check this point, but in some way we can come close to it, in the following sense.

The constituent quark model is rather crude, since it predicts a wrong value for g_A, 5/3 instead of 1.25. So the first question is: Can we believe in a sum rule based on this model? One could try to take a better quark model, but the result would still be model-dependent. What one would like to find instead is a model-independent sum rule. Such a sum rule exists in fact. It is the nuclear Adler-Weisberger (AW) relation which is based on the algebra of charges. The commutation relation of two axial charges follows from the quark model, but it has a validity which extends well beyond this model. While I do not want to enter into details, the AW relation also provides a link between the low- and high-energy axial strengths. This link is not along the line $\vec{q} = 0$ as our previous quark sum rule, but it is along the soft pion line where the energy and momentum are linked by the relation $|\vec{q}|^2 = \omega^2$. In the nuclear case the AW relation is written[24]:

$$N - Z = (N - Z) (g_A)^2_{\text{eff}} + \frac{f_\pi^2}{\pi} \int \frac{d\omega}{\omega} \left[\sigma_0^+(\omega) - \sigma_0^-(\omega) \right] . \qquad (10)$$

Here f_π is the pion decay constant, $\sigma_0^\pm$ are the total cross-sections for soft pions ($\pi^\pm$) on the nucleus; $(g_A)_{eff}^2$ $(N - Z)$ is the total GT strength at low energy [the difference of the two branches, the β^- branch being the quantity measured experimentally in the (p,n) reaction].

The pion cross-sections in the AW sum rule are not physical. They refer to soft pions, but this is not a long way off shell and we believe that the extrapolation from the physical cross-sections can be performed reliably. The dispersive integral is dominated by the Δ_{33} resonance. If only the Δ is retained and its width neglected, then the dispersive term is proportional to the square of $f_{\pi N\Delta}$, the $\pi N\Delta$ coupling constant. Through the Goldberger-Treiman relation the dispersive term is then related to the axial strength leading to the Δ. The superiority of the nuclear AW relation is that it takes into account the width of the Δ and the contribution of the other resonances in a model-independent way. The physics however, is basically the same as that of our quark sum rule.

In addition, the AW sum rule involves (nearly) measurable cross-sections and has therefore a predictive power about $(g_A)_{eff}$, which our quark sum rule did not have.

Can we understand on these new grounds the quenching of the low-energy sector and the blocking of the high-energy one? It is remarkable that this last effect comes very naturally in this description. The Δ region is dominant in the dispersive integral. In that region the cross-sections which are large are also strongly shadowed. In particular, the difference $\sigma^- - \sigma^+$ does not increase linearly with the neutron excess. There is an A-dependent quenching factor, which reaches zero in heavy nuclei[25-27]. We believe that the same quenching factor should hold for soft pions at least in the Δ resonance region. This quenching has implications for the low-energy sector, due to the existence of the AW sum rule.

In a heavy nucleus, if we suppress completely the dispersive integral on the argument of total shadowing, then we obtain the result that the low-energy strength by itself saturates the sum rule, while it represents 156% of the sum in the nucleon case (the dispersive contribution is negative). Therefore, in the nucleus $g_{A eff}^2$ is reduced, reaching the value 1: $g_{A eff}^2/g_A^2 = 0.64$. If one assumes total shadowing in the Δ region only and no shadow above, one obtains $g_{A eff}^2/g_A^2 = 0.48$. In all cases the AW relation can account for the observed quenching which comes naturally in this description, and was the subject of an early prediction[24]. Subsequently a detailed investigation of the experimental data then available, by Wilkinson[28], indicated the existence of a quenching.

In the AW relation the quenching is obviously A-dependent since there is no shadow in very light nuclei. What is not yet clear is the link between the shadow and the microscopic description of the LL effect. Maybe if we understand this link, we can obtain an independent determination of the force which mixes NN^{-1} and ΔN^{-1} states.

MISSING STRENGTH: WHERE HAS IT GONE?

Let me now come to another question, which comes up naturally in relation to the missing strength. The question is: Where has the strength gone? A statement is commonly made that the strength has gone in the high-energy (Δ-resonance) region. This statement is misleading, since it evokes, at least to me, the following picture. When the interaction between NN^{-1} and ΔN^{-1} is switched on, the low-energy sector is quenched, but the high-energy one should be enhanced accordingly to conserve the total strength. This is indeed what we had found with our quark sum rule, but it applied only to the difference of the β^- and β^+ branches. In each branch separately there is no conservation of strength. Let me illustrate this by the example of symmetric nuclear matter. Here you can see immediately that the mixing with NN^{-1} states cannot influence the high energy strength for the following reason. The quantity $\pi^N(\omega)$ which determines this influence in the RPA expression vanishes in the high-energy limit since its crossing properties are such that

$$\pi^N(\omega) \propto \frac{1}{\omega + \omega_{exc}} - \frac{1}{\omega - \omega_{exc}} \,,$$

where ω_{exc} is a typical nuclear excitation energy. In the Δ region, which is well above the nuclear energies $\omega_\Delta \gg \omega_{exc}$, π^N becomes very small. We have here an example where the strength missing at low energy is not at all compensated at high energy and is totally lost. With a neutron excess there may be some compensation but there is absolutely no reason for a strict conservation, since it is well known to be destroyed by the RPA correlations.

This feature is illustrated in Fig. 7 which displays the strength $S_T(q)$ in the nuclear sector of the response (including the collective state) as a function of the momentum in the free case and in the ring approximation, with and without Δ. Even when the Δ excitation is absent, i.e. when all the axial strength is concentrated in the low-energy nuclear sector, there is a quenching of the strength due to the RPA correlations. The quenching effect is enhanced by the introduction of the Δ(LL effect).

We will now follow the evolution of the response when the momentum increases.

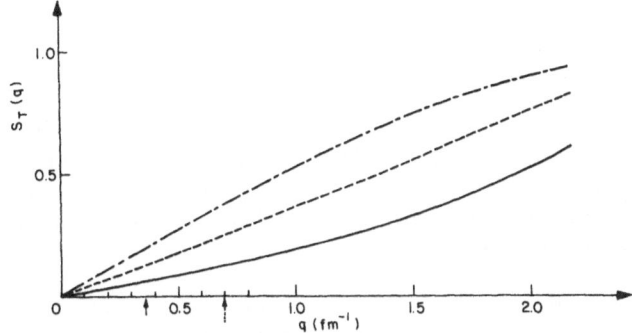

Fig. 7. Strength $S_T(q)$ in the transverse channel for a free Fermi gas (dot-dashed line) and in the ring approximation, without (dotted line) and with (continuous line) Δ. The arrows indicate the momentum at which the collective state disappears. From Ref. 29.

RESPONSE AT MODERATE MOMENTA (0.3 fm^{-1} $\lesssim$ q $\lesssim$ 2 fm^{-1})

We have now to distinguish between the two channels, since the corresponding forces are quite different.

Transverse channel[29,30]

The force remains appreciably repulsive. The collective features of the response displayed in Fig. 8 are still pronounced. There is no longer a collective state in the strict sense of condition (3). It has dissolved in the p.h. continuum. But it survives in the form of a hardening of the response[29-31]. The high-energy side of the response is enhanced, while the low-energy states are depopulated. In addition, the response is quenched by the repulsive interaction, as was shown in Fig. 7.

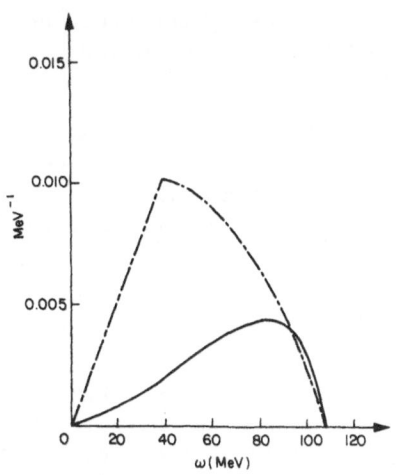

Fig. 8. The imaginary part of the response function of infinite nuclear matter for the transverse channel at $q = 1.3$ fm^{-1} as a function of ω for a free Fermi gas (dot-dashed line) and in the ring approximation (continuous line). From Ref. 29.

We have investigated whether this collective character of the response appears in the experimental data or not. We have explored in particular the inelastic electron scattering process which provides the isovector transverse spin response. Indeed the separation between the longitudinal (charge) and the transverse (magnetic) responses has been performed for ^{12}C [32] and ^{56}Fe [33]. For the magnetic part the coupling is $\vec{\sigma} \times \vec{q}$, a transverse one. It is essentially isovector due to the nearly perfect cancellation between proton and neutron magnetic moments.

Figures 9a and 9b show the experimental cross-section on ^{56}Fe as a function of the energy transfer for two values of the momentum $q = 1.45$ fm^{-1} and 1.65 fm^{-1}. The curves are the theoretical predictions. The upper one corresponds to the case of a free Fermi gas with $k_F = 1.15$ fm^{-1} and the lower one is our ring approximation. The agreement between the last curve and the experiments is satisfactory, apart from the region of high excitation where our description totally fails. The experimental data show a filling of the no man's land between the low-energy nuclear sector and the Δ sector. This filling is believed[34,35] to be due to the 2p.2h. contribution not incorporated in our description. At low energy, where the omission of the 2p.2h. contribution is less crucial, our description reproduces well the slope at the origin (as well as the position of the maximum).

Although encouraging, this agreement should not be taken as conclusive evidence for the collective character of the response. Our theory applies to infinite nuclear matter and in addition it neglects the antisymmetrization and Hartree-Fock effects. It is quite possible that a single-particle description which incorporates these properly accounts for the experimental points, without collectivity. In this case we would have to understand why the collective features, which are so pronounced at low momentum, have already faded away at $q \approx$ ≈ 1 fm^{-1}.

In order to settle this question we suggest using the proposed experimental signature: the contrast between the transverse and longitudinal responses which we will discuss shortly.

Another place where we believe[12] that collectivity may show up is the μ-capture process that we are investigating again at present. The total rate is dominated by the axial contribution. The momentum transfer is small, $q \approx 0.5$ fm^{-1}, and corresponds to the strongly repulsive region of the p.h. force. The quenching of the axial strength should influence the total rate.

Response in the longitudinal channel[36]

Below 1 fm^{-1} the longitudinal force is still repulsive, but less so than the transverse one. It produces some quenching and hardening of the response. Beyond $q \geq 1$ fm^{-1} the force turns attractive. It

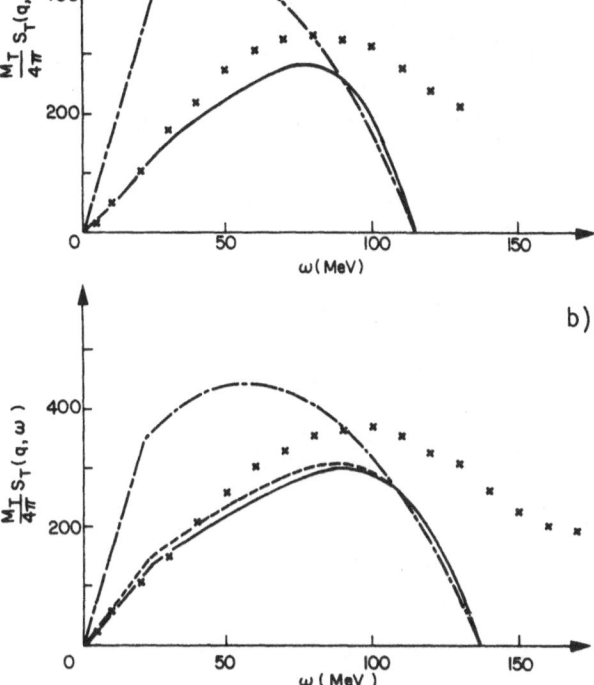

Fig. 9. The separated transverse magnetic response in ^{56}Fe at q = = 1.45 fm^{-1} (Fig. 9a) and 1.65 fm^{-1} (Fig. 9b) as a function of ω. The dot-dashed line refers to the free Fermi gas response with k_F = = 1.15 fm^{-1}. The continuous line is the ring approximation. From Ref. 29.

is only mildly attractive owing to the strong repulsion of the Migdal force. We will discuss the collective features associated with this attraction.

It had been suggested[37] that the attraction can produce a collective state, called a pion condensate. If it exists, the attractive force pulls it below the p.h. continuum, which starts itself at ω = 0. The collective state thus has to occur at ω = 0 (Im π(q,0) = 0. This makes it of a different nature from the zero sound. It corresponds to a phase transition, a reordering of the system.

The condition for its realization would be

$$1 - V(q,0)\pi^0(q,0) = 0 . \tag{11}$$

However, this condition cannot be met at the ordinary density, for the force is not sufficiently attractive. One has to increase $|\pi^0|$, i.e. the density (the estimates for the critical density are $\rho_c \gtrsim 3\rho_0$).

Suppose that we are at ρ_c. The condition (11) for pion condensation is fulfilled for a value q_c of the momentum. Then at $q = q_c$ the single-particle states are depopulated in favour of the collective one at $\omega = 0$. If the density is decreased below ρ_c the collective state disappears but it survives in the form of a softening of the response: the high-energy states are depopulated in favour of the low-energy ones as shown in Fig. 10. Here the value of g' has been set at $g' = 0.7$, for which the softening is mild. A lower value, such as $g' = 0.6$, considerably enhances the collective character of the response and the softening effect.

In addition the strength $S_L(q)$ is enhanced in the vicinity of q_c by the attractive force as shown in Fig. 11.

These two effects, enhancement and softening, are the precursors of pion condensation. At the critical density the softening would be total (the response becoming singular at $\omega = 0$). In addition the enhancement becomes infinite (critical opalescence[38,39]). This feature characterizes the occurrence of a second-order phase transitions. A normal collective state instead, the zero sound, does not affect the total strength. When it disappears nothing special happens as was shown in Fig. 7.

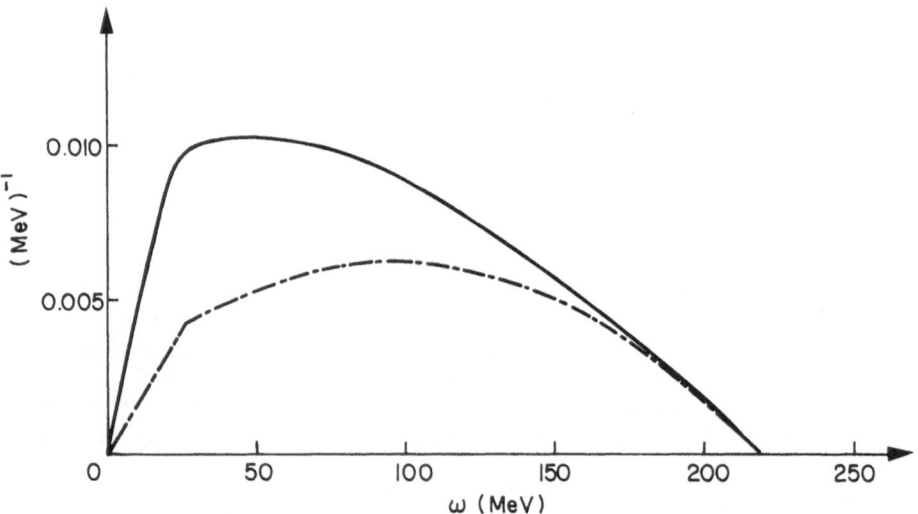

Fig. 10.   The imaginary part of the response function of infinite nuclear matter for the longitudinal channel at $q = 2.15$ fm^{-1} (critical value of the momentum) as a function of ω for a free Fermi gas (dot-dashed line) and in the ring approximation (continuous line). The value of g' is taken to be 0.7. From Ref. 36.

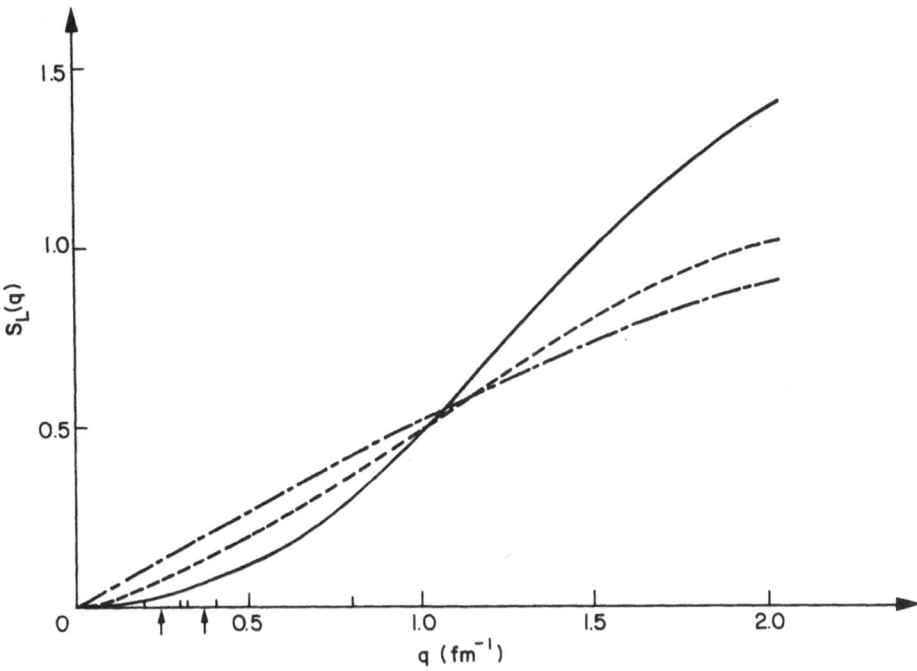

Fig. 11. Strength $S_L(q)$ in the longitudinal channel for a free Fermi
gas (dot-dashed line) and in the ring approximation, without (dotted line) and with (continuous line) Δ. The arrows
indicate the momentum at which the collective state disappears. From Ref. 29.

Between 1 and 2 fm^{-1} the contrast between the longitudinal and
transverse responses is striking, owing to the different character of
the forces. The spin longitudinal response functions are not yet
available experimentally. The difficulty is to find probes which explore the longitudinal spin density. The precursors have been searched
for only in discrete low-lying states. Softening in this case corresponds to the fact that states of unnatural parity 0^-, 1^+, etc., are
pulled down. This has been suggested and investigated from the very
beginning of the history of pion condensation[40]. However, no state
has been observed to be pulled down, at best they are at their unperturbed position. As for the critical opalescence it has been
searched for in the form factor of 1^+ transitions, with (p,p') and
(p,n) scattering[41-44]. The effect, if it is there, is mild (a factor
of two enhancement).

These negative findings should not discourage the exploration
of the longitudinal spin channel. The fact that no clear precursors
are seen in discrete levels does not exclude the possibility of a
detectable effect in the response function (see Fig. 10). Even in
the case where the softening and enhancement of the longitudinal re-

sponse would be too weak to be detectable by themselves the longitudinal response will serve as a contrast to the transverse one, as is shown in Fig. 12. An experimental display of this sharp contrast would provide indisputable evidence for the influence of the p.h. force and hence of the collective character of the response. This opportunity should be exploited. There are still exciting developments ahead of us in this field.

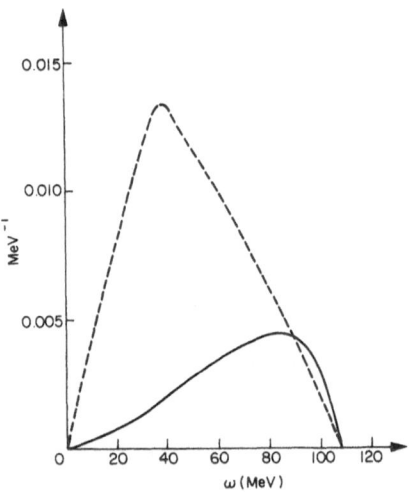

Fig. 12. Contrast between the imaginary parts of the response functions in the ring approximation at $q = 1.3$ fm^{-1} for the longitudinal channel (dotted line) and the transverse one (continuous line). From Ref. 29.

ACKNOWLEDGEMENT

Many of the ideas exposed in this talk have been developed jointly with Drs. W. Alberico, J. Delorme, A. Figureau, P. Guichon and Prof. A. Molinari. It is a great pleasure to thank them for their very stimulating collaboration.

REFERENCES

1. J. Speth, V. Klemt, J. Wambach and G. E. Brown, Nucl. Phys.
    A 343:382 (1980).
2. J. Meyer Ter Vehn, Symposium on Relativistic Heavy Ion Research,
    G.S.I. Darmstadt, 1978; Phys. Rep. 74:323 (1981).
3. G. E. Brown, S. O. Bäckmann, E. Oset and W. Weise, Nucl. Phys.
    A 286:191 (1976).
4. W. H. Dickhoff, A. Faessler, J. Meyer Ter Vehn and H. Müther,
    Phys. Rev. C23:1154 (1981).
5. K. Ikeda, S. Fujii and J. J. Fujita, Phys. Lett. 3:271 (1963).
6. D. J. Horen, C. D. Goodman, C. C. Foster, C. A. Goulding,
    M. B. Greenfield, J. Rapaport, D. E. Bainum, E. Sugarbaker,
    T. G. Masterson, F. Petrovich and W. G. Love, Phys. Lett.
    95B:27 (1980).
7. See the talk of C. Goodman at this conference.
8. See the talk of G. Crawley at this conference.
9. See the talk of C. Gaarde, J. S. Larsen and J. Rapaport at this
    conference.
10. W. Knupfer, R. Frey, A. Friebel, W. Mettner, D. Meuer, R. Richter,
    E. Spamer and O. Titze, Phys. Lett. 77B:367 (1978).
    A. Richter, Nucl. Phys. A 374:177C (1982).
11. M. Ericson, A. Figureau and C. Thévenet, Phys. Lett. 45B:19
    (1973).
12. J. Delorme, M. Ericson, A. Figureau and C. Thévenet, Ann. Phys.
    (USA) 102:273 (1976).
13. S. Barshay, G. E. Brown and M. Rho, Phys. Rev. Lett. 32:787
    (1974).
14. M. Rho, Nucl. Phys. A 231:493 (1974).
15. K. Ohta and M. Wakamatsu, Nucl. Phys. A 234:445 (1974).
16. E. Oset and M. Rho, Phys. Rev. Lett. 42:47 (1979).
17. I. S. Towner and F. C. Khanna, Nucl. Phys. A 372:331 (1981).
18. N. C. Mukhopadhyay, H. Toki and W. Weise, Phys. Lett. 84B:35
    (1979).
19. W. Knupfer, M. Dillig and A. Richter, Phys. Lett. 95B:349 (1980).
20. A. Bohr and B. M. Mottelson, Phys. Lett. 100B:10 (1981).
21. G. E. Brown and M. Rho, Nucl. Phys. A 372:397 (1981).
22. See the talk of A. Arima at this conference.
23. J. Delorme, M. Ericson and P. Guichon, Phys. Lett. 115B:86
    (1982).
24. M. Ericson, Ann. Phys. (USA) 63:p. 562 (1971).
25. C. Wilkin, C. R. Cox, J. J. Domingo, K. Gabathuler, E. Pedroni,
    J. Rohlin, P. Schwaller and N. W. Tanner, Nucl. Phys.
    B 62:61 (1973).
26. A. S. Clough, G. K. Turner, B. W. Allardyce, C. J. Batty,
    D. J. Baugh, W. J. McDonald, R. A. J. Riddle, L. H. Watson,
    M. E. Cage, G. J. Pyle and G. T. A. Squier, Phys. Lett.
    B 76:15 (1974).
27. J. Delorme, A. Figureau and N. Giraud, preprint Lycen, no. 8212,
    1982.

28. D. H. Wilkinson, Phys. Rev. C7:930 (1973).
29. W. Alberico, M. Ericson and A. Molinari, Nucl. Phys. A 379:429
     (1982).
30. W. Alberico, M. Ericson and A. Molinari, preprint CERN TH-3261
     (1982).
31. M. Kohno, Nucl. Phys. A 366:p. 245 (1981).
32. P. Barreau, M. Bernheim, M. Brussel, G. P. Capitani, J. Duclos,
     J. M. Finn, S. Frullani, F. Garibaldi, D. Isabelle, E. Jans,
     J. Morgenstern, J. Mougey, D. Royer, B. Saghai, E. de Sanctis,
     I. Sick, D. Tarnowski, S. Turck-Chieze and P. D. Zimmerman
     in: Perspectives in Electro- and Photo-nuclear physics,
     A. Gérard and C. Samour, ed., Saclay 1980, Nucl. Phys.
     A 358:287c (1981).
33. J. S. McCarthy, Nucl. Phys. A 335:27 (1980).
     R. M. Altemus, Thesis, Dept, of Physics, University of Virginia
     (1980).
34. J. M. Laget in: "Perspectives in Electro- and Photo-nuclear
     Physics, A. Gérard and C. Samour, ed., Saclay 1980, Nucl.
     Phys. A 358:275c (1981).
35. T. W. Donnelly and J. W. Van Orden, Ann. Phys. (USA) 131:451
     (1981).
36. W. M. Alberico, M. Ericson and A. Molinari, Phys. Lett. 92B:153
     (1980).
37. A. B. Migdal, Zh. Eksp. Teor. Fiz. 61:2210 (1971) [English trans.
     Sov. Phys. - JETP 34:1185 (1972)].
     F. Calogero, in: "The Nuclear Many-Body Problem", Proc. Int.
     Symposium on Present Status and Novel Developments in the
     Nuclear Many-Body Problem, Rome, 1972. F. Calogero and
     C. Ciofi degli Atti, ed. Editrice Compositori, Bologna
     (1972-73), Vol. 2, p. 535.
     F. Calogero and F. Palumbo, Lett. Nuovo Cimento 6:663 (1973).
     R. F. Sawyer, Phys. Rev. Lett, 29:382 (1972).
     D. J. Scalapino, Phys. Rev. Lett. 29:386 (1972).
     For a review, see A. B. Migdal, Revs. Mod. Phys. 50:107 (1978).
38. M. Gyulassi and W. Greiner, Ann. Phys. (USA) 109:485 (1977).
39. M. Ericson and J. Delorme, Phys. Lett. 76B:182 (1978).
40. S. Barshay and G. E. Brown, Phys. Lett. 47B:107 (1973).
41. J. R. Comfort, S. M. Austin, P. T. Debevec, G. L. Moake,
     R. W. Finlay and W. G. Love, Phys. Rev. C21:2147 (1980).
42. M. Haji-Saeid, C. Glashausser, G. Igo, W. Cornelius, M. Gazzaly,
     F. Irom, J. McClelland, J. M. Moss, G. Pauletta,
     H. A. Thiessen and C. A. Whitten Jr., Phys. Rev. Lett. 45:880
     (1980).
43. See the talk of J. Speth, S. Krewald and F. Osterfeld at this
     conference.
44. See the talk of W. Weise and A. Härting at this conference.

THE SEARCH FOR M1 STRENGTH

R.S. Hicks and G.A. Peterson

Department of Physics and Astronomy
University of Massachusetts
Amherst, MA, U.S.A. 01003

ABSTRACT

Current knowledge of M1 transition strength in nuclei is reviewed by studying selected examples. Attention is focused primarily on inelastic electron scattering, but information obtained using other techniques is also discussed. It appears that the utility of (e,e') as a spectroscopic tool for determining M1 strength is mainly restricted to nuclei with A<100. For nuclei below A≈40, the total measured M1 strength is in good accord with detailed shell model estimates, however, heavier nuclei show a strength deficit in comparison with model predictions.

INTRODUCTION

Arguments[1-3] usually advanced for studying M1 excitations in nuclei have been recently strengthened by speculations regarding the relevance of delta isobar-nucleon hole configurations. In the past, low momentum transfer electron scattering has been found to be an effective spectroscopic tool for the identification and measurement of M1 strength. The technique, pioneered by Barber[4] and his collaborators in the early sixties has now been exploited at a number of laboratories to study M1 excitations in some fifty nuclei. In the course of this development, the technique has come to transcend its traditional spectroscopic role, and the value of studying the M1 form factors at higher momentum transfers is now recognized. Since the (e,e') form factor is a simple Bessel transform of the transition density, measurements over an extended range of momentum transfers can provide rich insights into the nuclear dynamics of M1 excitations.

A tabulation of the nuclei studied by (e,e') shows that M1 strength has only been identified in three nuclei heavier than A=60: ^{88}Sr, ^{90}Zr, and ^{208}Pb. An important question to address concerns the failure to find the M1 excitation strength predicted in heavy nuclei. Is this the result of fundamental nuclear structure properties, or could it reflect the experimental limitations inherent in the (e,e') technique? Accordingly a brief critique of the experimental procedures will first be presented. We will then proceed to a discussion of M1 transitions in selected nuclei. In the case of medium and heavy nuclei, the (e,e') results occasionally appear to be at variance with other experiments, notably inelastic proton scattering at very forward angles which is reputed to have high sensitivity to M1 excitations. An example of such a discrepancy will be explored. Comprehensive reviews of the field have been given previously by Fagg[2] and Richter.[3]

DETERMINATION OF MULTIPOLE CHARACTER AND REDUCED TRANSITION PROBABILITY

In the plane-wave Born approximation, the (e,e') form factor for a magnetic transition of multipolarity L is defined in terms of the measured differential cross section by[5]

$$\frac{d(k,\theta)}{d\Omega} = \left[\frac{Z\alpha\cos\theta/2}{2k\sin^2\theta/2}\right]^2 \left[1 + \frac{2k}{M_N}\sin^2\theta/2\right]^{-1} F_{ML}^2(q),$$

where k is the incident electron momentum, θ is the scattering angle, and M_N is the nuclear mass. The form factor in turn is related to the reduced transition probability by

$$F_{ML}^2(q) \sim q^{2L} B(ML,q)\uparrow \tag{1}$$

For the case of small momentum transfer q, a convenient expansion is[5]

$$\left[\frac{B(ML,q)\uparrow}{B(ML,0)\uparrow}\right]^{1/2} = 1 - \frac{L+3}{L+1}\frac{q^2 \langle R_L^2 \rangle}{2(2L+3)} + \frac{L+5}{L+1}\frac{q^4 \langle R_L^4 \rangle}{8(2L+3)(2L+5)} \ldots \tag{2}$$

where the "effective transition radii" are defined in terms of the magnetic transition density $\rho_L(r)$ by

$$\langle R_L^\nu \rangle \sim \frac{\int \rho_L(r) r^{L+\nu} r^2 dr}{\int \rho_L(r) r^L r^2 dr}$$

Thus $\langle R_L^\nu \rangle$ varies from transition to transition, according to details of nuclear structure. Most existing information pertains to the $\langle R_1^2 \rangle^{1/2}$ radius determined for M1 transitions. The value of this radius is usually found to be close to 1.1 $A^{1/3}$ fm.[5]

When data are obtained at sufficiently low-momentum transfers, Eqs. (1) and (2) allow for a relatively model-independent determination of transition multipolarity and ground state radiation width

$$\Gamma_\gamma^o \sim E_x^{2L+1} \, B(ML, q = E_x/c)\! \uparrow \,\cong\, E_x^{2L+1} \, B(ML, q=0).$$

The usual procedure, as practiced by the Darmstadt group, for example, is to assume a definite multipolarity L and then plot Coulomb-distortion-corrected values of the deduced $B(ML,q)^{1/2} \sim q^{-L} F_{ML}(q)$ against q^2. If the multipolarity is correctly identified, then such a plot will show $B(ML,q)^{1/2}$ decreasing approximately linearly with q^2, as dictated by Eq. (2) for $q \to 0$. An erroneous multipole assignment will not only result in the $B(ML,q)^{1/2}$ versus q^2 plot having a conspicuous curvature,[3] but usually will also give values for Γ_γ^o and $<R_L^2>$ that are clearly unrealistic.

The $B(ML,q=0)\!\uparrow$ value is obtained by extrapolation of the (e,e') measurements to low-q. For this to be a reliable procedure we require the size of the q^4 term on the right hand side of Eq. (2) to be small compared to the difference of the two leading terms. In order to make some quantitative estimate, assume that this term should add less than 10% of the combined contribution of the lower-order terms. Making the further assumption that $<R_1^2>^{1/2} = <R_1^4>^{1/4} = 1.1 \, A^{1/3}$ fm, one finds that data are required such that $q^2 < 2A^{-2/3}$ if M1 strength is to be accurately determined. For ^{40}Ca, ^{90}Zr, and ^{208}Pb, respectively, measurements must be made at momentum transfers below 0.40, 0.30, and 0.25 fm^{-1}. Consideration of Coulomb distortion effects lowers these limits further, particularly for high-Z nuclei.

In practice, this largely model-independent technique is found to work well for nuclei below $A \simeq 60$. The main demands placed upon the experimental measurements are to resolve the states of interest, and to acquire sufficient statistics. However, for nuclei beyond $A \simeq 60$, data cannot be obtained at sufficiently low q values to permit the analysis described above. The principal difficulty here is the large "radiation tail" due to elastically scattered electrons which have lost energy by bremsstrahlung. For these nuclei it is necessary to determine transition strengths and multipolarities with the aid of a nuclear model. The technique reduces to the comparison of the measured data with a theoretical form factor, calculated in distorted-wave Born approximation from the transition density prescribed by the model. In the region of Zr^{90}, the results that emerge are relatively independent of the particular choice of model. However, for the case of ^{208}Pb, where the available data do not extend low enough to span the first form factor diffraction maximum, a strong model-dependence is observed. This particular example will be studied in more detail later on.

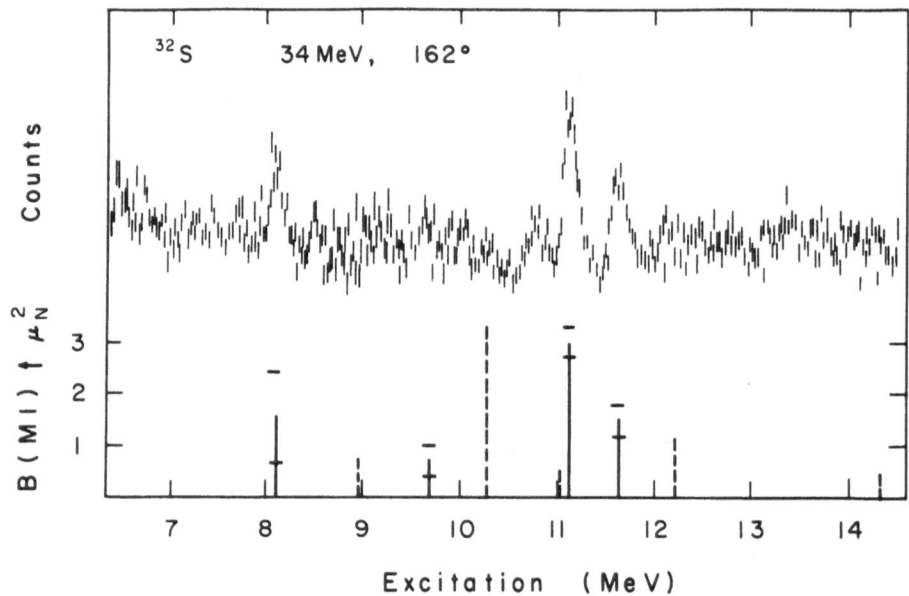

Fig. 1. ^{32}S(e,e') spectrum.[6] Derived B(M1)↑ values are indicated
          by the solid vertical lines below, with the associated
          errors represented by horizontal bars. The dashed lines
          show the strengths predicted by shell model calculations.[8]

SELECTED EXAMPLES

^{32}S and the s-d Shell

     Figure 1 shows a recent measurement by Burt et al.[6] of electrons
inelastically scattered at 162° from ^{32}S. The conspicuous peaks at
8.11, 11.12, and 11.63 MeV have been identified with M1 transitions,
and it now appears that the M1 strength in ^{32}S is more concentrated
than previously thought.[2] Thus ^{32}S displays the same feature
exhibited by other s-d shell and p-shell nuclei: the tendency for
the observed M1 strength to be concentrated into the lowest few
levels.[2]

     The summed B(M1,0)↑ strength for even-even self-conjugate nuclei
of the s-d shell is plotted in Fig. 2. As one proceeds from ^{16}O to
^{28}Si a systematic increase is observed in the B(M1,0)↑ value. Beyond
^{32}S, the measured strength decreases. The M1 cross sections observed
in these nuclei are mainly attributed to $1d_{5/2} \rightarrow 1d_{3/2}$ transitions,
and are thus expected to increase with increasing occupancy of the
$d_{5/2}$ shell. The reduction in strength beyond ^{32}S is due to blocking
by nucleons in the $d_{3/2}$ shell. In the simplest extreme single-

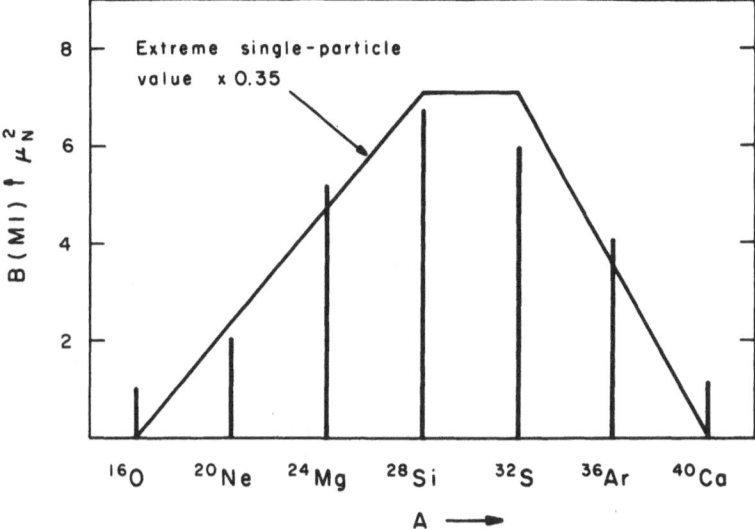

Fig. 2.   Total M1 strength measured for 4N nuclei in sd-shell, and
          comparison with extreme single-particle prediction.

particle model, zero M1 strength is expected in the closed shell
nuclei ^{16}O and ^{40}Ca. The observation of measurable M1 strength in
these nuclei attests to the presence of ground state correlations.
The envelope in Fig. 2 shows that, except in ^{16}O and ^{40}Ca, the
total observed M1 strength is approximately 35% of the prediction
of the extreme single-particle model.

     In view of the theme of this conference, it is relevant to
note that both spin-flip and non-spin-flip terms can contribute to
the cross sections observed in backward-angle (e,e') measurements.
Figure 3 indicates how, at low momentum transfers, the purely convective contributions interfere destructively with the $\Delta T = 1$ spin-
flip form factor calculated for ^{32}S. It is a general feature of
oscillator model calculations that the M1 transition strength between spin-orbit partners is reduced by 20% at low q due to orbital
convection currents.

     More general shell model calculations will also include other
M1 single-particle matrix elements. Within the s-d shell additional
contributions come from the $(d_{3/2}, s_{1/2})$, $(d_{5/2})^2$, $(d_{3/2})^2$, and
$(s_{1/2})^2$ terms. The q-dependences of these separate matrix elements
can vary markedly from that of the $(d_{3/2}, d_{5/2})$ term, as shown in
Fig. 3 for the $(d_{3/2}, s_{1/2})$ transition. Since $\Delta \ell = 2$, this latter
transition is mediated only by the spin-flip part of the M1 operator,
and the change in the nodal quantum number results in a strong
suppression of the $(d_{3/2}, s_{1/2})$ M1 strength at low q. Evidence for

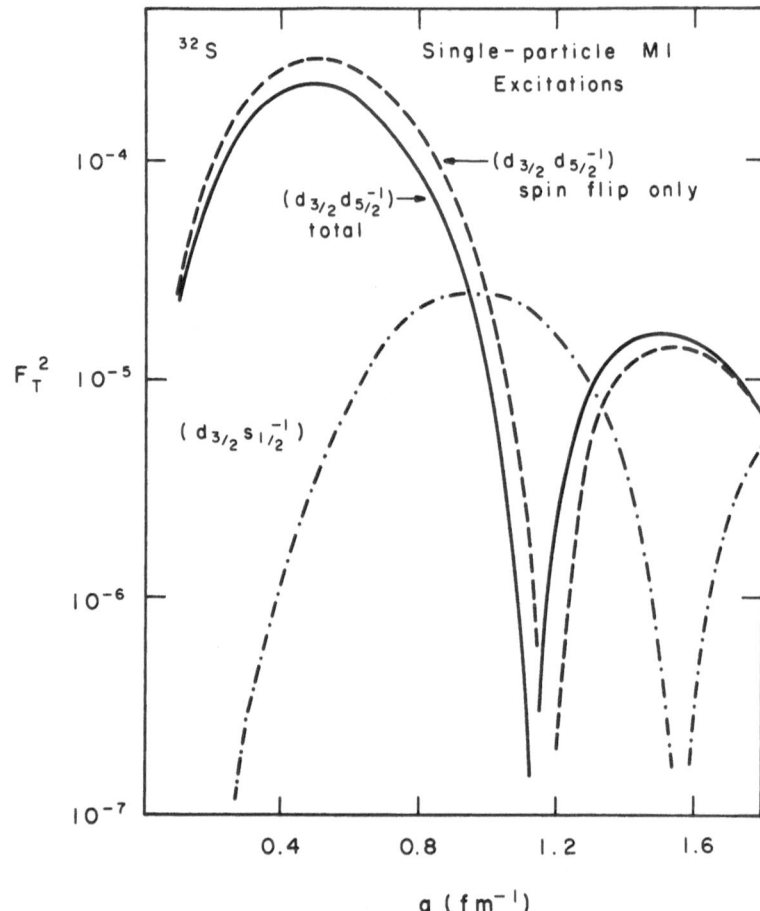

Fig. 3. Calculated (e,e') form factors for single-particle M1
         excitations in sd-shell. Harmonic oscillator radial wave
         functions were used with b = 1.82 fm.

transitions dominated by this so-called "ℓ-forbidden" matrix element
has been found[7] in ^{39}K and ^{41}K.

The spectrum of 1^+ states in ^{32}S has been calculated by Dubach[8]
in the complete sd-shell space using Kuo-Brown interactions.
Figure 1 shows that, although the excitation energies of the calculated states agree only approximately with the observed values, the
measured and predicted M1 strengths are of comparable magnitude. In
fact, the total M1 strength of 5.4 μ_N^2 calculated by Dubach lies in
excellent agreement with the measured 5.9 ± 0.9 μ_N^2. What is
important to note is that to reach this level of agreement one needs

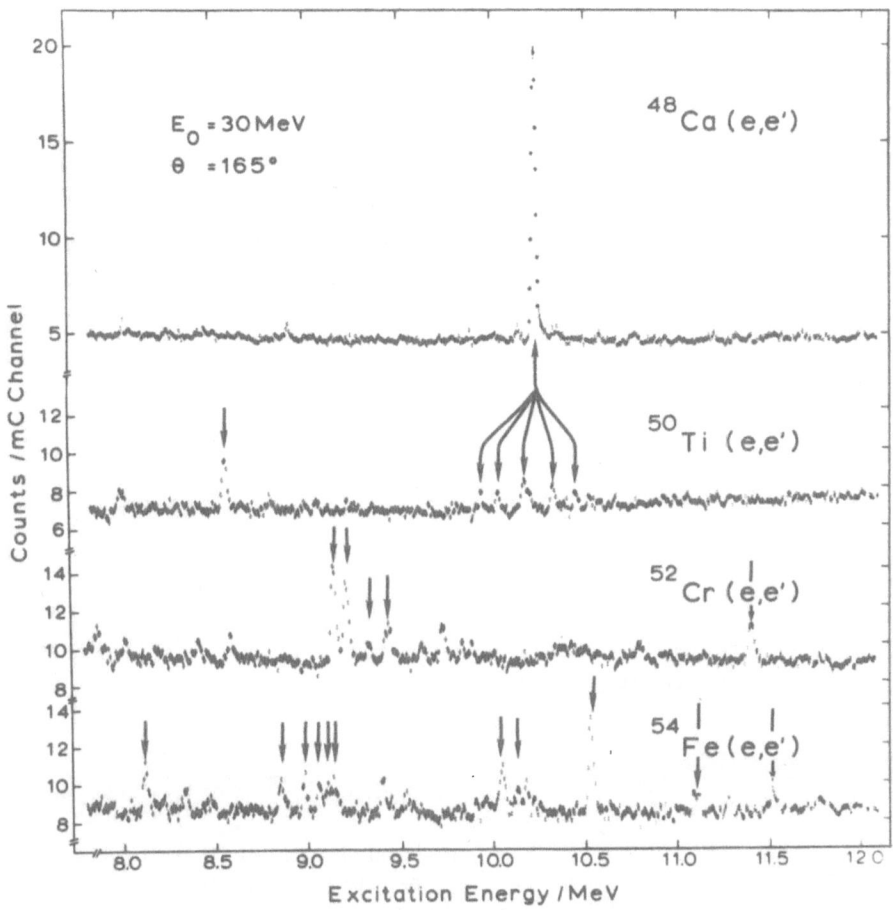

Fig. 4. Measured (e,e') spectra for N=28 isotones.[3] The arrows
denote M1 excitations.

only to treat the structure of ^{32}S on a realistic footing. When
this is done, the appreciable mixing of the sd-shell configurations
results in a sizeable reduction of the total expected M1 strength
from the high value (20.2 μ_N^2) predicted by the extreme single-
particle model.

^{48}Ca and Neighboring Nuclei

Recently, an (e,e') search for M1 transitions in $f_{7/2}$-shell
nuclei was undertaken at Darmstadt.[9] According to the simplest
model, the M1 strength of these nuclei would be mainly attributed
to $f_{7/2} \rightarrow f_{5/2}$ excitations. To date, the most striking result

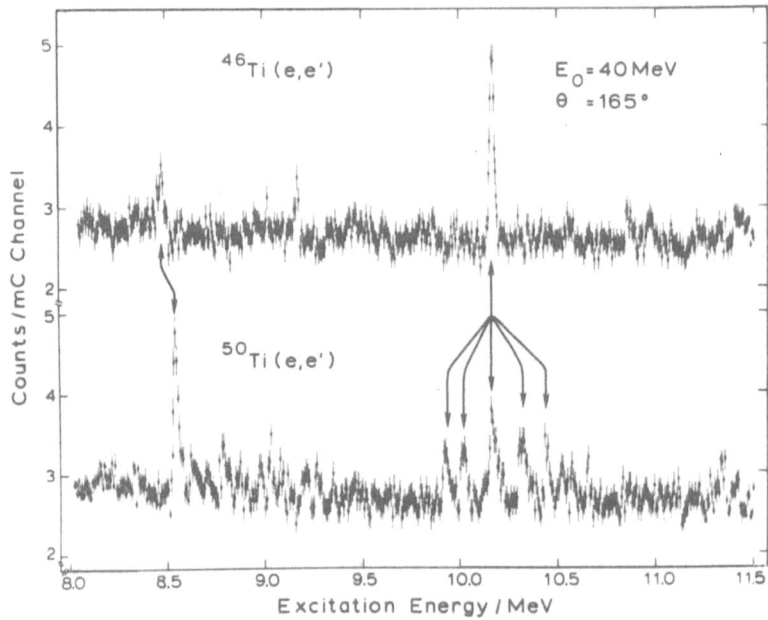

Fig. 5.   Inelastic (e,e') spectra for ^{46}Ti and ^{50}Ti, with M1 excitations indicated.[3]

has been the discovery[9] of a strong and isolated M1 transition at 10.23 MeV in ^{48}Ca, one of the best examples of a shell-model nucleus, with close to the limit of 8 $f_{7/2}$ neutrons and with protons well-confined in the closed-shell, A=40 core. The (e,e') spectrum indicated at the top of Fig. 4 appears to support this simple picture.

The addition of two protons to form ^{50}Ti leads to the result also shown in Fig. 4, where the M1 strength is fragmented into a tight cluster of states carrying about 75% of the strength observed in the 10.23 MeV transition of ^{48}Ca. The evenness of the splitting and distribution of strength is remarkable. An additional M1 peak appears at 8.56 MeV in ^{50}Ti. This is identified with the two $f_{7/2}$ protons, whereas the 10.2 MeV strength is associated with neutron excitations.

When four neutrons are omitted to form ^{46}Ti, the spectrum again takes on a very simple appearance with only two strong peaks being observed, as shown in Fig. 5. As in the case of ^{50}Ti, the lower peak is identified with proton excitations and the higher-lying strength with neutron promotions. Figure 4 shows the results of a systematic study of the four stable N=28 isotones. All spectra were measured at Darmstadt under the same kinematic conditions. The arrows indicate peaks of M1 character. The T=1 and T=2 isotones are seen to have a more complex 1^+ level structure than the T=3 and T=4 isotones. Presumably, this additional fragmentation arises largely from ground state correlations, which are known to play a more important role in the T=1 and T=2 nuclei.

In fact, the measured B(ML,0) strengths may already provide evidence for the existence of ground state correlations in ^{48}Ca, the nucleus which is assumed to have the most tightly closed shell structure. For example, the B(M1)↑ value of $4.0 \pm 0.3 \; \mu_N^2$ observed for the 10.23 MeV peak represents only one third of the $12 \; \mu_N^2$ given by the extreme single-particle model. The Darmstadt group have found a further $1.2 \; \mu_N^2$ distributed among states seen only weakly in the ^{48}Ca spectrum.[10] With a detection sensitivity of about $0.1 \; \mu_N^2$, any remaining strength would have to be very extensively fragmented in order to escape experimental observation. A reduction in the expected M1 strength is also seen in the low momentum transfer scattering of 160 MeV protons,[11] as well as in the related (p,n) reaction[12] to Gamow-Teller resonances in ^{48}Sc.

Open shell calculations by McGrory and Wildenthal[13] reduce the predicted M1 strength to about $8.6 \; \mu_N^2$. However, this is still 65% larger than the experimental results. Two recent calculations which include Δ_{33} isobar-nucleon hole configurations in addition to the conventional particle-hole pairs, give M1 strengths very close to the observed $5.2 \; \mu_N^2$. The first, by Harting et al.,[10] employs an effective operator formalism and finds a total M1 strength of about $5.6 \; \mu_N^2$. The second calculation[14] includes particle-hole and Δ-hole configurations in $6\hbar\omega$ random-phase wave functions, and yields a predicted M1 strength of $5.3 \; \mu_N^2$.

That Δ-hole configurations may play such an important role, even in relatively low energy processes, is an idea that warrants thorough investigation. The mass region around A=48 would appear a felicitous choice for such studies. For example, the comparatively simple systematics observed for the M1 strength distribution suggests that certain nuclear structure aspects may be more straightforward than is usual. In addition, high multipole M7 and M8 transitions have also been observed.[15,16] The description of these transitions in terms of customary models is very restricted, and so the opportunity exists to explore the multipole dependence of the proposed Δ-hole quenching of magnetic excitation strength. Finally, the A=48 region is one that can be conveniently studied by many different reactions, enabling one to extract more detailed information by exploiting the different sensitivities of the various probes.

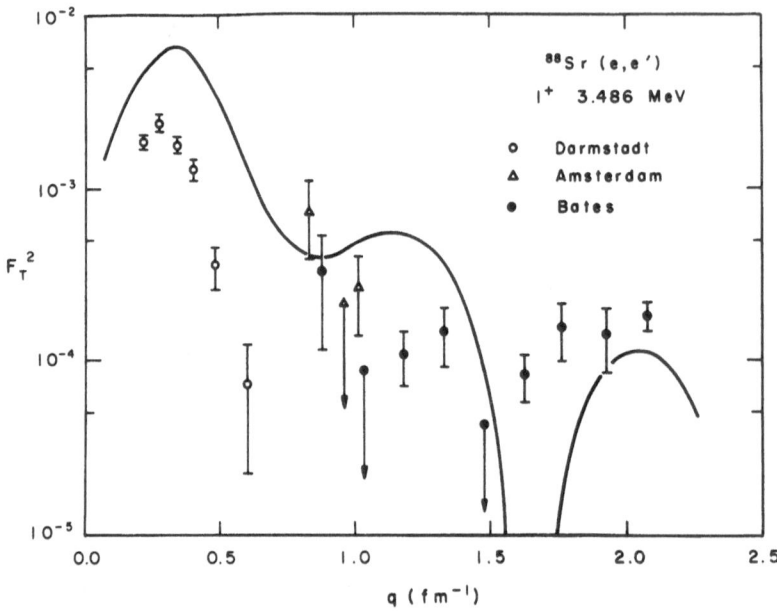

Fig. 6. Measured M1 form factor[17] for the 3.486 MeV state in ^{88}Sr, and comparison with BCS 'broken-pair' calculation.[18]

^{88}Sr and ^{90}Zr

Among heavier nuclei, the bound 1^+ level at 3.486 MeV ^{88}Sr provides the only unequivocal example of M1 strength observed in (e,e'). Blok and collaborators[17] have measured the corresponding form factor up to $q = 2.1$ fm^{-1}, as shown in Fig. 6. BCS 'broken-pair' calculations suggest that the structure of this state is determined by an almost pure $2p_{3/2} \rightarrow 2p_{1/2}$ proton excitation, with a small $g_{9/2} \rightarrow g_{7/2}$ neutron admixture.[18] The B(M1)↑ value predicted by this calculation exceeds by about a factor of 2 the 0.92 ± 0.15 μ_N^2 measured in γ-ray fluorescence. In keeping with this result, the (e,e') form factor measured at low-q lies well below the BCS prediction. One means of reducing the theoretical prediction is to apply a 'quenching' factor to the nucleon spin-g factor g_s used in the calculations. However, Blok et al. have shown that this procedure leads a result which does not reproduce the q-dependence observed above the first diffraction maximum. Thus, to account for the data by this method, it is necessary that any reduction of g_s inside the nucleus be q-dependent.

Recently, Weise and Härting[19] have attributed the quenching of the isovector part of g_s to the virtual excitation of Δ-hole components. An evaluation of this mechanism leads to a curve which

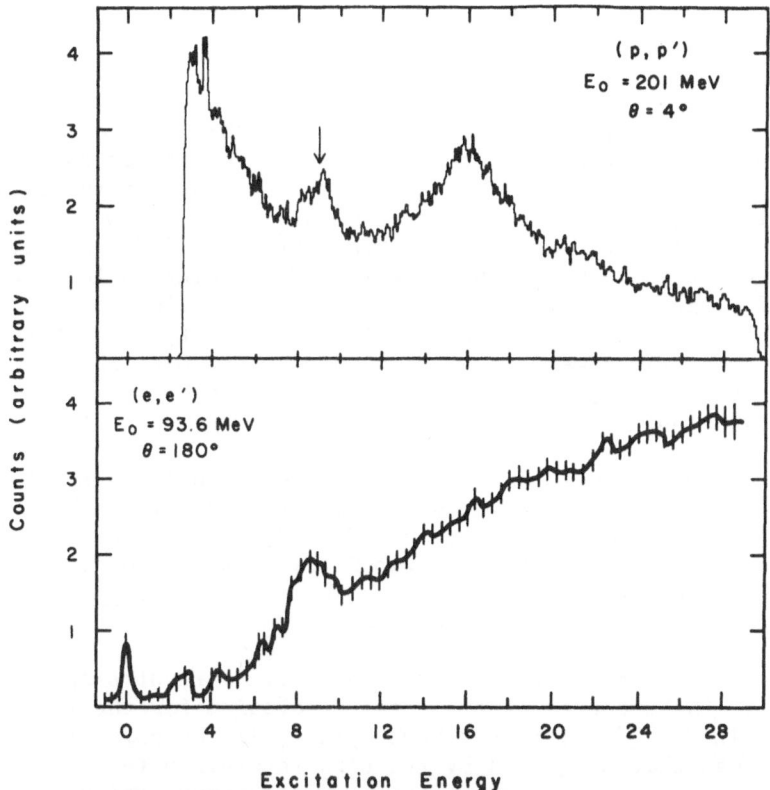

Fig. 7. Comparison of forward-angle (p,p') and 180° (e,e')
spectra[20,22] measured for ^{90}Zr.

faithfully reproduces the high-q diffraction pattern seen in the
(e,e') measurements. The uniqueness of this interpretation is of
course unknown. Whether or not the data can also be more con-
ventionally described by improving the standard nuclear structure
treatment remains to be seen.

    In recent years several inelastic scattering experiments have
been conducted to search for M1 strength in ^{90}Zr. For example,
forward-angle (p,p') measurements by Anantaraman et al.[20] and
Horen et al.[21] both reveal a broad peak at 8.9 MeV with a width of
1.7 ± 0.2 MeV. This resonance follows an L=0 angular distribution
and has been identified as being of M1 character. In Fig. 7, the
4° spectrum of Anantaraman et al. is compared with a 180° (e,e')
measurement by Ogino and coworkers.[22] The striking similarity of
the broad 9 MeV structure observed in the two spectra suggests that
the same states are being excited, however, the strength shown in

the (e,e') spectrum is almost certainly not of M1 character, since the momentum transfer is unfavorable for M1 excitation.

The nature of the strength observed in the (e,e') spectrum has been revealed by fine resolution (≈ 30 keV FWHM) electron scattering measurements performed at Darmstadt.[23] The results of this work, shown in Fig. 8, indicate that the level structure in the 8–10 MeV excitation region is complex and fragmented. It turns out that the bulk of the strength observed in this region is of M2 character. By searching for peaks which increase with decreasing incident electron energy E_o, three $J^\pi = 1^+$ states have been identified at 8.233, 9.000, and 9.371 MeV. There is some additional indication of further very weak M1 strength, however, the total dipole strength in the region is less than 2.5 μ_N^2, i.e., less than 1/4 of the 11 μ_N^2 predicted by RPA calculations.[24]

The Darmstadt work carries several valuable lessons. Firstly, to convincingly determine the distribution of M1 strength in ^{90}Zr, very high resolution, probably <5 keV FWHM, is required, Experimental techniques which suffer from intrinsic resolutions worse than a few tens of keV need to have extremely high multipole selectivity to safeguard against possible M2 contamination.

At the present time the (p,p') and (e,e') results appear to conflict. The measured experimental resolution for the (p,p') spectrum in Fig. 7 was 80 keV FWHM. Thus, if the only M1 strength measured was that suggested by the high resolution (e,e') experiment, then one would expect to see fragmented structure near 9 MeV, not a single broad peak. Secondly, the M1 strength measured in (p,p') exceeds the (e,e') estimate by a factor of two or more,[25] although practical and theoretical difficulties[25] presently hinder the accurate determination of B(M1) values from forward-angle (p,p').

The 7.48 MeV Peak in ^{208}Pb

In the simplest shell model of ^{208}Pb, only two low-lying states can be constructed:

$$\psi_1 = A\,\pi\,(h_{9/2},\,h_{11/2}^{-1}) + \sqrt{1 - A^2}\;\nu(i_{11/2},\,i_{13/2}^{-1})$$

$$\psi_2 = \sqrt{1 - A^2}\;\pi(h_{9/2}, h_{11/2}^{-1}) - A\nu(i_{11/2}, i_{13/2}^{-1}),$$

where the parameter A is a mixing parameter. The first-state, of isoscalar character, is predicted to be weak. It is sometimes identified with one member of a postulated doublet of 4.843 MeV. For now, however, attention is focussed on the second state. This is commonly considered to exist as a doorway state for the

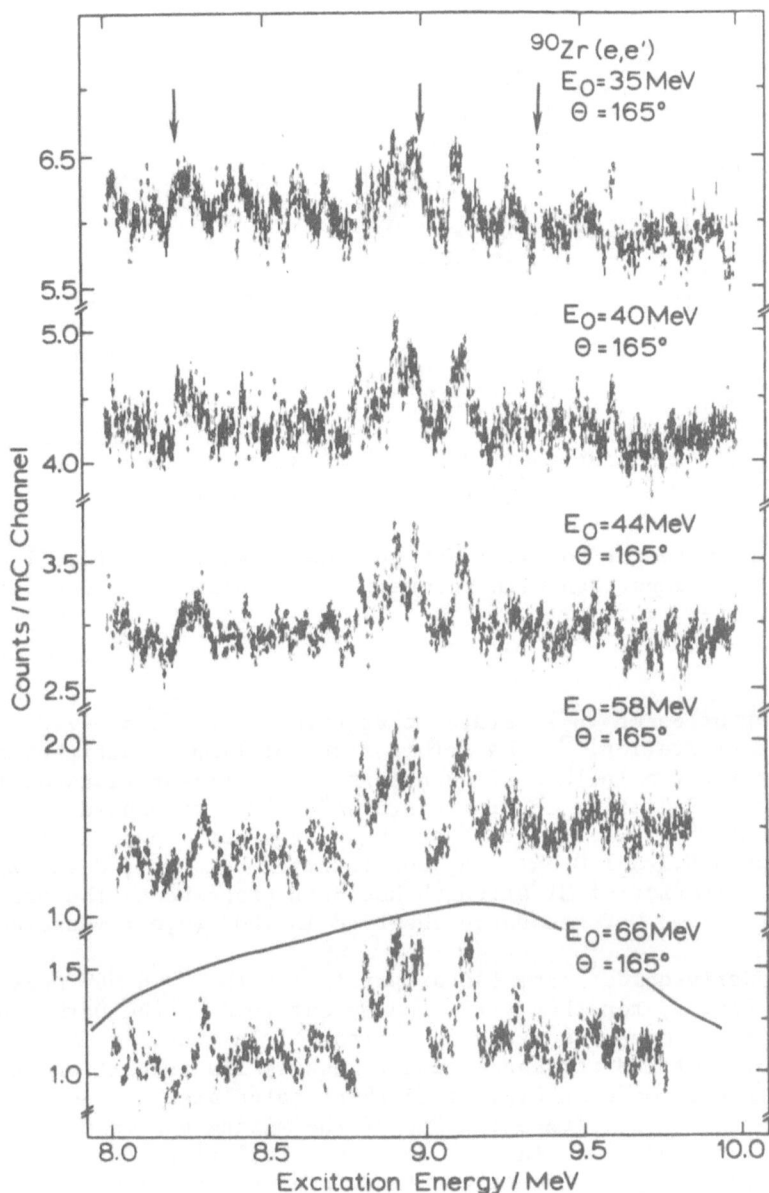

Fig. 8. Fine resolution Darmstadt data,[23] with M1 excitation
strength arrowed. The curve towards the bottom of the
figure indicates the envelope of the peak measured in the
(p,p') experiment of Anantaraman et al.[20]

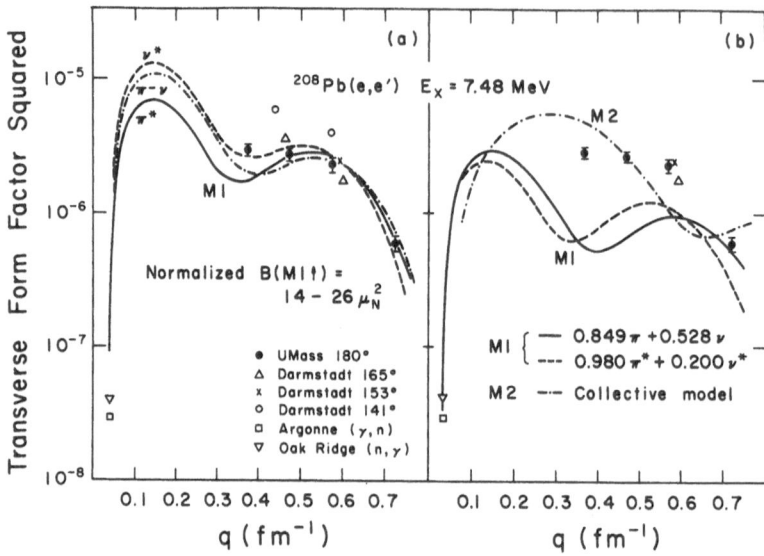

Fig. 9.   Form factor for 7.48 MeV peak observed in 180°(e,e') data,[29] and comparison with theoretical calculations. For details, refer to text.

excitation of 35 known 1$^+$ states clustered in the 7.25 – 7.82 MeV region of excitation.[26]   The strong fragmentation is attributed to mixing of the two 1p-1h configurations with configurations of the 2p-2h type.   Radiative neutron capture[26,27] and threshold (γ, n) measurements[28] find a total M1 strength of about 7.5 μ_N^2 in the region, with 5.2 ± 1.0 μ_N^2 being concentrated between 7.40 and 7.55 MeV. This concentration of M1 strength has been proposed as the possible source of a 7.48 MeV resonance observed in 180° (e,e') measurements.[29]

The derived 180° form factor points for the 7.48 MeV peak are shown in Fig. 9, together with results obtained by the Darmstadt group at more forward scattering angles.[30]   In order to confirm the M1 character of the resonance, and to assess its transition strength, a comparison is made in Fig. 9a to three calculated curves, each corresponding to a different value of the mixing parameter in Ψ_2, A=1, A=2$^{-1/2}$, and A=0.   The A=1 and A=0 calculations for the pure neutron and pure proton configurations employ effective operators[31] with quenched spin g-factors.   All three curves reproduce the general trend of the 180° data, yet the corresponding B(M1)↑ values obtained by normalization range from 14 to 26 μ_N^2, i.e., 3 to 5 times larger than value given by the (n,γ) and (γ,n) measurements.   Fig. 9b shows that even the most favorable calculations underestimate the (e,e') data by not less than a factor of 2.5 when fixed to B(M1)↑ = 5.2 μ_N^2.

Possible reasons for this discrepancy are discussed in Ref. 29. For example, the (e,e') measurements may be inflated by transverse electric strength. A foremost observation, however, is that the existing electron data appear to be confined to a momentum transfer region lying well beyond the first diffraction maximum in the M1 form factor. Since this is a region where the form factor displays appreciable sensitivity to nuclear structure details, the hope that the multipole character and transition strength might be reliably determined seems overly optimistic.

It is not unthinkable that the (e,e') data may be entirely consistent with the (n,γ) and (γ,n) results. To visualize this consistency, one may, for example, assume a correlated ^{208}Pb ground state containing 2p–2h high -ℓ configurations.[32] The consequent angular momentum recoupling terms give rise to large orbital convection currents which enhance the (e,e') form factor in the vicinity of the available data. Near the q = 7.5 MeV/c photon point, however, strongly destructive interference between the orbital and spin currents leads to the diminished B(M1)↑ value measured in (n,γ) and (γ,n) experiments. What relationship such conjectures bear with reality remains to be seen.

It seems clear that, at its present stage of development, (e,e') does not represent a propitious tool for searching for M1 strength in the high level density regions of nuclei with A ≥ 100. If, however, 1^+ states can be identified by other means, extensive measurements of the (e,e') form factors impose strong constraints on possible model descriptions. An example of the utility of such endeavors is provided by the 3.487 MeV 1^+ state of ^{88}Sr, discussed earlier.

TOTAL M1 STRENGTH

The accompanying table compares the total M1 strength measured in various nuclei to theoretical predictions. Except for the ^{208}Pb results, the measured values are taken from (e,e') experiments. Theoretical estimates are based on detailed shell-model calculations which, in most cases, include multi-ℏω configurations. Reference is specifically constrained to calculations that employ bare nucleon charges and angular momentum g-factors.

The table shows that, providing the nuclear structure is realistically treated, good agreement is reached for A ≤ 40 nuclei. However, heavier nuclei show a systematic strength deficit in comparison to the theoretical predictions. In order to lessen this discrepancy it has become customary to reduce the theoretical results by introducing some mechanism that will directly or indirectly lead to a quenching of the nucleon magnetic moment in the nuclear environment. Configuration mixing with Δ-hole components represents one model for this effect.

Table I. Theoretically-predicted and Observed Total
M1 Strength in Various Nuclei

Nucleus	Theory	Observed	%Theory	Refs.
^{12}C	$2.47 \ \mu_N^2$	$2.79 \ \mu_N^2$	113	2,8
^{16}O	0.8	1.0 ± 0.1	125	3,33
^{28}Si	6.5	6.8 ± 0.3	105	3,34
^{32}S	5.4	5.9 ± 0.9	109	6,8
^{40}Ca	1.7	1.55 ± 0.1	91	3,35,36
^{48}Ca	8.6	5.2	61	10,13
^{58}Ni	8.5	5.7 ± 1.4	67	37,38
^{90}Zr	11	1.8 ± 0.4	16	23,24
^{208}Pb	30	9.6	32	23,39,40

Before the problem of the missing strength can be conclusively
answered, other possibilities must be addressed. For example, how
much fragmented M1 strength lies undetected in the inelastic continuum of heavy nuclei? One might also question the veracity of
current shell model descriptions for these nuclei.

ACKNOWLEDGMENTS

        For their unstinting support in the preparation of this con-
tribution, special acknowledgment is due to A. Richter and J.
Dubach. We are also indebted to H. Blok, L.W. Fagg, W.J. Gerace,
and R.A. Lindgren. Financial support was provided by the U.S.
Department of Energy.

REFERENCES

1.   S.S. Hanna, Lecture Notes in Physics 61, 275 (1977).
2.   L.W. Fagg, Rev. Mod. Phys. 47, 683 (1975).
3.   A.Richter, Nucl. Phys. A374, 177c (1982); 1980 Summer School on
         Nucl. Structure, Dronten, The Netherlands.
4.   W.C. Barber, Ann. Rev. Nucl. Sci. 12, 1 (1962).
5.   H. Uberall, Electron Scattering from Complex Nuclei (Academic,
         New York, 1971).

6.   P.E. Burt et al. private communication.
7.   Th. Grundey, A. Richter, G. Schrieder, E. Spamer, and W. Stock, Nucl. Phys. A357, 269 (1981).
8.   J. Dubach, private communication.
9.   W. Steffen, H.D. Graf, W. Gross, D. Meuer, A. Richter, E. Spamer, O. Titze, and W. Knupfer, Phys. Lett. 95B, 23 (1980).
10. A. Harting, W. Weise, H. Toki, and A. Richter, Phys. Lett. 104B, 261 (1981).
11. K. Rehm, P. Kienle, D.W. Miller, R.E. Segel, and J.R. Comfort, Phys. Lett. 114B, 15 (1982).
12. B.D. Anderson, J.N. Knudson, P.C. Tandy, J.W. Watson, R. Madey, and C.C. Foster, Phys. Rev. Lett. 45, 699 (1980).
13. J.B. McGrory and B.H. Wildenthal, Phys. Lett. 103B, 174 (1981).
14. T. Suzuki, S. Krewald, and J. Speth, Phys. Letts. 107B, 9 (1981).
15. B.D. Anderson, proceedings of this conference.
16. J. Wise et al., private communication.
17. L.T. van der Bijl, H.P. Blok, R. Frey, D. Meuer, A. Richter, and P.K.A. de Witt Huberts, Z. Physik 305, 235 (1982).
18. K. Allaart and E. Boeker, Nucl. Phys. A198, 33 (1972).
19. W. Weise, proceedings of this conference.
20. N. Anantaraman, G.M. Crawley, A. Golonsky, C. Djalali, N. Marty, M. Morlet, A. Willis, and J.-C. Jourdain, Phys. Rev. Letts. 46, 1318 (1981).
21. F.E. Bertrand, E.E. Gross, D.J. Horen, J.R. Wu, J. Tinsley, D.K. McDaniels, L.W. Swenson, and R.J. Liljestrand, Phys. Lett. 103B, 326 (1981).
22. H. Ogino and A. Hotta, private communication.
23. D. Meuer, R. Frey, D.H.H. Hoffmann, A. Richter, E. Spamer, O. Titze, and W. Knupfer, Nucl. Phys. A349, 309 (1980).
24. S. Krewald and J. Speth, Phys. Lett. 52B, 295 (1974).
25. T.N. Taddeucci, J. Rapaport, and C. Goodman, Phys. Rev. C25, 2130 (1982).
26. S. Raman, in Neutron Capture Gamma Ray Spectroscopy, ed. R.E. Chrien and W.R. Kane (Plenum, New York, 1979) p. 193.
27. S. Raman, M. Mizumoto, and R.L. Macklin, Phys. Rev. Lett. 39, 598 (1977).
28. R.J. Holt, H.E. Jackson, R.M. Laszewski, and R.H. Specht, Phys. Rev. C20, 93 (1979).
29. R.S. Hicks, R.L. Huffman, R.A. Lindgren, B. Parker, G.A. Peterson, S. Raman, and C.P. Sargent, Phys. Rev. C26, 920 (1982).
30. R. Frey, A. Richter, A. Schwierczinski, E. Spamer, O. Titze, and W. Knupfer, Phys. Lett. 74B, 45 (1978).
31. W. Knupfer, R. Frey, A. Friebel, W. Mettner, D. Meuer, A. Richter, E. Spamer, and O. Titze, Phys. Lett. 77B, 367 (1978).
32. R.A. Lindgren, private communication.
33. A. Arima and D. Strottman, Phys. Lett. 96B, 23 (1980).
34. R. Schneider, A. Richter, A. Schwierczinski, E. Spamer, O. Titze, and W. Knupfer, Nucl. Phys. A323, 13 (1979).
35. W. Gross, D. Meuer, A. Richter, E. Spamer, O. Titze, and W. Knupfer, Phys. Lett. 84B, 296 (1979).

36. P.E. Burt, L.W. Fagg, H. Crannell, D.J. Sober, W. Stapor,
         J.T. O'Brien, X.K. Maruyama, J.W. Lightbody, and R.A.
         Lindgren, Phys. Rev. C25, 2805 (1982).
37. R.A. Lindgren, W.L. Bendel, E.C. Jones, Jr., L.W. Fagg, X.K.
         Maruyama, J.W. Lightbody, and S.P. Fivozinsky, Phys. Rev.
         14, 1789 (1976).
38. J. Rapaport, T. Taddeucci, T.P. Welch, D.J. Horen, J.B.
         McGrory, C. Gaarde, J. Larsen, E. Sugarbaker, P. Koncz,
         C.C. Foster, C.D. Goodman, C.A. Goulding, and T.
         Masterson, Phys. Lett. 119B, 61 (1982).
39. U.E.P. Berg, private communication.
40. J.S. Delesa, J. Speth, and A. Faessler, Phys. Rev. Lett. 38,
         208 (1977).

CHARGE EXCHANGE REACTIONS AND SPIN MODES

C. Gaarde, J.S. Larsen and J. Rapaport[*]

Niels Bohr Institute, University of Copenhagen
DK 2100 Copenhagen, Denmark
*Permanent address: Ohio University, Athens
OH 45701, USA

1. INTRODUCTION

The recent (p,n) experiments at intermediate energy at the Indiana University Cyclotron[1-6] have led to a major breakthrough in the understanding of spin correlations in nuclei. The (p,n) reaction for bombarding energies between 100 and 200 MeV has proven to be a unique tool for the study of spin excitations in nuclei. Fig.1 shows zero degree neutron spectra for E_p=200 MeV.[5] The spectra are seen to be dominated by a single peak which is interpreted as a collective 1^+-state, the giant Gamow-Teller state, the isospin-spin nuclear sound. The collective state carries a large fraction of the total Gamow-Teller strength, i.e. strength associated with the operator $\sigma_\mu t_-$ that governs Gamow-Teller β-decay. The concentration of strength into a collective state was first discussed by K. Ikeda, S. Fujii and J.I. Fujita as early as 1963.[7] In a shell model picture of e.g. ^{208}Pb the G.T. operator has non-zero matrix elements with all 44 excess neutrons (and the 12 $h_{11/2}$ neutrons in the next lower shell). The collective state then corresponds to a transition where all these nucleons coherently change direction of spin and isospin. In a field description the giant corresponds to a vibration in $\vec{\sigma\tau}$-space. The field in which the vibration takes place is on the microscopic level generated by $\sigma\sigma\tau\tau$ particle-hole interactions.

The impressive "signal to noise" ratio as displayed in fig.1 is characteristic for (p,n) reactions at intermediate energies. The high energy part of the inclusive neutron spectra is dominated by single step processes in contrast to spectra at lower energies where the continuum, the "background", is dominated by multistep processes. For bombarding energies corresponding to a momentum

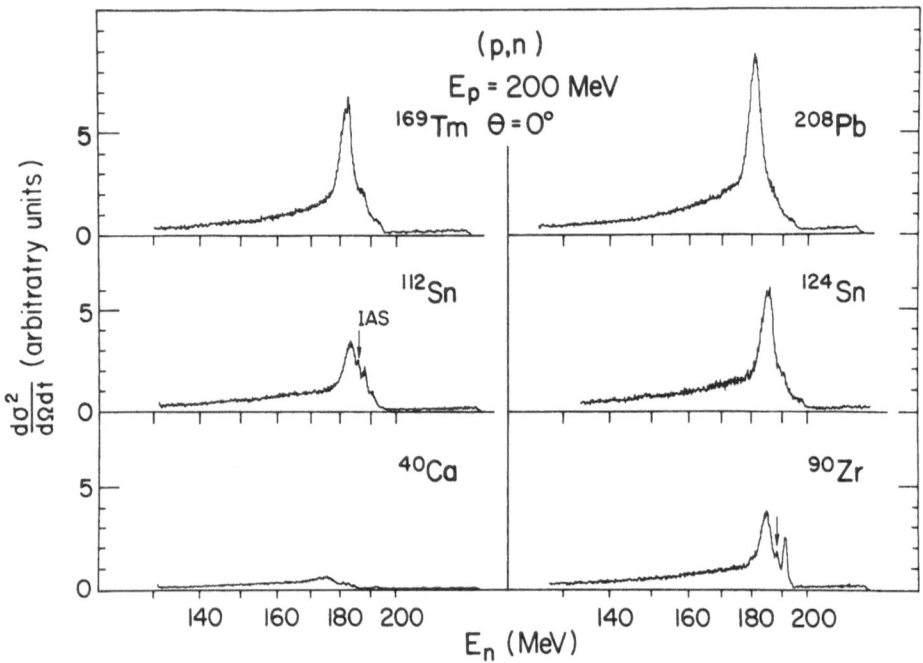

Fig. 1. Neutron time of flight spectra at 200 MeV. The spectra
         are normalized to show relative cross sections.

significantly larger than the Fermi-momentum the probability is
small for getting a nucleon out at small angles with energy (or
momentum) close to the incoming after collision with 2 (or more)
nucleons.   The high energy part of the neutron spectra is therefore a result of one step processes and can then be calculated in
microscopic models.[8] Another aspect of the selectivity is that at
intermediate energies the non spin transfer excitations are strongly suppressed whereas at energies below 50 MeV they are comparable
to the spin transfer excitations.[9]

2. EXPERIMENTAL SETUP

The data discussed are mainly coming from (p,n) experiments
at the Indiana University Cyclotron Facility, and we shall therefore briefly describe the experimental setup.

The data are obtained with the time of flight facility[10]which
consists of a beam swinger and 2 neutron detector stations. Time
compensated large volume plastic scintillators with dimensions

$15 \times 15 \times 100$ cm^3 are used as neutron detectors[11], and they are
tilted at the appropriate angle to obtain subnanosecond resolution.
At e.g. 200 MeV the tilt angle is around 10°, the 15×15 cm^2 end
is (almost) facing the target. Typical overall time resolution is
700-900 psec, and to obtain a reasonable energy resolution the
flight path has to be long. For 120 MeV and 100 m flight path an
energy resolution of 350 keV is obtained, whereas f. ex. 70 m path
and 200 MeV gives around 1 MeV resolution.

The efficiency x solid angle is typically 5×10^{-6} sr per
detector, and even with 2 or 3 detectors running in parallel at
each detector station the count rate would seem small. This is
however not the case. The large cross sections encountered in
these reactions and the use of targets of thickness 30-150 mg/cm^2
give rates that allow neutron spectra to be obtained in 10-40 min.

3. REACTION MECHANISM

The data from the (p,n) experiments consist of time of flight
spectra at angles from 0° to about 50° at several bombarding energies. Asymmetries have been obtained with polarized proton beams
at 120 and 160 MeV. We shall not discuss the polarization data
here. Targets ranging from ^{6}Li to ^{238}U have been used.

3.1. DWIA

At bombarding energies between 100-200 MeV we expect the
impulse approximation to apply. The effective interaction between
the incoming nucleon and each of the target nucleons is taken as
the free N-N t-matrix. We have specifically used a parametrization by Love and Franey[12] of the t-matrix into local potentials.
The parametrization is based on experimental phase-shift data for
N-N scattering. Exchange effects (between the incoming and the
struck nucleon) give major contributions to the cross section and
are taken into account. The interaction strengths in the different channels as parametrized by Love and Franey are then used in
DW codes like DWUCK4[13] and DWBA70[14] to calculate (p,n) cross sections for single particle transitions. In fig.2 are shown angular
distributions for a number of single particle transitions with
^{90}Zr as the target.

Such DWIA calculations show that the zero degree cross sec-
tion for $\Delta \ell = 0$ $\Delta s = 1$ transitions is proportional to the Gamow-Teller
strength for the single particle transition considered. In ^{90}Zr
e.g. $d\sigma/d\Omega(\theta=0)/B(GT)$ for the $g_{9/2} \to g_{9/2}$ and $g_{9/2} \to g_{7/2}$ transitions
differs by less than 4%. That means that a cross section unit
for GT strength can be calculated as a function of A for a given
bombarding energy, i.e. $d\sigma/d\Omega(\theta=0^\circ)/B(GT)=k(Q,A)$. The results of
such calculations are given in fig.5 for $E_p=160$ MeV. A similar

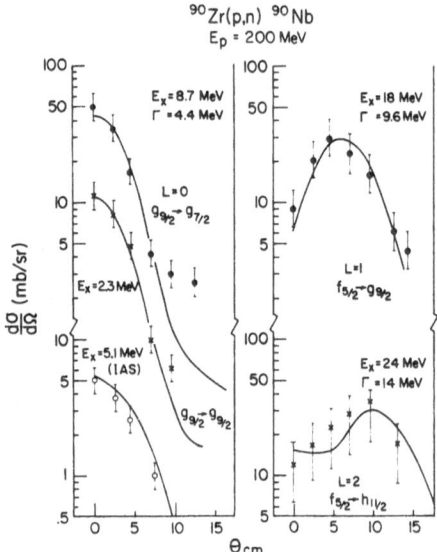

Fig. 2.   Angular distributions from the ^{90}Zr(p,n)^{90}Nb reaction.
The curves are angular distributions with arbitrary
normalization for typical single particle transitions
calculated in DWIA with interaction strength from ref.[12]

calculation at 200 MeV is described in ref.[5]

The following analysis assumes that the cross section to a
final state can be calculated from a coherent sum of amplitudes
of single particle transitions. For the (p,n) reaction to measure
the GT strength for a given transition it is therefore not only
necessary that the above mentioned condition for single transitions is fulfilled but also that the relative phase between the
different amplitudes is retained.

In the low momentum transfer limit the zero degree cross
section in the impulse approximation reduces to

$$\frac{d\sigma}{d\Omega}(\theta=0^{o}) = (\frac{\mu}{\pi\hbar^2})^2 \frac{k_f}{k_i} N_{\sigma\tau} J_{\sigma\tau}{}^2 B(GT)$$

$$= (\frac{\mu}{\pi\hbar^2})^2 \frac{k_f}{k_i} N_{\tau} J_{\tau}{}^2 B(F)$$

for GT and Fermi transitions respectively. The B-values are
defined as

$$B(GT) = \sum_{\mu} <f| \sum_{k=1}^{A} \sigma_{\mu}(k)t_{-}(k)|i>|^2$$

$$B(F) = <f| \sum_{k=1}^{A} t_{-}(k)|i>^2$$

N_{τ} and $N_{\sigma\tau}$ are distortion factors by which the cross sections are reduced due to absorption of the incoming and outgoing waves. J_{τ} and $J_{\sigma\tau}$ are volume integrals of the effective interactions.

We note that in this limit we obtain the proportionality between cross section and strength.

3.2. Energy Dependence

The ratio of cross sections for spin transfer- to non spin transfer-transitions changes dramatically between 80 and 200 MeV. In fig.3 spectra are shown for the $^{90}Zr(p,n)$ reaction at 120 and 200 MeV.

The isobaric analogue state, the 0^+ state with $\Delta\ell=0$ $\Delta S=0$, is seen to become smaller with increasing energy relative to the 1^+-states which are formed with $\Delta\ell=0$ $\Delta S=1$ transfers.

In fig.4 we have plotted experimental values for $(J_{\sigma\tau}/J_{\tau})^2$ using the formulas given above. For a number of nuclei where cross sections are measured for both GT- and Fermi-transitions and where the corresponding B-values are known from β-decay, the values for $J_{\sigma\tau}$ and J_{τ} are extracted. In the same figure is shown the calculated ratio between squares of t-matrix elements in the $\sigma\tau$- and τ-channels from the parametrization by Love and Franey. The figure shows that at 200 MeV transitions with isospin-spin ($\sigma\tau$) transfer will have one order of magnitude larger cross sections than transitions with only isospin (τ) transfer.

3.3. Normalization to β-decay

For a number of GT β-transitions throughout the periodic table we have measured the zero-degree (p,n) cross section for the corresponding transitions. In fig.5 we have plotted $d\sigma/d\Omega(\theta=0^\circ)/B(GT)$ as a function of $A^{1/3}$. We note that B(GT) for the transition in $^{42}Ca-^{42}Sc$ is 2.57 whereas the transition in $^{51}V-^{51}Cr$ has B(GT)= 0.016. The figure, however, shows that also the cross sections differ by a factor more than 100. It is the results as shown in fig.5 that lead us to conclude that the (p,n) reactions at these energies indeed measure the GT strength.

Also shown in the figure are calculated values for $d\sigma/d\Omega/B(GT)$. For ^{42}Ca e.g. the experimental numbers are $d\sigma/d\Omega(\theta=0^\circ)/B(GT)=5.9\pm1.0$

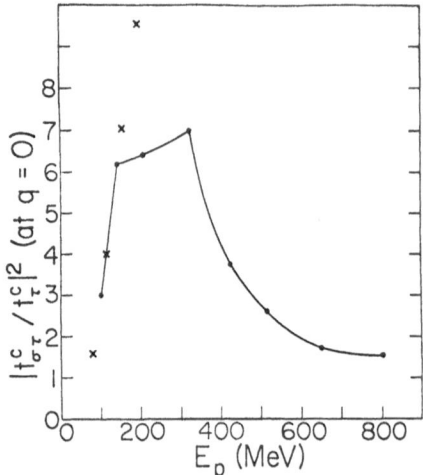

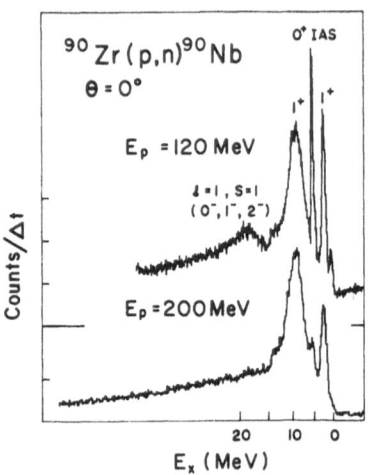

Fig. 3. Neutron spectra for ^{90}Zr at 120 and 200 MeV. The $\ell=1$ resonance is only seen at 120 MeV because the angular distribution is narrower (in angle) at 200 MeV (see figs. 2 and 17).

Fig. 4. The ratio between the squares of t-matrix elements for the central part of the interactions in the $\sigma\tau$- and τ-channels is plotted versus bombarding energy.[12] The crosses are experimental values[15] using the expression for the zero degree cross section in the q=0 limit.

mb/sr (^{42}Sc; E_x=0.61 MeV, E_p=160 MeV) whereas the DWIA calculations give 6.8 mb/sr. It is also seen that the DWIA calculations give the correct slope with $A^{1/3}$, i.e. the absorption is well described by the calculations. For the nuclei heavier than A=144 we have used the calculated cross section unit renormalized by 15% to be in accordance with the measured quantities in the region A=26-144.

In fig. 6 we have used the same data together with the formula in the q=0 limit given above to extract $J_{\sigma\tau}$. It is seen that the data are well described with a value of $J_{\sigma\tau}$=151±5 MeV fm^3; independent of A.

3.4. Fermi Strength

In fig. 7 we have plotted the zero degree cross section for a number of Fermi transitions, i.e. transitions to isobaric analogue states. The cross sections are divided by $N_\tau(k_f/k_i) \cdot (\mu/\pi\hbar^2)^2$ and plotted versus (N-Z). We see that the data seem to be described

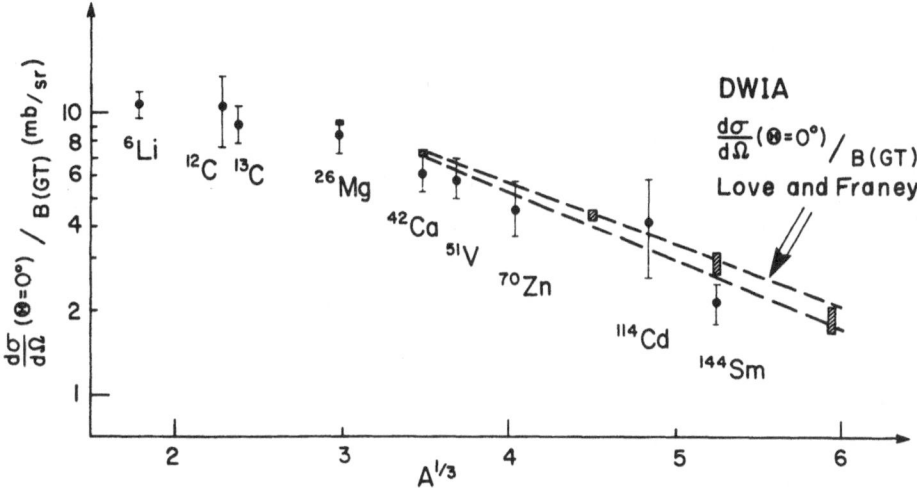

Fig. 5. For a number of GT transitions where B(GT) values are known
        from β-decay the (p,n) zero degree cross section at 160 MeV
        is divided by B(GT) and plotted versus $A^{1/3}$. Also shown is
        the same quantity calculated in DWIA with interaction
        strength as parametrized by Love and Franey. Different
        single particle transitions give slightly different values
        for $d\sigma/d\Omega(\theta=0^\circ)/B(GT)$, e.g. $f_{7/2} \to f_{7/2}$ or $i_{13/2} \to i_{11/2}$ tran-
        sitions in ^{208}Pb.

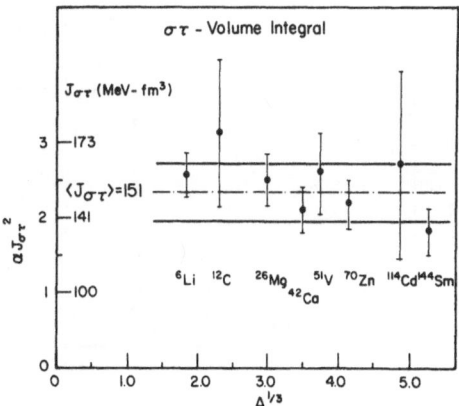

Fig. 6. The data given in fig.5 are analyzed using the expression
        for the zero degree cross section in the q=0 limit. The
        distortions $N_{\sigma\tau}$ are calculated as described in ref.[2]. The
        $J_{\sigma\tau}^2$'s are plotted versus A on a linear scale; the corre-
        sponding values for $J_{\sigma\tau}$ are indicated.

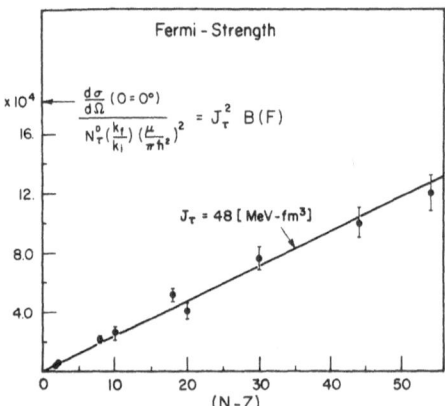

Fig. 7. The zero degree (p,n) cross sections at 160 MeV for transi-
        tions to analogue states are corrected for kinematic and
        distortion effects and plotted versus (N-Z).   The line
        corresponds to $J_T^2(N-Z)$ with J_T=48 MeV fm^3.   We note that
        ^{54}Fe is one of the targets and for this nucleus it is known
        from β-decay that B_F=(N-Z).

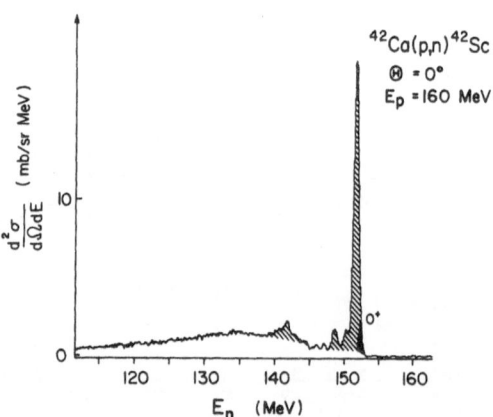

Fig. 8. Neutron spectrum from ^{42}Ca(p,n)^{42}Sc.   The strongly excited
        state is a 1^+-state (E_x=0.61 MeV) for which the B(GT) for
        the corresponding β-transition is known.   The hatched
        regions indicate other 1^+-states (or groups of states).

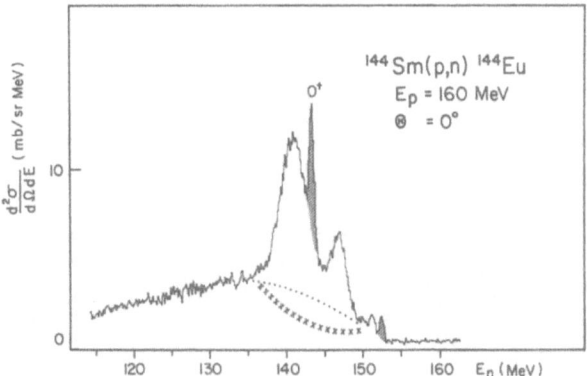

Fig. 9. Neutron spectrum from ^{144}Sm(p,n)^{144}Eu. The g.s. transition (shown hatched) has a known B(GT) value and this transition is in heaviest nucleus where such normalization to β-decay has been possible. The dotted line shows "an experimentalist's background". The crosses show a background more in line with calculations in ref.[8].

by the simple formula given above with J_T=48 MeV fm^3, independent of A. We further note that ^{54}Fe is one of the targets and for this nucleus the Fermi strength is known from β-decay to be N-Z=2. From the figure we would therefore conclude that all the Fermi strength is found in the analogue states, also for the very heavy nuclei.

4. SUMRULES

For charge exchange modes very general sumrules can be derived from commutator relations for operators involving t_+ and t_-.[16] For GT transitions we find

$$S_{\beta-}-S_{\beta+} = \sum_f B(GT,i\xrightarrow{\beta-}f) - \sum_{f'} B(GT,i\xrightarrow{\beta+}f')$$

$$= \sum_{f\mu}|<f|\sum_{k=1}^{A}\sigma_\mu(k)t_-(k)|i>|^2$$

$$- \sum_{f'\mu}|<f'|\sum_{k=1}^{A}\sigma_\mu(k)t_+(k)|i>|^2$$

$$= 3(N-Z)$$

The equation gives the difference between β_- and β_+ GT strength from an initial state $|i\rangle$ to all final states. The sumrule is model independent since it follows directly from commutators of onebody operators t_- and t_+. The only assumption is that σ_μ and $t_\pm$ are onebody operators that can only change direction of spin and isospin of nucleons.

To use this sumrule for comparison with experiment, we would have to know $S_{\beta+}$. For nuclei with a large neutron excess $S_{\beta+}$ will be small since all final states for transferring a proton into a neutron will be blocked. For e.g. ^{208}Pb described as a doubly closed shell $S_{\beta+}$ is zero.

The expression for the sumrule assumes that the nucleons are the building blocks in our system. However, inclusion of internal degrees of freedom of the nucleon changes the sumrule. The coupling to the Δ-isobars (M=1232 MeV, T=M=3/2) is very important for the isospin-spin modes.[18,19]

In the constituent quark model for the nucleon the reduced (in spin and isospin) matrix element for the $N \rightarrow \Delta$ transition is[19]

$$\langle\Delta|||\sigma\tau|||N\rangle = \frac{24}{5}\sqrt{2} \; ;$$

the corresponding one for the N–N transition is

$$\langle N|||\sigma\tau|||N\rangle = 6 \; .$$

The $p \rightarrow \Delta^{++}$ transition e.g. then corresponds to a $d_{quark} \rightarrow u_{quark}$ transition as in the $n \rightarrow p$ decay, and we could think of the Δ being the spin flip of the nucleon.

The GT-strength connected with the $N \rightarrow \Delta$ transitions is very large because all A nucleons in the nucleus can contribute, whereas only the excess neutrons are active in the $N \rightarrow N$ transitions, because of the Pauli-blocking mentioned above. In fig.10 the (unperturbed) GT-strength distribution for ^{208}Pb is shown including the Δ-isobars. In the particle-hole region the strength is 3(N–Z) = 132 and in the Δ region, 800 in the same units.

Fig.11 shows the Gamow-Teller strength for a number of heavy nuclei. The strength is obtained from the measured zero degree cross sections using the strength unit from fig.5 (corrected for Q-value effects by DWIA calculations). The strength is given for the collective state separately as well as the total strength, i.e. the low lying GT-strength is included. In certain nuclei the low lying strength is collected in a well defined peak (see figs.1, 3 and 9) whereas in others (see fig.1) it is a broader structure. The numbers for the collective state depend on how the "background"

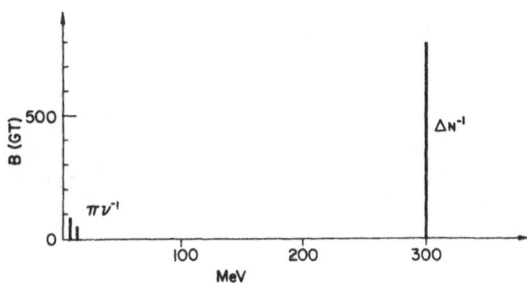

Fig. 10. GT-strength distribution for ^{208}Pb including the ΔN^{-1}
excitation. The strength in the proton-particle neutron-
hole states is collected in 2 states with strength
$3(N-Z)\cdot(1-f)$ and $3(N-Z)f$; $f=0.36$ (see text).

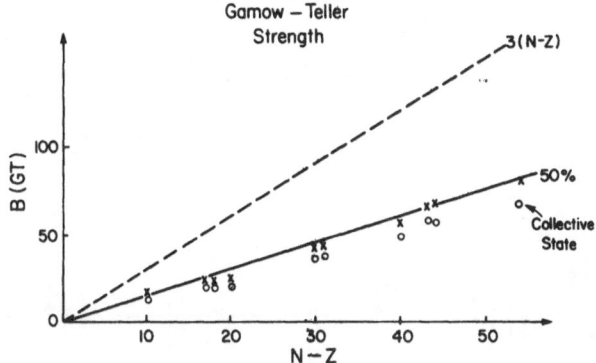

Fig. 11. The GT-strength extracted from the zero degree (p,n)
cross sections using the cross section unit (corrected
for Q-value effects) given in fig.5. The calculated
$d\sigma/d\Omega/B(GT)$(DWIA)-values (properly renormalized) are
used for inter- and extrapolations. The strength is
given for the collective state separately as well as
for the total strength. The strength given in this
figure corresponds to using a background exemplified
in fig.9 as "the experimentalist's background".

is defined. The numbers given in fig.11 are extracted using "an experimentalist's background" indicated in fig.9 for the ^{144}Sm spectrum. Errorbars are not given in the figure but the extracted data demonstrate that the analysis is performed in a systematic way. For the ^{90}Zr(p,n)^{90}Nb reaction we have rather systematic data and a detailed analysis of the background problem has led us to the following (preliminary) conclusion on the fraction of 3(N-Z) GT-strength observed for ^{90}Zr: i) (52±9)% with "experimentalist background" (dotted line in fig.9); ii) (61±10)% with a "background" based on angular distributions and in line with a calculated background[8] (crosses in fig.9); iii) (67^{+7}_{-10})% including everything under the GT-peaks.

In fig.12 is shown a summary of results for the GT-strength extracted from 160 MeV data. For the lighter nuclei the strength is often in a few peaks and the background problem is less severe. For ^{14}C,[20] ^{19}F[21] and ^{42}Ca,[22] most of the strength is in states where the B(GT) is known from β-decay, and this reduces the errorbars on the total GT-strength extracted from the (p,n) data. In all cases we have attempted to estimate $S_{\beta+}$ in order to use the sumrule.

We conclude that Gamow-Teller strength is missing in the E_x<50 MeV part of the spectra. This is consistent with a model where the internal degrees of freedom for the nucleon are involved in the isospin-spin vibrations.

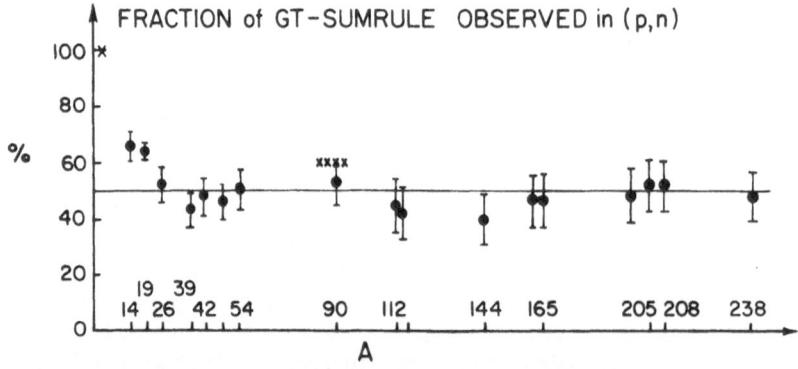

Fig. 12. Fraction of Gamow-Teller sumrule strength observed in (p,n) reactions. For the heavy nuclei the numbers correspond to fig.11. For the lighter nuclei the strength is often in a few sharp peaks. In all cases the $S_{\beta+}$ have been estimated. For ^{90}Zr we give 3 numbers for the total strength: i) 52% with "experimentalist background"; ii) 61% with calculated background; iii) 67% taking all the background under the GT-peaks as GT-strength.

5. STRUCTURE

5.1. GT-Resonance

The spectra in fig.1 show that for heavy nuclei a significant part of the GT-strength is collected in a single state, and we shall describe this state as a coherent particle-hole state. In a field description the collective state corresponds to a vibration in $\vec{\sigma}\vec{\tau}$ space. For the GT-state e.g. the vibration corresponds to a change of direction (in spin- and isospin-space) of many nucleons.

The coherent state is in such a field description generated by an oscillating average field proportional to $\vec{\sigma}\vec{\tau}$. This is equivalent to expressing the twobody interaction in separable form. Here we shall specifically take the particle-hole interaction as $V_{12} = \kappa\vec{\sigma}_1\vec{\sigma}_2\vec{\tau}_1\vec{\tau}_2$, corresponding to all radial matrix elements being equal. If we further assume a volume interaction we get a $1/A$ dependence for the coupling constant κ.

The selfconsistency condition for the oscillating potential leads to a dispersion relation for the energy of the collective state. This relation is equivalent to the RPA equation for a separable force

$$\sum_i \frac{<(\nu^{-1}\pi)_i,1^+,M=0|\sigma_o\tau_{-1})0>^2}{(\varepsilon_\pi-\varepsilon_\nu)_i-\varepsilon} + \sum_j \frac{<(\pi^{-1}\nu)_j,1^+,M=0|\sigma_o\tau_{+1}|0>^2}{(\varepsilon_\nu-\varepsilon_\pi)_j+\varepsilon}$$

$$= -\frac{1}{\kappa} \tag{5.1}$$

The equation relates to the M=0 states and σ_o and $\tau_{\pm1}$ are spherical tensor components of the operators $\vec{\sigma}$ and $\vec{\tau}$. ε_π and ε_ν are proton-particle and neutron-hole energies.

For nuclei with a large neutron excess the second sum of terms will be approximately zero because the relevant final states, for the β_+-decay of the ground state, are blocked. The equation then corresponds to the Tamm-Dancoff approximation.

The equation has been solved for ^{208}Pb using experimental values for particle and hole energies, and if we adjust the coupling constant to reproduce the observed energy of the collective GT-resonance (19.2 MeV relative to ^{208}Pb g.s.), we find

$$\kappa_{\sigma\tau} = \frac{23}{A} \text{ MeV.}$$

The coupling constant so determined is an effective coupling constant relating to the energy of the collective state but including e.g. the coupling to the Δ-resonance. The sum of strength is however still conserved, and we should not use the above equation to extract the absolute GT-strength.

We can treat the isobaric analogue state (IAS) in the same approximation. The collective state is then a result of $V_{12} = \kappa_\tau \vec{\tau}_1 \vec{\tau}_2$ particle-hole interactions, and the dispersion relation now refers to 0^+-states with matrix elements $<(\nu^{-1}\pi)_{0+}|\tau_{-1}|0>$. If we solve the equation for ^{208}Pb again, using the same particle and hole energies as for the GT-state, and adjust the coupling constant to fit the experimentally observed energy (18.8 MeV relative to the ^{208}Pb g.s.) we find

$$\kappa_\tau = \frac{28}{A} \text{ MeV}.$$

This corresponds to a symmetry energy of $V_1 = 4 \times 28$ MeV = 112 MeV. We note that for ^{238}U the GT-resonance lies below the IAS; the experimental evidence suggests that $\kappa_{\sigma\tau} < \kappa_\tau$.

The collective state can also be described by particle-hole interactions of quite different form, e.g. a δ-force in r-space.

$$V_{12} = g_o' \delta(\vec{r}_1 - \vec{r}_2) \vec{\sigma}_1 \vec{\sigma}_2 \vec{\tau}_1 \vec{\tau}_2$$

This is a good approximation for small momentum transfer. If we adjust g_o' to reproduce the GT-resonance energy in ^{208}Pb as before, we find

$$g_o' = 245 \text{ MeV fm}^3 .$$

This quantity should be compared with the volume integral obtained in the field approximation

$$\kappa \cdot \text{volume} = \frac{23}{A} \cdot \frac{4\pi}{3} r_o^3 A = 168 \text{ MeV fm}^3 (r_o = 1.2 \text{ fm})$$

The two approximations for the particle-hole interactions give rather different results for the volume integrals.

5.2. Effect of Δ in Field Approximation

In the field approximation it is straightforward to include the coupling to the Δ-isobars.

$$\sum_i \frac{|<(\nu^{-1}\pi)_i; J=1 \ M=0|\sigma_o \tau_{-1}|0>^2}{(\varepsilon_\pi - \varepsilon_\nu)_i - \varepsilon} + \sum_{N=1}^{A} \frac{<N^{-1}\Delta; J=1 \ M=0|\sigma_o \tau_{-1}|0>}{(\varepsilon_\Delta - \varepsilon_N) - \varepsilon}$$

$$+ \sum_{N=1}^{A} \frac{<N^{-1}\Delta; J=1 \ M=0|\sigma_o \tau_{+1}|0>^2}{(\varepsilon_\Delta - \varepsilon_N) + \varepsilon} = -\frac{1}{\kappa} \qquad (5.2)$$

The first sum of terms comes from the nuclear proton-particle neutron-hole excitations. We have neglected contributions from $\sigma\tau_{+1}$ particle-hole excitations which is a good approximation when $T_o \gg 1$. The next two sums come from nucleon-hole Δ-particle excitations, and the sum is over all nucleons. In the constituent quark model the matrix elements are given above and the dispersion relation becomes

$$\sum_i \frac{<(\nu^{-1}\pi)_i|\sigma_o\tau_{-1}|0>^2}{(\varepsilon_\pi - \varepsilon_\nu) - \varepsilon} + \frac{32}{25} \cdot \frac{3}{2} \left(\frac{Z+N/3}{300-\varepsilon} + \frac{Z/3+N}{300+\varepsilon}\right) = -\frac{1}{\kappa} \qquad (5.3)$$

The effect of inclusion of the Δ-resonance is then to renormalize all nuclear matrix elements involving $\sigma\tau$. We can therefore first solve the equation for the nuclear particle-hole states.

$$\sum_i \frac{<(\nu^{-1}\pi)_i|\sigma_o\tau_{-1}|0>^2}{(\varepsilon_\pi - \varepsilon_\nu)_i - \varepsilon} = -\frac{1}{\kappa_N} \qquad (5.4)$$

For ^{208}Pb we find $\kappa_N = 23/A$ MeV, and it is shown below that this value accounts well for the resonance energies of the collective GT-state for all heavy nuclei. From equation (5.3) we then get a value of $\kappa = 29/A$ MeV. The same value is obtained from an analysis of the reduction of the spin part of the magnetic moments around ^{208}Pb arising from the coupling to the Δ-resonance.[19]

In eq.(5.3) the first sum of terms can be replaced by one term (energy $\varepsilon_{\pi\nu}$) with all the strength $2(N-Z)$ (in the M=0 states for the operator $\sigma\tau_{-1}$)

$$\frac{2(N-Z)}{\tilde{\varepsilon}_{\pi\nu} - \varepsilon} + \frac{48}{25}\left(\frac{Z+N/3}{300-\varepsilon} + \frac{Z/3+N}{300+\varepsilon}\right) = -\frac{1}{\frac{29}{A}} \qquad (5.5)$$

This equation can now be solved for any specific nucleus, but we see that the reduction of strength in the nuclear particle-hole region is almost independent of A. We find e.g. for ^{208}Pb that 36% x 3(N-Z) of the GT-strength is shifted to the Δ energy region.

The coupling to the Δ-isobars also means an admixture of Δ-particle N-hole configurations into the GT wave function. The amplitude with the coupling constant used above will be around 8%: $0.997|ph> - 0.080|\Delta N^{-1}>$.

5.3. GT-Energy Systematics

The energy systematics for the GT-resonances can be discussed in a simple model which contains all the important aspects of the problem. We shall assume that the unperturbed GT-strength is clustered in 2 groups of states with energies ε_i and $\varepsilon_i + \Delta_{\ell s}$ and the

total strength divided as $3(N-Z)(1-f)$ and $3(N-Z) \cdot f$. In any specific case the value of f can be calculated. For ^{208}Pb $f = 0.36$ and
for ^{90}Zr $f = 0.57$. The dispersion relation then reads

$$\frac{2(N-Z)(1-f)}{\varepsilon_i - \varepsilon} + \frac{2(N-Z)f}{\varepsilon_i + \Delta_{\ell s} - \varepsilon} = - \frac{1}{23/A} \qquad (5.6)$$

For $\Delta_{\ell s}$ we take 5.6 MeV which applies to ^{90}Zr and ^{208}Pb but also to
^{48}Ca where $\Delta_{\ell s}(\pi f_{5/2} - \pi f_{7/2}) \sim 5.7$ MeV.

The solution to the equation is plotted in fig.13 relative to
the straight line $\varepsilon - \varepsilon_i = 2 \times 28(N-Z)/A$.

We see that this simple model accounts well for the energy
systematics of the collective state. This model places around 80%
of the strength (in the nuclear region) in the collective state,
and this is then in agreement with the experimental findings that
around 1/4 of the strength in the collective state is found at lower
excitation energies. The ratio is, however, rather uncertain being
dependent on how the continuum is defined in the high energy end of
the spectra.

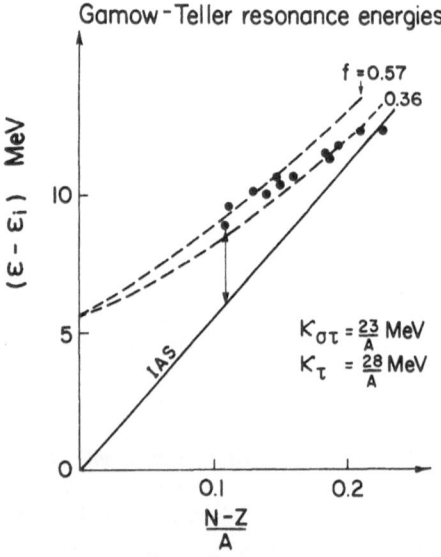

Fig. 13. GT-resonance energy systematics. The experimental ener-
          gies are plotted relative to the IAS as explained in the
          text. The dashed lines correspond to a model where the
          unperturbed particle-hole energies lie in 2 groups with
          $\varepsilon = \varepsilon_i$ and $\varepsilon = \varepsilon_i + \Delta_{\ell s}$ and strength $3(N-Z)(1-f)$ and $3(N-Z)f$,
          respectively. The IAS has (degenerate) particle-hole
          energies $\varepsilon = \varepsilon_i$.

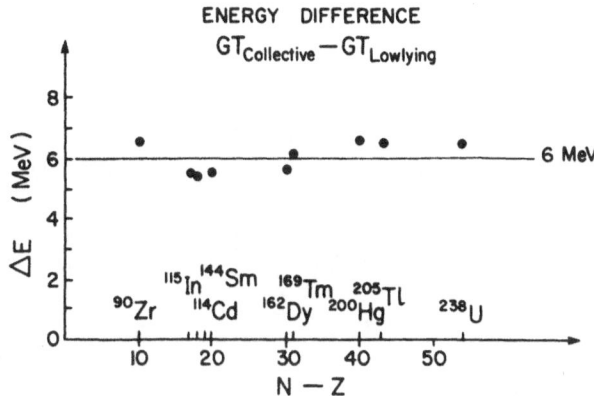

Fig. 14.   The energy difference between the collective GT-state and
           the low lying GT-state (see figs. 1, 3, 9) is plotted
           versus (N-Z). For the very heavy nuclei the low lying
           strength is not developed into a peak.

     In fig.14 the energy difference between the collective GT-
state and the low lying GT-state is plotted, and we observe a
rather constant difference. We note that eq.(5.6) gives significantly larger values for this quantity. Eq.(5.6) is however only
expected to apply to the collective state.

5.4. Isospin Structure of GT-strength Function

     Fig.15 shows spectra for ^{54}Fe and ^{58}Ni together with a shell-
model calculation of the β_--strength function for ^{54}Fe.[17] The isospin of ^{54}Fe(g.s.) is $T_0=1$ and the GT-strength can be on states
with T=0, 1 or 2. The shell-model calculation divides the strength
in the ratio 5.7 : 6.8 : 1.3. The sum of β_--strength is 13.8 and
the β_+ strength is 6x1.3 = 7.8, the double analogue of the T=2
β_--strength.[17] We note that ^{54}Fe and ^{58}Ni are examples where the β_+-
strength is large and has to be considered in using the $S_{\beta_-}-S_{\beta_+}$ =
3(N-Z) sumrule. The ^{54}Fe nucleus has $T_0=1$, like ^{42}Ca, and for both
therefore $S_{\beta_-}-S_{\beta_+}$ = 6. The GT-strength is however twice as large
for ^{54}Fe.[23] We find, however, in both cases about 50% of the strength.
Fig.15 shows that ^{54}Fe and ^{58}Ni have very similar GT-strength distributions, and furthermore that the shell-model calculation accounts
for the gross features of the spectra.

     Fig.16 gives spectra for the (p,n) reactions on ^{58}Ni and ^{60}Ni.
Also shown are relative B(M1) values from (e,e') measurements.[24] The
comparison shows that the (e,e') experiments seem not to be sensitive to the major part of the isospin-spin excitations in nuclei.

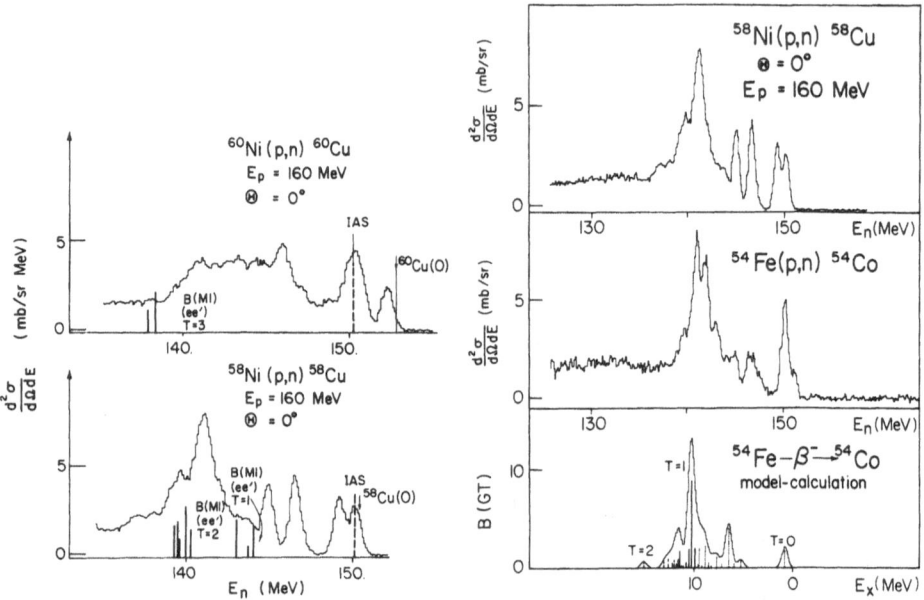

Fig. 15. Neutron spectra for ^{54}Fe and ^{58}Ni are shown together with a shell-model calculation of the GT-strength distribution for ^{54}Fe $\beta^- \to ^{54}$Co.[17] The states in the calculated spectrum are folded with 600 keV (FWHM) Gaussian shapes.

Fig. 16. Neutron spectra for ^{58}Ni and ^{60}Ni are shown together with published B(M1)-values from (e,e') experiments[24]

6. SPIN DIPOLE RESONANCE

In (p,n) spectra at intermediate energies a broad resonance characterized by an angular momentum transfer of 1 is observed for all heavy nuclei. The energy dependence of the cross section for the resonance shows a spin transfer, $\Delta S=1$. The form factor for the resonance is therefore $r(Y_1\sigma)_\lambda \tau_{-1}, \lambda^\pi = 0^-, 1^-, 2^-$, and we would describe the excitation as a spin dipole (charge exchange) resonance. The dipole resonances have widths of around 10 MeV and are interpreted as the envelopes of 3 collective states 0^-, 1^-, and 2^-. Simple RPA calculations do support such interpretations,[5,25] but so far it has not been possible from the (p,n) experiments to resolve the broad resonance into different spin components.

Some very general sumrules apply for the dipole (charge ex-change) strength[5] but an analysis of the data with uncertainties as described above for the GT-resonances is at present not possible. We shall not go into any detail here, but only mention some of the problems in such an analysis i) the extracted cross section is rather uncertain because of the question of how to define the con-tinuum; ii) the relation between cross section and strength is not tested yet by β-decay. There are several unique first forbidden β-decays where the corresponding transition can be measured in (p,n) reactions, but in most cases the resolution has not been sufficient to isolate the transition in question. We shall below describe an analysis of the dipole strength in ^{40}Ca where the nor-malization is obtained from (e,e') measurements.

6.1. Dipole Energy Systematics

The dipole resonance is described as a $1\,\hbar\omega$ excitation, and we can calculate the energy of the collective states in a simple model by assuming that the unperturbed particle-hole energies are degenerate with an energy (relative to the target g.s.)

$$1\,\hbar\omega_o \mp (\frac{V_1(N-Z)}{2A} - E_c)$$

for β_- and β_+ excitations, respectively. V_1 is the symmetry energy. E_c is the Coulomb energy. The dispersion relation in the field approximation (for each λ)

$$\sum_i \frac{|<(ph)_i|r(Y_1\sigma)_\lambda \tau_{-1}|0>|^2}{\varepsilon_i - \varepsilon} + \sum_j \frac{|<(ph)_j|r(Y_1\sigma)_\lambda \tau_{+1}|0>|^2}{\varepsilon_j + \varepsilon}$$

$$= -\frac{1}{\kappa_{\ell=1}} \tag{6.1}$$

then reduces to

$$\frac{\frac{1}{2\lambda+1}S^{(\lambda)}_{-1}}{\hbar\omega_o + (\frac{V_1(N-Z)}{2A} - E_c) - \varepsilon} + \frac{\frac{1}{2\lambda+1}S^{(\lambda)}_{+1}}{\hbar\omega_o + (\frac{V_1(N-Z)}{2A} - E_c) + \varepsilon} = -\frac{1}{\kappa_{\ell=1}} \tag{6.2}$$

where $S^{(\lambda)}_{-1}$ is the total strength.

With $x = (\varepsilon + V_1\frac{N-Z}{2A} - E_c)/\hbar\omega_o$ \hfill (6.3)

we find

$$X^2 - \frac{\kappa_{\ell=1}}{(2\lambda+1)\hbar\omega_0} (S_{-1}^{(\lambda)} - S_{+1}^{(\lambda)})X + \frac{\kappa_{\ell=1}}{(2\lambda+1)\hbar\omega_0} (S_{-1}^{(\lambda)} + S_{+1}^{(\lambda)}) - 1 = 0$$

$$(6.4)$$

The quantities $(S_{-1} - S_{+1})$ and $(S_{-1} + S_{+1})$ can be obtained from sum-rules[5]

$$S_{-1}^{(\lambda)} - S_{+1}^{(\lambda)} = (2\lambda+1) \cdot 2 \frac{N-Z}{4\pi} <r^2>_{\text{neutron excess}} \qquad (6.5)$$

$$\sum_f \varepsilon_f |<f|r(Y_1\sigma)_\lambda \tau_0|0>|^2 = \frac{3Ah^2}{8\pi M} (2\lambda+1) \qquad (6.6)$$

$$\hbar\omega_0 S_0^{(\lambda)} = \hbar\omega_0 1/2(S_{-1}^{(\lambda)} + S_{+1}^{(\lambda)}) = \frac{3Ah^2}{8\pi M} (2\lambda+1)$$

where in the last expression the assumption about degeneracy has been introduced. With

$$\kappa_{\ell=1} = \frac{4\pi}{<r^2>} \kappa_{\ell=0} = \frac{4\pi}{3/5 \, r_0^2 A^{2/3}} \frac{23}{A} \text{ MeV fm}^{-2} \quad ,$$

$$<r^2>_{N-Z} = 1.34<r^2>_A \text{ and } \hbar\omega_0 = 41A^{-1/3} \text{ MeV}$$

we find

$$X^2 - 2.0 \, \nu X - 2.95 = 0 \quad (\nu \equiv 0.76(N-Z)A^{-2/3}) \quad . \qquad (6.7)$$

We note that in this approximation the energy is independent of spin.

The solution corresponding to the collective state is plotted in fig.18 as a function of ν, the neutron excess measured in units of shells; ν is 0.92 for ^{208}Pb.

The ordinate X measures the energy in units of $\hbar\omega_0$ so X=1 corresponds to the unperturbed energy of the dipole transitions. We see that this simple model accounts for the gross features of the energy systematics. We note that the $\sigma\tau$-correlations are very strong in shifting the strength (for ^{208}Pb) almost 2 $\hbar\omega_0$ (from X=1 $\hbar\omega_0$ to X$\sim$3 $\hbar\omega_0$).

The same equations apply for the energy of the charge exchange dipole resonance with $\Delta S=0$. The formfactor is in this case $r Y_1\tau_{-1}$, and the spin-parity is 1^-.

A recent (p,n) experiment at E_p=45 MeV[26] identifies resonances

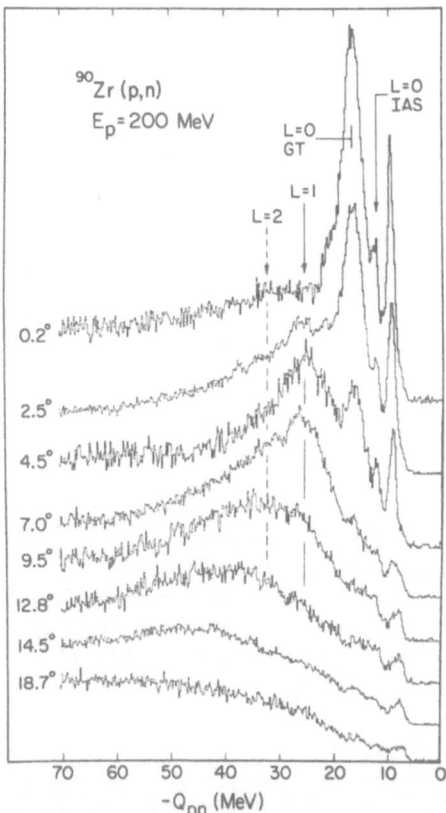

Fig. 17.   Neutron spectra at E_p=200 MeV for the ^{90}Zr(p,n)^{90}Nb reaction at different angles.[5] Angular distributions are given for some of the peaks in fig.2.

observed in Zr, Sn and Pb as collective 1$^-$-states. For ^{40}Ca we have used the value from a (γ,n) experiment[27] (shifted by the Coulomb energy).

In fig.18 the observed resonance energies are plotted using the equations given above. The equation (6.4) becomes

$$X^2 - 2.4 \ \nu X - 3.4 = 0 \qquad\qquad (6.8)$$

using $\kappa_\tau(\ell=0) = 28/A$ MeV. The solution corresponding to the collective state is also plotted in fig.18.

As for the spin dipole resonance we would conclude that the simple model accounts for the general trend of the data. We note that spin orbit effects are neglected in these models for the dipole resonances.

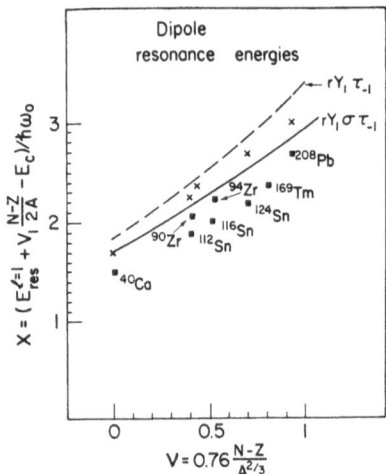

Fig. 18. Spin dipole energy systematics. The experimental energies for the spin transfer dipole[4],[5] (squares) and (non)
spin transfer) dipole resonances[26] (crosses) are plotted
following eq.(6.3). The symmetry energy is taken as
V_1=112 MeV. The curves correspond to eqs.(6.7) and
(6.8). The unperturbed particle-hole energies correspond to 1 $\hbar\omega_0$ excitations, i.e. x=1.

     We see from fig.18 that for both types of dipole (charge
exchange) resonances the coupling constant extracted from the ℓ=0
resonances (GT and IAR) seems to be too large to account for the
energies observed.

     In a more detailed analysis of the dipole resonance energies
in ^{40}Ca and ^{208}Pb we find in the field approximation
$\kappa_{\ell=1} \sim 0.75 \times 4\pi/\langle r^2 \rangle$ $\kappa_{\ell=0}$ or g_0' = 225 MeV fm^3 with a δ-force[5]
for the particle-hole interaction.

6.2. Spin Dipole Strength

     For ^{40}Ca we have estimated the total spin dipole strength.
Fig.19 shows the spectrum from the ^{40}Ca(p,n)^{40}Sc reaction at θ=5°
and E_p=160 MeV. The broad resonance has an angular distribution
corresponding to $\Delta\ell$=1; so has the peak labeled as 2⁻. The estimate
of the strength is then obtained from the ratio of areas (or cross
sections) for the resonance and the 2⁻-state together with a value
for the spin dipole strength for the 2⁻-state. This number is
calculated from the B(M2)-value from (e,e') data for the 2⁻ analogue state in ^{40}Ca. The main configuration for the 2⁻-state
(in e.g. ^{40}Sc) is $\pi f_{7/2}\nu d_{3/2}^{-1}$; the RPA calculation for the dipole

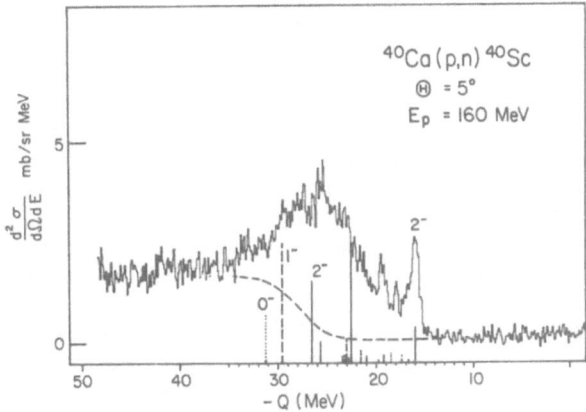

Fig. 19. Neutron spectrum at θ=5° for the ^{40}Ca(p,n)^{40}Sc reaction.
The dotted line shows the assumed background in extracting
the total dipole strength. The result of an RPA calculation of the spin dipole strength on the different final
spins is shown.

strength shown in the figure gives an amplitude of 0.95 for this
configuration. The B(M2) value is   (235±20) μ_0^2 fm^2, whereas a pure
$f_{7/2}d_{3/2}^{-1}$ state has 1487 μ_0^2 fm^2. If we only ascribe the polarization effect to the spin part (and not the orbital part) we find
the strength is reduced a factor of 5.2, i.e. the repulsive $\vec{\sigma}\vec{\sigma}\vec{\tau}\vec{\tau}$
interactions have removed strength from the low lying 2$^-$-state.

    The cross section for the spin dipole resonance depends on
how the background is defined. In the figure we have indicated
the background used to extract the resonance cross section.

    The result of the analysis of the total strength is that we
observe about 80% of the sumrule strength. We note that the sumrule
for ^{40}Ca (since T=0) as given in eq.(6.6) is an energy weighted
sumrule and therefore not as general as the sumrules discussed above
for the GT-strength.

    In the field approximation we can include the quenching effect
from the Δ-isobars. In the constituent quark model we find that 27%
of the spin dipole strength is removed from the nuclear region.[29]The
corresponding number for the GT-strength was 36%.

7. SUMMARY

    The (p,n) reaction at intermediate energies is a unique tool
for the study of isospin-spin correlations in nuclei. The specta-

cular "signal to noise" ratio observed in the spectra is related
to the continuum, "the background", being of the same nature as
the peaks, i.e. the result of one step processes. The (p,n)
experiments have shown that there are strong isospin-spin correlations in nuclei that build up collective states carrying a significant part of the sumrule strength. Simple one particle-one
hole models account for the energy systematics for both the ℓ=0
and ℓ=1 resonances. By normalizing to β-decay it has been possible
to relate the zero degree cross section for ℓ=0 transitions to
Gamow-Teller or Fermi strength. An analysis of the total Gamow-
Teller strength has shown that we consistently find less than the
sumrule strength. The very general nature of the sumrules for
charge-exchange modes suggests that internal degrees of freedom
for the nucleon are involved in the isospin-spin vibrations.

ACKNOWLEDGEMENTS

     The results discussed here are a result of a collaboration
of many people. We shall here specifically give the names of
people involved in most of the (p,n) measurements at Indiana:
C.C. Foster, C.D. Goodman, C.A. Goulding, D.J. Horen, T. Masterson,
E. Sugarbaker, and T. Taddeucci. In addition, we would like to
acknowledge the IUCF staff for enthusiastic support.

     This work was funded in part by U.S. National Science Founda-
tion and the Danish Natural Science Research Council.

REFERENCES

1. D. E. Bainum, J. Rapaport, C. D. Goodman, D. J. Horen, C. C.
   Foster, M. B. Greenfield, and C. A. Goulding, Phys. Rev. Lett.
   44:1751 (1980).
2. C. D. Goodman, C. A. Goulding, M. B. Greenfield, J. Rapaport,
   D. E. Bainum, C. C. Foster, W. G. Love, and F. Petrovich, Phys.
   Rev. Lett. 44:1755 (1980).
3. D. J. Horen, C. D. Goodman, C. C. Foster, C. A. Goulding, M. G.
   Greenfield, J. Rapaport, E. Sugarbaker, T. G. Masterson, F.
   Petrovich, and W. G. Love, Phys. Lett. 95B:27 (1980).
4. D. J. Horen, C. D. Goodman, D. E. Bainum, C. C. Foster, C.
   Gaarde, C. A. Goulding, M. B. Greenfield, J. Rapaport, T. N.
   Taddeucci, E. Sugarbaker, T. Masterson, S. M. Austin, A. Galon-
   sky, and W. Sterrenburg, Phys. Lett. 99B:383 (1981).
5. C. Gaarde, J. Rapaport, T. N. Taddeucci, C. D. Goodman, C. C.
   Foster, D. E. Bainum, C. A. Goulding, M. B. Greenfield, D. J.
   Horen, and E. Sugarbaker, Nucl. Phys. A369:258 (1981).
6. J. Rapaport, T. Taddeucci, C. Gaarde, C. D. Goodman, C. C.
   Foster, C. A. Goulding, D. J. Horen, E. Sugarbaker, T. G.
   Masterson, and D. Lind, Phys. Rev. C24:335 (1981).
7. K. I. Ikeda, S. Fujii, and J. I. Fujita, Phys. Lett. 3:271 (1963).

8. F. Osterfeld, these proceedings.

9. R. R. Doering, A. Galonsky, D. M. Patterson, and G. F. Bertsch, Phys. Rev. Lett. 35:1691 (1975).

10. C. D. Goodman, C. C. Foster, M. B. Greenfield, C. A. Goulding, D. A. Lind, and J. Rapaport, IEEE Trans. Nucl. Sci. 26:2248 (1979).

11. C. D. Goodman, J. Rapaport, D. E. Bainum, and C. E. Brient, Nucl. Instr. and Meth. 151:125 (1978).

12. W. G. Love and M. A. Franey, Phys. Rev. C24:1073 (1981).

13. P. D. Kunz, unpublished. Present version extended by J. R. Comfort.

14. R. Schaeffer and J. Raynal, unpublished.

15. T. N. Taddeucci, J. Rapaport, D. E. Bainum, C. D. Goodman, C. C. Foster, C. Gaarde, J. S. Larsen, D. J. Horen, T. Masterson, and E. Sugarbaker, Phys. Rev. C25:1094 (1982).

16. C. Gaarde, J. S. Larsen, M. N. Harakeh, S. Y. van der Werf, M. Igarashi, and A. Müller-Arnke, Nucl. Phys. A334:248 (1980).

17. C. Gaarde, J. S. Larsen, A. G. Drentje, M. N. Harakeh, S. Y. van der Werf, and A. Müller-Arnke, Nucl. Phys. A346:497 (1980).

18. M. Ericsson, Ann. of Phys. 63:562 (1971); M. Ericsson, A. Figureau, and C. Thévenet, Phys. Lett. 45B:19 (1973); E. Oset and M. Rho, Phys. Rev. Lett. 42:47 (1979); W. Knüpfer, M. Dillig, and A. Richter, Phys. Lett. 95B:349 (1980); H. Toki and W. Weise, Phys. Lett. 97B:12 (1980); G. Bertsch, Nucl. Phys. A354:157 (1981); G. E. Brown and M. Rho, Nucl. Phys. A372:397 (1981).

19. A. Bohr and B. Mottelson, Phys. Lett. 100B:10 (1981).

20. C. D. Goodman, C. C. Foster, D. E. Bainum, C. Gaarde, J. S. Larsen, C. A. Goulding, D. J. Horen, T. Masterson, J. Rapaport, T. N. Taddeucci, and E. Sugarbaker, Contributed papers to IX Intern. Conf. on High Engergy Physics and Nuclear Structure, Versailles 6-10 July (1981); C. D. Goodman and S. Bloom, this conference.

21. J. Rapaport et al., to be published.

22. C. D. Goodman, C. C. Foster, D. E. Bainum, S. D. Bloom, C. Gaarde, J. S. Larsen, C. A. Goulding, D. J. Horen, T. Masterson, S. Grimes, J. Rapaport, and T. N. Taddeucci, Phys. Lett. 107B: 406 (1981).

23. S. Krewald, F. Osterfeld, J. Speth, and G. E. Brown, Phys. Rev. Lett. 46:103 (1981), and references therein.

24. R. A. Lindgren, W. L. Bendel, E. C. Jones Jr., L. W. Fagg, X. K. Maruyawa, J. W. Lightbody,Jr., and S. P. Fivozinsky, Phys. Rev. C14:1789 (1976).

25. F. Osterfeld, S. Krewald, H. Dermawan, and J. Speth, Phys. Lett. 105B:257 (1981).

26. W. A. Sterrenburg, S. M. Austin, R. D. DeVito, and A. Galonsky, Phys. Rev. Lett. 45:1839 (1981).

27. J. Miller et al., J. de Phys. 27:8 (1966).

28. A. Richter, private communication.

29. T. Suzuki, H. Sagawa and C. Gaarde, Phys. Lett. 116B:91 (1982).

OBSERVATION OF M1 STRENGTH IN MEDIUM-HEAVY NUCLEI

VIA THE (p,p') REACTION

G.M. Crawley, N. Anantaraman and A. Galonsky

Cyclotron Lab, Michigan State University
E. Lansing, MI 48824

and

C. Djalali, N. Marty, M. Morlet, A. Willis, J.C. Jourdain
and P. Kitching

Institut de Physique Nucleáire
Orsay, FRANCE

ABSTRACT

A broad resonance has been observed by inelastic scattering of 200 MeV protons from ^{51}V, ^{58,60,62}Ni, ^{68}Zn, ^{90,92,94,96}Zr, ^{92,94,96,98,100}Mo, ^{120,124}Sn and ^{140}Ce. The resonance occurs between 8 and 9 MeV in most of the nuclei and has a width of around 2 MeV. In all cases, the angular distribution is very sharply forward peaked and is consistent with an orbital angular momentum transfer of zero. The excitation energy, angular distribution and strength of the resonance suggest that it is the giant M1 resonance. In the nickel isotopes, and in ^{51}V, both the T_0 and T_0+1 components of the resonance are observed.

I. INTRODUCTION

The study of spin excitations in nuclei has been pursued by a number of different yet complementary techniques including electron scattering, (γ,n) reactions, β-decay and, more recently, (p,n), (^{3}He,t) and (p,p') reactions. The program of this conference is proof of the richness and diversity of this field. Both the beautiful (p,n) results showing the excitation of the Gamow-Teller (G-T) resonance in many nuclei[1,2] and the very nice high resolution inelastic electron scattering work[3,4,5] have stimulated renewed

91

interest in this subject. In addition, improvements in the theoretical formulation of the effective interaction used in direct
reaction calculations at high energy[6] and the resulting ability to
predict the absolute magnitudes of the cross sections, confirm that
magnetic transitions are substantially quenched.[7] The quenching has
apparently two components, one being configuration mixing and the
other more exotic mesonic effects such as the mixing of the M1 state
with a Δ particle-nucleon hole configuration excited through the Δ-
nucleon interaction[8,9].

     In spite of the strong excitation of the G-T resonance
in the (p,n) reaction, until last year there was no corresponding
clear observation of the M1 state in the parent nucleus either by
inelastic proton scattering[10] or inelastic electron scattering[5,11,12] in nuclei heavier than the nickel isotopes. However, the
(p,n) data did suggest to us that the appropriate kinematic region to
search for the M1 state in (p,p') would be at high bombarding
energies (>120 MeV) and at very forward angles. The (p,n) reaction
to the G-T state is very forward peaked and the similarity of the
reaction mechanism suggested a similar behaviour for the (p,p')
reaction. Of course, the measurement of inelastic scattering cross
sections at very forward angles is experimentally challenging,
primarily because of the background from scattered particles and the
Landau tail of the elastic peak.

II. EXPERIMENTAL METHOD

     The experiments reported here were carried out using 201 MeV
protons from the synchocyclotron at the IPN, Orsay. The spectrometer
attached to this facility is ideal for forward angle measurements
because of its large size. In addition, a counter system consisting
of two multiwire detectors and two plastic scintillators measures
the trajectory of the particles emerging from the spectrometer.[13]
Such measurements serve to minimize the background from scattered
particles. However one problem with this counter was that it had
small differential non linearities which gave rise to spurious fine
structure in the spectra. For this reason each spectrum was taken
with two slightly different settings of the magnetic field to clearly
identify any spurious structure. An example of such overlapping
spectra is shown for ^{51}V in Fig. 3a. In addition, since the fine
structure was found to be very stable during a week long run, a
correction function was obtained which could then be applied to all
the spectra to further reduce the effect of the non linearities in
the counter.

     The absolute cross section was obtained in two ways. First, the
cross section was measured by a comparison with the known p-p
scattering cross section and second by a comparison with elastic
scattering calculations at angles forward of $10°$. The two methods
agreed to better than 10%.

III. RESULTS

Since the T_o component of the G-T strength had been resolved from the T_o-1 component in (p,n) experiments on the Zr isotopes at 120 MeV,[1,14] a reasonably accurate prediction could be made of the position of the M1 parent state in these nuclei. The energetics are displayed in Fig. 1. The T=5, 1^+ state is about 8.5 MeV above the 0^+ I.A.S. in ^{90}Nb suggesting that the excitation energy of the corresponding M1 state in ^{90}Zr will also be about 8.5 MeV. Theoretical predictions of the excitation energy of the M1 state gave similar values.

Energetics of 1^+ states in ^{90}Zr, ^{90}Nb

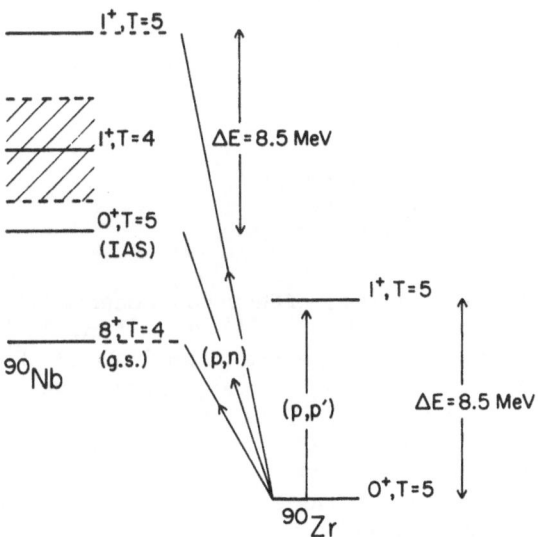

FIG. 1. Energies of 1^+ states excited in ^{90}Zr and ^{90}Nb by the (p,p') and (p,n) reactions. The excitation energy of the 1^+ T=5 state in ^{90}Zr is similar to the energy of the 1^+ T=5 state above the 0^+ IAS in ^{90}Nb.

Our first measurements were therefore made on ^{90}Zr, ^{92}Zr and ^{94}Zr.[15] In all these isotopes a broad peak was observed close to the expected excitation energy of between 8 and 9 MeV (Fig. 2).   A similar feature has also been observed in a (p,p') experiment on ^{90}Zr at TRIUMF.[16]

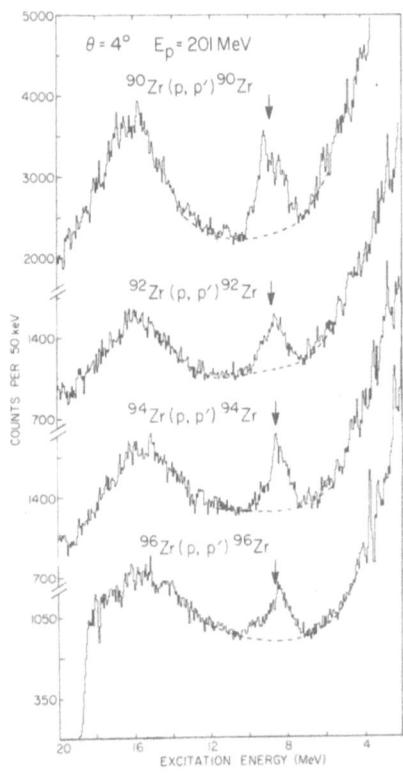

FIG. 2.   Spectra of protons inelastically scattered from ^{90}Zr, ^{92}Zr, ^{94}Zr and ^{96}Zr at 4°. The arrows indicate the centroids of the M1 resonance.

In later experiments this effect was observed in ^{96}Zr[17] and in thirteen other nuclei, viz. ^{51}V, ^{58,60,62}Ni, ^{68}Zn, ^{92,94,96,98,100}Mo, ^{120,124}Sn and ^{140}Ce.[18] Two spectra from ^{51}V, at 3° and at 8°, are shown in Fig. 3.   The bump is very prominent at 3° but has practically disappeared by 8°. This feature of a very forward peaked angular distribution is characteristic of all the broad peaks observed. Unfortunately, the tail from the elastic peak increases in the heavier nuclei so that in the Sn isotopes and particularly in the Ce case, it is very difficult to extract the broad bump from the underlying smooth background even at 4°.   (See Fig. 4).

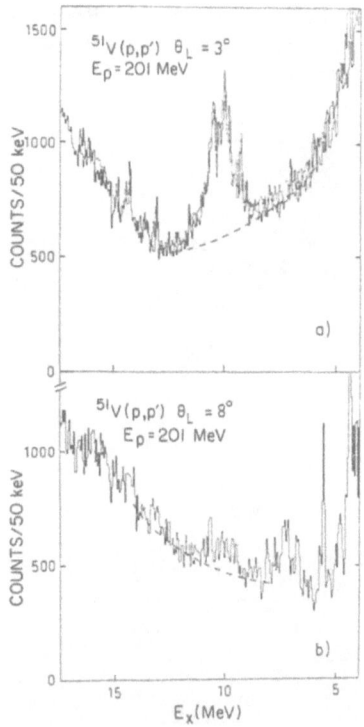

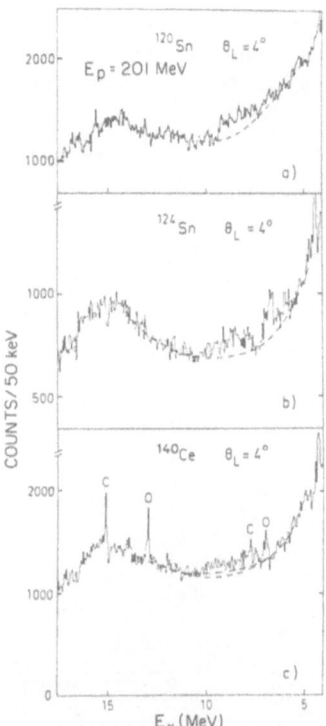

FIG. 3. Spectra of protons inelastically scattered from ^{51}V. a) at 3° and b) at 8° in the laboratory. At 3° the overlap of two spectra taken with different magnetic fields is shown. The dashed lines indicate background subtracted to obtain peak areas.

FIG. 4. Spectra of protons inelastically scattered from a) ^{120}Sn b) ^{124}Sn and c) ^{140}Ce at 4°. The dashed lines indicate backgrounds assumed in extracting peak areas. Two choices are shown for ^{140}Ce.

Angular distributions of the broad bump were measured for many of the nuclei. Some examples for the Zr isotopes are shown in Fig. 5. These angular distributions are fitted rather well by a macrosopic DWBA calculation[19] for a $\Delta L=0$ transition as would be expected for a 1$^+$ state excited by a $\Delta L=0$ spin-flip transition. An additional check is shown in Fig. 5, where a comparison is made with the ^{90}Zr(p,n)^{90}Nb angular distribution measured at a bombarding energy of 200 MeV.[20] The agreement between the (p,p') and (p,n) angular distributions is excellent back to 8° where the cross sections in both (p,p') and (p,n) reactions are down

by about a factor of ten from the value at 4°. Microscopic
calculations of the (p,p') reaction are also shown in Fig. 5 and
these will be discussed in a later section.

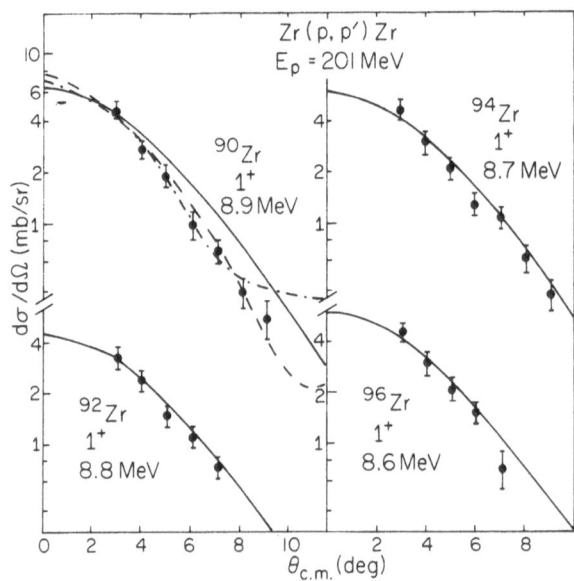

FIG. 5. Angular distributions for the M1 state in ^{90}Zr, ^{92}Zr, ^{94}Zr
and ^{96}Zr. The solid (DWBA70) and dashed (RESEDA) curves are DWIA
predictions, normalized to the data. The curve shown for ^{96}Zr is the
calculation for ^{90}Zr. The dot-dashed curve is from a ^{90}Zr(p,n)
measurement (ref. 20).

Thus both the excitation energy and angular distribution
suggest that the broad bump observed systematically in a fairly large
range of nuclei is indeed an M1 state. Let us now turn to a further
interesting aspect of these measurements which is illustrated in the
nickel isotopes.

IV. HIGHER ISOSPIN COMPONENT OF THE M1 STATE

In a heavy nucleus such as Zr, where the M1 transition proceeds
by neutron excitation, the isospin of the excited state is the same
as the isospin, T_o, of the ground state. However, in a nucleus with
both neutron and proton valence orbits, the particle and hole couple
to isospin 0 and 1, the latter then coupling to the core to produce
M1 states of isospin T_o and T_o+1. The energy splitting of these two
isospin components is related to the depth of the symmetry potential
V_1 by the relation

$$\Delta E = E_{T_o+1} - E_{T_o} = \frac{V_1}{A} (T_o+1).$$

The spectra observed for the ^{58}Ni, ^{60}Ni and ^{62}Ni(p,p') reactions at $4°$ are shown in Fig. 6. The spectra have two components, a broad peak centered near 8-9 MeV excitation energy and a strong sharp peak which moves to increasing excitation energy with increasing mass number. The broad peak is identified as the T=T$_o$ component and the sharp peak as part of the T=T$_o$+1 component of the

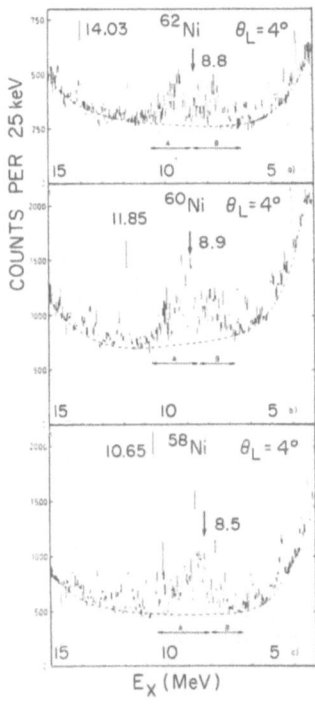

FIG. 6. Spectra of proton inelastic scattering at $4°$ from a) ^{62}Ni, b) ^{60}Ni, c) ^{58}Ni. The centroids of the broad bumps are shown by arrows and excitation energies are indicated in MeV.

M1 resonance. The energetics for ^{60}Ni are displayed in Fig. 7. While the 1^+,T=3 state at 11.85 MeV is above the neutron threshold, this state is isospin forbidden to decay by neutron emission. Proton decay of the state is isospin allowed but with an energy of only 2.32 MeV. The coulomb and centrifugal barriers ensure that the decay width of the state is small. Since the level density of T=3 states near 12 MeV is not large, the spreading width of the 1^+,T=3 state is also small so that the state has a narrow total width. The width measured (~70 keV) is determined by the experimental resolution.

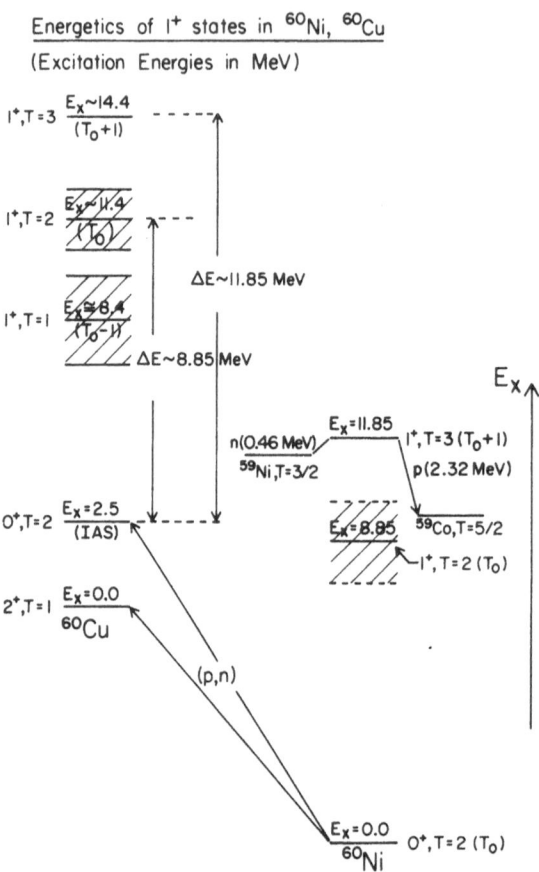

FIG. 7. Energetics of 1$^+$ states excited by (p,p') in ^{60}Ni and
(p,n) in ^{60}Cu. The excitation energies are given in MeV. Cross
hatching indicates a broad state. The decay energies for proton and
neutron emission from the T=3 state in ^{60}Ni are also shown.

     In order to check on these assignments, the broad peak was
divided approximately in half. Angular distributions for the high
energy piece (A), the low energy piece (B), the total broad resonance
(Σ) and the sharp states are shown in Fig. 8. The solid lines in
this figure are macroscopic DWBA calculations for ΔL=0. Forward of
6°, the theoretical curves match well the slopes of the experimental
angular distributions, but at larger angles the experimental cross
sections are larger than the calculations, presumably because of
contributions from states of higher J$^\pi$. This effect is particularly
apparent for the low excitation energy component (B) in ^{58}Ni.

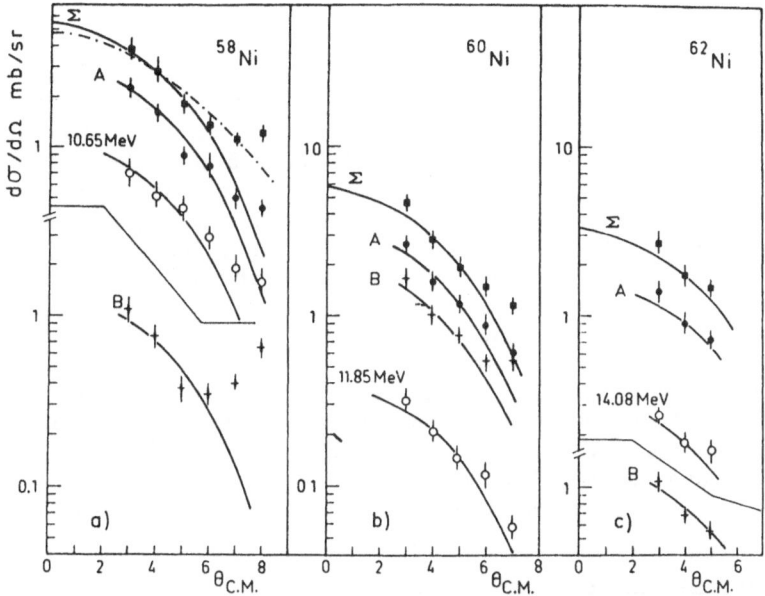

FIG. 8. Experimental angular distributions from some peaks and broad resonances in the Ni isotopes compared with macroscopic DWBA, $\Delta L=0$ calculations. a) For ^{58}Ni, Σ which is the sum of regions A and B, is also compared with a DWIA RESEDA calculation.

If we assume that the T_0+1 component is given by the single sharp peak and that the centroid of the broad structure gives the excitation energy of the T_0 component, the values of V_1 extracted are 62.4 MeV for ^{58}Ni, 59.0 MeV for ^{60}Ni and 81.1 MeV for ^{62}Ni. If all the smaller fine structure at high excitation is included in the T_0+1 component of the M1 resonance in ^{58}Ni and ^{60}Ni, the values of V_1 obtained are 82.5 MeV for ^{58}Ni and 74.5 MeV for ^{60}Ni. These values are in reasonable agreement with the estimate of $V=85$ MeV obtained from (p,n) measurements in heavier nuclei[21] of E_{T_0-1} and the assumption that $E_{T_0}=40A^{-1/3}$ MeV.

The relative excitation of the various isospin components of the 1^+ states excited in (p,p') and (p,n) reactions can be predicted using simple isospin coupling algebra. These predictions have been given recently for the Ni isotopes[22]. The energetics of the three components, T_0, T_0+1 and T_0-1, produced by (p,n) in ^{60}Cu are also displayed in Fig. 7. A recent (p,n) collaboration between Kent State and MSU on the 4 even-even Ni isotopes at IUCF[23] confirms that it is possible to observe all 3 isospin components at least in ^{58}Cu, ^{60}Cu and ^{62}Cu. The isospin splitting in these nuclei is similar to that observed in the Ni isotopes.

V. DISCUSSION

A. Excitation Energies of M1 States

A summary of the excitation energies of the M1 states observed is given in Table 1. Apart from the value in ^{51}V, the excitation energies are all very similar lying between 8 and 9 MeV. This is not too surprising since the excitation energy is expected to be determined primarily by the splitting of the spin orbit partner levels, (j=ℓ±½), involved in the transition, and increased by the repulsive particle-hole interaction which stays fairly constant with mass number. The spin orbit splitting also does not change very much with mass number since the effect of decreasing ℓ.s splitting with mass is partially compensated for by the larger ℓ values involved in M1 transitions in the heavier nuclei. For example, Bertsch predicts[24] that the excitation energy of the M1 state only changes from 10.6 MeV in ^{48}Ca to 7.4 MeV in ^{208}Pb. A simple shell model calculation by Toki, Cha and Bertsch[25] also finds good agreement with the observed excitation energies of the broad resonance observed in the Zr isotopes.

B. Strength of the M1 Excitation

While the shape of the angular distribution for the broad bump can be well fitted by a macroscopic ΔL=0 DWBA calculation, microscopic calculations are needed in order to obtain the strength of the transition. Two different microscopic calculations have been made. For ^{90}Zr, ^{92}Zr and ^{94}Zr, microscopic calculations have been carried out using the code DWBA70. For ^{90}Zr, a simple $(\nu g_{7/2} g_{9/2}^{-1})1^+$ configuration was assumed. The wavefunctions used for the calculations in ^{92}Zr and ^{94}Zr were taken from a shell model calculation of Anantaraman and Wildenthal. Further details of the DWIA calculations are given in ref. 17. The ratios of experimental to predicted cross section are about 1/4 to 1/5 for the three Zr isotopes and are given in Table 1.

Independent calculations have also been carried out using the code RESEDA for the nuclei ^{58}Ni, ^{90}Zr, ^{92}Mo, ^{120}Sn and ^{140}Ce assuming simple wavefunctions in each case. Details of these calculations are given in ref. 18. The ratio of experimental to theoretical cross sections ranges from 0.36 for ^{90}Zr to 0.22 for ^{140}Ce and is also listed in Table 1. The values obtained from the RESEDA calculations are probably less reliable than from DWBA70. Exchange effects, which are known to be very significant in the calculations, are treated exactly in DWBA70 but are treated only approximately in RESEDA.

The uncertainties in these ratios are large, both due to the uncertainties in the theoretical calculations and to the lack of a perfect fit to the experimental angular distributions. The ratios do not appear to be very dependent on the mass of the target. Calculations with more detailed wave functions are needed before any predicted by Toki and Weise[26] for the quenching due to coupling with the Δ-resonance with a finite range interaction, can be definitively established.

Table 1.   Excitation Energies and Comparison with Distorted Waves
Calculations for M1 States.

Nucleus	E_x (MeV) T_0	E_x (MeV) T_0+1	$N=\dfrac{\sigma(exp)}{\sigma(RESEDA)}$	$N=\dfrac{\sigma(exp)}{\sigma(DWBA70)}$
^{51}V	10.15±0.15	13.08		
^{58}Ni	8.5 ±0.1	10.65 11.36[a]	0.23	
^{60}Ni	8.9 ±0.1	11.85 12.58[a]		
^{62}Ni	8.8 ±0.1	14.03		
^{68}Ni	9.6±0.1 8.6±0.1			
^{90}Zr	8.9±0.2		0.36	0.26
^{92}Zr	8.8±0.2			0.19
^{94}Zr	8.7±0.2			0.26
^{96}Zr	8.6±0.2			
^{92}Mo	9.0±0.1 7.95±0.1		0.30	
^{94}Mo	8.6±0.2			
^{96}Mo	8.4±0.2			
^{98}Mo	8.5±0.2			
^{100}Mo	8.5±0.2			
^{120}Sn	8.4±0.2		0.23	
^{124}Sn	8.7±0.3			
^{140}Ce	8.6±0.3		0.22	

a) E_x for T_0+1 state if high excitation fine structure is included.

C. Comparison with (p,n) Measurements

In a nucleus like ^{90}Zr where we assume that the $j_>$ state is completely filled with neutrons and empty of protons, if we further assume that the (p,p') reaction to the 1^+ state and the (p,n) reaction to its analogue are mediated only by $V_{\sigma\tau}$, then taking the ground state isospin to be T_0, the (p,p') cross section to the 1^+ state can be compared with the (p,n) cross section to the T_0 component of the G-T transition by

$$\sigma(p,p')(M1) = \frac{T_0}{2}\ \sigma(p,n)\ (G\text{-}T,T_0)$$

Since the (p,n) cross sections to the T_0 and T_0-1 components of the G-T transition are in the ratio of $(1/(2T_0-1))$ in this model, then,

$$\sigma(p,p')(M1) = \frac{1}{4}\ \sigma(p,n)\ (G\text{-}T,\ T_0+(T_0-1))$$

In ^{90}Zr, the value predicted for the (p,p') cross section at 0° using the measured 0° (p,n) cross section at 200 MeV[20] is 12.5 mb/sr. Of course, there are a number of known corrections to this very simple comparison including the mixing of the (T_0-1) G-T state with lower excitation energy 1^+ states of different configurations, the contribution of the V_σ term in the (p,p') reaction and the different distortions for the outgoing proton and neutron. These effects gave about a 30% decrease (to 9 mb/sr) in the predicted cross section. An extrapolation using a DWBA calculation give a "measured" 0° (p,p') cross section of 7 mb/sr.

Thus in this simple model, there is reasonable agreement between the measured (p,p') cross section and the value expected from the (p,n) measurements.

Since the ^{208}Pb(p,n) cross section has been measured at 200 MeV,[20] it would be tempting to apply the same argument to predict the ^{208}Pb(p,p') cross section to the M1 state. However, since many neutron orbits contribute in the Pb case and also since proton excitations also contribute to the M1 excitations, the simple relationship between (p,n) and (p,p') cross sections will no longer apply. At present, inelastic proton scattering on ^{208}Pb at 200 MeV and at angles as small as 5° has not given any clear indication of M1 strength.

D. Comparison with Electron Scattering

While one might naively expect that direct comparisons could be made between (e,e') and (p,p') reactions on the same target, there are a number of difficulties both practical and theoretical with such an expectation. A comparison with earlier low resolution inelastic electron scattering on Zr[28] and Ce[29] would suggest that similar structures were excited in the two reactions. However, more detailed (e,e') experiments with better energy resolution[11,12] showed that the main multipolarity observed by backward electron scattering in these heavy nuclei was M2. There still appears to be a discrepancy between the present (p,p') results and the high resolution (e,e') data,[12] particularly on ^{90}Zr where a direct comparison can be made. While the overall strength observed in both (e,e') and (p,p') may not be inconsistent, considering the uncertainties in the DWIA predictions for the (p,p') reaction, the detailed features are quite different. A broad structure is observed in (p,p'), but only a few discrete 1^+ states are seen in (e,e'). These states would have been readily resolvable with the 70 keV resolution of the (p,p') experiment.

In lighter nuclei such as in the calcium and nickel isotopes where sharp states can be identified, the situation is still not completely clear. Let me first turn to the case of ^{58}Ni.

The excitation energies of the 1^+ states observed in two different electron scattering experiments on ^{58}Ni[11,30,31] are given in Table 2. At excitation energies above 9.5 MeV, there is overall agreement between these two experiments. In addition, they both agree with the energies found for the sharp states in the (p,p') spectra. The (e,e') studies assign an isospin $T=T_o+1=2$ to these states. However, the two (e,e') measurements differ for states at lower excitation energies. For example, in ref. 31, a single 1^+ state is found at 7.7 MeV which agrees in excitation energy with a strong state seen in (p,p'). But in ref. 30, three 1^+ states are reported at 6.05, 6.41 and 7.09 MeV. The main difference between the (e,e') and (p,p') results is that very little strength is seen in (e,e') in the region of the broad structure observed in (p,p').

For ^{60}Ni, there is general agreement in the energies of the high lying states observed in (e,e')[30] and in the present experiment. However in the electron scattering experiment, no 1^+ state is reported in the region from 6.75 to 10.75 MeV where considerable strength is seen in the (p,p') reaction.

In principle, one can obtain information on the wavefunctions of states excited in the two reactions by comparing the ratio of the observed strengths. The operators for exciting an M1 state with protons and electrons are similar particularly for neutron excitations where there is no orbital contribution to the M1 operator. In fact Crawley et al.[17] reported that the ratio of the (p,p') cross sections calculated for the M1 transitions in the Zr isotopes was very similar to the ratio of B(M1)'s calculated for the same states. These transitions were all assumed to be pure neutron transitions. Calculations using a microscopic DWIA code also show[18] that the cross sections for exciting proton or neutron particle hole configurations are very similar for the (p,p') reaction. However, this is not true in inelastic electron scattering. For example, if we assume that in ^{58}Ni the two states at 9.85 MeV and 10.65 MeV are both M1 transitions, the ratio of their strengths in (e,e') is about 1:1.3, whereas in (p,p'), the ratio of strengths is about 1:5 at 4°.

These different relative strengths suggest therefore a different neutron-proton structure for these two states.

Table 2. Comparison of 1^+ States in ^{58}Ni and ^{60}Ni observed in (e,e') and (p,p') reactions. Energies are in MeV.

^{58}Ni			^{60}Ni	
(e,e') (ref. 30)	(e,e') (ref. 31)	(p,p') present work	(e,e') (ref. 30)	(p,p') present work
6.05			11.87	11.85
6.41			12.34	12.20
7.09				
	7.7	7.7*		12.73
9.85	9.852	9.82	13.11	13.25
10.18	10.224	10.18	13.35	13.55
10.55	10.515	10.48	13.84	13.99
10.66	10.676	10.65		
11.03	11.020	10.98		
	11.92	11.84		
	12.00	12.25		
		12.70		
		13.25		
	14.18			

*Only the strongest 1^+ state of the broad resonance is given in the Table.

There are also differences observed in the calcium isotopes. In ^{40}Ca, an (e,e') experiment[31],[32] notes 1^+ states at 10.319 MeV and possibly at 9.868 MeV where there is probably a close doublet. In our (p,p') experiment we observe the 10.31 MeV state clearly but also observe a state with a $\Delta L=0$ angular distribution at 12.03 MeV. (See Fig. 9) The analogue of this 12.03 MeV state has also been seen in a ^{40}Ca(p,n) experiment[33] suggesting that it has $J^{\pi}=1^{+}$. Differences between (p,n) and (e,e') are also observed in ^{42}Ca where a single 1^{+} state was reported at 11.235 MeV in (e,e') and a state at about 10.2 MeV, suggested to be 1^{-}, was seen in (p,n).[34] The resolution of these differences in cases where individual levels can apparently be resolved may cast light on further comparisons in heavier nuclei.

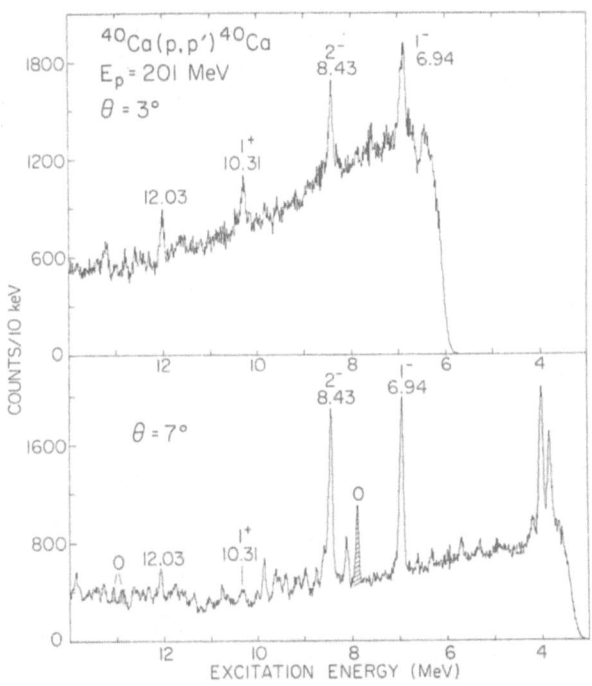

FIG. 9. Spectra of protons inelastically scattered from ^{40}Ca at 3° and 7°. Peaks due to an oxygen contaminant are shown hatched.

As a final comparison with electron scattering, a recent measurement of ^{48}Ca(p,p') at 200 MeV shows the very strong excitation of the 10.2 MeV 1^{+} state at forward angles similar to the observation in (e,e') at backward angles.[3,5] The high selectivity of the (p,p') reaction is seen in Fig. 10 where the spectra at 2°, 4°, 6° and 8° are displayed. At 2° only the 1^{+} state is visible, but at 8° many other levels are clearly seen. The angular distribution for the 10.2 MeV 1^{+} state, which is very sharply forward peaked, is shown in Fig. 11.

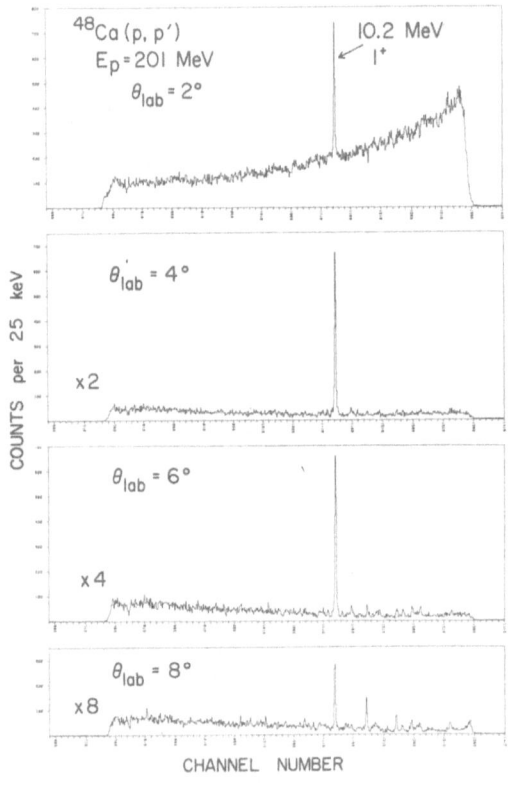

FIG. 10. Spectra from the ^{48}Ca(p,p') reaction at 201 MeV at laboratory angles of 2°, 4°, 6° and 8°.

FIG. 11. Preliminary angular distributions (in arbitrary units) for the 10.2 MeV 1$^+$ state observed in ^{48}Ca(p,p') at 201 MeV.

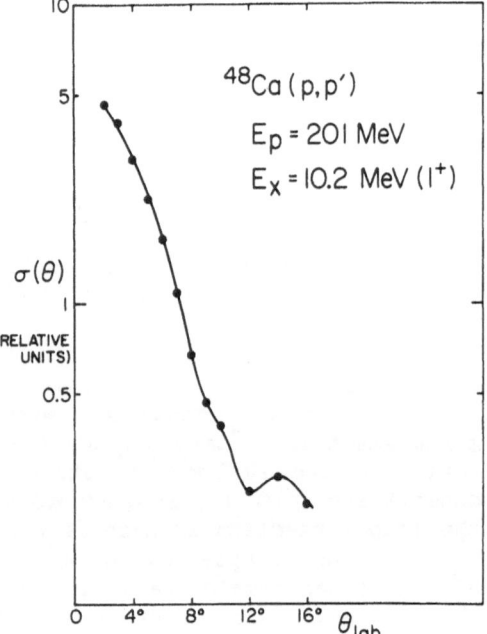

VI.   CONCLUSIONS AND FUTURE DIRECTIONS

Following the clue provided by the (p,n) results on G-T resonances, we have found broad structures with widths of a few MeV in seventeen nuclei from ^{51}V to ^{140}Ce using the (p,p') reaction at 200 MeV. The excitation energies of these broad bumps are consistent with those expected for an M1 transition. The angular distribution which is very forward peaked, and the strength, which is comparable to that expected from (p,n) measurements using a simple model, also support this assignment. Comparison with microscopic calculations in a few cases using simple wave functions, implies that the transitions are only about 30% of the predicted values. The precise value of the "quenching" and particularly the mass dependence await more sophisticad calculations.

Apart from the quenching problem, there are a number of other interesting avenues which could be pursued using the (p,p') reaction at high energy.

(1)   The measurement of analyzing power and particularly spin-flip probability in the region of the M1 transition would be very interesting. One might expect that the spin-flip probability would be high for an M1 transition and thus serve as an additional tool in extracting the resonance from the background.

(2)   The decay of the state may also provide a means of enhancing the resonance relative to the background. There is considerable interest in finding the M1 resonance in ^{208}Pb and a coincidence measurement using the γ-decay of the state may help in such a search.

To recapitulate, the interaction between different experimental methods has proved extremely fruitful in studying M1 transitions. Both (e,e') and (p,n) data provided stimulation for the present (p,p') work. This has in turn led to other (p,n) studies searching for higher isospin components of the G-T strength and to more detailed comparisons between (e,e'), (p,n) and (p,p').

ACKNOWLEDGEMENTS

This work was partly supported by the U.S. National Science Foundation under Grant No. Phy-78-22696. Three of the authors (GMC, NA and AG) received travel support from the INT Division of the National Science Foundation. One of the authors (GMC) would also like to thank the University of Paris XI for financial support and warm hospitality during his visit as Professeur d'echange. Finally, the authors are indebted to H. Toki, W.G. Love, N. Auerbach, Nguyen Van Giai, D. Gogny and J. Decharge for many enlightening discussions.

REFERENCES

1. D.E. Bainum, J. Rapaport, C.D. Goodman, D.J. Horen, C.C. Foster, M.B. Greenfield and C.A. Goulding, Phys. Rev. Letters 44, 1751 (1980) and C.D. Goodman, The (p,n) Reaction and the Nucleon-Nucleon Force, Proceedings of the Telluride Conference March 29-30 (1979), Plenum Press, New York (1980) p. 149.

2. R.R. Doering, A. Galonsky, D.M. Patterson and G.F. Bertsch, Phys. Rev. Letters 35, 1691 (1975).

3. W. Steffan, H.D. Graf, W. Gross, D. Meuer, A. Richter, E. Spamer, O. Titze and W. Knupfer, Phys. Lett. 95B, 23 (1980).

4. W. Knupfer, R. Frey, A. Friebel, W. Mettner, D. Meuer, A. Richter, E. Spamer and O. Titze, Phys. Lett. 77B, 367 (1978).

5. A. Richter, International School on Nuclear Structure, Alushta, USSR (1980).

6. W.G. Love and M.A. Franey, Phys. Rev. C24, 1073 (1981).

7. T. Yamazaki, "Mesons in Nuclei" edited by M. Rho and D. Wilkinson (North Holland, Amsterdam 1979) Vol. II, p. 651.

8. M. Ericson, A. Figureau and C. Thevenet, Phys. Lett. 45B, 19 (1973) and J. Delorme, M. Ericson, A. Figureau and C. Thevenet, Ann. Phys. 102, 273 (1976).

9. M. Rho, Nucl. Phys. A231, 493 (1974) and A. Bohr and B. Mottelson, Phys. Lett 100B, 10 (1981).

10. F.E. Cecil, G.T. Garvey and W.J. Braithwaite, Nucl. Phys. A232, 22 (1974).

11. A. Richter, in Nuclear Physics with Electromagnetic Interactions, edited by H. Arenhovel and D. Drechsel, Lecture notes in Physics Vol. 108 (Springer, Berlin 1979) p. 19.

12. D. Meuer, R. Frey, D.H.H. Hoffman, A. Richter, E. Spamer, O. Titze and W. Knupfer, Nucl. Phys. A349, 309 (1980).

13. A. Willis, M. Morlet, N. Marty, R. Frascaria, C. Djalali, V. Comparat and P. Kitching, Nucl. Phys. A344, 137 (1980).

14. W. Sterrenberg, S. Austin, A. Galonsky, T. Nees, D. Bainum, J. Rappaport, C. Foster, C. Goodman, D. Horen, C. Goulding and M. Greenfield, Proceedings of the Inter. Conf. on Nucl. Phys., Aug. 1980, Berkeley, (Abstracts) p. 176.

15. N. Anantaraman, G.M. Crawley, A. Galonsky, C. Djalali, N. Marty, M. Morlet, A. Willis and J.C. Jourdain, Phys. Rev. Letters 46, 1318 (1981).

16. F.E. Bertrand, E.E. Gross, D.J. Horen, J.R. Wu, J. Tinsley, D.K. McDaniels, L.W. Swenson and R. Liljestrand, Phys. Lett. 103B, 326 (1981).

17. G. Crawley, N. Anantaraman, A. Galonsky, C. Djalali, N. Marty, M. Morlet, A. Willis, J.C. Jourdain and P. Kitching, Phys. Rev. C 26, 87(1982).

18. C. Djalali, N. Marty, M. Morlet, A. Willis, J.C. Jourdain, N. Anantaraman, G.M. Crawley, A. Galonsky and P. Kitching, Nucl. Phys. A388, 1(1982).

19. Code DWUCK-4, P.D. Kunz, University of Colorado Report (unpublished).
20. C. Gaarde, J. Rapaport, T.N. Taddeucci, C.D. Goodman, C.C. Foster, D.E. Bainum, C.A. Goulding, M.B. Greenfield, D.J. Horen, E. Sugarbaker, Nuclear Phys. A369, 258 (1981).
21. W.A. Sterrenburg, S.M. Austin, R.P. Devito and A. Galonsky, Phys. Rev. Letters 45, 1839 (1980).
22. H. Toki - preprint 1982.
23. N. Anantaraman et al., unpublished.
24. G.F. Bertsch, Nucl. Phys. A354, 157 (1981).
25. H. Toki, D. Cha and G. Bertsch, Phys. Rev. C24, 1371 (1981).
26. H. Toki and W. Weise, Phys. Lett. 97B, 12 (1980).
27. C. Djalali, N. Marty, M. Morlet and A. Willis, Nucl. Phys. A380, 42 (1982).
28. L.W. Fagg, Rev. Mod. Physics 47, 683 (1975).
29. R. Pitthan, Z. Phys. 260, 283 (1973).
30. R.A. Lindgren, W.L. Bendel, E.C. Jones, L.W. Fagg, X.K. Muruyama, J.W. Lightbody and P.P. Fivozinsky, Phys. Rev. C14, 1789 (1978).
31. R. Frey et al., Int. Conference on Nucl. Structure, Tokyo 1977.
32. P.E. Burt et al., Phys. Rev. C25, 2805 (1982).
33. T.N. Taddeucci et al., - preprint 1981.
34. C.D. Goodman et al., Phys. Lett. 107B, 406 (1981).

NUCLEUS AS A CHIRAL FILTER: THE ROLE OF THE Δ (1232)[†]

Mannque Rho

Service de Physique Théorique
CEN SACLAY
91191 Gif-sur-Yvette Cedex, France

ABSTRACT

I describe how two different modes of chiral symmetry can be seen in nuclei. In particular, it is shown that the nuclear axial charge or more precisely the $0^+ \leftrightarrow 0^-$, $\Delta T=1$ transition at zero momentum transfer probes the nuclear configuration wherein the axial charge g_A is effectively enhanced in nuclear medium due to soft pions, symptomatic of the Goldstone realization of chiral symmetry in the medium while the Gamow-Teller resonances probe the configuration wherein soft pions are no longer operative, suggesting an approach toward the Wigner realization of chiral symmetry. Using the celebrated Adler-Weisberger relation, it is argued that the observed ~ 50 % quenching of the Gamow-Teller strength reflects the possibility that the Gamow-Teller operator sees the quarks inside the bag, blind to the Goldstone vacuum outside. Some implications on chiral phase transitions are also discussed.

I. INTRODUCTION

The recent experimental developments on nuclear axial charges and giant Gamow-Teller resonances bring into focus the interplay of chiral symmetry in nuclear many-body system and highlights some fascinating aspect of nuclear structure that is now emerging as one begins to understand better the underlying theory of strong interactions, namely QCD[1]. In this talk, I would like to discuss in a unified way how to correlate the experimental observations

---

[†]Invited talk given at the International Conference on Spin Excitations in Nuclei, Telluride, Colorado (March 25-27, 1982)

with what is now believed to be a complex structure of the vacuum
wherein nuclear matter provides a convenient laboratory with controlled conditions.

The crux of the matter is the question : what is a nucleus in
the quark-gluon picture of QCD ? Nobody knows the answer. But I
claim that we have some beginning of understanding at least in one
aspect of QCD, namely chiral symmetry. In hadrons made up of up
(u) and down (d) quarks whose current masses are nearly zero, there
is an almost perfect global symmetry $SU(2) \times SU(2)$ called chirality:
strictly zero mass quarks conserve their helicity in their equation
of motion. Thus the QCD Lagrangian for the u and d quarks is almost
invariant under this symmetry transformation. [For convenience, let
us forget the tiny masses (known approximately to be $m_u \approx$ 4 MeV,
$m_d \approx$ 7 MeV) whenever no ambiguity arises]. The vacuum, on the other
hand, is known to break this symmetry, presumably due to strong
coupling which induces quark-antiquark pair condensation, as a consequence of which zero-energy Goldstone bosons are excited. Nature
confirms this picture, the relic of the Goldstone bosons being the
triplet of pions $\pi^\pm, \pi^0$. Thus in nature, chiral symmetry is realized
in the Goldstone mode rather than in the alternative mode, Wigner-
Weyl mode. If QCD is to be a correct theory, this feature must form
an ingredient as essential as quark confinement. At the moment,
nobody has a very clear idea how one can *derive* these two features
but there is no reason to doubt them either. The sensible thing to
do at least for the moment seems to assume them and see how well
things fit in.

A phenomenological way of looking at a hadron is to consider
it as a bubble in a complex medium or simply as a bag[2]. The bag
may be visualized as a means of confining quarks and gluons, but
it also delineates the inside and outside of the bag as two regions
of chiral symmetry realization. Inside it is in a Wigner mode realized with zero mass spin 1/2 objects (spin $\geq$ 3/2 cannot be zero
mass) ; outside it is in a Goldstone mode realized with zero mass
scalar mesons, e.g., pions. Classically pions do not penetrate into
the inside region though fluctuations of the same quantum number
will occur and propagate as $q\bar{q}$ pairs. Is the confinement size the
same as the size of the Wigner bubble ? Probably not as I will
argue later on.

What is the order parameter for the two regions if considered
as two phases in statistical mechanics ? It is actually $<\bar{\psi}\psi>$, but
for our discussion, the relevant quantity will be the axial-vector
coupling constant g_A. It is well known[3] that because of pions,
$g_A \neq 1$ in the Goldstone mode (experimentally $\sim$ 1.26), while in the
Wigner mode $g_A = g_V = 1$.

What is a nucleus in this picture ? It seems reasonable to
consider it as a collection of bags. If the bag size is not too

big, then those bags can be thought to be surrounded by the exact
vacuum infested with Goldstone pions. Two nucleons in a nucleus
must be in various configurations, namely those in Fig.1 with appropriate probabilities. The wave function is made up of linear combinations of these configurations. In what follows I will discuss
how one can set a filter to *see* the configuration I or the configuration II. The quantity g_A will in principle tell us which configuration is observed.

2. PROBING WITH E.M. AND WEAK CURRENTS

When a nucleus reponds to a current, say the vector current
V_μ^i or the axial-vector current A_μ^i where μ is the Lorentz index and
i the isospin index (i=1,2,3), the current samples all configurations with certain weights associated with their probabilities.
The wave function should of course take care of the latter. In developing the idea, I shall use the following which is well established
in nuclear physics although its origin is not understood at all :
that two nucleons feel a repulsive core at short distances. (Where
does this repulsive core come from is an open question in nuclear
physics, usually sought in quark-gluon dynamics[4]). As a consequence,
two nucleons like to stay apart, given kinematical and symmetry

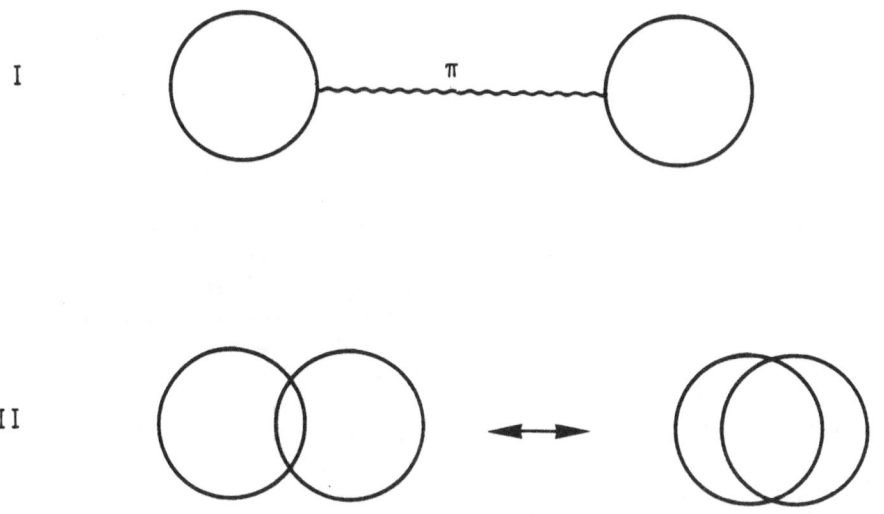

Fig. 1. Configuration I where soft pions influence the nuclear
response and configuration II where soft pions "evaporate".

conditions to allow them to do so. This phenomenon, usually referred to as short-range anticorrelation, is assumed to play the role of
a *filter* to block the second configuration in Fig.2.

   Let me first present a qualitative argument to set the stage
for the quantitative discussion to be made later. Consider two
nucleons in interaction responding to the current. The two nucleons,
when not overlapping, are surrounded by the true vacuum infested
with the Goldstone excitation (e.g. pions). I can then make an expansion in terms of number of pions exchanged between the two which
is equivalent to an expansion in terms of *configurations*. This is

Fig. 2. Chiral expansion of the amplitude (the cross indicating
        where the current acts) in terms of number of pions exchan-
        ged. The first term (a) gives the soft-pion limit; the rest
        corrections to the soft-pion limit.

quite similar to Weinberg's discussion of the $\pi\pi$ interaction in
terms of a chiral Lagrangian[5], so I shall pattern my argument[6] after
his, although our case is more complex and much less rigorous than
his. In diagrams, the amplitude looks like Fig.2. I will generical-
ly designate a characteristic energy or momentum scale involved in
the process by Q (in a nucleus, it is actually the three momentum
carried by the pions). To make this expansion meaningful, Q should
not be much larger than m_π. Chiral symmetry arguments allow one to
write[6]

$$M \sim Q^{-1} \left[Z + \left(\frac{Q}{f_\pi}\right)^2 b \left(\frac{Q}{\mu}\right) + \ldots \right] \quad . \tag{1}$$

Here Z is the one soft-pion exchange term corresponding to Fig. 2a, independent of Q, f_π the pion decay constant and μ the renormalization scale parameter needed to define loop diagrams. Beyond Q^2 order, there are lots of other terms that come in, but the nice thing is that one can have some ideas on those higher order terms from quite general considerations, as Weinberg demonstrated for the $\pi\pi$ scattering[5].

Now the crucial assumption is that whenever the soft-pion term is kinematically allowed, it dominates over the higher Q^2 terms, the reason behind this being that the short-range anti-correlation keeps those terms small. (So far this assumption has been checked to be valid in light nuclei by model calculations, though the fundamental mechanism is still to be discovered). Conversely whenever the Z term is suppressed kinematically or by symmetry, the non-soft-pion terms *need not* be negligible. Furthermore processes not appearing in the chiral expansion (1) can contribute. This corresponds to filtering off of the soft-pion exchange configuration I, allowing the configuration II to manifest.

Let us see how this argument works[7]. For two bags (i.e. two nucleons) that are widely separated, the exchanged pion is soft, so the relevant vertex to concentrate on in Fig.2a is that one where the current is acting. There the current algebra[3] gives the relevant matrix elements

$$[V_\mu^i, Q_A^j] = i\,\epsilon_{ijk}\,A_\mu^k \quad , $$
$$[A_\mu^i, Q_A^j] = i\,\epsilon_{ijk}\,V_\mu^k \quad . \tag{2}$$

Here Q_A^i is the axial charge. Since nucleons in nuclei are non-relativistic, one can look at the space and time components of the currents separately in non-relativistic limit and discovers that the two components behave quite differently : for the probes V_μ^i, A_μ^i (those appearing on the left-hand side of Eq.(2)), the vertex matrix elements have the properties :

	$\mu=0$ (Charge)	$\mu=1,2,3$
Vector current	O(P/M) Charge form factor	O(1) M1
Axial vector current	O(1) Axial charge	O(P/M) Gamow-Teller

From this follows our first conclusion : the M1 and the axial charges should sample primarily the soft-pion configuration I whereas the vector charge and the Gamow-Teller transition should "see" the short-range configuration II.

3. "SEEING" SOFT PIONS

I now discuss how this prediction fares with experiments. The well-known case where this argument works beautifully is the electro-disintegration of deuterons with small energy transfer and large momentum transfer[8] (see Fig.3). This has been discussed extensively in recent conferences[9] so I will not repeat it. I will just say that it is rather surprising that soft pions survive up to a large momentum transfer. This indicates rather persuasively that pions do not evaporate until small two-nucleon distances and if one argues on the basis of the two-phase nucleon picture discussed above, then the chiral bag must be considerably smaller than that implied by the usual hadronic size.

A more recent exciting confirmation of our simple picture comes from the axial charges, $<0^-|A_o^i|0^+>$ or in general the matrix element $<f|A_o^i|i>$. This information can be accurately extracted from the mass-12 triplets, since these nuclei have been thoroughly studied both experimentally and theoretically. The essential idea is to measure various angular correlations of the $\beta^\pm$ decays and, supplemented with the $1^+ \rightarrow 0^+ \gamma(M1)$ decay and μ-capture processes, to separate out the time component. A detailed theoretical analysis has recently been done by Guichon and Samour[10]. They have been able to show a clear evidence of the soft-pion presence in both Γ_γ and A_o matrix elements. The results are summarized in Fig.4.

The most convincing evidence comes from the axial charge

$$<0^-|Q_A^i|0^+> \qquad . \tag{3}$$

Suppose I write this quantity in terms of a single-particle charge operator $Q_{As.p.}^i$:

$$<0^-|Q_A^i|0^+> = \eta<0^-|Q_{As.p.}^i|0^+> \quad . \tag{4}$$

One can easily calculate η from the soft-pion term [i.e. Eq.(2) and Fig.2a] and the prediction was[7]

$$1.4 \lesssim \eta \lesssim 1.6 \tag{5}$$

where the lower limit corresponds to light nuclei like ^{16}O and the upper limit to the nuclear matter. One notes the insensitive density dependence and the extraordinarily large contribution from the soft-pion presence. One can express Eq.(4) in terms of an "effective charge" g_A^π ;

$$g_A^\pi = \eta \, g_A \approx 1.75\text{-}2 \quad . \tag{6}$$

One can interpret this as saying that in the presence of soft pions,

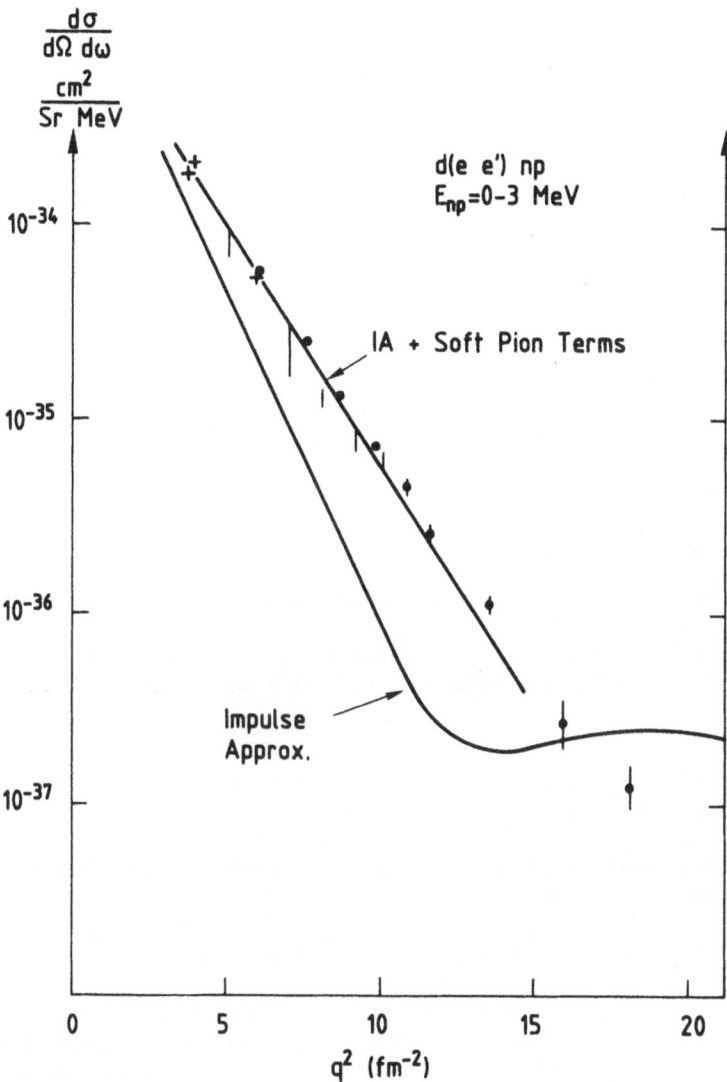

Fig. 3. Electro-disintegration of the deuteron. The data are from
Saclay[8] and the theoretical prediction with soft-pion
corrections from Ref.1.

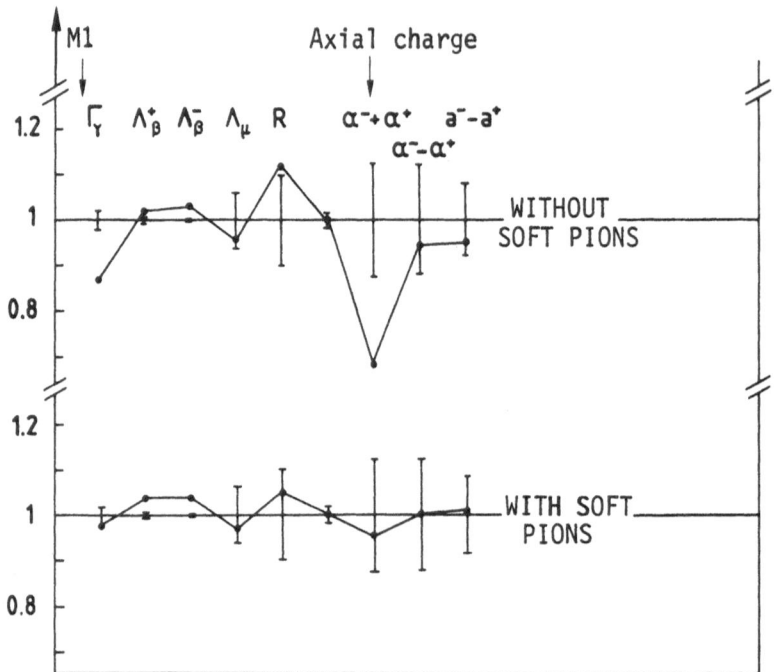

Fig. 4. Ensemble of the theoretical results normalized to the experimental data for A=12 system. The quantities most sensitive to the soft-pion influence are the gamma width $\Gamma\gamma$, and
the angular correlation $(\alpha^- + \alpha^+)$ of $\beta^\pm$ decays. See Ref.10.

the effective g_A is *enhanced*. I will discuss the meaning of this
enhanced g_A^π more carefully later on.

It now appears that this large enhancement is clearly pin-
pointed in the process

$$^{16}N(0^-,T=1) \rightarrow {}^{16}0(0^+,T=0) + e^- + \tilde{\nu}_e \quad . \tag{7}$$

Because of selection rules, the time component shows up significantly. Because of cancellations between the single-particle matrix
elements, the two-body process with a soft-pion comes in with ven-
geance. A model-independent prediction is that the decay rate Λ_β [11]
would be enhanced by a factor of more than three. Towner and Khanna,
in a careful calculation, confirmed this expectation and predicted
(using the force which they had constructed for understanding of
Gamow-Teller transitions, see later)

$$\Lambda_\beta^{Th} = 0.42 \text{ sec}^{-1} \tag{8}$$

which was in agreement with the only data available then[12]

$$\Lambda_\beta = (0.43 \pm 0.10) \text{ sec}^{-1} \quad . \tag{9}$$

The recent beautiful re-measurement by Gagliardi, Garvey, Wrobel and Freedman[13] corroborated this result and provided a convincing case by further reducing the error bar

$$\Lambda_\beta = (0.39 \pm 0.05) \text{ sec}^{-1} \quad . \tag{10}$$

The important point is that Λ_β^{Th} is consistent with the inverse process, μ-capture data and that if the soft pions were not present, the theory would undershoot the experiment by a factor of more than three.

The Ml process (e+d → e'+n+p) and the axial charge provide a common information : that the soft-pion presence is *seen*. Furthermore the $0^- \rightarrow 0^+$ β transition shows that the axial charge associated with non-relativistic nucleons is enhanced *when soft pions contribute* ; and totally blind to the quark degrees of freedom !

One additional remark : consider PCAÇ,

$$\partial_\mu A_\mu^i(x) = C \, \phi^i \quad .$$

If I take a matrix element between nuclear states i & f and integrate over the space, then

$$(E_f - E_i) < f|Q_A^i|i> = C \int d^3x \, <f|\phi^i|i> \quad .$$

The right-hand side is just the s-wave πfi vertex. Assuming that things extrapolate smoothly as required by PCAC, then the enhanced axial charge (i.e. the left-hand side) implies that the two-body contribution to the s-wave pion absorption is precisely (η-1) times the single-particle term or 40 to 60 % increase. This should be a strong constraint on the models built to describe s-wave pion nuclear interaction.

4. FILTERING OFF SOFT PIONS

Let me now turn to the situation where soft pions are (by kinematical constraints) screened. There is a dramatic change of physics in this case. As noted in section 2, the relevant observables could be e.m. charge form factors of nuclei and Gamow-Teller transitions. Because of charge conservation, there cannot be multibody corrections to the charge and hence one has to look at form factors to uncover new degrees of freedom. Although there are numerous attempts to calculate mesonic corrections to charge form fac-

tors (all of which are model-dependent) in such light nuclei as
^{2}H and ^{3}He, there is at present no reliable theory whatsoever*.
We know very little of what sorts of physics are involved there.
This remains an open problem.

The situation with the Gamow-Teller transition is quite different and in my opinion opens up an extremely interesting window
to a new area of physics. The recent beautiful experimental results
on the giant Gamow-Teller resonances[14] attest to this aspect.

The specific feature that is relevant to my discussion is the
quenched Gamow-Teller strength that Goodman talked about in this
conference. In fact the quenching of the Gamow-Teller strength has
rather a long history, so rather than describing the quenching argument in detail which is now fairly well known, I shall summarize
the main development and then discuss the quenching phenomenon in
a new light, namely in terms of the chiral structure of the vacuum.

As far as I know, Migdal[15] was the first to propose calculating
Gamow-Teller matrix elements in a way presaging the recent development, although he was not exactly on the right track. He assumed
that the axial current matrix element for nuclear transitions $i \to f$
can be written as

$$e_A <f|\underset{\sim}{A}^{\pm s.p.}|i>_{g_o'} \qquad (11)$$

where $\underset{\sim}{A}^{s.p.}$ is the *single-particle* axial current operator, e_A the
effective charge associated with the current $\underset{\sim}{A}^{\pm}$ and the nuclear
states i and f are to be calculated with the (g_o') Landau-Migdal
effective force[15]. For the Gamow-Teller transitions, say $0^+(T=0) \to$
$1^+(T=1)$, the relevant Landau-Migdal force is

$$\sim g_o' \vec{\tau}_i \cdot \vec{\tau}_j \sigma_i \cdot \sigma_j \ \delta(r_i - r_j) \qquad . \qquad (12)$$

Migdal did not know how to calculate e_A (since he did not think
that e_A was a conserved charge), so he simply assumed it to be the
same as the renormalization of magnetic moments and set $e_A \approx 0.9$.
This Migdal mechanism was then later invoked[16] to explain the factor of more than two discrepancy between the usual shell-model
calculation in RPA and the experiment (which was quenched) in the
process $\mu^- + {}^{16}O(0^+) \to \nu_\mu + {}^{16}N(2^-, T=1)$. It was found[16] that with
$g_o' = 0.5$ and $e_A \lesssim 0.9$, the capture rate could be easily quenched by
a factor of three or more : Here e_A was still taken to be a parameter < 1. The first attempt[17] to calculate e_A was made in 1970 in
terms of soft-pion exchange currents. As we now know, this was not
the right way to get the quenching and led instead to an enhancement.

---

*
Of course this does not prevent some people from calculating
lots of Feynman diagrams without knowing what they are doing.

(As demonstrated in Sec.3, soft pions are *screened* for this phenomenon).

Shortly afterwards, Wilkinson[18] showed by a careful analysis
of allowed Gamow-Teller transitions that the axial vector coupling
constant g_A is systematically quenched by about 10 % in light nuclei.
The understanding of this phenomenon came from two independent developments. The first was the proposal by M.Ericson and her colleagues[19] at Lyon that the g_A can be quenched in nuclei due to the
Lorentz-Lorenz effect already known in pion-nuclear interaction.
(The original derivation of the quenching $\gamma = g_A^{eff}/g_A$ was not quite
right ; this was corrected in their later works). The second was
the simple description[20] of the Lorentz-Lorenz effect in terms of
Δ-hole configurations where the Δ was treated as "elementary" in
the sense that it has a three-quark wave function that differs from
the nucleon only in its spin-isospin content. The combination of
the two ideas led to the prediction[21] (for uniform nuclear matter)

$$\gamma = e_A = \frac{1}{1+\alpha} \tag{13}$$

with

$$\alpha = \frac{1}{3} \left(\frac{8}{9} f^{*2} \frac{\rho}{m_\pi^2 \omega_R} \right) \tag{14}$$

where f^* is the $\pi N\Delta$ coupling [$(f^*/f)^2 = 72/25$ in constituent quark
model, $=2$ in the Chew-Low model], ρ is the nuclear matter density,
$\omega_R \simeq 300$ MeV the Δ-N mass difference.

The next step was then to realize that the classical Lorentz-
Lorenz factor 1/3 in Eq.(14) should be replaced by a Landau-Migdal
parameter g_o' appropriate in the Δ-nucleon sector. The simplest
thing to do was to assume a universality-type relation[22], namely
that

$$g_o' = \bar{g}_o' = \bar{\bar{g}}_o' \tag{15}$$

when the Landau-Migdal effective forces are parametrized as in
Fig.5. The upshot of all these is that Eq.(11) now reads

$$e_A(g_o') \ <f| \underset{\sim}{A}^{\pm s \cdot p \cdot} |i>_{g_o} \quad , \tag{16}$$

with $e_A(g_o')$ given by Eq.(13), but g_o' replacing the 1/3 factor in
Eq.(14). Thus the same g_o' is seen to govern the "effective charge"
and the "effective matrix element". I must say that this is not
universally accepted. For instance, Migdal has argued that while
$g_o' \approx 0.5$, both $\bar{g}_o'$ and $\bar{\bar{g}}_o'$ are negligibly small. In fact, he has set
$\bar{g}_o' = \bar{\bar{g}}_o' = 0*$.

---
*As pointed out by several people (e.g. Meyer-ter-Vehn[23]), this led
Migdal to predict pion condensation at a too low nuclear density.

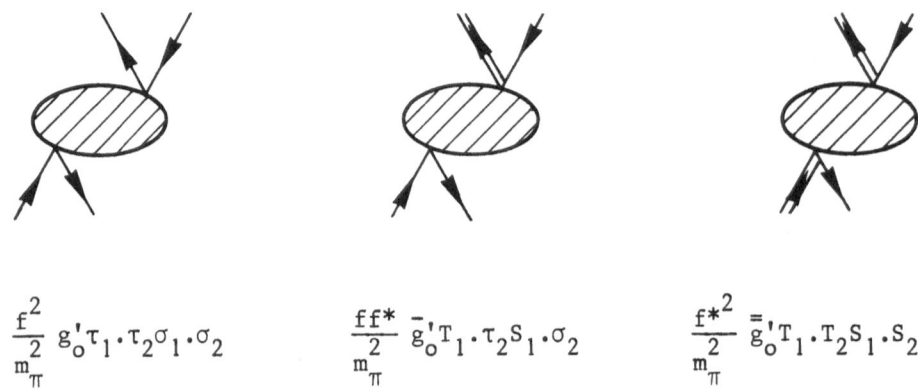

$$\frac{f^2}{m_\pi^2} g_0' \tau_1 \cdot \tau_2 \sigma_1 \cdot \sigma_2 \qquad \frac{ff^*}{m_\pi^2} \bar{g}_0' T_1 \cdot \tau_2 S_1 \cdot \sigma_2 \qquad \frac{f^{*2}}{m_\pi^2} \bar{\bar{g}}_0' T_1 \cdot T_2 S_1 \cdot S_2$$

Fig. 5. The Landau-Migdal interactions between N-hole and Δ-hole states. S and T are transition operators.

Although they did not discuss this in the context of the Gamow-Teller quenching, Brown, Weise, Bäckman, Oset and others[24] were predating this development in their studies of pion condensation by calculating

$$g_0' = \bar{\bar{g}}_0' \tag{17}$$

using π- and ρ-exchange mechanisms. Equation (15) was implicit in all their work**. g_0' was found to be dominated by the ρ-exchange – and *the associated force very short-ranged* – with the numerical values ranging

$$g_0' = 0.5 \sim 0.8 \qquad . \tag{18}$$

The Δ-hole interaction through the g_0' force, appropriately modified for finite nuclei, was found to correctly describe the observed Gamow-Teller quenching in light nuclei[25] and seems to account for much of the missing strength in the Gamow-Teller resonances[26]. This is roughly what had happened before ; anything further is the subject of this topical meeting.

_____

** This is a natural consequence of the π- and ρ-exchange mechanism. See Ref.26.

Let me now turn to a novel interpretation of the Gamow-Teller quenching. I am sure that there will be a lot of calculations in various different approximations which will support or refute the mechanism based on the Δ-hole configuration and the g_0' force. The controversy is bound to last for some time. Here I am going to present another argument, a more fundamental one, why the quenching mechanism and the magnitude of the quenching must be correct.

The amount of quenching that we are talking about is (in heavy nuclei or nuclear matter)

$$e_A \approx 0.70 \tag{19}$$

or

$$g_A^{eff} \approx e_A g_A \approx 0.88 \quad . \tag{20}$$

Note that this is considerably smaller than the soft-pion-modified quantity g_A^{π}, Eq.(6). We will return to the relation between the two later.

I am going to make somewhat sweeping assumptions to simplify the matter enormously. I will assume that the quenching arises when two or more nucleons interact at a very short distance (in the meson-exchange language, g_0' comes from the ρ-exchange), in fact *when the bags overlap.* In other words, Goldstone pions are banished from the consideration. Then the two overlapping *bags can be in either of the possible* configurations described in Fig.6. The two configurations differ by simultaneous spin-isospin flip of two quarks only and as far as the vacuum outside is concerned cannot be distinguished. [The mass difference $(M_\Delta - M_N)$ in free space is presumably due to gluon exchanges within the bag]. So when one piece of the bag beta-decays, *we cannot say that it is a pure nucleon that decays.* It must represent a decay of some object which cannot be labelled as a pure nucleon or a pure Δ. Imagine that we have an SU(4) commutation and derive a sum rule from it :

$$\left[\sum_i \tau_i^+ \sigma_i^3 \, , \, \sum_j \tau_j^- \sigma_j^3 \right] = \sum_i \tau_i^3 \quad . \tag{21}$$

Take an expectation value with a nucleon state and saturate the intermediate sum (on the left) by the N and the Δ state. Separating out the nucleon term and defining g_A as

$$g_A = \langle P_\uparrow | \sum_i \tau_i^+ \sigma_i^3 | N_\uparrow \rangle \tag{22}$$

and evaluating the N-Δ transition explicitly in terms of the constituent (non-relativistic) quark model, one gets the usual SU(6)

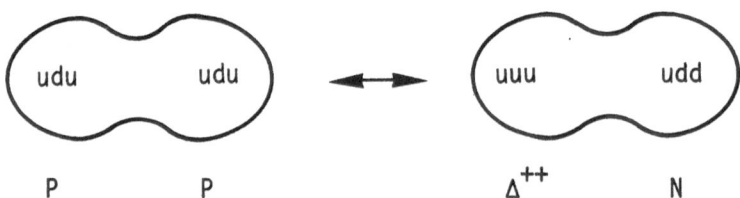

Fig. 6. The six-quark structure of two overlapping baryons probed
        by the Gamow-Teller operator. The latter seems blind to the
        Goldstone mode outside.

value :

$$g_A = 5/3 \quad .$$

(23)

Now assume that the hadronic state put in the intermediate sum is
something like a left piece of Fig.6*. Then the change in the nucleon wave function (in the intermediate state which is probed by
the GT operator) eats up the contribution from the would-be Δ, so
when finished one would get

$$g_A^{eff} = 1 \quad .$$

(24)

At this point, one would have

$$e_A = 3/5 = 0.6 \quad .$$

(25)

But the statement that $g_A = 1$ is the same as that the hadron is made
up of free quark plasma. In terms of the constituent (non-relativistic) quark picture, it is not obvious whether (23) and (24) represent two symmetry realizations of chiral symmetry (since the role

---

*The same argument goes through for a nucleus of Z protons and N
neutrons. One just has to divide by Z-N on both sides.

of chiral symmetry in such a model is not clear*). In terms of the quark bag model, however, $g_A \approx 1$ can arise only if the Wigner mode is sampled or approached.

Now to a more realistic case. Consider the Adler-Weisberger derivation of g_A, which comes from the chiral charge algebra

$$[Q_A^i, Q_A^j] = i\epsilon_{ijk}Q^k \tag{26}$$

by sandwiching between nucleon states, writing the commutators by inserting a complete set of intermediate states and going to infinite momentum frame[3] (this to obtain a Lorentz invariant quantity**). Imagine saturating the intermediate sum by the nucleon, the $\Delta(1232)$, the N*'s and the Δ*'s. The nucleon contribution is separated out to define g_A and the rest evaluated with experimental input. This calculation was done long ago and by many people. Let me quote the result of Della Selva and Masperi[27] as an exemplary calculation :

$$1 = 1/g_A^2 - \sum_{n=\Delta,\Delta*,N*} C_n \tag{27}$$

with the numerical values (for baryon masses $\lesssim$ 3 GeV)

$$C_{\Delta(1232)} = -0.675$$
$$\sum_{n=\Delta*} C_n = -0.064 \tag{28}$$
$$\sum_{n=N*} C_n = 0.341 \quad .$$

Equation (27) gives, with (28),

$$g_A = 1.29 \tag{29}$$

close to the experimental value 1.26.

Imagine now the configuration of Fig.6 as an intermediate state. As demonstrated above with the simple model, this just means that the N and the Δ states are no longer distinguishable and we may assume this as before to mean that the change in g_A^{eff} eats up completely the $\Delta(1232)$ contribution, i.e.,

$$1 = (g_A^{eff})^{-2} - \sum_{n\neq\Delta(1232)} C_n \quad . \tag{30}$$

---

* Equation (21) may be viewed as the axial charge algebra in infinite momentum frame. In this sense, there is something to it from chiral symmetry point of view.

** For alternative ways of deriving this and subtleties involved therein, see Ref.3.

Using (28), we see that

$$g_A^{eff} = 0.88 \tag{31}$$

$$e_A \approx 0.68$$

close to the value obtained in the "traditional" way, i.e. the results (19) and (20). The fact that g_A^{eff} is less than 1 means that higher-energy states (N*) still contribute as distinct states, so if one assumes that at normal nuclear densities, the Δ(1232) loses its identity in Gamow-Teller transitions, one would expect that at some higher density, say $\rho \gtrsim 2\rho_0$ (ρ_0 = nuclear matter density$\approx 0.48 m_\pi^3$), all the other Δ*'s and N*'s lose their identities at which point one would have

$$g_A \approx 1 \quad . \tag{32}$$

This means that as far as Gamow-Teller transitions are concerned, quarks behave as if there were no vacuum with Goldstone phase. Put differently the probe is blind to the true vacuum surrounding the overlapping bags. In an infinite system, this must be like a percolation phenomenon Gordon Baym likes to talk about[28] but probably not the true Wigner phase in which pions no longer exist. (See below).

5. CHIRAL PHASE TRANSITION

In nucleus, both the soft-pion-dominated configuration and the overlapping configuration with different effective axial charges can be probed with a judicious choice of probes. If the bag is small as in the "Little Bag" model[29], then the latter configuration must occur with small probability. But as the temperature or the density is increased, its probability must increase, g_A will approach closer to unity and at some point there will be a phase transition in which the Goldstone phase vacuum will disappear. This is an interesting problem we would like to get at its solution without subjecting the nucleus to an extreme condition such as collisions at relativistic energies. We are still far from answering this question.

I can, however, speculate a little here. The present lore (backed by lattice QCD calculations[30]) is that if a nuclear matter is heated to a temperature

$$T_c \approx 200 \text{ MeV} \tag{33}$$

quarks get deconfined and a quark plasma phase is formed. Translated roughly into the confinement size, this corresponds to a sphere of radius R_c ,

$$R_c \approx 1 \text{ fm} \quad . \tag{34}$$

This is about the M.I.T. bag size. Now according to the "Little Bag" picture[29] based on chiral symmetry, a nucleon has a "pion skin" which may be as thick as 1/2 fm. This structure implies another phase transition at a higher temperature, a phase transition in which pions evaporate, thus resulting in an overall Wigner mode. If the size of the core is one-half of the confinement size, this must occur at

$$T_{Chiral} \sim 2T_c \sim 400 \text{ MeV} \quad . \tag{35}$$

In terms of density, the latter phase transition must be harder to get to. If the deconfinement transition occurs, say, at ρ_c, then the chiral phase transition is expected to occur at

$$\rho_{Chiral} \approx 2^3 \rho_c = 8\rho_c \quad . \tag{36}$$

At the moment, there is no reliable estimate of ρ_c.

What we learned from the above discussion is that looked at in terms of g_A^{eff}, the transition from $g_A^{eff} \neq 1$ to $g_A^{eff} \approx 1$ cannot be abrupt. All ranges of g_A^{eff} are sampled *even in finite nuclei*. Clearly a different signal would be needed. This is a subject of greatest interest in relativistic heavy ion physics. Very little can be said of it at the moment, but let me illustrate a point by an argument due to Pisarski.

Pisarski[31] points out that while g_A changes rather undramatically from ~ 1.26 to ~ 1, the $\rho \to 2\pi$ width Γ_ρ (for instance) would have basically different properties depending upon whether or not the "pion skin" is present. If there is no "pion skin" as in the M.I.T. bag which also means $T_c \approx T_{Chiral}$, then as the temperature approaches T_c, the width diverges like

$$\Gamma_\rho \sim \left(\frac{T-T_c}{T_c}\right)^{-1/4} \quad , \tag{37}$$

whereas if there is a "pion skin" so that $T_{Chiral} > T_c$, then as $T \to T_c$, the width goes to zero,

$$\Gamma_\rho \sim \left(\frac{T-T_c}{T_c}\right)^{3/4} \quad . \tag{38}$$

Equations (37) and (38) result from the behavior of the ρ mass and f_π (the pion decay constant) near $T \sim T_c$ deduced from soft-pion theorems. It remains to be seen whether such a dramatic difference can be detected in experiments.

6. CONCLUSION

As a way of conclusion, let me clarify one point which was postponed up to here. The question is : why does the axial charge

measured in the $0^+ \leftrightarrow 0^-$ transition look much bigger than that seen in
the Gamow-Teller resonances ? This may look more puzzling at first
sight if one recalls that what goes into the Adler-Weisberger relation is the charge operator. Now the answer to this question is
as follows. The axial charge (or the axial charge matrix element)
is not a Lorentz invariant quantity. So in the frame where the
nucleons are moving slowly, i.e., Fig.7a, soft pions influence the
charge in an important way. Thus the enhancement. However one should
actually go into the frame where the nucleus moves with infinite
momentum (∞ momentum frame) to define a Lorentz invariant axial charge.

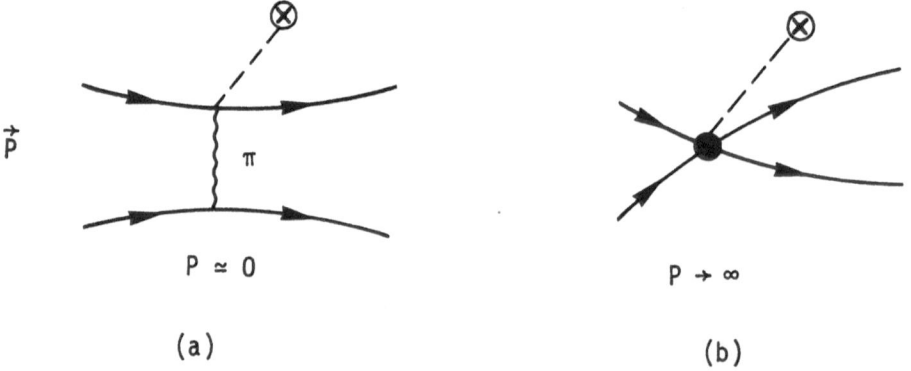

Fig. 7. Axial charges "seen" in $0^+ \leftrightarrow 0^-$ nuclear transitions (a)
and in infinite momentum frame (b). The g_0' interaction quenches the charge in (b).

This is where the Adler-Weisberger relation should be derived to get
at the g_A. Each of the nucleon in the nucleus moves with the momentum $\sim \vec{P}/A$ (A the number of nucleons) ; therefore in this frame the
soft pions evaporate (or decouple) as

$O(q/P)$

for q fixed and $P \to \infty$. But the contact interaction (g_0') which was
negligible in the case of Fig.7a persists in the case of Fig.7b,
inducing the *effective quenching*. An important point here is then

that the charge measured in the Gamow-Teller transition is the invariant charge given by the Adler-Weisberger relation, not the soft-pion-infested charge seen in the matrix element of A_0 in nuclei.

My discussion so far was at best very heuristic ; although it smells right, it demands a more convincing treatment. I have been trying to construct a field theory model incorporating the general features discussed in this note, so far without much success. Also it would be nice to devise a systematic way to calculate the corrections to the main terms found here. Whatever it may be, such a model must arise from a better understanding of QCD. Or conversely, a model of this type may lead to a more profound understanding of how QCD works in complex nuclei.

REFERENCES

1. For review, M. Rho and G.E. Brown, Comments in Nuclear and Particle Physics 10, 201 (1981).
2. T. De Grand, R.L. Jaffe, K. Johnson and J. Kiskis, Phys. Rev. D12, 2060 (1975); C.G. Callan, R.F. Dashen and D.J. Gross, Phys. Rev. D19, 1826 (1979); G.E. Brown and M. Rho, Phys. Lett. 82B, 177 (1979).
3. See e.g. V. De Alfaro, S. Fubini, G. Furlan and C. Rossetti, Currents in Hadron Physics, North-Holland Publishing Co. (1973) p. 468 ff.
4. G.E. Brown, 1981 Erice lectures on Nuclear Physics, Erice, Italy; 21-30 April 1981.
5. S. Weinberg, Physica (Utrecht) 96A, 327 (1979).
6. M. Rho, 1981 Erice Lectures on Nuclear Physics, Erice, Italy; 21-30 April 1981.
7. K. Kubodera, J. Delorme and M. Rho, Phys. Rev. Lett. 40, 755 (1978).
8. M. Bernheim et al., Phys. Rev. Lett. 46, 402 (1981).
9. See e.g. G.E. Brown, Nucl. Phys. A374, 63C (1982).
10. P.A.M. Guichon and C. Samour, Nucl. Phys. A382, 461 (1982).
11. I.S. Towner and F.C. Khanna, Nucl. Phys. A372, 331 (1981).
12. L. Palffy et al., Phys. Rev. Lett. 34, 212 (1975).
13. C.A. Cagliardi, G.T. Garvey, J.R. Wrobel and S.J. Freedman, Phys. Rev. Lett., 48, 914 (1982).
14. For review see C.D. Goodman, Nucl. Phys. A374, 241C (1982).
15. A.B. Migdal, Theory of Finite Fermi Systems and Applications to Atomic Nuclei (Interscience Publishers; New-York 1967).
16. M. Rho, Phys. Rev. 161, 955 (1967).
17. M. Chemtob and M. Rho, Nucl. Phys. A163, 1 (1971).
18. D.H. Wilkinson, Phys. Rev. C7, 930 (1973); for review, 1977 Les Houches Lectures, ed. by R. Baian, M. Rho and G. Ripka (North Holland Publishing Co., Amsterdam, 1978) Vol. II, p. 877.

19.   M. Ericson, A. Figureau and C. Thevenet, Phys. Lett. $\underline{47}$B, 381
      (1973); J. Delorme et al., Ann. of Phys. (N.Y.) $\underline{102}$, 272 (1976).
20.   S. Barshay, G.E. Brown and M. Rho, Phys. Rev. Lett. $\underline{32}$, 787
      (1974).
21.   M. Rho, Nucl. Phys. A$\underline{231}$, 493 (1974); K. Ohta and M. Wakamatsu,
      Nucl. Phys. A$\underline{234}$, 445 (1974).
22.   M. Rho, 1976 Erice Lectures on Nuclear Physics, in Prog. in
      Particle and Nucl. Phys. $\underline{1}$, 105 (1978).
23.   J. Meyer-ter-Vehn, Physics Reports $\underline{74}$, 323 (1981).
24.   G.E. Brown, S.O. Bäckman, E. Oset and W. Weise, Nucl. Phys.
      A$\underline{286}$, 191 (1977).
25.   E. Oset and M. Rho, Phys. Rev. Lett. $\underline{42}$, 42 (1979); I.S.
      Towner and F.C. Khanna, Phys. Rev. Lett. $\underline{42}$, 51 (1979).
26.   A. Bohr and B.R. Mottelson, Phys. Lett. $\underline{100}$B, 10 (1981); G.E.
      Brown and M. Rho, Nucl. Phys. A$\underline{372}$, 397 (1981).
27.   A. Della Selva and L. Masperi, Nuovo Cimento $\underline{50}$, 997 (1967).
28.   See G. Baym, 1981 Erice Lectures on Nuclear Physics, Erice,
      Italy; 21-30 April 1981.
29.   G.E. Brown and M. Rho, Ref. 2; G.E. Brown, M. Rho and V. Vento,
      Phys. Lett. $\underline{84}$B, 383 (1979).
30.   L.D. McLerran and B. Svetitsky, Phys. Rev. D$\underline{24}$, 450 (1981) and
      references given there.
31.   R.D. Pisarski, Phys. Lett. $\underline{110}$B, 155 (1982).

NOTE added : After this paper was presented at the meeting, a pre-
print by J.Kogut, M.Stone, H.W.Wyld, J.Shigemitsu, S.H.Shenker and
D.K.Sinclair (Illinois preprint, March 1982) appeared in which a
Monte-Carlo Simulation on lattice QCD for $SU(2)_c$ gave T_{chiral}/T_c =
1.60 ± 0.20. The Little Bag prediction (35) is close to it.

MAGNETIC MOMENTS REVISITED

I.S. Towner and F.C. Khanna

Chalk River Nuclear Laboratories
Atomic Energy of Canada Limited
Chalk River, Ontario, Canada K0J 1J0

ABSTRACT

Consideration of core polarization, isobar currents and meson-exchange processes gives a satisfactory understanding of the ground-state magnetic moments in closed-shell-plus (or minus)-one nuclei, A = 3, 15, 17, 39 and 41.

Ever since the earliest days of the nuclear shell model the understanding of magnetic moments of nuclear states of supposedly simple configurations, such as doubly closed LS shells ±1 nucleon, has been a challenge for theorists. The experimental moments, which in most cases are known with extraordinary precision, show a small yet significant departure from the single-particle Schmidt values. The departure, however, is difficult to evaluate precisely since, as will be seen, it results from a sensitive cancellation between several competing corrections each of which can be as large as the observed discrepancy. This, then, is the continuing fascination of magnetic moments.

In this contribution, we revisit the subject principally to identify the role played by isobar currents, which are of much concern at this conference. But in so doing we warn quite strongly of the dangers of considering just isobar currents in isolation; equal consideration must be given to competing processes which in this context are the mundane nuclear structure effects, such as core polarization, and the more popular meson-exchange currents. In particular we attempt to provide answers to the following questions:

(a) How does the quenching of spin operators from isobar currents

grow with the size of the nucleus? Does the quenching reach an
asymptotic limit for large mass?

(b) How good are soft-pion theorems? The assertion[1] is that as long
as the virtual momentum of the exchanged pion is small, the two-body
M1 operator should be dominated by single-pion exchange: all other
meson exchanges and corrections to the soft-pion theorem should be
suppressed. This statement is well established in the lightest of
systems, but how effective is the assertion in heavier systems and
how useful is it in the discussion of magnetic moments?

(c) Second-order core polarization involving the pion-tensor force
in the residual interaction couples strongly to intermediate states
of high excitation and gives a considerable correction to magnetic
moments. First-order core polarization involving the two-body meson-
exchange operator also couples strongly to highly excited states but,
being first order, its contribution to the magnetic moment comes with
the opposite sign. Thus there is a degree of cancellation which in
light nuclei has been demonstrated to be nearly complete. Is this
still true in heavier systems?

Isobar Currents

        Our calculation is the standard first-order correction to mag-
netic moments (but summed here to all orders in the RPA series)
involving isobar-hole states. We use a nonrelativistic model in
which the isobar is described as a baryon bound in a harmonic-oscil-
lator potential of the same characteristic frequency as that used
for nucleons. Two-body interactions between nucleons and between
nucleons and isobars are described by one-boson-exchange transition
potentials. The coordinate-space representation for the case of
one-pion exchange between nucleons (in the static limit with the
δ-function term removed) is

$$V(\underset{\sim}{r}) = \frac{1}{3} m_\pi f^2_{\pi NN}(\underset{\sim}{\tau}_1 \cdot \underset{\sim}{\tau}_2) \{(\underset{\sim}{\sigma}_1 \cdot \underset{\sim}{\sigma}_2)Y_0(x_\pi) + S_{12}Y_2(x_\pi)\}$$

where $S_{12} = 3(\underset{\sim}{\sigma}_1 \cdot \underset{\sim}{r})(\underset{\sim}{\sigma}_2 \cdot \underset{\sim}{r}) - \underset{\sim}{\sigma}_1 \cdot \underset{\sim}{\sigma}_2$; $Y_0(x) = e^{-x}/x$; $Y_2(x) = (1+3/x +$
$3/x^2)Y_0(x)$ and $x = m_\pi r$. The transition potential between nucleons
and isobars is obtained by substituting[2] $f_{\pi NN} \to f_{\pi N\Delta}$, $\underset{\sim}{\sigma} \to \underset{\sim}{S}$, $\underset{\sim}{\tau} \to \underset{\sim}{T}$
at the appropriate vertices where $\underset{\sim}{S}$ and $\underset{\sim}{T}$ are generalizations of the
Pauli spin and isospin matrices acting as transition operators
between 1/2 and 3/2 spinors. The coupling constants are $f^2_{\pi NN} = 0.08$
and $f_{\pi N\Delta}/f_{\pi NN} = 6\sqrt{2}/5$ from the quark model[2]. Similar expressions
and substitutions occur for the ρ-exchange potential, for which we
use the strong coupling version, $f^2_{\rho NN}(1+K)^2 = 5.31$ with $K = 6.6$ as
advocated by Höhler and Pietarinen[3]. Short-range behaviour and
wave-function correlation difficulties are avoided by introducing,
purely phenomenologically, a sharp cut-off at 0.5 $\hbar/m_\pi c$ (0.71 fm)
as the lower limit for all radial integrals.

Other forms for the residual interaction have been used in the literature. For example, in Landau-Migdal theory[4] a simple δ-function is used

$$V(\underset{\sim}{r}) = C_o\, g_o'(\underset{\sim}{\sigma}_1\cdot\underset{\sim}{\sigma}_2)(\underset{\sim}{\tau}_1\cdot\underset{\sim}{\tau}_2)\,\delta(\underset{\sim}{r})$$

with $C_o \sim 400$ MeV fm^3 and $g_o' \sim 0.6$. The Bohr-Mottelson interaction[5] is of similar form. Again the Pauli spin operators are replaced by transition spin operators for isobars. Only direct matrix elements of the δ-function are included, the exchange matrix elements are assumed to have been summed in defining the Landau parameter g_o'. Other authors use a combination of finite-range boson exchange and zero-range δ-function, e.g. $\pi+\delta$ used by Oset and Rho[6], $\pi+\rho+\delta$ used by Speth et al.[7]

In fig. 1a we display the calculated correction from isobar currents to the isovector magnetic moment of a 0s-orbital in a closed-shell-minus-one configuration expressed as a percentage of the Schmidt value for various sizes of the LS closed shell. The correction is negative (quenching) and for all interactions considered reaches an asymptotic limit for large nuclei. Of the interactions considered the δ-function gives the largest correction, but the correction scales with g_o', the larger g_o', the bigger the quenching. Adding the finite range π-exchange to the δ-function, as in Oset and Rho[6], reduces the quenching. Similar results are obtained if instead of the δ-function the full one-boson-exchange potential (OBEP) is used (with a radial cut-off), when only the direct matrix elements are retained. Adding form factors (FF) at the πNN-vertices and keeping exchange matrix elements each reduces the degree of quenching still further.

The quenching is largest for the 0s-orbital; increasing the orbital angular momentum of the single-particle state reduces the quenching. To make contact with the experimentally known magnetic moments, we plot in fig. 1b the quenching of the moment in closed-shell-plus-one configurations for the nodeless unoccupied state outside the LS closed shell, (viz. 0p orbit at $A = 4$, 0d at $A = 16$, 0f at $A = 40$ etc.) We note that asymptotically there is a quenching of 12% in the model of Oset and Rho[6] for valence orbitals with $j = \ell + \frac{1}{2}$ (15% for $j = \ell - \frac{1}{2}$; not shown) which is less than the 30% value deduced empirically from experiment by Knüpfer et al.[8] Our preferred model of (OBEP)$^{\text{dir+exch}}$ gives even less quenching.

Meson-exchange Currents (MEC)

As is now well known, the standard one-body orbital and spin operators describing magnetic moments must be augmented by two-body operators representing the interaction of an exchanged meson with

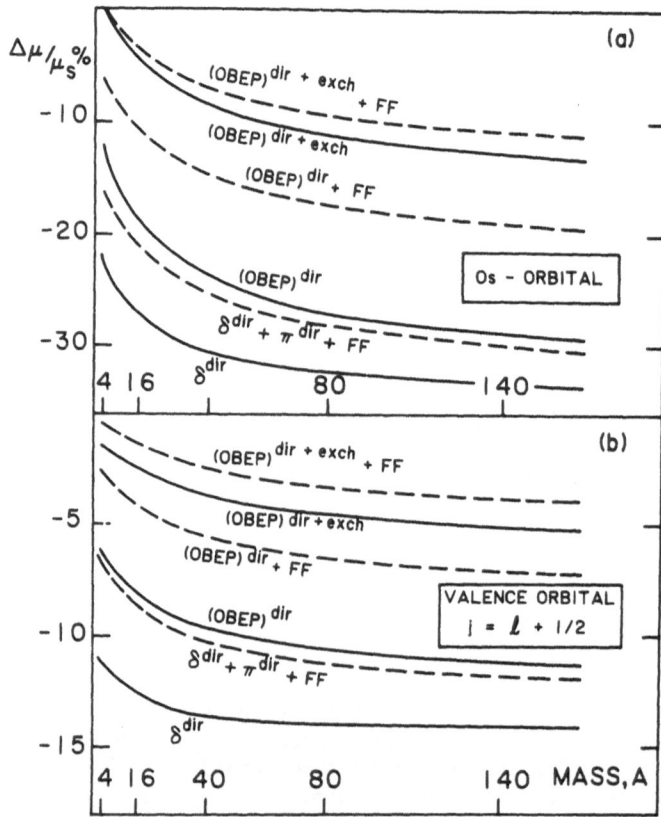

Fig. 1.   Calculated correction from (Δ-h) states to the isovector
          spin magnetic moment expressed as a percentage of the
          Schmidt spin value in closed-shell-plus-one nuclei for
          various sizes of the closed-shell core.

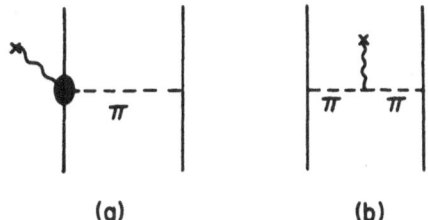

(a)   **(b)**

Fig. 2. (a) photo-pion production, and (b) pionic current meson-
         exchange diagrams. These are the only relevant diagrams
         in the soft-pion limit.

the external electromagnetic probe. For simple shell-model configurations such as the closed-shell-plus-one system, the MEC calculation involves two-body matrix elements evaluated between the valence
nucleon and a nucleon in the closed-shell core and summed over all
the nucleons in the core. Thus for larger systems, larger MEC
effects can be anticipated. The pion, being the lightest mass meson,
gives by far the longest range to the MEC operator it generates and
thus is likely the most important component. Other components
(multipion or meson resonances, e.g. ρ-meson) have much shorter
range and will be strongly suppressed by the repulsion which pre-
vents nucleons coming close together. (Short-range correlations
are simulated in our calculations by a lower cut-off in the radial
integrals).

     It is of interest to see to what extent pion terms dominate
the MEC calculation. Indeed the low mass of the pion can be traced
back to a practically exact symmetry, the $SU(2) \times SU(2)$ chiral symmetry[9]. As a consequence, the production amplitude of pions by an
external electromagnetic current on a nucleon in the zero-momentum
limit can be written down in a model independent way simply from the
knowledge of the algebra governing this underlying symmetry. The
finite momentum corrections correspond to shorter range operators
(including tensors) and fall as before in the more uncertain domain
of short distance phenomena. The pion-production amplitude is thus
fixed in the soft-pion limit and this leads to the repeatedly emphasized conclusion[10] that pion exchange should be important for the
magnetic part of the vector current (and hence magnetic moments).

     To the pion-production diagram (shown on the left in Fig. 2)
must be added the pion-current diagram (on the right) which though
formally of higher order in the pion momentum contains an extra pion
propagator and thus contributes in the chiral limit $m_\pi \to 0$. To get

an idea of the effects of these terms on magnetic moments in complex nuclei, Delorme[11] in a Fermi gas model has derived expressions for the correction to the orbital and spin isovector g-factors:

$$\delta g_\ell / g_\ell = \frac{g^2}{4\pi^2} \frac{m_\pi^2}{2Mp_F} \left[(1 + \frac{m_\pi^2}{2p_F^2}) \; \ell n \; (1 + \frac{4p_F^2}{m_\pi^2}) \; -2 \right]$$

$$\delta g_s / g_s = \frac{g^2}{4\pi^2} \frac{1}{4.71} \frac{m_\pi}{M} \left[2tg^{-1}(2p_F/m_\pi) + \frac{m_\pi}{p_F} - \frac{3}{2} \frac{m_\pi}{p_F} (1 + \frac{m_\pi^2}{6p_F^2}) \; \ell n \; (1 + \frac{4p_F^2}{m_\pi^2}) \right]$$

where the valence nucleons have been supposed to stay at the Fermi surface (p_F = Fermi momentum, $g^2/4\pi = 14.4$). Both graphs make important contributions which separately survive but cancel each other in the chiral limit ($m_\pi = 0$). This cancellation is especially severe (up to order m_π^2) for δg_ℓ and leaves room for a sizeable rho-meson contribution. Delorme[11] estimates, for a value of p_F corresponding to nuclear matter density, $\delta g_s/g_s$ of 13.6% and $\delta g_\ell/g_\ell$ of 20.2%.

In Fig. 3, we try to make contact with these results by computing MEC corrections in finite nuclei (of simple closed-shell-plus-one configuration) for various closed LS shells and extrapolating to heavy nuclei. The correction is positive (enhancement), increases with mass value and reaches an asymptotic limit for large nuclei. For a deeply-bound orbital such as the 0s, pionic terms lead asymptotically to an 11% enhancement to the spin magnetic moment (compared to Delorme's estimate of 13.6%) while heavy mesons are estimated to add a further 2%. Likewise pionic terms lead asymptotically to a 27% enhancement to the orbital magnetic moment (Delorme estimates 20.2%) for a 0p orbit, but there is in addition a huge enhancement from heavy mesons. Here we have used 'strong' coupling[3] for the ρ-meson.

As was the case with isobar currents, the computed MEC correction depends on the angular momentum of the valence nucleon, the correction decreasing as the orbital angular momentum of the nucleon goes up. Thus on the right side of Fig. 3, we plot the same quantities this time for the valence nucleon in the first unoccupied nodeless orbital outside the closed shell. The enhancement is less, and for the spin part of MEC it is quite small. It is clear that for the MEC calculation of magnetic moments in complex nuclei terms beyond just those given by soft-pion theorems must be considered even though they cannot be computed in a model independent way. The success of the soft-pion theorems in explaining the electrodisintegration of deuteron data at small energy transfer and large momentum[12] is due in part to the dominance of S-states in the calculation.

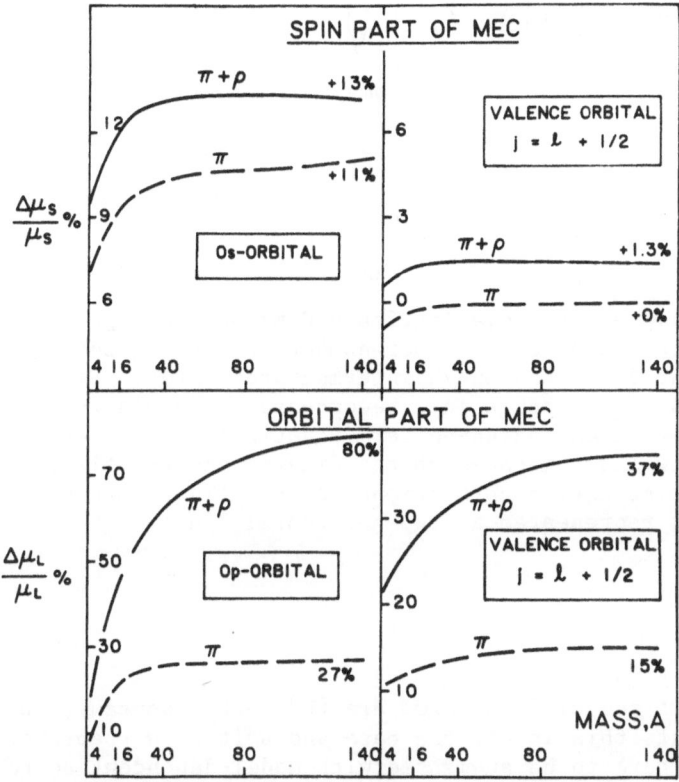

Fig. 3. Calculated correction from MEC to the isovector spin magne-
         tic moment (upper half of figure) and to the isovector
         orbital magnetic moment (lower half) expressed as a percen-
         tage of the respective Schmidt value in closed-shell-plus-
         one nuclei for various sizes of the closed-shell core.

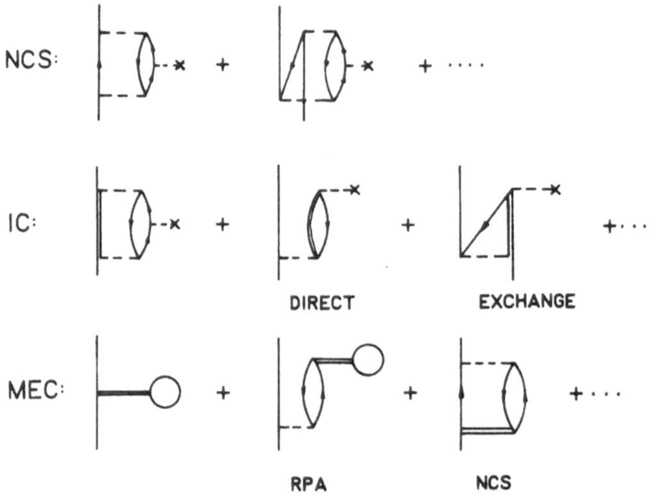

Fig. 4. Sample core-polarization and meson-exchange diagrams. First
         row shows second-order number-conserving sets (NCS) involv-
         ing 2p-1h and 3p-2h intermediate states. Second row shows
         isobar currents in a second-order NCS diagram, and the
         direct and exchange term starting the RPA series. Third
         row shows meson-exchange diagram and the first-order core-
         polarization corrections to MEC. The two-body MEC operator
         is represented as two horizontal lines.

Tensor terms and orbital terms are then not important, but for magnetic moments this is not the case and soft-pion theorems, as we
have seen, have to be augmented with model-dependent corrections.

Core Polarization

Finally, we turn briefly to a consideration of nuclear structure corrections. By limiting our discussion to closed LS shells ±1
nucleon, there is no first-order core-polarization correction of the
Arima-Horie[13] type. In second order, we need only consider the number-conserving sets[14] (NCS), a sample of which is shown in the first
row of Fig. 4. These diagrams have previously been calculated by
Shimizu, Ichimura and Arima[15] who show that intermediate states of
high excitation make important contributions. We have recomputed
these diagrams explicitly summing all states up to an excitation

Table: Calculated correction to the isovector magnetic moment from various core-polarization, isobar-current and meson-exchange effects expressed as a percentage of the single-particle Schmidt value.

		A = 3 $(0s_{1/2})^{-1}$	A = 15 $(0p_{1/2})^{-1}$	A = 17 $(0d_{5/2})$	A = 39 $(0d_{3/2})^{-1}$	A = 41 $(0f_{7/2})$
	Core polarization					
1	NCS(2ħω)	−4.8	+13.1	−6.2	+25.1	− 7.5
2	NCS(4→∞ħω)	−5.0	− 3.9	−7.9	+12.4	−12.0
	Isobar currents					
3	NCS	+1.5	+ 5.0	+0.6	+ 7.6	+ 0.8
4	RPA direct	−4.7	+ 0.5	−2.5	−14.3	− 3.1
5	RPA exchange	+5.3	+10.8	+1.6	+11.8	+ 1.4
	Meson currents, π					
6	MEC	+6.5	−18.4	+1.8	−24.9	+ 2.6
7	MEC−RPA	+0.7	+ 1.7	+0.0	+ 0.4	− 0.1
8	MEC−NCS	+4.9	− 4.6	+7.0	−22.1	+ 9.9
	Meson currents, ρ					
9	MEC	+1.8	− 2.2	+3.3	−17.9	+ 4.6
10	MEC−RPA	−0.2	+ 1.3	+0.0	+ 0.9	+ 0.0
11	MEC−NCS	+0.8	+ 6.2	−0.3	+12.1	− 0.2
12	Sum	+6.8	+ 9.3	−2.7	− 9.1	− 3.6
13	Expt	+8.5	+11.1	−1.3	−38.4	−8.9±0.3

energy of 12 ħω above the ground state and geometrically extrapolating to infinite energy. Calculations use harmonic oscillator wavefunctions, a one-boson-exchange residual interaction including π, ρ, ω and the fictitious σ meson (mass 500 MeV, coupling constant adjusted so that a Hartree-Fock calculation reproduced the ground-state binding energy of the closed-shell nucleus), and a radial cut-off at 0.5 ħ/m_πc to simulate short-range correlations. Monopole form factors at the boson-nucleon-nucleon vertices are now included of range Λ_π = 1 GeV for pions and Λ = 1.44 GeV for the other bosons. The results are given in the first two rows of the Table where the contributions from intermediate states of low excitation (arbitrarily set at 2 ħω) are separated from contributions from high excitation.

In lines four and five, we quote the results for isobar currents discussed earlier giving both the direct and exchange

contributions for a one-boson-exchange potential. We also give in
line three, the second-order contribution from the NCS diagram when
one of the intermediate particle lines is replaced by an isobar.
Finally in lines six to eleven, we give the meson-exchange contributions, where we have separated the contribution from the pion diagrams shown in Fig. 2 from all other contributions involving heavy
mesons (loosely categorised in the Table as ρ). As was stressed by
Arima and Hyuga[16], it is also necessary to consider the first-order
core-polarization correction for the two-body meson-exchange operator. There are two types, called RPA and NCS (see row three in Fig.
4), of which the former is negligible but the latter is very
important.

To see this, consider all contributions in the table labelled
NCS (lines 2, 3, 8 and 11) coming from highly excited intermediate
states. Each of these involves extensive computation because the
pion and rho tensor forces couple strongly to high lying states.
Yet the group, taken as whole, show appreciable mutual cancellation.
Thus the conjecture of Rho[17] that this group taken together could be
omitted in a first estimation seems a reasonable compromise and probably it would be necessary to make this approximation in heavier
nuclear systems.

Note that the summed contribution from isobar currents (with
the exception of mass $A = 15$) is only a small piece of the total and
is not necessarily a quenching effect. Lastly we turn to the bottom
lines in the Table where the sum of all these contributions are compared with experiment. Certainly the agreement is not marvellous,
but considering the large amount of cancellation going on the trends
are well reproduced by the calculation.

REFERENCES

1. M. Rho, Nucl. Phys. A354, 3c (1981)
2. G.E. Brown and W. Weise, Phys. Rep. 22C, 279 (1975); W. Weise,
       Nucl. Phys. A278, 402(1977)
3. G. Höhler and E. Pietarinen, Nucl. Phys. B95, 210 (1975)
4. A.B. Migdal, Rev. Mod. Phys. 50, 107 (1978); J. Speth, E. Warner
       and W. Wild, Phys. Rep. 33C, 127 (1977); G.E. Brown,
       S.O. Backman, E. Oset and W. Weise, Nucl. Phys. A286, 191 (1977)
5. A. Bohr and B.R. Mottelson, Phys. Lett. 100B, 10 (1981).
6. E. Oset and M. Rho, Phys. Rev. Lett. 42, 47 (1979); E. Oset and
       W. Weise, Phys. Lett. 77B, 159 (1978); M. Rho in Proc. of the
       NATO advanced institute in theoretical physics, Banff, 1978,
       ed. by B. Castel, F. Khanna and B. Goulard (Plenum, New York,
       1979)
7. J. Speth, V. Klemt, J. Wambach and G.E. Brown, Nucl. Phys. A343,
       382 (1980)
8. W. Knüpfer, M. Dillig and A. Richter, Phys. Lett. 95B, 349 (1980)

9. M. Rho and G.E. Brown, Comments Nucl. Part. Phys. 10, 201 (1981);
    M. Rho, Erice Lectures, 1976; M. Chemtob and M. Rho, Nucl.
    Phys. A163, 1 (1971)

10. K. Kubodera, J. Delorme and M. Rho, Phys. Rev. Lett. 40, 755
    (1978)

11. J. Delorme, Nucl. Phys. A374, 541c (1982)

12. M. Rho and G.E. Brown, ref. 9; G.E. Brown, Nucl. Phys. A374, 63c
    (1982); J. Hockert, D.O. Riska, M. Gari and A. Huffman, Nucl.
    Phys. A217, 14 (1973)

13. A. Arima and H. Horie, Prog. Theor. Phys. 12, 623 (1954)

14. P.J. Ellis and S. Siegel, Phys. Lett. 34B, 177 (1971)

15. K. Shimizu, M. Ichimura and A. Arima, Nucl. Phys. A226, 282
    (1974)

16. A. Arima and H. Hyuga, in Mesons in Nuclei, ed. D.H. Wilkinson
    and M. Rho (North-Holland, Amsterdam, 1979); H. Hyuga,
    A. Arima and K. Shimizu, Nucl. Phys. A336, 363 (1980)

17. M. Rho, in Progress in particle and nuclear physics, ed. by
    D.H. Wilkinson (Pergamon, New York, 1976); A.M. Green and
    T.H. Schucan, Nucl. Phys. A188, 289 (1972); M. Ichimura,
    H. Hyuga and G.E. Brown, Nucl. Phys. A196, 17 (1972); a simi-
    lar cancellation occurs in the Gamow-Teller matrix element,
    see I.S. Towner and F.C. Khanna, Phys. Rev. Lett. 42, 51 (1979)

THE EXPERIMENTAL EVIDENCE FOR GAMOW-TELLER QUENCHING

C. D. Goodman
Indiana University
Bloomington, IN 47405

S. D. Bloom
Lawrence Livermore National Laboratory
  and Department of Applied Science
University of California
Livermore, CA 94450

I. INTRODUCTION

The question of whether there is Gamow-Teller (GT) quenching relates in a fundamental way to our understanding of the nucleus and, thus, has been a subject of discussion for more than a decade.[1-6] The question has acquired considerable new interest following the publication of new (p,n) data which allow us to get a view of "complete" GT strength functions[7,8]—complete in the sense of covering the excitation energy range containing all states with GT strength on the basis of conventional shell model calculations. Although, strictly speaking, GT quenching simply means that less spin-isospin transition strength is observed than is calculated on the basis of some model, the special significance of GT quenching is that it deals with intrinsic nucleon degrees of freedom, and the strength discrepancy violates the very essence of the model rather than fine details of the model.

Gamow-Teller transitions are mediated by a spin-isospin product operator of the form $\sigma\tau$. In beta decay in the non-relativistic limit, the GT operator is essentially,[9]

$$GT = g_A \, \vec{\sigma} \, t^{\pm} \tag{1}$$

where g_A is the axial vector coupling constant. If, for a beta

143

decay transition, the matrix element of the GT operator is known, a
value for the coupling constant can be deduced from the measured
decay rate. If a process exists outside the model used to calculate the matrix elements that systematically alters the beta decay
of bound nucleons, then we should find that the value of the coupling constant deduced from nuclear decays will systematically differ from the value deduced from the free neutron decay. We, therefore, define a quenching factor that serves as a single parameter
for describing any suppression of the bound nucleon decay rate when
compared to the free nucleon decay rate. We write,

$$G_A = \eta_A g_A \qquad (2)$$

where G_A is the apparent coupling constant for nuclear GT decay and
g_A is the axial vector coupling constant that describes the decay
rate of the free nucleon.

Spin-isospin transitions may be observed with probes other
than beta decay, but we shall view the transitions as though they
occurred in beta decay, and we shall convert the strength into
units that allow us to use the quenching factor as defined above.
Among the probes available are $\beta-$, $\beta+$, (p,n), (n,p), and M1. Table
1 lists the allowed isospin transfers for several probes. For each
probe the value of ΔT_z is given. Thus, starting from a given target, different probes reach states of different T_z. Consistent
with the selection rules for a given probe, states of several different values of total isospin may be reached. Thus, for example,
with the (n,p) reaction from a target with (T,T) (where the first
argument is the total isospin and the second is its projection),
one can reach a state with (T+1,T+1). With an M1 excitation one
can reach the same isospin multiplet in its (T+1,T) projection, and
with the (p,n) reaction one can reach the same isospin multiplet in
its (T+1,T-1) projection. If these transitions are mediated by
different isospin projections of the same operator, leading to different isospin projections of the same isospin multiplet, then the
relative strengths are governed by isospin coupling coefficients.
The relative strengths are then as follows: in each case it is
understood that the initial state is (T,T). The transitions in the
denominator are $\Delta T_z=-1$ as in (p,n).

B(GT) [(T,T) → (T',M)]/B(GT)[(T,T) → (T',T-1)]

$$= T \quad \text{(for } T'=T,\ M=T) \qquad (3a)$$

$$= 2T+1 \quad \text{(for } T'=T+1,\ M=T) \qquad (3b)$$

$$= (T+1)(2T+1) \quad \text{(for } T'=T+1,\ M=T+1) \qquad (3c)$$

It is apparent from these ratios that if T is large it is over-
whelmingly advantageous to observe the strength to a given isospin
in the channel with the highest T_z. Looking at the ratios in-
versely, it is apparent that for large T the (p,n) spectrum will
be dominated by the isospin component T-1, which, in fact, includes
the collective Gamow-Teller mode.[7,8,10] This effect is further
reinforced by the essentially complete Pauli blocking of the T →
T+1 transitions for A≥80. When we reach the lead region the T → T
transitions are completely swamped by the T → T-1 resonances in the
(p,n) mode, though they can and have been seen in the M1 mode as
discussed below.

Table 1. Allowed isospin (ΔT) transfers for hadronic, electromag-
          netic, and leptonic GT reaction modes. We use the symbol
          N for nucleon. We assume $T=T_3$ for the parent state. The
          spin transfer selection rule is Δs = 0, ±1 for all reac-
          tions.

T of parent state →	0	1/2	>1/2
(N,N') or M1	1	0, +1	0, +1
(p,n) or β⁻	1	0, +1	-1, 0, +1
(n,p) or β⁺	1	0, +1	+1

    All of the above bears directly on deducing the quenching fac-
tor, η_A, from the various experimental measurements. Using either
beta decay or M1 exictation modes, one is generally limited to particular state-to-state transitions or a narrow band of final states
out of the total range possible. In these cases the model dependence is far more critical than for the (p,n) mode (see below). In
the case of M1 transitions one has two additional difficulties:
first, there is the complicating presence of the orbital component,
which contributes (for protons) a large amount to the observed
M1 strength; second, there is the difficulty of disentangling the
M1 strength from other strongly competing modes--e.g., E2. In a
few cases, like the calcium isotopes, there is a fair assurance
that the orbital components can be neglected. It is only in yet
fewer cases, e.g., ^{48}Ca,[11,12] that we can rely on the simple shell

model for a wave function, and in ^{48}Ca the Ml mode is strong enough to be clearly distinguishable from other excitations,[13] but this is indeed an exceptional case.

Only in the case of the (p,n) reaction does there exist the possibility of scanning the whole GT strength function for all these spin and isospin transfer modes. Thus, one can take a global look at the quenching question, which has the advantage that the total strength is much less model-dependent than the strength to a particular level. This is largely due to the operation of a very powerful sum rule. In the case of GT transitions the sum rule is very simple and useful, as well as rigorous,

$$B_{TOT}^{-}(GT) - B_{TOT}^{+}(GT) = 3(N-Z) \tag{4a}$$

where the subscript TOT implies a summation over all final states. For a single final state

$$B_{if}^{\pm}(GT) \equiv |\langle f \| \sigma t^{\pm} \| i \rangle|^2/(2J_i+1). \tag{4b}$$

We will discuss applications of this sum rule in Section III.

For Ml transitions there is no general sum rule such as Eq. (4a). For the rather special case of n neutrons in a single j orbit, $0 < n < 2j+1$, there is a relatively simple rule, however, which would be applicable, e.g., to the calcium isotopes, at least in first order,[14]

$$B_{TOT}(M1) = n_j(2j-1)/2j \tag{5}$$

where we use the same units as in Eqs. (4a,b); these units are 1/2 the size of the usual Ml units due to the presence of the T_z operator; see Eq. (1). It is assumed in Eq. (4) that the effective charge of the neutron is zero.

It is clear that the Ml sum rule is really not very useful because of the highly restrictive character of its model assumptions. The GT sum rule, on the other hand, is quite powerful and, as we shall see in Section III, affords considerable help in estimating total strengths, particularly if both (p,n) and (n,p) GT-excitation strength functions are known. The (n,p) strength is essentially zero in heavy nuclei, however. The sum rule applies to the strength summed over spin and isospin components, and the

partial sums are not subject to simple rules. The partial sums
require "realistic" model calculations.

II. EVIDENCE FROM BETA DECAY AND M1 TRANSITIONS

Because the theory of beta decay is so well developed, it
holds a unique place as a probe of spin-isospin excitations. Once
the half-life and end-point energy of an allowed beta transition
have been measured, one can reliably extract the corresponding
nuclear transition matrix element. For all other probes the connection between the experimentally measured quantities and the
nuclear matrix elements involve far more uncertainties. Thus, one
might think that beta decay would provide the most definitive information on GT quenching. The vitiating circumstance is that the
connection between the transition matrix element and the quenching
factor can be made only through a nuclear structure model, and, as
shall be discussed further in Section III, the matrix elements for
the particular transitions that can be observed in beta decay exhibit a high degree of model sensitivity with few exceptions.

Four notable exceptions are the LS closed shell ±1 nuclei,
A=15, 17, 39, and 41. For these cases the calculated values of the
GT transition matrix elements depend only on the model assertion
that these are LS closed shells ±1 nucleon and do not depend on
values of the spin-orbit splitting or the two body residual force.
These cases are listed among many others in the summary Table 2,
which gives quenching factors, η_A, in the last column. The LS
closed shell ±1 nucleon cases should be singled out as the most
reliable quenching information that can be obtained from beta
decay.

The remainder of the entries in Table 2 suffer from varying
and considerable degrees of structure model sensitivities. The
data shown are certainly not exhaustive; many more GT ft values
have been measured, and the corresponding matrix elements can be
calculated, but Table 2 includes a fair representation of what can
be learned about GT quenching from beta decay, M1 static moments,
and in two cases, from M1 transitions moments for (putatively) pure
neutron transitions of the $j=\ell+1/2 \rightarrow j'=\ell-1/2$ type. Table 2 does
not take into account the relativisitic corrections (Ref. 23) which
would increase the tabulated quenching factors by 2%-5%. To date
these corrections have been published only for A<40.

Table 2. Gamow-Teller quenching factors (η_A) deduced from M1 and β-decay moments; see Sect. II. The relativistic corrections have not been included; they should raise the quoted values for η_A by ≤5%. The quenching factors deduced from (p,n) data are given in Table 3. All the transitions are of the T→T type except ^{6}He, ^{10}C, ^{26}Mg, and ^{42}Ca, which are T → T −1. In two cases two β-decay branches have been summed.

Transition	Experimental Data Type	B(GT) Theory		Quenching Factor, η_A
^{3}H → ^{3}He	β	3		0.95
^{6}He → ^{6}Li	β	5.51	ref. 22	0.94
^{7}Be → ^{7}Li(g.s. + 0.5)	β(2)	2.87	ref. 16	0.91
^{10}C → ^{10}B	β	4.89	"	0.85
^{13}N → ^{13}C	β	0.32	"	0.90
^{15}O → ^{15}N	β	1/3	ref. 15	0.91 ± .01
^{17}F → ^{17}O	β	3/5	ref. 15	0.90 ± .01
^{19}F → ^{19}Ne	β(2)	2.85 2.95	ref. 21	0.76
^{26}Mg → ^{26}Al(0.6+1.1)	β(2)	2.7 2.5	ref. 20	0.80 ± .02
^{39}Ca → ^{39}K	β	3/5	ref. 15	0.74 ± .01
^{41}Ca	M1(static)	9/28		0.83
^{41}Sc → ^{41}Ca	β	9/7	ref. 15	0.81 ± .01
A = 17-39	M1(static)	—	ref. 17	0.87 ± .03
A = 16-40	M1(static)	—	ref. 18	0.85 ± .05
^{42}Ti → ^{42}Ca	β	4.92	ref. 19	0.74
^{42}Ca	M1(e,e')^c	(6/7)x0.62	ref. 13	0.81
^{48}Ca	M1(e,e')^{c,d}	(24/7)x0.75	refs. 12, 13	0.78
A = 90	M1(static)	—	ref. 18	1.1 ± 0.1
A = 208	M1(static)	—	ref. 18	1.1 ± 0.1

^{c}See ref. 13. ^{d}W. Steffen et al., private comm., quoted in refs. 13, 14.

In the case of ^{6}He the assumption that it can be described as
two particles in the p shell is sufficient to fix the total GT
strength at B(GT)=6. Supermultiplet symmetry would place all that
strength in the transition to the ground state of ^{6}Li. Inasmuch
as 80% of the total expected strength is seen in the ground state
transition, the quenching cannot be very great for any model assumption. A realistic p-shell model calculation[22] yields very good
agreement with the beta decay measurement without quenching, as
shown in Table 2.

In the case of ^{7}Be the calculation shows that most of the
strength should be in the first two states, both of which are observable in beta decay.[16] Thus, we expect that the total observed
strength should give a good measure of the actual total strength,
and the quenching factor deduced therefrom should be reliable. In
the case of ^{13}N, by contrast, the fractional strength in the observable transition is small, and the deduced quenching factor is,
therefore, quite model-dependent. This case is discussed again
with additional data on ^{13}C(p,n)^{13}N in Section III.

The analyses of static M1 moments for the range A=16-40 done
in Refs. 17 and 18 are very different in character but nonetheless
yield about the same quenching factors of about 0.85. This is in
fair agreement with the beta-decay results for A<17, the static M1
moment result for ^{41}Ca, and the beta-decay result for ^{41}Ca. There
is, however, an apparent contradiction with the results from the
beta decay of ^{39}Ca and the (e,e') results for ^{48}Ca. For these
cases the derived quenching factors are about 0.75.

For heavier nuclei the analyses of static M1 moments in the
region A ~ 90 and A ~ 208 yield a slight enhancement, albeit with
some uncertainty, as noted in Ref. 18. As will be seen, this is in
sharp contradiction with (p,n) results discussed in the next
section.

There is an important case for which an M1 transition of the
$j=\ell+1/2$ to $j=\ell-1/2$ type has been observed which is not included in
Table 2, ^{17}O (Refs. 24 and 25). In this case the transition is
observed to be 1/3 the size of the single-particle prediction. The
fact that the static $d_{5/2} \to d_{5/2}$ M1 moment is very weakly quenched,
whereas the $d_{5/2} \to d_{3/2}$ transition is very strongly quenched, could
be explained by the unobserved fragmentation of the 3/2+ state.
This is qualitatively in agreement with the observed fragmentation
of the 5/2+ hole state in ^{39}Ca with both (p,d)[26] and (p,n); the
latter is described below. Both the mass 17 and the mass 39 cases
illustrate the difficulty in trying to elicit a total B(GT)
strength of the T$\to$T type, at least via electromagnetic probes
alone.

In summary, the analysis of the beta-decay and M1 moments by themselves do indeed offer evidence of the GT quenching phenomenon, but the evidence is quite ambiguous. In particular, the very large quenching in total strength (~50%) predicted in 1973 and 1974 (Refs. 2 and 6) and recently (Ref. 27) is poorly supported, if at all, by these data alone. As we shall see, the situation is much changed by adding in the results from the (p,n) experiments.

III. EVIDENCE FOR QUENCHING FROM THE (p,n) REACTION

When a proton is injected into the nucleus and a neutron is ejected, the change in the intrinsic nucleon quantum numbers in the projectile-ejectile system is at least consistent with a GT transition. If we could further establish that the process took place with no momentum transfer in a direct, one-step reaction, we could be fairly sure that a corresponding transition in intrinsic nucleon quantum numbers also took place in the nucleus. While we cannot work at exactly zero momentum transfer, it has been shown that experimental conditions can be set up under which GT and Fermi transitions are strongly emphasized.[7,28] Theoretical discussions of this point are given in some detail by Petrovich[29] and by Love[30]. When the projectile and ejectile spins are not measured, the experiment does not distinguish Fermi from Gamow-Teller transitions. However, the Fermi strength can be identified in the spectrum as the sharp state with Q-value equal to the negative of the coulomb displacement energy. In this regard we accept the CVC hypothesis, which tells us that the Fermi strength is not renormalized, and we use the empirically demonstrated fact that the isobaric analog of a stable ground state is narrow. The IAS then becomes a useful calibration.

The (p,n) reaction as a probe relieves us of the limitation of observing only transitions whose energetics happen to permit beta decay to occur. Thus, we can explore the broader and less model-dependent aspects of GT strength functions, as well as look into details not accessible to beta decay. On the other hand, were we to use only (p,n) data to answer the question of whether there is quenching, we would find that we have traded an uncertain structure calculation with an uncertain reaction calculation. We can, however, combine the best of both worlds. We can cross-calibrate between beta decay and the (p,n) reaction. The essence of the approach is to measure the same or analogous transitions with both probes.

Of the parameters by which to measure GT quenching, the least model-sensitive is the total strength. To illustrate this point we can describe schematically a shell model calculation of GT strength The steps involved are to specify the model space in which the calculation is to be carried out (e.g., the complete s-d shell), to specify the one-body interaction energies, and to specify the two-body residual interaction energies. One then calculates the state vector for the ground state of the target and operates on it with the GT operator, creating a new vector which we may call the collective GT (CGT) vector. It is apparent that the scalar product $\langle CGT|CGT \rangle$ is the total GT strength. This number is a property of the target ground state, and the smallest value that this quantity can assume is $3(N-Z)$, which occurs when the protons are so arranged as to maximize the Pauli blocking of neutron-to-proton transitions (see Eq. 4(a)). This value is essentially the unblockable residue of strength. An attractive two-body force will tend to cause the protons and neutron to fill the same orbits as far as possible, so there is a tendency for $\langle CGT|CGT \rangle$ to be close to the minimum value, although this condition is not always fulfilled in light nuclei. Small changes in the one-body and two-body energies may shift some of the levels slightly and may cause some redistribution of GT strength among the possible final states, but will have very little effect on the toal strength. By comparison, the strength contained in a single state that collects only a small fraction of the total strength may undergo a large relative change due to a small change in the residual interaction.

The expression for the GT strength in a single final state is $|\langle f|CGT \rangle|^2$. This depends on both the target state and the final state, and through that argument alone is more model-dependent than the total strength. One can demonstrate further that quenching factors deduced from weak states are more model-dependent than those deduced from strong states. Imagine a fictitious case where 90% of the strength goes to final state A and 10% goes to final state B. Suppose that a small change in the residual force redistributes the strength so that only 89% of the strength goes to the strong state. The quenching factor deduced from that transition changes by hardly more than 1%, but the quenching factor deduced from the weak state changes by 10%. Therefore, we should choose the total strength or strong states to deduce quenching factors from comparisons between shell model calculations and measured values of B(GT). We can find the strong states with the (p,n) reaction.

We can assert quite generally that the calculated value of B(GT) for transitions that account for large fractions of the sum

rule strength are model-insensitive, while the values of B(GT) for
weak transitions are model-sensitive, where sensitivity is measured
by relative changes with respect to small changes in model parameters. Since M1 and (n,p) transitions always account for a small
fraction of the total GT strength, these transitions are always
model-sensitive.

For mass 14 nuclei (p,n) data, beta decay data, electron scattering M1 measurements, and shell model calculations are all available. This case not only constitutes an example of a reasonably
reliable determination of a GT quenching factor, but also provides
an illustration of how the different kinds of data can be used in
combination.

From the beta decay of ^{14}C to the ^{14}N g.s. we find B(GT)=10^{-6}.
According to the criteria stated above, this value is extremely
model-sensitive, as is well-known from attempts to fit it with
structure calculations. From our point of view B(GT) for that
transition is essentially zero and contains no useful information
concerning GT quenching. In the beta decay of ^{14}O the transition
to the ground state again yields no GT strength, but the transition
to the 3.95 MeV state yields B(GT)=2.8 ± 0.1.[31] This is nearly
half of the total expected GT strength. Thus, the calculation
should be reasonably insensitive to model parameters. A model
calculation yields B(GT)=4.84. Comparing the measured with the
calculated beta decay strength we find a ratio of 0.58 yielding a
value of η_A=0.76.

The calculation also places the remaining portion of the sum
strength in a T=1,1+ state of ^{14}N. The correctness of this prediction cannot be verified with beta decay data alone. The deduction of the GT quenching factor could be significantly wrong if the
model predicted the wrong distribution of strength. The (p,n) data
at once yield a picture of the distribution of the GT strength
among the several states, and the (p,n) spectrum in fact verifies
that the calculated distribution of strength is quite correct. The
several pieces of information about GT strength in mass 14 are
summarized in Table 3. A new value for the quenching factor can
now be found by comparing the total strength seen in the (p,n) data
with the total expected model strength of 6. This yields 0.60 of
the model strength observed and η_A=0.78.

The picture one finds here is that the shell model calculations give a good representation of the distribution of strength
but that a substantial fraction of the model strength simply does
not appear. This picture is repeated in all cases studied.

Table 3. Gamow-Teller strengths for mass-14 0^+(T=1) to 1^+(T=0,1) transitions.

Final State in ^{14}N (J,T, excitation in MeV)		B(GT)		
		Calculation[16]	Beta Decay[31]	(p,n)
1,0	0.0	0.02	33×10^{-6}	0.03
1,0	3.95	4.84	2.8 ± 0.13	2.8(normalized)
1,0	11.5 (experiment)			0.07
1,1	13.75	1.13		0.72
1,0	~15 (theory)	0.01		

The general survey of cases is summarized in Table 4. Some cases warrant special comments:

In the case of mass 19 nearly all of the strength is contained in the ground state mirror transition, and the normalization relies on a well measured ft value.

Table 4. GT quenching factors deduced from (p,n) measurements.

A	Expected B(GT)$_{sum}$	Observed fraction	η_A
7	3.04	0.78	0.88
13	3.95	0.5	0.71
14	6.0	0.60	0.78
19	3.0	0.55	0.74
26	7.5	0.51	0.71
39	3.0	0.55	0.74
42	6.0	0.54	0.73
48	24.0	0.5	0.71
90	30.0	0.5	0.71

For mass 26 the shell model calculation is quite complicated but is generally in agreement with the measurement. In this case T=0,1, and 2 final states separate clearly in the calculation. Inasmuch as there is good general agreement between the data and the calculation shown in Fig. 1, there is no reason to assume that the quenching is T-dependent. However, since the T=1 and 2 components account for only small fractions of the strength, conclusions about T dependence are model-dependent.

Mass 39 is a case discussed above in connection with model in-sensitivity of the beta decay data. What the (p,n) data add here is a view of the $d3/2^{-1} \rightarrow d5/2^{-1}$ transition. This is observed to be highly fragmented, and thus an accurate determination of whether there might be a j-dependence of quenching is impossible. It is apparent from the (p,n) spectrum shown in Fig. 2 that the lowest $d5/2$ hole state contains only a small part of the $d5/2$ hole strength. Thus, any deduction of a quenching factor based on the M1 transition from the ground state to the first $d5/2$ state is sus-pect. An analogous situation arises in mass 17, as mentioned in Section II, and fragmentation of the $d3/2$ particle state could ac-count for the apparent excessive quenching of the $d5/2 \rightarrow d3/2$ tran-sition in that case.

It is very difficult to define the total GT strength from the $^{39}K(p,n)^{39}Ca$ spectrum, but the spectrum is consistent with the 4:1 $d5/2{:}d3/2$ strength ratio expected from the LS closed shell -1 nucleon model for the ground state of ^{39}K. Thus, we adopt the quenching factor derived from the ground state transition.

The mass 42 case provides a good example of the background subtraction problem encountered in trying to determine the total strength from (p,n) spectra in medium-and heavy-weight nuclei. In Ref. 19 the contribution to the zero degree $^{42}Ca(p,n)^{42}Sc$ spectrum from other than GT processes was estimated by subtracting the $^{40}Ca(p,n)^{40}Sc$ spectrum. Here we take a slightly different approach which yields the same conclusion. We assume that all of the strength that has an $\Delta L=0$ angular distribution is GT strength. Thus, the problem is to subtract the non $\Delta L=0$ contributions from the 0 degree spectrum. We do this by first subtracting the 0 degree spectrum from the 2.5 degree spectrum scaled by the angular distribution for the L=0 part so that the strong GT peak subtracts out. The result is taken to be an approximation of the non $\Delta L=0$ spectral shape as seen at zero degrees. This is then scaled by the cross section ratio between the 2.5 and 0 degree spectra in the region of the L=1 resonance and subtracted from the 0 degree

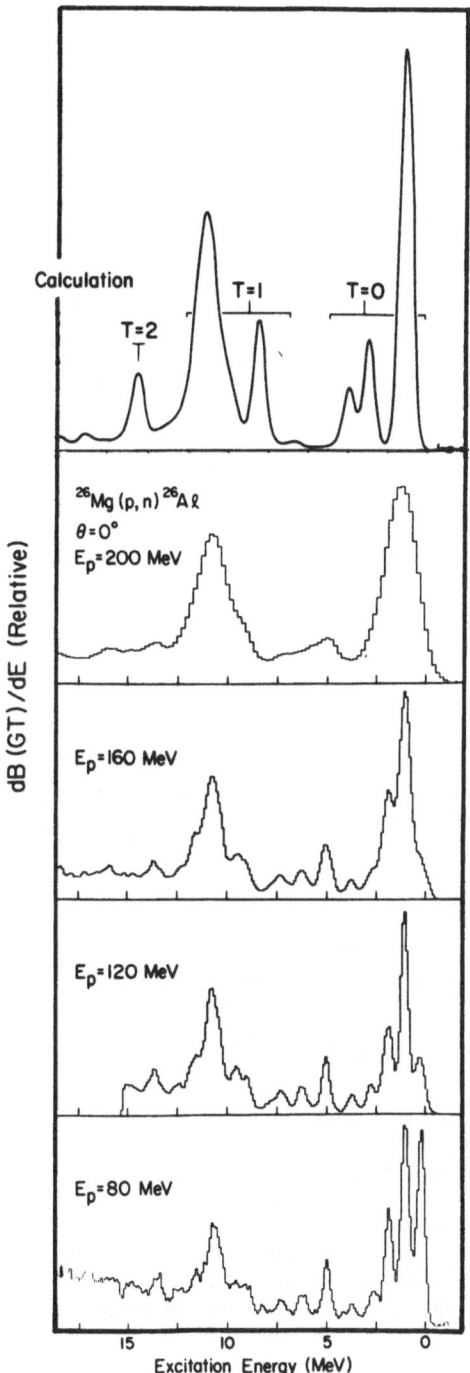

Fig. 1. Calculated GT strength function and 0° (p,n) spectra at several energies. No attempt has been made to subtract background from the experimental spectra. Note that the IAS transition is quite strong at 80 MeV and is progressively weaker in the spectra obtained with higher bombarding energies.

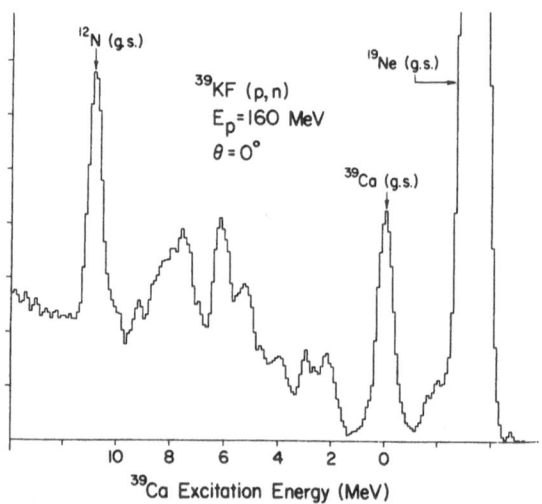

Fig. 2a.   shows the raw spectrum from a ^{39}KF target.

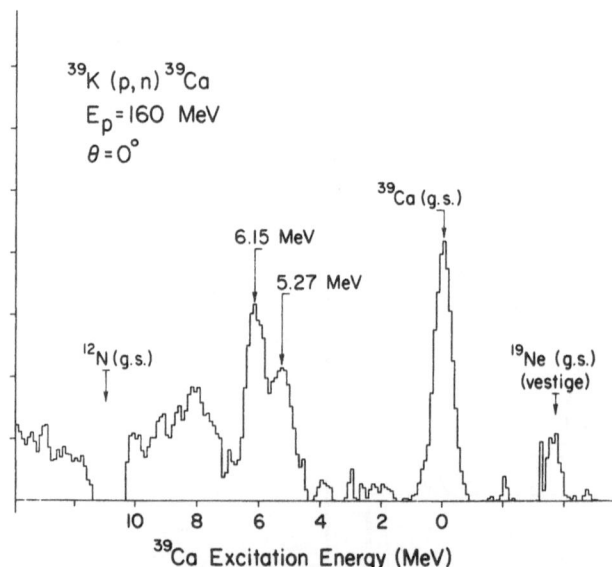

Fig. 2b.   shows the spectrum after subtraction of the
^{19}F(p,n)^{19}Ne contribution taken from a Teflon target.
Since Teflon contains relatively more carbon than the
mylar cover of the KF target, the ^{12}C(p,n)^{12}N peak is
over subtracted.

spectrum. The result is shown in Fig. 3. Our conclusion from this exercise is that the "background" in the high excitation does not constitute GT strength, in agreement with the conclusions of ref. 19.

This analysis indicates that the primary background in the 0 degree spectra is from the putative L=1 resonance, and background estimates obtained by projecting the high excitation background as a smooth curve under the giant GT resonance are probably incorrect and overestimate the background. This leads to less quenching for ^{48}Ca and ^{90}Zr than previously reported. (See also papers by F. Osterfeld and by C. Gaarde, J. S. Larsen, and J. Rapaport, in these proceedings.)

IV. CONCLUSIONS

We have examined the evidence for quenching in spin-isospin transitions and found that analyses of M1 moments and beta decay yield evidence of quenching. However, detailed model calculations are required to extract quenching factors from the data, leaving considerable ambiguity in the interpretation. The introduction of (p,n) data allows us to use the summed transition strength to all levels as the point of comparison, removing the sensitivity to the fine details of the model.

With the combined use of beta decay and (p,n) data we find that appreciable GT quenching occurs as a general feature of nuclear structure. With the exception of a few light nuclei, the quenching appears to be independent of A and can be described by a quenching factor $\eta_A = 0.74 \pm 0.05$. There is also no apparent isospin dependence of the quenching factor. That is, in cases where two or three isospin components of the GT strength function can be observed, the model calculations reproduce the relative strengths between the different isospin components quite well.

The general features of the quenching are in agreement with the theoretical considerations of Refs. (15, 28). A common feature of these theoretical treatments is a coupling of nucleon particle-hole states with Δ-isobar, N-hole states. From this point of view, Δ states are involved in spin-isospin degrees of freedom. The strength missing in ordinary GT transitions must then appear in high-lying Δ excitations. A broad peak has been observed in the Δ region in both high energy (p,n) and (^{3}He,t) experiments which might well include the missing GT strength. While no direct connection has been established between the high excitation peak and GT strength, there is a good deal of plausibility in the association.

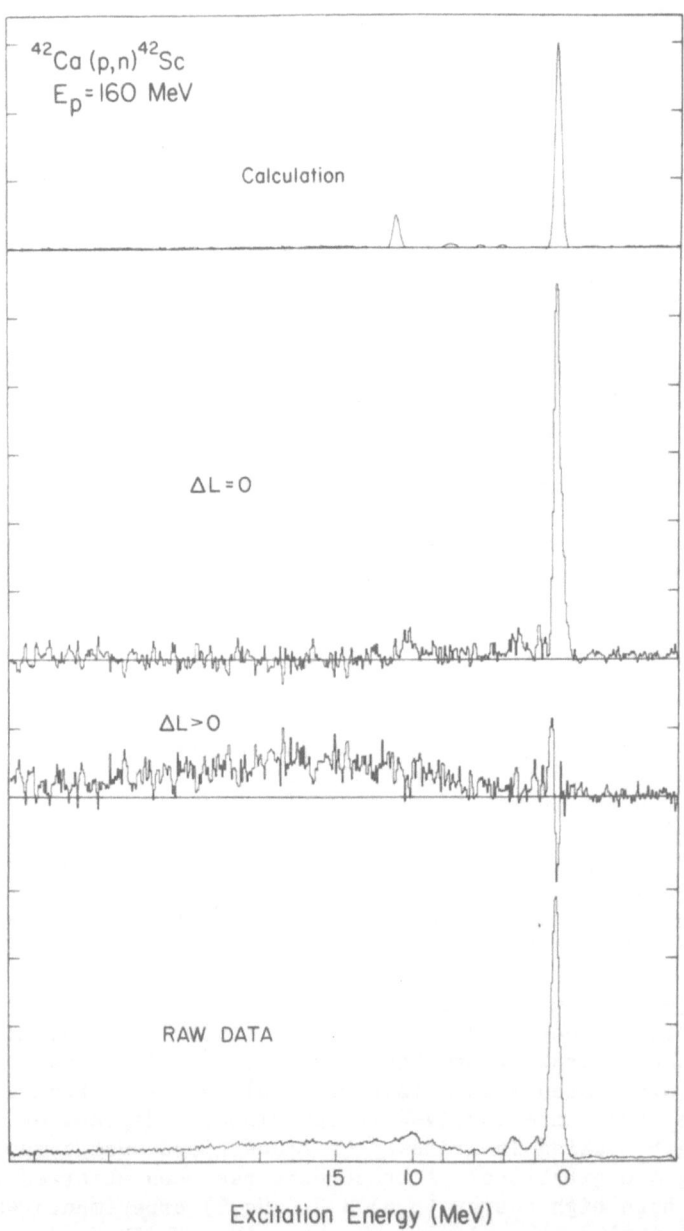

Fig. 3. Subtraction of ΔL > 0 components in ^{42}Ca(p,n)^{42}Sc as
described in text.

Even though no direct connection has been established experimentally between the missing strength and Δ couplings, that explanation is at present the preferred explanation for the missing GT strength, and it has the appealing feature of tying together nuclear structure and particle physics concepts.

V. ACKNOWLEDGMENTS

The data referred to in this paper, where not expressly referenced otherwise, are the results of a collaborative effort by a group consisting of the following people in addition to the co-authors: C.C. Foster, C. Gaarde, C.A. Goulding, D.J. Horen, J.S. Larsen, T.G. Masterson, J. Rapaport, E. Sugarbaker, and T.N. Taddeucci.

This work has been supported in part by the National Science Foundation, the U. S. Department of Energy, and the Danish Natural Science Research Council.

REFERENCES

1. R. J. Blin-Stoyle and M. Tint, Phys. Rev. 160:803 (1967).
2. M. Ericson, A. Figureau, and C. Thevenet, Phys. Lett. 45B:19 (1973).
3. D. H. Wilkinson, Phys. Rev. C 7:930 (1973).
4. D. H. Wilkinson, Nucl. Phys. A209:470 (1973).
5. D. H. Wilkinson, Nucl. Phys. A231:365 (1974).
6. M. Rho, Nucl. Phys. A231:493 (1974).
7. C. D. Goodman, in: "The (p,n) Reaction and the Nucleon-Nucleon Force," C. D. Goodman, S. M. Austin, S. D. Bloom, J. Rapaport, and G. R. Satchler, eds., Plenum, New York (1980).
8. C. D. Goodman, Nucl. Phys. A374:241C (1982).
9. R. J. Blin-Stoyle, in: "Mesons in Nuclei," M. Rho and D. H. Wilkinson, eds., North-Holland, Amsterdam (1979), p. 5.
10. D. J. Horen, C. D. Goodman, C. C. Foster, C. A. Goulding, M. B. Greenfield, J. Rapaport, D. E. Bainum, E. Sugarbaker, T. G. Masterson, F. Petrovich, and W. G. Love, Phys. Lett. 95B:27 (1980).
11. A. Harting, W. Weise, H. Toki, and A. Richter, Phys. Lett. 104B:261 (1981).
12. J. B. McGrory and B. H. Wildenthal, Phys. Lett. 103B:173 (1981).
13. W. Steffen, H.-D. Graf, W. Gross, D. Meuer, A. Richter, E. Spamer, O. Titze, and W. Knupfer, Phys. Lett. 95B:23 (1980).

14. W. G. Love, private communication.
15. E. Oset and M. Rho, Phys. Rev. Lett. 42:47 (1979).
16. D. Kurath, private communication.
17. B. H. Wildenthal and W. Chung, in: "Mesons in Nuclei," op.
     cit. Ref. 9, p. 721.
18. T. Yamazaki, in: "Mesons in Nuclei," op. cit. Ref. 9, p. 651.
19. C. D. Goodman, C. C. Foster, D. E. Bainum, S. D. Bloom, C.
     Gaarde, J. Larsen, C. A. Goulding, D. J. Horen, T.
     Masterson, S. Grimes, J. Rapaport, T. N. Taddeucci, and E.
     Sugarbaker, Phys. Lett. 107B:406 (1982).
20. S. D. Bloom, C. D. Goodman, S. M. Grimes, and R. F. Hausman,
     Jr., Phys. Lett. 107B:336 (1981).
21. S. D. Bloom and C. D. Goodman, unpublished.
22. S. Cohen and D. Kurath, Nucl. Phys. 73:1 (1965); see also
     erratum, Nucl. Phys. 89:707 (1966).
23. A. Barroso and R. J. Blin-Stoyle, Nucl. Phys. A251:446 (1975);
     R. J. Blin-Stoyle, Nucl. Phys. A254:353 (1975).
24. R. J. Holt, H. E. Jackson, R. M. Laszewski, J. E. Monahan, and
     J. R. Specht, Phys. Rev. C 18:1962 (1978).
25. R. G. Johnson, B. L. Berman, K. G. McNeill, J. G. Woodworth,
     and J. W. Jury, Phys. Rev. C20:27 (1979).
26. P. Doll, G. J. Wagner, K. T. Knopfle, and G. Mairle,
     Nucl. Phys. A263:210
     (1976); D. W. Devins, D. L. Friesel, W. P. Jones, A. C.
     Attard, S. F. Collins, G. G. Shute, B. M. Spicer, V. C.
     Officer, I. D. Svalbe, R. S. Henderson, and W. E. Dollhopf,
     Phys. Rev. C 24:59 (1981).
27. A. Bohr and B.R. Mottelson, Phys. Lett. 100B:10 (1981).
28. T. N. Taddeucci, J. Rapaport, D. E. Bainum, C. D. Goodman, C.
     C. Foster, C. Gaarde, J. Larsen, C. A. Goulding, D. J.
     Horen, T. Masterson, and E. Sugarbaker, Phys. Rev. C
     25:1094 (1982).
29. F. Petrovich in "the (p,n) Reaction and the Nucleon-Nucleon
     Force" op. cit. Ref. 7, p. 115.
30. G. Love in "The (p,n) Reaction and the Nucleon-Nucleon Force"
     op. cit. Ref. 7, p. 23.
31. This value is the weighted mean of two measurements: H. S.
     Wilson, R. W. Kavanagh and F. M. Mann, Phys. Rev. C 22:1696
     (1980); A. M Hernandez and W. W. Daehnick, Phys. Rev. C
     24:2235 (1981).

EVIDENCE FOR Δ-h COMPONENTS FROM PION REACTIONS

C. L. Morris

Los Alamos National Laboratory
Los Alamos, New Mexico 87545

INTRODUCTION

The role of excited baryon resonances (such as the $\Delta_{3,3}$) in nuclear matter is an important problem which is not yet fully understood. Although much theoretical work has been done in this field,[1-5] there are few experimental tests. The most convincing experimental confirmation of the influence of deltas on the properties of low lying nuclear states is the quenching of the Gamow-Teller strength recently observed in (p,n) reactions.[6-7] Because of uncertainties in the Gamow-Teller sum rule this quenching does not lead to direct estimates of the Δ-nucleus admixtures.

Also, enhancements in cross sections to M1 states at large momentum transfer observed in both electron[8] and proton[9] inelastic scattering have been interpreted as evidence for a renormalization of the N-N force due to Δ-nucleus admixtures. However, because of the many theoretical uncertainties in both the reaction mechanism at large q and the nuclear structure, this evidence is not convincing.

Recently, some anomalies observed in pion inelastic scattering have been interpreted as due to $\Delta_{3,3}$ admixtures in low-lying nuclear states.[10,11] If this interpretation is correct, these experiments may provide the first quantitative measurement of the size of these $\Delta_{3,3}$ admixtures. These data and their interpretation are the subject of this talk.

161

THE PION-NUCLEUS INTERACTION

At kinetic energies of 100-300 MeV the pion is unique among nuclear probes because it interacts with the nucleus through an intermediate pion nucleon resonance, the $\Delta_{3,3}$. At energies where this resonance is dominant, the pion-nucleon interaction has a simple form with the size of the central, spin dependent, and isospin dependent pieces fixed relative to each other by the quantum numbers of the $\Delta_{3,3}$ ($T=3/2$, $J=3/2$). The difficulties in understanding the pion-nucleus interaction arise largely from the resonant nature of the interaction. Since the $\Delta_{3,3}$ has a lifetime, it can propagate through nuclear matter. Interactions between the $\Delta_{3,3}$ and the remaining nucleons in the nucleus can lead to higher order effects such as changes in the position and width of the resonance from its free values. Some of these effects can be empirically accounted for in the impulse approximation by shifting the energy of the 3-3 resonance when evaluating the pion-nucleon t-matrix.[12]

Despite these difficulties a large body of resonance-energy pion-nucleus inelastic-scattering data is well described by the distorted-wave-impulse approximation (DWIA) using free pion-nucleon phase shifts, and with conventional nuclear structure.[13-16] The agreement of these simple calculations with the data for most of the cases studied indicate that the ratios between the pieces of the pion-nucleus force predicted from pion-nucleon scattering are correct.

Among the cases for which this simple model does not work are inelastic scattering to the 1^+ $T=0,1$ doublet in ^{12}C and pion double charge exchange (DCX). These are discussed below.

1^+ DOUBLET in ^{12}C

One of the predictions which can be made from the relative sizes of various parts of the pion-nucleus interaction at resonance energies is that the ratio of cross sections for members of an analog-antianalog pair of states (states with the same space-spin wave functions but different isospin) is $\sigma(T=0)/\sigma(T=1)=4$. This ratio is rigorous if only the $\Delta_{3,3}$ resonance contributes to the interaction. The 1^+ doublet in ^{12}C should provide an ideal test of this prediction.

The possibility that this ratio is not 4 and that it varies with incident pion energy for these states was first reported by Peterson et al.[17] Both of these features, if correct, are inconsistent with our present understanding of either the pion-nucleus interaction or with the structure of these states.

Cross sections for exciting the 1^+ states have been remeasured with better energy resolution and lower backgrounds in a recent experiment[10] using the EPICS spectrometer at the Clinton P. Anderson Meson Physics Facility (LAMPF). Resulting spectra at three different pion energies and at a constant q near the peak of the 1^+ angular distribution are shown in fig. 1. The T=0 1^+ state at 12.71 MeV and the T=1 1^+ state at 15.11 MeV are visible in all three spectra. From these data it is clear that while at low energies the cross section ratio is near the predicted value of 4, at higher energies it approaches 1.

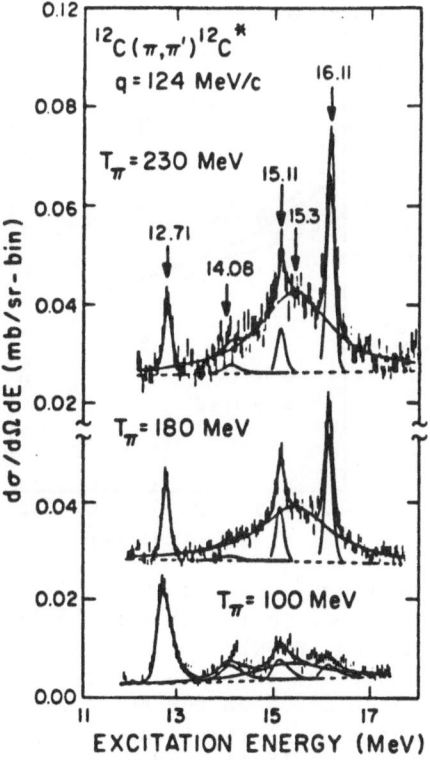

Fig. 1. The average of π^+ and π^- spectra for inelastic pion scattering from ^{12}C, at incident energies of 100 (bottom), 180 (middle), and 230 MeV (top).

The quantity $[\sigma(\pi^+) + \sigma(\pi^-)]/2$ evaluated at q = 124 MeV/c is plotted in fig. 2 along with the ratio R = $4\sigma(15.11$ MeV$)/\sigma(12.71$ MeV$)$.   Not only does R deviate significantly from unity, but also it shows an unexpected energy dependence with the maximum deviation occurring at 180 MeV where the assumption of Δ_{3-3} dominance is best. The energy dependence of the ratio resembles that of the elementary <u>pion-nucleon</u> reaction rather than the <u>pion-nucleus</u> interaction.

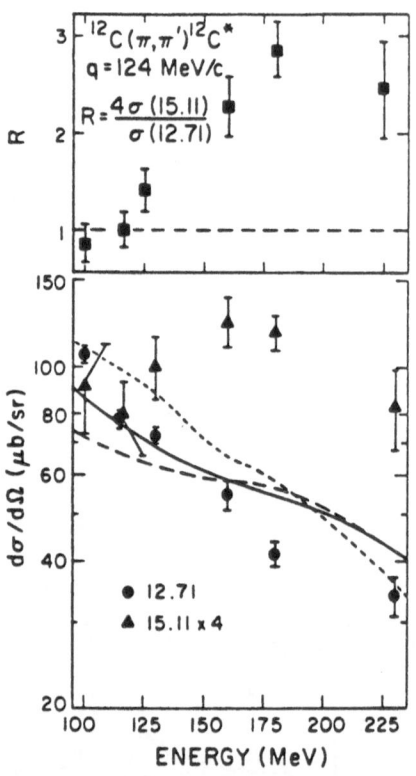

Fig. 2. Bottom: Averaged π^+ and π^- inelastic pion scattering cross sections to the 12.71 (circles) and 15.11 MeV (triangles) states in ^{12}C compared to DWIA predictions. The solid curve is for the 12.71-MeV state, and the dashed curve is for the 15.11-MeV state.   The short dash curve is for the 12.71 MeV state with the inclusion of Fermi-momentum corrections. Top: Ratio of cross sections for populating the 15.11 (x4) and 12.71 MeV levels versus incident energy.

Also plotted in fig. 2 are absolute DWIA calculations[18] made with microscopic transition densities derived from Cohen-Kurath wave functions[19] for the 12.71-MeV state (solid line) and the 15.11-MeV state (dashed line). Although these calculations predict the 12.71-MeV cross section within 20% at all energies, the cross section predicted for the 15.11-MeV level is too low by more than a factor of 2 at energies near the peak of the resonance.

A comparison of the data for these two states suggests a mechanism which enhances the cross section to the T=1 level but not the T=0 level, and which retains the intrinsic width of the $\Delta_{3,3}$ resonance. Some mechanisms which have been calculated or can be estimated and ruled out are:

1) Two-step effects which involve $\Delta T \neq 0$ or $\Delta T_z \neq 0$ in each step. These cross sections should be similar to DCX cross sections and are too small to explain the observed data.

2) Meson exchange current corrections to the cross sections for the 15.11-MeV T=1 state have been calculated and shown to be small.[20]

3) Fermi momentum corrections (current corrections) have been shown to contribute equally to the T=0 and T=1 cross sections.[21]

4) Delta propagation effects. Cross sections to these states have been calculated in the Δ-h formalism including the effects of binding, Pauli corrections and pion absorption.[22] These higher order terms affect the T=0 and T=1 states similarly, and do not change the cross section ratio significantly.

One mechanism which may explain the data is shown in the diagram in fig. 3. In this diagram the pion excites a nucleon to a $\Delta_{3,3}$, creating a Δ-N^{-1} state. Because of the intermediate Δ this amplitude is expected to be resonant. Since the isospin of the Δ (T=3/2) and the hole (T=1/2) can couple only to T=1 or 2, this mechanism cannot contribute to the T=0 cross section.

A semi-quantitative estimate of the amount of Δ-N^{-1} admixture in the 15.11-MeV state wave function can be made. First we note that the ratio of the coupling constant $f_{\Delta\Delta\pi}$ for the second vertex in the diagram in fig. 3 to $f_{\Delta N\pi}$ can be obtained from the quark model and is $\sqrt{25/8}$,[23] favoring the decay $\Delta \rightarrow \Delta + \pi$ over $\Delta \rightarrow N + \pi$. Using this ratio and collecting together the isospin-dependent terms for pion nucleus scattering as in ref. 3, but generalized to include a Δ in the final state, one arrives at the relative cross sections for states with the same space-spin wave functions but different isospin given in table 1.

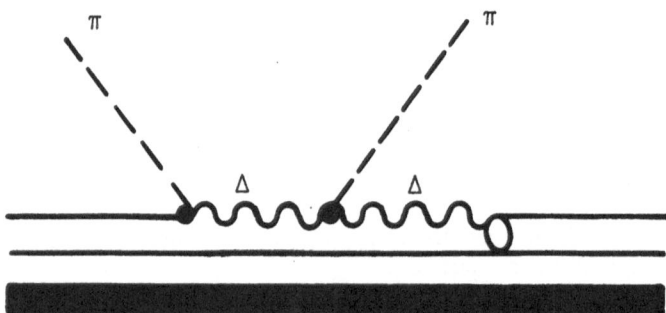

Fig. 3.   Diagram proposed to explain the observed enhancement in
           the 15.11 MeV cross section.

Table 1

Isospin dependence of the cross sections
for $T_i, \Lambda_i = 0,0 \rightarrow T_f, \Lambda_f$ transitions

T_f	Λ_f	Final state wave function NN^{-1}	ΔN^{-1}
0	0	1.000	0.000
1	0	0.250	1.953
1	1	0.250	1.953
2	0	0.000	0.078
2	1	0.000	0.234
2	2	0.000	0.468

Denoting the amplitude of the $\Delta-N^{-1}$ component of the 15.11-MeV state as β, the cross section is given as the square of the sum of two amplitudes:

$$\sigma_{15.11}(E,\theta) = |A(E,\theta) + \beta B(E,\theta)|^2 \tag{1}$$

where $A(E,\theta)$ is the amplitude for exciting the $N-N^{-1}$ component of the wave function, and $B(E,\theta)$ is the amplitude for exciting the $\Delta-N^{-1}$ component. $A(E,\theta)$ can be estimated from the cross section for the 12.71-MeV level and using the results in table 1 as:

$$|A(E,\theta)| = \sqrt{\sigma(12.71)/4} \tag{2}$$

This gives $|A| = 3.2\ \sqrt{\mu b/sr}$ at 180 MeV.

In a naive model of ^{12}C (i.e. a closed $1p_{3/2}$ shell) the 1^+ states are a simple $(1p_{1/2}\boxtimes 1p_{3/2}^{-1})_{J=1,T=0,1}$ configuration with respect to the ground state. Since the $\Delta_{3,3}$ components which can mix with this configuration are not inhibited by the Pauli exclusion principle the important admixtures are expected to be $(1p_{1/2},1p_{3/2}^{-1})$, $(1p_{3/2},1p_{3/2}^{-1})$, and $(1p_{5/2},1p_{3/2}^{-1})$, where the particle is a $\Delta_{3,3}$. All of these wave function components can be reached from the ground state by a one-body operator.

Cohen and Kurath wave functions lead to a quenching of the cross sections for both 1^+ states of a factor of 6 from that expected for a pure $p_{3/2} \rightarrow p_{1/2}$ transition. If we assume this quenching is not present for the $\Delta-N^{-1}$ components, that the three possible transitions $p_{3/2} \rightarrow p_{5/2,3/2,1/2}$ contribute coherently and are weighted as $\sqrt{2j_\Delta+1}$ and using the cross section ratios from table 1) we find:

$$|B(E,\theta)| = \frac{\sqrt{6}+\sqrt{4}+\sqrt{2}}{\sqrt{2}}\ 1.40\ \sqrt{6\sigma(12.71)} \tag{3}$$

This gives $|B(E,\theta)| = 91\ \sqrt{\mu b/sr}$. A similar result is obtained using DWIA calculations. Using the results of eq. 2 and 3 to solve eq. 1 for β gives $0.096 < \beta < 0.026$ depending upon the phase between the two terms. The lack of an interference dip in the excitation function lends more credence to the lower number.

Recently, $\Delta-N^{-1}$ components of the 15.11 MeV state have been calculated in an extended particle hole model by Grecksh, Dillig and Knüpfer.[1] The admixture coefficients they obtain for $1p_{1/2},1p_{3/2}^{-1}$, $1p_{3/2},1p_{3/2}^{-1}$, and $1p_{5/2},1p_{3/2}^{-1}$ are 0.043-0.061, 0.023-0.035, and 0.027-0.068 respectively. The agreement between these calculations and the rather crude experimental determination outlined above is encouraging.

DOUBLE CHARGE EXCHANGE

From table 1 we see that although the coupling from T=0 to T=2 final states in pion inelastic scattering is small, the coupling in pion double charge exchange is large. Since DCX to nucleon components of the final state must involve two steps (two single nucleon charge exchanges and two intermediate deltas) whereas it can proceed in one step (two single quark charge exchanges on the same nucleon with one intermediate delta) to Δ components of the final state, one might expect to see a large signature of these Δ components in this reaction.

In fact the data obtained for DCX present several puzzles which cannot be explained with a simple model in which this reaction is pictured as proceeding by two sequential single-charge exchanges through an intermediate analog state to the final double analog state. This model predicts a smooth energy dependence, and predicts zero cross section for $\Delta T \neq 0$ transitions.[24] These are:

1) At energies near 160 MeV forward angle differential cross sections for T=0 targets ($\Delta T \neq 0$ transitions) are comparable to those for T=1 targets[25] ($\Delta T = 0$).

2) DCX cross sections on both T=0 and T=1 targets are strongly energy dependent, both showing peaks near 160 MeV.[25]

3) Angular distributions for the allowed cases ($\Delta T = 0$) are diffractive at an incident pion energy of 292 MeV but are not simply diffractive at 164 MeV.[25]

4) Angular distributions for T=0 targets appear to be simply diffractive at 164 MeV.[26]

5) Although at 292 MeV DCX cross sections follow a simple target mass (A) dependence, $(N-Z)(N-Z-1)A^{-10/3}$, at 164 MeV they do not. DCX cross sections on T=0 targets follow an $A^{-4/3}$ dependence.[11]

With the present model there are two amplitudes for DCX on $T \geqslant 1$ targets. These are the conventional amplitude $\pi^+ + n + n \to \pi^0 + n + p \to \pi^- + p + p$ and the process $\pi^+ + n \to \pi^- + \Delta^{++}$, where the arrows represent intermediate deltas. For T=0 targets only the latter amplitude can contribute. The non-diffractive angular distribution at resonance energies, where the second amplitude has a maximum, can be understood as arising from an interference between these two amplitudes. Since only the second amplitude can contribute to DCX on T=0 targets one would expect these angular distributions to be diffractive, as is the case.

With respect to 5) above the A dependence of DCX on T=0 targets (arising from the second process) should be very similar to that for single charge exchange. In a simple two state model for the mixing of the Δ-N state into the low lying state the admixture amplitude can be obtained as:

$$\beta = \frac{\langle \Delta N | H' | NN \rangle}{300 \text{ MeV}} \qquad (4)$$

where the matrix element is the interaction which causes $NN \rightarrow \Delta N$ evaluated between the NN state and the ΔN state, and the energy denominator is the excitation energy of the state which is predominately ΔN. Ignoring details this matrix element should scale as the volume of the two-body interaction over the nuclear volume, i. e. A^{-1}. In evaluating a cross section the square of this matrix element multiplies a fundamental cross section which scales as the nuclear radius squared, $A^{2/3}$. Thus the cross

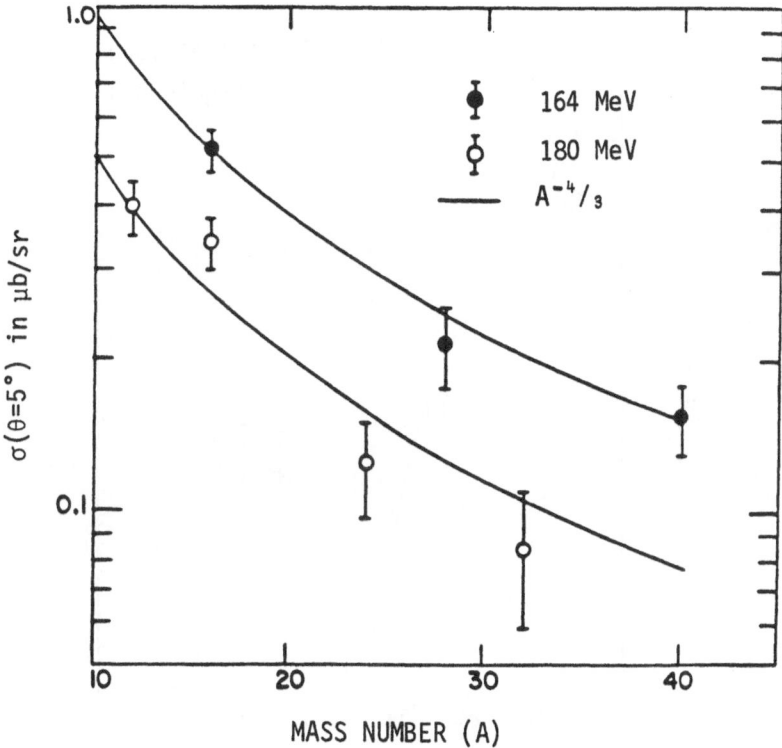

Fig. 4.   Cross sections for DCX between ground states at 5° and 164 MeV (filled circles) and 180 MeV (open circles) for (π^+, π^-) reactions on T=0 targets, plotted versus target mass (A). The curves represent an $A^{-4/3}$ dependence.

section should scale as $A^{-4/3}$. As shown in fig. 4 this fits the data quite well.

In order to calculate an admixture coefficient for the Δ^{++} component of the ground state of ^{16}Ne the momentum space DWIA code ARPIN[27] was run for a $(1p_{1/2} \boxtimes 1p_{1/2}^{-1})_{J=0,T=0}$ transition in ^{16}O. The resulting cross section was scaled by the appropriate factor from table 1 and by β^2 to fit the angular distribution for ^{16}O$(\pi^+,\pi^-)^{16}$Ne. The resulting fit, shown in fig. 5, is remarkably good. This results in value for the Δ^{++} admixture coefficient of $\beta = 0.017$.

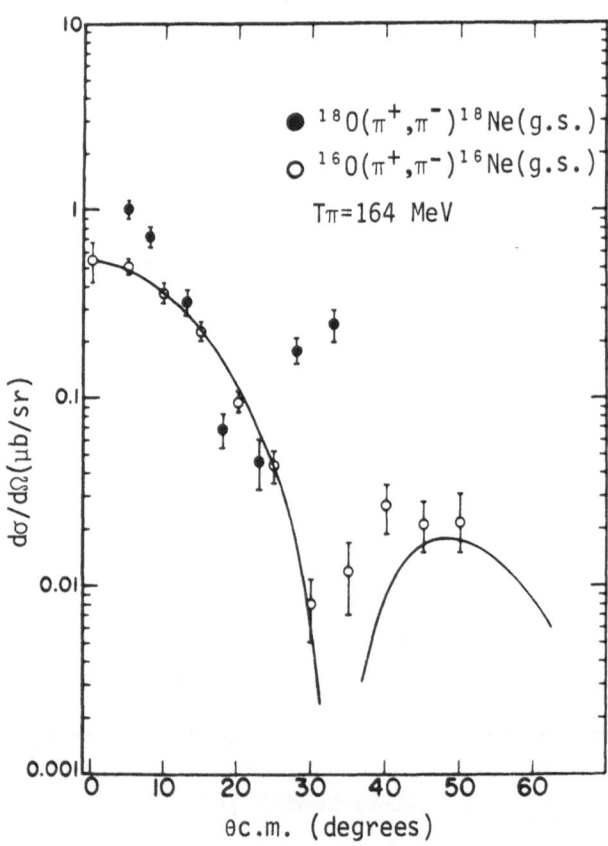

Fig. 5. Cross section angular distributions measured for DCX reactions on ^{16}O and ^{18}O at 164 MeV. The curve is the result of the calculated angular distribution for ^{16}O described in the text.

CONCLUSION

Some anomalies observed in pion-induced reactions have been qualitatively explained with a model which includes $\Delta_{3,3}$ admixtures in low lying nuclear states. Semi-quantitative analysis of these effects indicates the amplitudes for the $\Delta_{3,3}$ admixtures necessary to explain these effects are on the order of a few percent. Although a more rigorous theoretical treatment of this problem is necessary, it appears that pion-induced reactions may provide a tool with which the spectroscopy of these $\Delta_{3,3}$-admixtures can be studied.

ACKNOWLEDGEMENT

This work was supported in part by the Department of Energy

REFERENCES

1. E. Grecksch, M. Dillig, and W. Knüpfer, Z. Phys. A. 302 (1981) 247.
2. A. Bohr, and B. R. Mottelson, Phys. Lett. 100B (1981) 10.
3. A. K. Kerman and L. S. Kisslinger, Phys. Rev. 180 (1969) 1483.
4. A. M. Green, Rep. Prog. Phys. 39 (1976) 1109.
5. H. J. Weber and H. Arenhövel, Phys. Rep. 36C (1977) 280.
6. C. D. Goodman et al., Bull. Am. Phys. Soc. 26 (1981) 634.
7. C. D. Goodman, contribution to this conference.
8. H. Toki and W. Weise, Phys. Lett. 97B (1980) 12.
9. J. R. Comfort et al., Phys Rev C 21 (1980) 2147.
10. C. L. Morris, W. B. Cottingame, S. J. Greene, C. J. Harvey, C. Fred Moore, D. B. Holtkamp, S. J. Seestrom-Morris, and H. T. Fortune, Phys. Lett. 108B (1982) 172.
11. C. L. Morris, H. T. Fortune, L. C. Bland, R. Gilman, S. J. Greene, W. B. Cottingame, D. B. Holtkamp, G. R. Burleson, and C. Fred Moore, Phys. Rev. C 25 (1982) 3218.
12. L. S. Kisslinger, and W. L. Wang, Phys. Rev. Lett. 30 (1973) 1071; also see W. B. Cottingame and D. B. Holtkamp, Phys. Rev. Lett. 45 (1980) 1828.
13. T.-S. H. Lee and R. D. Lawson, Phys. Rev. C 21 (1980) 679; T.-S. H. Lee and D. Kurath, Phys. Rev. C 21 (1980) 293; and Phys. Rev. C 22 (1980) 1670.
14. C. L. Morris et al., Phys. Rev. C 24 (1981) 231.
15. K. G. Boyer et al., Phys. Rev. C 24 (1981) 598.
16. S. J. Seestrom-Morris et al., Phys. Rev. C26 (1982) 594 and Los Alamos National Laboratory Report LA-8916-T (1981).
17. R. J. Peterson et al., Phys. Rev. C 21 (1980) 1030.
18. F. Petrovich and J. A. Carr, to be published.
19. S. Cohen and D. Kurath, Nucl. Phys. 73 (1965) 1.

20. J. Cohen and J. M. Eisenberg, J. Phys. G: Nucl. Phys. 7 (1981) 881.
21. E. R. Siciliano and G. E. Walker, Nucl. Phys. A256 (1976) 444.
22. F. Lenz, M. Thies, and Y. Horikawa, Ann. Phys. (N. Y.) 149 (1982) 266.
23. P. Mulders, and A. Aerts, private communication; also P. J. Mulders, A. T. Aerts, and J. J. de Swart, Phys. Rev. D 19 (1979) 2635.
24. G. Miller, and J. E. Spencer, Phys. Lett. 53B (1974) 329; Ann. Phys. (N. Y.) 100 (1976) 562.
25. S. J. Greene et al., Phys. Rev. C 25 (1982) 924.
26. L. C. Bland et al., to be published.
27. The computer code ARPIN was written by T. S. H. Lee and was adapted by M. A. Franey to run on a CDC7600 computer.

CONSTRAINTS ON PION CONDENSATES AND PRECURSORS FROM NUCLEAR SPIN-

ISOSPIN EXCITATIONS[+]

W. Weise
Department of Physics, SUNY, Stony Brook, NY 11794
  and
Institute of Theoretical Physics, University of Regensburg
D-8400 Regensburg, W. Germany[++)]
  and
A. Härting
Institute of Theoretical Physics, University of Regensburg
D-8400 Regensburg, W. Germany

ABSTRACT

We discuss connections between the quenching of spin-isospin transitions at low momentum transfer, the absence of pion conden-sates, and the structure of nuclear spin-isospin correlations from an intermediate energy physics point of view in a picture where nuclei consist of nucleons, $\Delta(1232)$ isobars and mesons.

1. INTRODUCTION AND MOTIVATION

Investigations of nuclear spin-isospin excitations have become a substantial branch of current low and intermediate energy nuclear physics activities. The majority of workers in this area is con-vinced that studies of the nuclear response to external fields of the type $\sigma_i \tau_\lambda e^{i\vec{q}\cdot\vec{r}}$ offers exceptional possibilities to explore the borderlines between the traditional picture of nuclei consis-ting of structureless nucleons, and its opening towards mesonic and $\Delta(1232)$ degrees of freedom.

Pions and $\Delta(1232)$ have become well established ingredients of the nuclear many-body problem at intermediate energies [1,2] (i.e. at nuclear excitation energies $\omega > m_\pi = 140$ MeV). In order

[+)]Work supported in part by Deutsche Forschungsgemeinschaft (contract We 655/7-4) and by USDOE Contract No. DE-AC02-76ER13001.
[++)]permanent address

173

to see their effects on nuclear properties even at low energy
($\omega \ll m_\pi$), spin-isospin probes provide the necessary selectivity
in preparing the appropriate quantum numbers. For example, the
$\Delta(1232)$ is reached from the nucleon ground state by a $\Delta S = \Delta T = 1$ transition. This transition is strong because it proceeds by
spin-isospin flip at the level of quarks without changing their
orbital $(1s_{1/2})^3$ configuration, the 300 MeV ΔN mass difference
being essentially related to chromomagnetic fine structure splitting.

The response to operators of the type $\vec{\sigma} \cdot \vec{q} \, \tau_\lambda$ directly involves
the nuclear pion field. Much of the developments in recent years
have been influenced by Migdal's original suggestion [3] that at
momentum transfers q of the order of the Fermi momentum, the static
pion field (or equivalently, the static one-pion exchange interaction) might develop sufficient attractive strength to form a pion
condensate [3,4]. Pion condensates, as well as their precursors
[5], appear now to be ruled out [10], both by the available spectroscopic information on unnatural parity states [6] and by inelastic
proton scattering into unnatural parity states at high momentum
transfers [7-9]. The absence of a tendency towards pion condensation has been related to repulsive screening effects accompanying
the pion field. To the extent that such screening effects are of
short-ranged origin, their magnitude should not vary much on a
momentum transfer scale $0 \leq q \leq 2m_\pi$. The assertion has therefore
been made that this screening finds its counterpart at long wavelength in a quenching of isovector spin transitions [11], the $\Delta(1232)$
isobar playing an important role in this procedure.

Substantial experimental progress has been made in the past few
years, by the exploration of Gamow-Teller (GT) strength distributions
in (p,n) reactions [12-14], and by investigations of magnetic transitions in a variety of nuclei using high resolution inelastic electron
scattering [15,16]. Both the GT strength [12-14] and the magnetic
transitions of low multipolarity [15,17] are observed to be systematically quenched [18]. Part of the quenching can be attributed to
ground state correlations and "standard" nuclear core polarization,
but there remain sizeable reductions, as will be shown, which are
not easily accommodated within standard models. Their explanation
follows quite naturally if one accepts the presence of polarization
effects involving virtual excitations of nucleons into $\Delta(1232)$ isobars by repulsive interactions essentially identical to those which
screen nuclei against pion condensation.

The plan of this presentation will be to discuss connections
between the quenching of spin-isospin transitions at low momentum
transfer, the absence of pion condensates, and the structure of
nuclear spin-isospin correlations from an "intermediate energy"
point of view, i.e. in a picture where nuclei consist of nucleons,
$\Delta(1232)$ isobars and mesons. On the way, two specific examples of

magnetic quenching (of 1^+ transitions in ^{48}Ca and ^{88}Sr) will be
discussed. Gamow-Teller excitations will be touched only briefly,
since they are covered extensively by several speakers at this
conference. We shall review some of the recent theoretical developments to obtain the magnitude of the parameter g', the one multi-
plying the $\vec{\sigma}_1 \cdot \vec{\sigma}_2 \vec{\tau}_1 \cdot \vec{\tau}_2$ part of the Landau-Migdal interaction. This
g' is held responsible for at least part of the observed spin-
isospin screening. Finally, the fraction of g' related to the
classical Lorentz-Lorenz effect [42] will be reexamined from a
chiral quark bag model point of view.

2. BASIC INGREDIENTS OF NUCLEAR SPIN-ISOSPIN RESPONSE

We shall consider spin-isospin excitations first for nucleons
only and later including $\Delta(1232)$ isobars, following developments
in ref. [2]. There are two basic types of excitations represented
by operators

$$O_\ell = \vec{\sigma} \cdot \hat{q} \; \tau_\lambda \; e^{i\vec{q} \cdot \vec{r}} \quad \text{(longitudinal)} , \tag{1a}$$

and

$$O_t = (\vec{\sigma} \times \hat{q})_z \; \tau_\lambda \; e^{i\vec{q} \cdot \vec{r}} \quad \text{(transverse)} . \tag{1b}$$

Here $\hat{q}$ is the direction of momentum transfer $\vec{q}$. The terms "longitudinal" and "transverse" indicate the preferred alignment between
spin and momentum transfer. Excitations driven by $O_\ell(\vec{q})$ will be
called "pion-like", because of their pseudoscalar-isovector
structure. Transverse excitations via $O_t(\vec{q})$ are encountered for
example in magnetic isovector-spin transitions. The (p,n) and
(p,p') reactions involve combinations of both O_ℓ and O_t.

In response to these operators, a nuclear medium will be
polarized, primarily by particle-hole excitations coupled to the
appropriate quantum numbers. Here O_ℓ excites unnatural parity
states (pion like states) while O_t excites both natural and unnatural parity states. The <u>direct</u> particle-hole interaction which
drives the response must be of the general form[+)]

$$V_{\sigma\tau}(\vec{q}) = [V_\ell(q^2)\vec{\sigma}_1 \cdot \hat{q} \; \vec{\sigma}_2 \cdot \hat{q} + V_t(q^2)(\vec{\sigma}_1 \times \hat{q}) \cdot (\vec{\sigma}_2 \times \hat{q})] \; \vec{\tau}_1 \cdot \vec{\tau}_2 . \tag{2}$$

Note that $\vec{\sigma}_1 \cdot \hat{q} \; \vec{\sigma}_2 \cdot \hat{q} = \frac{1}{3} \vec{\sigma}_1 \cdot \vec{\sigma}_2 + \frac{1}{3} S_{12}(\hat{q})$,

and $(\vec{\sigma}_1 \times \hat{q}) \cdot (\vec{\sigma}_2 \times \hat{q}) = \frac{2}{3} \vec{\sigma}_1 \cdot \vec{\sigma}_2 - \frac{1}{3} S_{12}(\hat{q})$

[+)] We ignore possible spin-orbit interactions which are empirically
small in isovector channels.

with $S_{12}(\hat{q}) = 3\vec{\sigma}_1 \cdot \hat{q}\ \vec{\sigma}_2 \cdot \hat{q} - \vec{\sigma}_1 \cdot \vec{\sigma}_2$. The prototype interaction for V_ℓ is one-pion exchange (OPE). In its simplest form without cutoff factors,

$$V_\pi(q^2) = -\frac{f^2}{m_\pi^2}\ \frac{q^2}{q^2 + m_\pi^2}\ ,\quad \frac{f^2}{4\pi} = 0.08\ . \tag{3}$$

The prototype for the transverse coupling interaction V_t is isovector two-pion exchange. If the 2π mass spectrum is reduced to a single δ-function at the ρ meson mass $m_\rho = 770$ MeV, one obtains

$$V_\rho(q^2) = -\frac{f_\rho^2}{m_\rho^2}\ \frac{q^2}{q^2 + m_\rho^2}\ ,\quad \frac{f_\rho^2}{m_\rho^2} \simeq 2\ \frac{f^2}{m_\pi^2}\ . \tag{4}$$

One should keep in mind that ρ exchange is really the exchange of a continuous mass spectrum of two interacting pions. In fact, while the tensor part from ρ exchange produces just about the right amount of destructive interference with the OPE tensor force to account for the isovector-tensor force of, say, the Paris NN potential [19] over a momentum transfer range $0 \le q \lesssim 2m_\pi$, the spin-spin part turns out to be subject to strong cutoff corrections at high momentum transfers. In any case, one expects the prototype interactions to be accompanied by screening effects at short distances from several possible sources.

(i)   short-range repulsive correlations;
(ii) many-body vertex corrections (e.g. from exchange terms);
(iii) possible effects from the exclusion of meson fields out of the quark core volume, a situation familiar from chiral bag models [20], and to be discussed later.

We assume in the following that screening effects are introduced phenomenologically by a repulsive Fermi liquid type interaction, $g'\ \vec{\sigma}_1 \cdot \vec{\sigma}_2\ \vec{\tau}_1 \cdot \vec{\tau}_2$, to be added to OPE and ρ exchange. Note that g' acts identically in longitudinal and transverse channels, since $\vec{\sigma}_1 \cdot \vec{\sigma}_2 = \vec{\sigma}_1 \cdot \hat{q}\ \vec{\sigma}_2 \cdot \hat{q} + (\vec{\sigma}_1 \times \hat{q}) \cdot (\vec{\sigma}_2 \times q)$. Hence the ansatz

$$V_\ell(q^2) = g' - \frac{f^2}{m_\pi^2}\ \frac{q^2}{q^2 + m_\pi^2}\ ;\quad V_t(q^2) = g' - \frac{f_\rho^2}{m_\rho^2}\ \frac{q^2}{q^2 + m_\rho^2} \tag{5}$$

results. In the long wavelength limit, $V_{\sigma\tau}(q^2=0) = g'\ \vec{\sigma}_1 \cdot \vec{\sigma}_2 \vec{\tau}_1 \cdot \vec{\tau}_2$, and g' is related, up to a constant, to the Landau-Migdal Fermi liquid parameter G'_0. For infinite nuclear matter, the response to longitudinal operators O_ℓ is determined by $V_\ell(q^2)$, while the response to O_t is determined by $V_t(q^2)$, each of them acting in the <u>direct</u> particle-hole channel only, following the Fermi liquid concept.

The lowest order response of symmetric nuclear matter to a $\sigma\tau$-dependent static probe involves the nucleon-hole spin-isospin susceptibility

$$\chi_N(q^2) = 2\sum_{ph} \frac{|<ph|\sigma_i\tau_\lambda e^{i\vec{q}\cdot\vec{r}}|0>|^2}{E_p - E_h} \tag{6}$$

which is proportional to the familiar Lindhard function. It is useful to introduce diamesic functions, in complete analogy with the dielectric function describing the response of a polarizable medium to an electric field. For example, a pion field ϕ_π couples to nucleons via a longitudinal source function proportional to O_ℓ. A pion field in free space, $\phi_\pi^{(o)}$, will be modified by to the polarization of the medium according to

$$\phi_\pi(\vec{q}) = \frac{\phi_\pi^{(o)}(\vec{q})}{\varepsilon_\ell(q^2)} \quad , \tag{7}$$

where ε_ℓ is the longitudinal diamesic function,

$$\varepsilon_\ell(q^2) = 1 + V_\ell(q^2)\chi_N(q^2) \quad . \tag{8}$$

Similarly, the response to fields carrying the symmetry of the transverse operator, O_t of eq. (1b), will be determined by a transverse diamesic function,

$$\varepsilon_t(q^2) = 1 + V_t(q^2)\chi_N(q^2) \quad . \tag{9}$$

Note that the longitudinal response is a matter of balance between a repulsive part weakly dependent on q^2 (the g' term) and attraction (from OPE) rapidly varying with q^2. If the attraction wins at high q (for sufficiently small g'), the enverse of ε_ℓ acts to amplify the pion field in the medium. The amplification becomes critical as $\varepsilon_\ell \to 0$. This is the threshold for pion condensation in this model.

For sufficiently large g', the transverse diamesic function will always be larger than unity as long as momentum transfers are small compared to the ρ meson mass. In the long wave length limit, both ε_ℓ and ε_t reduce to

$$\varepsilon(q^2=0) = 1 + g' \chi_N(q\to0) \tag{10}$$

For symmetric nuclear matter, $\chi_N(q\to 0) = \dfrac{2M^* k_F}{\pi^2}$, where M^* is the nucleon effective mass and k_F the Fermi momentum. Hence any spin-isospin operator (longitudinal or transverse) will be quenched by a factor $\varepsilon^{-1}(q^2=0)$ in this limit.

So far, only nucleon-hole excitations have been considered, and the procedure outlined above is equivalent to RPA in a large model space of nucleons only. Next, Δ-hole components have to be added. The total spin-isospin susceptibility in lowest order becomes

$$\chi = \chi_N + \chi_\Delta \ ,$$

where

$$\chi_\Delta(q^2) = 2\delta \sum_{\Delta h} \frac{|<\Delta h|S_z^+ T_\lambda^+ e^{i\vec{q}\cdot\vec{r}}|0>|^2}{E_\Delta - E_h} \ , \qquad (11)$$

and we have used $N\Delta$ spin and isospin transition operators $\vec{S}^+$ and $\vec{T}^+$ connecting nucleon and Δ states. The factor δ relates the spin-isospin coupling strength for $N\to\Delta$ transitions to that for nucleons. The SU(4) quark model gives $\delta = 72/25$, while the empirical number is closer to the Chew-Low value, $\delta=4$. We use the latter one in all applications. Similar factors appear in the OPE and ρ exchange $N\Delta\to\Delta N$ interactions.

In the static limit, $\chi_\Delta = \dfrac{8}{9}\delta \dfrac{\rho}{M_\Delta-M}$, where ρ is the density and $M_\Delta-M \simeq 300$ MeV the ΔN mass difference. Note that at nuclear matter density in the long wavelength limit, a substantial fraction of the static spin-isospin susceptibility in lowest order is due to virtual Δ-hole excitations, the ratio of χ_Δ to χ_N at $q=0$ being about 0.35 at nuclear matter density.

An important issue is to understand the Δ-isobar analogue of the Fermi liquid interaction, $g'_\Delta \vec{S}_1\cdot\vec{S}_2^+ \vec{T}_1\cdot\vec{T}_2^+$, which introduces a repulsive screening parameter g'_Δ again in the strict sense of Landau-Migdal theory, namely such that it incorporates all vertex corrections from exchange terms (see Fig. 1).

We shall return to a discussion of this point later on. For the moment we assume that there is a universal screening parameter g' for both nucleons and Δ's. In this case, the longitudinal and transverse diamesic functions including Δ-hole components are obtained from eqs. (8,9) upon replacing χ_N by $\chi = \chi_N + \chi_\Delta$. The Δ components play a substantial role in the pion condensation problem.

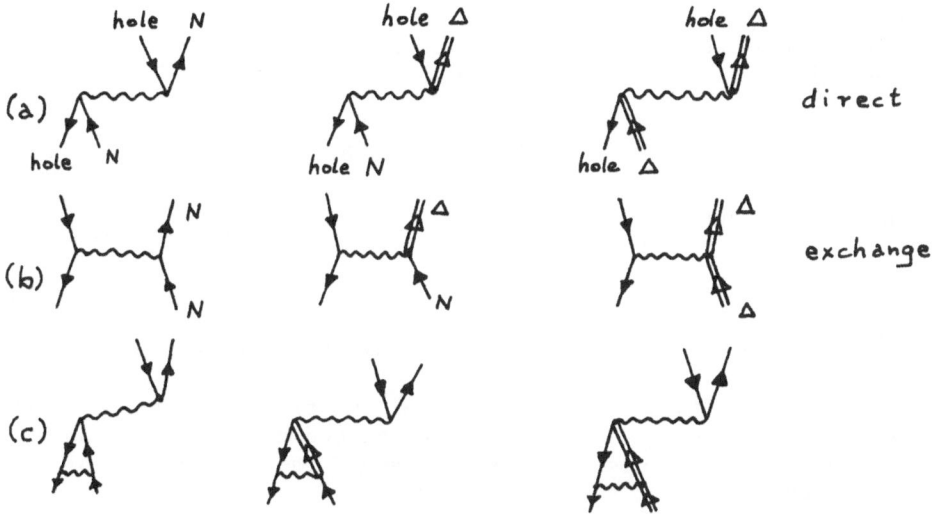

Figure 1: (a)   direct nucleon-hole, nucleon-hole→Δ-hole and
Δ-hole→ Δ-hole interaction and (b) corresponding
exchange terms.   (c) shows examples how exchange
terms enter as vertex corrections in the direct
particle-hole interactions.   These corrections are
already incorporated in the phenomenological g'.

If no screening ($g'_\Lambda = 0$) were allowed for Δ's, the critical density
ρ_c would come down close to nuclear matter density [3], even if
$g' = 0.6\ m_\pi^{-2}$ for nucleons.   A universal $g' > 0.5\ m_\pi^{-2}$ for both
nucleons and Δ's moves ρ_c well above twice the density of normal
nuclear matter.

   If g' for Δ-isobars is finite and positive, virtual Δ-hole
excitations add to the quenching of both longitudinal and transverse
operators in the long-wavelength limit.   In fact, in dealing with
finite nuclei, the nucleonic part of the susceptibility, χ_N, is
usually treated to a large extent by an explicit RPA or shell model
calculation in a given model space.   Effective operators and interactions are then introduced to account for the remaining parts
outside the model space, not treated explicitly.   It is convenient

and practical to treat the Δ-hole components that way, because of their large energy denominators of about 300 MeV separating them from the low-lying nucleonic particle-hole spectrum. The remaining Δ-induced quenching at q=0 applies to the isovector spin g-factor g_s^V as well as to the axial vector coupling constant g_A and is expressed as

$$g_{s,eff}^V = \gamma \, g_s^V \, , \qquad\qquad g_{A,eff} = \gamma \, g_A \qquad\qquad (12)$$

with a common quenching factor

$$\gamma = [1 + g' \chi_\Delta(q=0)]^{-1} \, . \qquad\qquad (13)$$

Note that $\gamma^2 \simeq 0.5$ at nuclear matter density for $g' = 0.5 \, m_\pi^{-2}$.

At the same time, the longitudinal and transverse parts of the effective interaction, operating in a reduced nucleon-hole space only, have to be renormalized, $V_\ell(q^2)$ and $V_t(q^2)$ of eq. (5) being replaced by

$$V_\ell^{eff}(q) = \frac{V_\ell(q)}{1 + V_\ell(q)\chi_\Delta} \qquad V_t^{\ eff}(q) = \frac{V_t(q)}{1 + V_t(q)\chi_\Delta} \, , \qquad (14)$$

respectively. The procedure outlined here is essentially the one used by Brown and Rho [21] in their treatment of Gamow-Teller resonances.

In applications to finite nuclei [2], the following modifications have to be made:

(i)   The particle-hole states are classified by their total angular momentum J and, for unnatural parity states, by two possible orbital angular momenta, L=J±1. The direct particle-hole interaction has matrix elements

$$<(p'h')J|V_{\sigma\tau}|(ph)J> = \sum_{L'L} \int_o^\infty \frac{dq \, q^2}{(2\pi)^3} \, \rho_{p'h'}^{*JL'}(q) \, V_{L'L}^J(q) \, \rho_{ph}^{JL}(q) \qquad (15)$$

where $\rho_{ph}(q)$ are particle-hole transition formfactors specified in detail in ref. [2], and

$$V_{L'L}^J(q^2) = a_{JL'} V_\ell(q^2) \, a_{JL} + b_{JL'} \, V_t(q^2) b_{JL}. \qquad\qquad (16)$$

The coefficients a_J and b_J project onto the longitudinal and transverse parts of the interaction for each particle-hole angular momentum J. Explicitly,

$$a_{J,J+1} = - \sqrt{\frac{J+1}{2J+1}} \qquad\qquad b_{J,J+1} = \sqrt{\frac{J}{2J+1}}$$

$$a_{J,J-1} = \sqrt{\frac{J}{2J+1}} \qquad\qquad b_{J,J-1} = \sqrt{\frac{J+1}{2J+1}}$$

Note that the a_J and b_J are normalized and orthogonal, in the sense that

$$\sum_L a_{JL}^2 = \sum_L b_{JL}^2 = 1 \; , \quad \sum_L a_{JL} b_{JL} = 0.$$

(ii)   For each particle-hole channel of given J, the diamesic functions are replaced by matrices $\varepsilon_{L',L}^{J}(q',q)$ which are non-local in momentum space due to the nuclear finite size. The procedure for obtaining spin-isospin response functions involves a matrix inversion in L,L' together with the solution of a momentum space integral equation, rather than the simple geometric series obtained for infinite nuclear matter. Details can be found in ref. [2]. The non-locality of the diamesic function becomes important for small and medium heavy nuclei.

As a consequence, the quenching of spin-isospin transitions becomes both angular momentum and A-dependent [22]. In particular, the $\Delta(1232)$-induced quenching for N=Z nuclei decreases with increasing J and increases with increasing mass number A. The systematics for N≠Z is discussed in ref. [23].

3.   TWO EXAMPLES OF $\Delta(1232)$-INDUCED MAGNETIC QUENCHING

We return now to the basic question about the participation of $\Delta(1232)$ degrees of freedom in low energy, low momentum transfer spin-isospin response problems. We recall from eq. (13) that if g'=0 for Δ-isobars, then no Δ-induced quenching appears, but the pion condensation threshold, or at least a signature of its precursors, would have to be expecred at densities close to normal nuclear matter. On the other band, a finite g' of about 0.5 (in pion mass units) introduces quenching factors of about 1/2 for M1 and GT transition probabilities and at the same time moves the pion condensation threshold essentially out of any experimentally accessible range.

The Δ-induced quenching introduced in ref. [11] has been discussed recently by various authors [21,22,24-30]. The main difficulty in isolating Δ(1232) effects lies in discriminating them
from conventional core polarization [31] and ground state correlations [29,32]. In addition, pionic exchange current contributions
[29] have to be considered for M1 transitions. In a situation like
this, the emphasis will be on the systematics of the Δ-quenching in
as many nuclei as possible, as seen by various complementary probes
spin-isospin probes, like (e,e')[15,16],(p,p')[7-9], (p,n)[12-14]
and (π,γ)[33]. A necessary condition is that the unnatural parity
states under consideration should be as simple as possible from the
conventional nuclear structure point of view. There are not many
"clean" cases of the kind for which experimental data covering
finite momentum transfers exist. We shall investigate here two
examples of M1 transitions in ^{48}Ca and ^{88}Sr.

3.1 ^{48}Ca(1^+; 10.2 MeV)

According to ref. [34], the 1^+ state in ^{48}Ca at 10.23 MeV has
a relatively simple shell model structure dominated by a $f_{5/2}f_{7/2}^{-1}$
neutron-hole configuration. The proton core seems to be well
closed. The wave function following ref. [34] obtained from a
full fp-shell model calculation is:

$$|^{48}Ca; 1^+> = 0.89 \; \nu|f_{5/2}f_{7/2}^{-1}; \; 1^+> + 0.11 \; \nu|f_{7/2}f_{5/2}^{-1}; \; 1^+> \quad (17)$$

plus additional small amounts of more complicated configurations.
The dominant neutron-hole component makes this a favorable case for
studying renormalization effects of the spin g factor. A pure
$f_{7/2} \to f_{5/2}$ single particle transition using the free space value
for g_s gives B(M1)↑ = 12 μ_k^2, whereas the experimental value is
4 μ_k^2 (5.2 μ_k^2 if all the 1^+ strength in the vicinity of the
10.23 MeV state is collected) [35]. Using the wave function, eq.
(17), the B(M1) value comes down to 7.3 μ_k^2. A major fraction of
this comes from 2p-2h ground state correlations of the type shown
in Fig. 2 (see also ref. [29]). Such 2p2h-correlations are

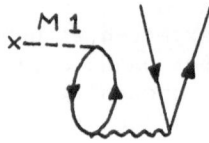

Fig. 2: Quenching of B(M1) by 2p-2h ground state correlations.

incorporated in standard RPA calculations [28], where $B(M1)=8.2 \ \mu_k^2$ is found. The importance of 2p-2h correlations has also been emphasized in ref. [32]. It is obvious, however, by comparison with the McGrory and Wildenthal [34] calculation, that an additional quenching by about 1 μ_k^2 is due to more complicated many particle-many hole configurations. Pionic and pair exchange current contributions have been investigated by Kohno and Sprung [29]. The pionic current adds to the quenching, but the pair current works in opposite direction, such that the total exchange current contribution increases the B(M1) value by about 0.5 μ_k^2. Altogether, it is therefore difficult to obtain a B(M1) much less than 8 μ_k^2 from a combination of ground state correlations, nucleonic core polarization and exchange currents. Another factor of 1.5 or 2 reduction seems to be needed.

Now, one would expect that Δ-hole induced screening of the type shown in Fig. 3 is a good candidate for supplying a major fraction of the remaining quenching, by processes of the type shown in Fig. 3.

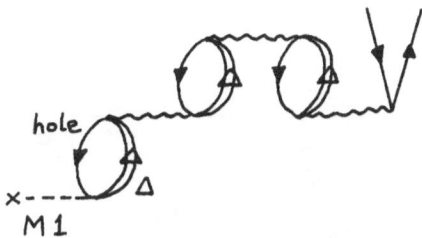

Figure 3:   Δ-hole induced screening of M1 transitions.

The calculation has been carried out using the diamesic function method fully taking into account non-localities and the proper angular momentum projections as described in refs. [27,2]. The direct Δ-hole interaction used here [27] is of the type shown in eq. (5), with π and ρ exchange plus a Δ-hole Fermi liquid interaction proportional to g'. The parameter g' essentially controls the amount of Δ-induced quenching. The result is shown in Fig. 4. Selecting g' between 0.5 and 0.6 gives roughly a factor of 2 reduction over and beyond the McGrory and Wildenthal calculation. This corresponds to an effective isovector-spin g factor $g_s^{eff} \approx 0.7 \ g_s^{free}$. Note again that pionic and pair current contributions add an amount of about 0.5 μ_k^2 to curve (c) of Fig. 4. This contribution is absent in the $^{48}Ca(p,n)^{48}Sc(1^+; 16.8 \text{ MeV})$ GT transition, but on the other hand, 2p2h correlations are also

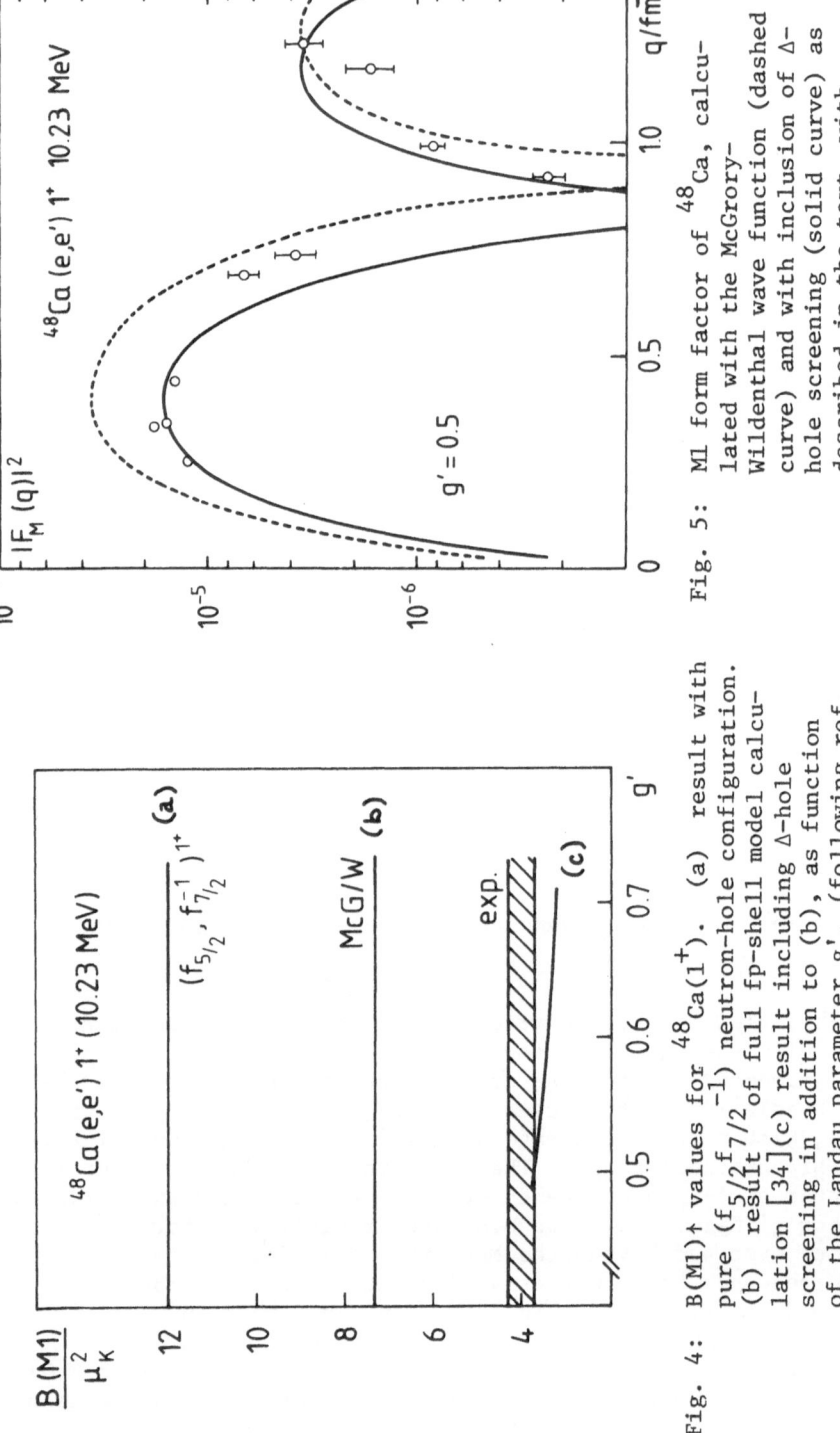

Fig. 5:   M1 form factor of ^{48}Ca, calculated with the McGrory-
Wildenthal wave function (dashed
curve) and with inclusion of Δ-
hole screening (solid curve) as
described in the text, with
g'=0.5. Exp. data are taken
from ref. [15] .

Fig. 4:   B(M1)↑ values for ^{48}Ca(1^+).   (a) result with
pure ($f_{5/2} f_{7/2}^{-1}$) neutron-hole configuration.
(b) result of full fp-shell model calcu-
lation [34](c) result including Δ-hole
screening in addition to (b), as function
of the Landau parameter g' (following ref.
[27]). Exp. numbers taken from ref. [35].

suppressed in this process. A Δ-induced quenching of GT strength
by a factor of two would give about half of the reduction required
by experimental data at 160 MeV [36].

   It should be noted that a proper treatment of momentum space
non-localities in the Δ-hole polarization process, Fig. 3, is
important for nuclei as small as ^{48}Ca [27]. Consequently, there is
a mixing of transverse and longitudinal parts of the Δ-hole interaction in finite nuclei, even though the probing M1 field is purely
transverse. As a consequence, the attraction from OPE and (to lesser
extent) ρ exchange reduces the quenching obtained from g' alone by
about 10% (at q=0 and for g' = 0.5; this effect is included in
curve (c) of Fig. 4).

   At finite momentum transfers, the Δ-induced quenching helps to
close the gap between the McGrory-Wildenthal calculation and the
experimental M1 form factor, as Fig. 5 shows. It is to be expected,
however, that at momentum transfers q $\sim$ 1 fm^{-1} core polarization
effects outside the fp-shell model space will be of some importance.

3.2 ^{88}Sr(1^+; 3.48 MeV)

   It might appear that the low-lying 1^+ state in ^{88}Sr is an
exceedingly complicated case from the nuclear structure point of
view. On the other hand, extended broden-pair (BP) calculations
[37] reveal that the leading component of that state is a simple
proton $(p_{3/2}p_{1/2})$-particle-hole configuration accompanied by an
admixture of $(g_{7/2}g_{9/2}^{-1})$ neutron-hole pairs:

$$|^{88}Sr;1^+>=0.81\pi|2p_{1/2}2p_{3/2}^{-1};1^+>+0.27\pi|2p_{3/2}2p_{1/2}^{-1};1^+>$$

$$+ 0.02 \, \nu|g_{7/2}g_{9/2}^{-1}; 1^+> , \qquad\qquad (18)$$

additional more complicated components being negligibly small. The
magnetic form factor for this state has recently been measured [38]
and found to be systematically smaller than the expectation based on
the BP wave function. The measured B(M1) $\uparrow$ value, 0.92 ± 0.15 μ_k^2,
is a factor of two lower than the BP calculation result obtained
with the unrenormalized spin g-factor. We have calculated the effect
of Δ-hole screening using the same diamesic function technique as
for ^{48}Ca in order to see whether a consistent picture of the required quenching can be obtained. The situation here is different
from ^{48}Ca because the leading part of the M1 transition is now made
by the 2p-shell protons, where both orbital and spin g-factors
enter. (We leave g_ℓ unrenormalized (g_ℓ=1), but note that an
effective g_ℓ of 1.12 for protons and -0.03 for neutrons [39] does

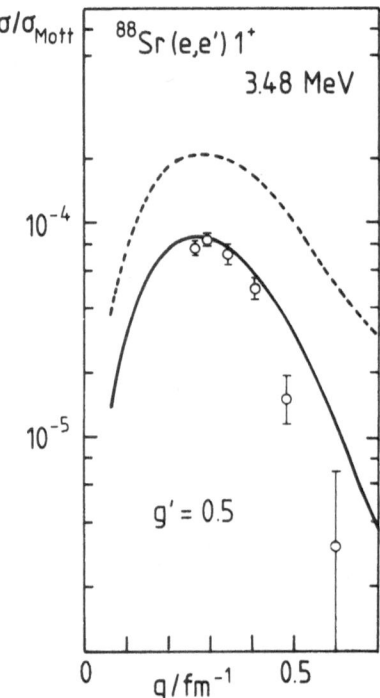

Figure 6: σ/σ_{MOTT} for the M1 transition to the 3.48 MeV
             state in ^{88}Sr, calculated with broken-pair wave
             functions following ref. [37] (dashed curve)
             and including in addition the effect of Δ-hole
             screening (solid curve), using a Landau parameter
             g' = 0.5 in the direct Δ-hole interaction. The
             experimental data are taken from ref. [38].
             DWBA corrections are included in the curves.

not change our conclusions very much). Results are shown in Fig. 6 for $g'=0.5$ m_π^{-2}. Again, a Δ-hole Fermi liquid parameter of this order gives the correct amount of quenching. The B(M1) value so obtained is 0.8 μ_k^2. One expects that pionic and pair current contributions will increase B(M1), so that in fact a somewhat larger Δ-induced quenching, with $g'=0.6$ m_π^{-2}, would be tolerable.

The Δ-hole screening gives the correct tendency for the steep decrease of the form factor at $q \sim 0.5$ fm^{-1}. In fact, the sensitivity to renormalization of g_s is greatly enhanced in the ^{88}Sr case because of destructive interference between orbital and spin parts of the M1 amplitude. The still remaining discrepancy can at least partly be explained by nucleon-hole polarization effects outside the model space implied by eq. (18).

So far, we can summarize the situation by saying that certainly a major fraction of the observed M1 quenching is due to ground state correlations and standard core polarization, but still an effective spin g-factor, typically $g_s^{eff} \simeq 0.7$ g_s, is needed to reproduce data at low momentum transfers. The combined contribution of pionic and pair currents does not help because it acts to increase B(M1). A natural source of quenching is the Δ-hole screening, provided g' for Δ-isobars is 0.5 m_π^{-2} or larger. Similar screening effects are needed to explain the reduction of GT strength in ^{208}Pb consistent with the location of the GT resonance in a schematic model [21].

4. HOW LARGE IS THE FERMI LIQUID PARAMETER g' FOR Δ(1232) ISOBARS?

The basic question is whether a g' of this size for Δ's can be understood in simple terms. In a model based on OPE and ρ exchange modified by short-range correlations, it has been pointed out that exchange terms of the NN↔NΔ interaction, Fig. (1b), tend to cancel a large part of the direct Δ-hole interaction responsible for the Δ-induced quenching [28,29][+]. If so, the effective g' for Δ isobars would have to be small, typically less than g'=0.25 following [29]. Cancellations from exchange terms are the reason why the authors of ref. [28], in a coupled Δ-hole and nucleon-hole RPA calculation, obtain less Δ-induced quenching. On the other hand, one must keep in mind that exchange terms involving the ΔN→ΔN interaction, Fig. 7 can certainly not be disregarded. Such terms receive contributions not only from π and ρ exchange, but also from isoscalar exchanges which have no counterpart in the direct Δ-hole channel. One expects that the latter give overall attraction in the particle-particle channel and increase g' through exchange of the type, Figure 7. Therefore statements about exchange terms based on π and ρ exchange alone must be regarded with some care.

+)see also Towner and Khanna in ref. [24], and A. Arima referred to in ref. [16].

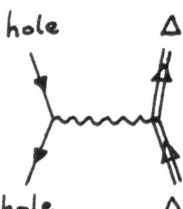

Figure: 7   Exchange term related to the direct Δ-hole
             interaction

It is obvious from the above that the problem of determining
all relevant exchange terms of the Δ-hole interaction is far from
being solved. In a situation like this, we find it useful to
strictly maintain the Landau Fermi liquid framework, for nucleons
as well as for Δ(1232) isobars, and constrain g' from as many
data as possible.

Some support comes from a coupled channel reaction matrix
calculation for both nucleons and Δ's [40], where the g' for Δ's
turns out to be only slightly smaller than that for nucleons, even
without the terms, Fig. 7. At the same time, reaction matrix
calculations of g' for nucleons [41] with realistic interactions
show that (a) a local approximation of the $\sigma\tau$-dependent G matrix
in momentum space, similar to the parametrization of $V_{\sigma\tau}$, eq. (5),
is reasonable and (b) a g' of about 0.6 m_π^{-2}, smoothly dependent
on energy and momentum transfer, is well within reach.

4.1   Chiral Bag Models and the Lorentz-Lorenz Correction

In trying to find common roots for at least part of the Δ-
induced screening and the size of Landau parameter g' for nucleons,
oneis naturally lead to quark models where nucleons and Δ's are
partners in their orbital three-quark structure, split only by
gluonic hyperfine interactions. In chiral bag models [20], the
space inside the bag is occupied by quarks, but no pions, while
the pion field exists outside the bag and joins to the quark
axial current at the bag boundary such as to make it continuous
across the boundary. The axial current in this model (we consider
the static limit, which is sufficient for our purposes here) is:

$$\vec{A}_\lambda(\vec{r}) = -\ \bar{\psi}\ \gamma_5\vec{\gamma}\ \frac{\tau_\lambda}{2}\ \psi(\vec{r})\ \theta(R-r) + f_\pi\ \vec{\nabla}\ \phi_\lambda(\vec{r})\ \theta(r-R) \qquad (19)$$

where the ψ are quark spinors and R is the bag radius. If the pion has zero mass, the axial current is conserved, $\vec{\nabla}\cdot\vec{A}_\lambda = 0$, and

$$\oint \vec{A}_\lambda \cdot d\vec{S} = 0 \, ,$$

where the integration is taken over the surface of the bag sphere. For each surface element $d\vec{S}$ normal to that sphere, we have:

$$-\bar{\psi}\, \gamma_5\, \vec{\gamma}\, \frac{\tau_\lambda}{2}\, \psi \, \cdot \, d\vec{S} \bigg|_{r=R-\epsilon} = f_\pi\, \vec{\nabla}\phi_\lambda \, \cdot \, d\vec{S} \bigg|_{r=R+\epsilon} \qquad (20)$$

Consider now a medium of well separated bags, where the density ρ is the number of bags per volume. We assume that the average distance between bags in large compared to the bag radius R and are interested in the <u>local</u> pion field ϕ^{loc} at the boundary of one particular bag. This field will differ from the pion field in free space because of the spin-isospin polarization of the medium surrounding the bag. The polarizability is just given by the spin-isospin susceptibility χ, see section 2, and the pion field in the medium satisfies the familiar equation $\vec{\nabla}(1-\chi)\vec{\nabla}\phi = 0$. Now, locally at the boundary of one bag, the condition eq. (20) in the presence of a polarizable medium is modified according to

$$\overset{\text{inside}}{-\frac{1}{f_\pi}\, \bar{\psi}\, \gamma_5\vec{\gamma}\, \frac{\tau_\lambda}{2}\, \psi\cdot d\vec{S} \bigg|_{r=R}} = \overset{\text{outside}}{(1-\chi)\vec{\nabla}\phi_\lambda\cdot d\vec{S} \bigg|_{r=R}} \qquad (21)$$

By comparison, one sees that there is a correction to the local field at the boundary due to the polarization of the surrounding medium,

$$\delta\, \vec{\nabla}\phi_\lambda \, \cdot \, d\vec{S} = -\chi\, \vec{\nabla}\phi_\lambda\cdot d\vec{S} \, ; \qquad (22)$$

this has to be averaged over angles which gives a factor 1/3. Performing the correction self-consistently yields for the local field at the bag boundary

$$\phi^{loc} = \phi^{(o)} - \frac{1}{3}\, \chi\, \phi^{loc} \qquad (23)$$

which can be solved to obtain

$$\phi^{loc} = \frac{\phi^{(o)}}{1 + 1/3\ \chi} \ . \tag{24}$$

Eq. (24) relates the pion field $\phi^{(o)}$ in the absence of a spin-isospin polarizable medium to the local field at the surface of a baryon in the presence of such a medium.

We have just repeated the textbook derivation of the classical Lorentz-Lorenz effect [42] under the assumption that the pion field is not allowed to penetrate the bag interior. This condition alone, even in the absence of any repulsive short-range correlation, introduces a screening effect which, by comparison with the development in section 2, can be represented by a Landau Fermi liquid parameter g'=1/3. This conclusion is obviously identical for nucleons and Δ(1232) isobars. A finite pion mass would not change much, as long as the bag radius R is small compared to the pion Compton wavelength. Thus the quark bag picture itself, a phenomenology based on the asymptotic freedom and confinement properties of QCD, gives a major fraction of a universal g' for both nucleons and Δ's, if the pion field is kept outside the bag.

5. SUMMARY

We have discussed screening mechanisms in nuclear spin-isospin response problems, those effects which prevent pion condensates and their precursors from developing. We found it most useful to treat these screening effects in terms of Landau Fermi liquid interactions for both nucleons and Δ(1232) isobars, assigning to them a Landau parameter g'. While g' is rather well established for nucleons, (values g' $\simeq$ 0.6 m_π^{-2} are commonly accepted), the corresponding parameters for Δ-isobars are subject to some debate. It appears that the observed quenching of Gamow-Teller and magnetic transitions (two examples of the latter kind we investigated in detail) finds a most natural explanation in terms of Δ-hole screening, if g' for Δ's is of about the same size as g' for nucleons. If this turns out to be true, it is tempting to assume that the dynamics of spin-isospin screening for nucleons and Δ's has a common root. Along these lines we presented a re-interpretation of the Lorentz-Lorenz correction, familiar from the pion-nucleus optical potential, in terms of chiral bag models which strictly prohibit pions from the bag interior. Such models give g'=1/3 for nucleons and Δ's, even in the absence of correlations.

ACKNOWLEDGEMENTS

One of us (W.W.) is grateful to G.E. Brown, S.O. Bäckman, W. Dickhoff, H. Toki and J. Wambach for many stimulating discussions. He wishes to thank G.E. Brown and the Physics Department at SUNY, Stony Brook, for their kind hospitality.

REFERENCES

[1]   M. Mirata, J. Koch, F. Lenz and E.J. Moniz, Ann. of Phys. 120
      (1979) 205; L.S. Kisslinger and W.L. Wang, Ann. of Phys.
      108 (1976) 314.
[2]   E. Oset, H. Toki and W. Weise, Phys. Reports 83 (1982)
      281; G. E. Brown and W. Weise, Phys. Reports 22 (1975)
      279.
[3]   A.B. Migdal, Rev. Mod. Phys. 50 (1978) 107.
[4]   G.E. Brown and W. Weise, Phys. Reports 27 (1976)1.
[5]   M. Ericson and J. Delorme, Phys. Lett. 76B (1978) 241;
      S.A. Fayans, E.E. Saperstein and V.E. Tolokonnikov, Nucl.
      Phys. A326 (1979) 463; H. Toki and W. Weise, Phys. Rev.
      Lett. 42 (1979) 1034.
[6]   J. Meyer-ter-Vehn, Phys. Reports 74 (1981) 323.
[7]   J.M. Moss et al., Phys. Rev. Lett. 44 (1980) 1189;
      M. Haji-Saeid et al., Phys. Rev. Lett. 45 (1980) 880;
      J.R. Comfort and W.G. Love, Phys. Rev. Lett.44 (1980)1656.
[8]   J.L. Escudie et al., Phys. Rev. C24 (1981) 792.
[9]   K. Rehm et al., Phys. Lett. 114B (1982) 15; M. Ejiri et al.,
      Phys. Rev. C24 (1981) 2001.
[10] W. Weise, Proc. 9 ICOHEPANS, Versailles 1981, Nucl. Phys.
      A374 (1982).
[11] M. Ericson, A. Figureau and C. Thévenet, Phys. Lett. 45B (1973)
      19; M. Rho, Nucl. Phys. A231 (1974) 493; K. Ohta and
      M. Wakamatsu, Nucl. Phys. A234 (1974) 445.
[12] C. Gaarde et al., Nucl. Phys. A369 (1981) 258.
[13] C.D. Goodman et al., Phys. Lett. 107B (1981) 406.
[14] C.D. Goodman, Proc. 9 ICOHEPANS, Versailles 1981, Nucl. Phys.
      A374 (1982) 241c.
[15] A. Richter, Proc. 9 ICOHEPANS, Nucl. Phys. A374 (1982) 177c.
[16] W. Bertozzi, Proc. 9 ICOHEPANS, Nucl. Phys. A374 (1982) 109c.
[17] H. Ejiri, private communication, and to be published.
[18] G. Bertsch, D. Cha and H. Toki, Phys. Rev. C24 (1981) 533,1371.
[19] M. Lacombe et al., Phys. Rev. C23 (1981) 2405.
[20] A. Chodos and C.B. Thorn, Phys. Rev. D12 (1975) 2733;
      G.E. Brown and M. Rho, Phys. Lett. 82B (1979) 383;
      V. Vento, M. Rho, E. Nyman, J. Jun and G.E. Brown, Nucl.
      Phys. A345 (1980) 413.
[21] G.E. Brown and M. Rho, Nucl. Phys. A372 (1981) 397.
[22] H. Toki and W. Weise, Phys. Lett. 97B (1980) 12.
[23] H. Toki, Phys. Rev. C26 (1982) 1256.
[24] E. Oset and M. Rho, Phys. Rev. Lett. 42 (1979) 47;
      I.J. Towner and F.C. Khanna, Phys. Rev. Lett. 42 (1979)51.
[25] W. Knüpfer, M. Dillig and A. Richter, Phys. Lett. 95B (1980)349.
[26] A. Bohr and B. Mottelson, Phys. Lett. 100B (1981) 10.
[27] A. Härting, W. Weise, H. Toki and A. Richter, Phys. Lett. 104B
      (1981) 261.
[28] T. Suzuki, S. Krewald and J. Speth, Phys. Lett. 107B (1981)9.
[29] M. Kohno and D. Sprung, preprints McMaster Univ. (1981/82).

[30] H.R. Fiebig and J. Wambach, Nucl. Phys. A386 (1982) 381 and
      proceedings of this conference.
[31] T. Suzuki, H. Hyuga, A. Arima and K. Yazaki, Phys. Rev. Lett.
      106B (1981) 19.
[32] I.S. Towner and F.C. Khanna, Nucl. Phys. A372 (1981) 331.
[33] C. Joseph et al. (Lausanne-Louvain-Zurich-Munich Collabora-
      tion), preprint (1982).
[34] J.B. McGrory and B.H. Wildenthal, Phys. Lett. 103B (1981) 173.
[35] W. Steffen et al., Phys. Lett. 95B (1980) 23;
      W. Steffen, R. Benz, H. Graf, A. Richter, E. Spamer and
      O. Titze, preprint.
[36] B.D. Anderson, et al., Phys. Rev. Lett. 45 (1980) 699.
[37] J. Akkermans, K. Allaart and G. Bonsignori,VU Amsterdam
      preprint, and private communication; K. Allaart and
      E. Boeker, Nucl. Phys. A198 (1972) 33.
[38] L.T. van der Bijl, H.P. Blok, R. Frey, D. Meuer, A. Richter,
      and P. de Witt Huberts, Z. Physik A305 (1982) 231.
[39] T. Yamazaki, in: Mesons in Nuclei, Vol. II, M. Rho and
      D.H. Wilkinson, eds., North-Holland (1979).
[40] W. Dickhoff, J. Meyer-ten-Vehn, M. Müther and A. Faessler,
      Phys. Rev. C23 (1981) 1154.
[41] S.O. Bäckman, A.D. Jackson and J. Niskanen, Stony Brook
      preprint (1979); W. Dickhoff, private communication.
[42] M. Ericson and T.E.O. Ericson, Ann. of Phys. 36 (1966) 383.

RENORMALIZATION OF AXIAL β AND MAGNETIC γ TRANSITIONS

WITH Δℓ=1 AND THE RELATED GIANT RESONANCES

Hiroyasu Ejiri

Dept. Physics, Osaka University
Toyonaka, Osaka 560
JAPAN

ABSTRACT

Renormalization factors for first forbidden β and analogous γ transitions between single particle states and giant resonances associated with those β and γ modes with Δℓ=1 are analyzed in terms of the polarization interaction $H=\chi\tau\sigma Y_1 \cdot \tau\sigma Y_1$. The renormalization factor g^{eff}/g and the giant resonance energy E^{GR} for the vector mode are found to be consistent with each other, while those for the axial vector mode are inconsistent, in a framework of the spin-isospin polarization of the nucleon particle-hole type only. A possible effect of the Δ nucleon-hole polarization on the axial vector mode is deduced.

Spin-isospin interactions in nuclei give rise to the spin isospin polarizations and spin isospin giant resonances (GR). Consequently coupling constants g_α for spin-isospin mode operators T_α are renormalized by factors $g_\alpha^{eff}/g_\alpha \approx 1/1(1+\kappa_\alpha)$, where κ_α are the spin isospin susceptibilities. The giant resonance and the polarization are kinds of collective spin isospin excitations. The giant resonance absorbs most of the transition strength, resulting in reduction of the single particle transition strength. In other words the single particle transition is quenched because of the distructive interference of the giant resonance. These features are schematically illustrated in fig. 1. The collective features of the spin isospin modes may be described as the coherent particle-hole excitations. Since the Δ isobar has the spin isospin quantum number, the Δ particle nucleon-hole (N^{-1}) excitations may contribute to the spin isospin mode. It is interesting to study how the NN^{-1} nucleon mode and the ΔN^{-1} isobar mode contribute to the spin isospin polarizations and the spin-isospin giant resonances.

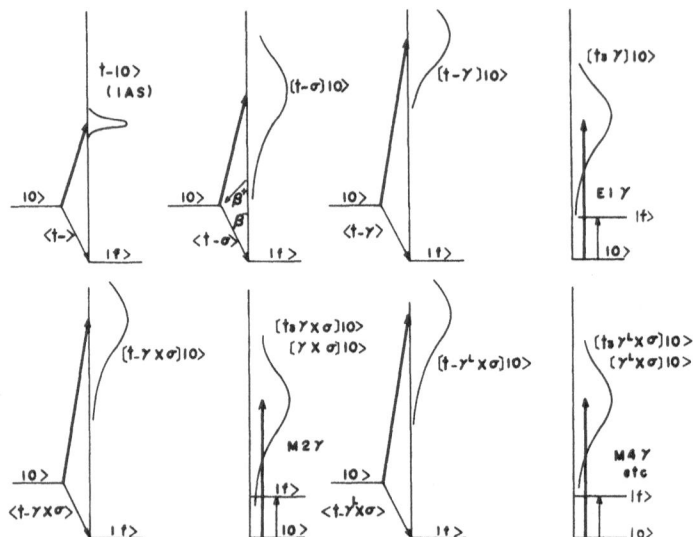

Fig. 1   Transition and level scheme showing giant resonances
          absorbing a large fraction of the transition strength
          and the reduced single particle transition (refs. 7,8).

        So far most studies of the spin-isospin modes are concentrated
on magnetic dipole (M1) and Gamow Teller (GT) operators with $\Delta\ell=0$.
Giant resonances (GR) associated with the GT transitions and the
quenching of the GT strength were theoretically predicted by J.I.
Fujita et al. in 1965[1]. The GT GR were found in (p, n) reactions[2].
Quenching of the M1 diagonal and transition moments[3-6] has been
discussed in terms of the Δ isobar effect and the core-polarization
effect.

        This report deals with the spin-isospin modes with $\Delta\ell=1$ such as
the first forbidden axial-vector β and magnetic quadrupole γ transitions in medium heavy nuclei. Merits of studying the $\Delta\ell=1$ spin
isospin modes are following. i) Nuclear particle-hole excitations
relevant to the $\Delta\ell=1$ mode are essentially $1\hbar\omega$ excitations, which
should be a uniform effect and hence insensitive to the details of
the valence nucleon configuration. ii) Unique first forbidden β transitions and the M2γ transitions from the unique parity state with
$j=\ell_{max}+\frac{1}{2}$ to the $j'=\ell_{max}-1-\frac{1}{2}$ are quite pure shell model transitions.
Comparison between the observed and shell-model transition rates
leads to the renormalization factor. iii) Renormalization factors of
the first forbidden β and analogous γ transitions for both vector

and axial vector modes have been studied by Ejiri and Fujita[7,8,9]. The giant resonances with $\Delta\ell=1$ were found in (p, n) reactions[10] as suggested by Ejiri and Fujita[7,8]. Thus one can now compare these observables for both vector and axial vector modes.

Experimentally the value g_α^{eff}/g_α for the $G_\alpha = t^T[i^L Y^L \sigma^S]_J$ mode with α=LST is obtained from the observed matrix element $<T_\alpha>_{exp}$ as

$$g_\alpha <T_\alpha>_{exp} = g_\alpha^{eff} <T_\alpha>_0 , \tag{1}$$

where $<T_\alpha>_0$ is the model calculation. The model used is the shell model+pairing, quadrupole and octupole interactions. Since the observed data for quite pure single quasi-particle transitions in near semi-magic nuclei with $N\approx82$ or $Z\approx50$ are used, the calculation error for the $<T_\alpha>_0$ is small. Then the g^{eff} stands for the effect of the spin isospin interaction which is left in the calculation. The renormalization factors g_α^{eff}/g_α for the vector and axial vector transitions are uniformly 0.3~0.2 as shown in fig. 2.

The spin-isospin interaction responsible for the spin-isospin collective mode T_α may simply be given as $H_\alpha = \chi_\sigma T_\sigma T_\alpha$. We first consider the NN^{-1} nucleon mode $1\hbar\omega$ excitations. Since the $1\hbar\omega$ jump energy is much larger than the single particle transition energy, one may use a schematic model, where the unperturbed NN^{-1} excitation energies E_{iJ} and the matrix elements G_{iJ} are represented by the mean values of $\hbar\omega$ and G, respectively. Then we get for $\beta^\pm$ decays

$$\frac{f(1\mp k)}{E_\pm^{GR}/\hbar\omega - 1} - \frac{f(1\pm k)}{E_\pm^{GR}/\hbar\omega + 1} = 1, \tag{2}$$

$$E_\pm^{GR}/\hbar\omega = \mp kf + (k^2 f^2 + 1 + 2f)^{1/2}, \tag{3}$$

$$\kappa = \frac{f(1+k)}{1 - E_1/\hbar\omega} + \frac{f(1-k)}{1 + E_1/\hbar\omega}, \tag{4}$$

where $k \equiv (N-Z)/n$, $f = n\chi G^2/\hbar\omega$, n is the number of nucleons in one $\hbar\omega$ shell. The unique first forbidden β and $M2\gamma$ transitions have large matrix elements for the stretched transitions of $(j = \ell_{max} + \frac{1}{2}) \ddagger (j = \ell_{max} - 1 - \frac{1}{2})$. Their energies are shifted by the spin orbit energy $\Delta_{s\ell}$. Then, taking into accounts this shift explicitly, we get

$$E_-^{GR}/\hbar\omega = f'k + (f'^2 k^2 + 2f' + 1)^{1/2}, \tag{5}$$

$$\kappa = (\frac{(1+k)-\eta}{\hbar\omega - E_1} + \frac{(1-k)}{\hbar\omega + E_1} + \frac{\eta}{\hbar\omega - \Delta_{s\ell} - E_1}) f\hbar\omega , \tag{6}$$

$$\frac{1}{f'} = \frac{1}{f} + \frac{1}{f}(\frac{\eta}{E_-^{GR}/\hbar\omega - 1} - \frac{\eta}{E_-^{GR}/\hbar\omega - 1 + \Delta_{s\ell}/\hbar\omega}) \tag{7}$$

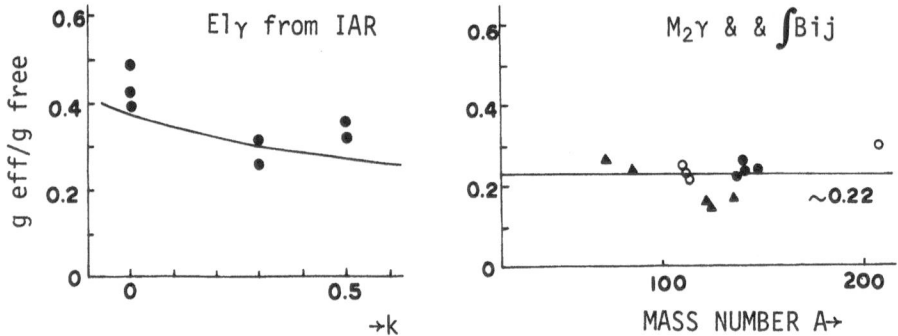

Fig. 2 Renormalization factors for the first forbidden vector tran-
         sitions obtained from the Elγ decays from IAR and those for
         the first forbidden axial-vector transitions (refs. 8,9,11,12).

The η is the ratio of the strength with $\hbar\omega-\Delta_{s\ell}$ to the total sum
strength. It is given as $\eta\approx0.41-0.14$ k. The interaction strength
χ_τ for the vector mode is obtained so as to reproduce the observed
Elγ giant resonance energy. We get $f_\tau\equiv n\chi_\tau G^2/\hbar\omega=1.15$. The value $\chi_{\tau\sigma}$
for the axial-vector mode can be evaluated by using the same ratio
$\chi_{\tau\sigma}/\chi_\tau\approx0.83$ derived from the GT giant resonance energy with respect
to the IAS energy[5]. Then we get $f_{\tau\sigma}\approx0.83 \times f_\tau\approx0.95$.

The calculated and observed values for the E^{GR} and g^{eff}/g are
shown in fig. 3. The renormalization factors g^{eff}/g for the vector
mode $(\tau_- rY_1)$ transitions are well reproduced by using the $f_\tau=1.15$
obtained from the Elγ giant resonance energies. The $\Delta\ell=1$ giant resonance energies in the (p, n) reactions are consistent with the spin
isospin GR with $f_{\tau\sigma}=0.95$ derived from the GT resonance. It is nice
to see that all data points fit our simple analytical line. The
values $g^{eff}_{\tau\sigma}/g_{\tau\sigma}$ for the axial vector mode, however, are smaller than
the prediction with the $f_{\tau\sigma}=0.95$. Similarly the M2γ transitions are
more quenched than evaluation based on the $f_{\tau\sigma}=0.95$ as shown in fig.4.

The inconsistency between the GR energies and the renormalization factors indicate some additional effect which is not taken
into account in the spin-isospin polarization interaction of the $1\hbar\omega$
NN^{-1} nucleon mode. The ΔN^{-1} mode bears some spin-isospin strength.
It lies in the very high excitation region. Assuming the similar
repulsive interaction as in the NN^{-1}, the ΔN^{-1} mode takes away the
transition strength from the low-lying single particle states as the
NN^{-1} nucleon mode does. On the other hand the ΔN^{-1} mode puses down
the GR lying near the $2\hbar\omega$ excitation region, while the NN^{-1} nucleon
mode pushes up the GR. These features are schematically illustrated
in fig. 5. Because of these different natures of the ΔN^{-1} and NN^{-1},
the ΔN^{-1} mode strength is necessary to be taken into account explicitly.

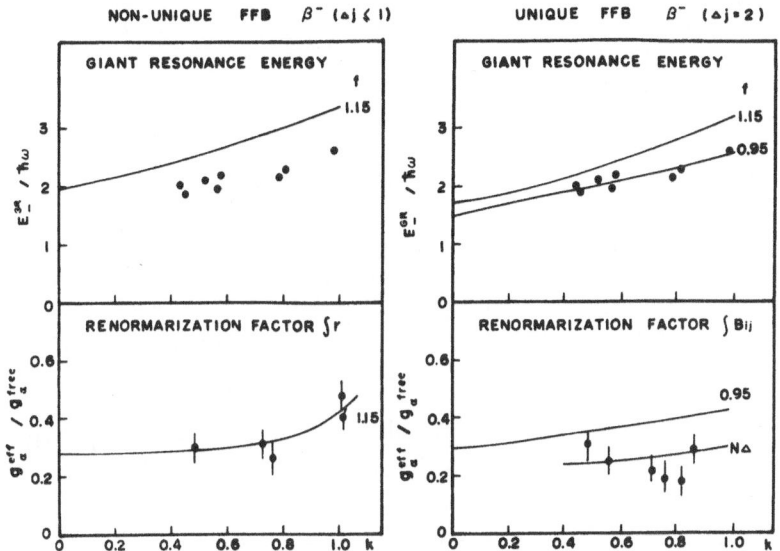

Fig. 3   Giant resonance energies E^{GR} and renormalization factors
         g^{eff}/g^{free}. $E^{GR}+E(IAS)-\Delta(SYMMETRY)$ gives the excitation
         energy. Left hand side: one degenerate level approxi-
         mation for non unique transitions. Right hand side: two
         degenerate level approximation for unique transitions.
         Experimental data E^{GR}, g^{eff}/g for $\int r$ and g^{eff} for $\int B_{ij}$
         are taken from refs. 8-12. The lines show the calculation
         for the NN^{-1} mode and that including the ΔN^{-1} mode also.

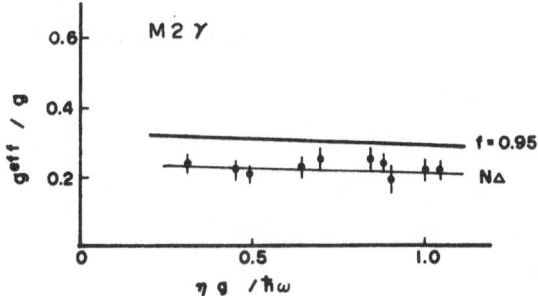

Fig. 4   Renormalization factors for the M2γ transitions between
         1h11/2 and 1g7/2 quasi-particle states in $Z\approx50$ nuclei
         (open circles) and $N\approx82$ nuclei (closed circles). The
         data are taken from refs. 8, 9, 12. Solid lines give the
         calculation (f=0.95) for the NN^{-1} nucleon mode only with
         $f_{\tau\sigma}$=0.95 and that (NΔ) for both the NN^{-1} and ΔN^{-1} modes.

Now let's consider the X mode with $\chi_X G_X^2$ at the high excitation region of E_X. The dispersion equation for the GR energy for the axial vector mode is given as

$$f\left(\frac{1+k-\eta}{E^{GR}/\hbar\omega-1} - \frac{1-k}{E^{GR}/\hbar\omega+1} + \frac{\eta}{E^{GR}/\hbar\omega-1+\Delta_{s\ell}/\hbar\omega}\right)-h = 1, \qquad (8)$$

where $h=2\chi_X G_X^2/E_X$ is the X mode contribution. The susceptibilities κ for the unique forbidden β and the M2γ transitions are

$$\kappa = f\left(\frac{1+k-\eta}{1-E_1/\hbar\omega} + \frac{1-k}{1+E_1/\hbar\omega} + \frac{\eta}{1-\Delta_{s\ell}/\hbar\omega-E_1/\hbar\omega}\right)+h, \qquad (9)$$

$$\kappa = f\left(\frac{2(1-\eta)}{1-E_1^2/\hbar\omega^2} + \frac{2(1-\Delta_{\ell s}/\hbar\omega)\eta}{(1-\Delta_{\ell s}/\hbar\omega)^2-E_1^2/\hbar\omega^2}\right)+h \qquad (10)$$

The nucleon mode contribution f and those due to the X mode h are deduced from the two different type observations of the $g^{eff}/g=1/(1+\kappa)$ and the E^{GR}. The observed values shown in figs. 3 and 4 lead to the values $f_{\tau\sigma}\approx1.3$ and $h_{\tau\sigma}\approx0.3{\sim}0.4$. Similar analysis for the vector mode gives $f_{\tau}\approx1.15$ and $h_{\tau}\approx0$ as expected from the consistency of the g_{τ}^{eff}/g_{τ} with the E^{GR} in the NN^{-1} frame. Thus the X mode with the strength $h_{\tau\sigma}/f_{\tau\sigma}\sim0.3$ is the spin-isospin mode at the high excitation region. The ΔN^{-1} is considered to be just such a mode. Using the $E_X\approx300$ MeV for the ΔN^{-1} mode, one gets $\chi_\Delta G_\Delta^2/\chi_N G_N^2\approx6$. This is the same order of magnitude as the Δ mode suggested for the M1 moment[3,5]. Another contribution may be due to the tensor interaction as emphasized for M1 moments[13].

ACKNOWLEDGEMENT

The author thanks Prof. M. Ishimura and Prof. H. Ohtsubo for valuable discussions.

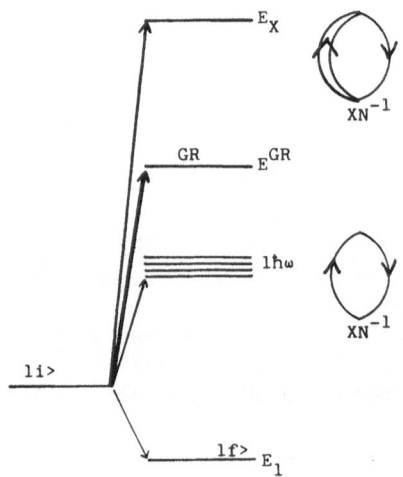

Fig. 5.
Schematic picture of the β^- transition scheme, showing absorption of transition strength from the single particle state E_1 and the E^{GR} energy shifts due to the NN^{-1} and XN^{-1} modes.

REFERENCES

1. J.I. Fujita and K. Ikeda, Nucl. Phys. 67 (1965) 145;
    K. Ikeda, S. Fujii and J.I. Fujita, Phys. Letters 3 (1963) 271;
    K. Ikeda, S. Fujii and J.I. Fujita, Phys. Letters 2 (1962) 169.
2. R.R. Doering, A. Galonsky, D.M. Patterson and C.F. Bertch, Phys.
    Rev. Letters 35 (1975) 1961.
3. A. Bohr and B. Mottelson, Phys. Letters, 100B (1981) 10.
4. E. Oset, H. Toki and W. Weise, Phys. Reports 83 (1982) 281.
    E. Oset and M. Rho Phys. Rev. Letters 42 (1979) 51.
5. T. Suzuke, private communication.
6. A. Arima and H. Horie, Prog. Theor. Phys. (Kyoto) 11 (1954) 509.
7. H. Ejiri and J.I. Fujita, Phys. Rev. 176 (1968) 1277.
    H. Ejiri, Nucl. Phys. A166 (1971) 594; ibid. A211 (1973) 232.
8. H. Ejiri and J.I. Fujita, Phys. Reports 38C (1978) 85.
    H. Ejiri and T. Shibata, Phys. Rev. Letters 35 (1975) 148.
9. H. Ejiri, Nucl. Phys. A178 (1972) 350.
10. D.J. Horen, et al., Phys. Letters 99B (1981) 383.
    D.J. Horen, Proc. RCNP Symp. Highly Excited States in Nuclear
    Reaction, Osaka, (1980) 223.
    D.E. Bainum, et al., Phys. Rev. Letters 95B (1980) 27.
    D.J. Horen, et al., Phys. Letters 95B (1980) 27.
11. H. Ejiri, P. Richard, S. Ferguson, R. Heffner and D. Perry,
    Phys. Rev. Letters 21 (1968) 373; Nucl. Phys. A128 (1969) 388;
    K.A. Snover, J.F. Armann, W. Hering and P. Paul, Phys. Letters
    37B (1971) 29.
12. R. Brenn, S.K. Bhattacherjee, C.D. Sprouse and L.E. Young,
    Phys. Rev. C10 (1974) 1414.
13. A. Arima, private communication.

LORENTZ-LORENZ QUENCHING FOR THE GAMOW-TELLER SUM RULES

J. Delorme, M. Ericson, A. Figureau and N. Giraud

Institut de Physique Nucléaire
Université Claude Bernard, Lyon I, 43 Bvd du 11 Novembre
69622 Villeurbanne Cedex, France

One of the most striking aspects of the recent detailed exploration of the Gamow-Teller resonances[1] is the systematic observation
that the summed axial vector strength is appreciably lower than that
expected from the Gamow-Teller sum rule. The latter is obtained un-
der the assumption of additivity of the free nucleon coupling $g_A \vec{\sigma}.\vec{\tau}$
(with $g_A = 1.25$) to the axial current ; indeed simple Pauli matrix
algebra gives then :

$$\sum_n \left\{ \left| \langle n | \sum_i^A g_A \sigma_i^z \tau_i^+ | 0 \rangle \right|^2 - \left| \langle n | \sum_i^A g_A \sigma_i^z \tau_i^- | 0 \rangle \right|^2 \right\} = (N-Z) g_A^2 \qquad (1)$$

In principle, both the positively and negatively charged branches
have to be measured ; in a nucleus with positive neutron excess however, equation (1) gives already a lower limit for the $\sigma \tau^+$ strength
which is the quantity accessible through (p,n) reactions and Pauli
blocking rapidly suppresses the τ^- branch as (N-Z) increases. Thus,if
one assumes that no part of the strength has escaped detection,a reduction such as that observed in a wide range of nuclei is strongly
suggestive of the intervention of new degrees of freedom.

An attractive mechanism is the now popular Lorentz-Lorenz effect
which was once proposed to explain the reduction of axial β decay rates (For a survey, see refs. (2) and (3)). It arises from the coupling between nucleon and Δ isobar degrees of freedom through a short
range spin-isospin dependent force. We present here some results of
a calculation in which the coupling is described in a Δ-hole polarizability picture ; the model we use is the same as that introduced
some years ago[4] to treat the modifications of the pion field in the
nuclear medium. This is in principle practically equivalent to currently developped R.P.A. treatments of Δ-hole propagation, the only
difference being the use of the static limit and of closure approxi-

mation on the Δ states (the Δ-hole polarizability becomes then local in configuration space).In our previous works it was shown that the coupling to Δ-hole excitations through a contact spin-isospin interaction characterized by the Landau-Migdal parameter g' induces a local renormalization of the axial (or pion) vertex (Lorentz-Lorenz effect) :

$$A^{\pm}(x) = g_A \sum_i \vec{\sigma}_i \tau_i^{\pm} \delta(x-x_i) \longrightarrow g_A/(1+g'\alpha(x)) \sum_i \vec{\sigma}_i \tau_i^{\pm} \delta(x-x_i) \qquad (2)$$

The axial polarizability $\alpha(x)$ is proportionnal to the nuclear density $\alpha(x) = 8 f^{*2}/9 m_\pi^2 \omega_\Delta \; \rho(x)$, ω_Δ and f^* being the excitation energy of the Δ resonance and the $\pi N\Delta$ coupling constant. The sum rule (1) is then simply modified to the following ground state expectation value, defining an effective axial coupling constant :

$$\sum_n \left\{ |\langle n|A_z^+(q=0)|0\rangle|^2 - |\langle n|A_z^-(q=0)|0\rangle|^2 \right\} = -\langle 0| \left[g_A/(1+g'\alpha(x)) \right]^2 \sum_i \tau_i^3 |0\rangle$$

$$= g_A^2 \int d^3x \; [\rho_n(x) - \rho_p(x)]/(1+g'\alpha(x))^2 = (N-Z)(g_A^{eff})^2 \qquad (3)$$

Actually the spin-isospin interaction is somewhat more complicated than the simple Landau-Migdal ansatz : in particular it comprises a one pion exchange part which, however, can be expected a priori to play little role here because it vanishes at zero momentum transfer (strictly speaking, it can have some influence for finite nuclear size, specially in the lightest nuclei). In this more general case, the derivation of the sum rule requires the knowledge of the full renormalization of the pion propagator. From P.C.A.C., one gets at fixed q :

$$\sum_n \left\{ |\langle n|A_z^+(q)|0\rangle|^2 - |\langle n|A_z^-(q)|0\rangle|^2 \right\} = \frac{g_A^2}{2f^2} (q^2+m_\pi^2)^2 \frac{m_\pi^2}{q^2} \langle 0|[\varphi(q),\varphi^+(q)]|0\rangle \qquad (4)$$

One checks easily that in a description limited to nucleon degrees of freedom $\varphi(q) = \frac{f_\pi \sqrt{2}}{m_\pi} \sum_i \vec{\sigma}_i \cdot \vec{q}/(q^2+m_\pi^2) \tau_i^{\pm}$, and the right hand side of eq .(4) becomes $(N-Z)g_A^2$, independent of q, so that one recovers the primitive sum rule (1).

Using methods developed elsewhere[5], we have computed the right hand side of (4) at q=0, which differs from the simplified sum rule (3) by the inclusion of pion propagation effects. As for the case of eq .(3), we need the neutron and proton densities, which are chosen with a Fermi profile for a series of nuclei covering the whole range of atomic numbers (a modified Gaussian shape is preferred for the lightest case, A=14). The ratio $(g_A^{eff}/g_A)^2$ is represented in fig.(1), for values of g' in the now favoured range 0.7 $\pm$ 0.1. The calculated quenching increases slowly with the mass number, following the average nuclear density (characteristic of a contact interaction). We have found, as expected, that the Lorentz-Lorenz quenching (eq .(3)) is only slightly altered by the consideration of pion propagation effects (these produce only a reduction of the effect ranging from 2.6 to 5.6% as the atomic number decreases). The smallness of the pion contribution is typical of the zero momentum limit ; it greatly contrasts with the situation which prevails e.g. at the

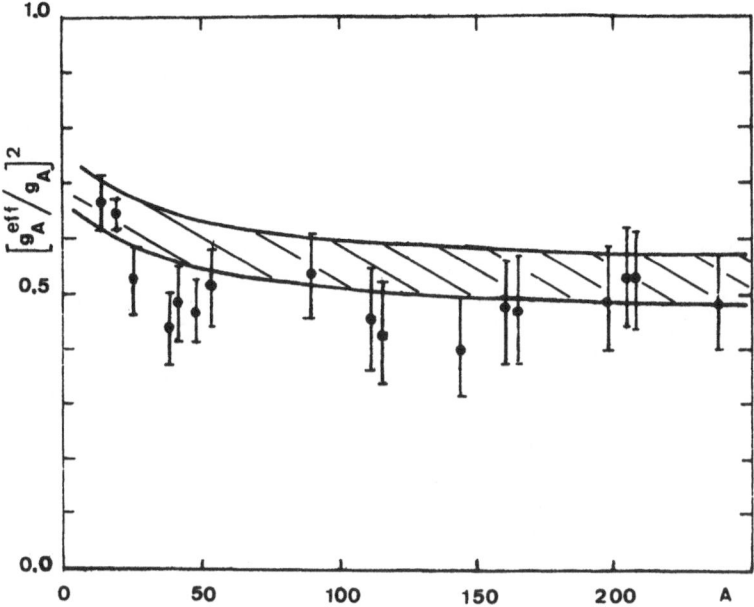

Fig. 1 : The reduction factor of the (squared) effective axial vector coupling constant with respect to the G-T sum rule value (1). The shaded area corresponds to values of g' between 0.6 (upper curve) and 0.8 (lower curve). The experimental points are taken from ref.(6).

pion pole ($q \to i\, m_\pi$) where the sum rule (4) defines an effective on-shell pion-nucleus coupling constant, a quantity of interest for π-nucleus dispersion relations : our previous works [4,5] show that besides the universal Lorentz-Lorenz quenching, a considerable decrease of the π-nucleus coupling constant is to be attributed to the pion propagation.

Here the quenching is less dramatic, being essentially due to the Lorentz-Lorenz effect : Though the reduction is moderate in very light nuclei, it attains a factor 2 in the heaviest ones ; the experimental data [6] follow approximately the same trend (see fig.(1)).It is then very satisfying that a simple mechanism like the Lorentz-Lorenz effect can explain the existing results. The value g' = 0.8 seems to be favoured ; one should add however two remarks concerning the validity of such an interpretation. First the relevant parameter is the product of g' by the polarizability. Our curves have been obtained with a ratio of $\pi N \Delta$ to πNN coupling constants f^{*2}/f^2 = 4.5 taken from the Chew-Low model with recoil terms. Furthermore, the axial polarizability is not saturated by the Δ alone [7], which means that other N* isobars can contribute, but we do not know whether their short-range interaction with the nucleons can be described by a universal value of g'. There is in fact no compelling proof (see e.g.ref.3,19 that the same g' applies even to the (N-hole)-(Δ-hole) and the (Δ-hole)-(Δ-hole) interactions as was implicitly assumed in eq .(2) ; the vertex renormalization should read more precisely $\left(1 - g'_{N\Delta}\,\alpha/(1 + g'_{\Delta\Delta}\,\alpha)\right)$ instead of $(1 + g'\alpha)^{-1}$. The second remark is that the interpretation of all the missing G.T. strength by a Lorentz-Lorenz quenching relies on the hypothesis that no strength can have been pushed at high excitation without appealing to the Δ isobar, as in the second order core polarization mechanism through a tensor force once suggested by the Arima group [8].

1. C.Gaarde, J.Rapaport, T.N.Taddeucci, C.D.Goodman, C.C.Foster, D.E.Bainum, C.A.Goulding, M.B.Greenfield, D.J.Horen and E.Sugarbaker : Nucl.Phys. A 369 (1981) 258.
   C.D. Goodman: Proc. of 9-Icohepans, Versailles 1981, Nucl. Phys. A 374 (1982) 241 C.
2. M.Ericson, this conference.
3. M.Rho, this conference.
4. J.Delorme, M.Ericson, A.Figureau and C.Thévenet : Annals of Physics 102 (1976) 273.
5. J.Delorme and A.Figureau : in 4ème Session d'Etudes Biennales de Physique Nucléaire, Report LYCEN 7702 (1977) p.C.7.1.
   J.Delorme, A.Figureau and N.Giraud:preprint LYCEN 8212(1982)
6. C.Gaarde, this conference.
7. M.Ericson and A.Figureau. J.Phys.G : Nucl.Phys.7(1981)1197.
8. K.Shimizu, M.Ichimura and A.Arima : Nucl.Phys. A 226(1974)282.
9. J. Speth, this conference.

THE NUCLEON-NUCLEON INTERACTION AND NUCLEON-NUCLEUS SCATTERING

W. G. Love

Department of Physics and Astronomy
University of Georgia
Athens, Georgia 30602

M. A. Franey

Department of Physics
University of Minnesota
Minneapolis, Minnesota 55455

F. Petrovich

Department of Physics
Florida State University
Tallahassee, Florida 32306

ABSTRACT

A derivation of a t-matrix interaction from the free nucleon-nucleon amplitudes is sketched and some of the most important characteristics of this operator for the excitation of spin modes are described. The nucleon and electromagnetic couplings to spin and current modes of the nucleus are cast in a longitudinal-transverse representation which displays explicitly the common and distinctive aspects of these interactions in the study of nuclear structure. A current ⊗ spin coupling present in the effective nucleon-nucleon interaction due to non-locality is described briefly and some of its possible implications for spin-observables are discussed. The derived nucleon-nucleon interaction is compared with the coupling predicted by a π+ρ exchange model and a few applications of the t-matrix interaction are made to transitions which illustrate some specific aspects of the coupling.

1.   INTRODUCTION

A knowledge of the relevant couplings between various probes and the nucleus is an essential element[1] for identifying and interpreting different modes of excitation of the nucleus. In the case of electron scattering, this coupling is weak and its characteristics are well understood so that it may be used in a relatively straightforward manner to study quantitatively a variety of types of nuclear excitations. Although nucleon scattering is potentially a richer probe of nuclear excitations due to the nucleon's spin and isospin structure, its coupling to the nucleus is not as well understood. In the absence of a complete and calculable theory of strong interactions, progress in interpreting nucleon-nucleus scattering has been made largely by using a few theoretical constraints as a guide to an otherwise phenomenological approach for representing the coupling between the projectile and target nucleons.

Two primary techniques have been used to derive an effective nucleon-nucleon (N-N) interaction for use in describing nucleon-nucleus scattering within a single-scattering framework. The first method[1-5] uses the free N-N t-matrix for the coupling. Below ~400 MeV a more sophisticated approach[6] has been attempted in which a density-dependent nuclear-matter G-matrix is constructed on the basis of a realistic N-N potential (Hamada-Johnston or Paris). To date, the largest differences between these two approaches[7] occur in those parts of the effective N-N interaction which excite non-spin (S = 0) modes; consequently, we focus primarily on the simpler t-matrix approach here. J. Kelly discusses density-dependent corrections[1,7] to the free t-matrix in some detail in a later paper in these proceedings.

In section 2 a description of the construction of the t-matrix interaction is presented and some of its most important characteristics are discussed. In section 3 the forms of the couplings for weak, electromagnetic and strong interactions are cast into a common representation which permits a direct comparison of these probes. This is followed by a brief discussion of current and current ⊗ spin couplings and their roles in the excitation of spin modes. In section 4 the S = 1 nucleon-nucleon t-matrix interaction is compared with that based on a one-pion-exchange and one-rho-exchange model. In section 5 a number of specific results for the excitation of spin modes at intermediate energies are discussed. These isolate and illustrate the roles of some of the most important features of the S = 1 nucleon-nucleon interaction and provide some measure of the uncertainties involved.

2. A t-MATRIX INTERACTION

The precise connection between N-N data (or phase shifts) and an effective interaction (V_{12}) for nucleon-nucleus scattering is neither simple nor unique. Here we outline one approximate scheme[5] for making this connection. This and other approaches are discussed in more detail elsewhere.[2-6]

The N-N scattering amplitude can be written

$$M(E_{cm},\theta) = A + B\vec{\sigma}_1\cdot\hat{n}\vec{\sigma}_2\cdot\hat{n} + C(\vec{\sigma}_1 + \vec{\sigma}_2)\cdot\hat{n} + E\vec{\sigma}_1\cdot\hat{q}\vec{\sigma}_2\cdot\hat{q} + F\vec{\sigma}_1\cdot\hat{Q}\vec{\sigma}_2\cdot\hat{Q} \quad (1)$$

where A,B,C,E and F are functions of the center-of-mass energy E_{cm}, the scattering angle θ and the two-body isospin. If $\vec{k}(\vec{k}')$ is the initial (final) momentum of either particle in the N-N c.m. system, the unit vectors $[\hat{q},\hat{Q},\hat{n}]$ form a right handed coordinate system with $\vec{q} = \vec{k} - \vec{k}'$ and $\vec{Q} = \vec{k} + \vec{k}'$. We see from Eq. (1) that only the A term cannot participate in transitions with S = 1 transfer; for S = 0 transfer only A and C contribute. For the practical purpose of identifying the spin and spatial ranks of the N-N interaction it is useful to rewrite[4,5] M as:

$$M(E_{cm},\theta) = A'P_S + B'P_T + C(\vec{\sigma}_1 + \vec{\sigma}_2)\cdot\hat{n} + E'S_{12}(\hat{q}) + F'S_{12}(\hat{Q}) \quad (2)$$

where S_{12} is the usual tensor operator, $P_S(P_T)$ is the singlet (triplet) spin-projection operator and

$$A' = A - (B+E+F), \quad B' = A + \frac{(B+E+F)}{3}, \quad E' = \frac{E-B}{3}, \quad F' = \frac{F-B}{3}. \quad (3)$$

Yet another form of M which is convenient for comparison with other probes is the longitudinal-transverse representation:

$$M = A + C(\vec{\sigma}_1 + \vec{\sigma}_2)\cdot\hat{n} + EO_\ell + (\frac{B+F}{2})O_t + (\frac{B-F}{2})O_u. \quad (4)$$

Here the one longitudinal and two transverse operators (O_ℓ, O_t, O_u) are defined by

$$O_\ell \equiv \vec{\sigma}_1\cdot\hat{q}\sigma_2\cdot\hat{q} = \frac{1}{3}[\vec{\sigma}_1\cdot\vec{\sigma}_2 + S_{12}(\hat{q})] \quad (5a)$$

$$O_t \equiv (\vec{\sigma}_1 \times \hat{q})\cdot(\vec{\sigma}_2 \times \hat{q}) = \sigma_{1Q}\sigma_{2Q} + \sigma_{1n}\sigma_{2n} = \frac{1}{3}[2\vec{\sigma}_1\cdot\vec{\sigma}_2 - S_{12}(\hat{q})] \quad (5b)$$

$$O_u \equiv (\vec{\sigma}_1 \times \hat{Q})\cdot(\vec{\sigma}_2 \times \hat{Q}) - (\vec{\sigma}_1 \times \hat{n})\cdot(\vec{\sigma}_2 \times \hat{n}) = \sigma_{1n}\sigma_{2n} - \sigma_{1Q}\sigma_{2Q} \quad (5c)$$

where $\sigma_{1n} = \vec{\sigma}_1\cdot\hat{n}$, etc. Apart from isospin considerations we note from the structure of Eq. (4) that for S = 1 excitations the π-N coupling[1] is similar to the C term whereas the corresponding e-N coupling[1] is similar to the static transverse term O_t. We will pursue these comparisons in more detail below.

A calculable form of the t-matrix interaction (V_{12}) is constructed[4,5] from the empirically determined amplitude (M) by taking

$$V_{12} = v^C(r_{12}) + v^{LS}(r_{12})\vec{L}\cdot\vec{S} + v^T(r_{12})S_{12}(r_{12}) \tag{6}$$

and adjusting its parameters until its <u>anti-symmetrized</u> N-N matrix
elements in momentum space

$$t^o_{NN} = \eta M = \int d^3r\, e^{-i\vec{k}'\cdot\vec{r}} V_{12}(1 - P_{12})e^{i\vec{k}\cdot\vec{r}} \ , \quad \eta = \frac{-4\pi(\hbar c)^2}{E_{cm}} \tag{7}$$

match those of the on-shell N-N t-matrix in each N-N spin and isospin
channel at a number of energies over the range of interest. Guided
by one-boson-exchange models and computational considerations, the
radial parts of the central and spin-orbit parts of V_{12} are taken to
be a sum of Yukawa terms; the radial form of the tensor part of V_{12}
is taken to be r^2 times a sum of Yukawa terms. For V^C_{12} the Yukawa
term of longest range is constrained to match the OPEP; the Yukawa
terms of shorter range are then adjusted to match t^o_{NN}. Because of
the static ansatz for V_{12}, the F' term in Eq. (2) is attributable
exclusively to the exchange terms which in the present procedure
must be included in the nucleon-nucleus problem as well.

Some of the most important characteristics of the t-matrix
interaction V_{12} can be illustrated by plotting the moduli of its
anti-symmetrized momentum space matrix elements t_{NN} in the nucleon-
nucleus system as a function of momentum transfer (q), projectile
kinetic energy (E_p) and spin and isospin transfers (S and T).[1,4,5,8]
For the immediate purpose the term anti-symmetrized matrix elements
implies that the effects of knock-on exchange are included in t_{NN}
by a suitable factorization approximation. See Ref. 2, 5 and 9
for details.

Figure 1 shows a plot for the central parts of the force at
q = 0 for $100 \leq E_p(MeV) \leq 800$. The subscripts σ and τ refer to spin and
isospin transfers of one unit. The most striking feature of these
curves is the strong dominance of the scalar-isoscalar part of t_{NN}
at all energies considered. The very small (and poorly determined)
t_σ suggests that the central part of the force is ineffective for
exciting isoscalar spin modes. In striking contrast, we see that for
isovector excitations $t_{\sigma\tau}$ dominates over t_τ at small q in this
energy regime. This is emphasized in Fig. 1 where the ratio
$|t_{\sigma\tau}/t_\tau|^2$ is plotted versus E_p at q = 0. It is now well
established[1,4,5,8,10] that this S = 1 dominance of t_{NN} renders proton
scattering an especially sensitive probe of isovector spin modes at
intermediate energies. This dominance, together with the fact that
$|t_{\sigma\tau}| \lesssim |t_\tau|$ below about 60 MeV has been especially important for the
identification and interpretation of Gamow-Teller(GT) resonances
using the (p,n) reaction.[1,4,5,8,10]
Figure 2 illustrates the relative importance of the spin-orbit

and tensor parts of the interaction as a function of bombarding
energy at a momentum transfer of $q = 1.5$ fm^{-1}. This value of q was
chosen since it represents a compromise of the peak positions of the
spin-orbit and tensor terms which, from Ref. 5, peak roughly near
2.0 and 1.0 fm^{-1} respectively. More importantly, these non-central
terms tend to dominate in this region of q. Most apparent from Fig.
2 is the near negligible size of the isovector spin-orbit term so
that isovector spin modes at large q are dominated by the tensor
force at all energies. This smallness of $t_{LS\tau}$ also implies a strong
sensitivity of isovector $S = 1$ analyzing powers to optical model
parameters since there is typically little source of asymmetry in
the inelastic reaction mechanism. Isoscalar $S = 1$ excitations are
expected to be excited competitively by the spin-orbit and tensor
parts of t_{NN} and this is borne out in detailed calculations.[1,4,5,8,11]

Figure 3 shows the $S = 1$ parts of t_{NN} at $E_p = 140$ MeV as a function of momentum transfer q. As described in Ref. 5, t_τ and $t_{\sigma\tau}$
should be multiplied by ~2.0 in order to compare directly with $\bar{t}^T_\tau$
and $\bar{t}^T_\tau$ for nucleon-nucleus scattering. These latter quantities ($\bar{t}^T$)
have already been normalized to represent the strength of the tensor
force in nucleon-nucleus collisions as described in the appendix of
Ref. 5. (The left hand sides of equations A5a, A5b and A5c of Ref.
5 should be $t^{T\beta}$, $t^{T\gamma}$ and $t^{T\alpha}$ respectively.) For both isoscalar and

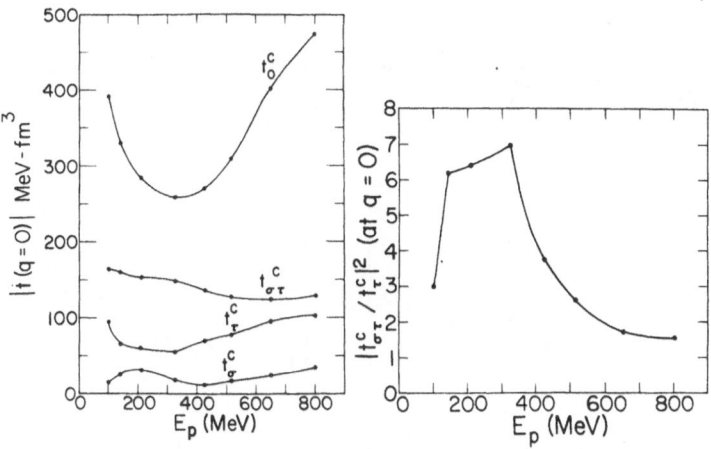

Fig. 1.   Energy dependence of the magnitudes of the central (direct
+ exchange) parts of the N-N t-matrix. On the right is
shown the energy dependence of the ratio $|t^C_{\sigma\tau}/t^C_\tau|^2$ at zero
momentum transfer.

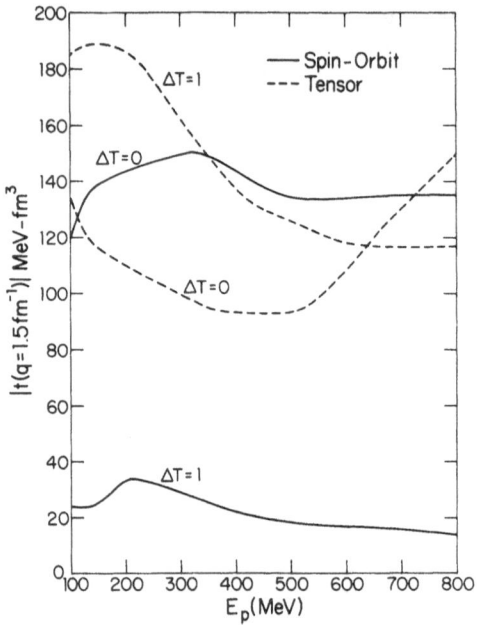

Fig. 2. Energy dependence of the spin-orbit and tensor parts of
the t-matrix interaction (t) at q = 1.5 fm^{-1}.

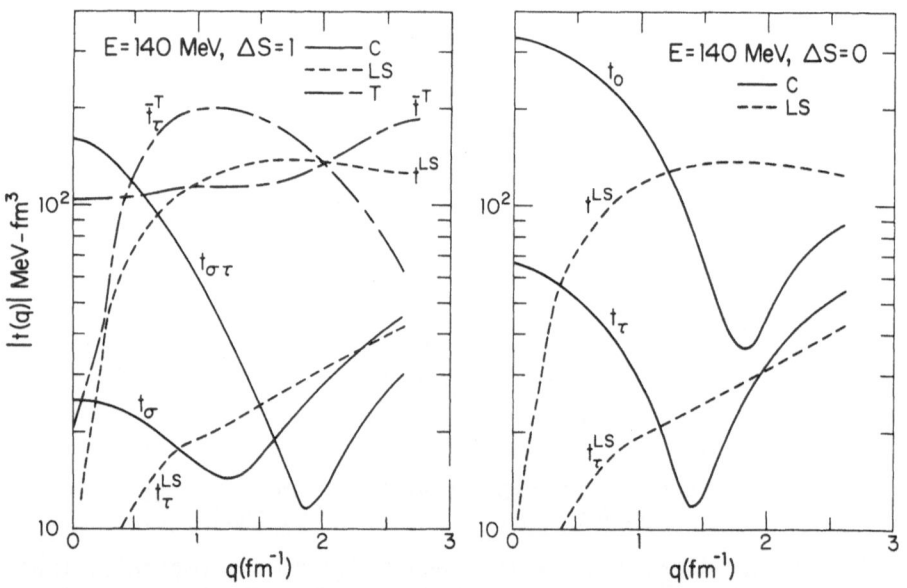

Fig. 3. Moduli of S = 1 and S = 0 parts of t as a function of q
at E_p = 140 MeV.

isovector transitions, it is seen to be important to include the tensor force in calculations of nucleon-nucleus scattering for essentially all values of momentum transfer. A significant exception is seen to occur for the excitation of $S = T = 1$ modes for $q \lesssim 0.5$ fm^{-1} and this is especially important[1,4,5,8,11] for interpreting GT excitations.

3. FORMS OF COUPLING TO SPIN EXCITATIONS

For comparison with nucleon-nucleus scattering data one typically[1,4,5,8] performs antisymmetrized distorted-wave-impulse-approximation (DWIA) calculations using a N-N interaction like that described by Eq. (6). Because of the complexities and uncertainties associated with V_{12} it is important when possible to compare the results of nucleon-nucleus scattering with the nuclear structure information obtained from other better understood excitation mechanisms such as electron scattering and β-decay. Although less definitive, comparisons between pion scattering and nucleon scattering can also be fruitful.[1] As we shall try to show, such comparisons can be used either to assess our level of understanding of nucleon-nucleus scattering or to obtain complementary information from the distinctive processes.

To make comparisons of the above type most transparent it is convenient to cast the couplings of the different probes to spin modes into a common form. Here, we will focus primarily on those couplings relevant for the excitation of unnatural parity states $(\Delta \pi = (-)^{J+1})$. For nucleon-nucleus scattering a necessary complication is the inclusion of the exchange terms associated with V_{12} of Eq. (6). Fortunately, at intermediate energies $(E_p \gtrsim 100$ MeV) there exist reliable short-range approximations[2,5,9] for the central and spin-orbit parts of V_{12}. The short-range factorization[2,5] approximation for the tensor part of the interaction is somewhat less satisfactory but suffices for present comparative purposes. The numerical DWIA calculations do not use these approximations. The net result of the factorization approximation[2,5] is, to lowest order, a local energy-dependent N-N interaction for which only <u>direct</u> type matrix elements need be calculated. We denote this interaction by

$$t_{ip} = t_0^C(r_{ip}) + t_1^C(r_{ip}) \vec{\sigma}_i \cdot \vec{\sigma}_p + t^{LS}(r_{ip}) \vec{L} \cdot \vec{S}$$

$$+ t^T(r_{ip}) S_{ip}(\hat{r}_{ip}) + X^T \delta(\vec{r}_{ip}) \qquad (8a)$$

where X^T denotes exchange terms arising from the tensor force. If t^T is taken to be V^T from Eq. (6) then (from Eq. (14) of Ref. 4 and Ref. 11)

$$\chi^T = -v^T(Q)S_{12}(\hat{Q}) \qquad (8b)$$

where Q is a vector of magnitude k_A in the direction $\vec{k}_A + \vec{k}_A'$ where $\vec{k}_A(\vec{k}_A')$ is the initial (final) momentum in the nucleon-nucleus system. Here p(i) denotes the projectile (target) nucleon and isospin variables have been suppressed for brevity. With these approximations the coupling terms for the various probes can be put in a common form by introducing the spin and current tensors defined by:

$$M_{LSJ}(q\vec{r},0_S) \equiv j_L(qr)(i^L Y_L(\hat{r}) \otimes 0_S)^J, \quad \text{spin } (S = 0,1) \qquad (9a)$$

and

$$P_J(q\vec{r},\vec{L}) \equiv \frac{j_J(qr)}{qr}(i^{J-1} Y_{J-1}(\hat{r}) \otimes \vec{L})^J, \quad \text{current} \qquad (9b)$$

where $0_0 = 1$, $0_1 = \vec{\sigma}$ and $\vec{L}$ is the orbital angular momentum operator. The longitudinal and transverse combinations of M_{L1J} are also useful for unnatural parity excitations and are given by

Longitudinal: $\quad M_J^\ell = \sqrt{\dfrac{J}{2J+1}}\, M_{J-11J} - \sqrt{\dfrac{J+1}{2J+1}}\, M_{J+11J}, \qquad (10a)$

Transverse: $\quad M_J^t = \sqrt{\dfrac{J+1}{2J+1}}\, M_{J-11J} + \sqrt{\dfrac{J}{2J+1}}\, M_{J+11J}. \qquad (10b)$

In terms of the above tensors t_{ip} can be written:

Central: $\quad t_{ip}^C = \dfrac{2}{\pi}\displaystyle\int_0^\infty q^2 dq \sum_{LSJ} (-)^{J+S} \tilde{t}_S^C(q) M_{LSJ}(i)\cdot M_{LSJ}(p) \qquad (11a)$

Spin-Orbit:

$$t_{ip}^{LS}\vec{L}\cdot\vec{S} = \dfrac{2}{\pi}\int_0^\infty q^2 dq\, \dfrac{q}{4}\,\tilde{t}^{LS}(q)\sum_J (-)^J \sqrt{\dfrac{2J+1}{J+1}}\; [P_J(i)\cdot M_J^t(p) + M_J^t(i)\cdot P_J(p)] \quad (11b)$$

Tensor:

$$t_{ip}^T S_{ip} = \dfrac{2}{\pi}\int_0^\infty q^2 dq\, \tilde{t}^T(q) \sum_{LL'J} i^{L+L'+2} Z_{LL',L'SJ}^J M(i)\cdot M_{LSJ}(p), \quad S=1 \quad (11c)$$

where the Fourier transforms ($\tilde{t}$) of the spatial forms of t_{ip} are defined in Ref. 5 and Eq. (14), the geometrical recoupling coefficients $Z_{LL'}^J$ are given in Ref. 12, and q is to be identified as the momentum transfer in the plane-wave Born approximation. Only the unnatural parity part of t_{ip} is given for t^{LS}.

Apart from kinematic factors[1] arising from the pion-nucleus mass difference, the coupling for the excitation of unnatural parity spin modes by pions is given in lowest order by the second term in t_{ip}^{LS}. Excluding coupling to the nuclear currents the spin-orbit part of the N-N coupling is (like (π,π')) only sensitive to the transverse part of the spin-density. This is also true for the magnetic multipole operator for the excitation of unnatural parity states by inelastic electron scattering which is given by[13]

$$\text{\underline{Electromagnetic}:}\quad \hat{T}_J^{\text{mag}}(q,i) = i^{2-J}\beta q\left[\frac{g_s}{2}M_J^t(i) + 2g_\ell\sqrt{\frac{2J+1}{J+1}}P_J(i)\right] \quad (12a)$$

where β is the nuclear magneton and $g_s(g_\ell)$ is the spin (orbital) g-factor. For natural parity excitations the spin dependent electric multipole operator[13] is

$$\text{\underline{Electromagnetic}:}\quad \hat{T}_J^{\text{el}}(q,i) = i^{-J}\beta q\frac{g_s}{2}M_{J1J}(i) . \quad (12b)$$

The more complicated convective part of T^{el} is not considered here. Finally, for allowed β-decay the operator in the nucleon space is

$$\text{\underline{β-decay}:}\quad M_\beta = \frac{t_-(i)}{\sqrt{4\pi}}[g_V + g_A\sigma(i)] \quad (13)$$

where $g_V(g_A)$ is the vector (axial-vector) coupling constant for Fermi (Gamow-Teller) transitions corresponding to $S = 0$ and $S = 1$ modes respectively, and t_- is the isospin lowering operator. The reduced nuclear matrix element of each of the above tensors defines a corresponding transition density[1] in momentum transfer space.

Aside from differences in isospin selectivity, one of the most notable dynamical differences between electromagnetic pionic and nucleonic couplings is that the (π,π') and the (e,e') reactions are only sensitive to the transverse part of the nuclear spin density for unnatural parity transitions while the (p,p') reaction is sensitive to both longitudinal and transverse parts. The first point follows from Eq. (11) and Eq. (12); the second point will be demonstrated in detail below. Much has been made[14] of this in the search for precursors to pion condensation.

To show how the central and tensor components of the N-N interaction couple to both longitudinal and transverse spin densities for unnatural parity transitions, we introduce the longitudinal and transverse components of the central and tensor parts of the interaction which in the N-N system would be η times the coefficients of O_ℓ, O_t and O_u. These are

$$v^\ell(q) = \tilde{t}_1^C(q) - 2\tilde{t}^T(q), \quad \text{longitudinal} \quad (14a)$$

$$v^t(q) = \tilde{t}_1^C(q) + \tilde{t}^T(q), \quad \text{transverse} \quad (14b)$$

$$v^u(Q) = \frac{3}{2}\tilde{v}_E^T(Q), \quad \text{exchange transverse} \quad (14c)$$

where

$$\tilde{t}_1^C(q) = \tilde{v}_D^C(q) + \tilde{v}_E^C(Q), \qquad \tilde{t}^T(q) = \tilde{v}_D^T(q) - \frac{1}{2}\tilde{v}_E^T(Q) \qquad (14d)$$

and $\tilde{v}_D^{C,T}$ is from Eq. (6) and $\tilde{v}_E^{C,T}$ is from Eq. (6) but with the signs of the odd-state parts of the V_{12} reversed;[9] Q is the momentum transfer (k_A) appropriate for exchange (Ref. 2, 5, 9 and Eq. (8b)). In this longitudinal-transverse representation some of the tensor exchange terms are included in v^t and v^ℓ and X^T of Eq. (8b) becomes $V^u O_u$ where the unit vectors in O_u are approximated by their asymptotic values. In terms of v^t and v^ℓ and the tensors from Eqs. (9) and (10) the static S = 1 central plus tensor parts of t_{ip} may be written as

natural parity:

$$t_1^C \vec{\sigma}_i \cdot \vec{\sigma}_p + t^T S_{ip} = \frac{2}{\pi} \int_0^\infty q^2 dq \sum_J (-)^{J+1} v^t(q) M_{J1J}(i) \cdot M_{J1J}(p) \qquad (15a)$$

unnatural parity:

$$t_1^C \vec{\sigma}_i \cdot \vec{\sigma}_p + t^T S_{ip} = \frac{2}{\pi} \int_0^\infty q^2 dq \sum_J (-)^{J+1} [v^t(q) M_J^t(i) \cdot M_J^t(p)$$

$$+ v^\ell(q) M_J^\ell(i) \cdot M_J^\ell(p)]. \qquad (15b)$$

The V^u term usually contributes to both natural and unnatural parity transitions (and is included in Figures 2 and 3). In the factorization approximation without distortion Moss[21] has shown that this term enters the differential cross section multiplied by the square of the transverse nuclear matrix element of Eq. (10b). Since the isovector part of V^u is weak (see Fig. 4) and it is without a precise counterpart in the pionic and electromagnetic couplings it will not be discussed further; it is of course included implicitly in the DWIA calculations.

Equation (15a) suggests that the N-N transverse coupling (for which ρ-exchange should be the leading isovector term) can best be studied empirically by examining S = 1 modes of natural parity; this is especially true when the spin density is available from measurements of the transverse electric[15] form factor. In practice this is most useful for T = 1 excitations because of the usual dominance (see Fig. 1) of the S = 0 terms for T = 0 excitations. In many ways the longitudinal coupling is even more difficult to isolate since from Eq. (15b) it appears with $v^t(q)$. A striking exception is the excitation of 0^- states where $M_0^t \equiv 0$ and only longitudinal excitation is possible. Unfortunately there is little reliable data for such excitations. We will examine one such transition[16] in a later section.

Another hybrid form of the N-N coupling can be useful for

considering unnatural parity excitations where more than one value
of L-transfer is usually allowed. This is:

$$t_1^C \vec{\sigma}_i \cdot \vec{\sigma}_p + t^T S_{ip} =$$

$$\frac{2}{\pi} \int_0^\infty q^2 dq \sum_J (-)^{J+1} \left\{ \left[\sqrt{\frac{J+1}{2J+1}} v^t(q) M_J^t(i) + \sqrt{\frac{J}{2J+1}} v^\ell(q) M_J^\ell(i) \right] \cdot M_{J-11J}(p) \right.$$

$$\left. + \left[\sqrt{\frac{J}{2J+1}} v^t(q) M_J^t(i) - \sqrt{\frac{J+1}{2J+1}} v^\ell(q) M_J^\ell(i) \right] \cdot M_{J+11J}(p) \right\}.$$

$$(16)$$

Although this form of the N-N coupling is less conventional than that
given in Eq. (11), it is often useful in DWIA calculations where the
different L-transfers to the projectile are largely incoherent. This
form also isolates the (longitudinal) nuclear structure information
inaccessible to the (e,e') and (π,π') reactions.

The practical limitation on the extraction of longitudinal
transition densities, search for precritical phenomena, etc., is
determined by the actual size of $v^\ell(q)$ relative to $v^t(q)$ and v^u. In
Fig. 4 we illustrate this limitation for $S = 1$ transitions in this
representation by showing the moduli of v^ℓ, v^t and v^u (the coefficient of 0_u) for the free t-matrix interaction[5] described above at
140 and 425 MeV. Surprisingly, the N-N isovector transverse coupling
v^t (believed to be dominated by ρ-exchange) is significantly larger
than the longitudinal coupling (dominated by π-exchange) for most
momentum transfers $q \lesssim 2 fm^{-1}$ at each of these energies. This suggests
that the really distinctive information obtainable from (p,p') may be
masked by the large transverse coupling. Figure 4 suggests that for
isoscalar $S = 1$ transitions where $v^\ell > v^t$ the situation is quite different. However, the isoscalar terms are more uncertain. In addition,
the non-static and less familiar v^u term (arising in the present model
entirely from the tensor exchange terms) is comparable to or larger
than v^ℓ for most of the relevant momentum transfers. For isovector
transitions v^u is quite small.

Although the dominant knock-on exchange terms are quite well
approximated above ~100 MeV by a local N-N interaction as in Eq. (8),
another class of smaller but interesting terms arises when explicit
account is taken of the non-locality (or velocity dependence) of an
effective interaction. The details of the derivation are given elsewhere;[17] here we simply give the (approximate) expression for
including the new type of terms which arise from the knock-on
exchange amplitudes associated with the central part of the interaction. The addition to Eq. (11a) is:

$$\delta t_{ip}^C = \frac{2}{\pi} \int_0^\infty q^2 dq \sum_{LSJ} (-)^{J+S} \hat{A}_{ST} q^2 [P_L(i) \otimes 0_S(i)]^J \cdot [P_L(p) \otimes 0_S(p)]^J \quad (17)$$

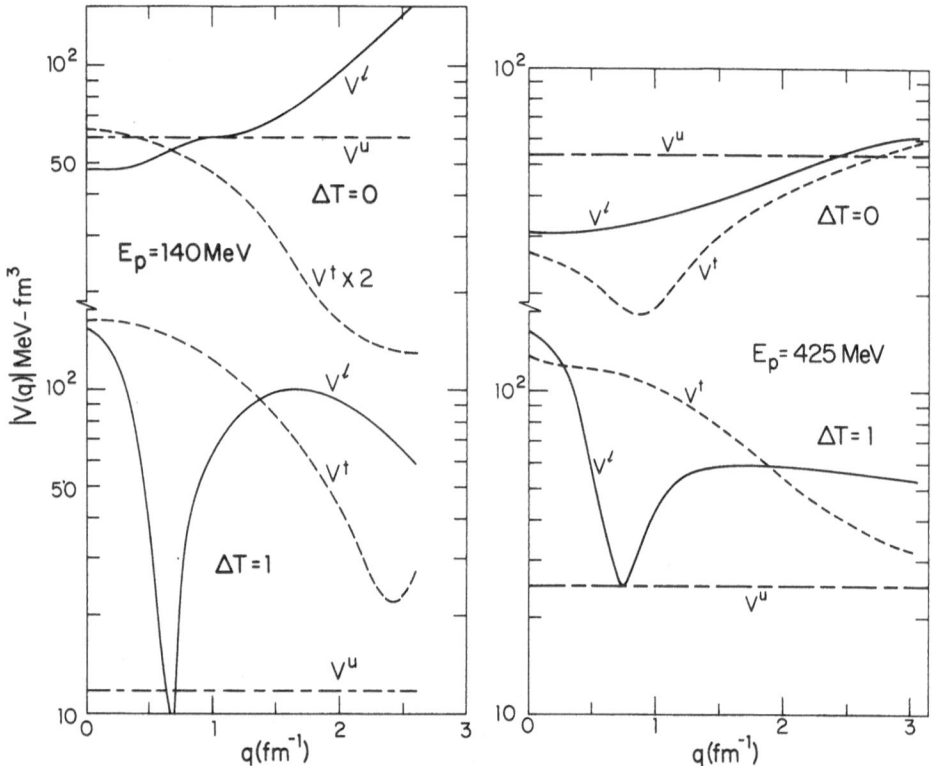

Fig. 4. Moduli of V^ℓ, V^t and V^u at E_p = 140 and 425 MeV.

where $\hat{A}_{ST} = 2(\partial \hat{v}^C_{ST}/\partial k^2)_{k^2}$. Here $\hat{v}^C_{ST}$ is the Fourier transform of
the central part of the exchange interaction of Eq. (8) appropriate
for spin (isospin) transfer S(T) evaluated at the forward asymptotic
momentum as discussed in conjunction with Eq. (8b). The isospin
operator (1 or $\tau_i \cdot \tau_p$) has been suppressed. Couplings like this are
of course already included in antisymmetrized DWIA calculations when
the knock-on exchange terms are treated exactly. The approximate
form above illustrates more clearly, however, the nature of the new
kinds of terms which can and do arise. δt^C_{ip} above represents a cur-
rent-current coupling between the target and projectile nucleons and
can excite unnatural parity states[17] even in the absence of spin
transfer.

Calculated cross sections are often slightly changed when δt^C_{ip}
and the larger δt^T_{ip} are included. Some specific examples of the
effect of t^C_{ip} for low nucleon energies appear in Ref. 18. However,
these terms can significantly affect polarization observables as is
illustrated in Fig. (5) for the excitation[19] of the 15.1 MeV 1^+
state in ^{12}C. By going to LS coupling and removing the L = S = J =
1 part of the transition density predicted by the Cohen-Kurath
wavefunctions[20] (CKWF) we see that the polarization P_y and analyzing
power A_y are qualitatively very similar whereas inclusion of this
(LSJ=111) term[21] yields $P_y \sim -A_y$. The transverse magnetic form factor
is insensitive to this term. This striking effect may be traced to
the relative sizes of the various parts of the transition "density"
(A_{LSJ}) in an LS representation where for the lowest $^{12}C(0^+ \rightarrow 1^+)$
transitions Lee and Kurath give[22]

Table 1. Amplitudes of Cohen-Kurath for 1^+ Levels in ^{12}C

LSJ	$E_x = 15.1$ MeV A_{LSJ}	$E_x = 12.7$ MeV A_{LSJ}
011	0.160	−0.152
211	0.096	−0.049
111	0.515	−0.537
101	−0.023	0.093

Although the normal parity <u>amplitudes</u> ($\Delta\pi = (-)^L$) are appreciable,
spectroscopically this transition might best be characterized as an
$\vec{L} \times \vec{\sigma}$ "mode." Even more dramatic results apply for the 1^+, T = 0 excitation at 12.7 MeV in ^{12}C where the inclusion of the LSJ = 111 terms
also increases the integrated cross section by a factor of 2.

Apart from sampling distinctly different transition densities
from those sampled by the static parts of the interaction in
unnatural excitations, δt^C_{ip} and especially δt^T_{ip} contribute[23] to one-
body optical and shell model spin-orbit potentials in those nuclei

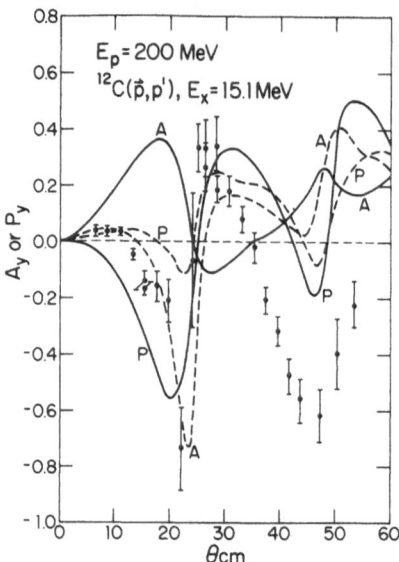

Fig. 5.   Polarization and analyzing power for excitation of the 1^+
state in ^{12}C at 15.1 MeV by 200 MeV protons.

having spin-unsaturated subshells (SUS). This is evident from Eq.
(17) where for J = 0, L = S = 1, the scalar product is proportional to
$\vec{L}(i)\cdot\vec{\sigma}(i)\vec{L}(p)\cdot\vec{\sigma}(p)$. It is the presence of such terms[23] which theoretically quenches the one-body spin-orbit potential in SUS nuclei. The
empirical absence of such quenching suggests that either the tensor
forces used for bound and scattering problems are incorrect or our
understanding of spin correlations in the ground state is incomplete.
This in turn can have a bearing on our understanding of those low-
lying M1 excitations which can only occur between spin-orbit partners.
The absence of quenching of the one-body spin-orbit potential in SUS
nuclei remains one of the long-standing unsolved problems[24] in nuclear
physics and provides further impetus for trying to understand $\delta t^{C,T}$ in
the excitation of unnatural parity states. Similar corrections to
t^{LS} are expected to be less important due its very short range. G.
Walker[25] discusses other sources of current terms in a later paper in
these proceedings.

4. N-N COUPLING: PHENOMENOLOGICAL vs $V_\pi + V_\rho$

It is interesting to compare the phenomenological N-N coupling
with that predicted on the basis of π- and ρ-exchange models. We
focus primarily on the isovector part of V_{12} where these processes
contribute large direct matrix elements. The basic π-exchange and
ρ-exchange couplings are given[26] in momentum space by

$$V_\pi(q) = -J_\pi \frac{x_\pi^2}{1+x_\pi^2} [S_{12}(\hat{q}) + \vec{\sigma}_1 \cdot \vec{\sigma}_2], \quad \text{longitudinal} \qquad (18a)$$

$$V_\rho(q) = J_\rho \frac{x_\rho^2}{1+x_\rho^2} [S_{12}(\hat{q}) - 2\vec{\sigma}_1 \cdot \vec{\sigma}_2], \quad \text{transverse} \qquad (18b)$$

$$J_i \equiv \frac{4\pi}{3} \frac{f_i^2}{m_i^2}, \quad x_i = qR_i, \qquad (18c)$$

with $R_\pi = 1.43$ fm, $R_\rho = 0.254$ fm, $J_\pi = 137.$ MeV $-$ fm^3, $J_\rho = 263.$ MeV $-$ fm^3. The isospin operator $\vec{\tau}_1 \cdot \vec{\tau}_2$ is suppressed for brevity. If, as is customary, we drop the (coordinate space) contact terms from Eq. (18) we obtain modified couplings $\hat{V}$ given by:

$$\hat{V}_\pi(q) = -J_\pi[\frac{x_\pi^2}{1+x_\pi^2} S_{12}(\hat{q}) - \frac{\vec{\sigma}_1 \cdot \vec{\sigma}_2}{1+x_\pi^2}] , \qquad (19a)$$

$$\hat{V}_\rho(q) = J_\rho[\frac{x_\rho^2}{1+x_\rho^2} S_{12}(\hat{q}) + 2\gamma \frac{\vec{\sigma}_1 \cdot \vec{\sigma}_2}{1+x_\rho^2}] \qquad (19b)$$

where $\gamma \simeq 0.4$ is said to account[26] roughly for short-range correlations most important for the ρ-exchange process. Independent of the value chosen for γ, these modified couplings $\hat{V}$ are from Eq. (14) no longer purely longitudinal and transverse. The longitudinal and transverse couplings (direct terms only) are now given by:

$$v^\ell(q) = \frac{J_\pi}{1+x_\pi^2} [1-2x_\pi^2] + \frac{J_\rho}{1+x_\rho^2} [2\gamma + 2x_\rho^2] , \qquad (20a)$$

$$v^t(q) = \frac{J_\pi}{1+x_\pi^2} [1 + x_\pi^2] + \frac{J_\rho}{1+x_\rho^2} [2\gamma - x_\rho^2]. \qquad (20b)$$

The isovector exchange terms arising from antisymmetrization can be included approximately by making the replacement:

$$v^{\ell,t}(q) \rightarrow v^{\ell,t}(q) - \frac{v^{\ell,t}(Q)}{4} = v_{D+E}^{\ell,t} , \quad \text{isovector} \qquad (21)$$

where $Q = k_A$ is the momentum transfer appropriate for the knock-on process. The contributions of $\hat{V}_\pi$ and $\hat{V}_\rho$ to the isoscalar longitudinal and transverse terms arising from nucleon exchange are given by

$$V_E^{\ell,t} = \frac{3}{4} V^{\ell,t}(Q) \; , \quad \text{isoscalar} \tag{22}$$

independent of momentum transfer q. Figure 4 shows a significant q dependence of the isoscalar part of V^ℓ and V^t in the phenomenological interaction suggesting the presence of a non-negligible direct isoscalar S = 1 interaction.

Figure 6 shows a comparison between the moduli of the phenomenological isovector parts of V^ℓ and V^t and their $\hat{V}_\pi$ plus $\hat{V}_\rho$ counterparts calculated from Eqs. (20) and (21) at $E_p = 140$ MeV. Although there is little resemblance between the phenomenological and calculated curves, we see that the inclusion of a single contact term

$$\delta V_{\sigma\tau} = -150 \text{ MeV} - \text{fm}^3 \; \vec{\sigma}_1 \cdot \vec{\sigma}_2 \; \vec{\tau}_1 \cdot \vec{\tau}_2 \tag{23}$$

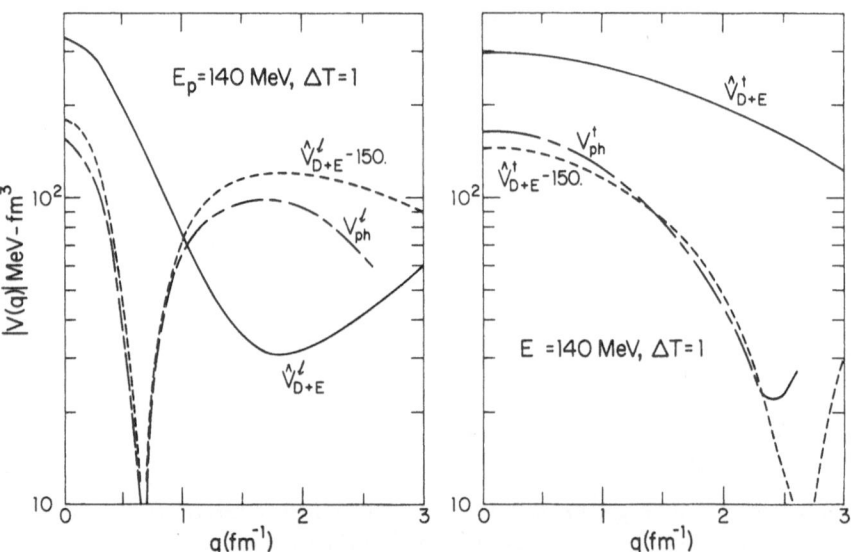

Fig. 6.   Comparison of phenomenological V^ℓ and V^t with $\hat{V}_\pi + \hat{V}_\rho$ at 140 MeV (see text).

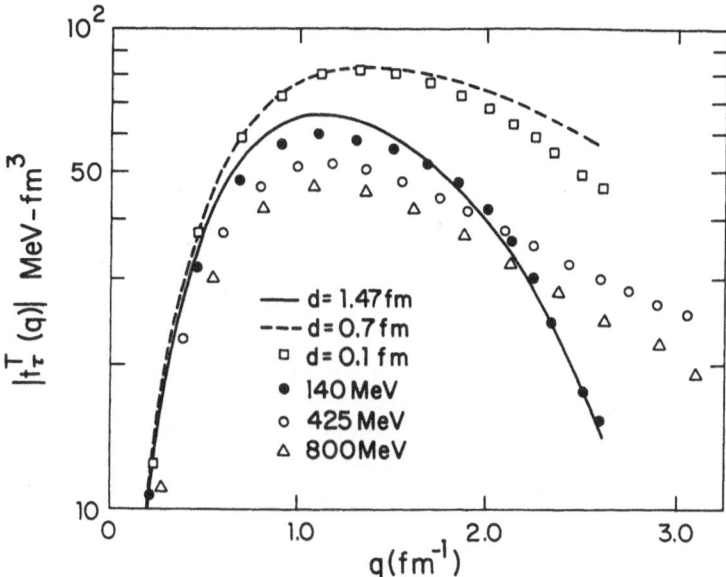

Fig. 7.   Phenomenological $V_\tau^T(q)$ vs cutoff $V_\pi^T + V_\rho^T$. Direct only,
d = inner cutoff.

in the theoretical momentum space interaction brings the phenomenological and "theoretical" curves into quite reasonable qualitative
agreement. Had we kept the full $V_\pi + V_\rho$ coupling, this same agreement
would have required $\delta V_{\sigma\tau} \simeq +87.$ MeV-fm^3. These corrections presumably
reflect the absence of a reliable procedure for including short-range
correlations and/or the presence of other terms in the bare interaction. The phenomenological isovector tensor force is almost
entirely real so that the above comparisons are meaningful.

It is interesting to isolate the isovector tensor force and
compare it with the predictions of $V_\pi + V_\rho$ especially since the tensor
part of the coupling remains unchanged under the particular manipula-
tions of the central part of the force described above. This is
shown in Fig. 7 where the direct part only of the tensor force is
compared with the corresponding part of $V_\pi + V_\rho$ at a number of bombarding energies. For $q \lesssim 2.0$ fm^{-1} there is little energy dependence
in the phenomenological interaction as predicted by the $V_\pi + V_\rho$ model.
There is, however, significant disagreement between the phenomenological and calculated tensor interactions, especially at the larger

momentum transfers. Radial cutoffs as large as 0.7 fm have a very small effect for $q \lesssim 2$ fm^{-1}. At 140 MeV a cutoff of $V_\pi + V_\rho$ of ~1.5 fm is needed to make the calculated interaction agree with the phenomenological one.

5.   APPLICATIONS

In this section we consider a few specific transitions which isolate particular parts of the N-N interaction.

Longitudinal Spin Modes: ^{16}O(p,p')

As noted in section 3, $0^+ \to 0^-$ excitations are especially inter- esting since they are only sensitive to V^ℓ, the longitudinal part of the N-N coupling. Consequently these transitions are forbidden in (π,π') and first-order forbidden in (e,e') reactions; isovector $0^+ \to 0^-$ transitions are seen in β-decay.[27] Similarly, this type of excitation is insensitive to the N-N spin-orbit coupling (see Eq. (11b)). Recently data have been obtained[16] for the T = 0, 0^- excita- tion at 10.95 MeV in ^{16}O using 135 MeV protons. Data for the higher lying T = 1 state could not be resolved from the background. Figure 8 shows a comparison between the calculated (T = 0 and T = 1) and experimental (T = 0) differential cross sections. Each transition was assumed to be described by a $1p_{1/2} \to 2s_{1/2}$ configuration. The T = 0 calcu- lated cross section is dominated by the exchange amplitudes arising from the tensor force and is larger than the data by a factor of 1.7. The calculated and observed shapes are in quite reasonable agreement. The relatively structureless differential cross section for this transition can be understood in terms of the relatively flat $V^\ell(q)$ for T = 0 excitations. Data for the ^{16}O(p,n) cross section for the 0^-, T = 1 state should be helpful for studying the more structured iso- vector part of the longitudinal force. Such data has been taken near 40 MeV and is being processed.[28]

Transverse Spin Modes: ^{12}C(p,p')

As indicated in section 3, the N-N transverse coupling can best be studied by considering natural parity modes having S = 1. The iso- vector 2^+ excitation in ^{12}C at 16.11 MeV is known from electron scat- tering[15] to be such a transition. In particular, both longitudinal and transverse electric form factors have been measured.[15] After scaling the Cohen-Kurath S = 0 and S = 1 transition densities to agree with the (e,e') measurements, the S = 1 transition density dominates. In addition, $|t_{\sigma\tau}| > |t_\tau|$ at intermediate energies while at larger q, t_τ^T should dominate over $t_{\sigma\tau}$. Figure 9 shows a comparison between the calculated[5] and measured[11,19,29] differential cross sections at 120 and 800 MeV using transition densities scaled to (e,e') measurements. The N-N spin-orbit contribution is negligible but as seen from the figure the tensor force is extremely important at each of these

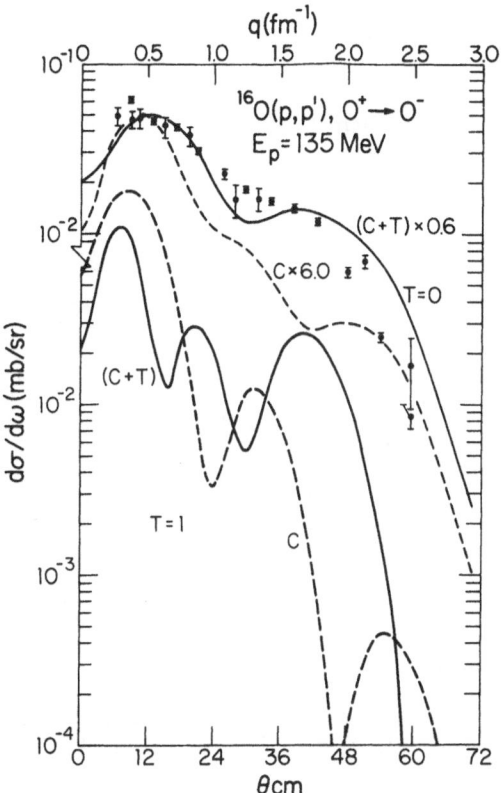

Fig. 8.   Calculated (T = 0 and T = 1) and measured (T = 0) cross sections
          for excitation of the 0⁻ states in ^{16}O at 10.95 (T = 0) and
          12.9 (T = 1) MeV.

energies and is apparently given correctly in the impulse approximation. More specifically, the transverse coupling predicted by the
free N–N t-matrix[5] appears adequate at the two energies considered.
There does appear (not shown) to be some difficulty[30,31] with this
transition near 200 and 400 MeV which we do not presently understand;
the data at 200 and 400 MeV are unpublished.

    The sensitivity of the (p,p') reaction to the S = 1 transition
density (largely because of the strong isovector tensor force) is in
striking contrast to the (π, π') reaction near the (3,3) resonance
which has been shown[22] to go almost entirely by S = 0. Consequently,
to the proton this excitation is predominantly a spin mode; to the
pion it is not.

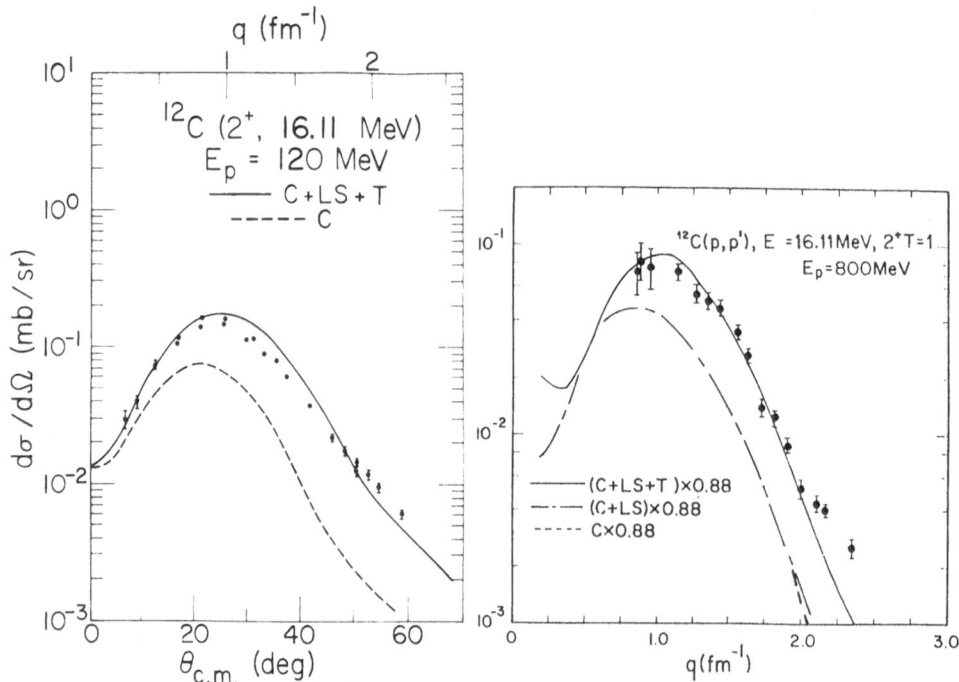

Fig. 9. Calculated and measured cross sections for the 2^+, T = 1
        state in ^{12}C at E_x = 16.1 MeV. See the text for details.

High Spin States: ^{28}Si(p,p')

The study[1,5,8,32] of stretched excitations (those having $j_p = \ell_p + \frac{1}{2}$, $j_h = \ell_h + \frac{1}{2}$ and $J = j_p + j_h$) in nuclei with different probes has proven especially helpful in understanding the high-momentum components of the spin dependent part of both the residual particle-hole interaction and the coupling of the target nucleons to the projectile. As can be seen from Eqs. (9–13), neither the current nor the $M_{J+11J}(i)$ terms contribute to this type of unnatural parity excitation. Only the single nuclear transition density

$$\rho_{JJ-1}^{S=1}(q) = <f||\sum_i M_{J-11J}(i)O_T(i)||i> ; \qquad (O_0 = 1,\ O_1 = \tau) \qquad (24)$$

is sampled in the (e,e'), (π,π') and p,p' reactions yielding a simple connection between the three reactions.

In ^{28}Si the stretched states have $J^\pi = 6^-$ and arise primarily from the $1f_{7/2} - 1d_{5/2}^{-1}$ particle-hole excitation.[32] For the 6^-, T = 1 excitation data is available for all three reactions; for T = 0 only (π,π') and (p,p') data are available.[32-34] Figure 10 shows a comparison between observed (p,p') differential cross sections at E_p = 134, 333 and 500 MeV, the (e,e') form factor and the corresponding calculated quantities. The $T = 1$ oscillator parameter ($b = 1.71$ fm) was determined[33] by fitting the (e,e') form factor without recoil corrections. The quantity S^2 is the factor by which the pure particle-hole calculated cross section is multiplied in order for the calculated and observed cross sections to agree. For the (p,p') reaction the average S^2 is $S_p^2 = 0.27$ which is ~15% smaller than that for (e,e') where $S_e^2 = 0.31$. The (π,π') reaction[1,34] gives $S_\pi^2 = 0.31$. As has been shown elsewhere,[1,5,8,32] the (p,p') cross section is strongly dominated by the tensor force. Therefore, the present results suggest that the isovector part of the tensor force may be ~5-10% too large for $1.5 \lesssim q(\text{fm}^{-1}) \lesssim 2.0$.

Results for the $T = 0$, 6^- excitation are less satisfactory. Particular problems are the energy dependent values of S_p^2 extracted and the difference in shapes of the calculated and observed cross sections forward of $q \approx 1.5$ fm^{-1}. At $E_\pi = 162$ MeV[1,34] $S_\pi^2 = 0.12$ which is in reasonable agreement with the average of S_p^2. Because of the relatively weak electromagnetic coupling to isoscalar modes, there

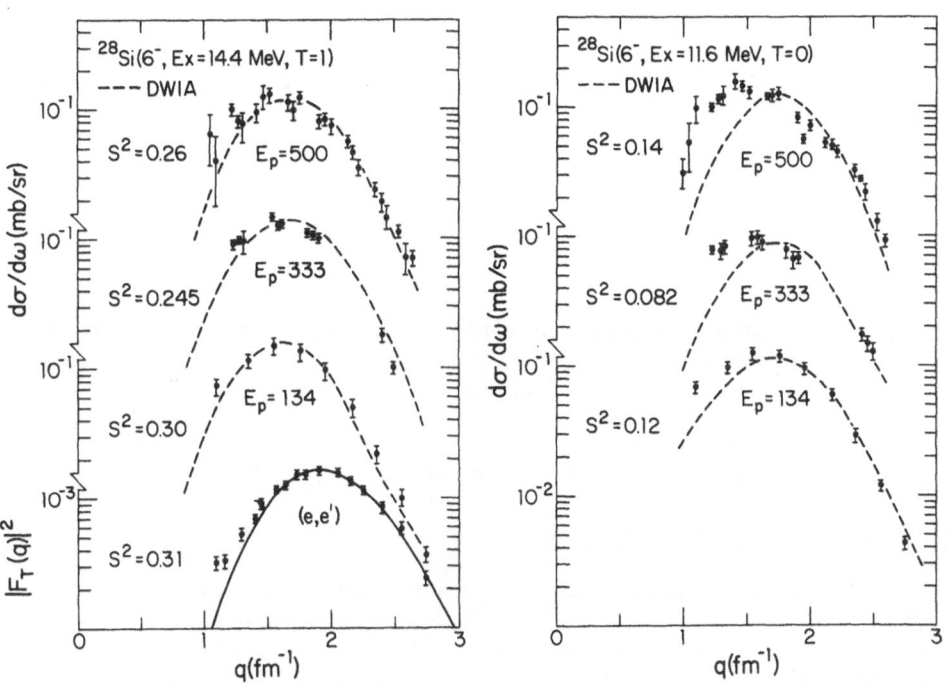

Fig. 10.   Calculated and observed cross sections (and form factor) for excitation of the 6^-, $T = 1$ and $T = 0$ states in ^{28}Si. The plotted $T = 1$ proton (electron) data are too small (large) by the factor 0.9223; the values of S^2 have been corrected for these renormalizations.

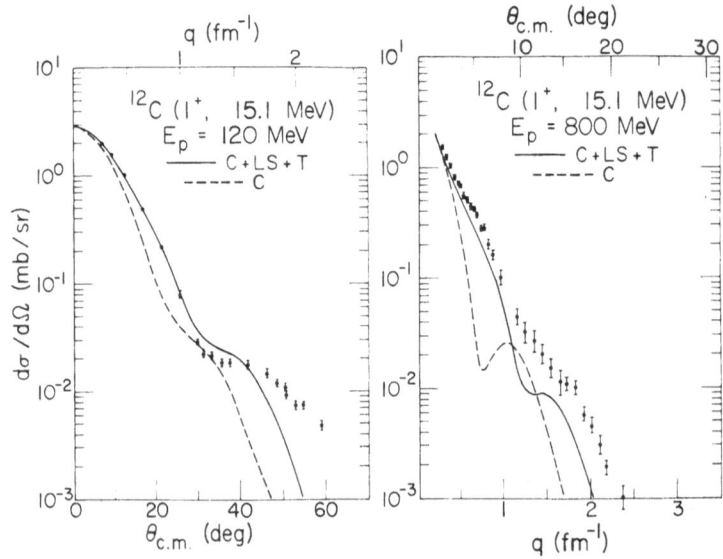

Fig. 11. Calculated and measured cross sections for the 1^+, $T = 1$ state in ^{12}C at $E_x = 15.1$ MeV by 120 and 800 MeV protons.

is no (e,e') data for this transition. For this excitation the (p,p') transition is mediated by nearly equal contributions[1,5,8,32] from t^{LS} and t^T at 134 MeV; at 333 and 500 MeV t^{LS} dominates (see Fig. 2). The t^C contribution is small at all energies. Recall that t^T arises primarily from exchange terms. Clearly, more work is required for a complete understanding of this transition.

1^+ Excitations: $^{12}C(p,p')$

Although 1^+ excitations are in some ways less selective than the transitions described above, the isovector ones can be correlated with (p,n) and β–decay measurements[1,4,5,8,10] of GT strength and have been generally helpful in establishing the DWIA as a quantitative procedure for extracting GT and M1 strengths.[1,4,5,8,10,35] Because of the largely direct nature of the isovector tensor interaction this type of isovector transition samples primarily the isovector part of t^C near $q = 0$ and this greatly simplifies the correlation with GT and

M1 excitations. The isoscalar 1^+ excitations are best compared with (π,π') scattering[1] where the isoscalar coupling is relatively strong although not forward peaked.

The isovector 1^+ excitation in ^{12}C at $E_x = 15.1$ MeV has now been studied at numerous bombarding energies[11,19,29,30,31] below 800 MeV by the (p,p') reaction. Part of the impetus for the abundance of data for this transition has been the search for precritical phenomena[14] associated with pion condensation. Above ~100 MeV the differential cross sections are nearly identical when plotted as a function of q. Data for the excitation of this state by the (e,e')[15] and (π,π')[36] reactions are also available. In Fig. 11 a comparison is shown between the observed and calculated (p,p') cross sections for the 1^+ state at $E_x = 15.1$ MeV and $E_p = 120$ and 800 MeV using the Cohen-Kurath wave functions which are known to reproduce[8] the corresponding β-decay rate and also provide a reasonable description of the magnetic form factor for $q \lesssim 1$ fm^{-1}. For larger q the wave functions are known (from (e,e')) to be completely inadequate.[37] For $q \lesssim 1$ fm^{-1} we see that the differential cross sections are given quite well in the DWIA. The tensor force is seen to be important for $q \lesssim 0.5$ fm^{-1}. An increase in the calculated cross section by ~10-20% would improve agreement with the experimental cross section at 800 MeV. Suggestions for the disagreement beyond 1 fm^{-1} abound [14,38] and are discussed elsewhere.

Data for the excitation of the isoscalar 1^+ state in ^{12}C by protons are available[11,19,29,30,31] for several proton bombarding energies below 800 MeV. Both (π,π')[22,36] and (e,e')[15] data are also available, with the (e,e') results being especially sensitive[15] to $T = 1$ impurities. Since (p,p') is much less sensitive to this admixture,[11] the (e,e') data are less helpful in pinning down the nuclear structure information sampled in the (p,p') reaction. Results from (π,π')[22] studies suggest that the Cohen-Kurath wave functions for this transition are adequate for $q \lesssim 2$ fm^{-1}. Figure 12 shows a comparison between measured and calculated (p,p') cross sections for this transition at 120 and 402 MeV. At $E_p = 120$ MeV we find very poor agreement. At 402 and 800 MeV (not shown),[29] the agreement between theory and experiment is much improved though still somewhat inferior to that found for the 15.1 MeV $T = 1$ transition. The overprediction of the forward angle cross section at 120 MeV can be traced[4,5] to the large tensor exchange amplitudes. This overprediction at small q is consistent with our results for the $0^+ \rightarrow 0^-$ transition in ^{16}O where the forward cross section (dominated by the isoscalar tensor force) is overpredicted by a factor of ~1.7. Recall from section 3 that the calculated cross section for this 1^+ transition at $E_p = 120$ MeV is quite sensitive to the current $\otimes$ spin terms which are not tested by the (e,e') data. At higher energies where the agreement is much better the calculated cross sections are much less sensitive to the current $\otimes$ spin terms.

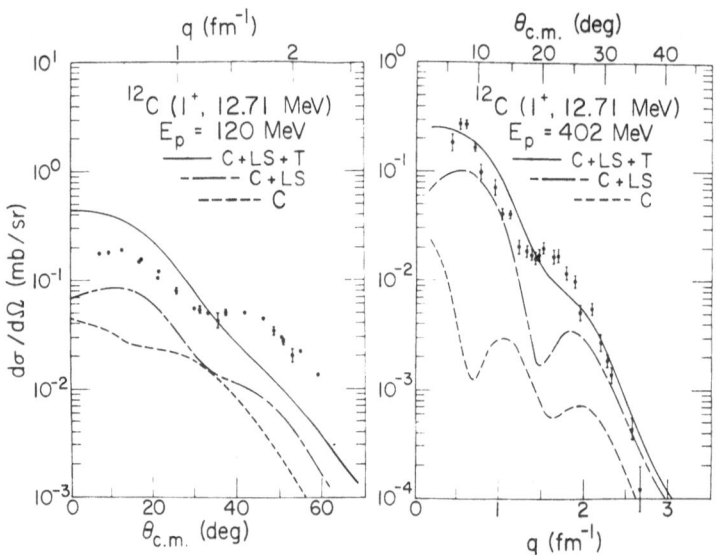

Fig. 12.   Calculated and observed cross sections for the 1^+, T = 0
state in ^{12}C at E_x = 12.7 MeV by 120 and 402 MeV protons.

6.   SUMMARY

We have outlined very briefly how a t-matrix interaction can be
derived from N–N data and have illustrated some of its most important
properties relevant to the study of spin modes. The N–N and electromagnetic couplings have been cast in the longitudinal-transverse form
to best identify those coupling characteristics for (e,e'), (π,π')
and (p,p') processes which are distinct and those which are the same
or similar. In particular, we have seen that the (π,π') and (e,e')
couple exclusively to transverse spin modes whereas (p,p') couples
to both longitudinal and transverse modes. The free t-matrix interaction implies that the transverse N–N coupling is comparable to or
larger than the longitudinal part at both 140 and 425 MeV except at
large q.

We have also indicated how non-locality corrections in the N–N
interaction lead to current ⊗ spin couplings not present in (e,e')
coupling nor in the simple[1] form of (π,π') coupling and these are
shown to have significant consequences on the calculation of spin
observables for those transitions which spectroscopically might most
accurately be regarded as current ⊗ spin modes.

A comparison of the S = 1 components of the N–N t-matrix with a
$\pi+\rho$ exchange model was made. It was found that at 140 MeV $V_\pi + V_\rho$
needs to be supplemented by a contact term in order to roughly agree

with the free t-matrix. In addition the amplitude of high momentum components of the tensor force is overestimated by $V_\pi + V_\rho$.

A number of applications of the t-matrix interaction to specific types of transitions have been made which suggest the following. The isoscalar tensor force given by the free t-matrix is too large at 140 MeV by as much as 40% at small q but is roughly correct for $E_p \gtrsim 400$ MeV. Near $q \simeq 1.7$ fm^{-1} the isoscalar tensor force appears correct near 135 MeV. The isoscalar spin-orbit force (t^{LS}) appears to be too strong at 333 and 500 MeV. From the ^{12}C(p,p') reaction ($E_x = 16.1$ MeV, $J^\pi, T = 2^+, 1$) at 120 and 800 MeV, the transverse isovector coupling appears correct. From the excitation of the isovector stretched 6^- state in ^{28}Si the tensor force near $q = 1.7$ fm^{-1} appears to be reasonably described by the t-matrix interaction.

In conclusion, a relatively simple phenomenological t-matrix interaction appears to give a reasonable first-order description of the coupling to spin-modes in nuclei, especially the isovector ones. Both the isoscalar central and tensor interactions are considerably less reliable, especially at small q where the exchange amplitudes arising from the tensor force dominate. The apparent deficiencies in the isoscalar $S = 1$ interaction can perhaps be delineated by a systematic study of a number of isoscalar unnatural parity transitions using both protons and pions coupled with an investigation of medium corrections[7] to the $S = 1$ coupling. Finally, although the (p,p') and (p,n) reactions are not completely understood they, like the (e,e') and (π,π') reactions, have proven to be extremely important for identifying and interpreting spin modes of excitation in nuclei and many of the subsequent ramifications.

ACKNOWLEDGMENTS

It is a pleasure to thank the Massachusetts Institute of Technology and the University of Minnesota groups for the use of data prior to publication. This work was supported in part by the National Science Foundation and the Department of Energy.

REFERENCES

1. F. Petrovich and W. G. Love, Nucl. Phys. A354: 499c (1981).

2. A. K. Kerman, H. McManus and R. M. Thaler, Ann. Phys. 8: 551 (1959); A. B. Clegg and G. R. Satchler, Nucl. Phys. 27: 431 (1961); M. Kawai, T. Terasawa and K. Izuma, Nucl. Phys. 59: 289 (1964).

3. A. Picklesimer and G. Walker, Phys. Rev. C17: 237 (1978).

4. W. G. Love, Alan Scott, F. Todd Baker, W. P. Jones and J. D. Wiggins, Jr., Phys. Lett. 73B: 277 (1978); W. G. Love, in "The (p,n) Reaction and the Nucleon-Nucleon Force," ed. C. D. Goodman et al., Plenum, New York (1980).

5. W. G. Love and M. A. Franey, Phys. Rev. C24: 1073 (1981).

6. F. A. Brieva and J. R. Rook, Nucl. Phys. A297: 206 (1978); H. V.
    von Geramb, F. A. Brieva and J. R. Rook, in: "Microscopic Opti-
    cal Potentials," ed. H. V. von Geramb, Springer-Verlag, Berlin,
    (1979); H. V. von Geramb, private communication; J. P. Jeukenne,
    A. Lejeune and C. Mahaux, Phys. Rep. 25: 83 (1976).

7. J. Kelly et al., Phys. Rev. Lett. 45: 2012 (1980) and Telluride
    Conf., 1982.

8. F. Petrovich, in: "The (p,n) Reaction and the Nucleon-Nucleon
    Force," ed. C. D. Goodman et al., Plenum, New York (1980); F.
    Petrovich, W. G. Love and R. J. McCarthy, Phys. Rev. C21: 1718
    (1980).

9. F. Petrovich, H. McManus, V. A. Madsen and J. Atkinson, Phys.
    Rev. Lett. 22: 895 (1969); W. G. Love, Nucl. Phys. A312: 160
    (1978).

10. C. D. Goodman et al., Phys. Rev. Lett. 44: 1755 (1980); B. D.
    Anderson et al., Phys. Rev. Lett. 45: 699 (1980); C. D. Goodman,
    Comments on Nucl. and Part. Phys. 10: 117 (1981); D. J. Horen et
    al., Phys. Lett. 95B: 27 (1980).

11. J. R. Comfort, S. M. Austin, P. T. Debevec, G. L. Moake, R. W.
    Finlay and W. G. Love, Phys. Rev. C21: 2147 (1980).

12. W. G. Love and L. J. Parish, Nucl. Phys. A157: 625 (1970).

13. T. de Forest, Jr. and J. D. Walecka, Adv. Phys. 15: 1 (1966).

14. W. Weise, Comments on Nucl. and Part. Phys. 10: 109 (1981) and
    references therein; J. Meyer-Ter-Vehn, Phys. Reports 74:323 (1981).

15. J. B. Flanz, Ph.D. Thesis, University of Massachusetts; J. B.
    Flanz et al., Phys. Rev. Lett. 41: 1642 (1978) and references
    therein.

16. J. Kelly, Ph.D. Thesis, M.I.T.

17. W. G. Love, Particles and Nuclei, 3: 318 (1972).

18. W. G. Love and G. R. Satchler, Nucl. Phys. A159: 1 (1970).

19. J. R. Comfort et al., Phys. Rev. C24: 1834 (1981).

20. S. Cohen and D. Kurath, Nucl. Phys. 73: 1 (1965).

21. J. Moss, Telluride Conf., 1982.

22. T. S. H. Lee and D. Kurath, Phys. Rev. C21: 293 (1980).

23. R. R. Scheerbaum, Phys. Lett. 73B: 381 (1976); W. G. Love,
    Phys. Rev. C20: 1638 (1979).

24. H. A. Bethe, in: "The Two-Body Force in Nuclei," ed. S. M.
    Austin and G. M. Crawley, Plenum, New York (1972); G. Bertsch,
    L. Zamick and A. Mekjian, in: "Nuclear Spectroscopy," ed. G. F.
    Bertsch and D. Kurath, Springer-Verlag, New York (1980).

25. G. Walker, Telluride Conf., 1982.

26. J. Speth, V. Klemt, J. Wambach and G. E. Brown, Nucl. Phys.
    A343: 282 (1980).

27. G. T. Garvey, Telluride Conf., 1982.

28. H. Orihara et al., Phys. Rev. Lett. 49: 1318 (1982) and pri-
    vate communication.

29. M. Haji-Saeid et al., Phys. Rev. C25: 3035 (1982).

30. J. R. Comfort, G.L. Moake, C.C. Foster, P. Schwandt, and W.G. Love, Phys. Rev. C26: 1800 (1982)

31. J. L. Escudie, private communication; J. L. Escudie et al., Phys. Rev. C24: 792 (1981).

32. P. F. Moffa and G. E. Walker, Nucl. Phys. A222: 140 (1974); G. S. Adams et al., Phys. Rev. Lett. 24: 1387 (1977); R. A. Lindgren, W. J. Gerace, A. D. Bacher, W. G. Love and F. Petrovich, Phys. Rev. Lett. 42: 1524 (1979); R. A. Lindgren, Telluride Conf., 1982; J. Lichtenstadt et al., Phys. Rev. Lett. 40: 1127 (1978); S. Krewald and J. Speth, Phys. Rev. Lett. 45: 417 (1980); S. Yen et al., Phys. Lett. 105B: 421 (1981) and to be published; F. C. Barker, R. Smith, I. Morrison and K. Amos, Jour. of Phys. G, Nucl. Phys. 7: 657 (1981).

33. M. Gazzaly, M. A. Franey and N. Hintz, private communication.

34. C. Olmer et al., Phys. Rev. Lett. 43: 612 (1979).

35. G. Bertsch, Comments on Nucl. and Part. Phys. 10: 91 (1981); N. Anantaraman et al., Phys. Rev. Lett. 46: 1318 (1981).

36. C. L. Morris, Telluride Conf., 1982; C. L. Morris et al., Phys. Lett. 108B: 172 (1982).

37. J. Dubach and W. C. Haxton, Phys. Rev. Lett. 41: 1453 (1978).

38. J. R. Comfort and W. G. Love, Phys. Rev. Lett. 44: 1656 (1980); J. R. Comfort, R. E. Segal, G. L. Moake, D. W. Miller and W. G. Love, Phys. Rev. C23: 1858 (1981) and references therein.

MESON EXCHANGE MODELS FOR EFFECTIVE INTERACTIONS

G.E. Brown*

State University of New York
Stony Brook, N.Y. 11794

ABSTRACT

The spin-isospin degrees of freedom; i.e., the giant Gamow-Teller resonances, are found to fit in well with the Boson exchange model of the nucleon-nucleon interaction. In particular, π- and ρ-meson exchange play major roles. Connection is made to the Ericson-Ericson-Lorentz-Lorenz correction in pionic atoms. Effects from the exchange of other mesons are discussed.

1. INTRODUCTION

The Boson exchange model is commonly used to describe nucleon-nucleon interactions. These, in turn, are the starting points for calculations of effective interactions in nuclei. The spin-isospin degrees of freedom have been of particular interest over the past years, chiefly because these are directly connected with pionic interactions. The tensor coupling of the ρ-meson involves the same degrees of freedom, and ρ-exchange enters in important ways to modify the short-range behavior of these interactions.

Many treatises have been written about the nucleon-nucleon interaction, and I shall refer to these for results, where needed. Considerable interest has been awakened recently in phenomena involving virtual $\Delta(1230)$ isobars, especially because of the missing strength in the giant Gamow-Teller resonances. In this talk, I wish to show that there is a coherent picture, known already for some years, connecting pion-nucleus scattering, pionic atoms, the giant

*Supported by the U.S.D.O.E. under contract No. DE-AC02-76ER13001.

Gamow-Teller resonances, and many other phenomena.

At the foundation of our discussion is the picture in which the nucleon and isobar are states of three quarks; i.e., they are members of the same family. This allows us to relate πNN and $\pi N\Delta$ couplings; in practice this is done in the following way, by the use of the transition-spin formalism

$$\delta\mathcal{L}_{\pi NN} = \frac{f}{m_\pi} \bar\psi \, \underset{\sim}{\sigma}\cdot\underset{\sim}{\nabla} \, (\vec\tau\cdot\vec\phi_\pi) \, \psi(x) \tag{1.1}$$

$$\delta\mathcal{L}_{\pi N\Delta} = \frac{f^*}{m_\pi} \bar\Delta \, \underset{\sim}{S}\cdot\underset{\sim}{\nabla} \, (\vec T\cdot\vec\phi_\pi)\psi(x) \tag{1.2}$$

where there are well-defined rules[1] for manipulating the transition spin $\underset{\sim}{S}$ and isospin $\underset{\sim}{T}$. The ratio f^*/f is given in the constituent quark model as

$$\frac{f^*}{f} = \sqrt{\frac{72}{25}} \quad, \tag{1.2}$$

the ratio of $SU(4)$ vector coupling coefficients. In fact, the empirical value of f^*/f, deduced from the width $\Gamma = 110$ MeV of the Δ, is

$$\frac{f^*}{f} \cong 2 \; . \tag{1.3}$$

The empirical enhancement is interesting; it has been suggested[2] that this results from D-state admixtures in the nucleon and in the isobar.

Similar arguments pertain to the tensor coupling of the ρ-meson:

$$\delta\mathcal{L}_{\rho NN} = \frac{f_\rho}{m_\rho} \bar\psi \, [\underset{\sim}{\sigma} \times \underset{\sim}{\nabla}] \, (\vec\tau\cdot\vec\rho) \, \psi(x)$$

$$\delta\mathcal{L}_{\rho N\Delta} = \frac{f_\rho^*}{m_\rho} \bar\Delta \, [\underset{\sim}{S} \times \underset{\sim}{\nabla}] \, (\vec T\cdot\vec\rho) \, \psi(x) \; . \tag{1.4}$$

Since the same $SU(4)$ vector coupling coefficients are involved here as in the pionic couplings, one would have

$$\frac{f_\rho^*}{f_\rho} = 2 \; . \tag{1.5}$$

From the above basic interactions we can obtain a unified quantitative picture of spin-isospin phenomena.

The one-pion and one-rho exchange interactions, fig. 1, are given by

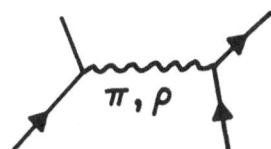

Fig. 1. One-pion or one-rho exchange interaction

$$V_{\pi} = -\frac{f^2}{m_{\pi}^2} (\vec{\tau}_1 \cdot \vec{\tau}_2) \frac{\underset{\sim}{\sigma}_1 \cdot \underset{\sim}{k} \, \underset{\sim}{\sigma}_2 \cdot \underset{\sim}{k}}{k^2 + m_{\pi}^2 - \omega^2} \qquad (1.6)$$

$$V_{\rho} = -\frac{f_{\rho}^2}{m_{\rho}^2} (\vec{\tau}_1 \cdot \vec{\tau}_2) \frac{[\underset{\sim}{\sigma}_1 \times \underset{\sim}{k}][\underset{\sim}{\sigma}_2 \times \underset{\sim}{k}]}{k^2 + m_{\rho}^2 - \omega^2} . \qquad (1.7)$$

It is convenient to decompose V_{π} and V_{ρ} into irreducible tensors

$$V_{\pi} = -\frac{f_{\pi}^2}{m_{\pi}^2}(\vec{\tau}_1 \cdot \vec{\tau}_2)\left[\frac{\underset{\sim}{\sigma}_1 \cdot k \underset{\sim}{\sigma}_2 \cdot k - \frac{1}{3}\underset{\sim}{\sigma}_1 \cdot \underset{\sim}{\sigma}_2 k^2}{k^2 + m_{\pi}^2 - \omega^2} + \frac{1}{3}\underset{\sim}{\sigma}_1 \cdot \underset{\sim}{\sigma}_2 - \frac{1}{3}\underset{\sim}{\sigma}_1 \cdot \underset{\sim}{\sigma}_2 \frac{m_{\pi}^2 - \omega^2}{k^2 + m_{\pi}^2 - \omega^2}\right]$$

$$(1.8)$$

$$V_{\rho} = -\frac{f_{\rho}^2}{m_{\rho}^2}(\vec{\tau}_1 \cdot \vec{\tau}_2)\left[-\frac{[\underset{\sim}{\sigma}_1 \cdot k \underset{\sim}{\sigma}_2 \cdot k - \frac{1}{3}\underset{\sim}{\sigma}_1 \cdot \underset{\sim}{\sigma}_2 k^2]}{k^2 + m_{\rho}^2 - \omega^2} + \frac{2}{3}\underset{\sim}{\sigma}_1 \cdot \underset{\sim}{\sigma}_2 - \frac{2}{3}\underset{\sim}{\sigma}_1 \cdot \underset{\sim}{\sigma}_2 \frac{m_{\rho}^2 - \omega^2}{k^2 + m_{\rho}^2 - \omega^2}\right]$$

$$(1.9)$$

The first term in each pair of brackets gives rise to a tensor-like interaction; the second term, when Fourier transformed, will behave like $\underset{\sim}{\sigma}_1 \cdot \underset{\sim}{\sigma}_2 (\vec{\tau}_1 \cdot \vec{\tau}_2) \, \delta(\underset{\sim}{r}_1 - \underset{\sim}{r}_2)$ in configuration space, and the third term is Yukawa in nature.

With inclusion of short-range correlations between nucleons, the

$\delta(\underset{\sim}{r}_1-\underset{\sim}{r}_2)$ term will become inoperative*; the two nucleons never come to the same point. So this term should be dropped; doing so is known as making the Ericson-Ericson-Lorentz-Lorenz correction. Rather than removing this term from the pion exchange, we find it convenient to leave the pion exchange expression unchanged, and modify V_ρ to

$$(V_\rho)_{mod} = V_\rho + \frac{1}{3} \frac{f^2}{m_\pi^2} (\vec{\tau}_1 \cdot \vec{\tau}_2)(\underset{\sim}{\sigma}_1 \cdot \underset{\sim}{\sigma}_2) \ . \tag{1.10}$$

The reason for grouping in this way is to include all short-range effects of spin-isospin nature in $(V_\rho)_{mod}$.

We have a further modification to make in V_ρ; namely, we have to include effects of short-range correlations (repulsions) originating from ω-meson exchange between nucleons. The ω-meson is coupled very strongly,[3]

$$\frac{g_\omega^2}{4\pi} = 10\text{-}12 \tag{1.11}$$

and exchange of ω-mesons leads to a highly repulsive interaction

$$V_\omega = \frac{g_\omega^2}{4\pi} e^{-m_\omega r}/r \tag{1.12}$$

which anticorrelates nucleons. Note that V_ω has a strength of several GeV at $r \stackrel{\sim}{=} \hbar/m_\omega c$.

Let us consider the Yukawa piece (last term) in V_ρ. Here ω can be neglected compared with m_ρ, since we shall be working with ω's which are $\leq m_\pi$. The Fourier transform of this part of V_ρ gives

$$(V_\rho)_{Yukawa} = \frac{2}{3} \frac{f_\rho^2}{4\pi} (\underset{\sim}{\sigma}_1 \cdot \underset{\sim}{\sigma}_2)(\vec{\tau}_1 \cdot \vec{\tau}_2) \frac{e^{-m_\rho r_{12}}}{r_{12}} \ . \tag{1.13}$$

Modification for effects from ω-exchange can be made in a fairly obvious way[4], defining

$$\hat{V}_\rho = \frac{2}{3} \frac{f_\rho^2}{4\pi} (\underset{\sim}{\sigma}_1 \cdot \underset{\sim}{\sigma}_2)(\vec{\tau}_1 \cdot \vec{\tau}_2) \frac{e^{-m_\rho r_{12}}}{r_{12}} g(r_{12}) \tag{1.14}$$

---

*Such a term follows from point sources for the pion field. As W. Weise will show (this meeting), the finiteness of the pion source dictated by confinement and asymptotic freedom in QCD means that the $\delta(\underset{\sim}{r}_1-\underset{\sim}{r}_2)$ term never occurs.

where g(r) is the two-body correlation function of the two nucleons calculated using (1.12). Since V_ρ and $\hat{V}_\rho$ are of short range, it is often a good approximation to take[5]

$$\hat{V}_\rho = \frac{2}{3} \frac{f_\rho^2}{4\pi} (\underset{\sim}{\sigma}_1 \cdot \underset{\sim}{\sigma}_2)(\vec{\tau}_1 \cdot \vec{\tau}_2) \frac{A}{m_\rho^3} \delta(\underset{\sim}{r}_1 - \underset{\sim}{r}_2) \tag{1.15}$$

with

$$\frac{A}{m_\rho^3} = \int \frac{e^{-m_\rho r}}{m_\rho r} [g(r) - g(o)] \, d^3r \tag{1.16}$$

In principle, one should modify the tensor part of V_ρ also to take into account short-range correlations. Since it chiefly connects relative s and relative d states in nuclei, and since the latter are suppressed at short distances by the centrifugal barrier, the modification is here relatively unimportant, and we neglect it.

Both before and after modification, V_ρ is of short range. Just as in the pion exchange, the $\delta(r_1-r_2)$ piece is inoperative. Modification by inclusion of g(r) amounts, in practice[5], to multiplication of the Yukawa piece by 0.4. Thus, we define an effective interaction

$$(\hat{V}_\rho)_{\text{mod}} = (\vec{\tau}_1 \cdot \vec{\tau}_2) \left\{ \frac{f_\rho^2}{m_\rho^2} [\underset{\sim}{\sigma}_1 \cdot \underset{\sim}{k} \ \underset{\sim}{\sigma}_2 \cdot \underset{\sim}{k} - \frac{1}{3} \underset{\sim}{\sigma}_1 \cdot \underset{\sim}{\sigma}_2 \ k^2]/(k^2+m_\rho^2) \right.$$

$$\left. + [\frac{2}{3} \frac{\hat{f}_\rho^2}{m_\rho^2} + \frac{1}{3} \frac{f^2}{m_\pi^2}] (\underset{\sim}{\sigma}_1 \cdot \underset{\sim}{\sigma}_2) \right\} \tag{1.17}$$

where $\hat{f}_\rho^2 \cong 0.4 \, f_\rho^2$. This interaction should be adequate for the regime of low $\omega \lesssim m_\pi$ and momenta, k << $m_\pi c$. This is essentially the interaction used by Speth (this meeting) in discussing effective spin-isospin forces in nuclei.

2. PION-NUCLEUS SCATTERING

The p-wave pion-nucleus scattering proceeds chiefly through virtual isobar excitation, as shown in fig. 2. As noted in eqs. (1.1) and (1.4), isobars can be included by generalizing $\underset{\sim}{\sigma}$ to $\underset{\sim}{S}$, $\underset{\sim}{\tau}$ to $\underset{\sim}{T}$. One finds easily that the p-wave scattering volume portrayed in fig. 2 is, in nuclear matter,

$$c_o = \frac{f^{*2}}{m_\pi^2} \frac{4}{9} \left[\frac{1}{\omega_R - \omega} + \frac{1}{\omega_R + \omega} \right] = 0.21 \, m_\pi^{-3} \tag{2.1}$$

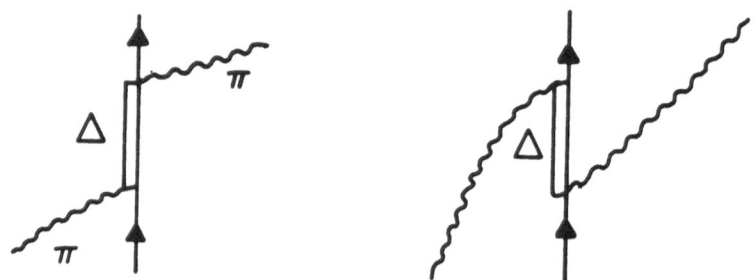

Fig. 2.   Isobar description of pion-nucleon scattering.

where we have used $f^* = 2f$, $\omega_R = 1.9\ m_\pi$. The factor 4/9 comes from
the sum over intermediate isobar states (See § 2.4 of ref. 1).   The
0.21 is the same value as follows from pion-nucleon scattering phase
shifts.

Now the optical-model potential is essentially the pion self-
energy. The $(V_\rho)_{mod}$ will affect this through vertex corrections
(See fig. 3).   One easily sees that this leads to an optical-

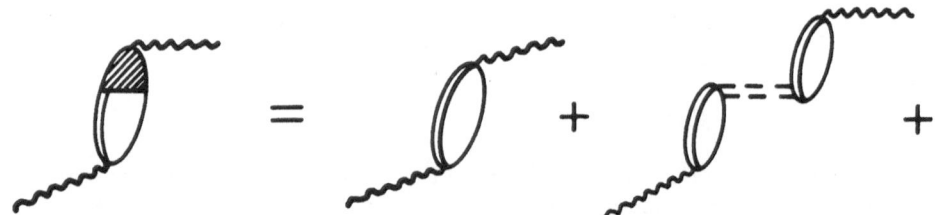

Fig. 3.   Modification of the pionic self energy through vertex
          corrections. The double dashed line represents the
          interaction $(V_\rho)_{mod}$.

model potential[4] for $\omega = m_\pi$ (to be used in pionic atoms)

$$2\omega \ V_{opt} = 4\pi \ \nabla \ \frac{c_o\rho}{1 + \frac{4\pi}{3} c_o\rho \ [1 + \frac{2f_\rho^2 m_\pi^2}{m_\rho^2 f^2}]} \ \nabla \qquad (2.2)$$

Keeping only the 1 in the square brackets in the denominator would amount to keeping the Ericson-Ericson Lorentz-Lorenz interaction. The effect of the $2\hat{f}_\rho^2 \ m_\pi^2/m_\rho^2 f^2$ is not small; this term is equal[4] to 1.6.* So we see that the square bracket is equal to 2.6. Unfortunately, this is much too large for empirical fits; the EELL value does better. But we realize through the above excercise that the second term in the denominator is a Fermi liquid parameter (as Migdal has often emphasized in his pion condensate work) and reflects the properties of all the short-range interactions. The analogy to the Lorentz-Lorenz interaction is made less direct by this realization.

By now we realize that we must sum higher-order terms in the pion self-energy; in particular, there are the isobar self-energy terms shown in fig. 4. A large contribution to the imaginary part

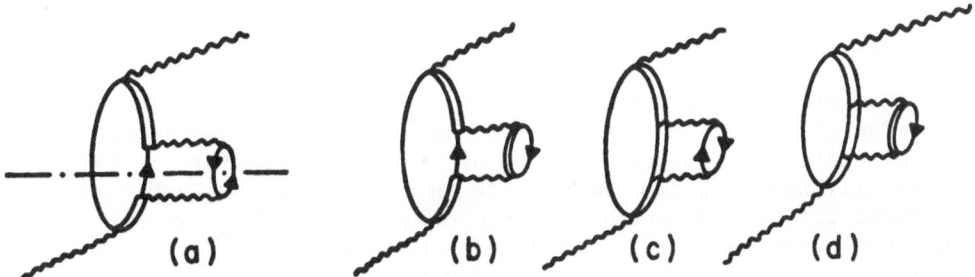

Fig. 4.   Terms in the isobar self-energy.

of the isobar self energy (to the isobar width) comes from the situation where the intermediate state, designated by the dash-dot line in fig. 4, is on shell. The off shell effects in a) and the contributions from b) - d) contribute real terms to the isobar self energy.

---

*We consider the "strong-ρ" coupling of Höhler and Pietarinen (Nucl. Phys. B95 (1975)210) to be well established. This gives $f_\rho^2/m_\rho^2 \cong 2 \ f^2/m_\pi^2$.

There are other important terms[6] contributing to both imaginary and real parts. Although long realized that processes b) – d) are large[1], they are seldom included in calculations. All of the above terms give attraction in the isobar self energy. They give convergent contributions only after cutoffs are used, usually in the form of vertex functions. It is clear that substantial contributions come from intermediate states of very high energy. The isobar self energies of fig. 4 form a geometrical series and can be summed, so that with their inclusion

$$2\omega \, V_{opt} = 4\pi \, \nabla \, \cfrac{c_o \rho}{1 + \dfrac{4\pi}{3} \, c_o \rho \, [1 + \dfrac{2\hat{f}_\rho^2 m_\pi^2}{m_\rho^2 f^2}] - \dfrac{C_o}{c_o} \rho} \, \nabla \tag{2.3}$$

where the C_o comes from the isobar self energy. Just because Re C_o is so cut off dependent, it is hard to obtain it reliably from theory, but we can turn matters around and note that empirically the denominator of (2.3) is roughly $1 + \frac{4\pi}{3} \, c_o \rho$, as would follow from the EELL[3].* This gives us a picture,[3] then, in which the effects from ρ-meson exchange cancel the dispersion corrections in the pionic atom work. It seems clear that these effects cannot be disentangled from each other in the pionic atom work alone; only the combination of EELL, ρ-meson effects and dispersion effects can be obtained.

As noted, processes 4c and 4d are seldom included. In fact, W. Weise and the author have noted that the process 4d) in which the pion is absorbed into a state of two $\Delta(1230)$ isobars is large, especially when the two isobars are in an I=2 state. Each of these isobars can decay into two nucleons, as shown in fig. 4a. This goes some way towards explaining the remarkable results of Schiffer and collaborators[8], who find that pion absorption tends to lead to states of $\sim$ 4 nucleons. Such processes of double isobar production also lead to large isotensor terms in the optical model potential. Just as one main theme of this conference is the importance of single-isobar excitation, I fully expect that future meetings will deal with multiple isobar excitations; certainly the results of ref. 8 seem to indicate this.

---

*From low-energy pion nucleus scattering one sees directly that this must be so. The additional term $(4\pi/3) \, c_o \rho$ in the denominator is just the term needed to remove the multiple scattering[8] for low $\omega \sim m_\pi$. The goodness of impulse approximation for light nuclei and low energies shows the multiple scattering to be small.

3. THE GIANT GAMOW-TELLER RESONANCE*

In working with GTR, we should note that the $\omega = E_{GTR}$ is quite different from that in the pionic atom discussion; in fact, as compared with m_π we can set E_{GTR} equal to zero. But our ω-dependence in equations like (2.1) is explicit, and this is easy to do.

The GTR, to first approximation, involves only spin-flip of the nucleons making it up, but no change in the orbital motion. Thus, to a good approximation, one can choose k=0 in the nucleonic particle-hole interaction making it up. In the way we've grouped interactions, this means $V_\pi \to 0$, and that only the k=0 limit of $(\hat{V}_\rho)_{mod}$ is involved. Now[4,9]

$$\left[\frac{2}{3} \frac{\hat{f}_\rho^2}{m_\rho^2} + \frac{1}{3} \frac{f^2}{m_\pi^2} \right] = 333 \text{ MeV fm}^3. \tag{3.1}$$

To this a screening factor γ must be applied, if we are to use the interaction in only the nucleon sector; this factor is[9] 0.72, so that the square brackets become

$$\left[\frac{2}{3} \frac{\hat{f}_\rho^2}{m_\rho^2} + \frac{1}{3} \frac{f_\pi^2}{m_\pi^2} \right] = 240 \text{ MeV fm}^3 \tag{3.2}$$

In calculating the position of the GTR, we can just as well use the δ-function potential, according to our above arguments. Thus

$$V = V_o \, \sigma_1 \cdot \sigma_2 (\vec{\tau}_1 \cdot \vec{\tau}_2) \, \delta(\underset{\sim}{r}_1 - \underset{\sim}{r}_2) \tag{3.3}$$

$$V_o = 240 \text{ MeV fm}^3$$

and this is the value of V_o needed to position the GTR correctly[10]. Thus, the GTR needs the full $\pi + \rho$ interaction term; the dispersion term in the isobar self energy would not enter here.

In ref. 9 corrections for going beyond a zero-range force are discussed; it is concluded that these corrections are largely cancelled by terms from the ω-dependence of the interaction.

It is amusing that the $\underset{\sim}{k}=0$ interaction is given by our $\underset{\sim}{r}=0$ effective interaction; usually small k means large distances. The connection is, of course, made by our effective interaction and

_____
* Following ref. 9.

results because $V_\pi \to 0$ as $k \to 0$. This should be much more general than our model, following from chiral invariance alone.

Mannque Rho discussed[11] already in 1974 the quenching of axial vector strength through coupling of the low-lying GTR to the high-lying Δ-hole states. The treatment has changed since then only through inclusion of the ρ-meson exchange interaction. In that paper he found

$$(g_A^{eff}/g_A)^\infty \sim 0.76 \tag{3.4}$$

where the superscript ∞ denotes the result for nuclear matter. It would appear from the foregoing that the damping would be much greater than this, since inclusion of the ρ-meson runs up the effect by a factor ~ 2.6; however, Brown and Rho[9] found the above ratio to be

$$\gamma = 0.72$$

for Pb^{208}. Part of this seeming discrepancy comes because Rho used somewhat too large parameters resulting in $c_0 = 0.24\ m_\pi^{-3}$ rather than $0.21\ m_\pi^{-3}$ (See eq. (2.1)). Part of the discrepancy is explained because Brown-Rho[9] took empirical energy shifts in Pb^{208} in order to estimate γ; evidently, Pb^{208} is still some distance from the infinite system. In any case, Brown-Rho[9] ended up with

$$(G_o')_{eff} = 1.68 \ ,$$

the value of the Fermi liquid parameter G_o' to be used when working in the nucleon sector. In the other much used notation this corresponds to

$$g' \stackrel{\sim}{=} 0.6 \ ,$$

a commonly accepted value these days. Note that our model here is the same as that of Speth (this conference).

4. ENERGY DEPENDENCE OF THE INTERACTIONS

One of the most exciting results connected with the GTR is that the spin-isospin interaction holds up well with increasing energy, whereas the $\vec{\tau}_1 \cdot \vec{\tau}_2$ interaction drops rapidly. Thus, energies of 100-200 MeV for (p,n) reactions give a real window through which to observe the spin-isospin degrees of freedom. These general features were foreseen in the works of Love[12] and Petrovich[13]. We shall give a simple model[14] for the energy dependence here. In this model, all of the $\vec{\tau}_1 \cdot \vec{\tau}_2$ interaction and only a small part of the $(\sigma_1 \cdot \sigma_2)\vec{\tau}_1 \cdot \vec{\tau}_2$ interaction come from second-order effects of the tensor force[15].

Let us look at 2nd order effects from V_π alone. Since the tensor interaction introduces mainly intermediate states of high momentum and energy when used in perturbation theory, we can choose some effective energy $\bar{E}$ for the energy of a typical intermediate state[16] and represent the higher-order effects of the tensor interaction by

$$V_{eff}(r) = - \frac{(V_{Tensor})^2}{\bar{E}}$$

$$= - \frac{1}{\bar{E}} (3-2\vec{\tau}_1 \cdot \vec{\tau}_2) [6 + 2\sigma_1 \cdot \sigma_2 - 2S_{12}] V^2_{t\pi}(r) \qquad (4.1)$$

where $V_t(r)$ is the radial part of the pion-exchange tensor force. Here the energies of the initial nucleons have been neglected in comparison with $\bar{E}$, which is taken to be $\sim$ 400 MeV. In actual calculations $V_{t\pi}$ is supplemented by $V_{t\rho}$, which has the opposite sign at short distances and acts as a cut off.

Now in application to (p,n) reactions, we must take into account the energy of the incoming proton (see fig. 5). The generalized expression is:

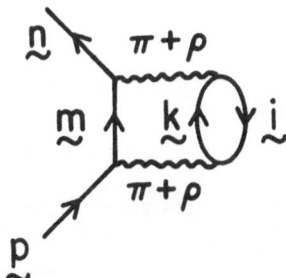

Fig. 5.   Second-order effects of the $V_{\tau\pi} + V_{\tau\rho}$ interaction as they enter into the (p,n) reaction.

$$M(\underset{\sim}{\varepsilon}_p) = \sum_{\underset{\sim}{k},\underset{\sim}{m},\underset{\sim}{i}} \frac{|\,(\underset{\sim}{k}\ \underset{\sim}{m}|V_{\tau\pi} + V_{\tau\rho}\,|\ \underset{\sim}{i}p)\,|^2}{\varepsilon_{\underset{\sim}{p}} - [\varepsilon_{\underset{\sim}{m}} + \varepsilon_{\underset{\sim}{k}} - \varepsilon_{\underset{\sim}{i}}]} \qquad (4.2)$$

where we made the approximation that $k_n = k_p$ which is approximately true for forward scattering. The ε_i, which refer to hole states, are relatively small. Thus, the energy denominator essentially

consists of the difference of the initial energy ε_p and the energy
of the two particles in intermediate states $\varepsilon_m + \varepsilon_k$. As soon as
the energy ε_p becomes comparable with typical intermediate-state
energies, the above sum will develop appreciable positive contributions, tending to cancel the negative ones, which predominate for
small ε_p (and give us the $V_{eff}(r)$ of eq. (4.1) in the limit $\varepsilon_p = 0$).
The above $M(\varepsilon_p)$ has been evaluated[14] by converting the sum
to a principal value integral, using plane-wave intermediate
states, and a simplified matrix element of $V_{\tau\pi}$ and $V_{\tau\rho}$. The latter
amounted to keeping the full $V_{\tau\pi}$ and $V_{\tau\rho}$ in the region where the
sum is negative. The interaction is set equal to zero inside .6 -
.7 fm., where the combined $\pi + \rho$ tensor interaction goes to zero.
The rationale for this is that in this inner region the repulsion
from ω-meson exchange strongly suppresses the interaction. The
resulting curve[14] is shown in fig. 6; comparison with the Love-
Petrovich results shows good agreement.

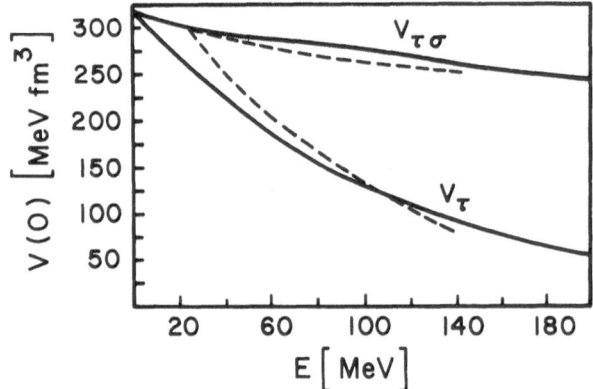

Fig. 6.   Dependence of the isovector coupling potentials
          $V_\tau(q=0)$ and $V_{\tau\sigma}(q=0)$ on the energy of the incoming
          particle energy. Dashed lines give the Love-Petrovich
          (refs. 12,13) interactions.

The effective interaction $V_{\tau\sigma}$ shown here is the one to be used in
the combined nucleon-isobar space. When we ask about excitation
of the GTR in (p,n) reactions, we work in only the nucleon space,
and we must apply the screening correction γ. Taking $\gamma = 0.72$, we
obtain for the ratio

$$R(E_p) = \frac{\gamma\, V_{\tau\sigma}}{V_\tau} \qquad\qquad (4.3)$$

the curves, as function of E_p, labelled by BSW in fig. 7. The
agreement between experiment and the simple model of ref. 14 is
sufficiently good to show that:

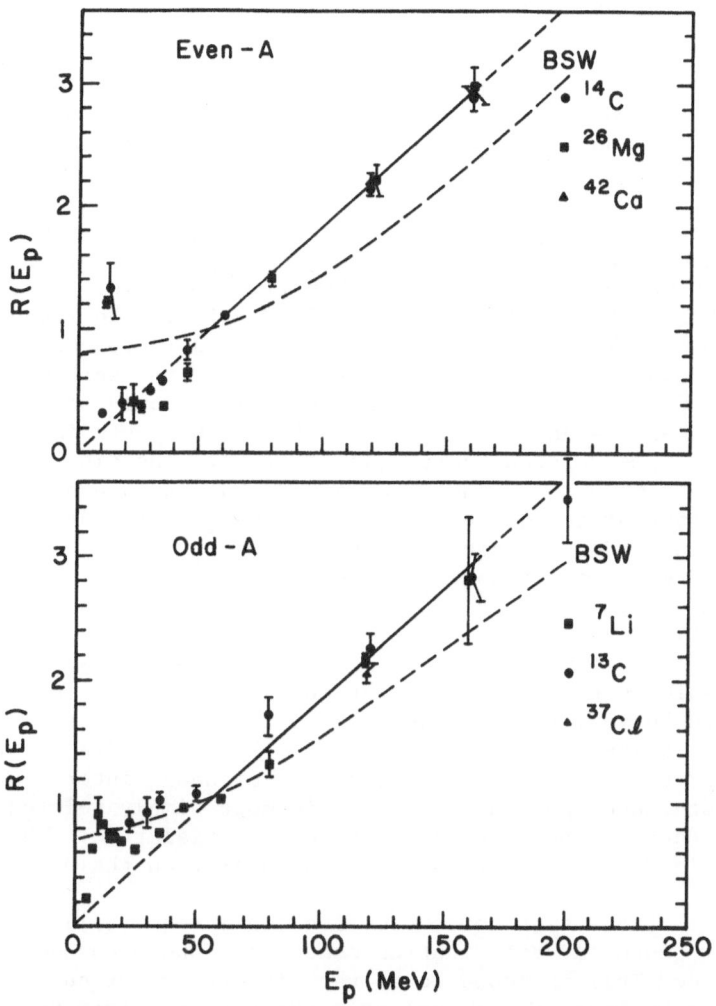

Fig. 7.   Comparison of experimental and theoretical values of
the ratio, eq. (4.3), of spin-isospin to isospin interactions. The experimental points are taken from
Taddeucci et al[17], the theoretical curve is that of
ref. 14, with $V_{\tau\sigma}$ multiplied by $\gamma=0.72$. (See eq.(4.3)).

   (i) The main part of $V_{\tau\sigma}$ comes from lowest-order processes,
π- and ρ- exchange in the model.

   (ii) The main part of V_{τ} comes from second (and possibly
higher) order processes.

In detail the model says that most of V_{τ} comes from second order
effects from the π- and ρ- exchange tensor forces. This is not so
surprising, because the deuteron, in which the tensor force is
operative, is bound, whereas the singlet S-wave nucleon-nucleon
system is not.

   Goodman et al[18] have deduced from an analysis of the experiments volume integrals of the interaction $J_{\sigma\tau}$. The volume integral
of the OPEP would be 122 MeV fm^3; they find for a proton energy
of 120 MeV that the empirical $J_{\sigma\tau} \cong 168$ MeV fm^3. The ratio to
OPEP is nothing like the factor 2.6 found upon introduction of the
ρ-meson exchange. We note, first of all, that the screening
factor γ = 0.72 for virtual isobar excitation should be applied.
Secondly, there is an ∿ 20% reduction in $V_{\tau\sigma}$ strength in going
from 0 to 120 MeV proton energy (see fig. 5). The discrepancy
between the simple theory here and experiment is then not so bad.
Note that most of the decrease in $V_{\tau\sigma}$ comes from the dropping out
of the second-order effects - there is some $\vec{\tau}_1 \cdot \vec{\tau}_2 \; \vec{g}_1 \cdot \vec{g}_2$ piece in
eq. (4.1) - with increasing proton energy.

5. DISCUSSION

   The spin-isospin degrees of freedom in effective interactions
are seen to be very directly related to the Boson exchange inter-
actions; to a lesser extent, so is the $\vec{\tau}_1 \cdot \vec{\tau}_2$ effective interaction.
One's first explanation is that the pion exchange interaction is
the longest-range one; therefore, it is most simply related. This
won't do, because the ρ-meson plays an essential role. Why do the
spin-isospin degrees of freedom seem simpler than the others?

   Part of the answer comes from looking at eq. (4.1). There a
fat scalar-isoscalar contribution comes from the second-order tensor
interaction. This is known to be very important for nuclear saturation, contributions to the nuclear binding energy from this term
dying out with increasing density, for the same reason as the V_{τ}
interaction dies out with increasing energy. But large, and oppos-
ing, contributions in the scalar-isoscalar channel come from the
ω-exchange and from the σ-degrees of freedom (mostly arising from
second-order processes involving virtual Δ(1230) isobars). It is
amusing that the net Fermi-liquid interaction in this channel seems
to come, after all of the above cancellations, from exchange terms
involving V_{π} and V_{ρ}[15]. This is, however, just a way of bookkeeping,
and it is clear that the situation is more complicated than in the
spin-isospin channel.

The most mysterious channel is the $\sigma_1 \cdot \sigma_2$ one. We know of no S=1, I=0 meson which enters into the nucleon-nucleon interaction in an important way. Thus, this interaction would be expected to come from higher-order effects (see eq. (4.1)), from ω-exchange involving small components of the Dirac wave function, exchange terms from other interactions, etc. The situation is clearly complicated. The $\sigma_1 \cdot \sigma_2$ component from (4.1) is of opposite sign to the $\sigma_1 \cdot \sigma_2 \ \vec{\tau}_1 \cdot \vec{\tau}_2$ one; yet at low energy, the effective interactions have the same sign. At higher energies, the $\sigma_1 \cdot \sigma_2$ effective interaction is weak.

There should be a strong $\sigma_1 \cdot \sigma_2$ interaction at large momentum transfers, coming from this component in ω-meson exchange. This is the Fermi-Breit interaction, now so prominent in quark spectroscopy where it results from gluon exchange. Because of the large mass of the ω-meson, this term is of short range, but it should show up at high energies and momenta.

All of our considerations have been within the context of the Boson exchange model. One can ask how quark degrees of freedom will change this picture. The answer, it seems to me, is that they cannot change it much, and if they do, in a given description, then the burden of proof on this description is to describe in the new language the phenomena described here.

COMMENTS

I would like to thank Mannque Rho for years of patient instruction on the problems considered here. I am also grateful to Wolfram Weise for criticism and help. I am thankful to Carl Gaarde, who "had faith" and bugged me for years to give him reasons why the GTR's were not found (I couldn't furnish them to him.). And finally one must remember that it was the experimentalists who turned these beautiful phenomena up, although one should also remember that Fujita[19] predicted the GTR in 1963. In any case, it's an incredibly beautiful story and there's enough glory for all!

REFERENCES

1. G.E. Brown and W. Weise, Phys. Reports 22C (1975) 281.
2. V. Vento, G. Baym and A.D. Jackson, Phys. Lett. 102B (1981) 97.
3. D.O. Riska and B. Verwest, Phys. Lett. 48B (1974) 7.
4. G.E. Brown, Meson-Nuclear Physics (Carnegie-Mellon Conference) AIP Conf. Proc. No. 33, (1976) p. 655.
5. M.R. Anastasio and G.E. Brown, Nucl. Phys. A285 (1977) 516.
6. K. Shimizu and A. Faessler, Nucl. Phys. A333 (1980) 495.
7. E. Oset, H. Toki and W. Weise, Phys. Reports, 83(1982) 281.
8. J.P. Schiffer, Comments on Nuclear and Particle Physics 10 (1981) 243.

9.   G.E. Brown and Mannque Rho, Nucl. Phys. $\underline{A372}$ (1981) 397.

10. C. Gaarde et al, Phys. Rev. Lett. $\underline{46}$ (1981) 902 find V_o =
     245 MeV fm^3.

11. Mannque Rho, Nucl. Phys. $\underline{A231}$ (1974) 493.

12. W.G. Love in The (p,n) Reaction and the Nucleon-Nucleon Force,
     eds. C.D. Goodman et al. (Plenum, New York, 1980), p. 30.

13. F. Petrovich in The (p,n) Reaction and the Nucleon-Nucleon
     Force, eds. C.D. Goodman et al. (Plenum, New York, 1980),
     p. 135.

14. G.E. Brown, J. Speth and J. Wambach, Phys. Rev. Lett. $\underline{46}$
     (1981) 1057.

15. G.E. Brown, in Lecture Notes of the Workshop on Nuclear
     Spectroscopy, Gull Lake, 1979, eds. G.F. Bertsch and
     D. Kurath (Springer-Verlag, 1980) p. 1.

16. T.S.S. Kuo and G.E. Brown, Phys. Lett. $\underline{18}$ (1966) 40;
     G.E. Brown, Unified Theory of Nuclear Models and Forces,
     North-Holland Publ. Co., 3rd Ed., 1971. A factor of 2
     error in the former reference is corrected in the latter.

17. T.N. Taddeucci, et al, Phys. Rev. $\underline{C25}$ (1982) 1094.

18. C.D. Goodman et al, Phys. Rev. Lett. $\underline{44}$ (1980) 1755.

19. K. Ikeda, S. Fujii and J.J. Fujita, Phys. Lett. $\underline{3}$ (1963) 271.

COMMENTS : Δ VERSUS CORE POLARIZATION

Mannque Rho

Service de Physique Théorique
CEN SACLAY
91191 Gif-sur-Yvette Cedex, France

Arima showed in his contribution to this conference the extent
to which a cancellation occurs between the direct and exchange terms
of the nucleon-hole-Δ-hole coupling potential derived from π and ρ
exchange. This seemed to indicate that the coupling to the Δ-hole
states brings about much smaller quenching of the Gamow-Teller
strength than the tensor-force-induced core polarization (call it
TCP in short). Although the issue was discussed *ad nauseam* at the
1976 Erice School on Nuclear Physics[1], I would like to present here
my argument again why the Δ *must be* the principal agent for shifting
a large amount of strength from the low-energy regime to the Δ region.

The most compelling theoretical argument why one should take
seriously the Δ-hole mechanism is its simplicity combined with its
wide range of predictiveness. The Landau-Migdal parameter g_0' in the
Δ-hole channel combined with the well-understood pion-exchange mechanism precludes any precocious pion condensation in nuclear matter, makes it obvious (at least to some of us) why no precursor
phenomenon (or critical opalescence) as recently discussed in the
literature can be relevant in nuclei, explains economically the
observed quenching of g_A in light nuclei and finally accounts for
most, if not all, of the massive shifting of the Gamow-Teller
strength observed in giant Gamow-Teller resonances. As Gerry Brown
discussed in his talk, there is also an intimate connection between
the g_0' mechanism and the Lorentz-Lorenz effect in π-mesic atoms.
Furthermore in this way of looking at things, the treatment can be
given a respectable field theoretic backing, an element clearly
necessary if one wants to go a step closer to being reconciled
with the notion of an ultimate field theory of strong interactions,
QCD. I discussed this aspect in my contribution to this conference:

249

The simple description is amenable to a simple interpretation in terms of quark degrees of freedom.

From a more practical point of view, it makes not much sense to attempt to calculate the g_0' effect in a *microscopic* way by taking only a simple π-and ρ-exchange force. This is a short-ranged interaction, so even in the framework of meson exchanges, many other correlated and uncorrelated mesons must contribute*. In terms of the bag picture, it may be close to overlapping bags. Furthermore there must be lots of cancellations. In particular, there is an almost complete cancellation between the tensor-induced core polarization (TCP) which involves excitation energies comparable to the Δ-N mass difference, the Δ-hole contribution involving (iterated) tensor operators and their normalization corrections as I discussed in the 1976 Erice School. The crucial point here is that one should consider all excitations of $E = M_\Delta - M_N$ *on the same footing* and it is simply inconsistent to treat one separately from others. This argument applies to all other exchange-current operators as well[2].

Gerry Brown discussed how one can get a reasonable value for g_0' by just taking the direct term of the π-and ρ-exchange force[3]. It is not impossible that this is just an accident. But it is more plausible that whatever other contributions there may be suffer a large cancellation. If this were not so, there would be no compelling reason why g_0' should be "universal", e.g., same in the ph-ph, Δh-ph and Δh-Δh channels, as demanded by the quark picture. (For instance, there is no way that the exchange terms can leave the "universality" relation intact). Whether such a cancellation does actually occur is not going to be easy to verify for the reason mentioned above. But to make the point, I will also use as Arima did the paper of Towner and Khanna[4], the only realistic microscopic treatment that I know of**. Specifically, consider the Gamow-Teller transitions for the A=39 and 41 nuclei. Referring to the Table 1 (quenching in percent) of Towner and Khanna, the relevant rows are 2 to 8. Arima combines the rows 4 and 5 to obtain what he considers to be the complete "Δ-hole" contribution. That is smaller than the row 2 which corresponds to the TCP (tensor-induced core polarization) ; i.e.

---

*
 Such as $2\pi, \pi\rho, A_1$, etc, particularly in the exchange term in microscopic derivations.
**The incompleteness of such a calculation cannot be overemphasized, however and one should look at the following illustration with a great deal of caution.

		A = 39	A = 41
TCP (Row 2)	:	−16.1	−11.4
"Δ-hole" (Rows 4&5)	:	− 3.1	− 4.1
Others (Rows 3,6,7,8)	:	7.7	6.3
Total	:	−11.5	− 9.2

One may look at this in a different way as proposed by Oset and Rho[5]. It is clear from Gerry Brown's discussion[3] that the row 4 corresponds roughly to the $g_o' \approx 0.6$, so I shall single it out and lump all the rest together* ; i.e.

		A = 39	A = 41
g_o' (Row 4)	:	−11.8	−9.8
Rest (Rows 2,3,5,6,7,8)	:	0.3	0.6
Total	:	−11.5	−9.2

Clearly what is left after the g_o' term is taken out is negligible and for heavy nuclei, can be simply dropped. The total is also quite close to the Δ-h contribution found by Oset and Rho for $g_o' \approx 0.6$.

That the TCP (tensor-force-induced core polarization) cannot be the main mechanism for quenching can be seen in the quenching pattern observed in experiments. One expects that TCP is fairly independent of mass number and particular transitions, which is not at all the case. To give one example, the triton β-decay would come out wrong if the TCP mechanism alone were invoked, as stressed in ref.4. Speth discusses more on this point in his contribution.

I am grateful for discussions with Gerry Brown from whom I first learned about the various cancellations, on which my arguments rely. He is of course not responsible for whatever wrong applications I may have made of his original ideas.

---

*
Note that they are all of same order from the point of view of the ranges involved.

REFERENCES

1.   See the lectures of A.Arima and M.Rho, in 1976 Erice Lectures
     on Nuclear Physics, Prog. in Particle and Nucl.Phys. $\underline{1}$, 105
     (1978).

2.   M.Rho, Nucl.Phys. A$\underline{354}$, 3c (1981).

3.   Also G.E.Brown and M.Rho, Nucl.Phys. A$\underline{372}$, 397 (1981).

4.   I.S.Towner and F.C.Khanna, Phys.Rev.Lett. $\underline{42}$, 51 (1979).

5.   E.Oset and M.Rho, Phys.Rev.Lett. $\underline{42}$, 42 (1979).

IMPLICATIONS OF DENSITY DEPENDENCE IN THE EFFECTIVE INTERACTION FOR THE EXCITATION OF SPIN MODES

James J. Kelly[1] and James A. Carr[2]

[1] Department of Physics and Laboratory for Nuclear Science
Massachusetts Institute of Technology
Cambridge, MA 02139

[2] Department of Physics
Florida State University
Tallahassee, FL 32306

INTRODUCTION

A substantial fraction of the research efforts dedicated to intermediate energy nucleon scattering has sought to ascertain the quantitative accuracy of a description of these phenomena in terms of one-body operators. Several important components of such a description are: 1) the properties of the free two-nucleon interaction; 2) the modifications of this interaction in the nuclear environment; and 3) the structure of the target nucleus. For the present discussion, we assume that the relevant aspects of the structure of the target are accurately known by independent means, such as electron scattering. We shall concentrate on the sensitivity of nucleon inelastic scattering and charge exchange to the properties of the free two-nucleon interaction and the modifications due to the nuclear density.

The importance of strong density dependence in the isoscalar spin-independent central component of the two-nucleon effective interaction near 150 MeV has already been demonstrated for normal parity isoscalar transitions whose spin and current contributions are negligible.[1,2] As shall be shown here, these effects are considerably larger than the ambiguities associated with the parameterization of the free two-nucleon interaction. The implications of density dependence in the effective interaction for the excitation of spin modes have not yet been as thoroughly studied. We shall show that the ambiguities in the free interaction are larger than the density dependent effects expected for spin excitations.

An important consideration in this study is the spin-isospin selectivity of nucleon-nucleus scattering. Inelastic scattering and charge exchange reactions allow one to study more varied, and generally more selective, linear combinations of scattering amplitudes than are sampled by nucleon-nucleon experiments. This selectivity is most useful in those cases where the observables are dominated by only one or two components of the effective interaction. Therefore, one may exploit the spin-isospin selectivity of nucleon-nucleus reactions to distinguish between alternative potentials which produce essentially equivalent descriptions of nucleon-nucleon scattering.

A second consideration is the greater range of momentum transfer available to nucleon-nucleus scattering. The center-of-mass momentum transfer available to the nucleon-nucleon system is limited to the incident laboratory momentum. Twice this momentum transfer is available to nucleon-nucleus scattering. Therefore, if the one-step reaction mechanism is reliable at larger momentum transfer, nucleon-nucleus scattering can be exploited to study the two-nucleon potential with finer spatial resolution.

Finally, the radial variation of the density of a finite nucleus can be exploited to study the detailed density dependence of the medium corrections to the two-nucleon effective interaction. We shall show that a reasonable description of these medium corrections is obtained from the local density approximation (LDA), in which the effective interaction in the finite system is assumed to be similar to that appropriate to infinite nuclear matter with the density in the vicinity of the interacting nucleons. Given this description, various nuclear excitations whose amplitudes are localized at different nuclear radii provide probes of the medium modifications appropriate to the nuclear densities at these radii. For example, the transition densities for monopole or isoscalar electric dipole transitions are concentrated in the high density nuclear interior where the medium corrections should be largest, while the transition densities for high spin states tend to be concentrated in the low density tails of the nuclear density distribution, where the effective interaction should be most similar to the free interaction.

MEDIUM CORRECTIONS

In principle, the two-nucleon effective interaction appropriate to scattering from an A-nucleon system is complex, nonlocal, energy dependent, and depends upon all the degrees of freedom of the (A+1) nucleons. Given the success of mean field theories, it is not unreasonable to suppose that the many-body modifications of the short-ranged two-nucleon interaction may be adequately described by the

local properties of the nuclear medium in the vicinity of the interacting pair. This is the essence of the local density approximation.
We assume that the effective interaction between an energetic nucleon
and a nucleon bound in a finite system is essentially the same as
that appropriate to infinite nuclear matter with the local density.
That is, we assume that in the intermediate energy regime, the
properties specific to the finite system are effectively decoupled
from the properties characteristic of the underlying nuclear medium.

In the formulation of Brieva and Rook,[3] the first step in the
construction of the effective interaction is the solution of a
Bethe-Goldstone equation modified so as to describe the correlated
pair wavefunction ψ of a continuum nucleon and a bound nucleon.
The density dependence arises through Pauli blocking. The Brueckner-
Hartree-Fock optical potential is included. The final solution is
self-consistent. The effective interaction is then constructed to
satisfy

$$t\phi = v\psi$$

where ϕ is the uncorrelated wavefunction and v is the two-nucleon
potential. Thus, observation of medium corrections to the effective
interaction may be interpreted as observation of the correlated pair
wavefunction in nuclear matter.

The effective interaction is then averaged over the Fermi distribution. The result is a local, density- and energy-dependent
effective interaction with central, spin-orbit, and tensor components.[4]

The original effective interaction constructed by Brieva and
Rook was based upon the Hamada-Johnston (HJ) potential.[3-5] They
and von Geramb parameterized this interaction in a form almost identical to that of the Love-Franey (LF) parameterization of the free
t-matrix.[6] A second G-matrix based upon the Paris potential[7] was
constructed by von Geramb.[8] The results presented in this work are
based upon these effective interactions.

That medium corrections to the two-nucleon effective interaction
can be vital to the interpretation of inelastic scattering data is
illustrated in Fig. 1, which compares, at 140 MeV, the isoscalar
spin-independent central components of the G-matrix based upon the
HJ potential with the LF free t-matrix. For momentum transfers less
than about 1.5 fm^{-1}, the low density limit of the HJ interaction is
similar to the LF interaction. As the density increases, the low-q
attraction is suppressed and the high-q repulsion is enhanced. At
saturation density, the forward scattering amplitude is reduced by
about a factor of three by the Pauli blocking. The high-q enhancement reflects the more pronounced wound in the correlated wavefunction at high density. This wound is simulated in the effective
interaction by enhanced repulsion.

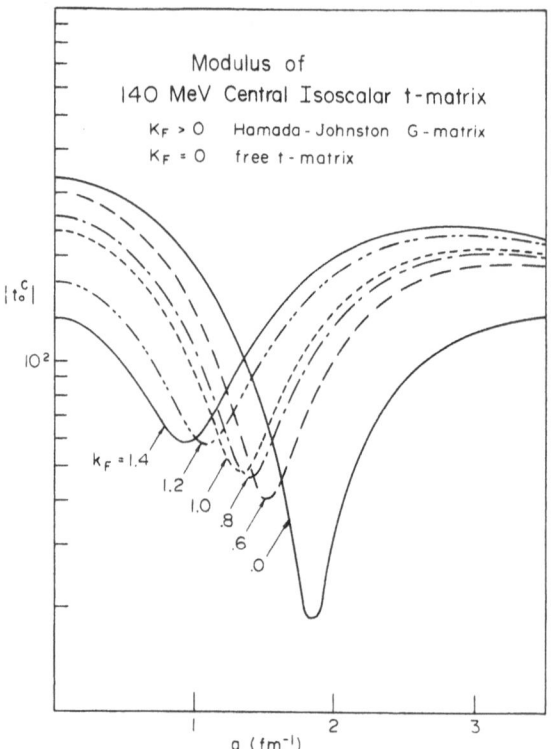

Fig. 1.   Isoscalar spin-independent central components of the
effective interaction at 140 MeV.

These phenomena are illustrated in Fig. 2, which compares calculations with data for the excitation of the first 1^- state of ^{16}O
by 135 MeV protons.[1,2] The impulse approximation (IA) uses the LF
interaction. The LDA calculations use the central HJ interaction
plus the LF spin-orbit interaction. Note that the isoscalar spin-
orbit interaction is predicted to show very little density dependence.[3] For both the IA and the LDA, the long-dashed curves use
distorted waves obtained from a phenomenological optical potential
fitted to elastic scattering data. The solid curves use consistent
distorted waves obtained from the microscopic optical potential
generated by the same interaction that induces the inelastic transition. The contribution of the central interaction, in the absence of the spin-orbit interaction, to the inelastic cross section
is given by the short-dashed curves.

The IA overestimates the low-q cross section by as much as a
factor of four, and then badly underestimates the high-q cross

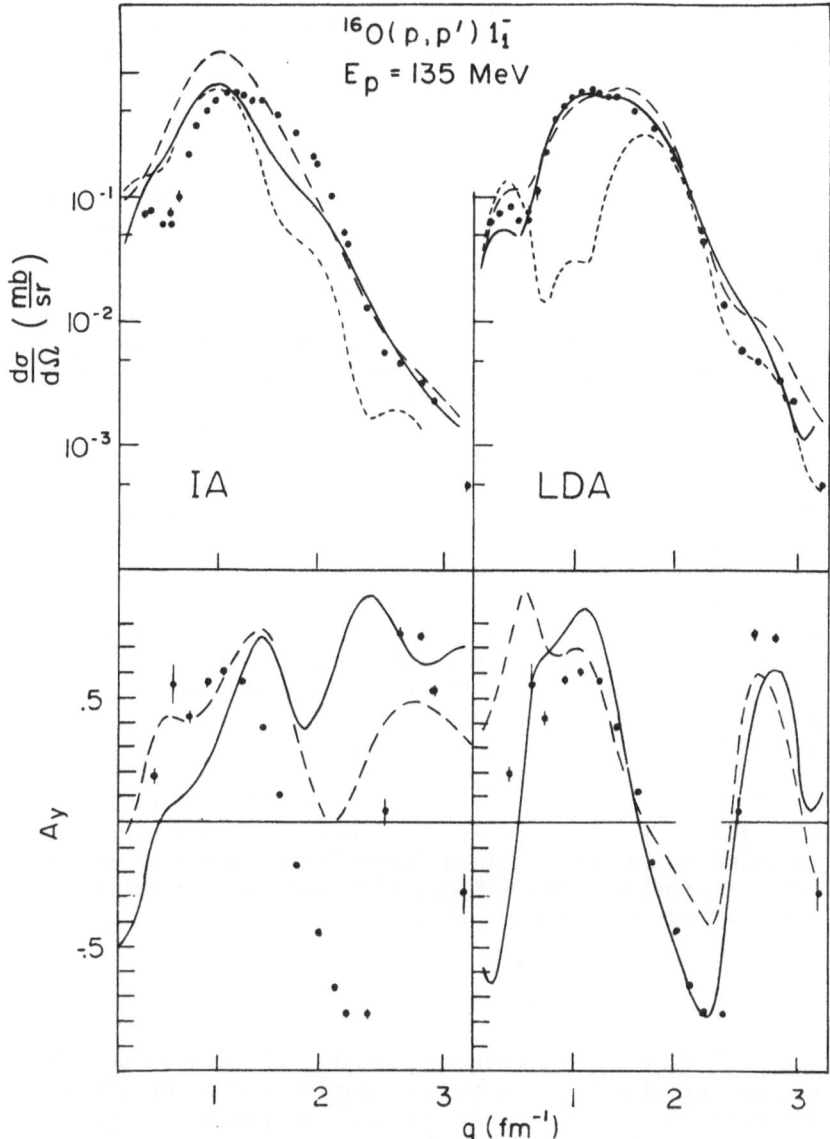

Fig. 2.   Calculations in the IA and LDA for excitation of 1^-, in
^{16}O with 135 MeV protons. Data are from Refs. 1 and 2.
Note that the horizontal scale is linear in angle, not
momentum transfer.

section. The IA analyzing power prediction completely fails to
reproduce the strong negative analyzing power observed near 2.4 fm^{-1}.
The central depth of the IA optical potential is 2-3 times that of
the phenomenological optical potential. The exaggerated absorption
present in the IA optical potential appears to improve the forward
angle cross section, but does nothing to improve the larger momentum transfer cross section while seriously degrading the analyzing
power prediction.

The LDA cross section and analyzing power predictions, on the
other hand, are both in fine agreement with the data. The central
contribution to the forward angle LDA cross section prediction is
greatly suppressed by Pauli blocking, in agreement with the data.
The repulsive maximum of the central interaction is greatly enhanced,
also in agreement with the data. The enhanced high-q repulsive maximum of the central interaction reproduces the strong negative analyzing power. Note that the LDA predictions are less sensitive to
the choice of optical potential because the microscopic optical potential based upon a density dependent effective interaction is more
similar to the phenomenological optical potential. In fact, the use
of a consistent microscopic optical potential improves the agreement
between the LDA predictions and the data.

These signatures of density dependence are characteristic features of the excitation of normal parity isoscalar states by intermediate energy nucleons and are by no means unique to this state or
nucleus.[1,2] The 1⁻ state was chosen for this comparison because its
transition density peaks in the high density nuclear interior where
the sensitivity to medium corrections is largest.

Therefore, we conclude that the data show definitive evidence
for strong density dependence in the isoscalar spin-independent central component of the two-nucleon effective interaction and that
this density dependence is well described by nuclear matter theory
and the local density approximation.

SENSITIVITY TO INPUT POTENTIAL

We recall that the particular formulation used to deduce the
effective interaction began with a potential model for the free two-
nucleon interaction.[3] How sensitive are the results to the initial
choice of potential?

The isoscalar spin-independent central (t_{oo}^{c}) component of the
140 MeV effective interactions resulting from the HJ[3,4] and Paris[8]
potentials are compared in Fig. 3. The low density limit of the
Paris interaction is similar to the LF interaction over a much
larger range of momentum transfer than is the low density limit of
the HJ interaction. However, this is not a strong argument in favor

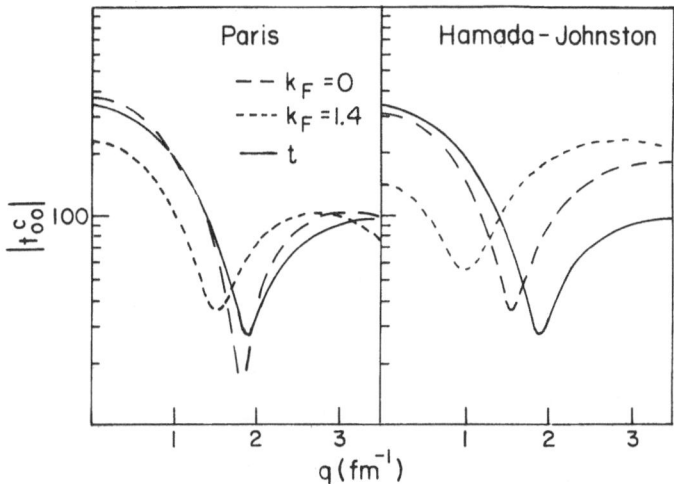

Fig. 3. Comparison of the isoscalar spin-independent central com-
         ponents of the effective interaction obtained from the
         HJ and Paris potentials at 140 MeV.

of the Paris potential because the LF interaction is not necessarily
reliable at these larger momentum transfers. Furthermore, the momentum transfer available to the nucleon-nucleon system is limited
to 2.5 fm^{-1}. We also note that the density dependence resulting
from the Paris potential is somewhat less pronounced than that arising from the HJ potential.

The difference between these interactions, particularly at
larger momentum transfer, reflects the difference between the short-
range repulsive cores of the input potentials. The hard core of the
HJ potential produces a more pronounced repulsive maximum than does
the soft core of the Paris potential.

The sensitivity of inelastic scattering predictions to the
difference between these potentials is illustrated in Fig. 4, which
compares the consistent LDA predictions for the excitation of the
first 1^- state of ^{16}O by 135 MeV protons using both the HJ and Paris
potentials. The high-q repulsive maximum of the HJ interaction
appears to be too strong, while that of the Paris interaction appears
to be too weak. That the high-q repulsion of the Paris interaction
is too weak is also evident in the analyzing power prediction, which
fails to become sufficiently negative at the momentum transfer corresponding to the repulsive maximum of the central interaction. For
surface transitions, it is even more apparent that the low density
limit of the HJ interaction is too repulsive at high-q. Therefore,
the data appear to favor a potential whose repulsive core is intermediate between the extreme hard core of the HJ potential and the
soft core of the Paris potential.

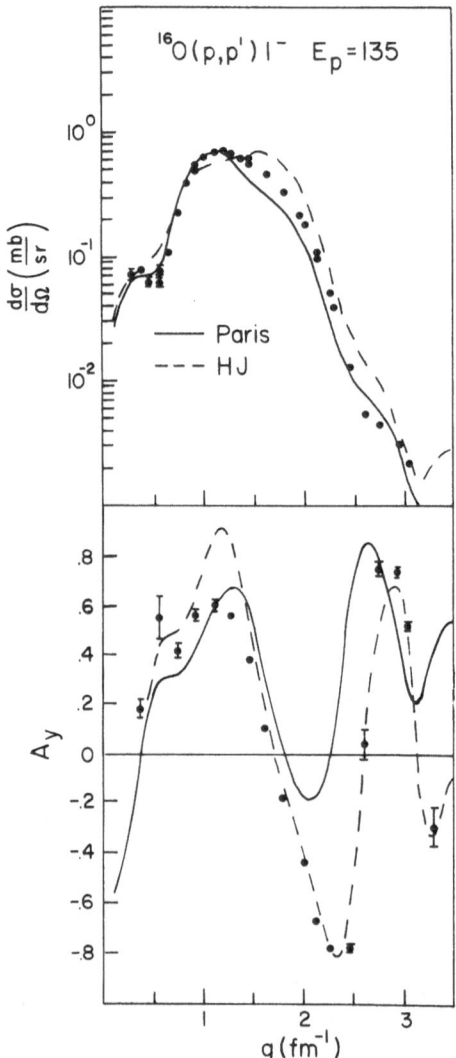

Fig. 4.   Calculations with the HJ and Paris effective interactions
for excitation of 1⁻, in ¹⁶O with 135 MeV protons.

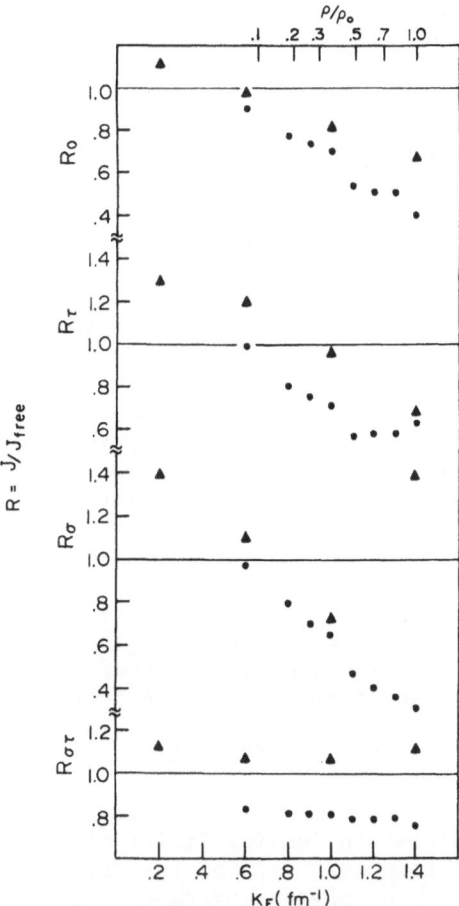

Fig. 5. Ratio of volume integrals of the central effective interaction to the free interaction for the HJ (circles) and
Paris (triangles) potentials shown as a function of
density.

This sensitivity may be further clarified by reformulating the construction of the nuclear matter effective interaction in terms of a t-matrix or in terms of scattering amplitudes fitted directly to the nucleon-nucleon data. In this way, the influence of medium corrections could be examined in a more direct manner, dispensing with the intermediate step of constructing a potential. The ambiguities associated with the choice of such a potential description would then be reduced, although the ambiguities associated with the off-shell extrapolation would remain.

The medium corrections to the central components of the effective interactions derived from these potentials are compared in Figs. 5 and 6. The ratio between the volume integrals of each central component of the effective interaction and the corresponding component of the LF free t-matrix is shown as a function of density in Fig. 5 for both the HJ and Paris potentials. Fig. 6 compares the momentum transfer dependence of the central components of the Paris and LF interactions. The spin-orbit components are shown in Fig. 7.

The isoscalar spin-dependent central term (t_σ^c) appears to have a strong, but irregular, density dependence. This particular component of the effective interaction is very weak, and was not well determined by the Love-Franey parameterization.[6] The nuclear matter calculation of this quantity is plagued by very severe cancellations. Therefore, the significance of the detailed behavior of this term is not clear. We also note that no intermediate energy scattering data reported to date are sufficiently sensitive to this term to discriminate between these various estimates. It appears that the most promising probes of this term are $0^+ \rightarrow 0^-$ T=0 transitions. At small q, the sign of the t_σ^c fitted by the LF interaction and that predicted by the Paris potential are opposite. For the LF interaction, the central and tensor contributions to the $0^+ \rightarrow 0^-$ cross section interfere constructively, while they interfere destructively for the Paris G-matrix. However, the tensor force is so dominant that the sensitivity to t_σ may not be sufficient, even in this case, to make a definitive statement.

The low and saturation density limits of the isovector spin-independent central component (t_τ^c) of the Paris G-matrix bracket the LF t-matrix at low-q, suggesting that their predictions for forward angle excitation of isobaric analog states will not differ substantially, despite the relatively strong density dependence. For momentum transfers larger than about 1.3 fm^{-1}, there is a substantial difference between the Paris and LF interactions. However, this difference is dominated by the difference between the low density limit of the Paris G-matrix and the free t-matrix, rather than by density dependence. A similar commentary applies to the isovector spin-orbit component. Therefore, large momentum transfer (p,n) data

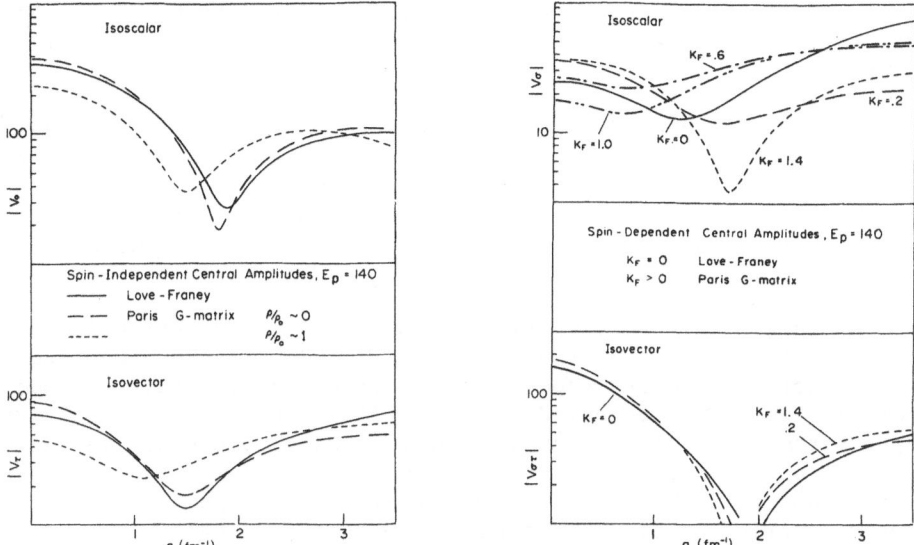

Fig. 6. Comparison of central spin-independent (left) and spin-
         dependent components of the Paris effective interaction
         to the LF t-matrix at 140 MeV.

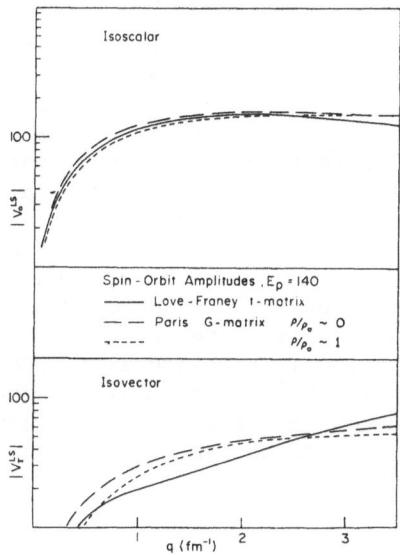

Fig. 7. Comparison of spin-orbit component of Paris effective
         interaction to the LF t-matrix at 140 MeV.

for isobaric analog states will be more sensitive to the input potential than to medium corrections. In this case, the ambiguity in the
input potential is more important than medium corrections.

The density dependent effects in the isovector spin-dependent
central component ($t_{\sigma\tau}^c$) are predicted to be small. However, there
is a large discrepancy between the volume integrals of the Paris
and HJ interactions. The 0° (p,n) data for Gamow-Teller transitions
favor a volume integral intermediate between the Paris and HJ interactions, similar to the LF interaction.[9]

At the time of this writing, a tabulated parameterization of
the tensor components of the G-matrix is not available. Suffice it
to say that the density dependence predicted for the tensor components is also small compared to the ambiguities associated with the
input potential.[8] As the tensor force tends to dominate abnormal
parity transitions, these small effects may be accessible to experiment if sufficiently accurate systematic comparisons between electron scattering, proton scattering, and charge exchange are performed.

SPIN MODES

Finally, we come to the ostensible topic of this talk, the
implications of density dependence for the excitation by nucleons
of spin modes in nuclei. For normal parity isoscalar transitions
whose spin and current contributions are negligible, it is very
important to include the medium modifications to the effective
interaction. In these cases, the medium modifications are so large
that they overshadow the ambiguities associated with the choice of
input potential. For isobaric analog transitions, the relative
importance of medium corrections is reduced, but still significant.
The excitation of spin modes, on the other hand, is dominated by
components of the effective interaction whose density dependence is
predicted to be much smaller than the ambiguities associated with
the initial assumptions concerning the free nucleon-nucleon interaction. Therefore, in studying the components of the two-nucleon
effective interaction applicable to the excitation of spin modes,
the dominant consideration is the nature of free interaction rather
than its modification in the nuclear medium. The spin-isospin
selectivity of nucleon-nucleus reactions provides a valuable method
to discriminate between alternative descriptions of the free two-
nucleon interaction.

CONCLUSIONS

Several components of the free two-nucleon effective interaction
are subject to substantial modifications in the nuclear environment.

The qualitative features of the strongest of these density dependent effects are well described by the local density approximation and are relatively insensitive to the initial assumptions concerning the free two-nucleon interaction. The medium modifications of those components of the interaction that are effective in the excitation of spin modes are smaller than the ambiguities in the free two-nucleon interaction. A large variety of potential models can describe nucleon-nucleon data equally well. There is no uniqueness to the potential description. It will be of great interest to exploit the spin-isospin selectivity of nucleon-nucleus reactions in a unified analysis of nucleon-nucleon and nucleon-nucleus data. The goals of this analysis would be to reduce the ambiguities associated with the description of nucleon-nucleon scattering and to distinguish between otherwise equivalent descriptions. It will, of course, be necessary to consider the medium modifications in the context of a local density approximation.

REFERENCES

1. J. Kelly, W. Bertozzi, T. N. Buti, F. W. Hersman, C. Hyde, M. V. Hynes, B. Norum, F. N. Rad, A. D. Bacher, G. T. Emery, C. C. Foster, W. P. Jones, D. W. Miller, B. L. Berman, W. G. Love, and F. Petrovich, Phys. Rev. Lett. 45:2012 (1980).
2. J. Kelly, Ph.D. thesis, MIT (1981).
3. F. A. Brieva and J. R. Rook, Nucl. Phys. A291:299, 317 (1977) and A297:206 (1978).
4. H. V. von Geramb, F. A. Brieva, and J. R. Rook, in: "Microscopic Optical Potentials," H. V. von Geramb, ed., Springer-Verlag, Berlin (1979).
5. T. Hamada and D. Johnston, Nucl. Phys. 34:382 (1962).
6. W. G. Love and M. A. Franey, Phys. Rev. C24:1973 (1981).
7. M. Lacombe, B. Loiseau, J. M. Richard, R. Vinh Mau, J. Cote, P. Pires, and R. de Tourreil, Phys. Rev. C21:861 (1980).
8. H. V. von Geramb, private communication.
9. C. Gaarde, these proceedings.

MODELS FOR PION-NUCLEUS REACTIONS

Frieder Lenz

SIN
CH-5234 Villigen
Switzerland

The theoretical concepts, to describe the pion-
nucleus dynamics, have been developed almost exclusively
within the context of investigations of elastic pion-
nucleus scattering. The large body of experimental data
on nuclear excitations in pion scattering has had little
impact on our theoretical understanding of the pion
nucleus interaction. First order theories, generally
used to describe these processes, are too poor in their
dynamical content. As is well known by now from the study
of elastic scattering, there is no theoretical basis for
a description of pion-nucleus scattering in terms of free
pion-nucleon scattering. Pion nucleus elastic scattering
can only be understood if one allows for large modifications
if the pion nucleon interaction takes place in the nuclear
environment. For pion scattering in the resonance region, the
Δ-h formalism has proved to be particularly useful to
account for these medium corrections. In the first part
of this talk, I shall give a brief summary of the results
obtained in applying the Δ-h formalism to elastic pion
nucleus scattering.

In the second part I shall present a modified form
of the distorted wave impulse approximation which is
based on the Δ-h formalism. In contrast to first order
theories, all the medium modifications present in elastic
scattering are incorporated into the transition operator.
Δ-propagation and Δ-nucleus interaction give rise to
excitation mechanisms which are specific for the pion-

nucleus system. Results for excitations in ^{12}C will be shown.

Although this description takes into account medium modifications of the one-body transition operator, the theory requires the presence of two (and many) body components in the transition operator which in contrast to elastic scattering cannot be taken into account, even on a phenomenological level, by an effective one body operator. Both,experimental evidence for and theoretical estimates of such two body pieces are discussed in the third part of this talk.

1. SUMMARY OF PION-NUCLEUS ELASTIC SCATTERING IN THE Δ-h FORMALISM

In all of the following qualitative discussions, I shall keep only the 3-3 part of the pion-nucleon inter-action. The numerical results will include scattering via π-N background amplitudes as well and the pion-nucleus Coulomb-interaction. In the Δ-h formalism, the pion-nucleus optical potential is written in the form

$$\mathcal{U}(\vec{p}',\vec{p}) = t_{\circ} \, \upsilon(p') \sum_{j_h \mu_h \tau} \int d^3x \, d^3x' \, \psi^*_{j_h \mu_h \tau}(\vec{x}) \, e^{-i\vec{p}'\vec{x}'} \, \vec{S}(\vec{p}' - \alpha\vec{\nabla}/i)$$

$$G_{\Delta h}(\vec{x}',\vec{x}) \, \vec{S}^+(\vec{p} + \alpha\overleftrightarrow{\nabla}/i) \, e^{i\vec{p}\vec{x}} \, \psi_{j_h \mu_h \tau}(\vec{x}) \, \upsilon(p) \qquad (1a)$$

and represented by the following diagram

$$\mathcal{U}(\vec{p}',\vec{p}) = \qquad \text{----} \bigcirc \text{----} \; . \qquad\qquad (1b)$$

The interpretation of the structure of eq. (1) is quite obvious. At the point $\vec{x}$, the incident pion of momentum $\vec{p}$ collides with the nucleon in the single particle orbit $(j_h \, \mu_h \, \tau)$ The formation of the 3-3 resonance is described by the π N-Δ vertex function $\vec{S}^+\vec{\kappa}$ v (κ), with κ the relative pion-nucleon momentum. In eq. (1), the pion nucleon relative momentum has been expressed explicitly in terms of pion and nucleon momentum in an approximate way with

$$\alpha = E_\pi / M_N \; . \qquad\qquad (1c)$$

The propagation of the Δ-h state from $\vec{x}$ to $\vec{x}'$ is described by $G_{\Delta h}(\vec{x}'\vec{x})$. At the point $\vec{x}'$ the Δ-h state decays back into pion and nuclear groundstate.

With the distorted waves $\Psi_{\vec{k}}^{(\pm)}(\vec{p})$ generated ·in the optical potential μ of eq. (1a), the elastic T-matrix is given by

$$T_o(\vec{k}',\vec{k}) = \int d^3p'd^3p \; \delta^3(\vec{k}'-\vec{p}') \; \mathcal{U}(\vec{p}',\vec{p}) \; \Psi_{\vec{k}}^{(+)}(\vec{p}) \qquad (2a)$$

which is represented in the following way

$$T_o(\vec{k}',\vec{k}) = \quad --\times--\langle\!\!\!\!\!\rangle------ \qquad (2b)$$

where the cross in the pion line denotes the distortion.

The central quantity which determines the optical potential is the Δ-h Greensfunction. In the context of elastic scattering any complicated pion-nucleus or equivalently Δ-nucleus interaction can be ultimately interpreted as a modification of the Δ-h Greensfunction.

The obviously simplest description results if the Δ-h Greensfunction is replaced by a constant $(G_{\Delta h}(\vec{x}',\vec{x}) \sim \delta(\vec{x}'-\vec{x}))$. The corresponding optical potential is the first order static pion nucleus potential. In the application of the Δ-h formalism to elastic scattering, the following form of the Δ-h Greensfunction has been adopted[1].

$$G_{\Delta h}^{-1} = E - R + i\Gamma/2 - H_{\Delta h} \qquad (3a)$$

$$H_{\Delta h} = T_\Delta + H_{A-1} + W_P + W \qquad (3b)$$

R and $\Gamma/2$ are position and width of the free πN resonance; for simplicity we have neglected the effects related to the energy dependence of these resonance parameters.

The Δ-h Hamiltonian of eq. (3b) accounts for Δ-propagation and Δ-nucleus interaction. H_{A-1} is the Hamiltonian of the (A-1) nucleus; W_p describes Pauli-quenching effects, i.e. modifications of the Δ self-energy arising from the dissociation into the π-N channel in the presence of the residual nucleus. W is the Δ-nucleus interaction which is parametrized as

$$W = W_o \, \rho(r)/\rho(0) \; + \; W_{LS}^\Delta \; \vec{L}_\Delta \cdot \vec{I}_\Delta \; f(r) \qquad (3c)$$

i.e. W contains a central interaction following the density of the nucleus and a spin orbit interaction with a surface peaked radial dependence. While Wp can be calculated microscopically, essentially nothing is known about the Δ-N interaction which would allow us to predict

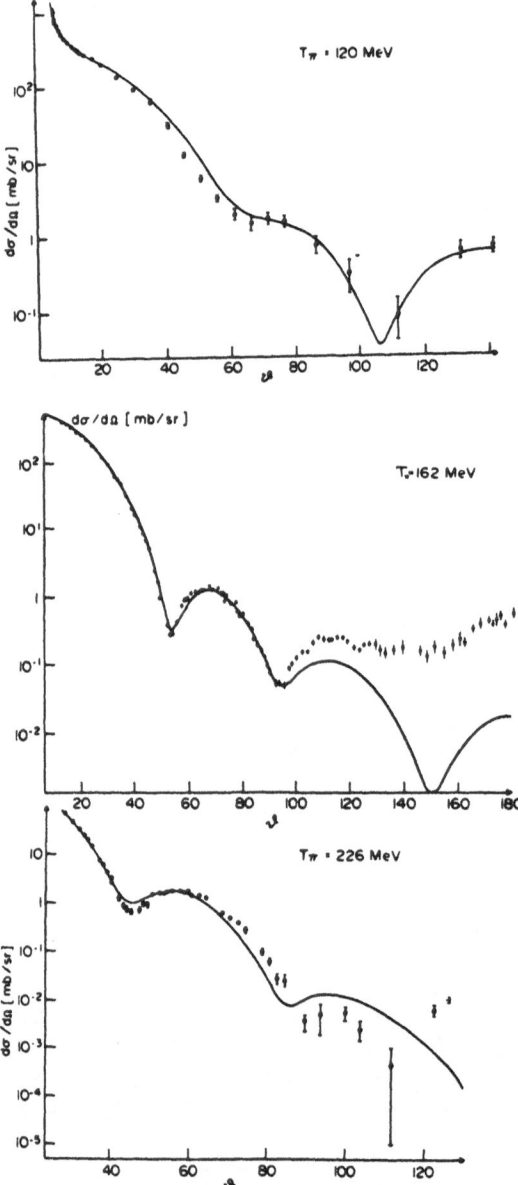

Fig. 1: π-^{12}C elastic angular distributions.
Data from ref. 2.

the Δ-nucleus interaction. The strength parameters W_0
and W_{LS}^Δ are determined from fits to pion-nucleus elastic
data. The average values of these strength parameters
obtained from analysis of pion scattering on ^{4}He, ^{12}C
and ^{16}O, in the resonance region are

$$W_0 \approx (-30 - i\,50)\,MeV \qquad\qquad (4a)$$

$$W_{LS}^\Delta \approx W_{LS}^N \qquad\qquad (4b)$$

i.e. the central Δ-nucleus interaction is attractive
with about half the strength of the nucleon shell model
potential and strongly absorptive. The Δ-nucleus spin
orbit force has the same sign and is of similar strength
as the nucleon-nucleus spin orbit force.

In Figure 1 we show the comparison between experimental and theoretical angular distributions for
π-^{12}C scattering in the resonance region. The Δ-h
formalism apparently provides a semi-quantitative
description of the data. With regard to the discrepancies,
I emphasize the order of magnitude disagreement in the
backward cross sections at 162 MeV and the missing
diffractive structure at the energies below the resonance.
I have no explanation for these discrepancies which are
very stable against variations in the phenomenolgically
determined strength parameters.

With the strength parameters (4) of the Δ-nucleus
interaction, pion nucleus elastic scattering has yielded
some qualitatively new data on the strong interaction.
For the interpretation of these strength parameters and
specifications of the Δ-nucleus interaction, both experimental data other than elastic ones and theoretical
models of the Δ-nucleus interaction are required. So far,
only the large imaginary part of the central Δ-nucleus
interaction has been explained phenomenolgically in a
satisfactory way, while both central attraction and
the spin orbit force has found no generally accepted
interpretation.

The large imaginary part of the Δ-nucleus interaction must be associated with true pion absorption via
inelastic Δ-N collisions ΔN ↔ NN. The other source of
inelasticity, nucleon knock-out in elastic Δ-N collisions
ΔN ↔ ΔN, cannot contribute more than by a few MeV to
the imaginary part of the Δ-nucleus potential for
energies at and below the resonance. With the assumption

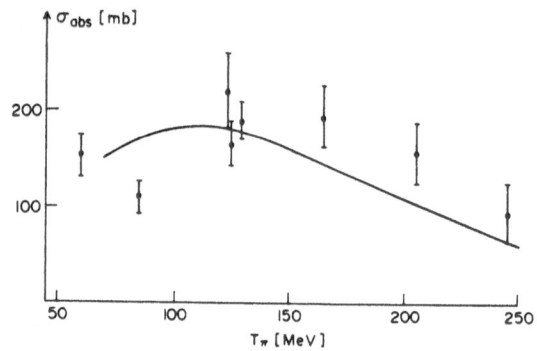

Fig. 2: $\pi-^{12}C$ absorption cross section as a function
of pion kinetic energy. Data from ref.3.

that the imaginary part of the Δ-nucleus is generated
completely through coupling to the absorption channel,
total pion absorption cross section can be predicted.
The theoretical results agree surprisingly well with the
experimental data as Figure 2 shows.

This result allows us to interpret the imaginary part
of the central interaction at a more fundamental level.
Assuming that Δ-N s-waves are most important for the
absorption process, which is supported experimentally
from absorption data on the deuteron and ^{4}He and
suggested theoretically by the small imaginary part of
the p-wave generated Δ-nucleus spin orbit force (eq.3c),
the imaginary part of W_0 in eq. (4a) can be interpreted
as arising from the Δ-N interaction

$$t = t_0 \, \delta(\vec{r}_\Delta - \vec{r}_N) \, P_{T=1} \, P_{S=2} \qquad (5a)$$

and the value of W_0 in (4c) translates after correction
for exchange processes into[4]

$$\mathrm{Im} \, t_0 \approx -1000 \ \mathrm{MeV} \ \mathrm{fm}^3 . \qquad (5b)$$

No direct experimental indication exists which would
allow to associate in a similar way the real part of the
Δ-nucleus interaction (4a,b) with definite channels in
the Δ-N interaction. I shall discuss in the last section,

how, in specific nuclear transitions, the interference
between first order processes and two step processes induced by the Δ-N interaction may be used to investigate
the dynamics which generates the real part of W.

2. PION NUCLEUS INELASTIC SCATTERING IN THE DISTORTED WAVE APPROXIMATION TO THE Δ-h FORMALISM

Nuclear transitions in pion nucleus scattering are
described in complete analogy to elastic scattering[5].
The same medium modified π-N t-matrix or equivalently
the same dressed Δ-h propagator appears in the
description of nuclear transitions. The distorted waves
are generated in the optical potential (1) and appear
now both in the initial and final states. Diagrammatically
a p-h excitation is given by

$$T_{fi}(\vec{k}',\vec{k}) = \quad \text{—×—} \quad \begin{array}{c} j_p \\ j_{\bar{k}'} \end{array} \tag{6a}$$

and corresponds to the following formal expression

$$T_{fi}(\vec{k}',\vec{k}) = \int d^3p' d^3p \; \psi_{\vec{k}'}^{(-)*}(\vec{p}') \, t_{fi}(\vec{p}',\vec{p}) \, \psi_{\vec{k}}^{(+)}(\vec{p}) \tag{6b}$$

with the undistorted amplitude

$$t_{fi}(\vec{p}',\vec{p}) = t_0 \, v(p') \sum_{\mu\tau} \int d^3x' d^3x \; \psi_{j_p,M+\mu,\tau}^{*}(\vec{x}') \, e^{-i\vec{p}'\vec{x}'} \, \vec{S}(\vec{p}' - \alpha\vec{\nabla}/i)$$

$$\cdot \, G_{\Delta h}(\vec{x}',\vec{x}) \, \vec{S}^{+}(\vec{p} + \alpha\vec{\nabla}/i) \, e^{i\vec{p}\vec{x}} \, (1+\tau_z/2) \, \psi_{j_h,\mu,\tau}(\vec{x})$$

$$(-1)^{j_h-\mu-\tau+1/2} \langle j_p, M+\mu; j_h-\mu \,|\, JM \rangle \langle \frac{1}{2}\tau; \frac{1}{2}-\tau \,|\, T0 \rangle .$$

$$\tag{6c}$$

Eq. (6c) describes the particle-hole excitation with
quantum number $(j_p \; j_h^{-1})$ and T in the scattering of π^+
on a closed shell nucleus. In contrast to the elastic
amplitude the Δ-h decays at the point $\vec{x}'$ to the particle-
hole state and pion rather than to the nuclear ground
state. To emphasize it again, the same medium modifications
which have been found to be important in elastic
scattering appear in the transition operator for nuclear

excitations as well and not only implicitly in the
distorted waves. This distinguishes essentially the
theoretically motivated Δ-h phenomenology from a purely
pragmatic phenomenological treatment of the pion-nucleus
optical potential. In the latter case, it is not possible
to infer from empirically determined optical potential
parameters the structure of the transition operator in
pion-nucleus reactions.

The physics content of eq. (6) will be displayed
both in analytical calculations and by results of
numerical evaluations of expression (6) for specific
nuclear transitions. The principal tool in the analytical
investigations is the non-locality expansion of the
nuclear density matrix in eq. (6c). In this approximation
it is assumed that the nuclear transition density matrix
$\Psi_f^*(\vec{x}')\,\psi_i(\vec{x})$ varies only little over the Δ propagation
distance which controls the behaviour of the Δ-h Greens-
function $G_{\Delta h}(\vec{x}',\vec{x})$. Using the coordinates

$$\vec{R} = \tfrac{1}{2}\,(\vec{x}+\vec{x}') \tag{7a}$$

$$\vec{r} = \vec{x}-\vec{x}' \tag{7b}$$

the non-locality expansion of the transition density
matrix reads

$$\psi_f^*(\vec{x}')\,\psi_i(\vec{x}) \approx \psi_f^*(\vec{R})\psi_i(\vec{R}) + \tfrac{\vec{r}}{2}\left[\psi_f^*(\vec{R})(\vec{\nabla}-\overleftarrow{\nabla})\psi_i(\vec{R})\right]+\dots \tag{8}$$

Only these first two terms of the expansion are kept
in the following.

With the momenta

$$\vec{q} = \vec{p}-\vec{p}' \tag{9a}$$

$$Q = \tfrac{1}{2}(\vec{p}+\vec{p}') \tag{9b}$$

conjugate to the variables (7), the amplitude (6c) is
expressed as

$$t_{fi}(\vec{p}'|\vec{p}) = \tfrac{2}{3}\,v(p')t_0 \int d^3R\,d^3r\; e^{i(\vec{q}\vec{R}+\vec{Q}\cdot\vec{r})}$$

$$\cdot\left\{\left[\rho_{fi}(\vec{R})+im\,\vec{r}\,\vec{J}_{fi}(\vec{R})\right]\left[\vec{p}\cdot\vec{p}'+\alpha/i(2\vec{Q}\vec{\nabla}_r-\tfrac{1}{2}\vec{q}\,\vec{\nabla}_R)\right]\right.$$

$$-\tfrac{i}{2}\left[\vec{S}_{fi}(\vec{R})+im\,\vec{r}\,\vec{U}_{fi}(\vec{R})\right]\left[\vec{p}'\times\vec{p}+\alpha/i(\vec{Q}\times\vec{\nabla}_R-\vec{q}\times\vec{\nabla}_r)\right]\Big\}$$

$$\cdot g_{\Delta h}(\vec{r},\vec{R})\,v(p). \tag{10}$$

At this level of the non-locality expansion, 4 different
nuclear structure form-factors appear as a result of
different modes of interaction in the pion nucleus
system. These basic reaction mechanisms are:

1. Coupling to the transition density ρ_{fi} (longitudinal
coupling)

$$\rho_{fi}(\vec{R}) = \langle f | \sum_{i}^{A} (1+ \tau_z(i)/2) \, \delta^3(\vec{R}-\vec{r}_i) | i \rangle \qquad (11\,a)$$

2. Coupling to the convection current $\vec{J}_{fi}$ (vector coupling)

$$\vec{J}_{fi}(\vec{R}) = \frac{1}{2m} \langle f | \sum_{i}^{A} [\vec{p}_i \delta^3(\vec{R}-\vec{r}_i) + \delta^3(\vec{R}-\vec{r}_i)\vec{p}_i](1+\tau_z(i)/2) | i \rangle \quad (11\,b)$$

3. Coupling to the spin density or magnetization current
$\vec{S}_{fi}$ (vector coupling)

$$\vec{S}_{fi}(\vec{R}) = \langle f | \sum_{i} (1+ \tau_z(i)/2) \, \vec{\sigma}_i \, \delta^3(\vec{R}-\vec{r}_i) | i \rangle \qquad (11\,c)$$

4. Coupling to the spin-flux tensor (tensor coupling)

$$\overleftrightarrow{U}_{fi}(\vec{R}) = \frac{1}{2m} \langle f | \sum_{i} [\vec{p}_i \vec{\sigma}_i \delta^3(\vec{R}-\vec{r}_i) + \delta^3(\vec{R}-\vec{r}_i) \vec{p}_i \vec{\sigma}_i](1+\tau_z(i)/2) | i \rangle \quad (11\,d)$$

In addition to the 3 standard form-factors also present
in electro-excitation, with the spin-flux tensor a new
nuclear structure form-factor appears here. This form-
factor arises from both spin and nucleon momentum dependent terms in the effective π-N interaction. It
characterizes the spatial changes of the magnetization
current in the nuclear transition and is related to
the spin-density in the same way as the convection
current is to the matter density. It satisfies an
analogous continuity equation

$$\vec{\nabla} \cdot \overleftrightarrow{U}_{fi}(\vec{R}) = -i\,(E_f - E_i)\, \vec{S}_{fi}(\vec{R}). \qquad (12)$$

This classification not only makes explicit the dependence
of the transition amplitude on nuclear structure, it also
yields some important selection rules for pion induced
nuclear excitations. The basis of this selection rules
are transformation properties of the transition amplitudes
corresponding to one of the 4 excitation mechanisms under
the two symmetry operations:

$$S_1 : \quad \vec{p} \leftrightarrow \vec{p}' \qquad \sim \qquad \begin{array}{c} \vec{q} \to -\vec{q} \\ \vec{Q} \to \vec{Q} \end{array} \qquad (13a)$$

$$S_2 : \quad \vec{p} \leftrightarrow -\vec{p}' \qquad \sim \qquad \begin{array}{c} \vec{q} \to \vec{q} \\ \vec{Q} \to -\vec{Q} \end{array} \qquad (13b)$$

Transformation properties of the 4 nuclear structure form-factor under the parity operation $\vec{R} \rightarrow -\vec{R}$ (e.g. $\rho_{fi}(-\vec{R}) = (-1)^J \rho_{fi}(\vec{R})$) imply definite transformation properties of the corresponding amplitudes in (10) under the operations $S_{1,2}$ (13).

Particularly simple results are obtained from these symmetry considerations for forward and backward amplitudes if differences in the initial and final distorted waves are neglected. In the following Table, the angular dependences around 0° and 180° are given for the amplitudes corresponding to the 4 coupling modes (11).

	J	T^ρ_{fi}, T^u_{fi}	T^j_{fi}, T^s_{fi}	T^u_{fi}	T^j_{fi}, T^s_{fi}
0°	even	$\cos\theta$	$\sin\theta$	$\sin\theta$	$\sin^2\theta$
	odd	$\sin\theta$	$\cos\theta$	$\sin^2\theta$	$\sin\theta$
180°		$\cos\theta$	$\sin\theta$	$\sin^2\theta$	$\sin\theta$
		natural parity		unnatural parity	

The same $S_{1/2}$ transformation properties of longitudinal T^ρ_{fi} and tensor coupling T^u_{fi} on the one hand and the vector couplings T^j_{fi} and T^s_{fi} on the other hand imply that in the unpolarized cross section, only amplitudes corresponding to tensor and longitudinal coupling interfere with each other as well as the amplitudes corresponding to vector coupling. There is no interference between T^j_{fi} and T^s_{fi} with either T^ρ_{fi} or T^u_{fi}. This implies that in an appropriate coordinate system, the two classes of coupling modes populate different magnetic substates.

So far only very general properties of the amplitude (10) have been used. To study in closer detail how the nuclear excitation via the various interaction modes (11) arises in pion scattering, additional approximations are necessary. In particular a simple model for the Δ-h Greensfunction $g_{\Delta h}(\vec{r},\vec{R})$ in (10) is needed. I assume that the Greensfunction $g_{\Delta h}$ is given by the Δ-propagator in a uniform medium

$$g_{\Delta h}(r, \vec{\mathcal{R}}) = -2M^*/4\pi \; e^{i\vec{\mathcal{R}}r}/r \qquad\qquad (14a)$$

with

$$\vec{\mathcal{R}}^2 = 2M^*(E - R + i\Gamma/2 - \overline{W})$$

where $\overline{W}$ is supposed to represent the medium corrections as described by the Δ-h Hamiltonian (3b). Because of the technical complexity, I will not discuss here any details of the coupling to the spin-flux tensor.

With this extremely simple model (14), the amplitude (10) can be evaluated in closed form

$$t_{fi}(\vec{P}'_1\vec{P}) \approx \tfrac{2}{3} t_0 \, \upsilon(P') \frac{2M^*}{\vec{\mathcal{R}}^2 - Q^2} \, \upsilon(P) \cdot$$
$$\cdot \left[\alpha^\ell \rho_{fi}(\vec{q}) + m \, \vec{Q} \, (\alpha^s \, \vec{J}^{\,s}_{fi}(\vec{q}) + \alpha^c \, j_{fi}(\vec{q})) \right] + t^{\,u}_{fi}(\vec{P}'_1\vec{P})$$
$$\qquad\qquad (15)$$

with the following expression for the parameters α^i characterizing the coupling strength to the various nuclear form-factors

$$\alpha^\ell(\vec{P}_1\vec{P}) = \vec{P}'\vec{P} - 2\alpha Q^2 \qquad\qquad (16a)$$

$$\alpha^s(\vec{P}'_1\vec{P}) = 1 - \alpha \qquad\qquad (16b)$$

$$\alpha^c(\vec{P}_1\vec{P}) = 2\left(\frac{\vec{P}'\vec{P} - 2\alpha Q^2}{\vec{\mathcal{R}}^2 - Q^2} - \alpha \right) \qquad\qquad (16c)$$

The magnetization current has been introduced in the standard way

$$2m \, i \, \vec{Q} \, \vec{J}^{\,s}_{fi}(\vec{q}) = (\vec{P}' \times \vec{P}) \, \vec{S}_{fi}(\vec{q}) . \qquad\qquad (17)$$

Longitudinal coupling arises from the non-spin flip part of the π-N amplitude. α^ℓ reflects the characteristic p-wave dependence. Similarly, the coupling to the magnetization current is due to the spin-flip piece of the pion-nucleon amplitude. The convection current coupling finally is generated through the dependence of initial and final πN-Δ vertices on the nucleon momenta as well as the dependence of the Δ-propagation on the initial nucleon momentum. Typical for these Δ-propagation induced couplings is the occurence of the square of the Δ-propagator in (15) and (16) or more generally, of the energy derivative of the pion-nucleon amplitude; this makes the effective coupling constant in convection current dominated excitations strongly energy dependent in the resonance region.

Formally, the structure of eq. (15) and (16) is very similar to the corresponding expression in electron scattering. Nuclear density and nuclear convection and magnetization currents couple to the pion density $\psi_{\vec{k}'}^{(-)*}(\vec{p}')\,\psi_{\vec{k}}^{(+)}(\vec{p})$ and pion current $\psi_{\vec{k}'}^{(-)*}(\vec{p}')[(\vec{p}+\vec{p}')/2]\psi_{\vec{k}}^{(+)}(\vec{p})$ respectively; the energy and momentum dependent form-factors of these couplings are proportional to $v(p')v(p)\alpha^1(\vec{p}',\vec{p})/(\bar{H}^2-Q^2)$ and therefore determined by the medium modified π-N amplitude. I emphasize difference in the vector coupling in π- nucleus scattering and electro-excitation. In electro-excitation, coupling to convection and magnetization current are comparable in strength for a $\Delta T = 0$ transition, $(\alpha^c = 1/2, \alpha^s = 1/2(\mu_p+\mu_n))$, while $\Delta T = 1$ transitions are dominated by coupling to the magnetization current $(\alpha^c = 1/2, \alpha^s = 1/2(\mu_p-\mu_n))$. In pion excitations, the strength of the two vector couplings are comparable and apart from an overall normalization, independent of the isospin of the excited state. The relative phase between the two couplings, however, varies strongly with the pion energy.

I now turn to a discussion of results for inelastic pion scattering on ^{12}C obtained from a direct numerical evaluation of the expressions (6b,c). In these calculations none of the approximations discussed above has been used and, in addition to the 3-3 part of the π-N interaction, excitation via background π-N partial waves is included. Note, that in contrast to elastic scattering, there is no free parameter in the description of the excitations. Still remaining uncertainties are due to the partly unknown nuclear structure of the transition. These nuclear structure uncertainties are minimal in ^{12}C, where longitudinal and transverse form-factors are separately known for most of the transitions[6]. It follows from the qualitative discussions above, that, in general, (π,π') cross sections will not exhibit simple patterns in their energy and angular dependences unless a specific excitation mechanism is singled out by the corresponding nuclear structure form-factor.

Examples of such simple excitations are provided by the 3 low-lying states in ^{12}C. They represent almost pure longitudinal excitations. Cross sections for pion scattering to the 3^- state at 9.6 MeV are shown in Figure 3. Here as in all of the following examples,

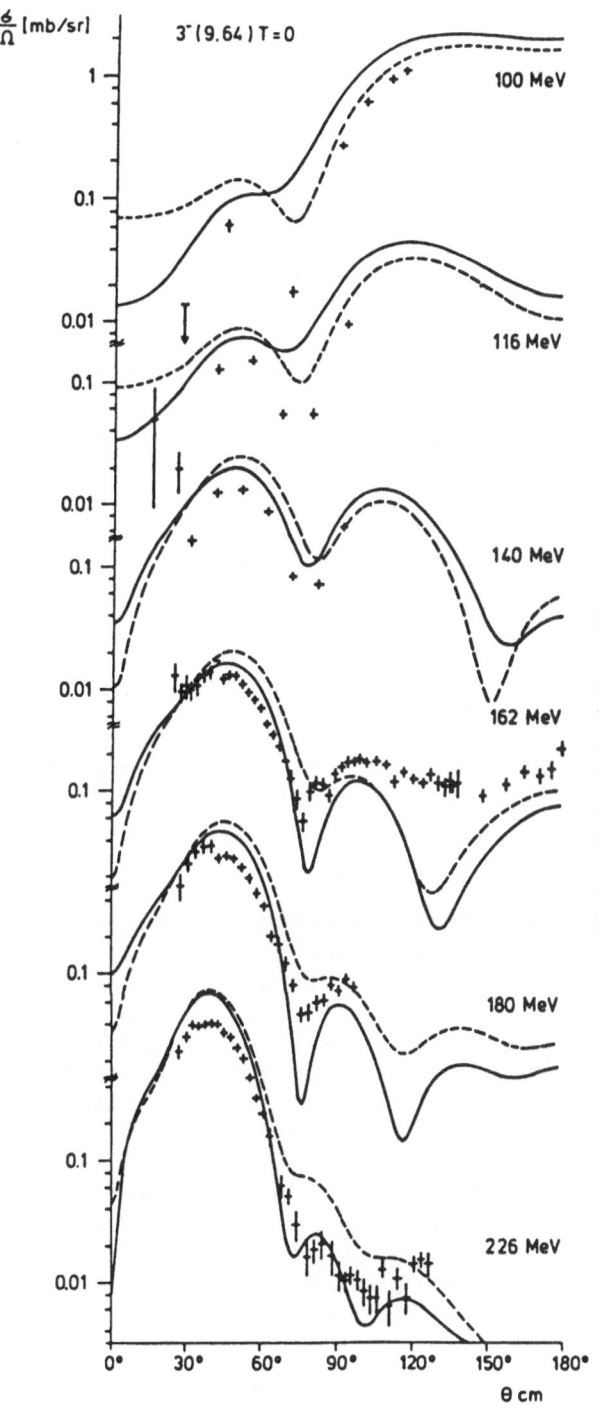

Fig. 3:

Angular distribu-
tions for the 3^-
excitation in ^{12}C.
Solid line:dominant
convection current;
dashed line: para-
metrization with
destructively inter-
fering convection
and magnetization
current contribu-
tions to the trans-
verse form factor.
Data from ref. 7.

the excited state is described by superposition of p-h
amplitudes with weights determined by fitting both, the
longitudinal and transverse form-factors. The two
theoretical curves in Figure 3 correspond to parametrizations which yield almost identical longitudinal
and transverse form-factors. In one case (solid line),
the transverse form-factor is dominated by convection
current coupling, in the other case the form-factor is
generated by destructive interference between convection
and magnetization currents.

Because of differences in the relative phase between
magnetization and convection current coupling in pion
reactions and electroexcitation respectively, no such
destructive interference occurs in pion scattering. Thus
uncertainties in the interpretation of the transverse
form-factors lead to non negligible effects in pion
scattering although the transverse form-factor is here
smaller by more than an order of magnitude than the
longitudinal one and as in electroexcitation, there is no
interference between longitudinal and vector couplings
in pion scattering. Qualitatively, the pion cross section
exhibits the characteristics of a longitudinal excitation,
the structure of the angular distribution and its
variation with energy can be understood as a result of
the interplay between the momentum transfer dependence
of the nuclear form-factor on the one hand and the
angle dependence ($\cos\theta_{CM}$) of the π-N amplitude on the
other hand.

Qualitatively similar results are obtained for the
two other low-lying excitations[5]. Within typically
20-30 %, the experimental cross sections are reproduced
in the first maximum. Like in elastic scattering (Fig. 1),
systematic discrepancies between theory and experiment
are observed in the backward cross sections at 162 MeV
where the theoretical values are, by up to an order of
magnitude, smaller than the experimental ones.

Very different patterns in both the size and the
angular dependence of the pion cross section are encountered, if the longitudinal excitation mechanism is
not any more dominant. Such an example is the 2^+ T = 1
excitation in ^{12}C at 16.1 MeV. In Figure 4 the two
curves correspond to p-h parametrizations of the
transverse form-factor in terms of magnetization current
(solid curve) only and convection current only (dashed

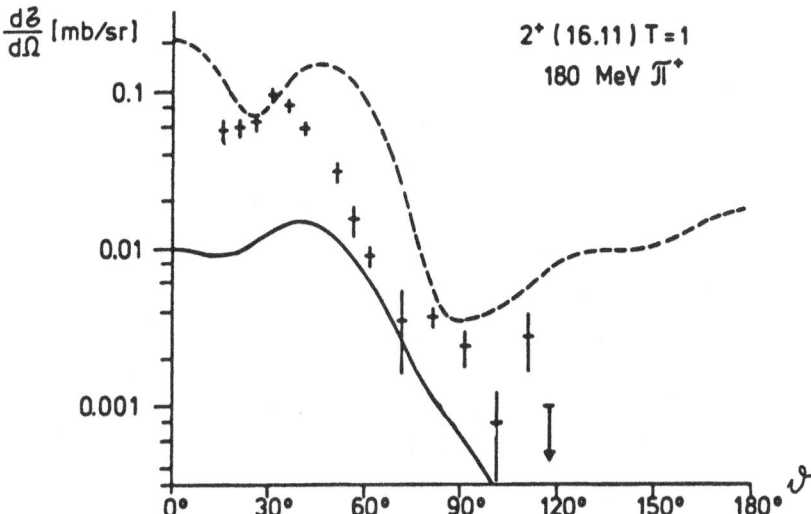

Fig. 4: Angular distributions for the $2^+T=1$ exciation in
^{12}C (16.1MeV). Prametrization of the transverse form-factor
in terms of magnetization current only (solid curve) and
convection current only (dashed curve). Data from ref. 7.

curve). The large differences obtained in pion scattering
reflect the order of magnitude difference in the relative
strength of the magnetization and convection current
coupling constants for $\Delta T = 1$ transitions in pion and
electron scattering respectively.

As a last example, Figure 5 shows the cross section
for the unnatural parity excitation of the $2^- T = 0$
state at 11.8 MeV in ^{12}C. Unpolarized cross sections,
as well as the magnetic substate populations are shown
for the two standard parametrizations of the electron
scattering transverse form-factor. Here, magnetization
current and convection current parametrizations yield
pion scattering cross sections of similar size due to
the similar relative strength of α^S and α^C for $\Delta T = 0$
transitions in pion and electron scattering. In the
two parametrizations the $|M| = \pm 1$ substate cross
sections are similar, but there are large differences
in the $|M| = 2$ cross sections. This difference arises
from coupling to the spin-flux tensor. It can be shown
that, with the momentum transfer as quantization axis,
vector couplings can populate only the $|M| = 1$ components
while the spin-flux tensor populates the $|M| = 2$
components. Here, as in other transitions, large contributions from this tensor coupling are obtained if

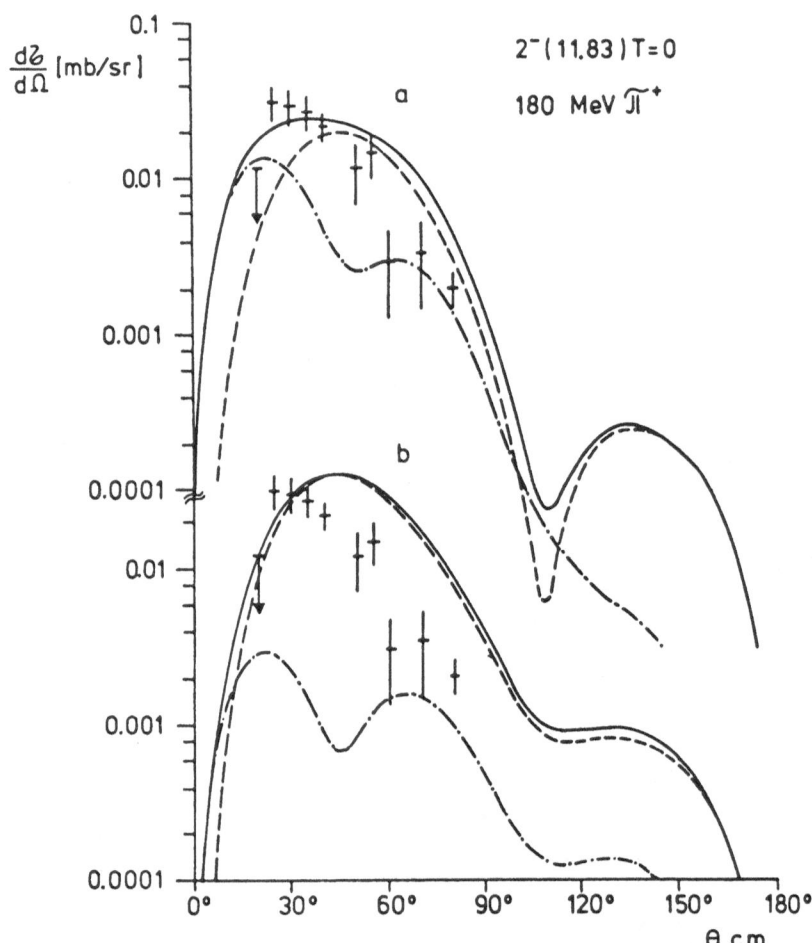

Fig. 5: Angular distributions (solid lines) for the 2^- excitation (11·83) in ^{12}C and magnetic substate population, $\sigma_{M=1} + \sigma_{M=-1}$ (dashed curves), $\sigma_{M=2} + \sigma_{M=-2}$ (dashed-dotted curves). Parametrization of the transverse form-factor in terms of convection current only (a) and magnetization current only (b). Data from ref. 7.

large convection currents are present. It would be of
interest to obtain experimental evidence for the presence
of such tensor couplings. Apart from measurements of
the magnetic substate population a detailed study of the
energy dependence of the cross section could be of help
in disentangling the tensor from the vector couplings.

3. BEYOND THE DISTORTED WAVE APPROXIMATION IN PION
NUCLEUS REACTIONS

        I introduce this section with a discussion of the
$\Delta T = 0$, 1 1^+ excitations in ^{12}C at 12.7 and 15.1 MeV
respectively. According to both, the analytical studies
and numerical evaluation of the Δ-h amplitude (6), this
pair of transitions is expected to be successfuly
described within the present formalism. As far as the
small momentum transfer behaviour of the form-factor is
concerned ($q \leq 1$ fm^{-1}), there seems to be little doubt
that these excitations correspond to transitions within
the p-shell. Furthermore, it can be shown from symmetry
arguments, that tensor coupling is forbidden in 1^+
transitions and therefore only vector couplings contribute
to the pion transition amplitude.

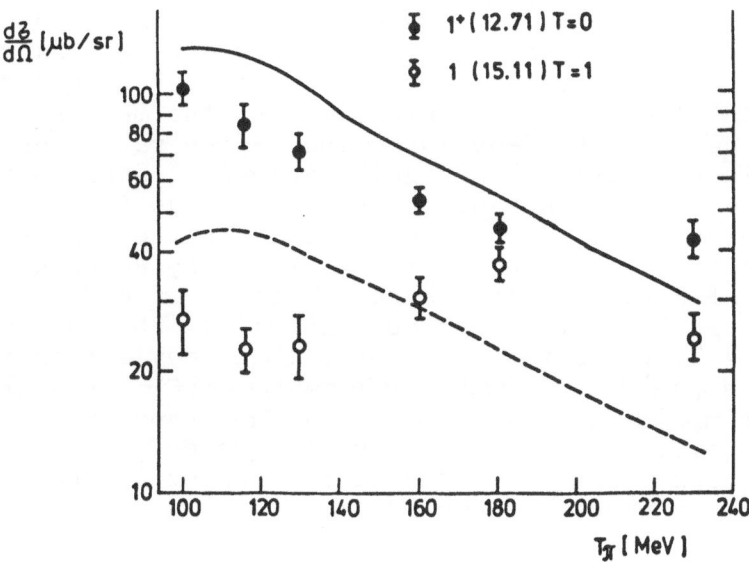

Fig. 6: Excitation functions for the 1^+states at 12.7
(T=0,solid curve) and 15.1MeV (T=1,dashed curve) at
q=120MeV/c. Data from ref. 8.

    In Figure 6 are shown the calculated excitation
functions for these transitions at fixed momentum
transfer (q≈120 MeV/c) in comparison with experiment.
Qualitative agreement between theory and experiment for
the $\Delta T = 0$ excitation, striking disagreement in the
$\Delta T = 1$ excitation are obtained. Allowing for even very
large convection current contributions to the $T = 1$
excitation which would not have been seen in electron
scattering, it is not possible to obtain such
qualitative changes in the energy dependence as required by the data. Obviously if one insists on the
identity of the wave functions in the two transitions
there is, in this formalism, no possibility to account
simultaneously for the $T = 0$ and $T = 1$ excitations.
Within the distorted wave approximation, the $T = 0$ and
$T = 1$ cross sections have a constant ratio of 4 for a 3-3
interaction only. Background π-N interaction, as well
as isospin impurities cannot change this ratio by more
than 20 - 30 %. Experimentally, however, this ratio is
of the order 1 - 2 around the resonance which indicates
the presence of reaction mechanisms beyond those contained in the distorted wave formalism.

    Further experimental evidence for the presence of
such processes has come from the study of the $(\pi,\pi N)$
reaction in ^{12}C[9] and ^{16}O[10] at 240 MeV incident energy.
In both experiments cross section ratios corresponding
to π^+ and π^- induced proton knock-out $\sigma(\pi^+,\pi^+p)/\sigma(\pi^-,\pi^-p)$
have been measured. For strict 3-3 dominance this ratio
has the value 9. In both experiments, the measured ratio
even exceed the already large quasi-free ratio for
specific kinematics. The "semi-exclusive" experiment on
^{12}C finds a 20 % increase close to the quasi-free peak
for a pion scattering angle of 130°, in the exclusive
^{16}O experiment, increase of this ratio by up to a
factor of 2 for a pion scattering angle of 60° and
quasi-free kinematics has been obtained. Similar to
the $\Delta T = 0, 1$ excitation in ^{12}C, these deviations from the
quasifree ratio require a coherent modification of the
impulse approximation by more complex processes.
In the following I indicate how the description
pion-nucleus scattering has to be changed in order to
account for these phenomena[4].

    The crucial point in the formal derivation of the
distorted wave approximation has been the assumption
that the medium correction to π-N scattering can be

accounted for by the Δ-nucleus optical potential W in
eq. (3b). With this assumption, the medium modified
π-N transition operator is as the free t-operator
a one body operator. As a consequence and independently
of any details, cross section ratios of the kind discussed
above are identical to the free ratios. Only if medium
corrections are described more fundamentally in terms
of the Δ-N interaction rather than the Δ-nucleus interaction, changes in these quasi-free ratios can be
described theoretically.

     Such a description accounts not only for elastic
scattering of the Δ by the residual nucleus but allows
as well the residual nucleus to be excited by Δ-N
collisions. The two basic processes contributing to p-h
excitations are shown in the following diagrams.

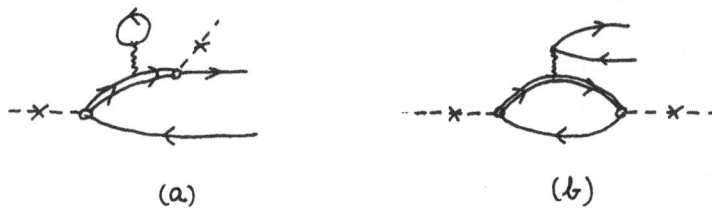

(a)   (b)

The process (a) represents the medium modification due
to elastic Δ-nucleus scattering. This process is
described by the Δ-nucleus optical potential. In process
(b), the nuclear transition actually occurs through
the Δ-N collision rather than in the decay of the Δ-h
state. This second process can only be described by a
two-body transition operator. It is important that this
change in the representation of the medium corrections
is relevant only for pion-nucleus reactions while elastic
scattering is not affected. Upon closing the particle-
hole lines, the two processes become identical and both
represent (after neglecting final state distortion) the
same medium modification of the π-N t matrix in elastic
scattering.

     In general, it is impossible to estimate the
importance of these Δ-N induced processes without
using a specific model for the Δ-N interaction. However,
changes in isospin ratios at resonance can be estimated
almost model-independently. First, it can be shown[4] that
modifications from quasi-free ratios can only occur
through the T = 1 part of the Δ-N interaction. Second,
at resonance where the Δ propagators are dominantly

imaginary, interference between first and second order
processes is possible only via the imaginary part of the
Δ-N interaction. This imaginary part of the T = 1
component has been deduced already from the Δ-nucleus
optical potential in accordance with the experimentally
determined true absorption cross sections (eq. 5). Thus
all the necessary ingredients for such an estimate are
available and the following form of the amplitude including those second order terms is obtained:

$$f_{T=0} - f^{\circ} - f^{1} \qquad\qquad f_{T=1} = \tfrac{1}{2}\left(f^{\circ} + f^{1}\right) \qquad\qquad (18a)$$

for the ΔT = 0,1   1^{+} excitations in ^{12}C and

$$g_{\pi^{+}p} = g^{\circ} - g^{1} \qquad\qquad g_{\pi^{-}p} = \tfrac{1}{3}g^{\circ} - 3g^{1} \qquad\qquad (18b)$$

for the π$^{\pm}$ induced proton knock-out reactions.

    Based on the value (5b) for the imaginary part of
the Δ-N interaction, the ratio of first (f^{0},g^{0}) to
second order amplitudes has been estimated to

$$f^{1}/f^{\circ} \approx 0.2 \qquad\qquad\qquad\qquad (19a)$$

and
$$g^{1}/g^{\circ} \approx 0.08 \qquad\qquad\qquad\qquad (19b)$$

which yields a value of about 1.8 for the ratio of the
ΔT = 0, 1 cross sections and an almost complete
suppression of the π^{-}p reaction. The sign of the
effect i.e. relative enhancement of the T = 1 amplitude
and the relatively stronger suppression of the π^{-}p
as compared to the π^{+}p reaction is essentially determined
by the sign of the imaginary part of t_{0} (eq. 5b) and by
the assignement of the channel quantum numbers of the
Δ-N interaction (eq. 5a). The size of the effect can be
related to modifications of the transition operator by
the Δ-nucleus potential which is known from both
approximate analytical estimates and detailed numerical
calculations.

    Clearly these estimates need to be refined in
detailed calculations, but they indicate that at the
level of about 15 - 30 % of inelastic cross sections,
deviations due to the two-body part of the transition
operator must be expected. The two body process of true
pion absorption induces a two body component in the
transition operator with the strength ultimately determined from the experimentally observed pion absorption

cross section. Away from the resonance, the real part
of the Δ-N interaction and therefore most likely other
processes than true pion absorption will become important.
This offers, in principle, the possibility to specify
further the Δ-N interaction and in particular to clarify
the dynamical origin of the attraction in the Δ-nucleus
potential.

CONCLUSIONS

   I have described the application of the Δ-h
formalism to nuclear excitations in pion-nucleus
scattering. As in elastic scattering, the central issue
in the theory of pion-nucleus reaction is the change of
the pion-nucleon interaction due to the surrounding
nucleons. Despite of the unavoidable phenomenological
ingredients of the Δ-h description of pion-nucleus
scattering, this formalism allows to incorporate in
the construction of the transition operator the same
medium modifications, which in elastic scattering have
been found to be important. In such a description, the
transition amplitude depends in a direct way on Δ-
propagation and Δ-nucleus interaction. This dependence
leads to a complex structure of the transition operator.

   Analytical insight into the excitation mechanisms,
implicitly described by this transition operator,has
been obtained within the non-locality expansion. Four
basic coupling modes of the pion-nucleus system have
been established. Longitudinal coupling to the transition
density,vector coupling to magnetization and convection
current,and tensor coupling to the spin-flux tensor.
Selection rules for forward and backward scattering,
interference properties and M-substate populations
have been discussed on the basis of symmetry properties
of these 4 excitation mechanisms.

   The interplay between these excitation mechanisms
has been demonstrated for some of the experimentally
observed nuclear excitations in ^{12}C. In the low-lying
natural parity excitations in ^{12}C one coupling mode,
the longitudinal one, prevails and comparatively simple
patterns in both energy and angular dependence arise.
Nuclear structure uncertainties have little effect.
Calculated cross sections agree with experiment at the
first maximum within typically 20 - 30 %. In most of
the other excitations, nuclear structure uncertainties

prohabit a similarly quantitative confrontation of theory with experiment. In particular, in $\Delta T = 1$ transitions, missing information on the convection current leads to large uncertainties in the theoretical predictions. Tensor coupling finally can be very important in specific transitions and nothing is known from any other source about the spin-flux tensor. This in turn may make pion scattering to a qualitative tool of nuclear structure. The very different energy dependences in the effective coupling constants as well as the population of different M-substates can be used to disentangle the different excitation mechanisms and thereby obtain qualitative information on the associated nuclear structure form-factors.

In the last section, I have discussed deviations from this description of pion-nucleus reactions which are due to the two-body component in the transition operator. Such two-body components are induced by the important two-body process of pion absorption. Inclusion of a two-body component requires to go beyond the description of the Δ-nucleus interaction. Here description of the medium modifications of π-N scattering in terms of the Δ-N interaction rather than in terms of the Δ-nucleus potential is required. I have discussed these effects and given some numerical estimates in the context of the observed violations of quasi-free isospin ratios which provide direct evidence for such two-body components.

REFERENCES

1) M. Hirata, F. Lenz and K. Yazaki, Ann. of Phys. 108 (1977) 116
   M. Hirata, J.H. Koch, F. Lenz and E.J. Moniz, Ann. of Phys. 120 (1979) 205
   Y. Horikawa, M. Thies and F. Lenz, Nucl. Phys. A345 (1980) 386

2) F. Binon, P. Duteil, J.P. Garron, J. Gorres, L. Hugon, J.P. Peigneux, C. Schmitt, M. Spighel and J.P. Stroot, Nucl. Phys. B17 (1970) 168

J. Piffaretti, R. Corfu, J.P. Egger, P. Gretillat,
C. Lunke and E. Schwarz, Phys. Lett. 67B (1977) 289
J. Piffaretti, R. Corfu, J.P. Egger, C. Lunke, E. Schwarz,
C. Perrin and B.M. Preedom, Phys. Lett. 71B (1977) 324;
E. Schwarz et al., private communication.

3) H. Byfield, J. Kessler and L.M. Lederman, Phys. Rev. 86
(1952) 17
E. Bellotti, D. Cavalli and C. Matteuzzi, Nuovo Cim.
18A (1973) 75
I. Navon, D. Ashery, G. Azuelos, H.J. Pfeiffer, H.K.
Walter and F.W. Schlepuetz, Phys. Rev. 22C (1979) 717
R.M. Edelstein, W.F. Baker and F. Rainwater, Phys. Rev.
122 (1961) 252

4) M. Hirata, F. Lenz and M. Thies, to be published

5) F. Lenz, M. Thies and Y. Horikawa, Ann. of Phys. 140
(1982) 266.

6) J. Flanz, Dissertation, Univ. of Massachusetts (1979),
and references therein.
J. Flanz, R.S. Hicks, R.A. Lindgren, G.A. Peterson,
J. Dubach and W.C. Haxton, Phys. Rev. Lett. 43 (1979)
1922

7) C.L. Morris, K.G. Boyer, C.F. Moore, C.J. Harvey,
K.J. Kallianpur, I.B. Moore, P.A. Seidl, S.J. Seestrom-
Morris, D.B. Holtkamp, S.J. Greene and W.B. Cottingame,
Phys. Rev. C24 (1981) 231
"Research in Experimental Nuclear Physics", W.J. Braith-
waite and C.F. Moore, LASL preprint distributed at the
Workshop on "Nuclear Structure with Intermediate Energy
probes" Jan. 14-16, 1980
J.P. Egger, private communication
8) C.L. Morris, W.B. Cottingame, S.J. Greene, C.J. Harvey,
C.F. Moore, D.B. Holtkamp, S.J. Seestrom-Morris and
T. Fortune, Phys. Lett. 108B (1982) 172

9) E. Piasetzky, D. Ashery, A. Altman, A.I. Yavin, F.W.
Schlepütz, R.J. Powers, W. Bertl, L. Felawka, H.K. Walter,
R.G. Winter and J.v.d.Pluym, Phys. Rev. 25C (1982) 2687

10) Th. S. Bauer, J.J. Domingo, Q. Ingram, J. Jansen, G. Kyle,
J. Zichy, R. Stamminger, F. Vogler, private communication.

SPIN EXCITATIONS IN PION INELASTIC SCATTERING

S. J. Seestrom-Morris[†], D. B. Holtkamp[††], and
W. B. Cottingame[†††]

[†]Los Alamos National Laboratory
Los Alamos, New Mexico, 87545

[††]University of Minnesota
Minneapolis, Mn. 55455

[†††]New Mexico State University
Las Cruces, NM

INTRODUCTION

During the last few years a large amount of pion-nucleus
(π-A) inelastic scattering data has been obtained with high-flux,
high-resolution facilities such as the SUSI spectrometer at SIN
and the EPICS spectrometer at LAMPF. The early studies for the
most part concentrated on strong transitions to low-lying
collective states. These studies have been reviewed in part in
Ref. 1 . One unanticipated feature of π-A inelastic scattering
that has become evident as the scope of studies has increased is
the strong excitation of high-spin stretched states in light
nuclei. By stretched states we mean those whose total angular
momentum is the maximum achievable in a $1\hbar\omega$ particle-hole
excitation. Transitions to stretched states have now been
observed in almost all p-shell nuclei that have been studied, as
well as in ^{28}Si.[2-8]

The strength of these transitions can be qualitatively
understood by examining the pion-nucleon (π-N) interaction.
Following Koltun[9], the π-N amplitude can be written as follows, if
the interaction is dominated by the [3,3] resonance:

$$f(k,k') = \alpha(k)[2\cos(\theta) + i\ \vec{\sigma}\cdot\hat{n}\sin(\theta)] \qquad (1)$$

where $\alpha(k)$ contains the energy dependence of the elementary π-N force, θ is the center-of-mass scattering angle, $\vec{\sigma}$ is the nucleon spin operator, and $\hat{n}$ is the normal to the scattering plane. Only the second term, which is proportional to the π-N spin-orbit operator, can induce a spin transfer. Petrovich and Love[10] have extracted the strength of the central and spin-orbit parts of the π-N interaction using the impulse approximation (IA) and the π-N phase shifts. Their results at 180 MeV, shown in Fig. 3 of Ref. 10, confirm the conclusions one would draw from the simple P_{33} result (eq. 1), namely that at small momentum transfers ($q < 1.2$ fm^{-1}) the central strength, t_c, is considerably larger than the spin-orbit strength, t_{LS}. At about $q = 1.4$ fm^{-1} ($\theta \simeq 70°$) the two strengths become comparable. The relative strengths of the central and spin-orbit interactions are the same for both isospin channels. The factor of two enhancement of the isoscalar over the isovector t-matrix that results from the isospin properties of the [3,3] resonance is also contained in the results of Petrovich and Love. It is this factor that gives pion inelastic scattering its unique sensitivity to the relative contributions of protons and neutrons to inelastic transitions.

EXCITATION FUNCTION MEASUREMENTS

$^{12}C(\pi,\pi')$

The strength with which stretched configurations are excited can be seen in Fig. 1 which shows a π^+, a π^-, and a difference ($\pi^- - \pi^+$) spectrum taken on ^{12}C at $T_\pi = 164$ MeV and $\theta_L = 70°$. The very strong groups near 19 MeV have been identified as two 4^- states, one an essentially pure proton transition (19.25 MeV) and the other a pure neutron transition (19.65 MeV). The large π^-/π^+ asymmetry is understood[11] as due to nearly maximal isospin mixing between a T = 0 and a T = 1 4^- state. The shapes of the angular distributions for these transitions are well described using a transition density having a total angular momentum transfer $\Delta J = 4$, orbital angular momentum transfer $\Delta L = 3$, and spin transfer $\Delta S = 1$.[2]

It was observed by Moore, et al.[12] that excitation functions measured for these transitions were quite different from those measured for low-lying natural parity transitions. When excitation functions measured at constant momentum transfers[13] (near the maximum in $d\sigma/d\Omega(\theta)$) for the 4^- (19.25) and 2^- (18.36) states were compared with those for the 2^+ (4.44), 3^- (9.56) and 0^+ (7.65) states (Fig. 2) it was found that the cross sections for

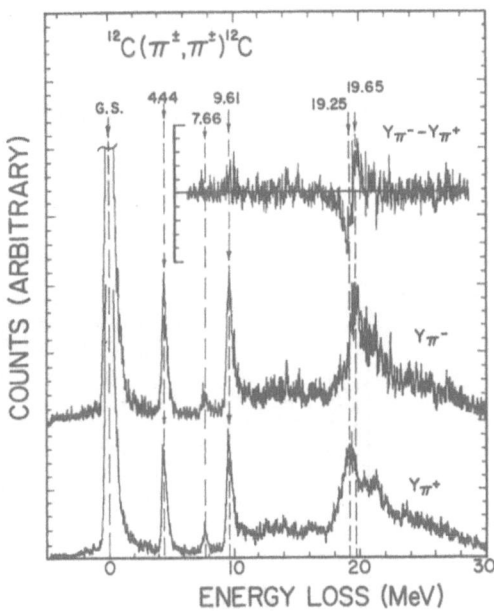

Fig. 1. π^+, π^-, and difference spectra for scattering from ^{12}C.

the unnatural-parity transitions decrease as the energy increases while the cross sections for the natural-parity transitions rise dramatically as the energy is raised.

Theoretical Interpretation of the Excitation Function Data

Siciliano and Walker[14] have provided a theoretical interpretation of the pion inelastic excitation-function data based on the assumptions of 1) the validity of the fixed-scatterer impulse approximation and 2) a one-step reaction mechanism. Their expression for the differential cross section for pion-nucleus inelastic scattering is:

$$\frac{d\sigma}{d\Omega} = \Gamma(E)[4M^2(q_0)\cos^2\theta + S^2(q_0)\sin^2\theta] \qquad (2)$$

where q_0 is a fixed momentum transfer near $d\sigma/d\Omega(q)_{max}$. In general only the S form factor can contribute to unnatural-parity transitions while both M and S can contribute to natural-parity transitions (although M usually dominates). The energy dependent factor $\Gamma(E)$ is a product of the distortion of the pion waves and

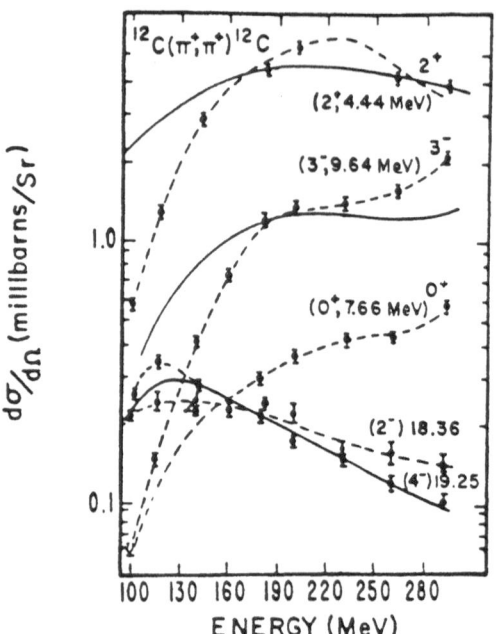

Fig. 2. Constant q excitation functions for (π,π') to states in ^{12}C.

the energy dependence of the elementary π-N force. Since the effect of attenuating the pion waves varies approximately as the inverse of the strength of the force, $\Gamma(E)$ is roughly constant in the vicinity of the [3,3] resonance. The energy dependence of pion-nucleus scattering is given by the $\cos^2\theta$ and $\sin^2\theta$ dependences of the pion-nucleon scattering amplitude. These angular dependences result in an energy dependence for constant q because the scattering angle must be adjusted as a function of energy to keep the momentum transfer constant. This results in $\Delta S = 1$ transitions decreasing with increasing energy since only the S form factor with its accompanying $\sin^2\theta$ dependence can contribute to unnatural-parity excitations. It should be noted that while in general Fermi motion corrections can allow the central part of the pion-nucleon interaction to contribute to unnatural-parity transitions they cannot contribute to the excitation of stretched configurations.

The solid lines in Fig. 2 are the simple $\cos^2\theta$ and $\sin^2\theta$ predictions of Siciliano and Walker. The 4^- and 2^- data are well represented by the $\sin^2\theta$ energy dependence. The $\cos^2\theta$ curves

reproduce   qualitatively   the   energy   dependence   of   the
natural-parity transitions. The discrepancies are due to the
non-constant nature of $\Gamma(E)$ which will be discussed in more detail
in the next section.

Application to Transitions in ^{13}C

In an odd mass nucleus more than one total angular momentum
transfer is generally allowed and hence there are few pure
unnatural-parity transitions. The main exceptions are transitions
to stretched states where a spin transfer is required to reach the
total angular momentum transfer necessary. One such state is the
9.5 MeV $9/2^+$ state in ^{13}C.   The shapes of the (π,π') angular
distributions for this state are characteristic of the $\Delta J = 4$,
$\Delta L = 3$, $\Delta S = 1$ transition density amplitude.[3] The ratio of
$\sigma(\pi^-)/\sigma(\pi^+) = 10\pm2$ has indicated that this state is reached by a
pure neutron particle-hole excitation[15]. The pure neutron nature
of the transition has been explained in a simple weak-coupling
model as well as in a DWIA calculation using Millener-Kurath wave
functions for their first predicted $9/2^+$ state.[3] (see Sec. III)
The shell-model wave functions predict a transition density that
is pure $\Delta S = 1$.

The constant-q excitation function for π^- scattering to this
state is shown in the lower part of Fig. 3 (diamonds). The energy
dependence is similar to that of the unnatural-parity transitions
in ^{12}C and is reproduced very well by the simple $\sin^2\theta$ dependence.
The upper data in Fig. 3 (solid circles) is for the
collectively-enhanced transition to the $3/2^-$ state (3.68 MeV)
which is understood to be predominantly $\Delta S = 0$. The solid curve
is the $\cos^2\theta$ prediction and the dashed curve is the result of a
DWIA calculation[16] using a collective form factor and normalized
to the data at 162 MeV. The energy dependence predicted by the
DWIA is in very good agreement with the data.

The excitation function for the $9/2^+$ state provides striking
confirmation of the $\Delta S = 1$ nature of the transition as deduced
from the angular distributions. The knowledge of the spin
transfer in a transition is very useful for testing wave functions
as well as for determining the spin and parity of states in nuclei
with spin-zero ground states. This method has been used to
identify four other M4 transitions in ^{13}C.[17]

This technique has proven very complementary to the
measurement of 180^0 electron scattering, for two reasons. The
first is that 180^0 (e,e') is sensitive essentially only to $\Delta T = 1$
transitions whereas pion scattering excites $\Delta T = 0$ and $\Delta T = 1$

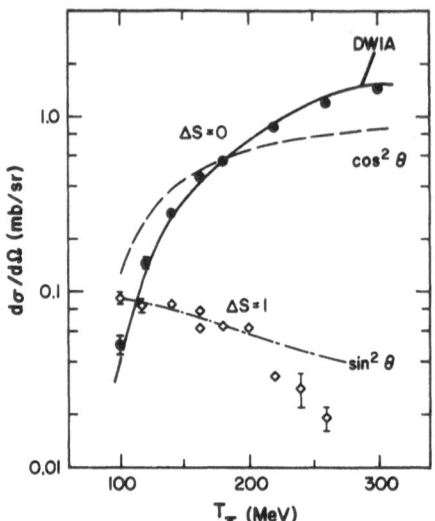

Fig. 3. Constant q excitation function for the 3/2⁻ (circles) and 9/2⁺ (diamonds) states in ^{13}C.

transitions in the ratio of 4/1. The second reason is the sensitivity of π^+ and π^- comparisons to the relative contributions of neutrons and protons. ^{13}C gives a good example of the utility of these comparisons. The first $9/2^+$ state (9.5 MeV) is reached by a pure neutron excitation while the second (16.1 MeV) is reached by a transition that involves mostly protons. Both of these transitions have been observed in 180° (e,e')[18]. The third M4 transition is excited about equally by π^+ and π^-, indicating a pure isospin transfer. This state is not seen in (e,e') indicating that the transition is probably pure $\Delta T = 0$.

EXCITATION OF STRETCHED STATES

P-Shell Nuclei

Stretched one-particle, one-hole states are those states having the maximum total angular momentum allowed in a single particle-hole excitation; i.e. $\Delta J = \ell_p + \ell_h + 1$, where ℓ_p (ℓ_h) is the particle (hole) orbital angular momentum and the orbital angular momentum transfer is $\Delta L = \ell_p + \ell_h$. Whenever ΔJ equals $\Delta L + 1$, states are reached by unnatural-parity transitions. In the p-shell, stretched states are made from $p_{3/2} \rightarrow d_{5/2}$ particle-hole excitations and are thus reached by M4 transitions ($\Delta J = 4$, $\Delta L = 3$, $\Delta S = 1$). Measurements of angular distributions are sensitive to the transferred J, L, and S, whereas excitation functions are sensitive to the spin transfer. These two together are very useful for locating stretched states. Because of the limited configurations that can make up stretched states they are ideal tests for the shell model. Fig. 4 shows the distribution of M4 transitions that have been measured in p-shell nuclei using π^+ and π^- inelastic scattering. M4 transitions in ^{11}B, ^{14}N, ^{14}C, and ^{16}O are shown along with those in ^{12}C and ^{13}C that have been previously discussed. Preliminary results for ^{15}N not contained in Fig. 4 will also be presented in this section.

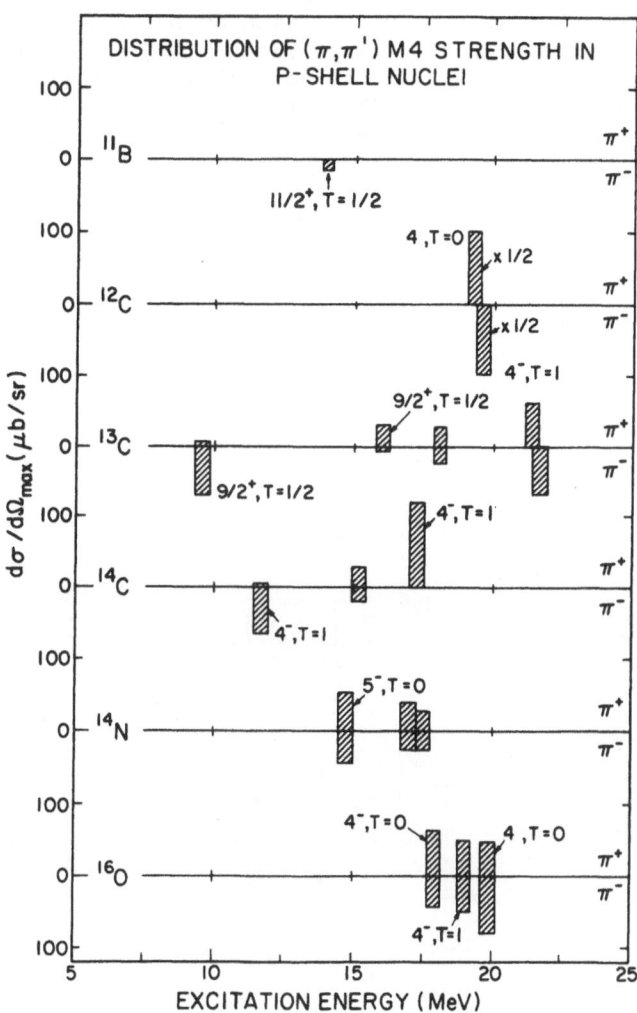

Fig. 4. Plot of (π,π') cross sections for M4 transitions in the p-shell:

The lightest nucleus in which an M4 transition has been observed is ^{11}B. The data of Zupransky, et al.[19] in Fig. 5 shows spectra for π^+ and π^- scattering from ^{11}B at T_π = 162 MeV and Θ_{lab} = 70°. The 11/2$^+$ state at 14.04 MeV can be seen to be excited much more strongly by π^- than by π^+, with a lower limit[19] of R($=\sigma(\pi^-)/\sigma(\pi^+)$) ≥5. An angular distribution and excitation function measured for π^- scattering indicate that this state is reached by an M4 transition. The large π^- enhancement is easily understood because the proton $d_{5/2}p_{3/2}^{-1}$ particle-hole excitations that can form 11/2$^+$ states in ^{11}B require a recoupling of the remaining $p_{3/2}$-shell protons. Such configurations cannot be reached in a single-step from the ground state. Microscopic DWIA calculations using the wave function of D. Kurath for the first 11/2$^+$ state in ^{11}B predict a large ratio $\sigma(\pi^-)/\sigma(\pi^+)$ and require an overall renormalization[8] factor of .22 in order to fit the absolute magnitude of the data. This renormalization factor is smaller than that required to fit the data for the ^{13}C 9/2$^+$ state using Millener-Kurath wave funcions.

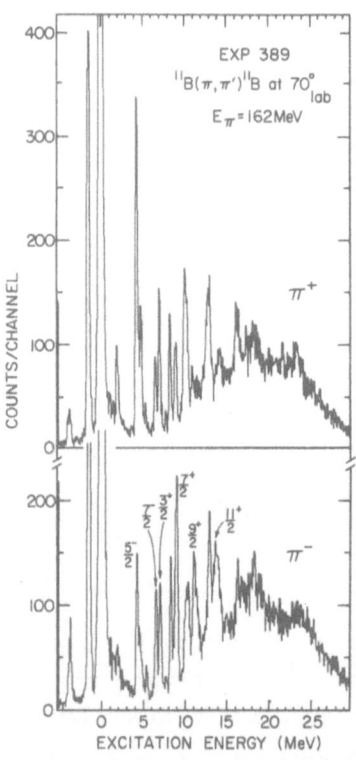

Fig. 5. Spectra for π^+ and π^- scattering from ^{11}B at T_π = 162 MeV.

As mentioned already, the M4 strength in ^{12}C is concentrated in two 4$^-$ states, which are strongly isospin mixed. In ^{13}C there is a group at slightly higher excitation energy which shows a similar π^-/π^+ asymmetry indicating that these states in ^{13}C (which are reached by M4 transitions) have a large parentage in ^{12}C(4$^-$) ⊠ $\nu(p_{1/2})$.

The π^+ and π^- angular distributions for the 9.5-MeV 9/2$^+$ state in ^{13}C are shown in Fig. 6. The solid curves are microscopic DWIA[20] calculations using the code ALLWRLD[21] to generate transition densities from the particle-hole amplitudes of Lee and Kurath with a harmonic oscillator parameter α = .632 fm^{-1}. The value of α required to reproduce the shape of the data is considerably smaller than that needed to fit[22] the transverse form factor measured in (e,e′), although it is nearly the same as the

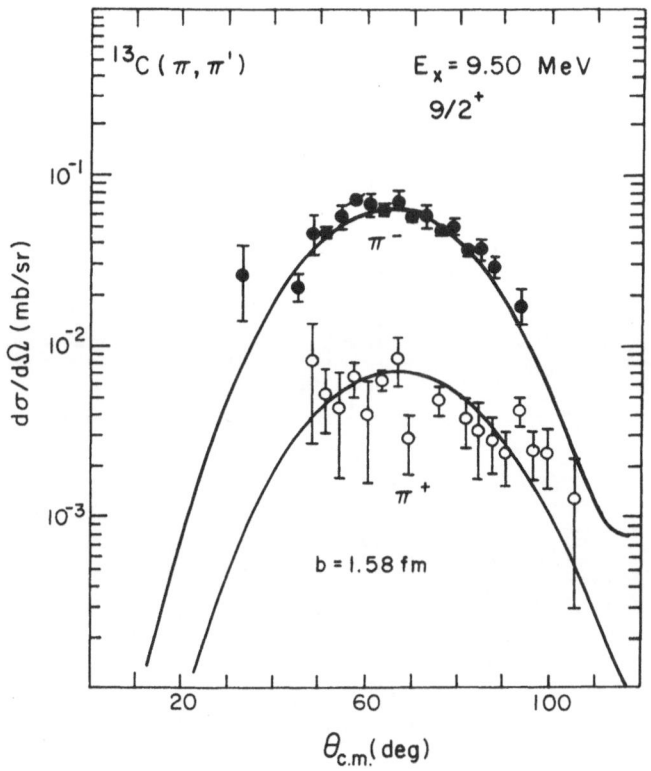

Fig. 6. Angular distributions for π^+ and π^- scattering to 9.5-MeV 9/2$^+$ state in ^{13}C.

value required to fit the 547 MeV (p,p′) data for this transition[23]. The overall renormalization required is .4, considerably smaller than that required to fit the transverse form factor from (e,e′) (.7).

Spectra for π^+ and π^- scattering[6] from ^{14}C, in which two 4^-, T = 1 states can be seen, are presented in Fig. 7 . The lower energy state (11.67 MeV) is π^- enhanced while the upper (17.26 MeV) is π^+ enhanced. Both ratios $\sigma(\pi^-)/\sigma(\pi^+)$ are larger than the free $\pi + n$ ($\pi + p$) values of 9 (1/9). These data have been compared with DWIA calculations[24] for $d_{5/2}p_{3/2}^{-1}$ particle-hole excitations to derive isoscalar and isovector spectroscopic amplitudes. The equation that must be solved to extract the relative isoscalar and isovector components from the measured ratio $\sigma(\pi^+)/\sigma(\pi^-)$,

$$R = \frac{\sigma(\pi^+)}{\sigma(\pi^-)} = \frac{(2S_0 - S_1)^2}{(2S_0 + S_1)^2}, \tag{3}$$

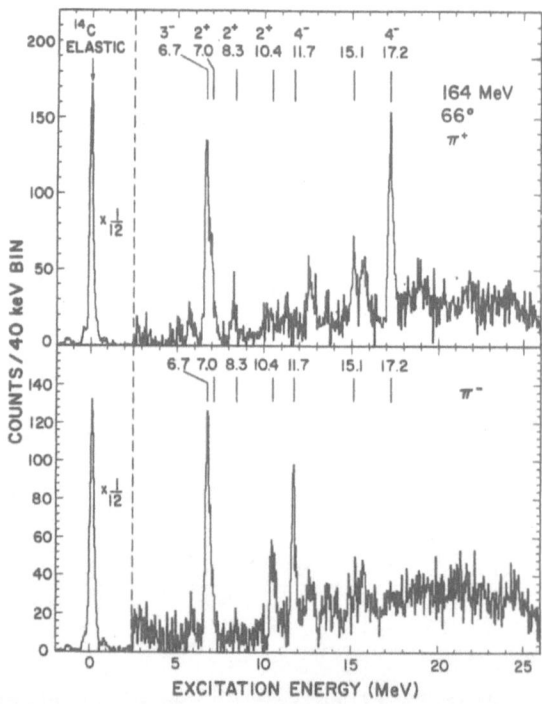

Fig. 7. Spectra for π^+ and π^- scattering from ^{14}C at T_π = 164 MeV.

has two solutions. This results in an ambiguity in the final
result for S_1 and S_0.   This must be eliminated through a
comparison with electron and/or proton scattering data.

        The strongest M4 transition observed in ^{14}N is to a 5^- state
at 14.7 MeV (Fig. 8, Geesaman[7] et al.).   The π^+ and π^- cross
sections for this transition are approximately   equal, indicating
that there is no significant isospin mixing.   A quenching factor
of .70 is needed to obtain agreement between the magnitude of the
data and DWIA calculations for the lowest 5^- state predicted by D.
Kurath[7]. These wave functions predict 51% of the 5^- strength to
be in the lowest 5^- with the remainder split between states at
17.3 and 18.0 MeV.   These model states have been tentatively
identified with groups seen at 16.86 and 17.46 MeV.   The angular
distributions and total strengths of these groups indicate that
they   contain   additional   unresolved   states   of   other
multipolarities.

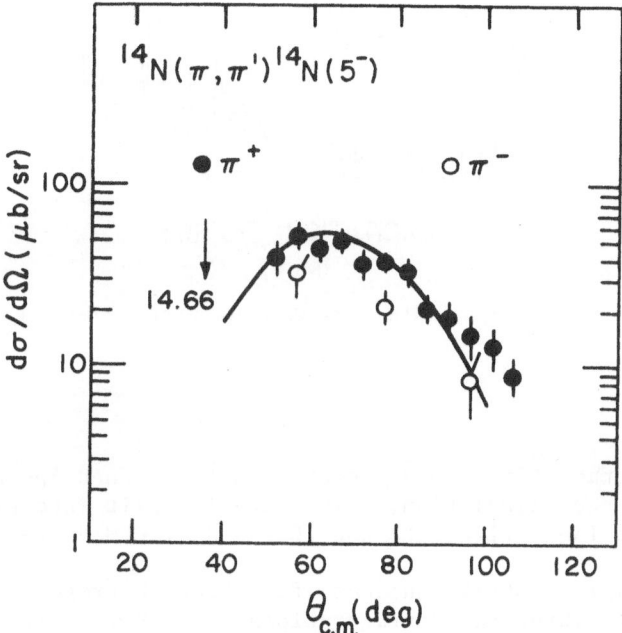

Fig. 8. Angular distributions for π^+ and π^- scattering to 14.0-MeV
5^- state in ^{14}N.

Very recently π^+ and π^- spectra for scattering from ^{15}N have been measured. Fig. 9 shows a π^+ and a π^- spectrum taken at $\theta_{lab} = 65^\circ$ and $T_\pi = 164$ MeV. Previously identified $9/2^+$ states at 10.7 and 12.6 MeV are labeled as well as a very strong state at 17.2 MeV which is possibly reached by an M4 transition. The two lowest $9/2^+$ states are very strongly π^+ enhanced, while the state at 17.2 MeV is slightly π^- enhanced. Shell model calculations of D. J. Millener[25] using a $1\hbar\omega$ basis predict a very large π^+/π^-

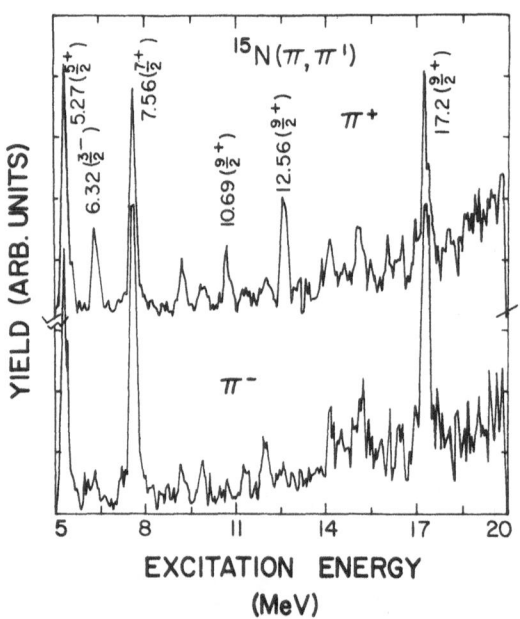

Fig. 9. Spectra for π^+ and π^- scattering from ^{15}N at $T_\pi = 164$ MeV.

ratio for the first $9/2^+$ state in ^{15}N. When 3p-4h terms are included in the calculation, this state is split into two states, which can be identified with the 10.7 and 12.6 MeV states.

The last p-shell nucleus for which we present data on the excitation of stretched configurations is ^{16}O. Fig. 10 shows angular distributions for three 4^- states excited in (π,π') on ^{16}O

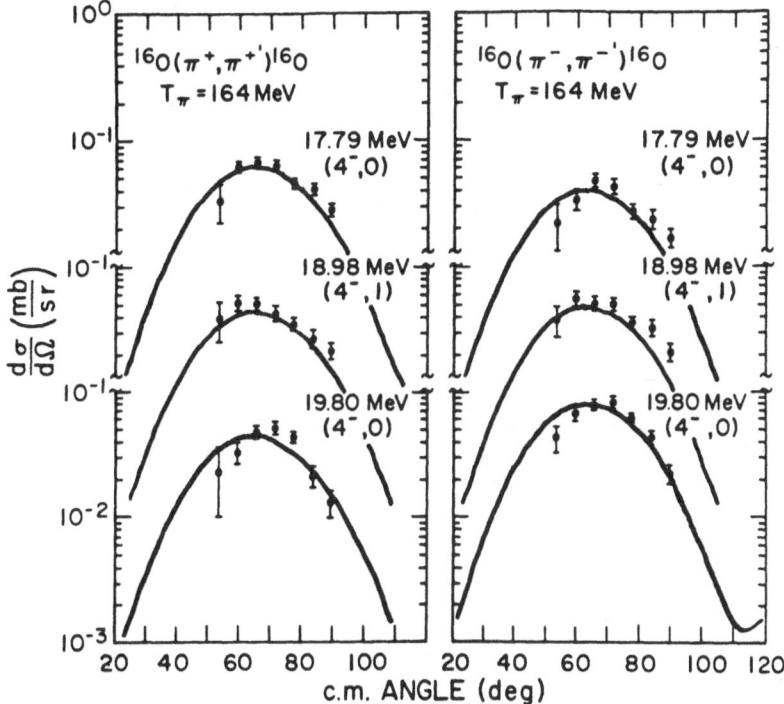

Fig. 10. Angular distributions for π^+ and π^- scattering to a triplet of isospin-mixed 4^- states in ^{16}O.

(T = 0 17.79 and 19.80 MeV, T = 1 18.98 MeV). The observed ratios $\sigma(\pi^+)/\sigma(\pi^-)$ indicate that these states are weakly isospin mixed. The solid curves are calculations using a $d_{5/2}p_{3/2}^{-1}$ transition density and a spectroscopic factor for the T = 1 state determined from electron scattering. The isospin mixing matrix elements derived from the pion scattering greatly improve the agreement of calculations with the electron scattering data to the lower T = 0 state.

Stretched States in Heavier Nuclei

^{28}Si is the only nucleus outside the p-shell for which excitations of stretched states have been measured with pion inelastic scattering. Fig. 11 shows angular distributions measured[4] for 6^- T = 0,1 and 5^-, T = 0 states. The solid curves for the 6^- states are calculated using a $f_{7/2}d_{5/2}^{-1}$ transition density and squared spectroscopic factors of .13 for the T = 0 state and .34 for the T = 1 state. The spectroscopic factor for

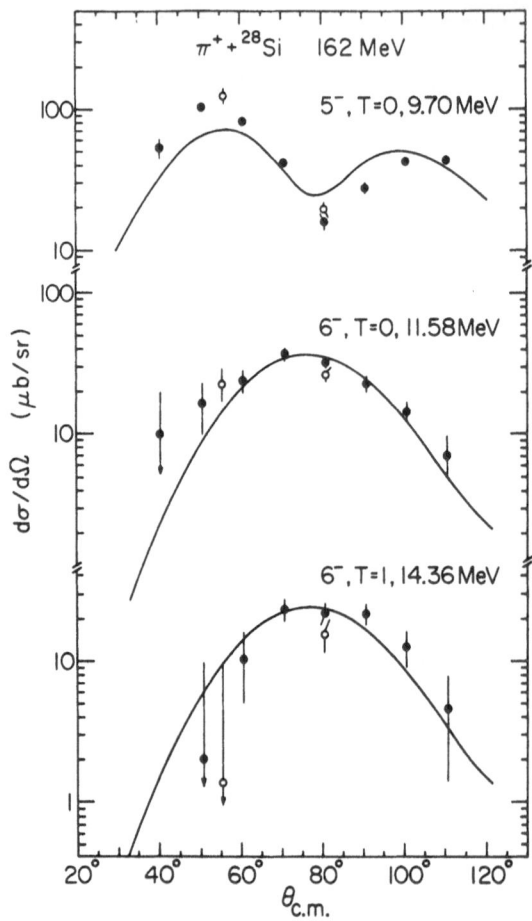

Fig. 11. Angular distributions for π^+ and π^- scattering to 5^- and 6^- states in ^{28}Si.

the T = 1 state is consistent with that necessary to reproduce proton and electron scattering data for this transition. Recent (p,p′) measurements[26] have indicated different structure for the T = 0 and T = 1 6^- states in ^{28}Si that is consistent with the isoscalar and isovector spectroscopic factors determined in the pion scattering measurements.

SPIN EXCITATIONS OF NON-STRETCHED CONFIGURATIONS

Although the majority of spin excitations that have been observed in pion inelastic scattering are transitions to stretched states, there are a few examples of $\Delta S = 1$ excitation of non-stretched states. The best studied of these have been the 1^+ T = 0 (12.71 MeV) and T = 1 (15.11 MeV) states in ^{12}C. Comparisons of π^+ and π^- cross sections for these states have indicated the presence of isospin mixing.[27] A constant-q excitation function measured for the T = 0 state shows the energy dependence expected for a pure $\Delta S = 1$ transition. The T = 1 state excitation function shows an unexpected bump near at 180 MeV. The interpretation of this anomolous excitation function as due to admixtures of ΔN^{-1} admixtures in the 15.11-MeV state wave function[28] will be discussed in another presentation.[29]

Most of the spin excitations seen in (π,π') have been transitions which are required by parity or angular momentum conservation to have a non-zero spin transfer. A natural-parity transition to a 1^- state (4.45 MeV) in ^{18}O recently observed[30] in (π,π') seems to be dominated by $\Delta S = 1$. The $\Delta S = 1$ assignment is based on the angular distribution, shown in Fig. 12. A number of

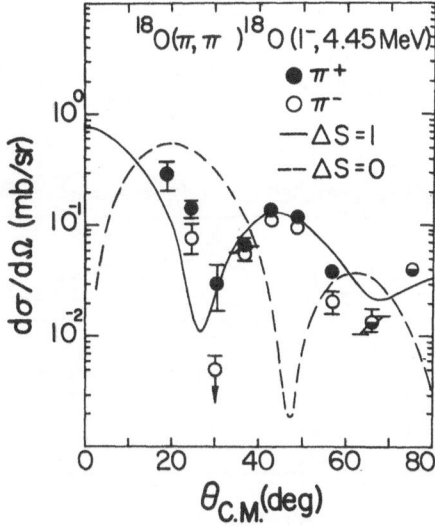

Fig. 12. Angular distributions for π^+ and π^- scattering to 4.45-MeV 1^- state in ^{18}O.

DWIA and eikonal model calculations have predicted dramatically different angular distributions for transitions to 1^-, $\Delta S = 0$ and 1^-, $\Delta S = 1$ states. The solid curve in Fig. 12, generated using the DWIA code ARPIN[30], is for a pure $\Delta L = 1$, $\Delta S = 1$ transition and the dashed curve is for pure $\Delta L = 1$, $\Delta S = 0$. The two calculations are completely out of phase and only the $\Delta S = 1$ calculation reproduces the data. This data represents the first observation in pion scattering of the different angular distributions predicted for $\Delta S = 0$ and $\Delta S = 1$ transitions.

A similar effect has been seen[32] in ^{12}C for the excitation of states near 25 MeV. Spectra taken at $T_\pi = 180$ MeV and $\theta_{lab} = 25^0$ for π^+, π^-, and the difference between them are shown in Fig. 13, along with a fit to the data. The large peak at 22.1 MeV is probably the well-known isovector giant dipole resonance and peaks at 18.3 and 19.3 MeV are an isospin-mixed doublet of 2^- states. The angular distributions for the 23.7 and 25.6 MeV groups are much less forward peaked than that of the 22.1-MeV state. In addition, these angular distributions resemble neither the

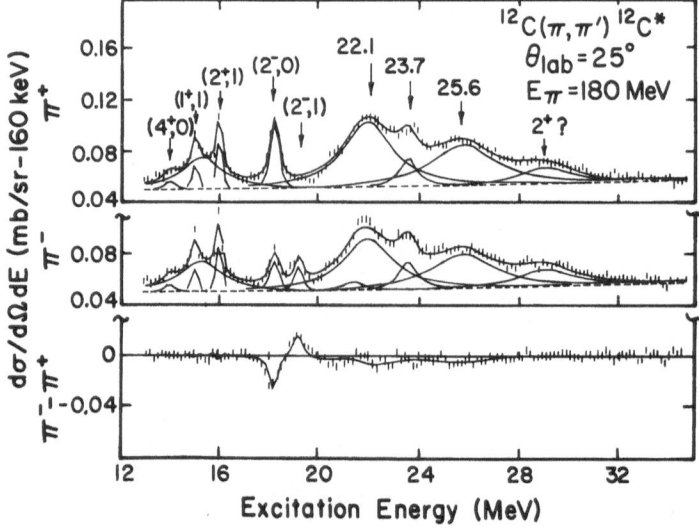

Fig. 13. π^+, π^-, and difference spectra for scattering to the giant dipole region of ^{12}C.

predicted shapes for $1^-, T = 1$ or 2^- states. (The 18 and 19 MeV 2^- data are well described by the 2^- shapes.) The experimental angular distributions can be fitted with a combination of 1^- and 2^- calculations, but this requires a total 2^- strength exceeding the sum rule limit[31]. This suggests the presence of some 1^-, $\Delta S = 1$ strength in this region. The summed cross sections for the 22.1, 23.7, and 25.6 MeV levels as well as the 18.3 and 19.3 MeV 2^- states can be fitted with a combination of 1^- $T = 1$ $\Delta S = 0$, 1^- $T = 0,1$ $\Delta S = 1$, and 2^- $T = 0,1$ for all allowed p to d excitations.

CONCLUSION

This paper has reviewed the data on spin excitations observed in pion inelastic scattering. A predominant feature of this process is the selectivity with which high-spin unnatural-parity states are excited. Constant-q excitation functions have proven valuable in identifying unnatural-parity states because of the unique signature of $\Delta S = 1$ transitions versus $\Delta S = 0$ transitions. It has recently also been shown that angular distributions measured for transitions to natural-parity states can be quite different for $\Delta S = 0$ and $\Delta S = 1$ transitions. Thus pion scattering should continue to prove useful in studying the spin structure of nuclear transitions because of the sensitivity of both excitation functions and angular distributions to the spin transfer to the nucleus. In particular, pion scattering measurements may be helpful in searches for spin-mode giant resonances.

REFERENCES

1. D. Dehnhard, Nucl. Phys. A374, 377c (1982).
2. W. B. Cottingame, to be published.
3. S. J. Seestrom-Morris,, D. Dehnhard, M. A. Franey, G. S. Kyle, C. L. Morris, R. L. Boudrie, J. Piffaretti, and H. A. Thiessen, Phys. Rev. C26, 594 (1982).
4. C. Olmer, B. Zeidman, D. F. Geesaman, T.-S. H. Lee, R. E. Segel, R. L. Boudrie, G. S. Blanpied, L. W. Swenson, H. A. Thiessen, C. L. Morris, and R. E. Anderson, Phys. Rev. Lett. 43, 612 (1979).
5. D. B. Holtkamp, W. J. Braithwaite, S. J. Greene, R. J. Joseph, C. F. Moore, C. L. Morris, J. Piffaretti, E. R. Siciliano, H. A. Thiessen, and D. Dehnhard, Phys. Rev. Lett. 45, 420 (1980).
6. D. B. Holtkamp, S. J. Seestrom-Morris, S. Chakravarti, D. Dehnhard, H. W. Baer, C. L. Morris, S. J. Greene, and C. J. Harvey, Phys. Rev. Lett. 47, 216 (1981).

7. D. F. Geesaman, K. Kurath, G. C. Morrison, C. Olmer, B. Zeidman,
    R. E. Anderson, R. L. Boudrie, H. A. Thiessen, G. S. Blanpied,
    G. R. Burleson, R. E. Segal, L.W. Swenson, Phys. Rev. C27, 1134
    (1983).
8. D. F. Geesaman, B. Zeidman, P. Zupransky, R. E. Segel, C. L.
    Morris, R. L. Boudrie, S. Greene, B. G. Ritchie, G. S.
    Blanpied, C. J. Harvey, L. W. Swenson, G. C. Morrison,
    Abstracts of Cont. Papers, IV ICOHEPANS, July 1981, 302.
9. D. S. Koltun, in Advances in Nuclear Science, Vol. 3, Ed. by
    M. Baranger and E. Vogt, pp.71,ff (Plenum Press, 1969).
10. F. Petrovich and W. G. Love, Nucl. Phys. A354, 499c (1981).
11. C. L. Morris, J. Piffaretti, H. A. Thiessen, W. B. Cottingame,
    W. J. Braithwaite, R. J. Joseph, I. B. Moore, D. B. Holtkamp,
    C. J. Harvey, S. J. Greene, C. F. Moore, R. L. Boudrie, and R.
    J. Peterson, Phys. Lett. 86B, 31 (1979).
12. C. Fred Moore, W. Cottingame, K. G. Boyer, L. E. Smith, C.
    Harvey, W. J. Braithewaite, C. L. Morris, H. A. Thiessen, J.
    F. Amann, M. Devereux, G. Blanpied, G. Burleson, A. W. Obst,
    S. Iversen, K. K. Seth, R. L. Boudrie, and R. J. Peterson,
    Phys. Lett. 80B, 38 (1978).
13. W. Cottingame, K. Allbright, S. Greene, C. Harvey, D. B.
    Holtkamp, R. J. Joseph, I. B. Moore, C. F. Moore, J.
    Piffaretti, C. L. Morris, N. King, R. L. Boudrie, J.
    Kraushaar, R. J. Peterson, B. Ristinen, and G. R. Smith, Bull.
    Am. Phys. Soc. 24, 821 (1979).
14. E. R. Siciliano and G. E. Walker, Phys. Rev. C23, 2661
    (1981).
15. D. Dehnhard, S. J. Tripp, M. A. Franey, G. S. Kyle, C. L.
    Morris, R. L. Boudrie, J. Piffaretti, and H. A. Thiessen,
    Phys. Rev. Lett. 43, 1091 (1979).
16. S. J. Seestrom-Morris, Ph.D. dissertation, University of
    Minnesota (1981) and Los Alamos Report LA-8916-T.
17. S. J. Seestrom-Morris, D. Dehnhard, D. B. Holtkamp, and C. L.
    Morris, Phys. Rev. Lett. 46, 1447 (1981).
18. R. S. Hicks, R. A. Lindgren, B. Parker, G. A. Peterson, H. A.
    Thiessen, H. Crannel, D. Sober, Proceedings of the Workshop on
    Nuclear Structure with Intermediate Energy Probes, Los Alamos,
    Ed. H. A. Thiessen, 292 (1980).
19. P. Zupransky, private communication.
20. R. A. Eisenstein and G. A. Miller, Comp. Phys. Comm. 11, 95
    (1976).
21. J. A. Carr, F. Petrovich, W. B. Cottingame, and D. B.
    Holtkamp, Phys. Rev. C27, 1636 (1983).
22. M. A. Franey, private communication.
23. S. J. Seestrom-Morris, M. A. Franey, D. Dehnhard, D. B.
    Holtkamp, J. F. Amann, R. L. Boudrie, and C. A. Goulding, to
    be published.
24. D. B. Holtkamp, S. J. Seestrom-Morris, D. Dehnhard, H. W.
    Baer, C. L. Morris, S. J. Greene, and C. J. Harvey, to be
    published.

25. D. J. Millener, private communication.

26. F. Petrovich, W. G. Love, A. Picklesimer, G. E. Walker, and E. R. Siciliano, Phys. Lett. 95B, 166 (1980).

27. C. L. Morris, R. L. Boudrie, J. Piffaretti, W. B. Cottingame, W. J. Braithwaite, S. J. Greene, C. J. Harvey, D. B. Holtkamp, C. Fred Moore, and S. J. Seestrom-Morris, Phys. Lett. 99B, 387 (1981).

28. C. L. Morris, W. B. Cottingame, S. J. Greene, C. J. Harvey, C. Fred Moore, D. B. Holtkamp, S. J. Seestrom-Morris, and H. T. Fortune, Phys. Lett. 108B, 172 (1982).

29. C. L. Morris, these proceedings.

30. S. J. Seestrom-Morris, C. L. Morris, D. B. Holtkamp, D. Dehnhard, C. Blilie, D. Gay, and H. T. Fortune, to be published.

31. The program ARPIN, written by T.-S. H. Lee, unpublished.

32. C. Fred Moore, private communication.

SPIN FLIP 1$^+$ EXCITATIONS BY INELASTIC HADRON SCATTERING

G.P.A. Berg, W. Hürlimann, I. Katayama[+], S.A. Martin,
J. Meissburger, F. Osterfeld, J.G.M. Römer, B. Styczen[++]
and J. Tain

Institut für Kernphysik, Kernforschungsanlage Jülich
D-5170 Jülich, West Germany

G. Gaul, R. Santo, and G. Sondermann

Institut für Kernphysik, Universität Münster
D-4400 Münster, West Germany

INTRODUCTION

In this contribution we report on inelastic hadron scattering
experiments to 1$^+$ spin flip states in ^{48}Ca, ^{58}Ni and ^{208}Pb using the
high resolution spectrometer BIG KARL at the Jülich cyclotron facility. Experimental and theoretical investigations of spin flip
excitations in nuclei are stimulated by the perspective to obtain
information on the nature of the spin ($\vec{\sigma} \cdot \vec{\sigma}$) and spin-isospin
($\vec{\sigma} \cdot \vec{\sigma} \vec{\tau} \cdot \vec{\tau}$) parts of the residual particle-hole (ph) interaction
which in the Landau-Migdal parametrisation[1,2], can be written as:

$$F_{ph} (\vec{r}_1 - \vec{r}_2) = C_o(g_o\vec{\sigma}_1 \cdot \vec{\sigma}_2 + g_o'\vec{\sigma}_1 \cdot \vec{\sigma}_2\vec{\tau}_1 \cdot \vec{\tau}_2)\delta(\vec{r}_1 - \vec{r}_2); \qquad (1)$$

$$\text{with } C_o = 302 \text{ m/m}^* \text{ (MeV fm}^3)$$

The magnitude of the force strength parameter g_o' which describes
the short range spin-isospin correlations is intimately related to
the experimentally observed quenching of the total M1- and Gamow-
Teller (GT) strength. For this reason spin excitations in nuclei
are presently investigated experimentally by many groups using pre-
ferably highly selective reactions like (p,n) charge exchange[3],

---

+ Permanent address: RCNP, Osaka University, Japan
++ Permanent address: Institute of Physics, Jagiellonian University
                      Cracow, Poland

inelastic electron scattering[4],[5] at backward angles and nuclear
resonance fluorescence with polarized photons $(\vec{\gamma}, \gamma')$[6].

There are two reasons why inelastic hadron scattering with
light projectiles (p, d, ^{3}He, α) is well suited to study spin flip
transitions. In table 1 projectiles and their selection rules for
allowed inelastic scattering processes are listed.

Table 1: Allowed inelastic scattering transitions

projectile	isospin: ΔT	spin: ΔS	
p, d, ^{3}He, α	0	0	(isoscalar, non-spin flip)
p, d, ^{3}He	0	1	(isoscalar, spin flip)
p, ^{3}He	1	0	(isovector, non-spin flip)
p, ^{3}He	1	1	(isovector, spin flip)

While protons and ^{3}He particles can induce all kinds of transitions
deuterons do not excite isovector states but they may induce spin
flip transitions. Alpha particles are even more selective, only
isoscalar non-spin flip transitions are allowed. These selection
rules are of course based on the assumption of direct reaction
mechanism. When these processes are dominant as e.g. at high indident
energies, the described selectivity can be used to determine the
spin-isospin mode of the investigated transition.

The second reason for the usefulness of inelastic scattering
is the energy dependence of the nucleon-nucleon interaction for
different spin-isospin modes which have been calculated by Petrovich
and Love[7]. The most favorable energy to study magnetic dipole (M1)
transition ranges from 150 to 400 MeV where the volume integral of
the spin-isospin interaction $V_{\sigma\tau}$ becomes relatively large compared
to the isoscalar spin – independent central component V_0. In addition to the particle and energy selectivity the shape of measured
angular distributions will help to identify and study 1$^+$ states.
Since in the excitation of the unnatural parity 1$^+$ spin flip state
no parity change is involved, the transfered orbital angular momentum must be even.

EXPERIMENTAL EQUIPMENT

In the present investigation of 1$^+$ states we used the energy
variable Jülich isochronous cyclotron JULIC [8] which provides light
ion particle beams (p, d, ^{3}He, α) from E = 22.5 to 45 MeV/A. The
required high resolution beam (E/ΔE = 5000 – 10000) was obtained
using the dispersive mode of the double analyzing magnet system. In
this mode the beam is reduced to about 100 nA. The inelastically
scattered particles are momentum analyzed in the high resolution
spectrometer BIG KARL[9]. Fig. 1 gives a schematic view of the QQDDQ
magnet system and some of the design values. The dispersion can be

varied using the last quadrupol Q3. Kinematical corrections are
accomplished with quadrupole Q2. The H_t correction coils[9],[10] in
both dipoles D1, D2 allow the correction of higher order aberrations.
Solid angles of up to 3.5 msr are used typically in high resolution
experiments where $E/\Delta E$ = 5000 - 10000 which is mostly limited by
the properties of the incoming beam.

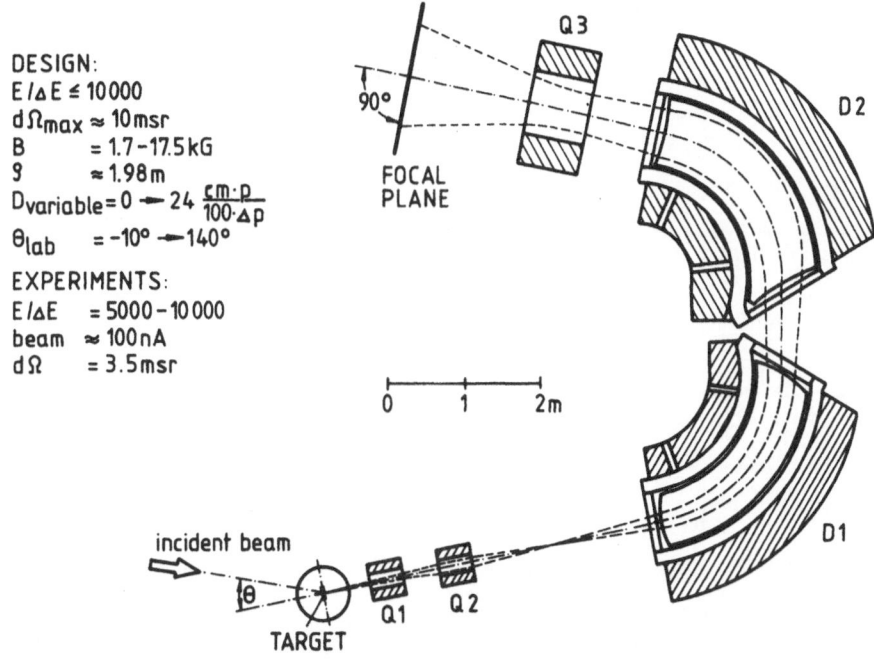

DESIGN:
$E/\Delta E \leq 10000$
$d\Omega_{max} \approx 10$ msr
B = 1.7-17.5 kG
9 ≈ 1.98 m
$D_{variable} = 0 \rightarrow 24 \frac{cm \cdot p}{100 \cdot \Delta p}$
θ_{lab} = $-10° \rightarrow 140°$

EXPERIMENTS:
$E/\Delta E$ = 5000-10000
beam ≈ 100 nA
$d\Omega$ = 3.5 msr

Fig. 1: Schematic view of the QQDDQ magnetic spectrometer BIG KARL.
        Design values and typical high resolution conditions during
        experiments are shown.

THE 1⁺ STATE IN ⁴⁸Ca AT 10.212 MEV

     In this experiment[11] we measured the 1⁺ state at
E_x = (10.212 ± 0.009) MeV which has been identified for the first
time by the Darmstadt group[4] at 10.227 MeV using inelastic electron
scattering at backward angles. The M1 strength is concentrated in
this state and only a very small fraction of the strength is fragmented into other states lying nearby. Fig. 2 shows some of the
high resolution spectra measured in our experiment[11]. At forward
angles the 1⁺ state shows a very dominant peak while at more backward angles many other states with high orbital angular momenta show
up. According to ref. 12 there are a few weakly excited 1⁺ states
identified in (e,e') measurement at E_x = 9.880, 10.138 and 10.350
MeV besides the strong 1⁺ state at 10.227 MeV. We find weakly excited states at E_x = 9.865, 10.126 and 10.345 MeV, too. The angular

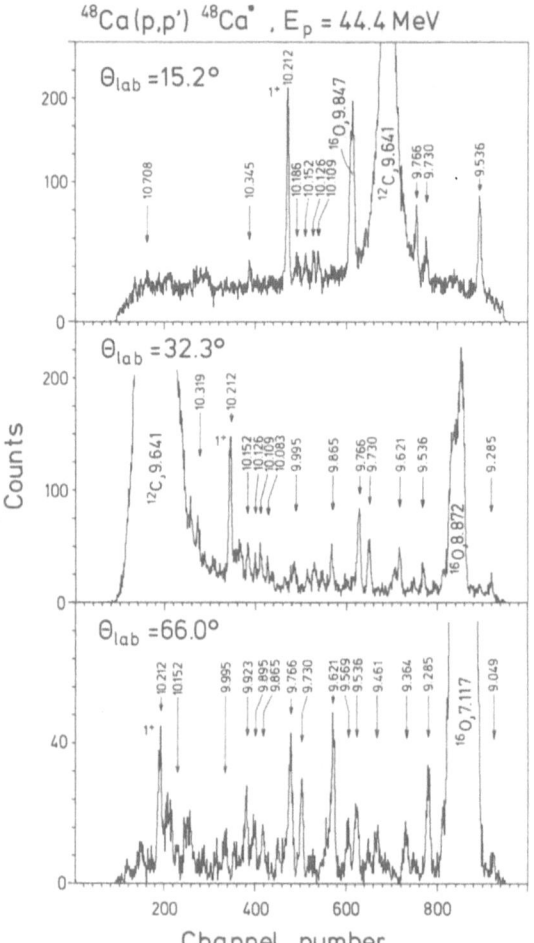

Fig. 2: Sample spectra of ^{48}Ca(p,p') from ref. 11. The resolution
of 7 - 10 keV allowed the identification of 50 new levels.
Note the strongly excited 1$^+$ state at E_x = 10.212 MeV.

distributions agree within the error limit with the data of the 1$^+$
state showing that some small amount of M1 strength is found in
these states. According to the Osaka group[13] the 1$^+$ state at 10.212
MeV is only very weakly excited in α-scattering measurements con-
firming that this state corresponds to a spin flip transition. Inelastic deuteron scattering measurements and also the comparison
with (p,n) data could help to determine the isospin mode of this
transition.

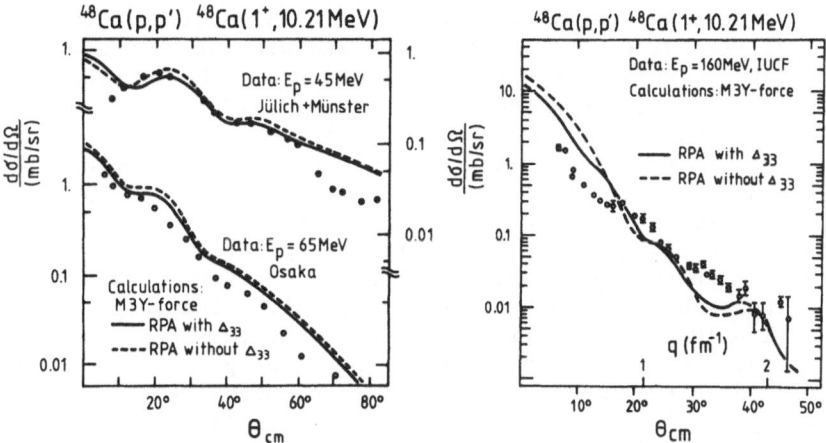

Fig. 3: Angular distributions of ^{48}Ca(p,p') leading to the 1^+ state
at E_x = 10.21 MeV at three different energies: 45 MeV
(ref. 11), 65 MeV (ref. 13) and 160 MeV (ref. 14). The
microscopic DWBA calculations with RPA wavefunctions
(for details see ref. 17) show the effects of the inclusion
of the Δ_{33}-resonance (solid line). The dashed line shows
the results without coupling to the Δ_{33}-resonance.

    In Fig. 3 (p,p')-data to the 1^+, 10.212 MeV state in ^{48}Ca are
shown for three different incident energies. The 160 MeV data have
been measured at the Indiana University cyclotron facility[14](IUCF),
the 65 MeV data are those of the Osaka group[13], and the 45 MeV data
have been measured in Jülich[11]. The experimental angular distributions are compared to microscopic cross section calculations which
treat the Δ(1232)-isobar degrees of freedom explicitly in both the
structure (random phase approximation RPA calculations of Suzuki,
Krewald and Speth[15]) and the DWBA reaction calculations[16] (see also
the contribution of Speth et al. to this conference[17]). For the
effective projectile-target nucleon interaction we used the M3Y-
force of Bertsch et al.[18] in case of the 45 and 65 MeV data, and
the G3Y-interaction of Love and Petrovich[7,19] in the case of the 160
MeV data. For each energy calculations are shown with inclusion of
the Δ(1232)-isobar effect (solid line) and without this effect
(dashed curve). While the explicit treatment of the Δ-isobar effect
results in a 10 % to 20 % reduction   (quenching) of the cross section at most scattering angles, there are some regions in the angular distributions where instead of a reduction an enhancement occurs.
This happens even at very forward angles for 45 MeV proton scattering
although one would expect quenching at these angles. This effect
has its origin in the hadronic transition operator the shape and
magnitude of which depends strongly on the incident projectile
energy. It has been shown by Osterfeld et al.[20] that high energy
protons (E > 100 MeV) penetrate much deeper into the nucleus and
are therefore much more sensitive to the inner part of the nuclear
transition density than low energy protons. Since the Δ-isobar nucleon-hole part of the M1-transition density is largest in the nuclear interior the Δ-isobar effect on the cross section should also
be largest when the incoming proton really probes the nuclear
interior. Note that all nucleons in the nucleus take part in the
ΔN^{-1} transitions while only nucleons around the Fermi edge contributes to the NN^{-1}-excitations. This happens in a most efficient
way at high incident energies (E > 100 MeV). This effect can
actually be seen in Fig. 3 by comparing the theoretical cross sections calculated with "RPA" and "RPA with Δ_{33}". The quenching is
larger at 160 MeV than at lower energies (45 and 65 MeV).

    Proton inelastic measurements have been performed recently
at 201 MeV[21] and measurements at 250 and 350 MeV using polarized
protons are planned in the near future using the MRS spectrometer
at TRIUMF.

SEARCH FOR 1^+ STATES IN ^{58}Ni

    According to shell model calculations[22] several isoscalar 1^+
states are expected in ^{58}Ni (E_x = 7.2, 7.6, 8.2, 8.4, 8.6 MeV).
From (γ,γ') and (e,e') measurements[5,23] J = 1 states were found at
E_x = 7.051, 7.710 and 8.240 MeV. In order to search for 1^+ states

in this energy region we measured angular distributions of ^{58}Ni(p,p')
at 45 MeV. Fig. 4 shows spectra at three different angles
with a resolution of 11 — 16 keV increasing slightly for larger
angles. More than 40 new levels have been identified in the range
from E_x = 7.1 to 8.4 MeV. The excitation energies shown in Fig. 4
are accurate within about ± 5 to 10 keV given by the uncertainty of the ^{48}Ca level energies[24] used as calibration standards. Measured angular distributions of the strongly excited levels are shown in Fig. 5. The solid lines represent macroscopic DWBA calculations using standard optical model parameters[25]. The transfered orbital angular momenta ℓ determined by comparing data and calculations are shown in Fig. 5. The dominantly excited state at E_x = 7.204 MeV displays clearly a ℓ = 3 shape. The state at 7.721 MeV agrees well with the ℓ = 2 calculation but a 1⁺ assignment cannot be excluded. In fact recent (γ,γ') and (e,e') measurements reveal a 1⁺ state at E_x = 7.710 MeV. Therefore we show in Fig. 5 for comparison also the DWBA calculation[11] for the transition to the 1⁺ state in ^{48}Ca showing the typical shape of a 1⁺ transition (see dotted line). Calculated and measured angular distributions are very similar.

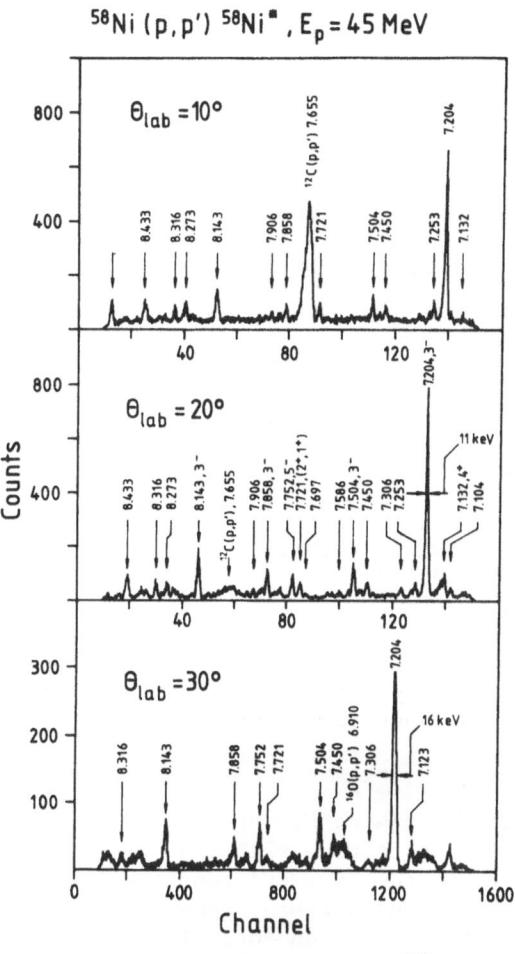

Fig. 4: Measured spectra of ^{58}Ni(p,p')
at three different angles.
For details see text.

The data of the weak states at 7.697 MeV do not allow discrimination between ℓ = 1 or ℓ = 2 but the shape is different from the 1⁺ state in ^{48}Ca. At 7.046 and 8.246 MeV close to the J = 1 states at 7.051 and 8.240 MeV from refs. 5, 23 we find two states but our data do not allow a determination of ℓ.

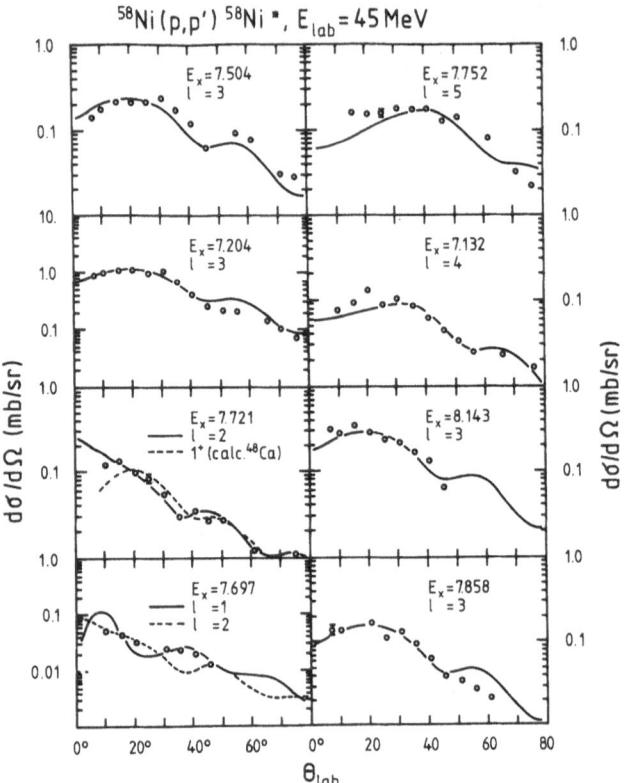

Fig. 5: Measured angular distributions of ^{58}Ni(p,p') and macroscopic
DWBA calculations. The absolute cross sections are accurate
within ± 30 % only due to the uncertainty of the target
thickness.

We conclude that in the measured range no strong 1^+
state is excited in inelastic proton scattering with the possible
exception of the 7.721 MeV state where the data are consistent with
the shape of a 1^+ state. We plan to extend our measurements to
higher energies where spin-flip states have been identified in
(e,e') measurements by Lindgren et al.[5]. Two strongly excited
states have been seen by the Orsay group[26] in ^{58}Ni(p,p') at
$E_x \sim 8.66$ MeV and 10.65 MeV at $\Theta_{lab} = 4^o$. While the 10.66 MeV level
is found to be a 1^+ state, the 8.65 MeV level is suggested to be
an E1 transition. Preliminary measurements of ^{58}Ni(p,p') at 45 MeV
show an extremly high level density at $E_x \sim 10$ MeV, so that the
identification of these states will be difficult.

THE ISOSCALAR 1$^+$ STATE IN ^{208}Pb AT 5.846 MEV

In ^{208}Pb the shell model predicts 1$^+$ states at the unperturbed energies of E_x = 5.86 and 5.55 MeV from simple one-particle one-hole configurations $\nu(i_{13/2}^{-1}, i_{11/2})$ and $\pi(h_{11/2}^{-1}, h_{9/2})$. In spite of many experimental investigations[27] only very few 1$^+$ assignments[28] have remained unchallenged. In a recent $(\vec{\gamma},\gamma')$ measurement[29] using polarized photons of 10 MeV a 1$^+$ state at E_x = (5.846 ± 0.001) MeV has been identified in ^{208}Pb. Spectra in this energy region[30] from ^{208}Pb(p,p') and ^{208}Pb(d,d') are shown in Fig. 6. At E_x = (5.841 + 0.002) MeV a weakly excited state is seen in (p,p'). In the (d,d^T) spectra a very weak excitation is possible at the same level energy

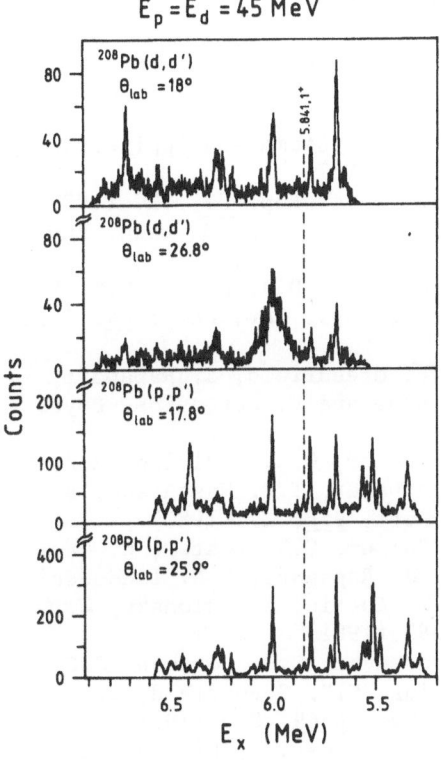

Fig. 6: Inelastic proton and deuteron spectra at 18° and 26° at 45 MeV in the range of E_x = 4.8 to 6.3 MeV. Note the weakly excited 1$^+$ state at 5.841 known from $(\vec{\gamma},\gamma')$ measurements[29]. The resolution was about 10 keV in the proton and slightly worse in the deuteron spectra.

which barely rises above the background in this measurement which has been carried out before the 1$^+$ state was identified. Clearly data with better statistics are needed before one can to draw quantitative conclusions on the isoscalar character of this state, which is believed to be the 1$^+$ state identified in the $(\vec{\gamma},\gamma')$ measurement. Since this transition involves a spin flip it cannot be excited in direct inelastic α-scattering. This measurement could be used as independent check of the spin-flip character of the transition.

These few examples show that high resolution inelastic hadron scattering using different particles and energies in connection with the information from other experiments is a promising tool to investigate spin flip 1$^+$ states in medium and heavy mass nuclei.

ACKNOWLEDGMENT

We would like to thank T. Suzuki, S. Krewald, and J. Speth for supplying us with their RPA wave functions and Profs. J. Speth and O.W.B. Schult for support for these investigations and valuable discussions.

REFERENCES

1. A.B. Migdal, Theory of Finite Fermi Systems and Applications to Atomic Nuclei (Wiley, New York, 1967)
2. J. Speth, E. Werner, and W. Wild, Phys. Rep. 33, 127 (1977)
3. R.R. Doering, A. Galonsky, D. Patterson, G.F. Bertsch, Phys. Rev. Lett. 35, 1961 (1975)
    D.E. Bainum, J. Rapaport, C.D. Goodman, D.J. Horen, C.C. Foster, M.B. Greenfield and C.A. Goulding, Phys. Rev. Lett. 44, 1751 (1980)
    C.D. Goodman, C.A. Goulding, M.B. Greenfield, J. Rapaport, D.E. Bainum, C.C. Foster, W.G. Love and F. Petrovich, Phys. Rev. Lett. 44, 1755 (1980)
    D.J. Horen, C.D. Goodman, C.C. Foster, C.A. Goulding, M.B. Greenfield, J. Rapaport, E. Sugarbaker, T.G. Masterson, F. Petrovich and W.G. Love, Phys. Lett. 95B, 27 (1980)
    D.J. Horen, C.D. Goodman, D.E. Bainum, C.C. Foster, C. Gaarde, C.A. Goulding, M.B. Greenfield, J. Rapaport, T.N. Taddeucci, E. Sugarbaker, T. Masterson, S.M. Austin, A. Galonsky, W. Sterrenburg, Phys. Lett. 99B, 383 (1981)
    C. Gaarde, J. Rapaport, T.N. Taddeucci, C.D. Goodman, C.C. Foster, D.E. Bainum, C.A. Goulding, M.B. Greenfield, D.J. Horen and E. Sugarbaker, Nucl. Phys. A369, 258 (1981)
4. W. Steffen, H.D. Gräf, W. Gross, D. Meuer, A. Richter, E. Spamer, O. Titze and W. Knüpfer, Phys. Lett. 95B, 23 (1980)
5. R.A. Lindgren, W.L. Bendel, E.C. Jones, L.W. Fagg, X.K. Maruyama, J.W. Lightbody and S.P. Fivozinsky, Phys. Rev. C14, 1789 (1976)

6. U.E.P. Berg, D. Rück, K. Ackermann, K. Bangert, C. Bläsing, K. Kobras, W. Naatz, R.K.M. Schneider, R. Stock and K. Wienhard, Phys. Lett. 103B, 301 (1981)

7. F. Petrovich and W.G. Love, Nucl. Phys. A354, 499c (1981).

8. J. Reich, S.A. Martin, D. Protic and G. Riepe, Seventh Int. Conf. on Cyclotrons and their Applications, Zürich, Switzerland, 1975

9. S. Martin, G. Berg, A. Hardt, W. Hürlimann, M. Köhler, J. Meissburger, T. Sagefka, O.W.B. Schult, Use of Magnetic Spectrometers in Nuclear Physics, Proc. of the Daresbury Study Weekend 10 - 11 March 1979, Ed. N.E. Sanderson, SRC Daresbury Laboratory 1979

10. K. Halbach, Nucl. Instr. Meth. 107, 515 (1973)

11. G.P.A. Berg, W. Hürlimann, I. Katayama, S.A. Martin, J. Meissburger, J. Römer, B. Styczen, F. Osterfeld, G. Gaul, R. Santo and G. Sondermann, Phys. Rev. C25, 2100 (1982)

12. A. Richter, private communication

13. Y. Fujita, M. Fujiwara, S. Morinobu, T. Yamazaki, T. Itahashi, S. Imanishi, H. Ikegami and S.I. Hayakawa, Phys. Rev. C25, 678 (1982)

14. P. Kienle, D.W. Miller, K.W. Rehm, and R.E. Segel, Verhandl. DPG (VI) 16, 742 (1981) J. Comfort, private communication.

15. T. Suzuki, S. Krewald, and J. Speth, Nucl. Phys. A306, 360 (1978)

16. T. Suzuki, S. Krewald, T. Suzuki, and J. Speth, to be published

17. J. Speth, S. Krewald, F. Osterfeld and T. Suzuki, see contribution to this conference

18. G. Bertsch, J. Broysowicz, H. McManus, and W.G. Love, Nucl. Phys. A284, 399 (1977)

19. W.G. Love and M.A. Franey, Phys. Rev. C24, 1073 (1981)

20. F. Osterfeld, J. Wambach, H. Lenske and J. Speth, Nucl. Phys. A318, 45 (1979)

21. G. Crawley, see contribution to this conference

22. J. McGrory and J. Speth, private communication

23. K. Ackermann, K. Bangert, U.E.P. Berg, G. Junghans, R.K.M. Schneider, R. Stock and K. Wienhard, Nucl. Phys. A372, 1 (1981)

24. Nuclear Data Sheets, Vol. 23, Nr. 1 (1978)

25. F.D. Becchetti and G.W. Greenless, Phys. Rev. 182, 1190 (1969)

26. N. Marty, private communication. C. Djalali, N. Marty, M. Morlet, A. Willis, J.C. Jourdain, N. Anantaraman, G.M. Crawley, A. Galonsky, and P. Kitching, Nucl. Phys. A388, 1 (1982).

27. S. Raman, Proc. of the Int. Symp. Neutron Capture Gamma-Ray Spectroscopy, Stony Brook 1978, 193

28. G.E. Brown and S. Raman, Comm. Nucl. Part. Phys. 9, 79 (1980)

29. U.E.P. Berg, private communication

30. G.P.A. Berg, P. Decowski, I. Katayama, S.A. Martin, J. Meissburger, H.P. Morsch, M. Rogge, B. Styczen and P. Turek, Verhandl. DPG (IV) 16, 617 (1981)

STUDIES OF SPIN EXCITATIONS WITH ELECTROMAGNETIC AND HADRONIC PROBES

Richard A. Lindgren

Department of Physics and Astronomy
University of Massachusetts
Amherst, MA 01003

Fred Petrovich

Department of Physics
Florida State University
Tallahassee, FL 32306

ABSTRACT

Excitation of unnatural parity states, predominantly of high spin, using electromagnetic and hadronic probes, is discussed. Spectroscopic strengths are deduced from studies of (e,e'), (p,p'), (π,π'), and (p,n) for states whose "doorway" is the "stretched" particle-hole configuration. These levels are excited primarily through the isovector electromagnetic-nucleon magnetization coupling, nucleon-nucleon tensor coupling, and pion-nucleon spin-orbit coupling. The extracted isovector spectroscopic strength is typically 38% of the extreme single particle-hole model and about 66% of that predicted by more realistic nuclear structure calculations. The observed isoscalar strength is only about one half of the isovector strength. The results obtained with the three different probes are quite consistent. The primary conclusion is that the "missing strength" for these high spin excitations is at least as large as for the low spin M1 and GT excitations. This implies the existence of other important "quenching" mechanisms since the Δ-N^{-1} mechanism involved in the discussion of the low spin excitation affects only the isovector transitions and contributes little to high spin excitations. A method for using (e,e') and π^{+}/π^{-} cross section ratios to separate and determine the absolute isoscalar and isovector spin densities for T_0 to T_0 transitions in $N \neq Z$ nuclei is also discussed and some comments on extracting information from (e,e') and (p,p') studies at high q on low spin 1^{+} and 2^{-} levels are presented.

INTRODUCTION

The development of medium energy accelerators and high-resolution spectrometers such as those at the Bates Linear Accelerator (Bates), Indiana University Cyclotron Facility (IUCF), Los Alamos Meson Physics Facility (LAMPF), and others have made it possible to significantly extend the measurement of nuclear observables to include the spin modes of nuclear excitations. Furthermore, the expectation that comparative spectroscopic analysis of electron and hadron scattering cross sections in this energy regime would be helpful in disentangling effects due to nuclear structure from those due to reaction dynamics in the study of hadron-nucleus scattering are being fulfilled. Up to this point most of the attention of this conference has been centered on the spin excitations with low angular momentum, i.e. GT and M1. In this presentation, I would like to highlight recent electron scattering data on high spin states that are predominantly excited via coupling to the nuclear magnetization current and illustrate how the complementary aspects of the electromagnetic and hadronic probes can be used to obtain important information on nuclear structure, effective interactions, and reaction models.

The high spin unnatural parity states, particularly those of "stretched" configurations are ideal for these proposed studies. In fact it was first noted by Donnelly and Walker[1] that in light nuclei the distribution of magnetic strength to high spin unnatural parity states as measured in (e,e'), was particularly sensitive to the strength of the spin-dependent parts of the effective interactions used in nuclear structure model calculations. Later, Moffa and Walker[2] reported that for special high spin unnatural states (e.g., "stretched" $(f_{7/2}d_{5/2}^{-1})$ 6^- particle-hole states in sd-shell nuclei) structure deduced from (e,e') could be used in (p,p') and (p,n) to extract information on the high momentum components of the effective interaction, in particular the tensor component. The first systematic comparison of the (e,e') and (p,p') transition strengths[3] for the excitation of unnatural parity states of "stretched" configurations verified that this observation was correct for proton scattering in self-conjugate nuclei. The tensor component is particularly interesting since its microscopic origin, in the low q limit, evolves naturally from one pion exchange considerations with significant contributions from one rho exchange in the high q region.[4] Since it is not known whether the most useful theory of strong interactions is described by nucleons interacting through pion and rho exchange or quarks interacting through gluon exchange, measurements on the tensor force in the nucleus are valuable in discriminating such models. Recently, Petrovich and Love[5] summarized the cross section formulae for (e,e'), (p,p'), (p,n) and also $(\pi^{\pm},\pi^{\pm})$ and illustrated how these processes are interrelated through nuclear matter, spin, and orbital transition densities.

First, a summary of relevant cross section formulae for electron, proton, and pion scattering to unnatural parity excitations is presented. Then a discussion of the special case of excitations of "stretched" configurations is given, highlighting the simplicity in the spin density resulting from the "stretched" angular momentum condition. A review of electron scattering data on selected high spin states and a comparison with nuclear model calculations follows. A comparison of (e,e'), (p,p'), (p,n), and $(\pi^{\pm},\pi^{\pm})$ results for selected transitions to states of "stretched" configurations is then given. Finally, similar comparisons are made for transitions to states of "nonstretched" configurations illustrating the additional considerations that arise when there are several independent contributions to the cross sections.

ELECTRON, PROTON, AND PION INELASTIC SCATTERING CROSS SECTIONS

Inelastic electron, proton, and pion scattering cross sections for unnatural parity excitations involve the magnetic multipoles of the electromagnetic interaction, the spin dependent central, tensor, and spin-orbit components of the nucleon-nucleon interaction, and the spin-orbit component of the pion-nucleon interaction and thus depend on the isoscalar and isovector spin and orbital transition densities of the target nucleus. Following closely the work of Ref. 5, we present here plane wave Born approximation expressions for the inelastic differential cross sections for $0^{+} \rightarrow J^{\pi}$ unnatural parity excitations in electron, proton, and pion-nucleus scattering to serve as a basis for the discussion to follow.

For the case of electrons[5,6]

$$\frac{d\sigma}{d\Omega} = \frac{z^2 \sigma_M}{\eta} \left(\frac{1}{2} + \tan^2 \frac{\theta}{2}\right) F_T^2(q) \tag{1}$$

where σ_M is the Mott cross section, η is a recoil factor, θ is the scattering angle, q is the momentum transfer, and $F_T^2(q)$ is the transverse form factor. The latter is given by

$$F_T^2(q) = \frac{4\pi}{z^2} (2J+1) \left|\frac{q\hbar}{2Mc} \sum_T \{\frac{1}{2} g_T^s \rho_{JT}^s(q) - 2g_T^\ell \rho_{JT}^\ell(q)\}\right|^2 \tag{2}$$

where T is an isospin index, g_T^s and g_T^ℓ are the spin and orbital g-factors, ρ_{JT} is the diagonal current transition density, and

$$\rho_{JT}^{s\perp}(q) = \left(\frac{J+1}{2J+1}\right)^{1/2} \rho_{JJ-1}^{sT}(q) - \left(\frac{J}{2J+1}\right)^{1/2} \rho_{JJ+1}^{sT}(q) \tag{3}$$

is the transverse spin transition density. In a shell model picture, the ρ_{JL}^{sT} and ρ_{JT}^ℓ transition densities can be written as a sum of a product of one body density matrix elements and single particle matrix elements as follows

$$\rho_{JL}^{sT}(q) = \sum_{j_a j_b} Z_{JT}(j_a j_b) \sqrt{2} \, \hat{j}_a \, \hat{J}^{-1} <j_a||j_L(qr)[Y_L(\hat{r}) x \bar{\sigma}]^J ||j_b> \qquad (4)$$

$$\rho_{JT}^{\ell}(q) = \sum_{j_a j_b} Z_{JT}(j_a j_b) \sqrt{2} \, \hat{j}_a \hat{J}^{-1} <j_a||q^{-1} j_J(qr)[Y_J(\hat{r}) x \nabla]^J ||j_b> \qquad (5)$$

where Z_{JT} is a spectroscopic amplitude proportional to the one body density matrix elements[7]

$$Z_{JT}(j_a j_b) = < J_f T_f ||[a_{j_a}^+ \, x \, \bar{a}_{j_b}]^{J,T} ||J_i = 0 \, T_i> \qquad (6)$$

which assumes the value unity for a simple isoscalar or isovector particle-hole excitation. When the transition densities are constructed from the shell model, i.e. eqs. (4) – (6), it is necessary to apply two corrections before using them in eq. (2).[6] One of these is the center of mass correction and the other is for the finite nucleon size. These may be indicated schematically as

$$\rho(q) = \rho_{SM}(q) \, \rho_{cm}(q) \, \rho_N(q). \qquad (7)$$

Only the center of mass correction is necessary for the hadronic reaction to be discussed next. The differential cross section for inelastic nucleon scattering depends on the longitudinal spin transition density

$$\rho_{JT}^{s,\|}(q) = (\frac{J}{2J+1})^{1/2} \rho_{JJ-1}^{sT}(q) + (\frac{J+1}{2J+1})^{1/2} \rho_{JJ+1}^{sT}(q) \qquad (8)$$

as well as the $\rho_{JT}^{s\perp}$ and ρ_{JT}^{ℓ} which appear in electron scattering. Specifically, in PWBA[5,8-10]

$$\frac{d\sigma^N}{d\Omega} = 4\pi \, (\frac{M_N}{2\pi\hbar^2})^2 \, (2J+1) \, [\,|\sum_T \bar{v}_T^\|(q) \, \rho_{JT}^{s\|}(q)^2$$

$$+ \frac{1}{2} k_N^4 \, q^{-4} \, \sin^2\theta \, |\sum_T \alpha_p \, \bar{v}_T^{LS}(q) \, \rho_{JT}^{s\perp}(q)|^2$$

$$+ \frac{1}{2} |\sum_T \{\alpha_t \bar{v}_T^{LS}(q) \, \rho_{JT}^{\ell}(q) + [\bar{v}_T^\perp(q) + \bar{v}_T(q)] \rho_{JT}^{s\perp}(q)\}|^2$$

$$+ \frac{1}{2} |\sum_T \{\alpha_t \bar{v}_T^{LS}(q) \, \rho_{JT}^{\ell}(q) + [\bar{v}_T^\perp(q) - \bar{v}_T(q)] \rho_{JT}^{s\perp}(q)\}|^2 \qquad (9)$$

where M_N (k_N) is the reduced energy (wave number) of the incident nucleon, the factor α_i arise from the decomposition of the relative nucleon-nucleon momentum $\bar{p} = \alpha_p \bar{P}_p - \alpha_t \bar{P}_t$, $\bar{v}_T^{LS}(q)$ refers to the spin orbit component of the effective nucleon-nucleon interaction, and

$$\bar{v}_T^{\parallel}(q) = \bar{v}_T^C(q) - 2\bar{v}_T^T(q) \tag{10}$$

$$\bar{v}_T^{\perp}(q) = \bar{v}_T^C(q) + \bar{v}_T^T(q) \tag{11}$$

are the longitudinal and transverse combinations of the central and tensor interaction components. The bars on the interaction components are to indicate that they are defined to include, approximately, contributions associated with knockout exchange.[8-10] The component $\bar{v}_T$ is solely due to tensor exchange.[8-10] To apply to charge exchange, eq. (9) must be multiplied by 2 and T is restricted to unity.

In the simplest picture of pion-nucleus scattering the excitation of unnatural parity levels proceeds entirely through the spin-orbit component of the effective pion-nucleon interaction. In PWBA the differential cross section is analogous to the second term in eq. (9) for nucleon-nucleus scattering[5,11,12]

$$\frac{d\sigma^{\pi}}{d\Omega} = 4\pi \left(\frac{M_{\pi}}{2\pi\hbar^2}\right)^2 (2J+1) \frac{1}{2} k_{\pi}^4 q^{-4} \sin^2\theta \left|\sum_T \alpha_{\pi} t_T^{LS}(q) \rho_{JT}^{s\perp}(q)\right|^2 \tag{12}$$

where M_{π} and k_{π} are the reduced energy and incident wave number of the incident pion, α_{π} is as α_t in eq. (9), and t_T^{LS} is the spin-orbit component of the effective pion-nucleon interaction. In this framework pion-nucleus scattering determines uniquely the transverse spin density. In the important incident energy region of the (3,3) resonance the pion-nucleon interaction is very non-local. This gives rise to corrections to eq. (12) which produce coupling to ρ_{JT}^{ℓ} and spin-current transition densities.[13-16] These will not be considered here.

In electron scattering, $g_1^s \gg g_0^s$, g_0^{ℓ}, and g_1^{ℓ} so this reaction provides mainly information on transitions where $\rho_{JT}^{s\perp}$ is large. For $q \simeq 2$ fm^{-1}, the approximate momentum transfer for observing high spin excitations $v_1^{\perp}$ is dominant,[5,8-10] so nucleon-nucleus scattering also "sees" isovector spin-flip states of high spin strongly. This reaction also provides information on isoscalar high spin states because both $\bar{v}_0^T$ and $\bar{v}_0^{LS}$ are appreciable at $q \sim 2$ fm^{-1} with most of $\bar{v}_0^T$ coming from knockout exchange.[5,8-10] For the important incident energy region corresponding to the (3,3) resonance, $t_0^{LS} = 2t_1^{LS}$, so pion-nucleus scattering favors isoscalar transitions.

STATES OF STRETCHED CONFIGURATION

The relationship between the electron, nucleon, and pion-nucleus scattering cross sections discussed above is greatly simplified for the excitation of states of stretched configurations. The "stretched" particle-hole configuration is of the form $(j_a j_b^{-1})_{J_{MAX}}$, where $J_{MAX} = j_a + j_b$, $j_a = \ell_a + 1/2$, $j_b = \ell_b + 1/2$, and j_a and j_b are the largest angular momentum found in the last filled shell and the first open shell, respectively. This configuration is unique in a space excluding 1p-1h excitations with $E > 3\hbar\omega$, so there is no mixing with other 1p-1h configurations. Of course, the "stretched" configurations can mix with multi-particle-multi-hole configurations within the same shell. Such mixing produces physical ground and excited states that are not pure closed shell and particle-hole wavefunctions, respectively; however, the additional components in the wave functions will not be connected by the one body spin and orbital current operators. The only effect is then a reduction of the transition strength with the cross sections remaining proportional to a single particle matrix element corresponding to the stretched configuration. An additional simplification that arises is that the ρ_{JT}^{ℓ} and ρ_{JJ+1}^{sT} transition densities vanish as a result of angular momentum restrictions on the single particle matrix element. Consequently, the longitudinal and transverse spin densities are related by

$$\rho_{JT}^{s\parallel}(q) = (\frac{J}{J+1})^{1/2} \rho_{JT}^{s\perp}(q) = (\frac{J}{2J+1})^{1/2} \rho_{JL}^{sT}(q), \tag{13}$$

and the electron, pion, and nucleon-nucleus scattering cross sections of eq. (2), eq. (9), and eq. (12) are connected through a common density.

Specifically, for transitions to pure isovector, isoscalar, neutron, or proton stretched states

$$F_T^2(q) = \frac{4\pi}{Z^2} (J+1) \left| \frac{q\hbar}{2Mc} \frac{1}{2} g_\alpha^s \rho_{JJ-1}^{s\alpha}(q) \right|^2 \tag{2'}$$

$$\frac{d\sigma^N}{d\Omega} = 4\pi(\frac{M_N}{2\pi\hbar^2})^2 [J|\bar{v}_\alpha^{\parallel}(q)|^2 + (J+1)\{|\bar{v}_\alpha^{\perp}(q)|^2 + |\bar{v}_\alpha(q)|^2\}$$

$$+ \frac{1}{2}(J+1)k_N^4 q^{-4} \sin^2\theta \; \alpha_p^2 |\bar{v}_\alpha^{LS}(q)|^2][\rho_{JJ-1}^{s\alpha}(q)]^2 \tag{9'}$$

$$\frac{d\sigma^\pi}{d\Omega} = 4\pi(\frac{M_\pi}{2\pi\hbar^2})^2 (J+1)\frac{1}{2} k_\pi^4 q^{-4}\sin^2\theta \; \alpha_\pi^2 \; |t^{LS}(q)\rho_{JJ-1}^{s\alpha}(q)|^2 \tag{12'}$$

where $\alpha = 0$, 1, p, or n, $g_{p,n} = g_0 \pm g_1$, $\bar{v}_{p,n} = \bar{v}_0 \pm \bar{v}_1$ for incident protons ($\mp$ for incident neutrons), $t_{p,n} = t_0 \pm t_1$ for incident π^+ ($\mp$ for incident π^-). An explicit expression for the transition density is

$$\rho_{JJ-1}^{s\alpha}(q) = (-1)^{\ell_b} (4\pi)^{-1/2} C_\alpha Z_{J\alpha} J^{-1/2} \hat{j}_a \hat{j}_b <j_a j_b \tfrac{1}{2} - \tfrac{1}{2}|J0> \quad x$$

$$<n_a \ell_a |j_{J-1}(qr)|n_b \ell_b> \qquad (14)$$

where $C_\alpha = \sqrt{2}$ for $\alpha = 0$ and 1, and $C_\alpha = 1$ for $\alpha = p$ and n, and $Z_{p,n} = (Z_0 \pm Z_1)/\sqrt{2}$ assume the value unity for pure proton and neutron particle-hole excitations. For harmonic oscillator radial wave functions and the stretched state conditions $n_a = n_b = 1$, $\ell_a = \ell_b + 1$, and $J = \ell_a + \ell_b + 1$, the radial overlap in eq. (14) is given by

$$<n_a \ell_a |j_{J-1}(qr)|n_b \ell_b> = \frac{(bq)^{J-1} \exp(-b^2 q^2/4)}{2^{(J-1)/2}[(J-1)!!(J+1)!!]^{1/2}} \qquad (15)$$

where b is the oscillator length parameter. A useful analytic expression for the transverse electron scattering form factor is obtained by substituting eq. (14) and eq. (15) into eq. (2').

$$F_T^2(q) = |\frac{\rho_{cm}(q)\rho_N(q)}{Z} \frac{q\hbar}{2Mc} (\frac{J+1}{J})^{1/2} \frac{1}{2} C_\alpha Z_{J\alpha} g_\alpha^s \quad x$$

$$\hat{j}_a \hat{j}_b <j_a j_b \tfrac{1}{2} - \tfrac{1}{2}|J0> \frac{(bq)^{J-1} \exp(-b^2 q^2/4)}{2^{(J-1)/2}[(J-1)!!(J+1)!!]^{1/2}}|^2 \qquad (2'')$$

Similar expressions can be constructed for the nucleon and pion inelastic cross sections, but will not be given here.

POPULATION OF STRETCHED STATES VIA ELECTRON SCATTERING

States of the $(d_{5/2}p_{3/2}^{-1})$ 4^- type were first observed at 19.5 MeV in ^{12}C and 18.98 MeV excitation in ^{16}O, and of the $(f_{7/2}d_{5/2}^{-1})$ 6^- type at 14.35 MeV in ^{28}Si over a decade ago using 0.5 MeV energy resolution electron scattering.[17-19] More recently, new data has been accumulated with considerably higher energy resolution. Specifically, new data[20,21,22] on ^{12}C, ^{16}O, and ^{28}Si have been obtained and additional levels[23,24] of the same configuration have been observed in ^{14}C and ^{24}Mg. Stretched levels have also been seen in heavier nuclei,[25,26,27,28,29] in particular, $(g_{9/2}f_{7/2}^{-1})$ 8^- states in ^{54,56}Fe and ^{58,60}Ni, $(h_{11/2}g_{9/2}^{-1})$ 10^- in ^{90}Zr, and $(i_{13/2}h_{11/2}^{-1})$ 12^- and $(j_{15/2}i_{13/2}^{-1})14^-$ in ^{208}Pb. Typical electron spectra taken at momentum transfers to enhance states of high spin and at sufficiently backward angles to retard non-magnetic transitions are shown in Fig. 1.

One important feature that has emerged from these studies is that in light self-conjugate nuclei, only one strong isovector "stretched" state is observed in (e,e'). In light and medium mass

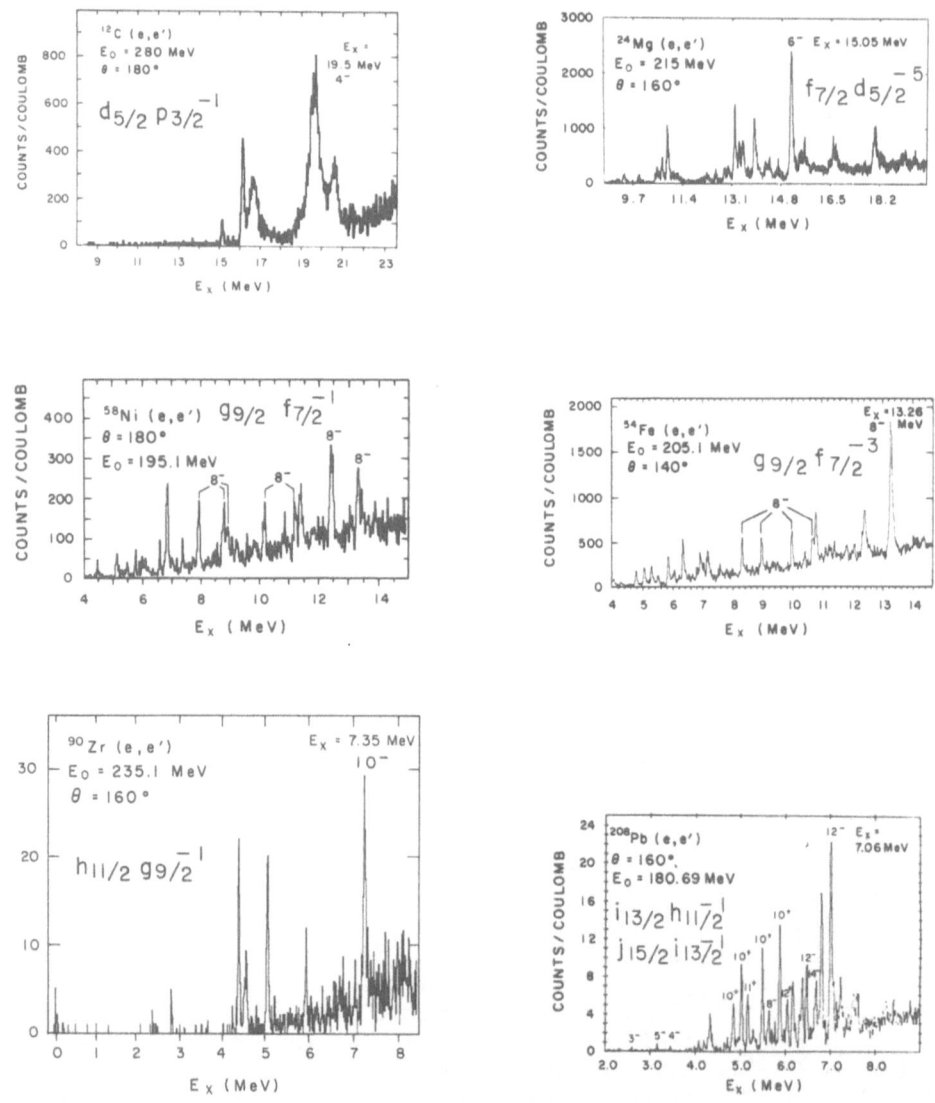

Fig. 1.   High resolution electron energy spectra illustrating
          stretched high spin states in several nuclei near
          closed shells.

neutron excess nuclei like ^{54,56}Fe and ^{58,60}Ni, the strength is
fragmented among several levels. The distribution of fragmented
strength is important to study as it provides information on the
spin dependent parts of the effective interaction in nuclei.

Although is it difficult to distinguish unambiguously between all
the T_0 and T_0+1 isospin components, it appears that most of the
fragmentation is among the T_0 components. For example the Millener-
Kurath $1\hbar\omega$ shell model calculation[30] for ^{14}C and the Lawson
$(g_{9/2}f_{7/2}^{-3})$ shell model calculation[26] on ^{54}Fe are in rough agreement
with experiment in this regard. In each case one strong T_0+1 level
is predicted and only one is observed experimentally. This appears
to be the $J^\pi = 4^-$, T=2 level at 24.3 MeV in ^{14}C and the $J^\pi = 8^-$,
T=2 level at 13.26 MeV in ^{54}Fe.

Because these stretched levels are located in the continuum, and
difficult to unravel from unresolved neighboring levels, some care
must be taken in identifying multipolarities. In addition many of
the levels are unbound to either proton or neutron emission so,
strictly, continuum wave functions in a realistic finite potential
well such as the Wood-Saxon form (WSWF) should be considered.[31,32]
As can be seen from Fig. 1 the states are narrow, so these problems
may not be severe and the question of realistic wells is to a large
part ignored here. Instead, the results of a systematic study of
most of the strong isovector levels observed in (e,e') based on the
use of harmonic oscillator radial wave functions (HOWF) is presented.
With this assumption the form factor for the $1\hbar\omega$ stretched particle-
hole configurations has the simple q dependence, indicated in
eq. (2''),

$$F_T^2 \propto q^{2J} \exp(-b^2q^2/2), \qquad\qquad\qquad (2''')$$

which peaks at $q = [2J/b]^{\frac{1}{2}}$.This simplicity is an important asset in
the identification of the multipolarity. The square of the transverse form factor F_T^2 is shown versus q_{eff} for several transitions
in various nuclei in Fig. 2. Distortion effects due to the mutual
electron-nucleus coulomb attractions are accounted for by correct-
ing the momentum transfer using the relation

$$q_{eff} = q(1 + \frac{3}{2} \frac{Z\alpha\hbar c}{E_0 R}) . \qquad\qquad\qquad (16)$$

The curves through the data points in Fig. 2 were determined by
a least square fit of the form factor calculated assuming the extreme single particle-hole (ESPHM) model. The oscillator parameter,
b, and a normalization factor,

$$s_\alpha^2(ESPHM) = z_\alpha^2(EXP)/z_\alpha^2(ESPHM),$$

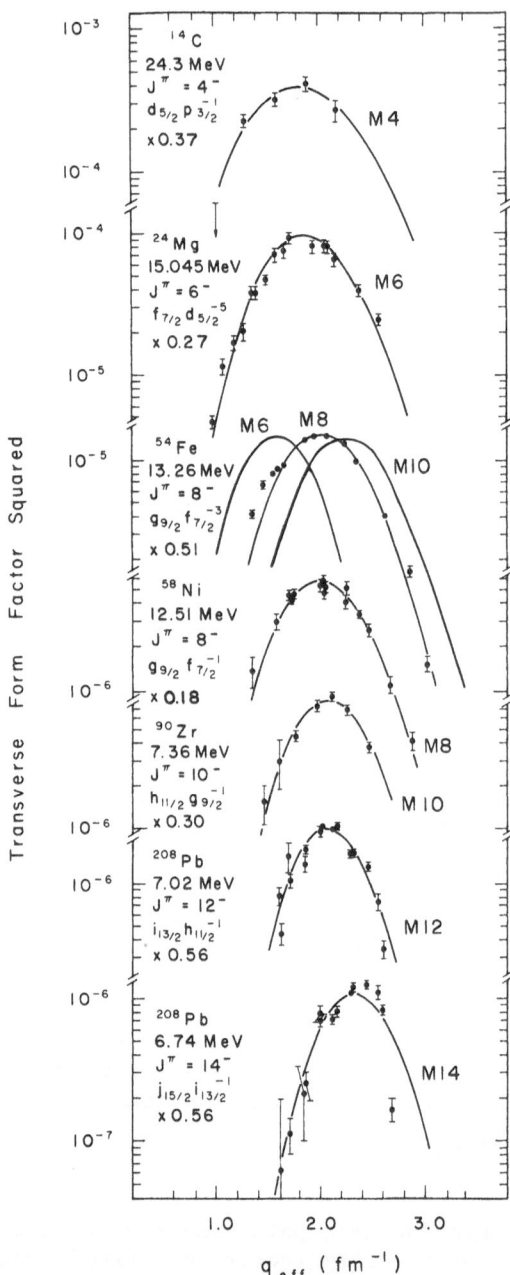

Fig. 2.   Transverse form factor squared for stretched high spin
states is plotted versus q_{eff} in PWBA.

Table 1. Deduced spectroscopic strength S^2 for stretched states
with strong isovector components from (e,e').

Nucleus	E_x (MeV)	J^π, T	CONF	b	S^2 (ESPHM)[++]	S^2 (THY)	Ref.
^{12}C	19.50	4^- 1	$d_{5/2}p_{3/2}^{-1}$	1.50	0.37±0.04	0.60	20
^{14}C	24.30	4^-, 2	$d_{5/2}p_{3/2}^{-1}$	1.50±0.08	0.37±0.07	0.42	23
^{16}O	18.98	4^- 1	$d_{5/2}p_{3/2}^{-1}$	1.63±0.03[+]	0.41±0.02	0.71	18,21
^{24}Mg	15.05	6^-, 1	$f_{7/2}d_{5/2}^{-5}$	1.85±0.04[+]	0.27±0.02	0.47	24
^{28}Si	14.36	6^-, 1	$f_{7/2}d_{5/2}^{-1}$	1.77±0.02[+]	0.31±0.01	0.55	22
^{54}Fe	13.26	8^-, 2	$g_{9/2}f_{7/2}^{-3}$	1.90±0.02[+]	0.51±0.02	0.72	26
^{58}Ni	12.50	8^-, 2	$g_{9/2}f_{7/2}^{-1}$	1.93±0.03[+]	0.18±0.01	0.31	25
^{90}Zr	7.36	10^-, 5	$h_{11/2}g_{9/2}^{-1}$	2.08±0.04	0.26±0.03	----	28
^{208}Pb	6.43	12^-,22	$j_{15/2}i_{13/2}^{-1}$	2.18±0.05	0.54±0.04	1.00	29
^{208}Pb	7.02	12^-,	$i_{13/2}h_{11/2}^{-1}$	2.30±0.03	0.56±0.02	1.00	29
^{208}Pb	6.74	14^-,22	$j_{15/2}i_{13/2}^{-1}$	2.22±0.04	0.56±0.03	1.00	29

+Harmonic oscillator parameter b and normalization S^2 were determined by a least square fit of the DWBA calculated cross section
to the data. The center of mass and finite size effects have been
included in the form factor during the fitting procedure.
++S^2(ESPHM) = Z_α^2(EXP)/Z_α^2(ESPHM), where Z_α^2(ESPHM) is defined in eqs.
(6) and (14). For ^{12}C, ^{16}O, ^{28}Si, $z_1^2 = 1$; For ^{90}Zr, ^{208}Pb, $z_{n(p)}^2 = 1$;
For ^{14}C, ^{58}Ni, $z_1^2 = 1/2$; for ^{24}Mg and ^{54}Fe, $z_1^2 = 2/3$ and
3/8, respectively.

were used as fitting parameters. The resulting values of b and the
normalizing factor S^2(ESPHM) are summarized for several transitions
in Table 1. The values obtained for b are close to that expected
from the $A^{1/6}$ rule. The factors S^2(ESPHM) are consistently much
less than one. These factors are intended to serve as a unit in
which to conveniently compare deduced spectroscopic strengths.
S^2(ESPHM) = 1 corresponds to the maximum isovector strength expected in a single level in the ESPHM. For example in ^{16}O the M4
($d_{5/2}2p_{3/2}^{-1}$) strength in the 18.98 MeV level is S^2(ESPHM) = 0.41 or
41% of the ESPHM prediction.

To test the sensitivity of the cross section to radial shapes, calculations for ^{16}O were performed with WSWF. The same shape and quality of fit were obtained, but with S^2(ESPHM) = 0.54. For ^{28}Si S^2(ESPHM) = 0.31 and 0.37 were obtained using HOWF and WSWF, respectively. In nuclei, ^{16}O, ^{24}Mg, ^{28}Si, and ^{54}Fe, DWBA calculations, using the code HEIMAG,[33] modified to use HOWF, were performed by fitting the cross sections instead of the form factors. The normalization factors were the same as those deduced in PWBA fits to within 3%, but the size parameter b was systematically larger by 2-4%.

There is ample evidence that the ground state orbits are not fully occupied as assumed in ESPHM. The column labeled S^2(THY) contains the results of various theoretical models that take into account core polarization effects and correlations in the ground state again normalized to the data. For example, Millener-Kurath[30] wave functions (MKWF) are used to calculate the M4 strength in ^{12}C, ^{14}C, and ^{16}C. In ^{12}C and ^{14}C the complete p shell space is used to calculate the ground state wave function. The space is further enlarged to allow one nucleon in the sd-shell to calculate the odd parity states. Some improvement is obtained in ^{12}C, but very little in ^{14}C. Clearly, a more realistic treatment of the excited state and ground state is required, for example, allowing for $3\hbar\omega$ configuration admixtures in the excited state as well as 2p-4h ground state admixtures. In ^{16}O where better agreement is obtained, multi-particle-multi-hole configurations are allowed both in the ground state and the excited state. In ^{24}Mg and ^{28}Si open shell RPA calculations[34] predict too much M6 strength by a factor of two. Again the ground state is treated realistically allowing for $1d_{5/2}$, $2s_{1/2}$, $1d_{3/2}$ admixtures while the excited state only allows for one nucleon in the $f_{7/2}$ orbit. Other shell model calculations on ^{28}Si that involve $1d_{5/2}$ and $2s_{1/2}$ orbits yield similar quenching factors.[35] In ^{54}Fe when the model space is enlarged[36] from $(g_{9/2}f_{7/2}^{-3})$ 8^- to include configurations of the type $[g_{9/2}^2 p_{3/2} f_{7/2}^{-4}]8^-$ and using the modified surface delta interaction (MSDI), the ratio of the experimental to theoretical cross section is 0.72. Calculations using MSDI involving the extra core nucleons in ^{58}Ni yield S^2(THY) = 0.31.[25]

In all of the above cases there is improved "agreement" between theory and experiment for the T_0+1 level when the model space is enlarged. The average S^2 is 0.38 for ESPHM and 0.66 for models that include realistic ground state wave functions. Two-body meson exchange currents when included in the calculation of the M4 strength in ^{14}C increase the cross section by 15%.[23,37] The S^2 in Table 1 for ^{14}C should be reduced by 15% to correct for this two body effect. Such corrections will tend to make the disagreement between experiment and theory even larger. The conclusion here is that there is no adequate model yet available to account for the (e,e') strength to the presumably simple high-spin stretched excitations in light and medium mass nuclei.

Although subnucleon degrees of freedom, in particular the quenching of the Gamow-Teller M1 strength in nuclei through Δ particle-nucleon hole admixtures in the wave functions have been discussed at length at this conference, this mechanism does not seem to be a reasonable possibility for quenching the strength of high spin states. Calculations using Δ particle-nucleon hole admixtures for the E_x = 18.98 MeV, J^π = 4$^-$, T=1 state in ^{16}O[38] show that the M4 strength is only reduced by 9% and, therefore, is not expected to play a major role in sapping strength away from the high-spin magnetic states. This is not to say that there could not be a higher multipole nucleon resonance that would more effectively couple with the high spin states. The situation with regard to the T_0 stretched magnetic excitations in non self-conjugate nuclei is more involved (see Refs. 23,25,26,30,36 regarding ^{14}C, ^{54}Fe, ^{58}Ni) and, with exception of ^{208}Pb, the disagreement between experiment and theory is larger than it is for the T_0+1 levels.

In ^{208}Pb, measurements of the (e,e') cross sections to the J^π = 12$^-$, E_x = 6.43 and 7.02 MeV states and to the J^π = 14$^-$, E_x = 6.74 MeV state are about half the values predicted by the ESPHM. Core polarization calculations by Hammamoto et al.,[39] which include a sum of 1p-1h, 1$\hbar\omega$, 2$\hbar\omega$, 3$\hbar\omega$, etc. configurations predict the correct strength. However, the calculations employ a zero range delta force (V_0 = 170 MeV/fm^3), which is known to be too strong[40] at large q in comparison to realistic effective interactions. Particle-vibration coupling calculations by Krewald et al.[41] show a 50% reduction in M12 and M14 strength to the E_x = 6.43, 7.02, and 6.74 MeV levels in agreement with experiment. However, the missing strength in these states appears as additional M12 and M14 strength in a group of states at 8.0-9.0 MeV excitation, which have not been identified experimentally. The model calculations predict that the J^π = 12$^-$ state at E_x = 6.43 MeV is predominantly a neutron state and the one at E_x = 7.02 MeV state is predominantly a proton state. It is encouraging that the best agreement between theory and experiment obtained to date for these high spin states occurs for ^{208}Pb where one might expect the model assumptions to be most adequate.

HIGH SPIN STRETCHED STATES AS BENCHMARKS: COMPARISON WITH HADRONIC PROBES

States of stretched configuration are being studied with a variety of probes besides electrons. As an example, Fig. 3 contains the spectra from a variety of reactions in which the 6$^-$, T=1 stretched level at E_x = 14.36 MeV has been observed. Although these pure spin transfer, stretched transitions are not yet completely understood in terms of current nuclear structure models, they are very useful as "benchmarks" in testing direct, one step inelastic scattering models for hadronic reactions such as (p,p') (p,n), and ($\pi^\pm$,$\pi^\pm$). The reason for this is that the same spin transition

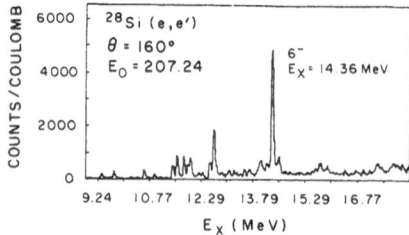

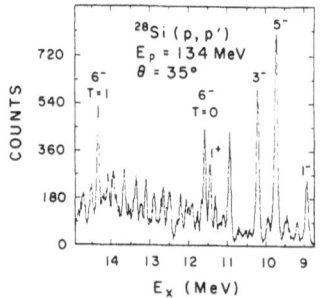

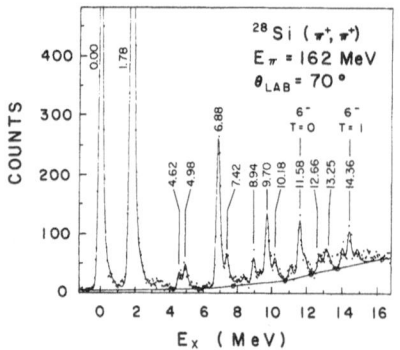

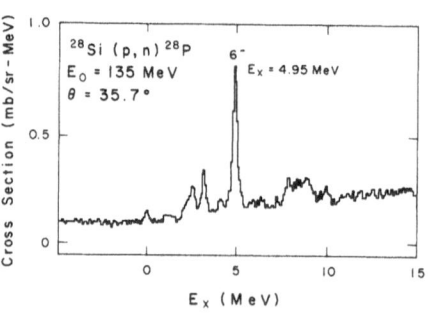

Fig. 3. Energy spectra illustrating inelastic electron,[22] proton,[42,46] and pion[48] excitation of the T=0 and T=1, J^π = 6$^-$ states in ^{28}Si and charge exchange[47] (p,n) excitation of the analog state in ^{28}P. The T=0, E_x = 11.58 and the T=1, E_x = 14.36 MeV are both observed in proton and pion scattering. In electron scattering, excitation of a known 3$^-$ state at E_x = 11.59 MeV obscures observing population of the T=0, 6$^-$ state at 11.58 MeV.[22]

density appears in the transition amplitude for all of these reactions as was displayed in eq. (2'), eq. (9'), and eq. (12') above. Experimental data and theoretical results on stretched excitations in (p,p'), (p,n), and (π,π') are discussed separately in several contributions to this conference, i.e. Ref. 42, 43, and 44, respectively. Here, some of these results are collected together for comparison much in the spirit of the discussion in Refs. 3 and 5.

Rather complete data on the 6^- stretched levels in ^{28}Si and the 4^- stretched levels in ^{16}O are available, i.e. (e,e'), (p,p'), (p,n), and (π,π'). Stretched levels in ^{24}Mg, ^{54}Fe, and ^{208}Pb have been seen in (e,e') and (p,p'). The stretched levels in ^{14}C have been recently studied simultaneously via (e,e') and (π,π'). These four subsets of results are discussed separately below in the light of available theoretical results. All of the theoretical calculations to be shown are in the sense of eq. (2'), eq. (9'), and eq. (12'), however, essential distortion effects are taken into account through the use of DWBA instead of PWBA and the exchange amplitudes for the nucleon-nucleus reactions are included exactly through the use of the code DWBA70. The nucleon-nucleus and pion-nucleus calculations use the local t matrix interactions of Ref. 8 and Ref. 11, 12, respectively, unless otherwise specified. An important feature of the interaction of Ref. 8 is that the radial dependence is given as sums of Yukawa terms with the long range parts of the central and tensor components matched to OPEP.

As was discussed above, the comparisons of interest here provide information on the isovector tensor part of the effective nucleon-nucleon interaction and the spin-orbit component of the effective pion-nucleon interaction. To illustrate the former, the modulus of $\bar{v}_1^C(q)$, $\bar{v}_1^{LS}(q)$ and $3\bar{v}_1^T(q)$ for 135 MeV protons as deduced from the free nucleon-nucleon amplitudes in Ref. 8, is shown in Fig. 4. Clearly, $3\bar{v}_1^T(q)$ is dominant in the range of q where the electron scattering form factors displayed in Fig. 2 peak. This shows very nicely that the comparison of (e,e') and (p,p') for isovector stretched state transitions provides information on the strength and q dependence of the isovector tensor interaction in the approximate range q = 1-2.5 fm^{-1}. Similar diagrams for the isoscalar parts of the effective nucleon-nucleon interaction[5,8-10] show clearly that isoscalar stretched levels are populated in (p,p') via exchange amplitudes associated with $v_1^T(q)$ and the isoscalar spin-orbit interactions $\bar{v}_0^{LS}$. For pictures of the pion-nucleon spin-orbit interaction the reader is referred to Ref. 5, 11, and 12.

Stretched 6^- Levels in ^{28}Si

Fig. 5 displays experimental differential cross section data and theoretical curves shown as a function of q for the population of the 6^-, T=1 stretched state in ^{28}Si via (e,e'), (p,p'), (p,n), and (π^+,$\pi^{+'}$). The (e,e') data are from Ref. 22, the (p,p') data taken at E_p = 135 MeV is from Ref. 42,46, the (p,n) data taken at

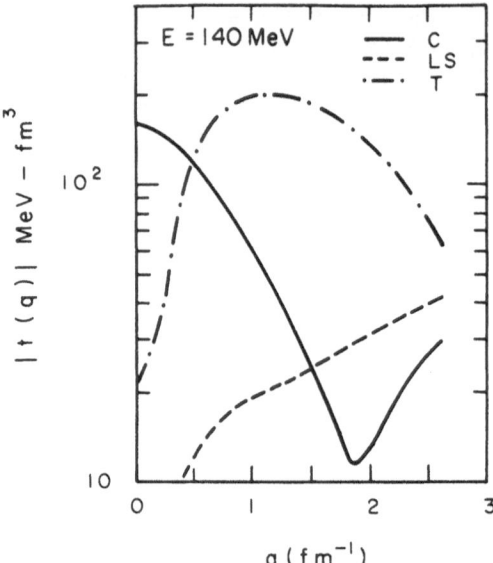

Fig. 4.   Modulus of the isovector spin dependent central, spin-
          orbit, and tensor components of the nucleon-nucleon
          t-matrix plotted in momentum space. Includes the effect
          of exchange forces. In addition the tensor force has
          been multiplied by 3 to account for the fact that the
          tensor force carries more weight in the cross section
          than the central force, i.e. see eq. 9', 10, and 11.

E_p = 135 MeV is from Ref. 43, 47, and the (π,π') data taken at
E_π = 162 MeV is from Ref. 48. The theoretical (e,e') results are
the same as those discussed in the preceding section of this paper,
the (p,p') results are from C. Olmer,[49] the (p,n) results are from
Ref. 43,47, and the (π,π') result is that of Carr et al.[12] The
(p,p') result is nearly identical to theoretical results reported
earlier[3,5,50,51] which used a somewhat different parameterization of
the nucleon-nucleon t matrix.[52] The transition density parameters,
b and S^2, corresponding to the theoretical results shown in Fig. 5
are summarized in Table 2.

     Overall the consistency of the results is quite satisfying.
The value of S^2 from the (p,n) and (p,p') reaction is about 15%
lower than the values of S^2 from (π,π') and (e,e'). Additional
results for the 6^-, T=1 level in ^{28}Si observed in (p,p') at
E_p = 333 and 500 MeV give S^2 = 0.25 and 0.29.[10,53] This indicates
the energy dependence of the isovector tensor interaction as given

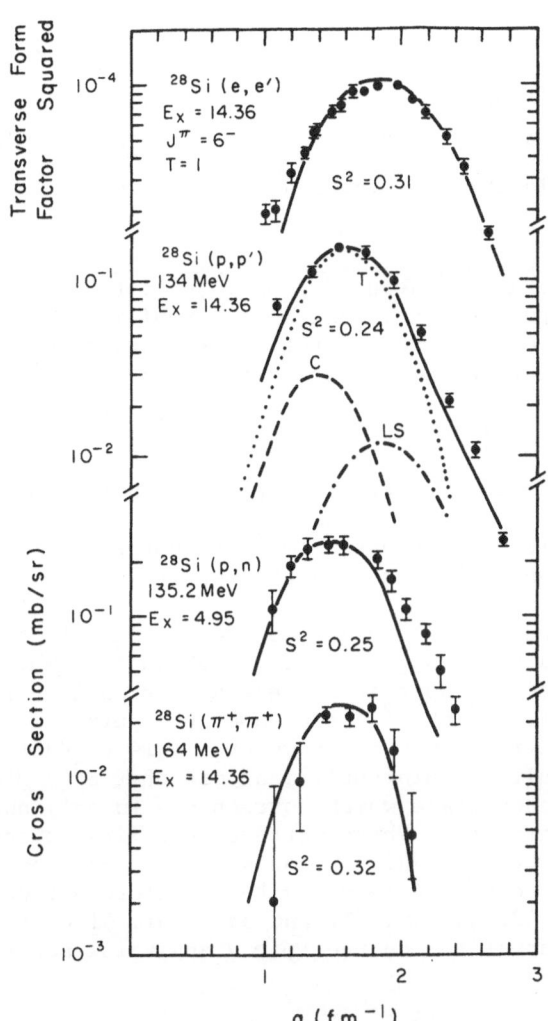

Fig. 5. A plot of (e,e'), (p,p'), (p,n), and (π,π') measured and calculated cross section for the $f_{7/2}d_{5/2}^{-1})6^-$ transition in ^{28}Si.

Table 2.   Values of S^2 for a $J^\pi = 6^-$, T=1 $(f_{7/2}d_{5/2}^{-1})$ transition.

Prove	E_x(MeV)	S^2 ESPHM	b^+	Ref.
^{28}Si (e,e')	14.36	0.31±0.01	1.77±0.02	22
(p,p')	14.36	0.24	1.78	46,49
(π^+,π^+)	14.36	0.32	1.76	48,12
(π^-,π^-)	14.36	0.32	1.76	48,12
(p,n)	4.95	0.25	1.74	47

+The values of b and S^2 deduced from (p,p'), (p,n), and (π,π') have been corrected for the fact that a center of mass correction was not made in the calculation of the cross section when fitting the data.   The correction factors are determined from[45]

$$b = b_{DWBA70}\left[\frac{A}{A-1}\right]^{\frac{1}{2}} \text{ and } S^2 = S^2_{DWBA70}\;(\frac{A-1}{A})^{J-1},$$

which are only valid for stretched particle-hole configurations.

by the free nucleon-nucleon t matrix is approximately correct in the q region sampled in these comparisons.

The (p,p') and $(\pi^\pm,\pi^\pm)$ reactions provide important information on the stretched 6^-, T=0, E_x = 11.577 MeV in ^{28}Si which is not observed in (e,e') because it is weak and unresolved from a 3^-,T=0 state at E_x = 11.585 MeV.[22]   Space does not permit a complete discussion of these results here; however, it is noted that the (p,p') and $(\pi^\pm,\pi^\pm)$ results consistently indicate that $S^2 \sim 0.12$ for the isoscalar stretched state which corresponds to only about 35% of the isovector strength.   These matters were first discussed in the second entry of Ref. 3, Ref. 5, and Ref. 50.   For more complete information the reader is referred to the aforementioned work and Ref. 8, 10, 12, 42, 44, 48, 51, and 53.   Ref. 51 contains some interesting observations on the (p,p') analyzing power for stretched states.

Stretched 4^- Levels in ^{16}O

Fig. 6 and Table 3 contains the information from (e,e'), (p,p'), (p,n), and $(\pi^\pm,\pi^\pm)$ on the 4^-, T=1 stretched state in ^{16}O organized in the same fashion as the results for the 6^-, T=1 level in ^{28}Si discussed above.   The (e,e') data are from Ref. 18, 21, the (p,p') data taken at E_p = 135 MeV are from Ref. 54, the (p,n) data taken at E_p = 135 MeV are from Ref. 43, and the (π,π') data taken at E_π = 162 MeV are from Ref. 55.   As in the case of ^{28}Si, the (e,e') theoretical curve is from the preceding section of this paper and

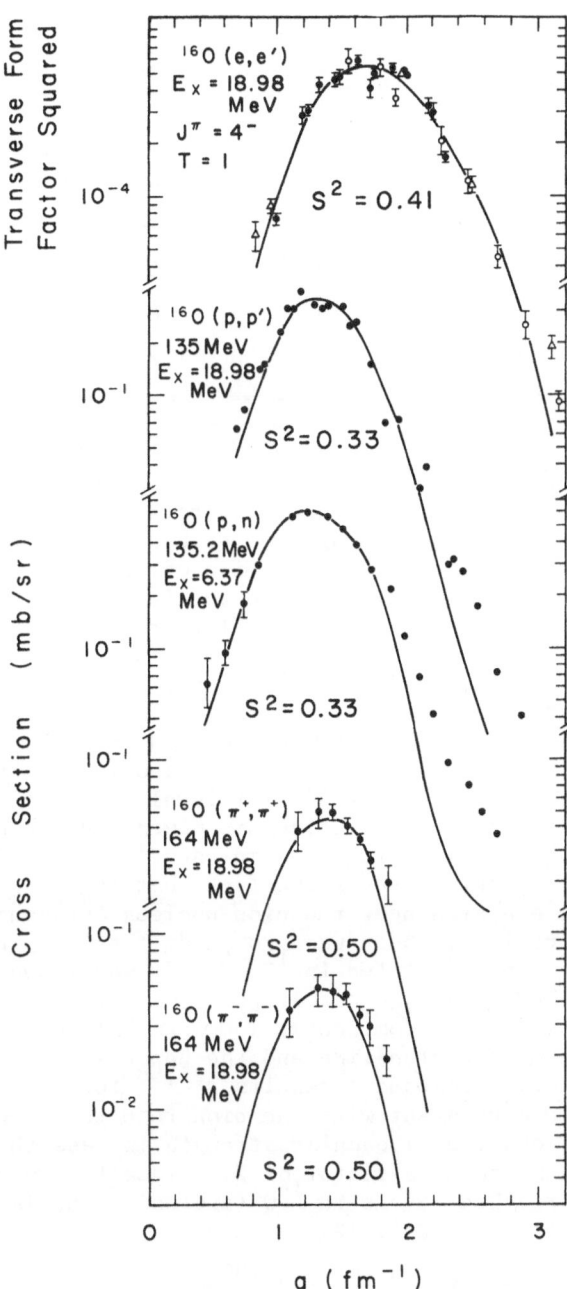

Fig. 6. A plot of (e,e'), (p,p'), (p,n), and (π,π') measured and calculated cross section for the $(d_{5/2}p_{3/2}^{-1})4^-$ transition in ^{16}O.

Table 3. Values of S^2 for a $J^{\pi} = 4^-$, T=1, $(d_{5/2}p_{3/2}^{-1})$ transition.

Probe	E_x(MeV)	S^2 ESPHM	b^+	Ref.
^{16}O (e,e')	18.98	0.41±0.02	1.63±0.03	18,21
(p,p')	18.98	0.33	1.70	35,54,56
(π^+,π^+)	18.98	0.50	1.62	12,55
(π^-,π^-)	18.98	0.50	1.62	12,55
(p,n)	6.37	0.33	1.72	43

+See Table 2.

the (p,p'), (p,n), and $(\pi^{\pm},\pi^{\pm})$ theoretical results are from Ref. 56, 43, and 12, respectively. Again the overall consistency of the results is satisfactory, although the values of S^2 deduced from the data show more scatter as a function of reaction than in the case of ^{28}Si. This is due in part to the fact that the (p,p') and (p,n) calculations assumed a somewhat larger value for b than is indicated by (e,e').

The isoscalar stretched 4^- strength in ^{16}O presents a very interesting picture which will only be touched on here. First, two 4^-, T=0 levels are observed at E_x = 17.79 and 19.80 MeV. These levels are above and below the 4^-, T=1 level which is observed at E_x = 18.98 MeV. The levels are isospin mixed and the mixing amplitudes were deduced from differences in the π^+ and π^- differential cross sections,[55] making use of the fact that $|t_{\pi^+p}/t_{\pi^-p}|=|t_{\pi^-n}/t_{\pi^+n}|$ =3 for incident energies near the pion-nucleus Δ-resonance. This technique had previously been used to study the isospin mixing of pairs of 1^+, 2^-, and 4^- states in ^{12}C.[57] It was found in Ref. 55 that the 18.98 MeV level was nearly pure T=1 while the 17.79 and 19.80 MeV levels were predominantly T=0 with T=1 admixtures making the lower state more proton-like and the upper state more neutron-like in character. Subsequent results[5,21,58] for (e,e') and (p,p') were found to be consistent with the pion results of Ref. 55. Again the total observed isoscalar strength is less than the isovector strength by about a factor of 2. Z coefficients which describe the stretched levels in ^{16}O (including the isospin mixing) may be found in Ref. 12.

Stretched Levels in ^{24}Mg, ^{54}Fe, and ^{208}Pb

The population of stretched states in ^{24}Mg, ^{54}Fe, and ^{208}Pb has been studied experimentally via (p,p')[46,59,60] as well as (e,e')[24,26,29] as discussed earlier with regard to Table 1.

Table 4. Comparison of S^2(p,p') with S^2(e,e').

Nucleus	E_x MeV	J^π,T	CONF	S^2(p,p') (ESPHM)	b^+	Ref.	$\frac{S^2(e,e')}{S^2(p,p')}^{++}$
^{24}Mg	15.05	6^-, 1	$f_{7/2}d_{5/2}^{-1}$	0.24	1.86	3,46	1.13±0.08
^{54}Fe	13.26	8^-, 2	$g_{9/2}f_{7/2}^{-1}$	0.45	1.92	59,61	1.13±0.04
^{208}Pb	6.74	14^-,22	$j_{15/2}i_{13/2}^{-1}$	0.50	2.30	3,60	0.89±0.05
^{208}Pb	6.43	12^-,22	$j_{15/2}i_{13/2}^{-1}$	0.80	2.30	3,60	1.50±0.11
^{208}Pb	7.02	12^-,22	$i_{13/2}g_{11/2}^{-1}$	0.20	2.30	3,60	0.36±0.11

+See Table 2.
++Error on ratios only include error in S^2(e,e').

Table 4 contains a summary of the essential results of theoretical
(p,p') calculations[3,60,61] based on the parametrization of the
nucleon-nucleon t matrix given in Ref. 52. The values of S^2(p,p')
and S^2(e,e') are compared in the Table. The (e,e') and (p,p')
results from the isovector 6^- and 8^- transitions in ^{24}Mg and ^{54}Fe
and the pure neutron 14^- transition in ^{208}Pb are in reasonable
accord.

The results for the two 12^- states in ^{208}Pb are not consistent.
One of these levels is primarily an $i_{13/2}h_{11/2}^{-1}$ proton configuration
and the other is primarily a $j_{15/2}i_{13/2}^{-1}$ neutron configuration.
Only the first is stretched and since there are two levels, this
is not a clean example of the <u>unique</u> stretched excitations which
have been the primary topic of discussion here. It was shown in
ref. 60 that a consistent explanation of the (e,e') and (p,p') data
for these 12^- levels could not be achieved by considering configuration mixing between the two levels alone. A possible problem is
that the proton cross section for the 6.43 MeV neutron 12^- level
may contain a contribution from an unresolved 13^- level.[60] Since
^{208}Pb is such an important shell model nucleus, it is essential
that it be examined more carefully with a more sophisticated
structure model such as the particle-vibration model of Krewald
et al.[41]

Finally, it is noted that an upper limit of the isoscalar 6^-
strength in ^{24}Mg has been set at S^2 = 0.11 in Ref. 50.

T=0 and T=1 Densities for Stretched Excitations in N≠Z Nuclei

For both (e,e') and (π,π') the probe-nucleus coupling is
sufficiently simple that information on the isoscalar and isovector

contributions to the transition amplitudes (equivalently the isoscalar and isovector spin transition densities) can be obtained
rather directly from cross section ratios in cases where both
contributions are present. This is essentially the idea applied
in earlier discussions of Coulomb mixing 1^+, 2^-, and 4^- states[57] in
^{12}C and in the 4^- levels[5,21,55,58] in ^{16}O. In these situations the
appropriate relations analogous to eq. (2') and eq. (12') are

$$F_T^2(q) = \frac{4\pi}{z^2} (J+1) \left| \frac{q\hbar}{2Mc} \frac{1}{2} \sum_T Z_{JT} g_T^s \rho_{JJ-1}^s(q) \right|^2 \tag{2'''}$$

$$\frac{d\sigma^\pi}{d\Omega} = 4\pi \left(\frac{M_\pi}{2\pi\hbar^2} \right)^2 (J+1) \frac{1}{2} k_\pi^4 q^{-4} \sin^2 \theta \alpha_\pi^2 \left| \sum_T Z_{JT} t_T^{LS}(q) \rho_{JJ-1}^s(q) \right|^2 \tag{12''}$$

where $\rho_{JJ-1}^s(q)$ is just $\rho_{JJ-1}^{sT}(q)$ of eq. (14) with the spectroscopic
amplitude divided out. In application of these relations, the
π^+/π^- cross section ratios mainly fix Z_0/Z_1 and the magnitude of Z_1
is primarily determined by the electron scattering cross section.
The redundant index J on Z has now been dropped. Nucleon scattering
is not so easily used in this way, since the transition amplitude
consists of several different incoherent terms with different isospin dependence even when $\rho_{JT}^\ell(q)$ vanishes (see eq. (9)). This was
explicitly demonstrated for the stretched 4^- levels in ^{16}O in Refs.
5 and 58.

As another example of the use of this technique, the case of
stretched $(d_{5/2}p_{3/2}^{-1})$ 4^- transitions in the N$\neq$Z nucleus ^{14}C will be
considered. In this nucleus, two T=1, 4^- states at E_x = 11.72 and
17.33 MeV and a T=2, 4^- state at 24.5 MeV are observed in (e,e')[23]
and (π,π')[62,63] as shown in Fig. 7. In a pure particle-hole model
for ^{14}C, the isoscalar transition strength, Z_0^2 has a value of unity,
and the isovector transition strength, Z_1^2, is equally divided between the T=1 and T=2 states, with a maximum value of 1/2 for a
given value of T. The value of Z_T deduced from the data are
summarized in Table 5. The large range of the value of the Z_0
coefficients are a result of the measured limits on the pion cross
section ratios.[62,63] The results in Table 5 imply that the observed transitions to the T=1 states exhaust 33 to 45% of the
total isovector transition strength, but only 1 to 15% of the
isoscalar transition strength. As suggested by the calculations
of Millener[30] and the (π,π') study[63], other 4^-, T=1 states reached
by strong isoscalar, but weak isovector transitions may be present
in this excitation region, but would not be seen in (e,e') because
magnetic electron scattering is most sensitive to isovector
transitions. A similar study[64] for stretched transitions in ^{54}Fe
is in progress.

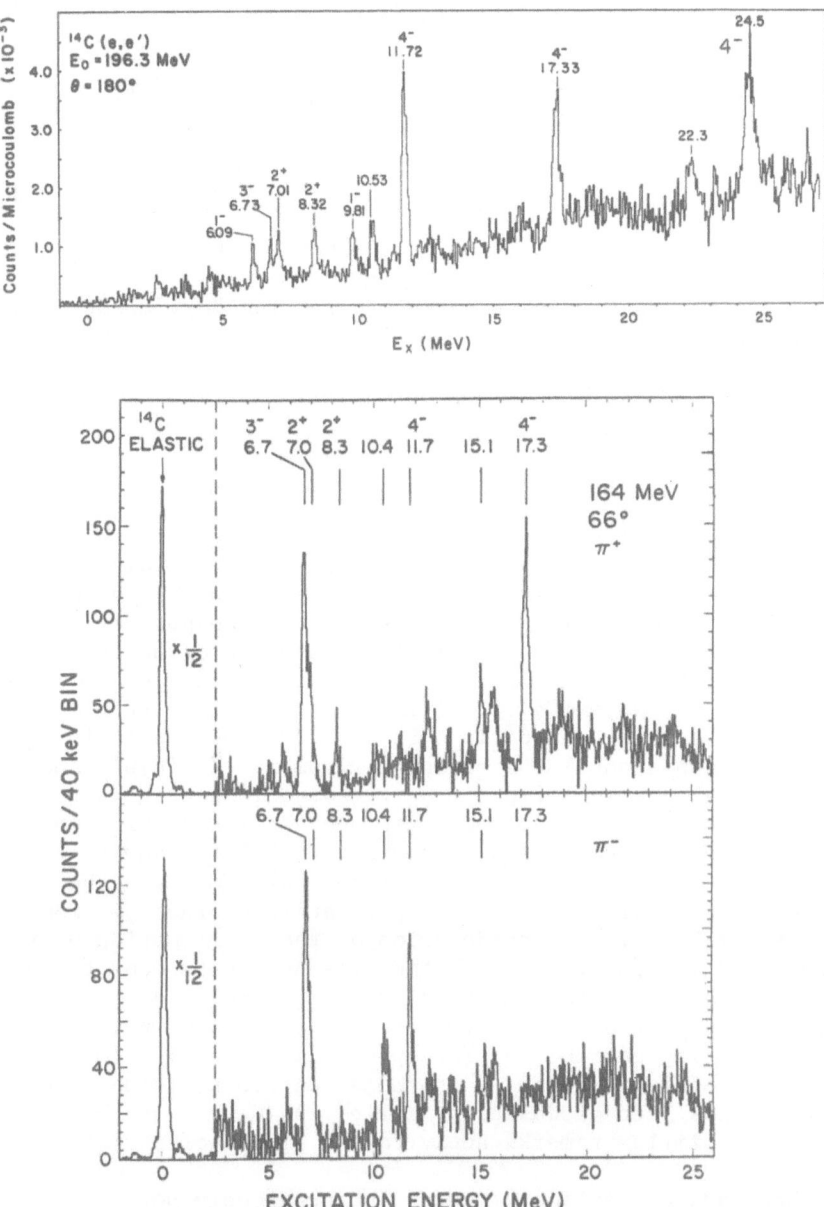

Fig. 7. An electron and pion spectrum illustrating excitations of
the 4⁻ states in ¹⁴C at E_x = 11.72, 17.33, and 24.5 MeV.

Table 5.   Deduced Z coefficients[+]from (e,e') and (π,π') on ^{14}C.

E_x MeV	J^{π}, T	CONF	Z_0	Z_1
11.7	4^-, 1	$d_{5/2}p_{3/2}^{-1}$	$0.092 \leq Z_0 \leq 0.29$	$0.30 \leq Z_1 \leq 0.36$
17.3	4^-, 1	$d_{5/2}p_{3/2}^{-1}$	$0.26 \leq Z_0 \leq -0.080$	$0.27 \leq Z_1 \leq 0.31$

+These coefficients have been corrected for two body meson exchange current contributions.

Discussion

The overall consistency of the results on stretched levels from
(e,e'), (p,p') and (π,π') over a wide range of multipoles and nuclei
is quite striking. This is strong support of the DWIA description
of the hadronic scattering process and a solid measure of the
strength of the tensor and spin-orbit components of nucleon- and
pion-nucleon interactions in the range of q near the peak cross
section. There are some difficulties in fitting the (p,p') and
(p,n) at higher q for the 6^- in ^{28}Si and ^{24}Mg and the 4^- in ^{16}O.
The possible appearance of a second maximum in ^{16}O (p,p') and (p,n)
suggest several possibilities. The tensor component gets rapidly
weaker at high q and the isovector spin-orbit interaction which is
not as well determined begins to dominate. Effects due to the shapes
of radial wave functions and the nuclear medium may also be important here. The local nucleon-nucleon t matrix excludes medium effects
such as Pauli blocking, consideration of the Fermi motion of target
nucleons, and the information on the off-shell nature of the nucleon-
nucleus interaction. Careful studies of the energy dependence of
the t-matrix and other effective interactions which include medium
effects are being carried out for spin excitations by Olmer et al.[65]
The question of medium effects in the effective nucleon-nucleon
interaction has been discussed by Kelley and Carr[66] at this
conference. Similar remarks apply to the pion-nucleus interaction.

Another word of caution concerns the previously noted correction
to the one body spin density extracted from (e,e') data due to two
body meson exchange currents.[37] This effect has not been included in the calculations in Tables 1-5. Meson exchange current
calculations[23] for ^{14}C using MK and ESPHM wave functions show that
the meson exchange current contribution has the effect of reducing
S^2(e,e') by about 15%. From this, one might conclude that the
spectroscopic factor S^2 deduced from (e,e') should be reduced before comparing with S^2 deduced from hadron scattering.

NON-STRETCHED 1^+ AND 2^- STATES IN ^{12}C

The low q cross section for the excitation of isovector states with low spin via (p,p') and (p,n) are determined primarily by the isovector central part of the effective interaction since the spin orbit and tensor components become negligible as q goes to zero.[3,5,8,10] See also Fig. 4. Consequently, the low q cross sections mainly provide a measure of $\rho^{s1}_{JJ-1}(q)$, which is generally larger than $\rho^{s1}_{JJ+1}(q)$ for low-lying states for low q. For the case J=1, i.e., 1^+ transitions, $\rho^{s1}_{10}(0)$ is determined by beta decay and a correlation between beta decay matrix elements and corresponding (p,n) and (p,p') cross sections is expected. This is the basis behind the studies of the GT and M1 strength with (p,p') and (p,n) which have been discussed in detail at this conference.[3,5,8,10,67-69] Comparative studies of isovector 1^+ states via (e,e') and (p,p') or (p,n) can provide information on both the spin and current densities[70] since (e,e') depends on ρ^{ℓ}_{11} and ρ^{s1}_{10} while the nucleon reactions depend mainly on ρ^{s1}_{10}. A discussion of M1 strength in nuclei via (e,e') at this conference[71] and comparisons of (e,e') and (p,n) or (p,p') data for 1^+ levels has been given in Refs. 69, 70, and 72.

The $0^+ \to 1^+$ T=1, E_x = 15.1 MeV transition in ^{12}C provides a specific illustration of these points. For this transition the wave functions of Cohen-Kurath (CKWF)[73] give $\rho^{s1}_{10}(0)$ = 0.22 and B(M1)↑ = $2.29\mu^2_o$. The corresponding experimental values from beta decay and electromagnetic studies are 0.22 and $2.79\mu^2_o$, respectively. The difference between the theoretical and experimental B(M1)↑ can be attributed to meson exchange current contributions. If CKWF are used including meson exchange current corrections,[74] then the calculated B(M1)↑ = $2.79\mu^2_o$ is in agreement with experiment. Inelastic proton scattering to the T=1, $J^\pi = 1^+$, E_x = 15.1 MeV state in ^{12}C at 120 MeV and electron scattering[76] are both in good agreement with the predictions of the CKWF at low q where the spin density dominates as is shown in Fig. 8a and 8b. At higher q in (e,e') and (p,p') there is no agreement. The full curve for (e,e'), HDWF,[77] which has the appropriate magnitude at the second maximum is merely a polynomial fit whose degree is constrained by only allowing for p shell wave functions. Clearly, it is more difficult to fit (e,e') and (p,p') at higher q's for 1^+ states because of the additional competing spin-orbit and tensor terms in (p,p'), and the probable increased importance of the interference of the spin and orbital densities in (e,e') and (p,p').

Although consistent results for the spin densities can be obtained at least at low q for 1^+ states using electrons and protons, this is not the case for the E_x = 16.58, $J^\pi = 2^-$, T=1 state in ^{12}C. The proton angular distribution[76] and electron scattering form factors are shown,[78] compared to predictions using the MKWF[30] in Figs. 8c and 8d. The shapes of the electron scattering form factor

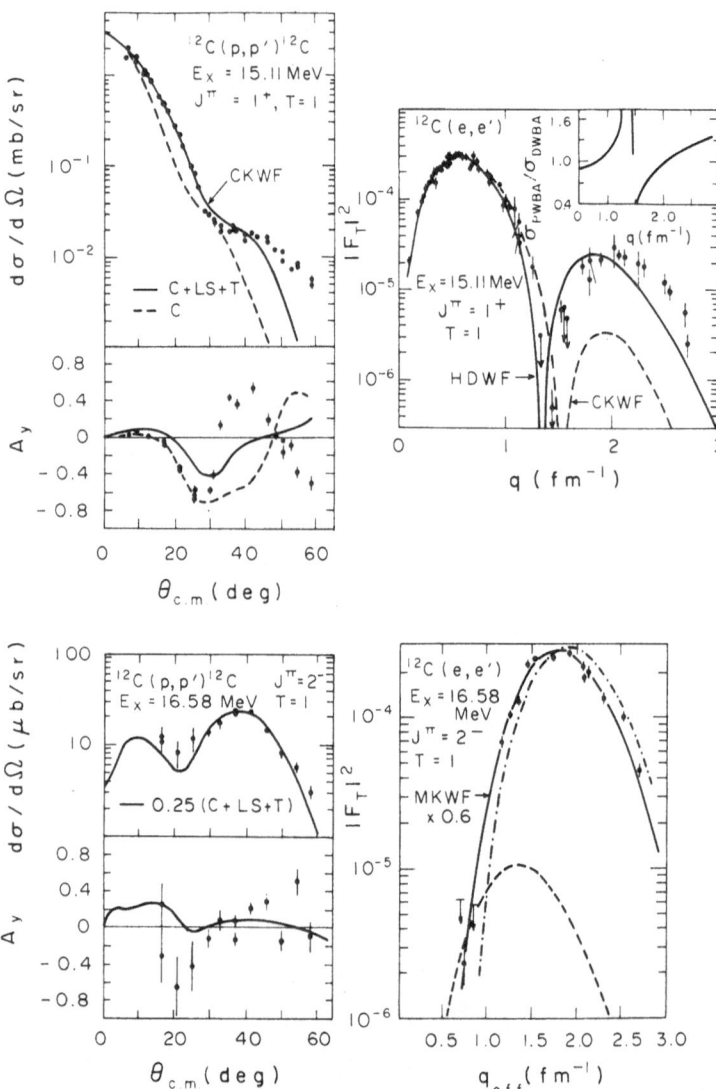

Fig. 8. A composite proton angular distribution and electron
         form factor compared to calculations for the isovector
         1^+ and 2^- states in ^{12}C.

and the proton angular distribution are in qualitative agreement with the MKWF predictions, but unlike the low q data for the 1^+ state, renormalization factors are required. More important is the fact that while the MKWF predictions for the electron form factor must be multiplied by 0.6, the predictions must be multiplied by 0.25 for the proton angular distributions. Because there are several competing terms contributing to the proton cross section (central, spin-orbit, and tensor) and several possible shell model configurations (($d_{5/2}p_{3/2}$), ($2s_{1/2}p_{3/2}$), etc.), contributing to the spin and orbital transition densities, it is quite difficult to resolve the inconsistencies between (e,e') and (p,p'). The analyzing powers which are sensitive to a different linear combination of potentials and densities are also in very poor agreement with theory. There is much work yet to be done for these more complicated states and new insights can be expected from such studies.

SUMMARY

Expressions for the electron, proton, and pion inelastic scattering cross sections have been written down in PWBA for transitions to high spin stretched states. The $\rho_{JJ-1}^{ST}(q)$ spin transition density appears as a common nuclear structure factor in each expression. Normalization of the theoretical cross sections to experimental cross sections provides a measure of the spectroscopic strength S^2 observed.

Where data is available, consistent results are obtained from (e,e'), (p,p'), (p,n), and (π,π'). Specific cases that have been considered here are the 4^- and 6^- levels in ^{16}O and ^{28}Si, respectively. These results indicate model effective interactions currently employed in nucleon and pion-nucleon scattering calculations are reasonable. In particular, the comparisons provide information primarily on the tensor component of the nucleon-nucleon interaction and the spin-orbit part of the pion-nucleon interaction. The results are particularly striking since several independent measurements are correlated. The extracted values of S^2 for isovector transitions are typically 38% of the ESPHM and 66% of that predicted by nuclear models that allow for configuration mixed ground state wave functions. The observed isoscalar strength is only about one half of the isovector strength. The origin of the quenching of the high spin magnetic strength is not known. The most likely possiblility is that it arises because of fragmentation of the single particle strength that is too weak to be observed and the use of reference model calculations that do not employ a large enough basis to sufficiently fragment the strength. Effects due to meson exchange currents (MEC) and isoscalar currents are small and affect only the isovector strength.

These conclusions are, thus far, based on a very selective data set and not over a very extended range of q. For example, in ^{12}C at low q, (e,e'), (p,p'), and (p,n) cross sections to the $J^\pi = 1^+$, $E_x = 15.1$ MeV state are reasonably well described by the CKWF including MEC corrections to (e,e'). At high q, near the second maximum, calculations using the same wave functions are too low by a factor of ten. It is clear that measurements on the high spin states must be extended to higher q at other incident proton and pion energies and on other nuclei in order to improve our understanding and substantiate the present findings. It should be kept in mind that the effective interactions and DWBA used in the description of the hadronic reactions represent a simplified treatment of a very complex situation. We should be cautious in our conclusions since we have either neglected or treated aprroximately the many body nature of the nucleus and undoubtedly many surprises are still awaiting us.

ACKNOWLEDGEMENTS

The authors would like to thank M.A. Plum, R.S. Hicks, J.A. Carr, C. Olmer, W.G. Love, and J. Millener for useful conversations and assistance concerning some of the specific results which have been presented.

REFERENCES

1. T.W. Donnelly and G.E. Walker, Annals of Physics 60, 209 (1970).
2. P.J. Moffa and G.E. Walker, Nucl. Phys. A222, 140 (1974).
3. R.A. Lindgren, W.J. Gerace, A.D. Bacher, W.G. Love, and F. Petrovich, Phys. Rev. Lett. 42, 1524 (1979). F. Petrovich and W.G. Love, Proc. LAMPF Workshop on Pion Single Charge Exchange, Los Alamos, New Mexico, (LA-7892C) 1979.
4. G.E. Brown, J. Speth, and J. Wambach, Phys. Rev. Lett. 46, 1057 (1981).
5. F. Petrovich and W.G. Love, Nucl. Phys. A354, 499c (1981).
6. T. DeForest and J.D. Walecka, Ad. Phys. 15, 1 (1966).
7. F. Petrovich, R.H. Howell, C.H. Poppe, S.M. Austin, and G.M. Crawley, Nucl. Phys. A383, 355 (1982).
8. W.G. Love and M.A. Franey, Phys. Rev. C24, 1073 (1981).
9. F. Petrovich, J.A. Carr, D. Halderson, W.G. Love, and J.J. Kelly, to be published.
10. W.G. Love, M.A. Franey, and F. Petrovich, these proceedings.
11. J.A. Carr, Proceedings of the Second LAMPF II Workshop, Los Alamos, New Mexico (LA-9572C) 1982.
12. J.A. Carr, F. Petrovich, D. Halderson, D.B. Holtkamp, and W.B. Cottingame, Phys. Rev. C27, 1636 (1983).
13. C. Wilkin, Nucl. Phys. A220, 621 (1974).
14. E.R. Sicilano and G.E. Walker, Phys. Rev. C23, 2661 (1981).

15. F. Lenz, these proceedings.
16. G.E. Walker, these proceedings.
17. T.W. Donnelly, J.D. Walecka, I. Sick, and E.B. Hughes, Phys. Rev. Lett. 21, 1196 (1968).
18. I. Sick, E.B. Hughes, T.W. Donnelly, J.D. Walecka, and G.E. Walker, Phys. Rev. Lett. 23, 1117 (1969).
19. T.W. Donnelly, J.D. Walecka, G.E. Walker and I. Sick, Phys. Lett. 328B, 545(1970).
20. R.S. Hicks, J.B. Flanz, R.A. Lindgren, G.A. Peterson, L.W. Fagg, and J. Millener, to be published.
21. C. Hyde, W. Bertozzi, T. Buti, M. Deady, W. Hersman, J. Kelly, S. Kowalski, R. Lourie, B. Pugh, C.P. Sargent, W. Turchinetz, B. Norum, B.L. Berman, M.V. Hynes, J. Lichtenstadt, F. Petrovich, D. Halderson, J.A. Carr, and W.G. Love, Bull Am. Phys. Soc. 26, 27 (1981) and to be published.
22. S. Yen, R. Sobie, H. Zarek, B.O. Pich, T.E. Drake, C.F. Williamson, S. Kowalski, and C.P. Sargent, Phys. Lett. 43B, 250 (1980); Phys. Rev. C27, 1934 (1983).
23. M.A. Plum, R.A. Lindgren, J. Dubach, R.S. Hicks, R.L. Huffman, B. Parker, G.A. Peterson, J. Alster, J. Lichtenstadt, and M.A. Moinster, to be published.
24. H. Zarek, B.O. Pich, T.E. Drake, D.J. Rowe, W. Bertozzi, C. Creswell, A. Hirsch, M.V. Hynes, S. Kowalski, B. Norum, F.N. Pad, C.P. Sargent, C.F. Williamson, and R.A. Lindgren, Phys. Rev. Lett. 38, 750 (1977).
25. R. A. Lindgren, C.F. Williamson, and S. Kowalski, Phys. Rev. Lett. 40, 504 (1978).
26. R. A. Lindgren, J.R. Flanz, R.S. Hicks, B. Parker, G.A. Peterson, R.D. Lawson, W. Teeters, C.F. Williamson, S. Kowalski, X.K. Maruyama, Phys. Rev. Lett. 46, 706 (1981).
27. R.A. Lindgren, M.A. Plum, R.S. Hicks, B. Parker, G.A. Peterson, R. Singhal, C.F. Williamson, and X.K. Maruyama, Phys. Rev. Lett. 47, 1266 (1981).
28. J. Heisenberg, to be published.
29. J. Lichtenstadt, C.N. Papanicolas, C.P. Sargent, J. Heisenberg, and J.S. McCarthy, Phys. Rev. Lett. 44, 858 (1980).
30. J. Millener and D. Kurath, Nucl. Phys. A255, 315 (1975); D.J. Millener, private communication.
31. E.R. Siciliano and D.L. Weiss, Phys. Lett. 93B, 371 (1980).
32. D. Halderson, R.J. Philpott, J.A. Carr, and F. Petrovich, Phys. Rev. C24, 1095 (1981).
33. HEIMAG, distorted wave code for magnetic states including high spin, obtained from J. Heisenberg. Modified by R.A. Lindgren to include HOWF.
34. D.J. Rowe, S.S.M. Wong and H. Chow, Nucl. Phys. A298, 21 (1978).
35. A. Amusa and R.D. Lawson, Phys. Rev. Lett. 51, 103 (1983).
36. B.C. Metsch, Ph.D. thesis, Utrecht.
37. J. Dubach, Nucl. Phys. A340, 271 (1980).
38. T. Suzuki, S. Krewald, and J. Speth, Phys. Rev. Lett. 107B, 9 (1981).

39. I. Hammamoto, J. Lichtenstadt, and G.F. Bertsch, Phys. Lett.
     96B, 249 (1980).
40. F. Petrovich, unpublished.'
41. S. Krewald and J. Speth, Phys. Rev. Lett. 45, 417 (1980).
42. G. Emery, these proceedings.
43. B.D. Anderson, these proceedings.
44. S. Joyce-Seestrom Morris, these proceedings.
45. J.R. Comfort, G.L. Moake, C.C. Foster, P. Schwandt, and
     W.G. Love, Phys. Rev. C26, 1800 (1982).
46. G.S. Adams, A.D. Bacher, G.T. Emery, W.P. Jones, R.S. Kouzes,
     D.W. Miller, A. Picklesimer, and G.E. Walker, Phys. Rev. Lett.
     38, 1387 (1977).
47. B. Anderson, A. Fazely, C. Lebo, J.W. Watson, and R. Madey,
     Phys. Rev. 25, 1715 (1982).
48. C. Olmer, B. Zeidman, D.F. Geesaman, T.S.H. Lee, R.E. Segal
     L.W. Swenson, R.L. Boudrie, G.S. Blanpied, H.A. Thiessen,
     C.L. Morris, and R.E. Anderson, Phys. Rev. Lett. 43, 612
     (1979).
49. C. Olmer, private communication.
50. F. Petrovich, W.G. Love, A. Picklesimer, G.E. Walker, and
     E.R. Siciliano, Phys. Lett. 95B, 166 (1980).
51. S. Yen, R.J. Sobie, T.E. Drake, A.D. Bacher, G.T. Emery, W.P.
     Jones, D.W. Miller, C. Olmer, P. Schwandt, W.G. Love, and F.
     Petrovich, Phys. Lett. 105B, 421 (1981).
52. W.G. Love, A. Scott, F.T. Baker, W.P. Jones, and J.P. Wiggins,
     Jr., Phys. Lett. 73B, 277 (1978); G. Bertsch, J. Børysowicz,
     H. McManus, and W.G. Love, Nucl. Phys. A284, 399 (1977).
53. M. Gazzaly, M.A. Franey, and N. Hintz, private communication.
54. R.S. Henderson, B.M. Spicer, J.D. Svalbe, V.C. Officer, G.G.
     Shute, D.W. Divine, D.L. Friesel, W.P. Jones, and A.C. Attard,
     Aust. J. Phys. 32, 411 (1979).
55. D.B. Holtkamp, W.J. Braithwaite, W. Cottingame, S.J. Greene,
     R.J. Joseph, C.F. Moore, C.L. Morris, J. Piffaretti, E.R.
     Sicilano, and H.A. Thiessen, Phys. Rev. Lett. 45, 420 (1980).
56. W.G. Love, private communication.
57. C.L. Morris, Workshop on Nuclear structure with Intermediate
     Energy Probes, Los Alamos, Ne Mexico (LA-8303C) 1980.
58. W. Bertozzi, Nucl. Phys. A374, 109c (1982).
59. C. Olmer, et al., to be published.
60. G.S. Adams, A.D. Bacher, G.T. Emery, W.P. Jones, D.W. Miller,
     W.G. Love, and F. Petrovich, Phys. Lett. 91B, 23 (1980).
61. J.A. Carr and F. Petrovich, unpublished.
62. D.B. Holtkamp, S.J. Seestrom-Morris, S. Chakravarti, D.
     Dehnhard, H.W. Baer, C.L. Morris, S.J. Greene and C.J. Harvey,
     Phys. Rev. Lett. 47, 216 (1981).
63. D.B. Holtkamp, S.J. Seestrom-Morris, D. Dehnhard, H.W. Baer,
     C.L. Morris, S.J. Greene, C.J. Harvey, D. Kurath, and J.A.
     Carr, to be published.
64. D.F. Geesaman, et al., to be published.
65. C. Olmer, Conference Proceedings, McCormicks Creek State
     Park, 1982.

66. J.J. Kelly and J.A. Carr, these proceedings.
67. C.D. Goodman, C.A. Goulding, M.B. Greenfield, J. Rapaport, D.E. Bainum, C.C. Foster, W.G. Love, and F. Petrovich, Phys. Rev. Lett. 44, 1755 (1980).
68. C. Gaarde, these proceedings.
69. G.M. Crawley, these proceedings.
70. F. Petrovich, W.G. Love, and R.J. McCarthy, Phys. Rev. C21 1718 (1980).
71. R.S. Hicks, these proceedings.
72. B.D. Anderson, R.J. McCarthy, M. Ahmad, A. Fazely, A.M. Kalenda, J. N. Knudson, J.W. Watson, R. Madey, and C.C. Foster, Phys. Rev. C26, 8 (1982).
73. S. Cohen and D. Kurath, Nucl. Phys. 73, 1 (1965); A101, 1 (1967).
74. P.A.M. Guichon and C. Samour, Nucl. Phys. A382, 461 (1982).
75. J.R. Comfort, G.L. Moake, C.C. Foster, P. Schwandt, C.D. Goodman, J. Rapaport and W.G. Love, Phys. Rev. C24, 1834 (1981).
76. J.B. Flanz, R.S. Hicks, R.A. Lindgren, G.A. Peterson, A. Hotta, B. Parker, and R.C. York, Phys. Rev. Lett. 41, 1642 (1978).
77. J.B. Flanz, R.S. Hicks, R.A. Lindgren, G.A. Peterson, J. Dubach, and W.C. Haxton, Phys. Rev. Lett. 43, 1922 (1979).
78. J.B. Flanz, Ph.D. thesis, University of Massachusetts (University Microfilms, Ann Arbor, 1979).

SPIN OBSERVABLES IN NUCLEON-NUCLEUS SCATTERING

J. M. Moss

Los Alamos National Laboratory

Los Alamos, New Mexico 87545

1.   INTRODUCTION

The curse of inelastic nucleon scattering and charge exchange has always been the enormous complexity of the nucleon-nucleon (N-N) interaction. This complexity, however, can also be viewed as the ultimate promise of nucleons as probes of nuclear structure. Given an adequate theoretical basis, inelastic nucleon scattering is capable of providing information not obtainable with other probes.

Recently a revolution of experimental technique has taken place that makes it desireable to re-examine the question of what physics is ultimately obtainable from inelastic nucleon scattering. It is now feasible to perform complete polarization transfer (PT) experiments for inelastic proton scattering with high efficiency and excellent energy resolution. Programs to measure PT observables are underway at several laboratories, and results are beginning to appear. Objectives of this talk are to examine how such experiments are done, and what physics is presently obtained and may ultimately be learned from them.

2.   POLARIMETRY WITH MEDIUM-ENERGY PROTONS

Availability of medium-energy protons is the crucial factor in the measurement of PT observables in the (p,p') reaction. The long range of protons with energies above 100 MeV makes feasible the design of polarimeters with scattering efficiencies in the range of 0.1% to 10%; this is several orders of magnitude larger

than is possible at low energies. When coupled to high-resolution
magnetic spectrometers these instruments are ideal for PT measurements. The most advanced system at present is the focal-plane
polarimeter on the high-resolution spectrometer (HRS) at LAMPF.[1]
A less ambitious polarimeter is attached to the focal plane of the
QDDM spectrometer at IUCF.[2] The latter system has the advantage
of a very high intensity (150 nA) polarized proton beam.

     The LAMPF-HRS polarimeter (Fig. 1) consists of a pair of planes
of x- and y-sensitive multiwire drift chambers (MWDC) and associated scintillators, which constitute the normal focal-plane array.
Following this the protons are scattered from 12 cm of carbon, and
detected by two additional planes of larger MWDCs and scintillators. Thus for each proton, the initial and final (after scattering
from the carbon block) trajectories are determined. From this
information the scattering angle in both planes perpendicular to the
outgoing momentum may be deduced. The data-acquisition system
includes a fast micro-processor front end, which rejects protons
that do not scatter in the carbon block. A flexible system of
initial polarization orientation in the LAMPF accelerator allows
one to measure all possible PT observables (because of spin
precession in the field of the HRS, not all observables can be
measured for all outgoing energies). Those consistent with parity
conservation are D_{NN}, D_{LL}', D_{SS}', D_{LS}', and D_{SL}'; where L, N, and S
are respectively in the direction of the incident momentum, $\vec{k}$,
normal to the reaction plane (along $\vec{k} \times \vec{k}'$), and normal to $\vec{k}$, in
the reaction plane ($\vec{N} \times \vec{L} = \vec{S}$). Final (primed) subscripts are
defined analogously with respect to the final momentum, $\vec{k}'$.

     One additional observable that will prove to be very interesting
is the polarization function, P, or more precisely, P-A, where A is
the analyzing power. Measurement of P is accomplished by measuring

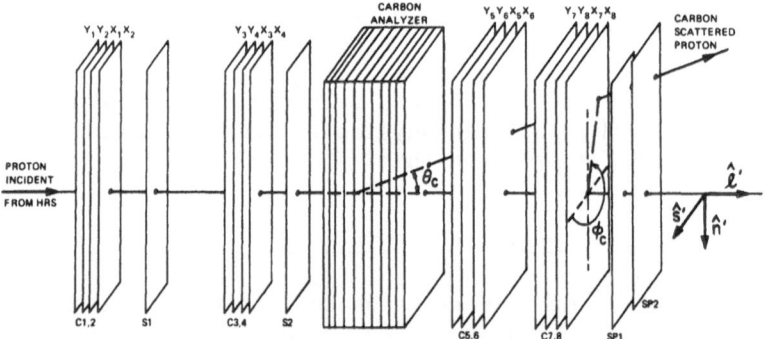

Fig. 1.   Schematic of the HRS focal-plane polarimiter.

the outgoing polarization in a reaction induced by an unpolarized beam.

3.   THEORETICAL FRAMEWORK

Now that one can measure these new observables, is it fair to ask, "What do they tell us?" We will address this question from a simple viewpoint that displays the physics involved in a fashion that is much more transparent than one gets from numerical calculations with either the distorted-waves impulse approximation (DWIA) or the Glauber model.

In the plane-wave impulse approximation (PWIA), the N-nucleus scattering amplitude is

$$\overline{M}_\mu(q) = < \mu \mid M(q) \, e^{-i\vec{q}\cdot\vec{r}} \mid 0> \ ,$$

where $M(q)$ is the N-N scattering amplitude, and μ is the projection of the total angular momentum transfer along the q axis. Following Kerman, McManus, and Thaler[3] (KMT)

$$M(q) = A + B \, \sigma_{1\hat{n}}\sigma_{2\hat{n}} + C(\sigma_{1\hat{n}}+\sigma_{2\hat{n}}) + E\sigma_{1\hat{q}}\sigma_{2\hat{q}} + F\sigma_{1\hat{p}}\sigma_{2\hat{p}} \ ,$$

with
$$\begin{aligned}
\hat{q} &= \vec{q}/|q| & \vec{q} &= \vec{k} - \vec{k}' \ , \\
\hat{n} &= \vec{n}/|n| & \vec{n} &= \vec{k} \times \vec{k}' \ , \\
\hat{p} &+ \vec{p}/|p| & \vec{p} &= \vec{n} \times \vec{q} \ .
\end{aligned}$$
(1)

3.1 Unnatural Parity States

First we consider the excitation of unnatural parity states. Now using the expression

$$D_{ij} = \sum_\mu \mathrm{Tr}(\overline{M}_\mu \sigma_i \overline{M}_\mu^+ \sigma_j) \, / \, \sum_\mu \mathrm{Tr}(\overline{M}_\mu \overline{M}_\mu^+) \ ,$$
(2)

one arrives at simple expressions for the PT observables in the $\hat{n}$, $\hat{p}$, $\hat{q}$ system[4] (n is identical to N)

$$\begin{aligned}
\sigma_o D_{\hat{n}\hat{n}} &= X_T^2 \, (C^2 + B^2 - F^2) - X_L^2 \, E^2 \ , \\
\sigma_o D_{\hat{p}\hat{p}} &= X_T^2 \, (C^2 - B^2 + F^2) - X_L^2 \, E^2 \ , \\
\sigma_o D_{\hat{q}\hat{q}} &= X_T^2 \, (C^2 - B^2 - F^2) + X_L^2 \, E^2 \ , \\
\sigma_o D_{\hat{q}\hat{p}} &= -\sigma_o D_{\hat{p}\hat{q}} = 2X_T^2 \, \mathrm{Im} \, (BC^*) \ ,
\end{aligned}$$
(3)

where the differential cross section, σ_o, is given by

$$\sigma_o = X_T^2 (C^2 + B^2 + F^2) + X_L^2 E^2 \ .$$

The transverse, X_T, and longitudinal form factors, X_L are defined by

$$X_T = \left(\frac{j+1}{2(2j-1)}\right)^{1/2} Q_{jj-1} + \left(\frac{j}{2(2j+3)}\right)^{1/2} Q_{jj+1} \ ,$$

$$X_L = \left(\frac{j}{(2j-1)}\right)^{1/2} Q_{jj-1} - \left(\frac{j+1}{(2j+3)}\right)^{1/2} Q_{jj+1} \ ,$$

$$(4)$$

where $Q_{j\ell}$ is a reduced matrix element defined in Appendix III of KMT.[3] Transformation of Eqs. (3) to the laboratory system (N, L, S, S', L') is straightforward[4]. Equations (3) may be inverted to yield

$$X_L^2 = \sigma_o/4E^2 (1 - D_{\hat{n}\hat{n}} + D_{\hat{q}\hat{q}} - D_{\hat{p}\hat{p}}) \ ,$$

$$X_T^2 = \sigma_o/4B^2 (1 + D_{\hat{n}\hat{n}} - D_{\hat{q}\hat{q}} - D_{\hat{p}\hat{p}}) \ ,$$

$$X_T^2 = \sigma_o/4C^2 (1 + D_{\hat{n}\hat{n}} + D_{\hat{q}\hat{q}} + D_{\hat{p}\hat{p}}) \ ,$$

$$(5)$$

$$X_T^2 = \sigma_o/4F^2 (1 - D_{\hat{n}\hat{n}} - D_{\hat{q}\hat{q}} + D_{\hat{p}\hat{p}}) \ ,$$

$$X_T^2 = \sigma_o/2Im (BC^*) D_{\hat{q}\hat{p}} \ .$$

Note that with the knowledge of the coefficients of the impulse approximation (IA) interaction, the PT observables may be used to[5] directly infer these two form factors. The transverse form factor is similar to that obtained from electron scattering. However, X_L is not present in (e,e') and thus represents a new aspect of nuclear structure obtainable in (p,p') experiments. Of course the separation of nuclear structure and reaction dynamics is not straightforward in the DWIA. However, the physics contained in Eqs. (3) and (5) must still be present.

It often occurs that a given transition is dominated by a single ℓ value, e.g., in stretched configurations. Then $X_L = X_T$ (apart from constants) and Eqs. (3) for the PT observables become independent of nuclear structure. In such cases the $D_{ii's}$ may be used to deduce the components of the effective N-N interaction. Evidence is accumulating that certain parts of this interaction[6] may differ considerably from the free N-N interaction.

By comparing to exact DWIA calculations we have found that the approximations that yield the simple equations are reasonably accurate for $q < 1$ fm^{-1}. Figure 2 shows such a comparison along with experimental data for the $^{12}C(p,p')^{12}C$ (15.11 MeV, 1^+, T=1) reaction at 500 MeV.[7]

3.2 Natural Parity Transitions

For natural parity transitions we consider only the case where a single j transfer is allowed, such as for transitions from a 0^+ ground state. Form factors with and without spin transfer are allowed; $Q_{j\ell}$ and Q_j respectively. Natural parity transitions are often dominated by a collective spin-independent amplitude. Intuition would say that in such cases, the effects on the PT observables from the spin-dependent form factor might be difficult to measure. Equations derived from (2) are best cast in the form of a spin-flip probability (SFP), where $S_{ij} = 1/2 \ (1-D_{ij})$.

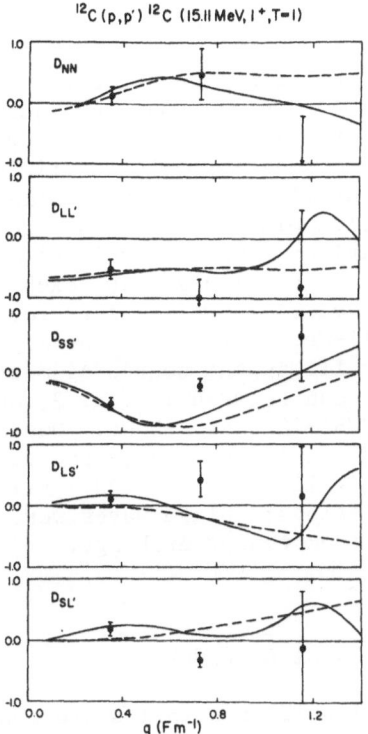

Fig. 2. Polarization transfer observables (preliminary analysis) at E_p = 500 MeV.

One finds

$$S_{\hat{n}\hat{n}} = Q_{j\ell}^2 \; F^2 \; /2\sigma_o \quad ,$$

$$S_{\hat{q}\hat{q}} = Q_{j\ell}^2 \; (B^2 + F^2) \; /2\sigma_o + Q_j^2 \; C^2/\sigma_o$$

$$S_{\hat{p}\hat{p}} = Q_{j\ell}^2 \; B^2 \; /2\sigma_o + Q_j^2 \; C^2/\sigma_o \tag{6}$$

$$S_{\hat{q}\hat{p}} = 1/2 \left[1 - 1/2 \; Q_{j\ell}^2 \; \text{Im} \; (BC^*) + 2Q_j^2 \; \text{Im} \; (AC^*) \right] \quad ,$$

$$\sigma_o = 1/2 \; Q_{j\ell}^2 \; (C^2 + B^2 + F^2) + Q_j^2 \; (A^2 + C^2) \quad .$$

Clearly $S_{\hat{n}\hat{n}}$ is different from zero only to the extent that the spin-transfer form factor, weighted by the spin-dependent terms of the N-N interactions, competes with the corresponding spin-independent factors.

3.3 Polarization and Analyzing Power

Evaluation of the polarization and analyzing power in the PWIA yields the result, P = A. As was shown by Squires[8] many years ago, this is a consequence of using a scattering amplitude, which depends only on q. Spin-orbit distortion effects eliminate this equality in the DWIA, but in general P and A have similar shapes unless one is close to a diffraction minimum. A much more interesting difference between P and A arises from the nonlocal/exchange nature of the N-N interaction.[9] In particular, the exchange amplitudes of the tensor interaction yield opposite signs for P and A. In the excitation of the 15.11-MeV state of ^{12}C at 150 MeV, the unnatural parity amplitude ℓsj = 111 is the source of most of the difference between P and A. This can be seen in Fig. 3, where PWIA calculations using the code DWBA-70[10] are shown. One set of curves employs the full Cohen-Kuruth[11] (CK) functions; the other uses the CK-wave functions with the ℓsj = 111 term removed. All exchange terms are present in both calculations; plane waves were used in order to isolate the P-A terms from tensor exchange.

4. EXPERIMENTS: PRESENT AND FUTURE

Experiments in which PT observables are measured are relatively new and as such few published results exist. We will discuss some of these experiments, often with preliminary data and interpretations, and speculate about areas of future interest.

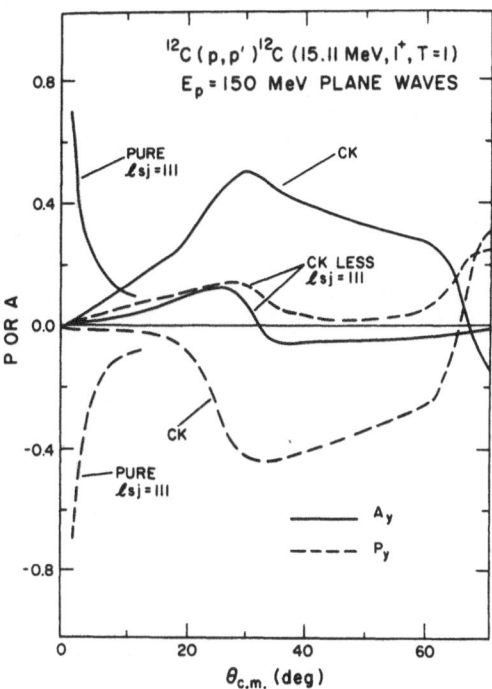

Fig. 3   Plane-wave calculations of P and A.   CK stands for the
Cohen-Kurath wave functions.

4.1   Complete Polarization Transfer Experiments for Discreet States

Recently at LAMPF we have measured the first complete set of
PT observables[7] for the excitation of unnatural parity states in
^{12}C.   At small momentum transfer where the data are most precise,
the two 1^+ states of ^{12}C are dominated by ℓsj = 011 transfer.   Thus
equations (3) can be reduced to

$$\bar{C}^2 = 1/4 \ (1 + D_{\hat{n}\hat{n}} + D_{\hat{p}\hat{p}} + D_{\hat{q}\hat{q}}) \quad ,$$

$$\bar{B}^2 = 1/4 \ (1 + D_{\hat{n}\hat{n}} - D_{\hat{p}\hat{p}} - D_{\hat{q}\hat{q}}) \quad ,$$

$$\bar{F}^2 = 1/4 \ (1 - D_{\hat{n}\hat{n}} + D_{\hat{p}\hat{p}} - D_{\hat{q}\hat{q}}) \quad ,$$

$$\bar{E}^2 = 1/4 \ (1 - D_{\hat{n}\hat{n}} - D_{\hat{p}\hat{p}} + D_{\hat{q}\hat{q}}) \quad ,$$

where

$$\bar{B}^2 = B^2 \ / \ (B^2 + C^2 + E^2 + F^2) \quad \text{etc.} \quad .$$

The magnitude of the cross section is not accounted for by the PWIA, hence, it is preferable to compare the experimental and theo-retical amplitudes[12] in terms of the normalized (barred) quantities. This comparison is shown in Figs. 4 and 5; experimental data for the 15.11-MeV state are shown in Fig. 2 along with the DWIA and PWIA curves calculated with the Love-Franey amplitudes. Data are still preliminary so it is not possible to draw firm conclusions. However, this is an indication of a possible problem with the isos-calar spin-orbit amplitude (C_0). Recently, independent evidence[13] for a need to increase the spin-orbit amplitude with respect to the IA value has been found in an elastic scattering experiment at 500 MeV.

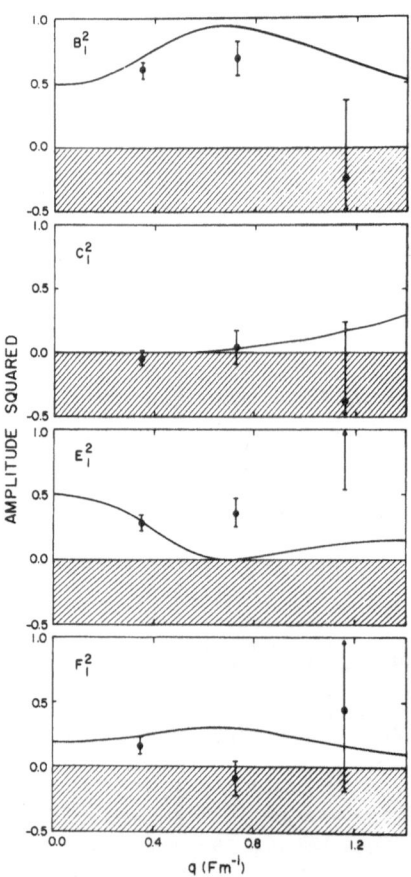

Fig. 4 Isovector impulse approximation amplitudes derived from polarization transfer data (preliminary analysis) at 500 MeV. The solid curves are the Love-Franey amplitudes.

Clearly, when experiments such as this become even more refined, current reaction theories will be put to severe tests. Our prejudice is that the real future of complete PT experiments is in testing models of nuclear structure at a level not previously possible.

4.2 Polarization and Analyzing Power

It is clear from Figs. 6 and 7 that large differences between P and A have been observed in the excitation of the 1^+ states in ^{12}C at 150 MeV.[14] At scattering angles smaller than 20°, P-A is dominated by the effects of tensor exchange discussed in section 3.3. The solid curve in Fig. 6 is a DWIA calculation using the Love effective interaction[15] and the CK-wave functions. The dashed

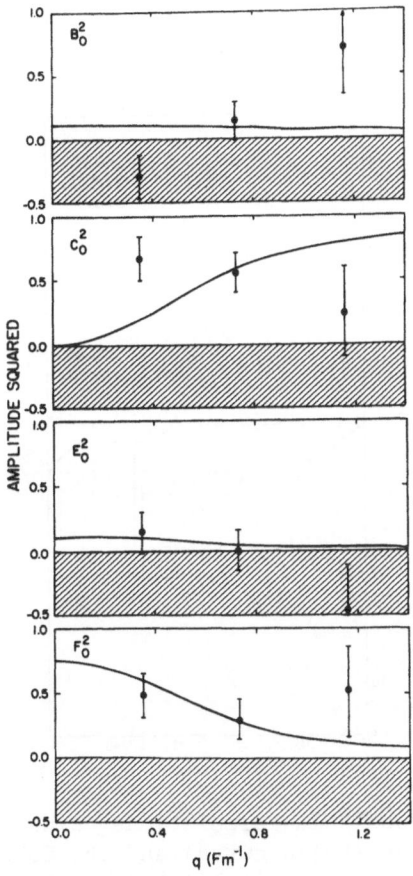

Fig. 5 Isoscalar impulse approximation amplitudes derived from polarization transfer data (preliminary analysis) at 500 MeV. The solid curves are the Love-Franey amplitudes.

curve is a similar calculation with the ℓsj = 111 term removed.
Our preliminary conclusion is that the ℓsj = 111 term in the CK-
wave functions is required to fit the small angle points. We consider these points more significant since the cross section is
large here. Additionally, variations in the magnitude of P-A at
larger angles are possible due to small changes in optical potential
distortions.

The ℓsj = 111 term determines the sum of the density matrix
elements, p1/2 p3/2 + p3/2 p1/2. As was pointed out by Dubach and
Haxton[16] this quantity is very poorly determined by electromagnetic
and weak interaction data on the 15.11-MeV state and its analogues.
Thus the (p,p') reaction is able to make a unique contribution to
the determination of the structure of this transition.

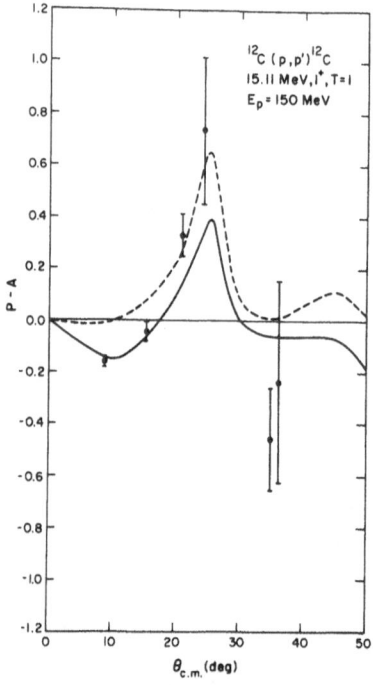

Fig. 6   P-A data versus DWIA calculations using the Cohen-Kurath
         wave functions (solid curve) and the Cohen-Kurath wave
         functions with the ℓsj = 111 term removed (dashed curve).

Figure 7 shows that large values of P-A are also seen in the
excitation of the 1^+, T = 0 state. The uncertainties in the knowledge of the interaction in the s = 1, T = 0 channel are such that
no definitive statement can be made at this time regarding the
large discrepancy between calculation and experiment. Further
study of the new polarization observables should eventually lead to
an understanding of the failure of the DWIA to reproduce the angular
distribution of this state in the energy range of 150 MeV.

4.3 Polarization Transfer and Spin Excitation

It is clear from the discussion in Section 3 that the spin
observables $D_{\hat{n}\hat{n}}$, $D_{\hat{q}\hat{q}}$, and $D_{\hat{p}\hat{p}}$ are different from unity (and the
corresponding SFPs are different from zero) only when spin excita-
tions are important. This rule has had considerable experimental
verification both at low[17] and intermediate energies in the case of
$D_{\hat{n}\hat{n}}$. Figure 8 shows $S_{\hat{n}\hat{n}}$ for several states in ^{12}C excited by 400-
MeV protons. Note that the collective 3^- state displays a SFP
close to zero.

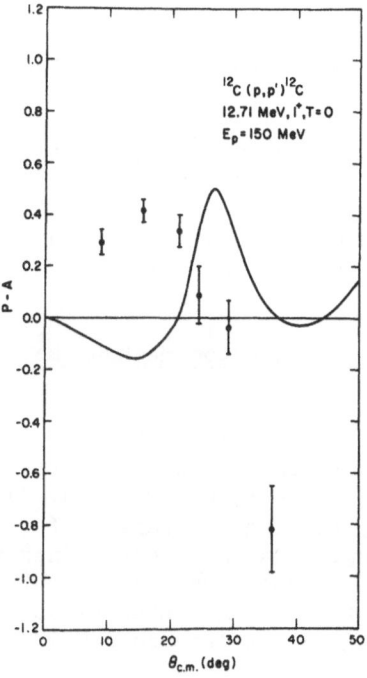

Fig. 7 P-A data versus DWIA calculations with the Cohen-Kurath
wave functions.

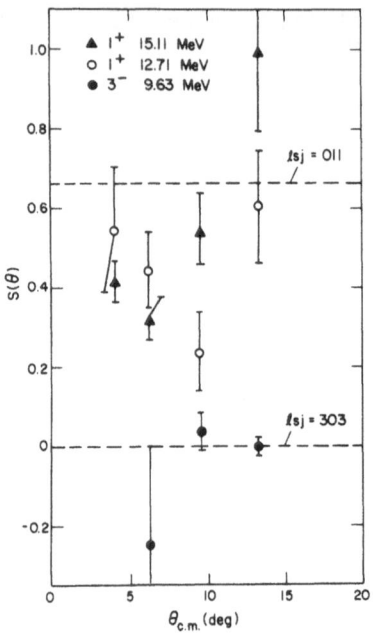

Fig. 8   Spin-flip probabilities for these states in ^{12}C at E_p = 500 MeV.

The simple connection between the PT observables and spin transfer means that they can be used to search for spin excitation in unexplored territory.   Although ^{12}C from E_x = 8 to 20 MeV can hardly be considered such a case, Fig. 9 indicates how a spectrum of spin flip might be used to "amplify" the signal for spin excitations.

4.4   Spin Observables and Reaction Mechanisms

It has been appreciated for some time that PT observables can be employed as probes of reaction mechanisms.[17]   For example, an $0^+ \rightarrow 2^-$ transition proceeding by ℓsj = 112 transfer from a purely central interaction (B = E = F, C = 0 in Eq. 3) yields $S_{\hat{n}\hat{n}}$ = 7/10 in the PWIA.   A purely two-step mechanism of the type $0^+ \rightarrow 3^- \rightarrow 2^-$ proceeding by two nonspin transfer scatterings yields $S_{\hat{n}\hat{n}}$ = 0. More realistic DWIA and coupled-channel calculations bear out the qualitative features of the PWIA.   Exceptions may occur in reactions where the two-body, spin-orbit interaction is of major importance.

4.5   Spin Observables and Meson Exchange

Bugg has pointed out[18] in connection with searches for precursors of pion condensation[19] that the dominance of certain meson

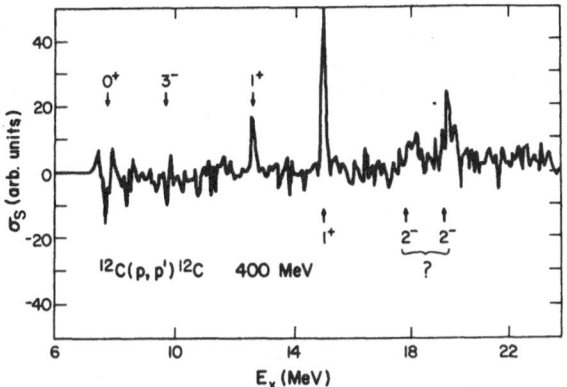

Fig. 9   Spin flip corss section for ^{12}C at 3.5°.

fields in the effective N-N interaction will lead to definite
signatures in the PT observables. Although no evidence of precursor phenomena[20] have been found, Bugg's ideas are interesting to
consider. As an example, pure one-pion exchange yields $D_{\hat{n}\hat{n}} = -1$,
$D_{SS'} = 1/2 \cos \theta$, and $D_{LL'}, = -1/2 \cos \theta$; ρ exchange results in a
different combination. It may be possible to select transitions
which, in selected regions of q, are dominated by nearly pure meson
exchange.

4.6 Longitudinal and Transverse Response Functions

One of the most exciting areas for future experiments concerns
spin-flip strength in the continuum. Where, for example, is the
missing Gamow-Teller strength? It is contained in the structureless region surrounding the Gamow-Teller resonance, or is in the
region of Δ-hole excitations? The answers to these questions have
an obvious impact on the understanding of some very important issues
in the nuclear structure.

A global view of the response of nuclei to spin perturbing
fields could be obtained if one could experimentally map out the
longitudinal (spin) and transverse nuclear response functions. The
importance of the longitudinal response functions has been emphasized
recently[21] in connection Gamow-Teller quenching and pion-condensation
precursor phenomena. This function is not measureable in backward-
angle electron scattering.

Fortunately the N-N interaction contains a longitudinal field
and to the extent that Eqs. (5) and (6) are valid the <u>longitudinal
response functions can be isolated</u> by the combination of observables
$\sigma_o (1- D_{\hat{n}\hat{n}} + D_{\hat{q}\hat{q}} - D_{\hat{p}\hat{p}})$. It is easy to verify from Eq. (6) that

this combination <u>always</u> yields zero for natural-parity excitations.
Even beyond the range where Eqs. (5) and (6) are clearly valid
($q < 1$ fm^{-1}), the above combination of observables will probably
be much more sensitive to the longitudinal than the transverse response functions. Thus it is not difficult to forsee the day when
inelastic scattering and charge-exchange reactions will provide
both longitudinal and transverse response functions over a wide
range of energy and momentum transfer.

5. CONCLUSION

I hope that it has become clearer that we are on the verge of
a new era in inelastic proton scattering and charge exchange.
Polarimetry has developed to the point where all of the allowed PT
observables may be measured with very high efficiency. The simple
expressions for these observables presented here make it clear that
in certain situations new nuclear structure information may be
obtained. In other cases reaction mechanism may be the dominant
effect. It is clear that in the future elucidations of spin excitations in nuclei the new spin observables will play an increasingly
important role.

REFERENCES

1. J. M. McClelland, J. F. Amman, W. Cornelius, and H. A. Thiessen
    Bull. Am. Phys. Soc. <u>26</u>, 1159 (1981).
2. T. A. Carey, J. M. Moss, D. W. Miller, H. Nann, C. Olmer,
    P. Schwandt, and E. J. Stephenson
    Bull. Am. Phys. Soc. <u>25</u>, 746 (1980).
3. A. K. Kerman, H. McManus, and R. M. Thaler, Ann. of Phys. <u>8</u>,
    551 (1959).
4. J. M. Moss, Phys. Rev. <u>C26</u>, 727 (1982).
5. E. Bleszinski, M. Bleszinski, and C. A. Whitten
    In Proc. of the 5th Int. Symposium on Polarization Phenomena
    In Nuclear Physics - 1980, eds. G. G. Ohlsen et al.,
    (Am. Inst. of Phys., New York), p. 556.
6. J. Kelley, W. Bertozzi, T. N. Buti, F. W. Hersman, C. Hyde,
    M. V. Hynes, B. Noram, F. N. Rad, A. D. Bacher, G. T. Emery,
    D. C. Foster, W. P. Jones, D. W. Miller, B. L. Berman, W. G.
    Love, and F. Petrovich, Phys. Rev. Lett. <u>45</u>, 2012 (1980).
7. J. B. McClelland, J. M. Moss, M. Bleszinski, G. Igo, B. Aas,
    A. Rahbar, J. Geaga, G. Weston, and M. Gazzaly
    to be published.
8. E. J. Squires, Nucl. Phys. 6, 504 (1958).
9. W. G. Love, private communication.
10. Computer code write by R. Schaeffer and J. Raynal, modified
    by M. A. Franey and W. G. Love.

11. S. Cohen and D. Kurath, Nucl. Phys. A101, 1 (1967).

12. W. G. Love and M. A. Franey, Phys. Rev. C24, 1073 (1981).

13. G. W. Hoffmann, L. Ray, M. L. Bartlett, R. Ferguson, J. McGill,
     E. C. Milnov, Kamal K. Seth, D. Barlow, M. Bosko, S. Iverson,
     K. Kaleta, A. Saha, and D. Smith,
     Phys. Rev. Lett. 47, 1436 (1981).

14. T. A. Carey, J. M. Moss, S. J. Seestrom-Morris, D. W. Miller,
     H. Nann, C. Olmer, P. Schwandt, E. J. Stephenson, and
     W. G. Love, Phys. Rev. C26, 266 (1982).

15. W. G. Love, in The (p,n) Reaction and the Nucleon-Nucleon Force
     eds. C. D. Goodman, (Plenum, New York, 1980) p. 23.

16. J. Dubach and W. C. Haxton, Phys. Rev. Lett. 41, 1453 (1980).

17. W. D. Cornelius, J. M. Moss, and T. Yamaya, Phys. Rev. C23,
     1364 (1981).

18. D. V. Bugg, J. of Phys. G 7, L141 (1981).

19. M. Ericson and J. Delorme, Phys. Lett. 76B, 182 (1978), eds.
     J. Delorme et al., Phys. Lett. 89B, 327 (1980).

20. M. Haji-Saeid, C. Glashausser, G. Igo, W. Cornelius,
     M. Gazzaly, F. Irom, J. McClelland, J. M. Moss, G. Pauletta,
     H. A. Thiessen, and C. A. Whitten,
     Phys. Rev. Lett. 45, 880 (1980).

21. W. M. Alberico, M. Ericson, and A. Molinari
     Phys. Lett. 92B, 153 (1980);
     M. Ericson, these preceedings.

STUDIES OF SPIN EXCITATIONS USING POLARIZED PROTONS

G. T. Emery, A. D. Bacher, and C. Olmer

Indiana University Cyclotron Facility
Bloomington, Indiana 47405

ABSTRACT

The energy dependence of ^{28}Si(p,p') has been studied using polarized proton beams at energies between 80 and 180 MeV. Differential cross sections and analyzing powers have been determined for elastic scattering and for inelastic excitation of the following states: 6^-, T=1 (14.35 MeV), 6^-, T=0 (11.58 MeV), and 5^-, T=0 (9.70 MeV). The results are compared with distorted-wave impulse approximation calculations using the Love-Franey t-matrix and an earlier effective interaction. Rather good agreement is found with the DWIA description of these transitions. Some refinements to the standard model are discussed.

INTRODUCTION

The use of inelastic nucleon scattering to study spin excitations in nuclei has both advantageous and fruitful aspects. From an experimental point of view, when compared to pion or electron measurements, proton scattering is characterized by larger count rates (as determined by cross section times beam intensity) and comparable or better resolution (as determined by beam preparation and/or radiative tails). A less trivial advantage involves the opportunity with nucleons to measure observables that depend directly on spin, such as analyzing power, spin-flip probability, and others.[1] Even more interesting, perhaps, is the way the various spin- and isospin-dependent terms in the interaction show some rather strong variations with bombarding energy and momentum transfer, such that spin-transfer reactions are strongly favored for selected regions of energy and momentum transfer. The

371

dominance of Gamow-Teller transitions over analog transitions near 200 MeV is a well-known example.[2] An additional example is the selective excitation of high-spin states, such as the "stretched" 6^-, T=0 and T=1 states in ^{28}Si, at large momentum transfer. As is evident from the comparison in Fig. 1, these states are prominent at intermediate energies (approximately 80 to 300 MeV), but at higher energies they are only barely visible in the presence of other more strongly-excited states.

The study of the excitation of these two 6^- states in ^{28}Si is especially useful. The excitation involves only one form factor, that of the magnetization density, which is the same form factor for proton, electron and pion inelastic scattering. One would like to use the excitation of these "stretched" states, which should have a simple structure, to test theoretical descriptions of inelastic scattering. The idea has been, for example, to compare spectroscopic factors for the excitation of the same state by different probes. Such comparisons (e.g., ref. 4) have so far relied on data at a single energy for each probe. We have felt that it would be useful to make a careful study of the reaction mechanism, over a range of energy where these states were relatively strong, as a test of the mechanism and its ingredients, with the idea of trying to establish the limits of validity of the nuclear structure information extracted. To investigate these effects we have measured the differential cross sections and analyzing powers, at 80, 100, 135, and 180 MeV, for these two 6^- transitions, and for others in ^{28}Si, though the only other transition we will mention here is that to the 5^-, T=0 state at 9.70 MeV.

THE DWIA

The standard description of these reactions, usually called the distorted-wave impulse approximation (DWIA), has been largely developed by Love and Petrovich,[5-7] based on the work of Kerman, McManus, and Thaler,[8] and it has an assumption and three ingredients. The assumption is that the excitation occurs in a single step; two-step contributions may have a somewhat different energy dependence and, although work from Osaka[9] indicates that two-step effects are present in the 40- to 65-MeV energy range, they seem to decrease in importance with increasing incident proton energy.

The first ingredient in the DWIA description is the form factor For these "stretched" states the form factor is nodeless and peaked near the nuclear surface. For the 6^-, T=1 state, the form factor is determined by the (e,e') results,[10] and for the 6^-, T=0 state it should be similar in shape. The situation is more complicated for the 5^- state, since in (e,e') one measures the longitudinal form

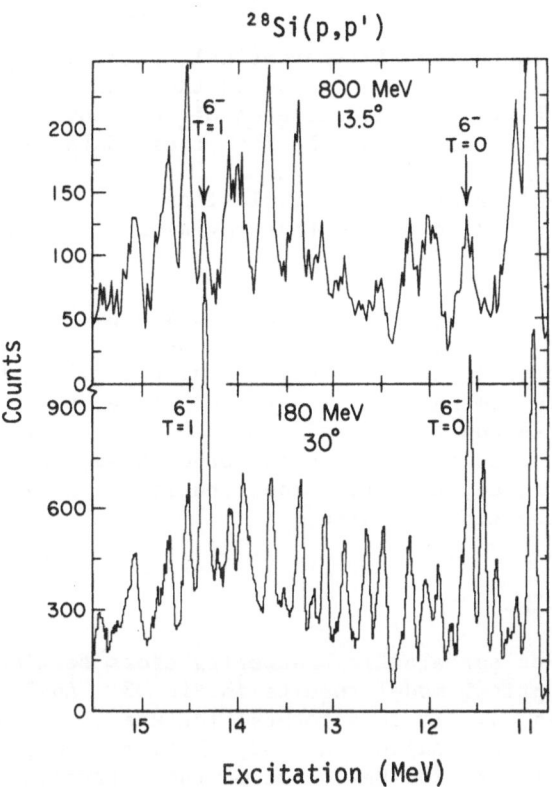

Fig. 1.   Inelastic proton spectra for the scattering of 180-MeV
          (bottom) and 800-MeV (top) protons from silicon.   The
          800-MeV data are from ref. 3.

factor and an upper limit on the magnitude of the transverse form
factor, but in (p,p') different combinations of form factors enter.
We have used a model wave function that is consistent with the
(e,e') data for this state.[11]

The second ingredient is the effective interaction, and we will
show results obtained from the Love-Franey interaction,[12] which is
explicitly energy-dependent, and from an older interaction,[13] whose
energy dependence comes only from interference between direct and
exchange terms.

The third ingredient is the optical optential used to generate
the distorted waves. We have measured the elastic scattering cross
sections and analyzing powers at each of the above energies, and
have extracted optical potentials from these data.

The inelastic scattering calculations were done using the code
DW81, Comfort's version of the Love revision of the standard
Raynal-Schaeffer code DWBA70.[14]

As a guide to what terms in the nucleon-nucleon interaction are
contributing to the excitation of these states, we reproduce in
Fig. 2 the results of some typical calculations.[15] The isovector
6^- excitation is dominated by the tensor force, while the isoscalar
6^- excitation has contributions from both the tensor and spin-orbit
terms. The 5^- cross section, on the other hand, is mostly due to
the spin-orbit force, with some contribution at smaller momentum
transfer from the central force.

ELASTIC SCATTERING

A portion of our elastic scattering cross section data is
compared with optical model results in Fig. 3. Analyzing powers
are shown in Fig. 4. It is a general feature of proton elastic
scattering from heavy nuclei (A > 40) that, as the proton energy
increases from 100 to 200 MeV, some of the diffractive oscillations
in the cross section are damped out,[16] while the corresponding
analyzing powers become more strongly oscillatory.[17] The silicon
data exhibit a similar trend.

Optical model fits to the elastic scattering cross section and
analyzing power data were made separately at 80, 100, 135 and 180
MeV.[18] These fits assumed the standard Woods-Saxon form. Recent
p+Si measurements[16] at larger angles lead to only small changes in
the potential parameters and do not indicate the necessity for
non-standard potential shapes, which had been found for $p+^{12}C$.[20]
The resulting optical potential parameters for p+Si exhibit a
smooth dependence on the proton bombarding energy. In particular,
both a decrease of V, V_{so} and W_{so} and an increase of W with proton

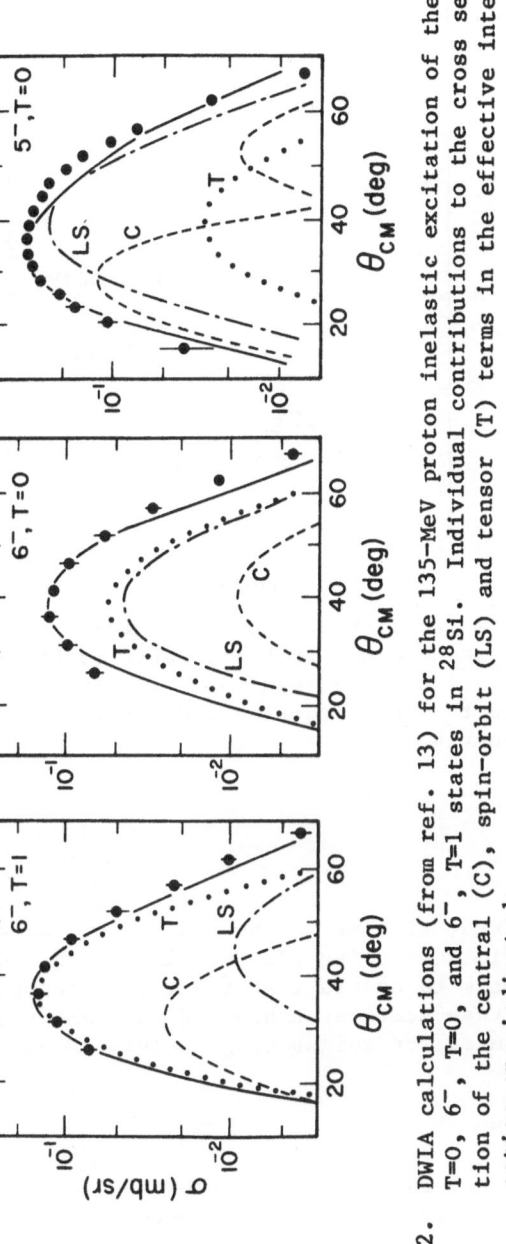

Fig. 2. DWIA calculations (from ref. 13) for the 135-MeV proton inelastic excitation of the 5^-, $T=0$, 6^-, $T=0$, and 6^-, $T=1$ states in ^{28}Si. Individual contributions to the cross section of the central (C), spin-orbit (LS) and tensor (T) terms in the effective interaction are indicated.

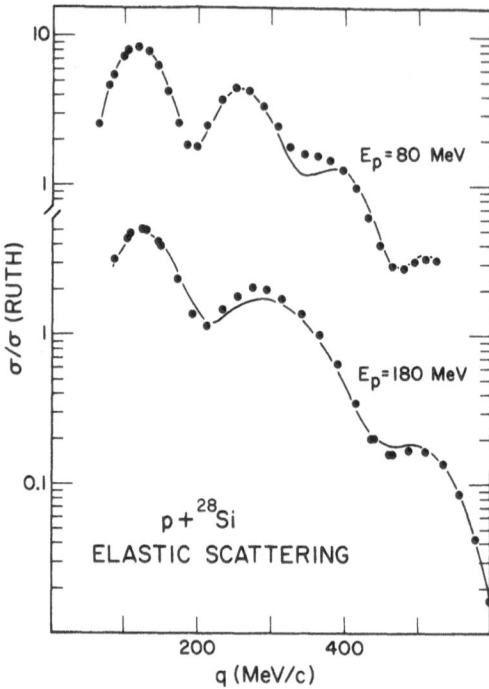

Fig. 3. Momentum transfer dependence of the cross sections for
         80-MeV and 180-MeV proton elastic scattering by silicon.
         The curves are the results of simultaneous optical-model
         fits to the cross-section and analyzing-power data at
         each energy.

energy are observed, similar to the results of Schwandt et al.[17]
for proton elastic scattering from Ca, Zr and Pb. Care was taken
in the present work to ensure that the resulting optical model
total reaction cross sections exhibited the same energy dependence
as has been measured for neighboring nuclei.

INELASTIC RESULTS

     In the next several figures we show the inelastic scattering
data at the four incident energies, together with the results of
DWIA calculations using the Love-Franey interaction. Each of these
calculations uses the newly-derived optical potential appropriate

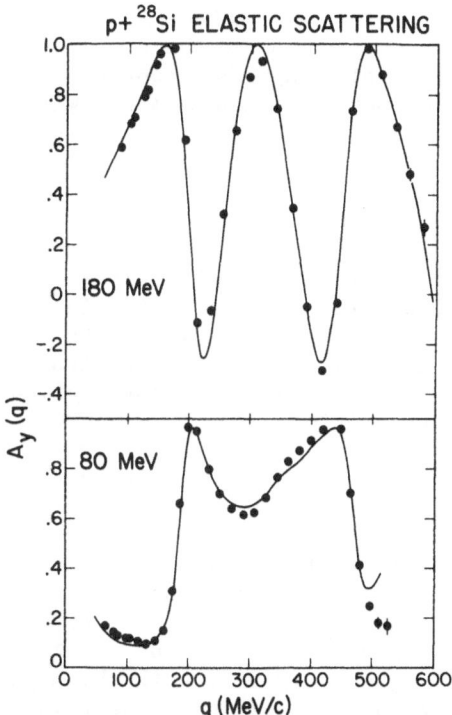

Fig. 4.   Momentum transfer dependence of the analyzing powers for
80-MeV and 180-MeV proton elastic scattering by silicon.
(See caption for Fig. 3.)

for that energy. For the "stretched" 6^- levels, the form factors
used are based on harmonic oscillator radial wave functions with
oscillator constants of 1.80 fm for the T=1 state and 1.91 fm for
the T=0 state, and the assumption that these are pure $f_{7/2}$ $(d_{5/2})^{-1}$
transitions starting with a filled $d_{5/2}$ shell. Oscillator radial
functions were also used for the 5^- state, with an oscillator
constant of 1.91 fm; the calculations used the RPA vector of ref.
11, including major contributions from the $f_{7/2}$ $(d_{3/2})^{-1}$ and $f_{7/2}$
$(d_{5/2})^{-1}$ configurations, and a smaller contribution from
$f_{5/2}(d_{5/2})^{-1}$. The calculated curves for each transition have been
multiplied by a normalization constant, independent of bombarding
energy, chosen to give an approximate fit to the data at the two
highest energies.

Fig. 5 shows the cross sections for the 6⁻, T=1 state. The
calculated results fit the data rather well at 180 and 135 MeV, but
do not reproduce the broadening of the cross section distribution
at the lower energies, although they continue to fit the location
and magnitude of the peak. The analyzing powers for this
transition are shown in Fig. 6. The experimental results show a
consistent increase from near zero at 200 MeV/c to positive values
at 400-500 MeV/c. At the two higher energies, the calculations
reproduce these qualitative features but are not completely
successful. It was noted earlier[15] that in the plane-wave limit,
the analyzing power for this transition is small, and that
spin-orbit distortions play an important role in producing the
calculated values. The present results, with a new optical
potential, are not significantly different from those given
earlier. Results with the older Love interaction are only slightly
different for this transition.

Cross sections for the 6⁻, T=0 state are shown in Fig. 7.
Again the calculations fit rather well at the two highest energies.
For this transition, however, the data at 100 and 80 MeV show a
movement of the peak to smaller momentum transfer, as well as a
broadening. The calculated results show only a much smaller shift.
The analyzing powers are shown in Fig. 8. At 180 and 135 MeV the
slope of A_y vs q is opposite to that for the T=1 transition. The
calculations do not agree very well with the higher-energy
analyzing powers for this transition. For this case, also, the
calculated analyzing powers are sensitive to spin-orbit
distortion.

Results for the normal-parity 5⁻, T=0 state are shown in Figs.
9 and 10. The shapes of the differential cross section distributions are reproduced rather well by the calculations. The
analyzing powers for this transition are already large in the
plane-wave limit and are only slightly changed by distortions, at
least at the higher energies. The crossing through zero at a
momentum transfer near 180 MeV/c is due to the change in sign of
the central term in the t-matrix at that value of q.[15] This
analyzing power is rather well given by the DWIA, with either of
the two interactions. Not only is the agreement with the data good
at the two highest energies, but the calculations also describe
very well the general decrease in magnitude of A_y as the energy
decreases.

ENERGY DEPENDENCE

In the DWIA description, the magnitude of the cross section
near the peak in the angular distribution will vary with energy
because (a) the effective interaction is energy dependent and (b)
the distortion effects are energy dependent. While the Love-Franey

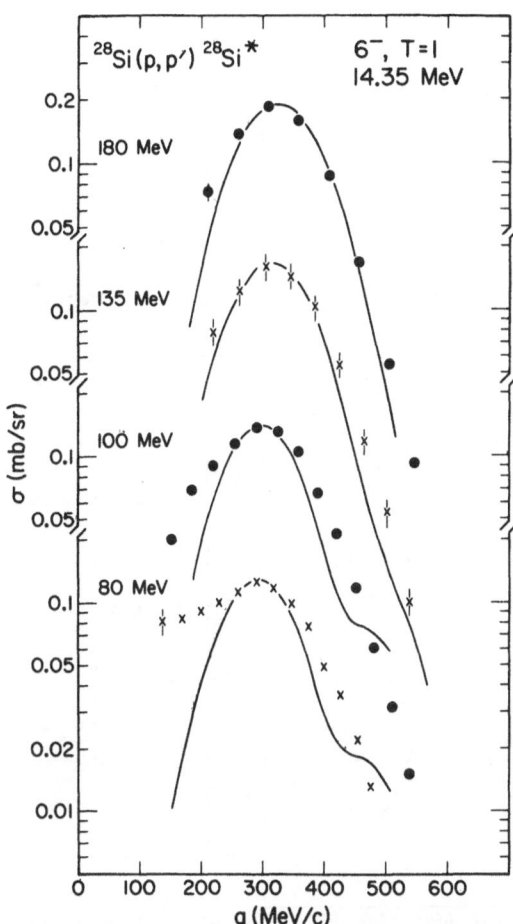

Fig. 5. Momentum transfer dependence of the cross sections for
         proton inelastic excitation of the 6⁻, T=1 state at 14.35
         MeV in ^{28}Si. The curves are DWIA calculations using the
         Love-Franey effective interaction and optical potentials
         appropriate for the various bombarding energies.

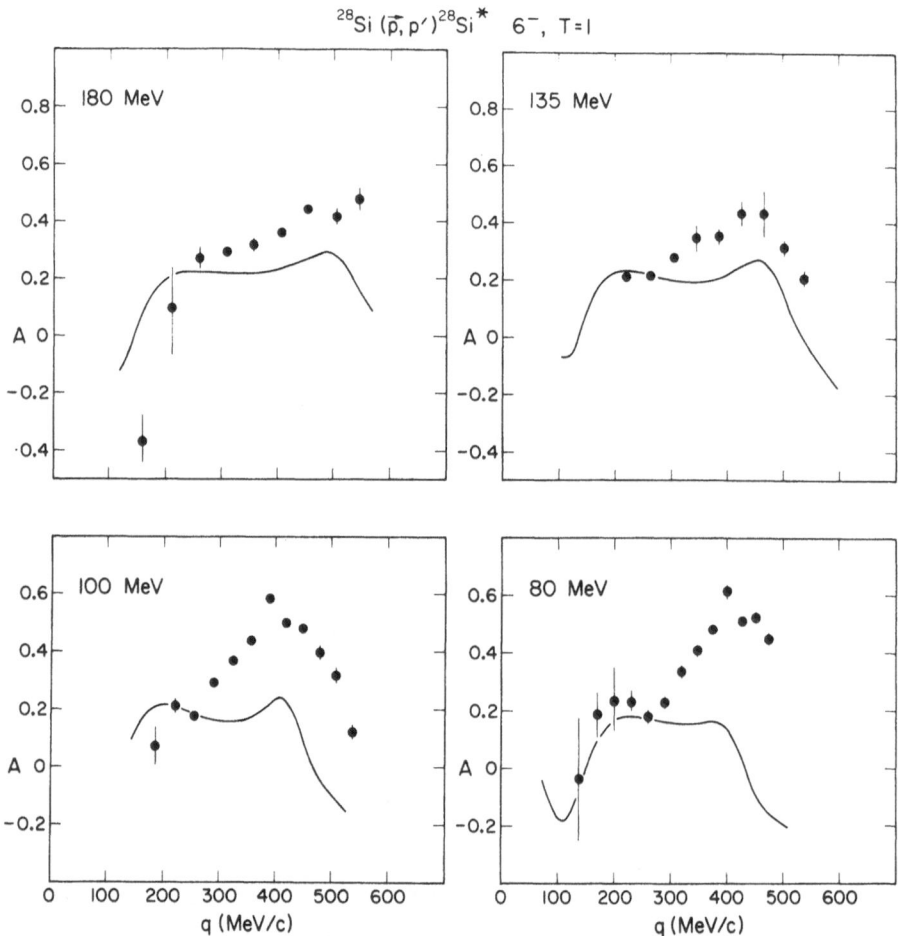

Fig. 6. Momentum transfer dependence of the analyzing powers for proton inelastic excitation of the 6^-, T=1 state at 14.35 MeV in ^{28}Si. (See caption for Fig. 5.)

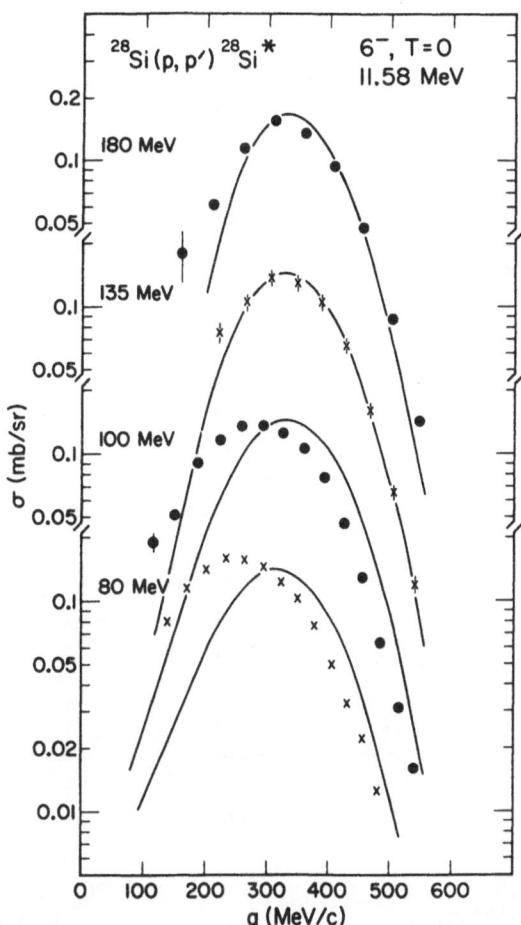

Fig. 7. Momentum transfer dependence of the cross sections for proton inelastic excitation of the 6⁻, T=0 state at 11.58 MeV in ²⁸Si. (See caption for Fig. 5.)

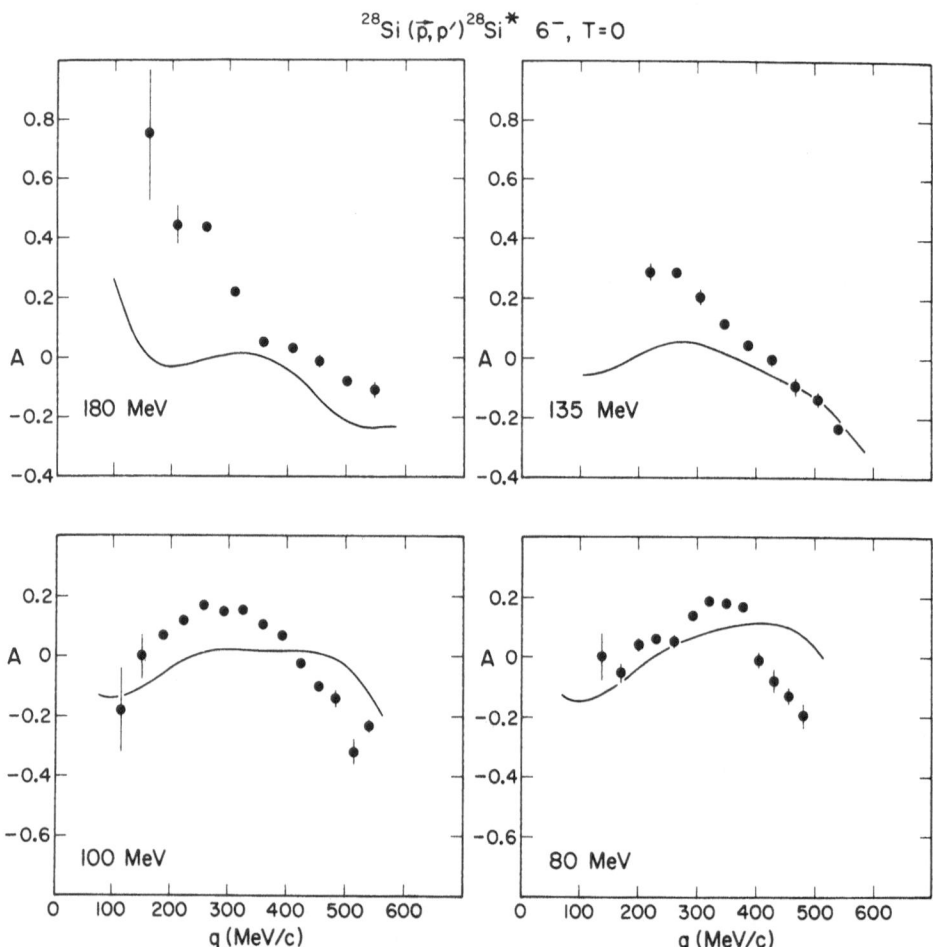

Fig. 8. Momentum transfer dependence of the analyzing powers for proton inelastic excitation of the 6^-, T=0 state at 11.58 MeV in ^{28}Si. (See caption for Fig. 5.)

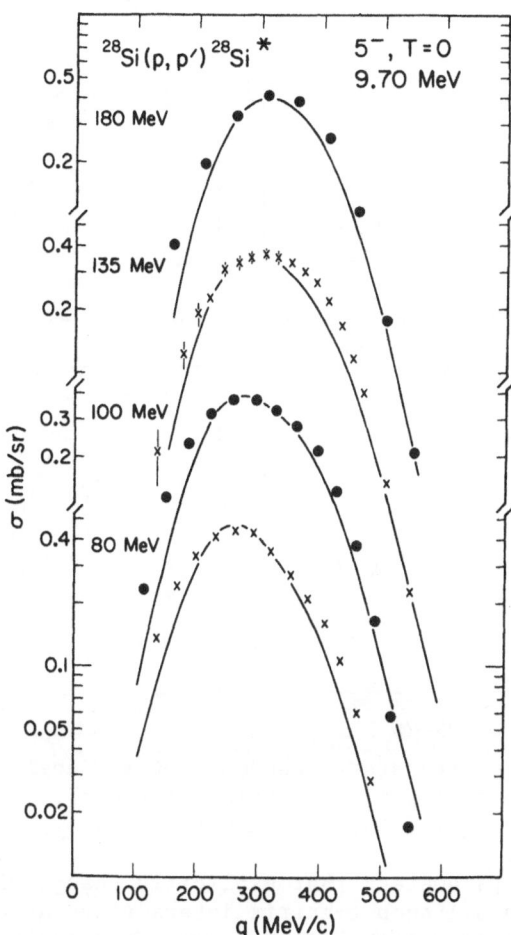

Fig. 9. Momentum transfer dependence of the cross sections for proton inelastic excitation of the 5⁻, T=0 state at 9.70 MeV in ^{28}Si. (See caption for Fig. 5.)

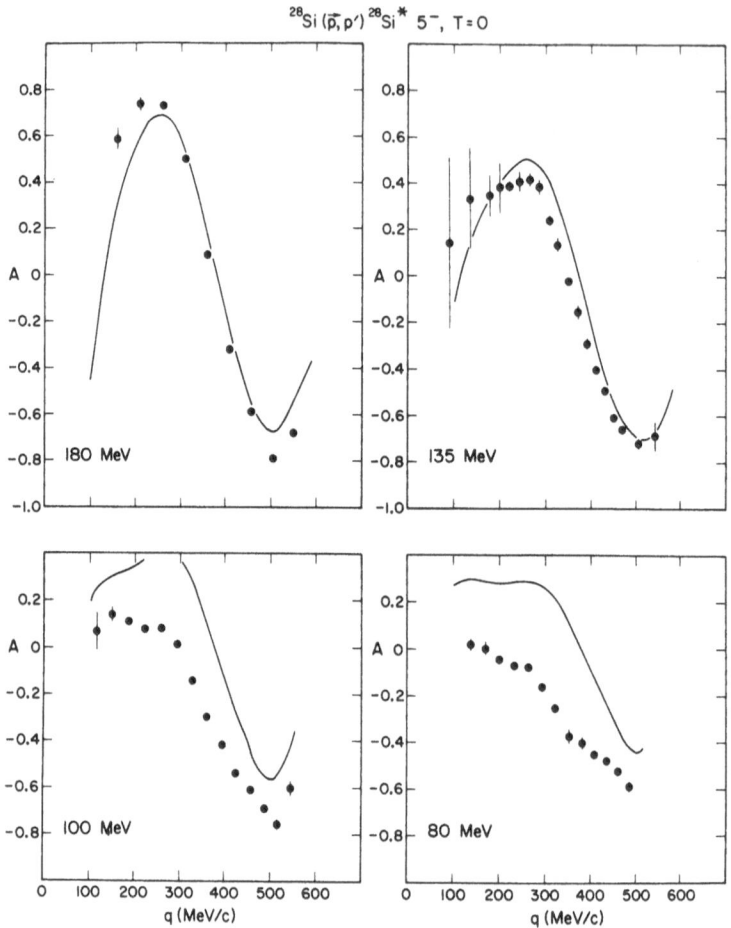

Fig. 10. Momentum transfer dependence of the analyzing powers for
          proton inelastic excitation of the 5⁻, T=0 state at 9.70
          MeV in ²⁸Si. (See caption for Fig. 5.)

interaction has parameters that explicitly change with energy,
there is also in all such t-matrix interactions an energy
dependence built in via the exchange term and the way it interferes
with the direct term. Similarly, the optical model parameters
derived from elastic scattering vary with bombarding energy, but
there would also be an energy dependence in the distortion effects
even if the parameters remained constant.

     As a way of comparing the energy dependence predicted by DWIA
calculations with that actually displayed by the data, we have
chosen a fixed momentum transfer of 300 MeV/c, near the peak of the
cross sections. Both experiment and theory are then easily
interpolated to give cross sections at that value of q. We show in

Fig. 11 the experimental cross sections for each of the states as a function of bombarding energy. The 6⁻, T=1 state exhibits a monotonic increase in cross section with energy, while the two T=0 states show a different, but common energy dependence, first level or dropping slightly, and then rising with increasing energy. It is clear from the ratio plotted in the bottom of the figure that the enhancement of the "stretched" isovector state relative to the natural-parity isoscalar state reaches a broad plateau in the 135- to 200-MeV region.

Comparisons with the DWIA calculations are shown in Fig. 12. Use of the Love-Franey force results in a description of the energy

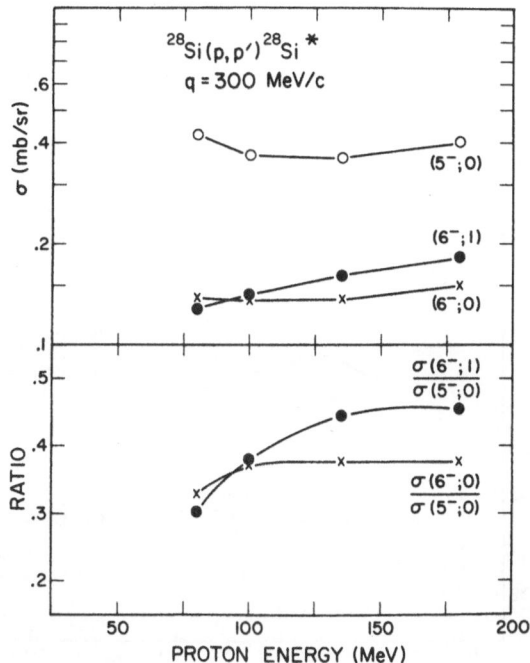

Fig. 11. Energy dependence of the experimental cross sections at a momentum transfer of 300 MeV/c for the inelastic excitation of high-spin states in ²⁸Si. The ratios of the 6⁻ cross sections to that for the 5⁻ transition are indicated at the bottom of the figure. The curves are drawn to guide the eye.

dependence of the 6⁻, T=1 state that is accurate to a few percent. For that transition the older Love force (not explicitly energy dependent) predicts a somewhat steeper increase of cross section with energy. Both interactions do reasonably well in tracking the energy dependence of the other two transitions.

We have also done plane-wave calculations to isolate the effects of distortions. That the distortion effects enter in a relatively subtle way is shown by the fact that the ratio of distorted-wave to plane-wave predictions is not only energy dependent but also transition dependent. For the 6⁻ transitions, both isovector and isoscalar, these ratios are approximately 3/4, while for the natural-parity 5⁻ transition the value is always near 1/2.

The DWIA calculations, especially with the Love-Franey force, provide an accurate description of the energy dependence of these cross sections, at least at the 10% level. The shapes of the cross section and analyzing power distributions are also reasonably well

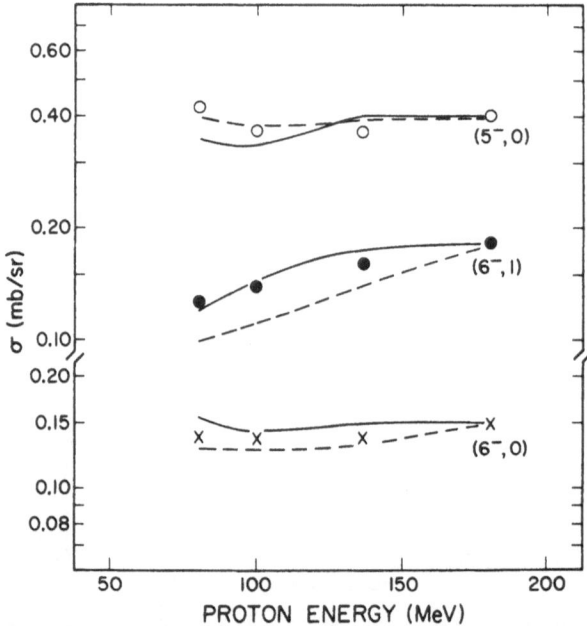

Fig. 12.   Energy dependence of the cross sections at a momentum
           transfer of 300 MeV/c. The solid (dashed) lines
           represent DWIA calculations using the Love-Franey
           (earlier Love) force. These calculations have been
           normalized to reproduce the data at 180 MeV.

described by the calculations, especially at 135 and 180 MeV. It
is instructive to compare the normalization factors for the
6^- states with those obtained with electron and pion inelastic
scattering, since they are sensitive to the same nuclear structure
information. While it would be premature to present final values,
for the 6^-,T=1 state our tentative value is 0.31, in good
agreement with earlier conclusions[4,15,21] and with results from
electron scattering[10] and pion scattering.[22,6] One must be more
cautious with the 6^-, T=0 state, since the data show a substantial
energy-dependent peak shift in momentum transfer with respect to
the calculations. We currently get about 0.15 for that transition,
somewhat larger than earlier,[15,21] but in agreement with the (π,π')
result[22,6] of 0.12 and still well below the value for its isovector
partner. On the other hand, several investigations[23-25] of the
particle-hole nature of these final states by the reaction
$^{27}Al(^3He,d)^{28}Si(6^-$, T=0 and 1) show approximately equal values of
the spectroscopic factor for transferring an $f_{7/2}$ proton starting
with a $5/2^+$ ground state.

We conclude that the DWIA description of the (p,p') excitation
of "stretched" states shows promise of becoming a quantitative
tool. The differential cross sections for ^{28}Si are given reasonably well over the range 80 to 180 MeV, though there is a shift in
the momentum transfer at which the 6^-,T=0 state peaks that is not
given by the calculations. Analyzing powers are distinctly different for the T=1 and T=0 states, and the general features are given
by DWIA. It is reassuring to note that the normalization constants
required to match the calculated results to the data are remarkably
independent of energy over this range.

FURTHER CONSIDERATIONS

There are still questions to be explored, however. The effects
of the density-dependence of the interaction, known to be important
for isoscalar natural-parity transitions of modest multipolarity,[26]
are being explored by Kelly, Geramb, and others. Fully
relativistic optical models must be explored further, and are being
explored by Clark, Schwandt, and their collaborators, and by
Shepard, Siciliano, and Rost.

An effect that has intrigued us for some time is connected with
the energy-dependence of the effective interaction, and the fact
that particular transitions emphasize particular regions of the
momentum-space wave function of the struck nucleon, and thus
correspond to different center of mass energies for the two
colliding nucleons. To participate in a "stretched" transition the
struck nucleon must have an initial momentum component opposite to
the momentum transfer to the nucleus -- the effect is then not so
much "Fermi averaging" as "Fermi-shifting." The shift in effective

beam energy for a "stretched" transition in silicon is shown in
Fig. 13 as a function of momentum transfer. The shift is about 40
MeV at 300 MeV/c and increases at larger q. While many of the
terms in the Love–Franey interaction vary slowly with energy, some
vary with moderate rapidity, at least at some momentum transfers.
Walker and Taddeucci have been exploring this same line of thought
in a slightly different way, and we plan to be working with them in
pursuing it further.

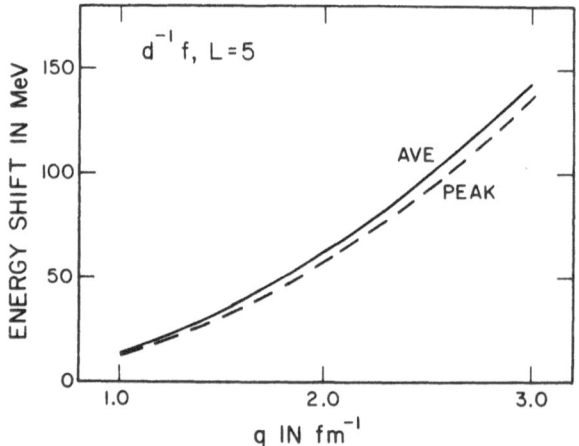

Fig. 13. Effective laboratory energy shift in MeV for a $d^{-1}f$, L=5
          transition in silicon, as a function of momentum
          transfer. AVE and PEAK denote two different ways of
          evaluating the shift. (Note that 300 MeV/c corresponds to
          a momentum transfer of ~1.5 fm^{-1}.)

ACKNOWLEDGMENTS

      We thank our experimental collaborators, C. W. Glover, W. P.
Jones, D. W. Miller, H. Nann, P. Schwandt, T. E. Drake, R. J.
Sobie, and S. Yen, and our theoretical collaborators, W. G. Love,
F. Petrovich, and G. E. Walker, for their cooperation. This work
was supported in part by the National Science Foundation.

REFERENCES

1. J. M. Moss, these proceedings.
2. C. Gaarde, J. Rapaport, T. N. Taddeucci, C. D. Goodman, C. C. Foster, D. E. Bainum, C. A. Goulding, M. B. Greenfield, D. J. Horen, and E. Sugarbaker, Nucl. Phys. A369:258 (1981); T. N. Taddeucci, J. Rapaport, D. E. Bainum, C. D. Goodman, C. C. Foster, C. Gaarde, J. Larsen, C. A. Goulding, D. J. Horen, T. Masterson, and E. Sugarbaker, Phys. Rev. C 25:1094 (1982).
3. N. Hintz, private communication.
4. R. A. Lindgren, in: "Giant Multipole Resonances," F. E. Bertrand, ed., Harwood, New York (1980), p. 399.
5. W. G. Love, in: "The (p,n) Reaction and the Nucleon-Nucleon Force," C. D. Goodman et al., eds., Plenum, New York (1980), p. 23; F. Petrovich, ibid., p. 115.
6. F. Petrovich and W. G. Love, Nucl. Phys. A354:499c (1981).
7. W. G. Love, M. A. Franey, and F. Petrovich, these proceedings.
8. A. K. Kerman, H. McManus, and R. M. Thaler, Ann. Phys. (NY) 8:551 (1959).
9. K. Hosono, M. Kondo, N. Matsuoka, T. Saito, K. Haranaka, T. Noro, H. Shimizu, S. Kato, K. Okada, K. Ogino, and Y. Kadota, RCNP Annual Report, Osaka, p. 26 (1979); H. Ejiri, private communication.
10. S. Yen, R. Sobie, H. Zarek, B. O. Pich, T. E. Drake, C. F. Williamson, S. Kowalski, and C.P. Sargent, Phys. Lett. 93B:250 (1980).
11. S. Yen, private communication.
12. W. G. Love and M. A. Franey, Phys. Rev. C 24:1073 (1981).
13. W. G. Love, first item cited in ref. 5.
14. J. Comfort, private communication.
15. S. Yen, R. J. Sobie, T. E. Drake, A. D. Bacher, G. T. Emery, W. P. Jones, D. W. Miller, C. Olmer, P. Schwandt, W. G. Love, and F. Petrovich, Phys. Lett. 105B:421 (1981).
16. A. Nadasen, P. Schwandt, P. P. Singh, W. W. Jacobs, A. D. Bacher, P. T. Debevec, M. D. Kaitchuck, and J. T. Meek, Phys. Rev. C 23:1023 (1981).
17. P. Schwandt, H. O. Meyer, W. W. Jacobs, A. D. Bacher, S. E. Vigdor, M. D. Kaitchuck, and T. R. Donoghue, to be published in Phys. Rev. C26:55 (1982).
18. C. Olmer, A. D. Bacher, G. T. Emery, W. P. Jones, D. W. Miller, P. Schwandt, T. E. Drake, R. J. Sobie, and S. Yen, to be published.
19. C. W. Glover, private communication.
20. H. O. Meyer, P. Schwandt, G. L. Moake, and P. P. Singh, Phys. Rev. C 23:616 (1981).
21. F. Petrovich, W. G. Love, A. Picklesimer, G. E. Walker, and E. R. Siciliano, Phys. Lett. 95B:166 (1980).

22. C. Olmer, B. Zeidman, D. F. Geesaman, T.-S. H. Lee, R. E. Segel, L. W. Swenson, R. L. Boudrie, G. S. Blanpied, H. A. Thiessen, C. L. Morris, and R. E. Anderson, Phys. Rev. Lett. 43:612 (1979).
23. S. Kato and K. Okada, J. Phys. Soc. Japan 50:1440 (1981).
24. H. Nann, Nucl. Phys. A376:61 (1982).
25. Y. Fujita, M. Fujiwara, M. N. Harakeh, K. Hosono, M. Noumachi, M. Sasao, and K. A. Snover, Ann. Rpt., Nucl. Phys. Lab., U. Washington (June 1981), p. 69.
26. J. Kelly, W. Bertozzi, T. N. Buti, F. W. Hersman, C. Hyde, M. V. Hynes, B. Norum, F. N. Rad, A. D. Bacher, G. T. Emery, C. C. Foster, W. P. Jones, D. W. Miller, B. L. Berman, W. G. Love, and F. Petrovich, Phys. Rev. Lett. 45:2012 (1980).

HIGH-SPIN STATES IN THE (p,n) REACTION

B.D. Anderson, A. Fazely, C. Lebo, J.W. Watson, and
R. Madey

Department of Physics
Kent State University
Kent, Ohio   44242

INTRODUCTION

In 1974, Moffa and Walker[1] stressed that certain "exotic
states" of relatively high spin should dominate medium-energy (p,p')
and (p,n) reactions at large momentum transfers. They predicted
that the study of the excitation of these states would be important
to determine the high-momentum transfer components of the nucleon-
nucleon (N-N) effective interaction at medium energies. Since the
discussion of Moffa and Walker, experimental studies of high-spin
states have been performed with both the (p,p') and (p,n) reactions.
These studies are providing not only information regarding the high-
momentum transfer components of the N-N effective interaction, but
are also beginning to provide nuclear structure information.[2]

The (p,n) measurements were performed at the Indiana University
Cyclotron Facility (IUCF) with the beam-swinger system. Three
separate neutron detector stations were used, providing measurements
in the angular region from 0° to about 70°. Large-volume, mean-
timed neutron detectors were used at flight paths up to 90 m. The
neutron detectors were fast plastic scintillators four inches thick
with frontal areas up to 2.25 m^2. The excellent timing characteristics of both the proton beam ($\sim$ 0.3 ns) and the mean-timed neutron
counters ($\lesssim$ 0.4 ns) resulted in overall timing resolutions of about
0.75 ns and energy resolutions varying from about 0.35 to 0.50 MeV,
depending on the flight path.

In Fig. 1, we show our measured neutron energy spectra at
three angles for the $^{16}O(p,n)^{16}F$ reaction at 135 MeV. Shown also

are the predicted spectra of Picklesimer and Walker,[3] whose calculations are an extension of the earlier predictions by Moffa and Walker. The calculations use shell-model wavefunctions in a microscopic Plane Wave Impulse Approximation (PWIA) calculation with an N-N effective interaction based on free N-N scattering parameters. The shell-model calculation assumes the ^{16}O target nucleus is a closed core and considers only one-particle, one-hole states of 1 and 2 $\hbar\omega$ excitation for the final states. The overall agreement between the predictions and the measurements is good. Note that the spectra reveal only a few strongly-excited states at each angle, and that the widest angle is dominated by a single transition to the 4^- high-spin state. This 4^- state is referred to as a 1 $\hbar\omega$ "stretched-state" because it is believed to be dominated by the $(\pi d_{5/2}, \nu p_{3/2}^{-1})$ configuration where the particle and hole are in different major shells and in subshells with $j = \ell + 1/2$ coupled to the maximum possible angular momentum. The excitation of this type of state must involve a change in the relative angular momentum plus a spin transfer, viz., $\Delta\ell = 3$ plus $\Delta s = 1$ in order to proceed from the 0^+ ground state of ^{16}O to this

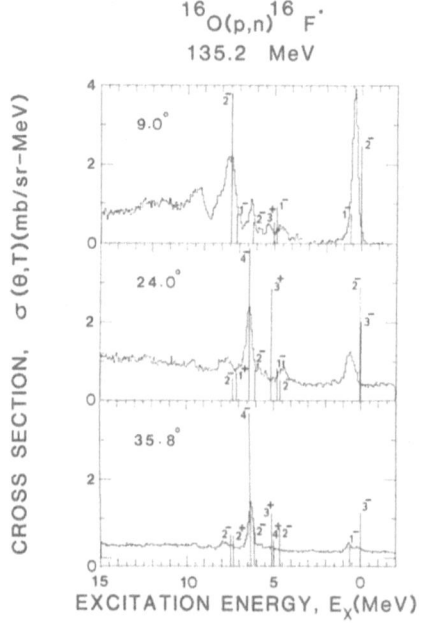

$^{16}O(p,n)^{16}F$
135.2 MeV

Fig. 1.   Comparison of excitation-energy spectra for the $^{16}O(p,n)^{16}F$ reaction with PWIA predictions by Picklesimer and Walker.[3]

$T = 1$, 4^- state in ^{16}F. Because this type of transition must involve spin transfer, it is logically part of the subject of this conference.

HIGH-SPIN STATES OF 1 $\hbar\omega$ EXCITATION

Let us consider in some detail the (p,n) excitation of the 4^- state in ^{16}F and also of the 6^-, $T = 1$ state in the $^{28}Si(p,n)^{28}P$

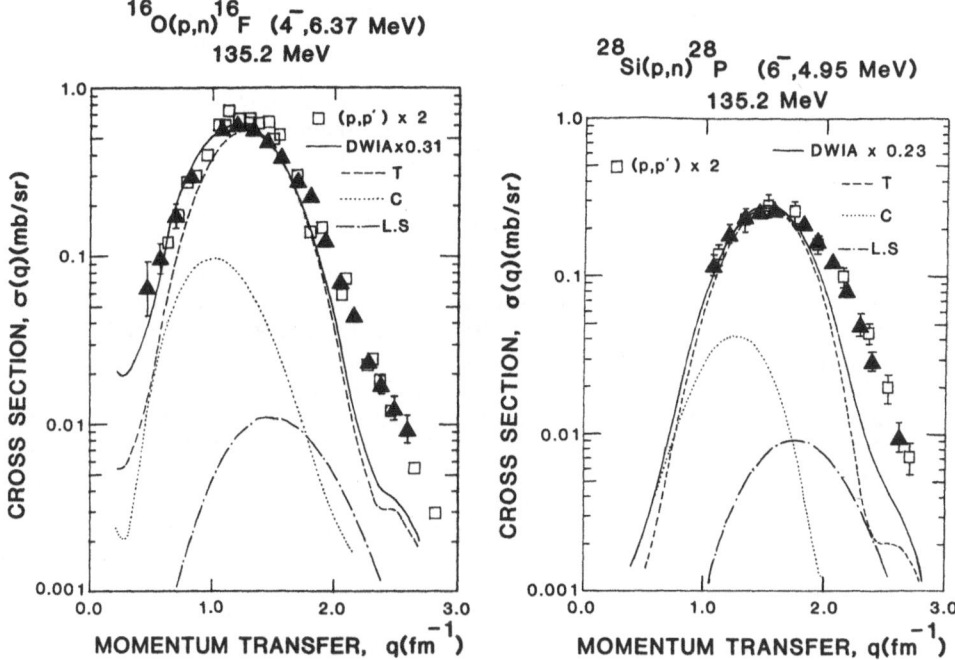

Fig. 2.   Differential cross sections for 1 ℏω stretched-state
         transitions from ^{16}O and ^{28}Si compared with DWIA
         calculations.

reaction. The latter state is another 1 ℏω stretched-state
believed to have the major configuration $(\pi f_{7/2}, \nu d\bar{5/2})$. Both of
these states were indicated by Moffa and Walker to be important
examples of these high-spin excitations.

    In Fig. 2, we show the measured angular distributions of the
cross sections for the ^{16}O(p,n)^{16}F (4$^-$, 6.37 MeV) and ^{28}Si(p,n)^{28}P
(6$^-$, 4.95 MeV) reactions. Shown also are the (p,p') cross sections
to the parent states in ^{16}O and ^{28}Si measured by Henderson et al[4]
and Yen et al,[5] respectively. The (p,p') cross sections are
multiplied by a factor of two for the ratio of isospin Clebsch-
Gordan coefficients for the (p,n) and (p,p') reactions. The agreement between the (p,n) and (p,p') measurements is seen to be quite
good. The solid lines represent DWIA calculations with the code
DWBA70.[6] Optical-model parameters are taken from Comfort and Karp[7]
for ^{12}C extrapolated to ^{16}O and from Schwandt et al[8] for ^{28}Si. The
N-N effective interaction is that of Love and Franey[9] at 140 MeV.
The calculation assumes that ^{16}O and ^{28}Si target nuclei are closed
cores and that the 4$^-$ and 6$^-$ final states have the one-particle,
one-hole stretched configurations indicated above.

The angular distributions are fit rather well, except at the widest angles. The contributions to the cross section from the individual terms in the effective interaction show that the tensor term clearly dominates. As indicated, the DWIA calculations require normalization factors of 0.31 and 0.23 in order to fit the measured cross sections on ^{16}O and ^{28}Si, respectively. The excitation of the analogs of these states in ^{16}O and ^{28}Si were studied with inelastic electron scattering,[10,11] where normalization factors of 0.44 and 0.33 are required, respectively, to make impulse approximation calculations agree with the experimental results. Since the interaction is well-known for inelastic electron scattering and since the impulse approximation should be reliable for high-energy electrons, the (e,e') results provide a measure of the nuclear structure overlap for these transitions. We note that the (p,n) and (p,p') normalization factors are only somewhat smaller than the (e,e') normalizations and that a decrease of only about 10% in the strength of the tensor term in the N-N effective interaction will bring the (p,n) and (p,p') results into agreement with the electron-scattering results. A 10% decrease in the tensor term is within the uncertainty for the strength of this term obtained from analyses of free nucleon-nucleon scattering data.

It is important to realize that the normalizations required for fitting these stretched-state transitions are only about 0.3 to 0.4 largely because of ground-state correlations in the target nuclei. Walker[12] estimated that consideration of correlations in ^{16}O would decrease theoretical calculations for inelastic-scattering transitions by approximately a factor of two. Recently, Halderson et al[13] pointed out that the extracted spectroscopic factor for the ^{27}Al (^{3}He,d)^{28}Si (g.s.) reaction indicates that ^{28}Si acts like a simple core, filled up through the $d_{5/2}$ single-particle levels only about 44% of the time. Thus, a normalization factor of 0.31 for the excitation of the T = 1, 6$^-$ state indicates that the final state looks like the assumed ($f_{7/2}$, $d_{5/2}^{-1}$) particle-hole state approximately 0.31/0.44 x 100 = 70% of the time. Halderson et al further showed that this result is consistent with the single-particle fraction obtained for this state (in ^{28}Si) from the measured width in low-energy ^{27}Al(p,p) scattering. Thus, these stretched-states are actually good one-particle, one-hole states.

HIGH-SPIN STATES OF 0 $\hbar\omega$ EXCITATION

In addition to these 1 $\hbar\omega$ stretched states, we have identified[14] another type of stretched state strongly excited in the (p,n) reaction. We refer to this kind of stretched state as a "0 $\hbar\omega$" stretched state because it does not involve the excitation of a particle from one major shell up to the next one. These 0 $\hbar\omega$ states can be excited by the (p,n) reaction on targets with N > Z and have a dominant configuration with the same quantum numbers n, ℓ, and j for both the proton particle and the neutron hole. These states have odd-integer spin, even parity, and must involve spin transfer to be excited. In Fig. 3, we show some of our measured neutron energy spectra for the ^{48}Ca(p,n)^{48}Sc reaction at 160 MeV. The wide-angle spectra are dominated by the excitation of a known 7^+ state at E_x = 1.10 MeV in ^{48}Sc. This state is the highest spin or "stretched" state of a 0^+ to 7^+ band formed from the ($\pi f_{7/2}$, $\nu f_{7/2}^{-1}$) configuration. While some of the lower spin members of this band can possibly mix with other available configurations, the 7^+ state can be formed by only this configuration (up to 2 $\hbar\omega$ of excitation). Note that this kind of stretched state can be excited only by a charge-exchange reaction on an N > Z target nucleus and thus cannot be excited in either (p,p') or (e,e') reactions.

In Fig. 3, we show also the extracted angular distribution for the 0 $\hbar\omega$ transition in the ^{48}Ca(p,n)^{48}Sc reaction at 135 MeV. The solid line represents a DWIA calculation similar to those discussed earlier for the 1 $\hbar\omega$ transitions. The optical-model parameters are from the global set of Schwandt et al,[15] the ^{48}Ca target nucleus is assumed to be closed core and the 7^+ state is assumed to be pure ($\pi f_{7/2}$, $\nu f_{7/2}$). The cross section angular distribution is fit well with a normalization factor significantly larger than those required for the two 1 $\hbar\omega$ transitions discussed earlier. This large normalization factor (viz., 0.61) is nearly as large as that determined for the transition in ^{28}Si after renormalizing for the ground-state correlations. We interpret this relatively large normalization factor to indicate that the ^{48}Ca target nucleus is a relatively good closed-shell nucleus in its ground state.

ANALYZING-POWER MEASUREMENTS

In addition to cross sections, analyzing powers were measured for the three reactions discussed above. The analyzing powers provide sensitive tests of the reaction mechanisms and nuclear structure assumed in the theoretical descriptions. In Fig. 4, we show the measured analyzing powers for the ^{16}O($\vec{p}$,n)^{16}F (4^-, 6.37 MeV), ^{28}Si($\vec{p}$,n)^{28}P (6^-, 4.95 MeV), and ^{48}Ca(p,n)^{48}Sc (7^+, 1.10 MeV) reactions at 135 MeV. The measurements are compared with the predicted analyzing powers from the DWIA calculations described earlier. The general shape of the analyzing-power measurements is approximately the same for all three reactions. The largest discrepancies between the measurements and the calculations are at large momentum transfers, where the calculations also do not fit the cross sections well. The predicted shape of these analyzing-power angular distributions is qualitatively correct even though, as reported by Yen et al[5] for ($\vec{p}$,p) measurements to stretched states in ^{28}Si and by Madey et al[16] for the ^{16}O($\vec{p}$,n)^{16}F (4^-, 6.37 MeV) reaction, the calculations are sensitive to distortion and to the

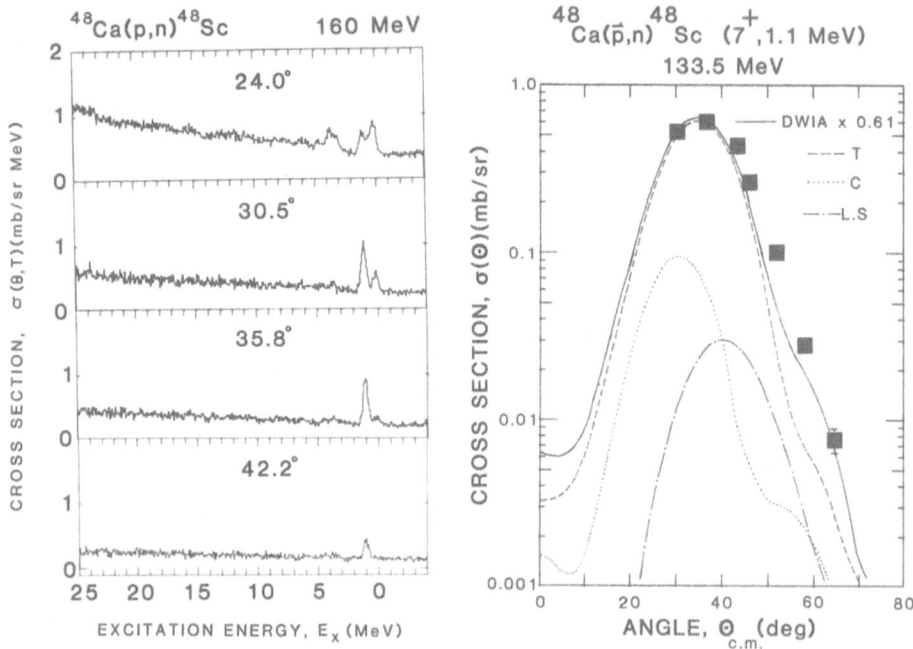

Fig. 3.   Excitation-energy spectra at wide angles for the
^{48}Ca(p,n)^{48}Sc reaction; and the cross section angular
distribution for the 0 ℏω stretched-state transition.

interference between the tensor term and both the central and spin-
orbit terms of the N-N effective interaction. The cross section
contains information on the interference between only the tensor
and central terms in the interaction. In fact, Madey et al noted
that the interference between the tensor and spin-orbit terms is
primarily responsible for the analyzing power of the ^{16}O(p,n)^{16}F
(4$^-$, 6.37 MeV) reaction; whereas the spin-orbit term is unimportant
for the cross section.

With regard to the sensitivity to the choice of optical-model
wavefunctions, we found that although distorted-wave calculations
seriously change the calculated analyzing powers from those obtained
in a plane-wave calculation, different "reasonable" optical-model
parameter sets all yield very similar analyzing-power angular
distributions. Specifically, we found for the ^{16}O($\vec{p}$,n)^{16}F (4$^-$, 6.37
MeV) reaction that two different parameter sets obtained by Kelly et
al[17] from analysis of ^{16}O proton elastic-scattering data, the
potential of Comfort and Karp[7] extrapolated from their analysis of
^{12}C proton elastic-scattering data, and two different parameter sets

from Schwandt et al[15] (also
extrapolated from ^{12}C data), all
yield similar results for the
calculated $(\vec{p},n)$ analyzing
powers.

The largest discrepancies
between the theoretical and
measured analyzing powers are
at large momentum transfers.
The cross sections are significantly underestimated in this
region also. These discrepancies may indicate incorrect
strengths in the large momentum-
transfer components of the
effective interaction of Love
and Franey. We note, for
example, that it is possible to
obtain a better fit to the
analyzing powers by significantly reducing the imaginary
tensor term, but such a step
would be a departure from the
interaction strengths deduced
from the nucleon-nucleon phase
shifts. Clearly, the possible
modification of these strengths
is a subject for further
investigation.

In summary, the comparison
of measured $(\vec{p},n)$ analyzing
powers with DWIA calculations
show some difficulties, but are
in overall qualitative agreement,
which indicates that the DWIA
description of these transitions
is largely correct.

(p,n) EXCITATION OF HIGH-SPIN
STATES AS A SPECTROSCOPIC TOOL

Even though there are
certainly several areas in the
theoretical description of high-
spin state transitions which
need more study, we think that

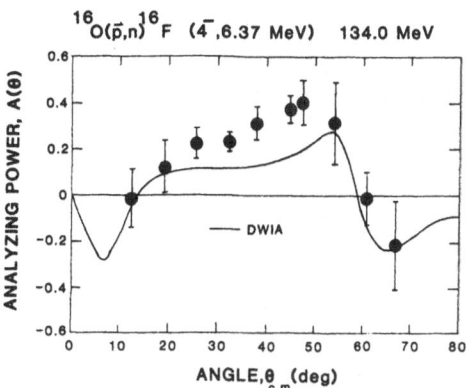

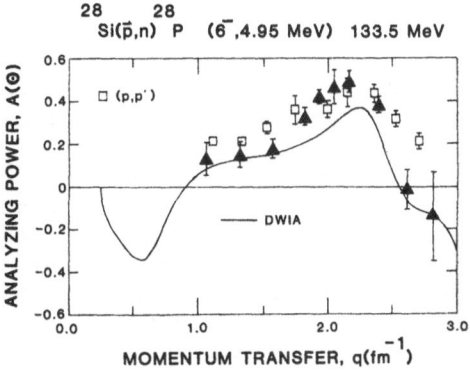

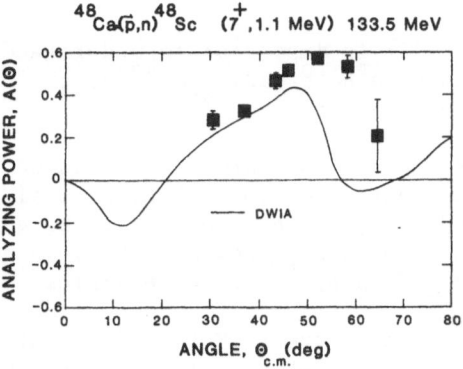

Fig. 4. Analyzing-power angular
distributions for $(\vec{p},n)$
stretched-state transi-
tions compared with DWIA
calculations.

the (p,n) excitation of stretched states can be used as a spectroscopic tool to study nuclear structure. We wish to consider a few
examples, starting with ^{40}Ca. In 1977, Adams et al[18] reported the
observation of strong excitations of the 6$^-$ stretched states in ^{24}Mg
and ^{28}Si(p,p') studies at 135 MeV. These excitations were seen to
be very narrow and clearly dominated the inelastic proton spectra
at large momentum transfers. These 6$^-$ states are believed to be
predominantly $(f_{7/2}, d^{-1}_{5/2})$ particle-hole states. At the same time,
they studied ^{40}Ca(p,p') and found no clear evidence for the excitation of this state, even though it is certainly still allowed in the
simple shell model. The (p,n) reaction is an ideal way to look for
this T = 1 state because, on a self-conjugate target like ^{40}Ca, the
(p,n) reaction will excite only T = 1 states and will eliminate
background from T = 0 states. An early attempt by us to look for
this state in ^{40}Sc was hampered by oxygen contamination in the
target which produced the strong 4$^-$ stretched state in ^{16}F exactly
in the excitation-energy region of interest; however, a more recent
experimental run with an oxygen-free ^{40}Ca target clearly reveals a
state in ^{40}Sc which dominates the wide-angle spectra. This state
is shown in Fig. 5, where it is seen to be a broad state, approximately 3 MeV wide. The excitation energy of this state is within
1 MeV of the location of the correct analog excitation energy of the
6$^-$ stretched state predicted in ^{40}Ca by Donnelly and Walker.[19] The
angular distribution for this state is shown in Fig. 5 and compared
to a DWIA calculation for a 6$^-$ state. Although the state is
unresolved from another state at angles less than about 30°, the
observed peak and wide-angle part of the angular distribution are
fit well by the calculation. The reason why this state was not
observed in the (p,p') reaction is that, because it is so broad, it
could not be easily seen against the background of T = 0 states in
^{40}Ca.

It is important to understand why this particular stretched
state is so broad. In ^{40}Ca, the $d_{5/2}$ state is no longer the valence
level, as it is in ^{24}Mg and ^{28}Si. As the $d_{5/2}$ level is pushed down
below the Fermi level by filling the $2s_{1/2}$ and $1d_{3/2}$ levels, it
broadens. This broadening can be seen directly in the ^{40}Ca(p,pn)
knockout spectrum of Watson et al,[20] shown in Fig. 5. The $1d_{3/2}$
and $2s_{1/2}$ states are seen to be quite sharp, whereas the $d_{5/2}$ state
is seen to be a few MeV wide, in good agreement with the state seen
in ^{40}Ca(p,n)^{40}Sc at wide angles.

In a recent (March 1982) run at the IUCF, we searched for 0 $\hbar\omega$
and 1 $\hbar\omega$ stretched states on several heavier nuclei. While the
quantitative results from this run are not yet available, they can
be summarized qualitatively by noting that we observed strong 0 $\hbar\omega$
excitations in most of the reactions, and only rarely did we see
any 1 $\hbar\omega$ strength; for example, while we see clear 9$^+$ and 13$^+$ 0 $\hbar\omega$
states in the ^{88}Sr and ^{208}Pb (p,n) reactions, respectively, we may

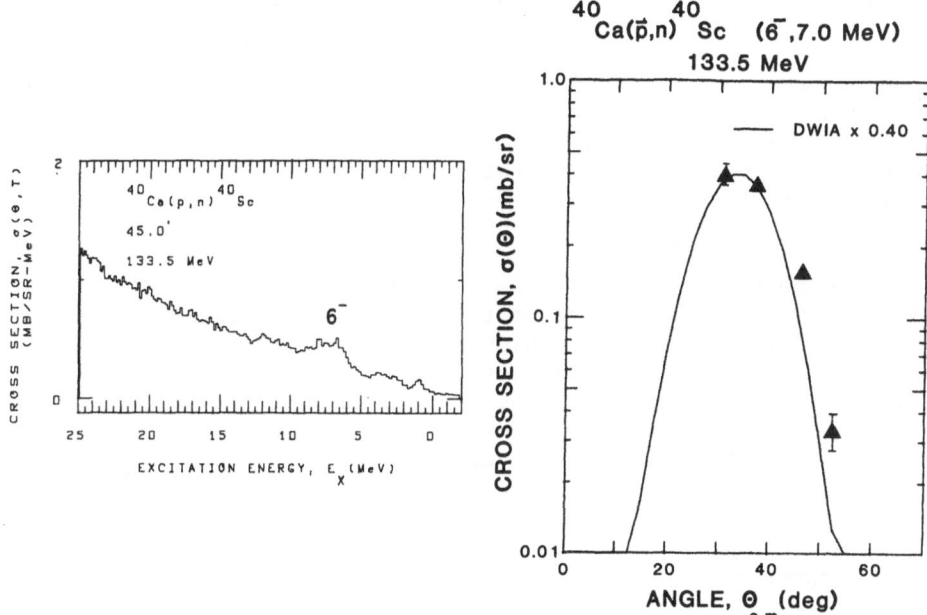

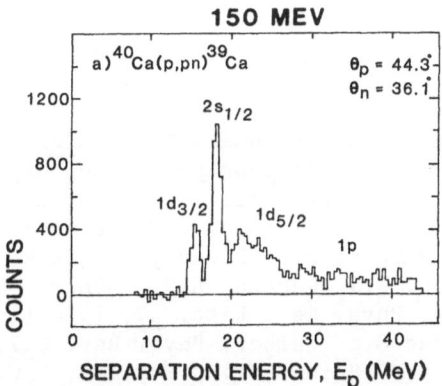

Fig. 5.   Excitation-energy spectrum at 45° for the ^{40}Ca(p,n)^{40}Sc
reaction; angular distribution for the transition to the
6$^-$ stretched state; and the ^{40}Ca(p,pn) neutron knockout
spectrum from Watson et al.[20]

see the 8^-, 1 $\hbar\omega$ strength in the ^{58}Ni(p,n)^{58}Cu reaction fragmented into several states, as expected from earlier (e,e') studies[21] of M8 strength in ^{58}Ni. These general results can be understood qualitatively in the following way: The 0 $\hbar\omega$ states usually involve particle-hole states in the valence shell of a nucleus. These states, at the Fermi level, are generally narrow states because they have low excitation energies. The 1 $\hbar\omega$ states in heavier nuclei increasingly involve placing the proton in a highly-excited particle state and/or removal of a neutron from a deep-hole state. These states far from the Fermi level have high excitation energies, and they are usually broad and fragmented because they can mix with other configurations with the same J^π.

In summary, studies of 0 $\hbar\omega$ stretched states provide significant information on the structure of target nuclei because they involve low-lying 1p-1h excitations which concentrate the strength into a single final state. Thus the normalizations required for theoretical predictions of these transitions provide information predominantly about the target ground-state wavefunctions. Detailed analysis of our measurements should provide important information for structure models of these nuclei.

ACKNOWLEDGEMENT

This work was supported in part by the National Science Foundation under grant PHY79-07790.

REFERENCES

1. P.J. Moffa and G.E. Walker, Nucl. Phys. A222, 140 (1974). Work
        motivated by early electron scattering experiments, e.g.
        Ref. 12 and 19, first called attention to these states.
2. F. Petrovich, The (p,n) Reaction and The Nucleon-Nucleon Force,
        ed. C.D. Goodman et al. (Plenum, New York, 1980) p. 115;
        R.A. Lindgren, W.J. Gerace, A.D. Bacher, W.G. Love, and
        F. Petrovich, Phys. Rev. Lett. 42, 1524 (1979).
3. A. Picklesimer and G.E. Walker, Phys. Rev. C17, 237 (1978).
4. R.S. Henderson, B.M. Spicer, I.D. Svalber, V.C. Officer, G.G.
        Shute, D.W. Devins, D.L. Friesel, W.P. Jones, and A.C.
        Attard, Aust. J. Phys. 32, 411 (1979).
5. S. Yen, R.J. Sobie, T.E. Drake, A.D. Bacher, G.T. Emery, E.P.
        Jones, D.W. Miller, C. Olmer, P. Schwandt, W.G. Love, and
        F. Petrovich, Phys. Lett. 105B, 421 (1981).
6. R. Schaeffer and J. Raynal (unpublished). The modified version
        supplied by W.G. Love.
7. J.R. Comfort and B.C. Karp, Phys. Rev. C21, 2162 (1980).
8. P. Schwandt, A. Nadasen, P.P. Singh, M.D. Kaitchuck, W.W.

Jacobs, J. Meek, A.D. Bacher, and P.T. Debevec, Indiana University Cyclotron Facility Prog. Rep. Jan. 31, 1978, p. 79.

9. W.G. Love and M.A. Franey, Phys. Rev. C24, 1073 (1981).

10. W. Bertozzi, Proc. LAMPF Users Group (1980), pp. 55-78; F. Petrovich and W.G. Love, Nucl. Phys. A354, 499 (1981).

11. S. Yen, R. Sobie, H. Zarek, B.O. Pich, T.E. Drake, C.F. Williamson, S. Kowalski, and C.P. Sargent, Phys. Lett. 93B, 250 (1980).

12. I. Sick, E.B. Hughes, T.W. Donnelly, J.D. Walecka, and G.E. Walker, Phys. Rev. Lett. 23, 1117 (1969).

13. D. Halderson, K.W. Kemper, J.D. Fox, R.O. Nelson, E.G. Bilpuch, C.R. Westerfeldt, and G.E. Mitchell, Phys. Rev. C24, 786 (1981).

14. J.W. Watson, M. Ahmad, B.D. Anderson, A.R. Baldwin, A. Fazely, P.C. Tandy, R. Madey, and C.C. Foster, Phys. Rev. C23, 2373 (1981).

15. P. Schwandt, private communication.

16. R. Madey, A. Fazely, B.D. Anderson, A.R. Baldwin, A.M. Kalenda, R.J. McCarthy, P.C. Tandy, J.W. Watson, W. Bertozzi, T. Buti, M. Finn, M. Kovash, B. Pugh, and C.C. Foster, Phys. Rev. C25, 1715 (1982).

17. J. Kelly, Ph.D. Dissertation, Mass. Inst. of Tech., p. 141 (1981).

18. G.S. Adams, A.D. Bacher, G.T. Emery, W.P. Jones, R.T. Kouzes, D.W. Miller, A. Picklesimer, and G.E. Walker, Phys. Rev. Lett. 38, 1387 (1977), and IUCF Prog. Rep. Jan. 31, 1977, p. 55.

19. T.W. Donnelly and G.E. Walker, An. Phys. (NY) 60, 209 (1970).

20. J.W. Watson, M. Ahmad, D.W. Devins, B.S. Flanders, D.L. Friesel, N.S. Chant, P.G. Roos, and J. Wastell, Phys. Rev. C26, 961 (1982).

21. R.A. Lindgren, C.W. Williamson, and S. Kowalski, Phys. Rev. Lett. 40, 594 (1978).

BACKGROUND CALCULATIONS FOR (p,n)-REACTIONS AT INTERMEDIATE

ENERGIES

Franz Osterfeld

Institut für Kernphysik
Kernforschungsanlage Jülich
D-5170 Jülich, West Germany

ABSTRACT

The background of ^{40}Ca(p,n)- and ^{48}Ca(p,n)-spectra have been calculated in a microscopic model for an incident proton energy of 160 MeV. The continuous parts of the spectra are reproduced within a factor of 1.3. It is shown that most of the background subtracted in the experimental analysis of the ^{48}Ca(p,n)-Gamow-Teller resonance is actually Gamow-Teller strength. The calculations predict a strong $\Delta L=2(3^+)$-resonance in ^{48}Ca(p,n) at 25 MeV excitation energy, but none in ^{40}Ca.

INTRODUCTION

The history of the "background problem" is as old as the history of giant resonances. These fundamental modes of nuclear excitation which involve the coherent motion of many nucleons in the nucleus appear energetically in the continuum region of the nuclear excitation spectrum. In inelastic hadron scattering[1] they are located on top of a large continuum (background) whose shape and magnitude is not known and whose nature depends on the probe used. Uncertainties in the decomposition of the spectra into resonance and background seriously limit the accuracy with which the amount of sum rule strength exhausted by the giant resonance states can be determined.

Several authors[2,3] have tried already to calculate the continuum below giant resonances within the DWBA. Most of these calculations were carried out for (p,p')-reactions at relatively low incident energies (E ≤ 62 MeV) with varying success. Bertsch and

Tsai[2] analyzed ^{208}Pb(p,p')-data[4] at E = 62 MeV assuming a one step
process for the reaction mechanism and using collective model form
factors for the nuclear transition potentials the strengths of
which were normalized to microscopic random phase approximation
(RPA)-strength distribution functions. They found that the one
step cross section could only account for $\sim$ 25 % of the experimental cross section. Similar results were obtained by Tamura and
Udagawa[3] who calculated, in addition to the one step contribution,
also the two step contributions to the cross section but still
underestimate the data at forward angles. A similar problem occurs
in the statistical theory of multistep direct reactions of Feshbach, Kerman and Koonin[5] who include in their analysis of 45 MeV
^{120}Sn(p,n)-data[6] multistep processes up to sixth order. Although
they vary the force strength of the effective projectile-target
nucleon interaction to obtain an overall best fit to the data,
they have difficulties reproducing the data at forward angles for
higher excitation energies.

All these calculations were carried out for relatively low
incident energies (E $\leq$ 62 MeV) where multistep processes obviously
make a large contribution to the cross section. The situation
should be much simpler, however, at high incident proton energies
(E > 100 MeV) where the reaction mechanism becomes dominantly
direct. In this paper we will show that this supposition is,
indeed, correct. In particular we investigate (p,n)-spectra[7,8] at
E = 160 MeV. One reason for choosing the (p,n)- and not the (p,p')-
reaction for such an investigation is that for (p,n) only a very
restricted class of nuclear excited states has to be considered at
E = 160 MeV. This is the class of charge exchange states with
unnatural parity. The second and actually most important reason
for choosing (p,n) is to investigate in detail the background
below the famous giant Gamow-Teller (GT) resonance which was
recently discovered in high energetic (p,n)-experiments[9-13] at the
Indiana University Cyclotron Facility. The GT-resonance is the
spin-isospin (ΔS=1, ΔT=1, ΔL=0) collective mode, which was already
predicted by Ikeda, Fujii and Fujita as early as 1963[14]. The
exciting thing about the GT-resonances is that only roughly 5o %
of the theoretically[14] expected total GT-strength is found in
these experiments. Several authors[15,16] have suggested that this
so-called quenching of the total GT-strength is due to the
admixture of Δ(1232) isobar-nucleon hole (ΔN^{-1}) excitations into
the proton particle-neutron hole (PN^{-1}) GT-state. For a quantitative understanding of this Δ isobar effect, however, it is of
utmost importance to calculate the background in a most reliable
way. Such a calculation is even more important for the ΔL=1-
resonances[12,13] and resonances of higher multipolarities[13] where
the "signal to background ratio" becomes rather bad. The ΔL=1-
resonance (and resonances with ΔL $\geq$ 2) should provide information
on the spin (J^{π}) dependence of the quenching.

THE BACKGROUND MODEL

     Our background model is chosen such that it describes the
discrete and the continuous parts of the spectrum as consistently
as possible, and that it also includes specific properties of the
target nucleus like neutron excess, collectivity, etc., in detail.
Our model assumptions are as follows:

     (1) For (p,n)-reactions at high incident energies (E ≥ 100
MeV) the reaction mechanism is direct, i.e. the whole spectrum
including peaks and continuum is a result of one step processes
only.

     (2) The effective projectile-target nucleon interaction can be
approximated by the free N-N t-matrix, i.e. by the G3Y-interaction
of Love and Petrovich[17].

     (3) The only nuclear states contributing to the (p,n)-back-
ground at E ≥ 100 MeV are spin flip ($\Delta S=1$, $\Delta T=1$) states. This
argument is based on the fact that the $\sigma\sigma\tau\tau$-part of the G3Y-inter-
action which excites spin flip states is nearly energy independent
while the $\tau\tau$-part which excites the non-spin-flip states gets
strongly reduced at E ≥ 100 MeV[17].

     (4) The final nuclear states are assumed to be of simple
proton particle-neutron hole doorway nature including bound,
quasibound and continuum states (see Fig. 1). The single particle
wave functions of the bound states are generated from a Woods-
Saxon potential which is chosen to reproduce the known experimental
single particle energies. The continuum states are generated from
the real part of the energy dependent Becchetti-Greenlees
potential[18]. The proton particle and neutron hole are coupled to
states of spin parity J^{π}. This is advantageous since to 0° cross
sections only states with low multipolarity can contribute.
Furthermore, by this procedure we obtain the contributions to the
background due to different final nucleus spins J^{π} separately.
This gives us the possibility to treat the nuclear structure of
states with different J^{π} in nuclear models of varying sophistica-
tion. For example, we may include[16] nuclear collectivity and
$\Delta(1232)$-isobar effects into the nuclear structure calculations for
the Gamow-Teller (1^+)-states while we use a simple particle-hole
model for the rest of the states. The latter form then the "background" for the GT-states. In addition, we can disentangle the
spectrum into the various multipolarities and discuss the strength
distribution for each J^{π}.

     (5) For the continuum wave functions we neglect the spin
orbit potential, so that the transitions to all final nuclear
states are completely incoherent. Then, the whole background is a

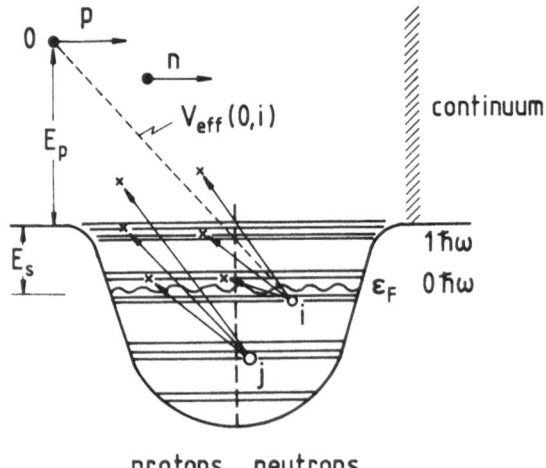

Fig. 1.   Schematic representation of the microscopic model used
for the background calculations. In the figure ε_F denotes
the Fermi energy, E_S the nucleon separation energy, and
E_p the incident projectile energy. For the effective
projectile target nucleon interaction V_{eff} the G3Y-inter-
action of Love and Petrovich[17] is used.

simple superposition of cross sections of inelastic excitations to
bound, quasibound and continuum states.

(6) The cross sections are calculated in the DWIA-approxima-
tion using the fast speed DWBA-code FROST-MARS[19] which includes
knock-out exchange amplitudes exactly.

The particle-hole doorway model discussed includes the
nuclear continuum exactly but treats nuclear collectivity explicitly only for certain selected states like the GTR or IAS. We
argue that for our purpose such a limited inclusion of nuclear
collectivity is sufficient. Our argument is based on the work of
Speth et al.[20] who have shown that for $\Delta S=1$, $\Delta T=1$-transitions
collectivity plays only a role for low multipolarities, i.e. for
0^-, 1^+, 1^- ($\Delta S=1$) and, maybe, 2^- states. This is simply an effect
of the finite range residual particle-hole (ph)-interaction in
the $\Delta S=1$, $\Delta T=1$-channel[20] which is strongly repulsive for low spin
states and weak for high spin states ($J^\pi \geq 2^-$). Therefore states
with large J^π are nearly unaffected by the residual ph-interaction
(see also ref. 21 for a detailed discussion).

RESULTS AND DISCUSSION

In the microscopic particle-hole model we have calculated the background at various scattering angles for the reactions ^{48}Ca(p,n) and ^{40}Ca(p,n). We chose these two target nuclei since a strong GT-resonance has been observed in ^{48}Ca[7] while only little 1^+-strength is found in ^{40}Ca[8]. (The latter is due to the presence of 2p-2h-components in the exact ground state of ^{40}Ca.) Moreover, these two nuclei have the same number of protons and differ only by the 8 neutrons which occupy the $1f_{7/2}$-neutron shell in ^{48}Ca. It is interesting to see to what extent these additional 8 neutrons modify the shape of the background or not.

In Figs. 2a and b we show 0° spectra for the reactions ^{48}Ca(p,n) and ^{40}Ca(p,n), respectively. The experimental data (thick full line) have been taken from ref. 7 in case of ^{48}Ca (using the normalization of ref. 8) and from ref. 8 in case of ^{40}Ca. The data are compared to the calculated spectra which are the incoherent sum of all cross sections with multipolarities $\Delta L=0$ through $\Delta L=3$ ($J^{\pi} = 0^-, 1^+, 1^-, 2^+, 2^-, 3^+, 3^-, 4^-$). The calculations reproduce the spectra at large Q-values within a factor of 1.33! This applies for both reactions, ^{40}Ca(p,n) and ^{48}Ca(p,n). The calculated continuum falls off sharply at Q $\sim$ -20 MeV for ^{48}Ca and at Q $\sim$ -25 MeV for ^{40}Ca. This falling off is a combined effect of the Coulomb and the centrifugal barrier which make the continuum wave function of the excited proton $|E_p, \ell_p j_p\rangle$ small in the nuclear surface region, especially for smaller energies ($E_p \leq 10$ MeV) and angular momenta $\ell_p \neq 0$. As a consequence the nuclear transition densities are small for these energies and therefore also the cross sections. The 0^- cross section at Q = -20 MeV in Fig. 2b is due to the $p_{3/2}$ proton single particle resonance which couples with the $d_{3/2}$ neutron hole to $J^{\pi} = 0^-$. This 0^- resonance would be shifted into the energy region around Q = -26 MeV if our model would include nuclear collectivity. The same is true for part of the theoretical 2^- strength at lower Q-values in ^{40}Ca.

By comparison of the experimental spectra for ^{40}Ca and ^{48}Ca (see Fig. 2), one immediately sees that for ^{40}Ca there is no cross section in the Q-value region from 0 to 15 MeV while for ^{48}Ca we have in this region the large cross sections due to the GT-states, due to the "background", and, to a small fraction, due to the IAS. Also our microscopic model gives zero cross section for ^{40}Ca(p,n) at $0 \geq Q \geq -15$ MeV. For ^{48}Ca(p,n) the background cross section below the GTR and IAS has then to be produced from all the transitions which promote a neutron from the 2s-1d-shell and $1f_{7/2}$-shell via charge exchange into the proton 2p-1f-shell (see Fig. 1) or into the continuum (the neutron separation energy is $\sim$ 10 MeV). The cross sections produced by these states are shown in Fig. 2a. (The discrete states, the cross sections to the GTR and IAS are not plotted.) Most of this background cross section is due to $\Delta L=1$

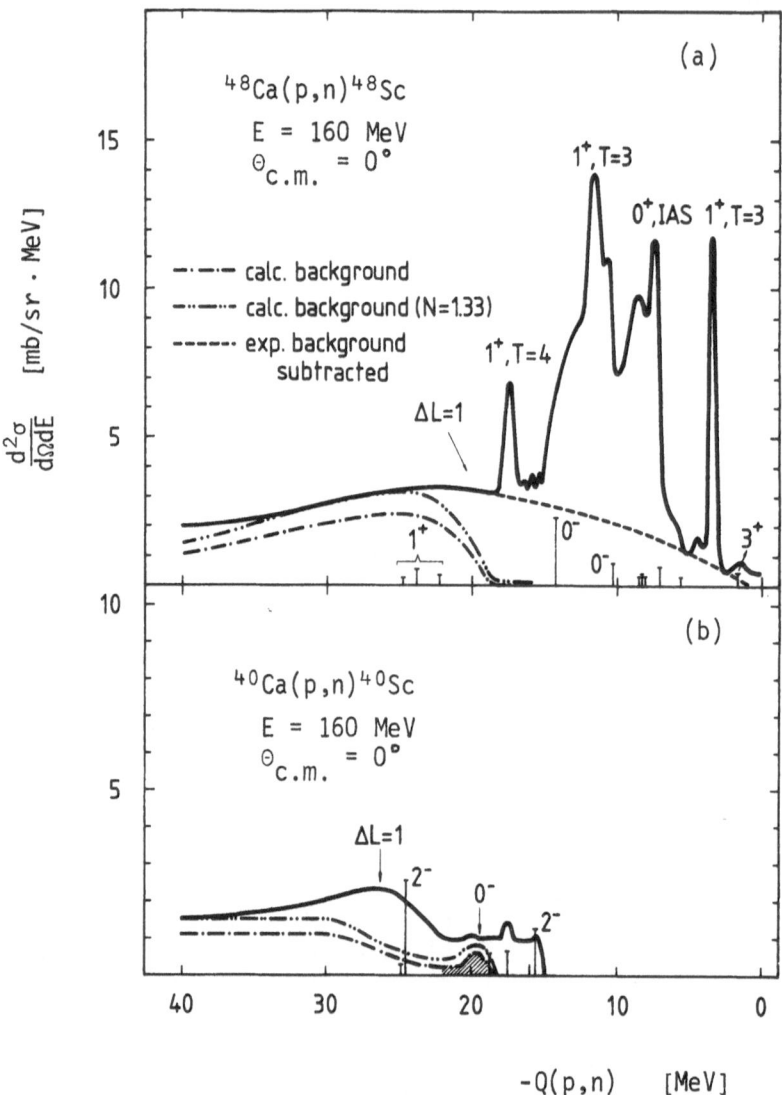

Fig. 2.   Zero degree spectra for the reactions ⁴⁸Ca(p,n) and
          ⁴⁰Ca(p,n). The data (thick full line) are taken from
          refs. 7 and 8 (see text). The discrete lines are calcu-
          lated cross sections due to bound and quasibound states.
          The arrow labelled with ΔL=1 indicates the location where
          the ΔL=1-resonance (0⁻,1⁻,2⁻) would occur if nuclear
          collectivity were included for these states. The
          theoretical cross sections due to the GTR and IAS are not
          plotted. The optical parameters for the cross section
          calculations have been taken from ref. 25.

$(0^-, 1^-, 2^-)$- and $\Delta L=2$ (3^+)-excitations. The sum of all cross sections in the Q-interval from 2 to 15 MeV amounts to 5.4 mb from which 2.5 mb are due to 0^-, 1.4 mb due to 2^-, and 1.0 mb due to 3^+ excitations. Note that most of the $\Delta L=1$ $(0^-, 1^-, 2^-)$-strength is shifted into the energy region around $Q = -22$ MeV (as indicated in the figure) when the residual ph-interaction is switched on. We emphasize that there exists a sum rule for $\Delta L=1$ charge exchange modes[14]. This sum rule tells us that when we consider a residual ph-interaction the strength is only redistributed, i.e. the strength is moved from the low to the high excitation energy region. Therefore the 5 mb calculated in our unperturbed ph-door-way model represent an upper limit for the background below the GTR in ^{48}Ca. In the experimental analysis, however, a background of roughly 17 mb is subtracted (see Fig. 2a). Our calculations show that at least 12 mb of this background are actually GT-strength. By adding this cross section of 12 mb to that of the 1^+, $T=3$, 11 MeV state the GT-cross section at 0° is changed from 48 to 60 mb which makes an effect of 25 %. This also means that the amount of GT-strength seen in ^{48}Ca is now increased from 43 % to 51 % of the total GT-strength.

In another contribution to this conference[22] (see also ref. 16) cross section calculations which treat the $\Delta(1232)$ isobar degrees of freedom explicitly in both the structure (RPA) and reaction calculations (DWIA) are shown for the GT-states in ^{48}Sc. It is found that the calculations describe the absolute magnitude of the experimental cross sections for the low-lying 1^+, $T=3$, 2.52 MeV and the high-lying 1^+, $T=4$, 16.8 MeV state both being only a little affected by the background subtraction. The experimental 0° cross section to the broad 1^+, $T=3$ state at $\sim$ 11 MeV, however, is overestimated by roughly 20 mb (the experimental cross section is 48 mb[8]). Part of this overestimate is due to isospin impurities in the RPA wave function (estimated error: 6 mb) whereas the rest of $\sim$ 14 mb is the amount of GT-strength which has been wrongly subtracted in the experimental analysis of the data.

In Figs. 3 and 4 we show calculated spectra for scattering angles of 6° and 12°, respectively. The striking point in these figures is that there appears a strong $\Delta L=2$ (3^+)-resonance in ^{48}Ca (p,n) (centered around $Q = -26$ MeV) while there is no $\Delta L=2$-resonance in ^{40}Ca. The 3^+-resonance in ^{48}Ca is a $2\hbar\omega$-excitation and is mainly built up by the $[\pi g_{9/2} \nu 1d_{3/2}^{-1}]_{3^+}$- and the $[\pi g_{7/2} \nu 2s_{1/2}^{-1}]_{3^+}$-configurations. Its width is $\sim$ 4.5 MeV at half maximum. Since our model includes only the escape width one would expect an additional broadening of the resonance due to 2p-2h admixtures (spreading width). Its location, however, should be roughly correct since the residual ph-interaction is weak for $J^\pi = 3^+$ so that no large energy shift should occur. Experimental evidence for a $\Delta L=2$-resonance has been found in the ^{90}Zr(p,n)- and ^{208}Pb(p,n)-reactions[13]. There is no experimental evidence for a

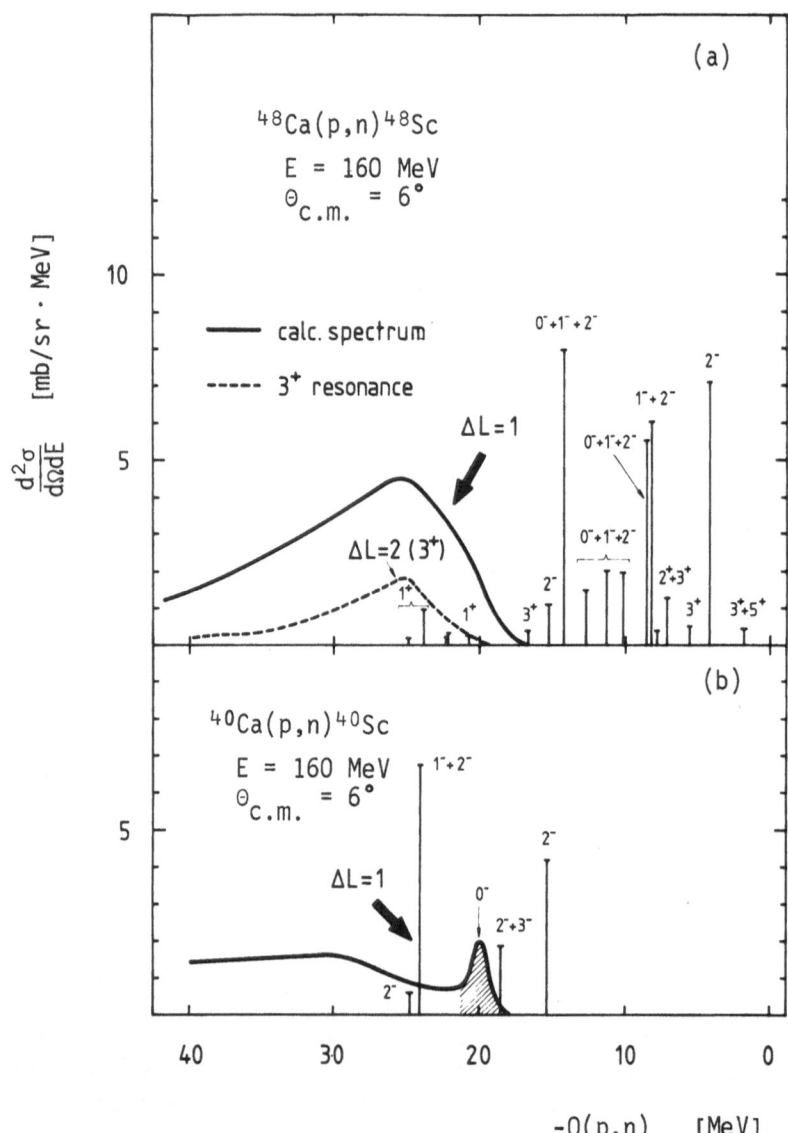

Fig. 3.   Same as in Fig. 2 but for $\Theta_{c.m.} = 6°$. No data are
          published yet for this scattering angle.

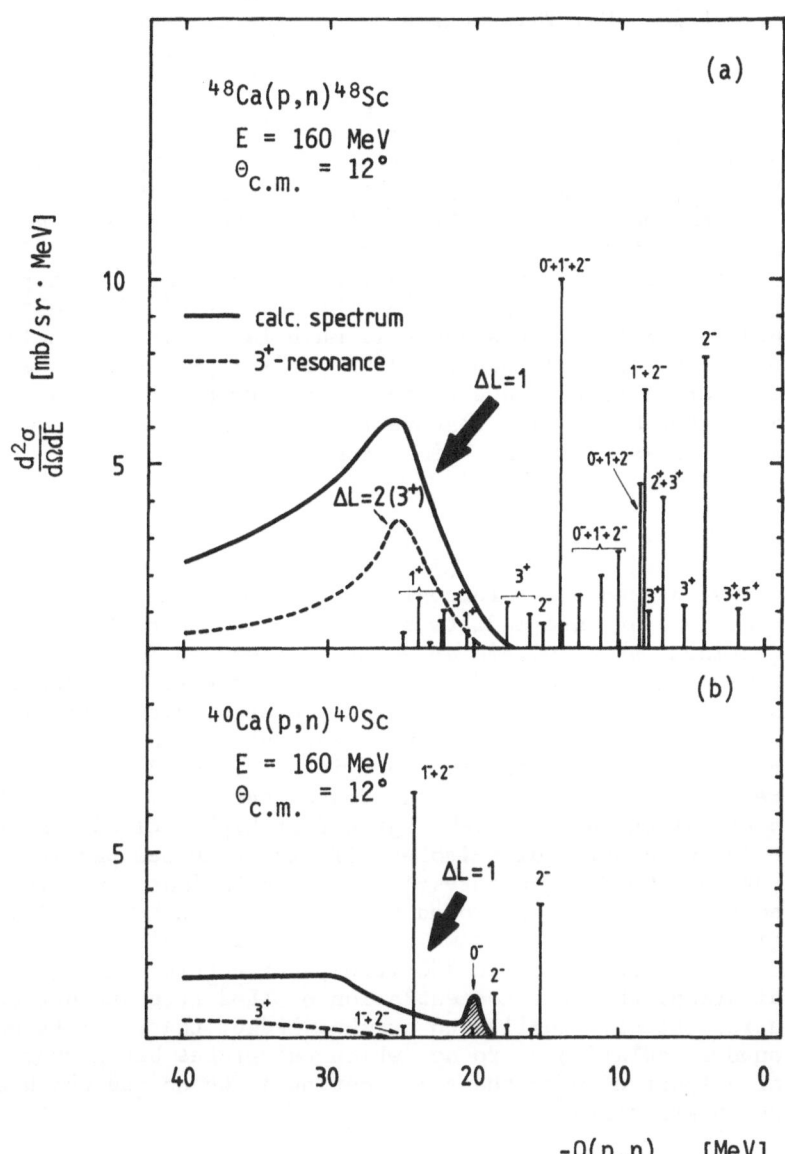

Fig. 4.   Same as in Fig. 2 but for $\Theta_{c.m.}$ = 12°. No data are
published yet for this scattering angle.

$\Delta L=2$-resonance in $^{40}Ca(p,n)^{23}$, and only weak evidence in $^{42}Ca(p,n)$ and $^{54}Fe(p,n)^{23}$. Our model calculations seem to explain this observation in a natural way: The $\Delta L=2$-strength is concentrated in ^{48}Ca but not in ^{40}Ca. This suggests that the concentration of $\Delta L=2$-strength depends on the neutron excess. At least for the nuclei ^{40}Ca and ^{48}Ca this explanation is right. An inspection of the effective potential $V_{eff} = V_\ell + V_C + V_N$ (V_ℓ: = centrifugal potential, V_C: = Coulomb potential, and V_N: = nuclear potential) felt by the $g_{9/2}$-proton wave function shows that the "potential pocket" in V_{eff} is much deeper for ^{48}Ca than for ^{40}Ca. This is simply due to the attractive $V_1 = -24$ N-Z/A term in the Becchetti-Greenlees potential[18] which is -5 MeV for ^{48}Ca but 0 for ^{40}Ca. Therefore the $g_{9/2}$-proton wave function which forms together with the $d_{3/2}$ neutron hole wave function the dominant $\left[\pi g_{9/2}\ \nu d_{3/2}^{-1}\right]_{3^+}$-configuration of the 3^+-resonance is much larger in the nuclear surface region for ^{48}Ca than for ^{40}Ca. This explains why there is a large 3^+-cross section due to this configuration in ^{48}Ca and not in ^{40}Ca. Furthermore, this feature is simply an effect of the neutron excess as has been claimed above.

SUMMARY

We have presented microscopic background calculations for (p,n)-reactions at intermediate energies which reproduce the ^{48}Ca (p,n)-continuum at $0°$ within an accuracy of 30 %. This good agreement between calculation and experiment shows that the background at high incident energies is dominantly made by direct processes only. The missing cross section might be due to two step processes, especially at higher excitation energies. This good result for (p,n)-reactions gives us some confidence that one might also be able to calculate the (p,p')-background at high incident energies. This is in our context of Δ-isobar effects in nuclei particularly important for the M1-resonances[24]. Our calculations for (p,n) show that the shape of the background below the GTR in $^{48}Ca(p,n)$ is quite different from that drawn by the experimentalists. We find a strong $\Delta L=2$, 3^+-resonance in $^{48}Ca(p,n)$, but not in $^{40}Ca(p,n)$. There are indications that the concentration of $\Delta L=2$-strength depends on the neutron excess. Finally, to our knowledge, there exists no background calculation up to now which calculates background and peaks of a spectrum with the same footing as we do and which comes so close to experiment.

REFERENCES

1. For reviews on the experimental and theoretical situation of giant resonances see for example:
    F.E. Bertrand, Ann. Rev. Nucl. Sci. 26:457 (1976);

G.R. Satchler, Proc. Int. School of Physics Enrico Fermi, LXIX
Corso, Varenna (1976);
J. Speth and A. van der Woude, Rep. Prog. Phys. 44:719 (1981).
2. G.F. Bertsch and S.F. Tsai, Phys. Rep. 18C:125 (1975);
S.F. Tsai and G.F. Bertsch, Phys. Rev. C11:1634 (1975); Phys.
Lett. 73B:248 (1978).
3. T. Tamura, T. Udagawa, D.H. Feng and K.K. Kan, Phys. Lett. 66B:
109 (1977);
T. Tamura and T. Udagawa, Phys. Lett. 78B:189 (1978); Proc.
1980 RCNP Int. Symp. on "Highly Excited States in Nuclear
Reactions", Osaka University (1980) p. 33.
4. F.E. Bertrand and R.W. Peelle, Phys. Rev. C8:1045 (1973); Oak
Ridge National Lab. Report ORNL-4638 (1971).
5. H. Feshbach, A.K. Kerman, S. Koonin, Ann. Phys. 125:429 (1980);
Proc. 1980 RCNP Int. Symp. on "Highly Excited States in Nuclear
Reactions", Osaka University (1980), p. 5.
6. L. Avaldi, R. Bennetti and L. Colli-Milazzo, to be published.
7. B.D. Anderson, J.N. Knudson, P.C. Tandy, J.W. Watson and R.
Madey, Phys. Rev. Lett. 45:699 (1980).
8. C. Gaarde, J.S. Larsen, C.D. Goodman, C.C. Foster, C.A.
Goulding, D.J. Horen, T. Masterson, E. Sugarbaker, preprint;
C. Gaarde, Trieste-Lectures, to be published (1981).
9. R.R. Doering, A. Galonsky, D. Patterson, G.F. Bertsch, Phys.
Rev. Lett. 35:1961 (1975).
10. D.E. Bainum, J. Rapaport, C.D. Goodman, D.J. Horen, C.C. Foster,
M.B. Greenfield and C.A. Goulding, Phys. Rev. Lett. 44:1751
(1980).
11. C.D. Goodman, C.A. Goulding, M.B. Greenfield, J. Rapaport,
D.E. Bainum, C.C. Foster, W.G. Love and F. Petrovich, Phys.
Rev. Lett. 44:1755 (1980).
12. D.J. Horen, C.D. Goodman, C.C. Foster, C.A. Goulding, M.B.
Greenfield, J. Rapaport, E. Sugarbaker, T.G. Masterson, F.
Petrovich and W.G. Love, Phys. Lett. 95B:27 (1980);
D.J. Horen, C.D. Goodman, D.E. Bainum, C.C. Foster, C. Gaarde,
C.A. Goulding, M.B. Greenfield, J. Rapaport, T.N. Taddeucci,
E. Sugarbaker, T. Masterson, S.M. Austin, A. Galonsky, W.
Sterrenburg, Phys. Lett. 99B:383 (1981).
13. C. Gaarde, J. Rapaport, T.N. Taddeucci, C.D. Goodman, C.C.
Foster, D.E. Bainum, C.A. Goulding, M.B. Greenfield, D.J.
Horen and E. Sugarbaker, Nucl. Phys. A369:258 (1981).
14. K.I. Ikeda, S. Fujii and J.I. Fujita, Phys. Lett. 3:271 (1963).
15. M. Rho, Nucl. Phys. A231:493 (1974);
K. Ohta and M. Wakamatsu, Nucl. Phys. A234:445 (1974);
J. Delorme, M. Ericson, A. Figureau and C. Thévenet, Ann. Phys.
102:273 (1976);
E. Oset and M. Rho, Phys. Rev. Lett. 42:42 (1979);
I.S. Towner and F.C. Khanna, Phys. Rev. Lett. 42:51 (1979);
W. Knüpfer, M. Dillig and A. Richter, Phys. Lett. 95B:349
(1980);

A. Härting, W. Weise, H. Toki and A. Richter, Phys. Lett. 104B: 261 (1981);
H. Toki and W. Weise, Phys. Lett. 97B:12 (1980);
S. Krewald, F. Osterfeld, J. Speth and G.E. Brown, Phys. Rev. Lett. 46:103 (1981);
A. Bohr and B.R. Mottelson, Phys. Lett. 100B:10 (1981);
G.E. Brown and M. Rho, Nucl. Phys. A327:397 (1981);
T. Suzuki, S. Krewald, J. Speth, Phys. Lett. 107B:9 (1981).
16. F. Osterfeld, S. Krewald, J. Speth, T. Suzuki, Phys. Rev. Lett. 49:11 (1982).
17. W.G. Love and M.A. Franey, Phys. Rev. C24:1073 (1981);
F. Petrovich, W.G. Love, Nucl. Phys. A354:499c (1981).
18. F. Becchetti and G.W. Greenlees, Phys. Rev. 182:1190 (1969).
19. F. Osterfeld, FROST-MARS-CODE, unpublished.
20. J. Speth, V. Klemt, J. Wambach and G.E. Brown, Nucl. Phys. A343:282 (1980).
21. F. Osterfeld, S. Krewald, H. Dermawan and J. Speth, Phys. Lett. 105B:257 (1981).
22. J. Speth, S. Krewald, F. Osterfeld, and T. Suzuki, these proceedings.
23. C. Gaarde, J. Rapaport, private communication.
24. N. Anantaraman, G.M. Crawley, A. Galonsky, C. Djalali, N. Marty, M. Morlet, A. Willis and J.-C. Jourdain, Phys. Rev. Lett. 46:1318 (1981).
25. A. Nadasen, P. Schwandt, P.P. Singh, W.W. Jacobs, A.D. Bacher, P.T. Debevec, M.D. Kaitchuck and J.T. Meek, Phys. Rev. C23:1023 (1981).

DAMPING OF SPIN RESONANCES

P.F. Bortignon

Istituto di Fisica, Università di Padova, Padova, Italy
INFN, L. N. Legnaro, Italy

R.A. Broglia

The Niels Bohr Institute, University of Copenhagen
DK-2100 Copenhagen Ø, Denmark

F. Zardi

INFN, Sezione di Padova, Padova, Italy

INTRODUCTION

In this conference we are learning how the Gamow-Teller giant resonance (GTR), after being a rather elusive vibrational mode[1,2], is becoming one of the best characterized resonant structures of the nuclear spectrum. It has quantum numbers $\lambda=0$, $\tau=1$ and $\sigma=1$, and it displays a forward peaked angular distribution in (p,n) reactions induced by protons with $E_p \sim 200$ MeV. The ratio of peak to background of these modes is strongly energy dependent and is determined by the energy dependence of the spin-isospin term $V(r)(\vec{\sigma}\vec{\sigma})(\vec{\tau}\vec{\tau})$ of the nucleon-nucleon interaction[3]. A basic difference between the GTR and other giant resonances is that only about 50% of the strength of the operator responsible for the excitation of the GT mode (i.e. $\sigma\tau_-$) is found in the main peak and low energy region of the spectrum[4].

Many of the talks address the question of the mechanisms through which the GTR are populated. Others are devoted to relating

the observed nuclear structure properties of these modes to the nucleon-nucleon interaction and to assess the role played by the nucleonic degrees of freedom.

In our contribution we investigate the couplings responsible
for the damping widths of the GTR. A detailed description of the
interplay of this mode and more complicated degrees of freedom may
prove useful in connection with the question of the missing strength
mentioned above. We present calculations of the spreading width $\Gamma^{\downarrow}$
for the GTR in ^{48}Ca, ^{90}Zr and ^{208}Pb. It is noted that the escape
widths $\Gamma^{\uparrow}$ are expected to be small in medium and heavy nuclei.
As reviewed in refs. 5 and 6, several models have been used to describe the coupling of a giant resonance to more complicated states.
Two of them which have been widely employed are based on the shell
model and on the elementary modes of excitation description of the
nuclear spectrum. In the first, the giant resonance is spread over
states which are obtained by the diagonalization of a residual interaction in an appropriately truncated many particle-many hole
space[7]. In the doorway state approach, the giant mode is coupled
to selected 2p-2h states containing an uncorrelated particle-hole
pair and a collective vibration[5,6].

As shown in Fig. 1, in the doorway state model the relation
between the widths of the particles and holes (graphs (a) and (b))
building the giant mode and those of the giant vibration itself
(graphs (c) and (d)) is rather explicit. Based on the systematic
analysis of Rapaport et al.[8] of elastic neutron scattering, it appears that at low excitation energies the imaginary part of the
single-particle potential is surface peaked (cf. Fig. 2). We can

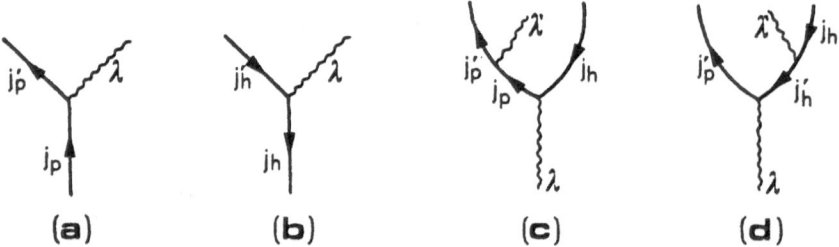

Fig. 1.   Basic couplings between particles and phonons which con-
          tribute to the damping widths of single-particles and of
          giant vibrations. A line marked by an arrow pointing up
          represents a particle, while one pointing down represents
          a hole. A wavy line stands for a surface vibration. In (a)
          and (b) a particle (hole) bounces inelastically off the
          surface and creates a surface vibration. In (c) and (d),
          a vibration decomposes into a particle above the Fermi
          surface and a hole in the Fermi sea. Subsequently, the par-
          ticle or the hole behaves as in (a) and (b) respectively.

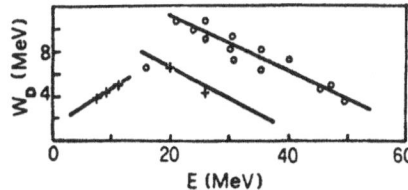 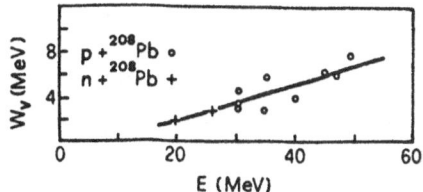

Fig. 2.   Energy dependence of the imaginary part W of the optical
          potential in the nucleon-^{208}Pb scattering (according to
          Rapaport et al.[8]). On the left side of the figure the e-
          nergy dependence of the surface peaked part of the imagi-
          nary potential is shown; on the right side the behaviour
          of the volume term is displayed. For more details see ref.
          8.

thus expect that the main contributions to the width of the single-
particle and single-hole states arise from the coupling to collective surface vibrations (wavy lines in the intermediate states of
Fig. 1). This is in fact the main conclusion of ref. 6, where the
width of different single-particle and single-hole states and giant
multipole resonances were calculated. Also discussed in refs. 5 and
6 is the mechanism by which in the doorway state model the correlations present in the giant mode between particles and holes reduce
its width as compared to the sum of the particle and of the hole
widths, a reduction which is displayed by the experimental data.

THE STRENGTH FUNCTION AND THE DOORWAY STATES

     The strength function is useful to describe the way a simple
state $|a>$ at energy E_a is spread over a set of doorway states $|\alpha>$
with energy E_α, by the coupling $V_{\alpha a}=<\alpha|V|a>$, V being the part of
the Hamiltonian which is not diagonal in the space $|a> \otimes |\alpha>$. The
strength function $P_a(E)$ which gives the probability to find the
state per unit energy can be expressed in the form[9]

$$P_a(E)=\frac{1}{2\pi}\frac{\Gamma_a(E)+\Delta}{\left[E_a+\Delta E_a(E)-E\right]^2+(1/4)\left[\Gamma_a(E)+\Delta\right]^2} \tag{1}$$

where

$$\Gamma_a(E)=\Delta\sum_\alpha\frac{V_{\alpha a}^2}{(E-E_\alpha)^2+(1/4)\Delta^2} \tag{2}$$

and

$$\Delta E_a(E) = \sum_\alpha \frac{V_{\alpha a}^2 (E-E_\alpha)}{(E-E_\alpha)^2 + (1/4)\Delta^2} \tag{3}$$

The quantity Δ denotes the energy interval around E over which ave-rages are carried out, and accounts in some way for the coupling of $|\alpha>$ to more complicated states[*].

We remark that only in the limiting conditions

$$E_\alpha = \alpha D \ , \qquad\qquad V_{\alpha a} = v >> D \ ,$$

i.e. of constant matrix elements and uniform distribution of states $|\alpha>$, the strength function assumes the simple Breit-Wigner form

$$P_a(E) = \frac{1}{2\pi} \frac{\Gamma^\downarrow}{(E_a - E)^2 + 1/4(\Gamma^\downarrow)^2} \quad .$$

The quantity

$$\Gamma^\downarrow = 2\pi v^2 / D \ ,$$

is in this case the full width at half maximum of the line shape (FWHM). In the general case, the parameters appropriate for the characterization of a resonant structure in a given energy inter-val $E_i \le E \le E_f$ are the centroid

$$<E> = \frac{1}{N} \int_{E_i}^{E_f} dE P_a(E) E \ ; N = \int_{E_i}^{E_f} dE P_a(E) \tag{4}$$

and the central multipole moments

$$\mu_n = <(E - <E>)^n> \ .$$

Instead of the second moment one usually introduces the standard deviation

---

[*]It is obvious that the matrix elements $<\alpha|V|a>$ do not exhaust the entire residual interaction V, which can couple states $|\alpha>$ with more complicated states.

$$\sigma=\sqrt{\mu_2} \qquad (5)$$

The spreading width is often related to σ by the relation

$$\Gamma^{\downarrow}=\sqrt{(8\ln2)\mu_2} \approx 2.35\sigma \qquad (6)$$

which is exact only for a Gaussian distribution.

For the calculation of the giant resonance strength function and the derived resonance parameters (4) and (6) we need the energies E_a and E_α and the matrix elements $V_{\alpha a}$. To obtain the energy E_a we use the Random Phase Approximation (see, e.g., ref. 10). The single-particle spectrum is calculated in the Hartree-Fock approximation. The residual particle-hole interaction used to generate the GTR mode is discussed in the following section.

The matrix elements $V_{\alpha a}$ coupling the giant mode $|a\rangle$ to the doorway states $|\alpha\rangle$ are schematically shown in Fig. 1, diagrams (c) and (d). The giant mode (of multipolarity λ) decays through the appropriate RPA amplitude $X(j_p j_h; \lambda)$ in the $(j_p - j_h)$ component. Afterward the particle j_p (the hole j_h) decays in a more complicated configuration consisting of a vibration and a fermion. The Hamiltonian responsible for the coupling of the particle to the vibration can be written as[11]

$$H_c = -R_0 \frac{\partial U}{\partial r} \sum_{\lambda\mu} \alpha_{\lambda\mu} Y^*_{\lambda\mu}(\hat{r}) \qquad (7)$$

where the matrix elements of $\alpha_{\lambda\mu}$ are calculated in RPA (see the next section). It follows that the functions $\Delta E_a(E)$ and $\Gamma_a(E)$ (cf. eqs. (2) and (3)) are the real and imaginary* part of the four diagrams shown in Fig. 3.

Graphs (a) and (b) represent single-particle renormalization processes while diagrams (c) and (d), often called vertex correction diagrams, lead to a change of the bare particle-hole interaction. The contributions (c) and (d) have in many cases opposite signs with respect to the graphs (a) and (b), reducing the coupling and eventually the width of the giant mode.

RESULTS AND COMPARISON WITH THE EXPERIMENTAL DATA

In what follows we calculate the energy and damping widths of the GTR in ^{90}Zr and ^{208}Pb, and quote results of a shell model calculation for ^{48}Ca[7]. According to the scheme we have outlined in the

---

*We have an imaginary part because of the averaging parameter Δ, the effect of which is to shift the energy into the complex plane: $E \rightarrow (E \pm i\frac{\Delta}{2})$.

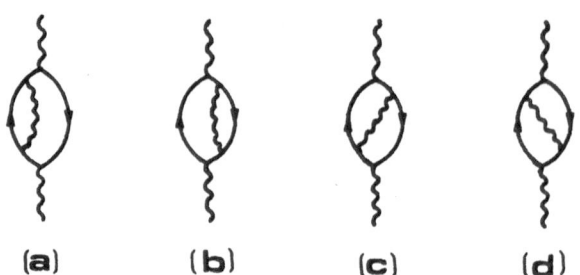

Fig. 3.   Graphs contributing to width of the giant resonance in the
           doorway state model; see the text for more details.

previous sections, the starting point for our calculations is to
find the energies and the eigenfunctions of the single-particle spectra. This was done using the Hartree-Fock approximation, with the
Skyrme III interaction, to which it is associated the effective mass
$m^*/m=0.76$.

To generate the GTR mode, the residual interaction

$$V=V_{\sigma\tau}\delta(\vec{r}_1-\vec{r}_2)\ (\vec{\sigma}_1\cdot\vec{\sigma}_2)\ (\vec{\tau}_1\cdot\vec{\tau}_2) \tag{8}$$

is diagonalized in the resulting single-particle basis using the
RPA. The value $V_{\sigma\tau}\approx200$ MeV fm^3 for the strength parameter leads to
a fair overall agreement with the observed centroids (see e.g. ref.
12). In Table 1 we show the RPA energies of the collective state and
the main components of the RPA wave functions together with the unperturbed particle-hole energies.

The collective vibrations λ' which are coupled to the single-
particle and the single-hole states (see Fig. 1) have also been cal-

Table 1

^{90}Zr		^{208}Pb	
$E_{GTR}=16.20$ MeV		$E_{GTR}=19.20$ MeV	
ε_{ph} (MeV)	X_{ph}	ε_{ph} (MeV)	X_{ph}
$5.65(\nu^{-1}g_{9/2}\pi g_{9/2})$	0.22	$13.1(\pi h_{9/2}\nu^{-1}h_{11/2})$	0.50
$13.29(\nu^{-1}g_{9/2}\pi g_{7/2})$	0.97	$14.7(\pi i_{11/2}\nu^{-1}i_{13/2})$	0.76

culated in the RPA. A schematic separable interaction with matrix elements

$$V_{php'h'}(\lambda)=k<j_p||r \frac{\partial U}{\partial r} Y_\lambda||j_h'><j_{p'}|r \frac{\partial U}{\partial r} Y_\lambda||j_{h'}> \qquad (9)$$

has been used; the constant k is determined by the condition that the isoscalar 1^- state is at zero energy. The properties of the vibrations are in overall agreement with the experimental data. This treatment leads to similar results as those one would obtain using as residual interaction the density derivative of the Hartree-Fock field (cf. ref. 10).

We have now all the quantities we need to evaluate the graphs of Fig. 3 and therefore the strength function and the derived parameters $<E>$ and $\Gamma^\downarrow$ (see eqs. (4) and (6)). The matrix element of the particle-vibration Hamiltonian H_c (see eq. (7)) is

$$<(j'\lambda)j|H_c|j>=\pm \frac{1}{\sqrt{(2j+1)(2\lambda+1)}} \beta_n(\lambda)<j||T_\lambda||j'>$$

where $\beta_n(\lambda)$ is obtained from the relation

$$<1_{n\lambda\mu}|\alpha_{n\lambda\mu}|0>=\sqrt{\frac{\hbar\omega_{n\lambda}}{2C_{n\lambda}}} = \beta_n(\lambda)/\sqrt{2\lambda+1}$$

and where $|1_{n\lambda\mu}>$ is the nth root of the RPA. The field $T_{\lambda\mu}$ is defined as

$$T_{\lambda\mu}=R_0\frac{\partial U}{\partial r} Y_{\lambda\mu}$$

The plus minus signs above correspond to a particle or to a hole scattering respectively.

By utilizing the above relations we obtain (cf. graphs (a),(c) of Fig. 3)

$$\Gamma_p(E,\lambda)_{GR}= \sum_{\substack{\lambda',n' \\ j_p j_p' j_h}} \frac{\beta_n^2(\lambda')X_n(j_p j_h^{-1};\lambda)}{(2\lambda'+1)(2j_p+1)} \cdot$$

$$\cdot \frac{<j_p||T_{\lambda'}||j_{p'}>^2\Delta}{\{E-[(\epsilon_{j_p'}-\epsilon_{j_h})+\omega_{n'}(\lambda')]\}^2+(1/4)\Delta^2} \qquad (10)$$

$$\Gamma_{ph}(E,\lambda)_{GR} = \sum_{\substack{\lambda' n' \\ j_p j_p' j_h j_h'}} (-1)^{j_h - j_p + \lambda} \beta_{n'}^2(\lambda') \left\{ \begin{matrix} j_p j_h \lambda \\ j_h' j_p' \lambda' \end{matrix} \right\}.$$

(10')

$$. \frac{X_n(j_p j_h^{-1};\lambda) X_n(j_p' j_h'^{-1};\lambda) <j_p'||T_\lambda'||j_p><j_h||T_\lambda'||j_h'>}{\{E - [(\varepsilon_{j_p'} - \varepsilon_{j_h}) + \omega_{n'}(\lambda')]\}^2 + (1/4)\Delta^2}$$

By substituting hole states with particle states, an expression similar to eq. (9) can be obtained for $\Gamma_h(E,\lambda)_{GR}$ (graph (b) of Fig. 2)). We can now calculate the basic quantity

$$\Gamma_a(E,\lambda) = \Gamma_p(E,\lambda)_{GR} + \Gamma_h(E,\lambda)_{GR} + 2\Gamma_{ph}(E,\lambda)_{GR}$$

(11)

Comparison of eq. (2) with eq. (3) suggests immediately the structure of $\Delta E_a(E)$.

A strong cancellation between $\Gamma_p + \Gamma_h$ and $2\Gamma_{ph}$ will occur provided: a) The mode λ is collective, b) the vibration λ' appearing in the doorway states is a density oscillation, c) $j \gg \lambda$ and thus the 6-j symbol can be replaced by its classical limit.

All the above conditions are approximately fulfilled in medium and heavy nuclei. The cancellation will thus be more complete for ^{90}Zr and ^{208}Pb than for ^{48}Ca. A more general discussion of the cancellation mechanism can be found in refs. 5 and 13.

In Table 2 the results for the centroids $<E>$ and the spreading widths $\Gamma^\downarrow$ are displayed in comparison with the experimental data. In Figs. 4(a) and 4(b) are shown the strength functions $P_a(E)$ for ^{208}Pb and ^{90}Zr respectively (an averaging parameter $\Delta = 1$ MeV has

Table 2

	ENERGY (MeV)		$\Gamma^\downarrow$ (MeV)	
	EXP.	THEOR.	EXP.	THEOR.
^{48}Ca	11[7]	10.6[7]	∿4[7]	4.5[7]
^{90}Zr	15.6[2]	14.5	4.4[2]	3.8
^{208}Pb	19.2[2]	18.8	4.2[2]	4.3

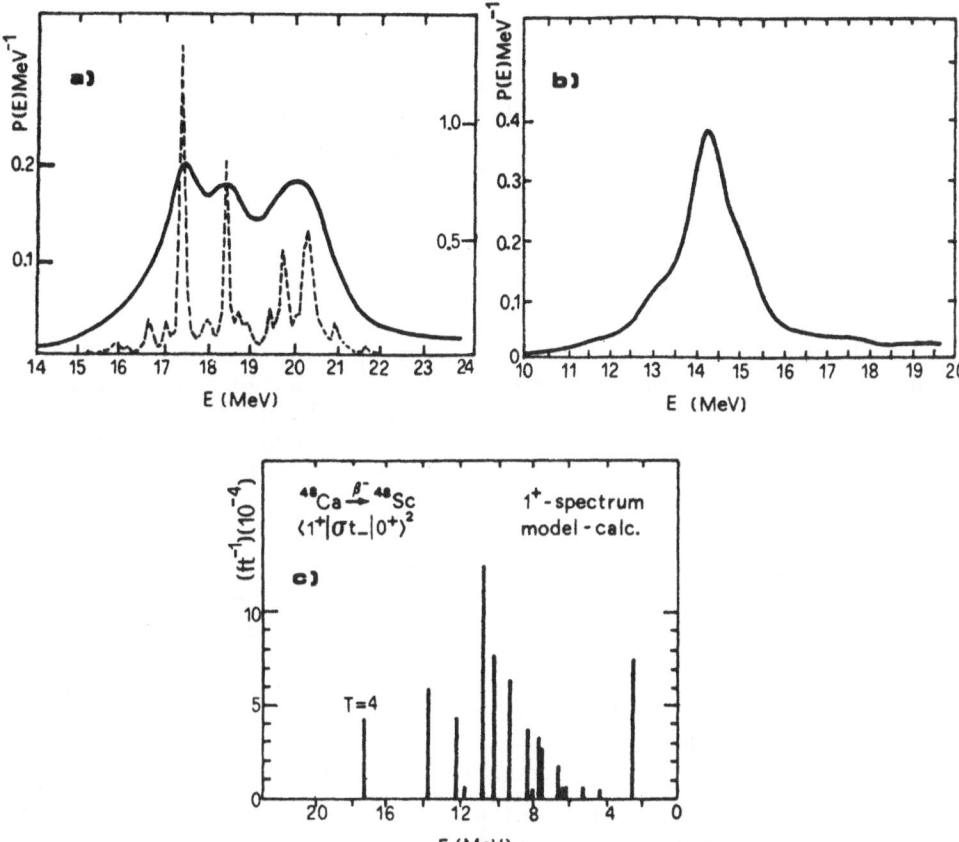

Fig. 4. a) Strength function for the Gamow-Teller giant resonance
            in the ^{208}Pb nucleus. The full line has been obtained
            with an averaging parameter $\Delta=1$ MeV; a calculation with
            $\Delta=50$ keV, which better evidences the underlying doorway
            state structure, is also displayed by the dotted line.
         b) The strength function for ^{90}Zr and $\Delta=1$ MeV.
         c) The shell model strength function for the nucleus ^{48}Ca
            evaluated by Gaarde et al.[7]

been used).In Fig. 4(a) the function $P_a(E)$ is also displayed with a
much smaller averaging parameter ($\Delta=50$ keV) to show the underlying
fine structure. The agreement between theory and experiment is satisfactory*.

*We do not expect the inclusion of the isovector modes in the inter-
 mediate states will change in a significant way the results for the
 widths, because of their high excitation energy. We have already
 noted that the escaping $\Gamma^{\uparrow}$ is expected to be very small [14].

For the ^{208}Pb nucleus, similar results have been obtained by Wambach et al.[15]. They used a residual interaction involving boson exchange terms to calculate the Gamow-Teller resonance in the RPA, explicitly including the coupling to the Δ_{33} nucleonic resonance. The particle-hole interaction diagonalized to obtain the collective vibrations in the intermediate state was parametrized according to the Landau-Migdal ansatz. In both the calculations the main contributions to the damping arises from the coupling to the quadrupole and octupole surface degrees of freedom. A strong cancellation, of the order of 50%, is found in both calculation, between diagrams (a, b) and (c, d) of Fig. 3.

Gaarde et al.[7] calculated the strength function for ^{48}Ca by the shell model approach. They assume the ^{48}Ca nucleus in the ground state to be a closed $(f_{7/2})^8$ configuration while, to obtain the 2p-2h space, a single nucleon is allowed to occupy the single-particle states of the 1f-2p shell. The single-particle energies and the two body matrix elements are taken from the experimental data. The results obtained are displayed in Fig. 4(c).

The surface excitation model has proved to give a good description of the damping widths of all the giant resonances observed in medium and heavy nuclei (cf. 5 and 6); in particular, of the GTR, as shown above. One can then conclude that, aside from background subtraction problems[16], the results shown in Table 2 can be interpreted as a further indication that the observed quenching of the $\sigma\tau_-$ strength has to be adscribed to mechanisms which shift GT strength to very high excitation energies (refs. 15, 17-19).

It is noted that the coupling to the Δ-isobars implies, e.g. for ^{208}Pb, an admixture of Δ-particle, N-hole configurations into the GT wavefunction of the order of 10%[15]. This small amplitude is enough to reduce the strength of the GT operator associated to the collective mode by 25%. On the other hand, due to the high excitation energy of the Δ-isobar, this admixture does not modify the results for the damping widths presented in this paper.

ACKNOWLEDGEMENT

Discussion with C. Gaarde is gratefully acknowledged.

REFERENCES

1. K. Ikeda, S. Fuji and J. I Fujita, Phys. Lett. 3:271 (1963).
2. C. Gaarde, these proceedings, and references therein.

3. F. Petrovich and W. G. Love, Nucl. Phys. A354:499c (1981); W. G. Love and M. A. Franey, Phys. Rev. C24:1073 (1981); and W. G. Love, F. Petrovich, and M. A. Franey, these proceedings.

4. C. D. Goodman, these proceedings.

5. G. F. Bertsch, P. F. Bortignon and R. A. Broglia, Rev. Mod. Phys. 55:287 (1983).

6. P. F. Bortignon and R. A. Broglia, Nuclear Physics A371:405 (1981) and references therein.

7. See, e.g. C. Gaarde, J. S. Larsen, M. N. Harakeh, S. Y. van der Werf, M. Igarashi and A. Müller-Arnke, Nucl. Phys. A334: 248 (1980).

8. J. Rapaport, T. S. Cheema, D. E. Bainun, R. W. Finlay and J. D. Carlson, Nucl. Phys. A296:95(1978); J. Rapaport, V. Kulkarni and R. W. Finlay, Nucl. Phys. A330:15(1979).

9. A. Bohr and B. R. Mottelson, Nuclear Structure, Vol. I app. 2D (Benjamin N. Y. 1969).

10. G. F. Bertsch and S. F. Tsai, Phys. Rep. C18:125(1975).

11. A. Bohr and B. R. Mottelson, Nuclear Structure, Vol. II (Benjamin, Reading Mass. 1975).

12. G. F. Bertsch, D. Cha and H. Toki, Phys. Rev. C24:533(1981); see also N. Van Giai and H. Sagawa, Phys. Lett. 106B:379 (1981).

13. R. A. Broglia, P. F. Bortignon and C. H. Dasso, to published.

14. C. Gaarde, J. S. Larsen, A. G. Drentje, M. N. Harakeh and S. Y. van der Werf, Phys. Rev. Lett. 46:902(1981); G. Wagner, in "Proceedings of the 1981 Nuclear Physics Trieste Workshop" (North-Holland), to be published.

15. J. Wambach and H. R. Fiebig, these proceedings and Nucl. Phys. A386:381 (1982).

16. F. Osterfeld, these proceedings.

17. G. F. Bertsch, Nucl. Phys. A354:157c(1981).

18. A. Bohr and B. R. Mottelson, Phys. Lett. 100B:10(1981).

19. G. E. Brown and M. Rho, Nucl. Phys. A372:397(1981) and these proceedings; T. Suzuki, S. Krewald and J. Speth, Phys. Lett. 46:1057(1981); J. Speth, S. Krewald, F. Osterfeld and T. Suzuki, these proceedings.

SPIN MODES IN THE LOW-ENERGY (p,n) REACTION

Hikonojo Orihara

Cyclotron and Radioisotope Center
Tohoku University
Sendai 980, Japan

ABSTRACT

Spin-mode excitation in nuclei has been investigated by observing the unnatural parity states with $J^\pi = 0^-$, 1^+, 2^-, ... and T=1 by means of (p,n) experiments at intermediate and low energies. Recent results for the stretched states of $1\,\hbar\omega$ character observed in the low-energy (p,n) reactions are discussed as the typical examples of spin excitations in nuclei. In addition, an isovector $0^+ \to 0^-$ transition studied in the $^{16}O(p,n)^{16}F$ reaction with high-resolution time-of-flight facilities is presented.

INTRODUCTION

The spin-isospin mode of excitation can be directly related to charge-exchange reactions such as (π^+,π_0), (p,n), $(^3He,t)$, etc. The magnetic excitations, in which spin-transfer takes place, is an area of high current interest. Quenching effects of nuclear magnetic properties have been extensively discussed[1-6] and it seems established[7-9] that the observed Gamow-Teller strength is usually (25-50%) of the sum-rule limit. One-pion exchange also gives rise to a strong tensor interaction in the $\sigma\tau$ channel. The tensor interaction can be studied in (p,n) reactions in which high-spin states are excited. Thus new data on high-spin unnatural-parity states, especially stretched particle-hole states, provide useful information on nuclear excitations formed by particle-hole interactions, since the number of particle-hole excitations which can contribute to these states is severely restricted. Stretched states have been studied so far in medium-energy electron scattering experiments at backward angles and (p,p') experiments.

427

Lindgren et al.[10] have reported a systematic comparison of the
(e,e') and (p,p') transition strengths for the excitation of
unnatural-parity states of stretched configurations. Recently,
such stretched states of $1\hbar\omega$ character have been observed in the
(p,n) reactions on ^{24}Mg (6^- state)[11] and on ^{16}O (4^- state)[12,13].

The $0^+ \to 0^-$, $\Delta T=1$ transition is of particular interest because
it carries the quantum numbers for symmetry properties of the pion
($I^\pi=0^-$, T=1). Experimentally the $0^\pm \to 0^\mp$ transitions so far have
been studied through β-decays and μ-capture in A=16 and A=18
nuclei.[14-16] It has been desired to carry out a larger momentum-
transfer study with the hadron-scattering experiment in which the
role of tensor interaction in one-pion exchange may be tested.

In this report we discuss the observation of stretched [$\pi d_{5/2}$,
$\nu p_{3/2}^{-1}$] 4^- states in ^{12}N and ^{16}F and a stretched [$\pi f_{7/2}, \nu d_{5/2}^{-1}$]
6^- state in ^{28}P by the (p,n) reaction at E_p = 40 and 35 MeV, res-
pectively, together with results described in Ref. 11. Results of
a high-resolution study of the [$\pi s_{1/2}, \nu p_{1/2}^{-1}$] 0^- state in ^{16}F at
E_p = 35 MeV are also given. The experiments were performed by using
proton beams from the AVF cyclotron and the time-of-flight facilities
at Cyclotron and Radioisotope Center, Tohoku University. We have
utilized a beam-swinger system and measured angular distributions
of emitted neutrons between 0° and 140°. A natural carbon foil of
a 2.47 mg/cm^2 thickness was prepared by rolling of graphite for ^{12}C.
Mylar foils of 3.5 and 5.3 mg/cm^2 in thickness were used for ^{16}O.
The carbon contributions were subtracted from the Mylar spectra. A
10 µm thick natural Si single crystal was used for ^{28}Si. Further
details of the experimental arrangements have been published elsewhere.[17]

RESULTS AND DISCUSSION

The ^{12}C(p,n)^{12}N Reaction

A typical neutron-energy spectrum for the ^{12}C(p,n)^{12}N
reaction obtained at E_p = 40 MeV is shown in Fig. 1. In addition
to the ground and first excited states, which are the isobaric
analogs to states in C at 15.110 MeV (1^+) and 16.107 MeV (2^+), a
prominent peak at E_x = 4.31 MeV in ^{12}N is seen to be excited. From
the energy systematics and the angular distribution of the cross
sections illustrated in Fig. 2, we tentatively assign this state to
be the analog of the 4^- state at E_x = 19.5 MeV in ^{12}C, which was
observed through M4 resonance in the (e,e') reaction.[18] The
DWBA predictions in Figs. 2 and 4 were obtained by the code DWBA-
70[19]. A set of effective interactions (M3Y)[20] is employed for the
p-n interaction. Pure ($\pi p3/2^4$, $\nu p3/2^4$) configuration is assumed for

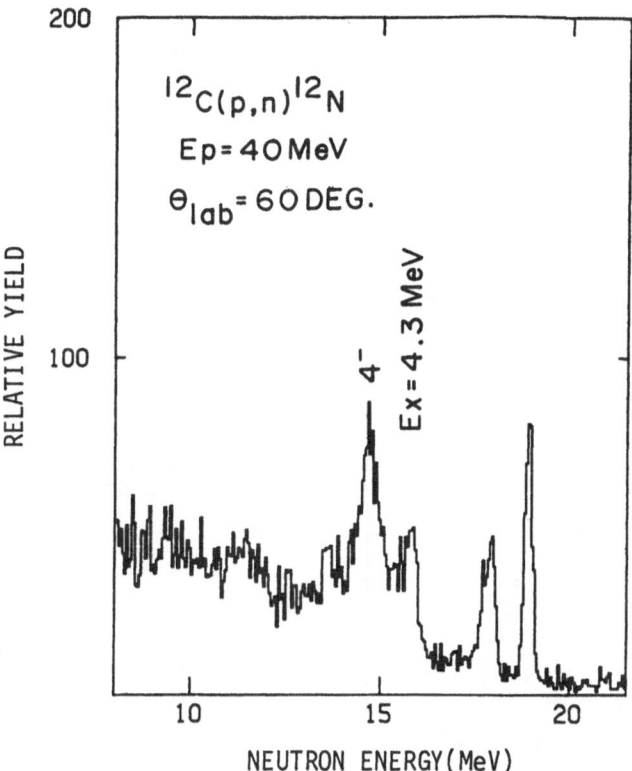

Fig. 1.   Neutron energy spectrum for the reaction ^{12}C(p,n)^{12}N at
θ(lab) = 60° measured with 40-MeV protons at a neutron
flight path of 24.6 m.   The ordinate is compensated for
the variation of the detector efficiencies with respect
to neutron energies.   Energy per bin is 50 keV.

the ground state of ^{12}C and $(\pi d5/2p3/2^4, \nu p3/2^3)4^-$ for the final
state.   The calculated angular distribution shape for this state is
in good agreement with the measurement, supporting the 4^- assignment.

The cross-section magnitude calculated for the 4^- state with
the pure configuration is much larger than the experimental value:
σ_{exp}/σ_{th} is found to be 0.2 as tabulated in Table 1.   It should be
worthwhile to point out that the ratio is in good agreement with
that from the (e,e') transition.   It has been shown that the tensor
force plays the most important role in the analyses of the 135-MeV
(p,p') data.[10,21,22]   In the present case about a half of the

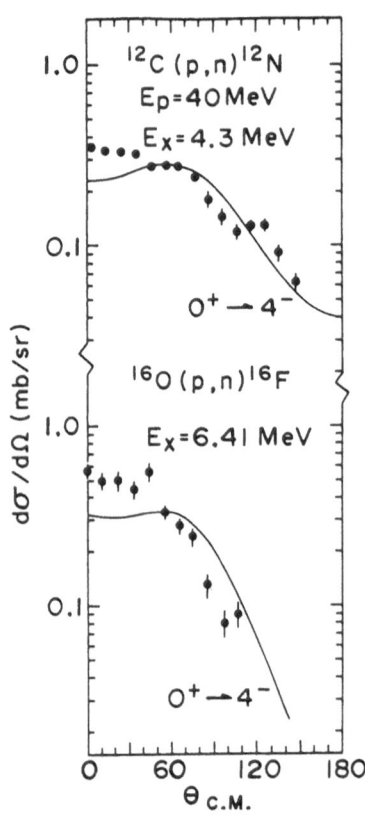

Fig. 2. Differential cross
sections for the 4.31-MeV
state in ^{12}N and 6.413-MeV
state in ^{16}F. The curves are
DWBA predictions calculated
with the M3Y interaction.

calculated cross section is due to the tensor force. Furthermore,
the exchange effect, which is correctly taken into account in the
code DWBA-70, is significant for the prediction of the cross section.

The ^{16}O(p,n)^{16}F Reaction

Many authors have discussed the stretched 4$^-$ states in A=16
nuclei as seen in (p,p') experiments[23-25], π^+ and π^- inelastic
scattering[26], the (p,n) reaction at E_p = 35 MeV[12], and $\sigma(\theta)$ and
$A(\theta)$ measurements for the (p,n) reaction at 134 MeV[12]. Our measured
angular distribution for the 4$^-$ excitation in the ^{16}O(p,n)^{16}F reaction at E_p = 40 MeV is not well reproduced as shown in Fig. 2.

Table 1

Excitation energies and ratios of σ_{exp}/σ_{th} (DWBA) for the stretched states in light N=Z nuclei obtained from the present study. A comparison for the excitation energies with the (e,e') and (p,p') experiments is also listed.

A	(e,e'), (p,p') Final state		(p,n) Final state		E_x in target(MeV)	$\dfrac{\sigma_{exp}}{\sigma_{th}}$
	J^{π}	E_x(MeV)	J^{π}	E_x(MeV)		
12	^{12}C, 4^-	19.5[a]	^{12}N, 4^-	4.31 ±0.08	19.41	0.2
16	^{16}O, 4^-	18.98[b]	^{16}F, 4^-	6.413±0.020	18.958	0.2–0.3
24	^{24}Mg, 6^-	15.137[c]	^{24}Al, 6^-	5.545±0.025	15.060	0.25
28	^{28}Si, 6^-	14.350[c]	^{28}P, 6^-	5.001±0.020	14.311	0.27

a) Ref. 15
b) Ref. 26
c) Ref. 27

This may be evidence for higher-order reaction processes, which may
involve virtual excitation of giant multipole resonances.

The ^{28}Si(p,n)^{28}P Reaction

Figure 3 shows the energy spectrum for the ^{28}Si(p,n)^{28}P
reaction at a large laboratory angle of 80°. A sharp prominent
peak is observed at E_x = 5.001 MeV. We estimate the width of this
state to be 45 keV by quadratically subtracting the contributions
of time spread and target thickness from the observed peak width.
The width is less than half that quoted in Ref. 11 for a 6⁻ stretched
state in ^{24}Al. The angular distribution shown in Fig. 4 and the excitation energy listed in Table 1, however, suggest a 6⁻ assignment.
Comparison of the spectroscopic factors for the 0⁺ → 6⁻ transition
in ^{24}Mg and ^{28}Si shows that the strength for ^{28}Si is 1.5 times larger
than that for ^{24}Mg. Present results indicate a very consistent value
of 1.6 for the ratio. Furthermore, it is noticeable that the

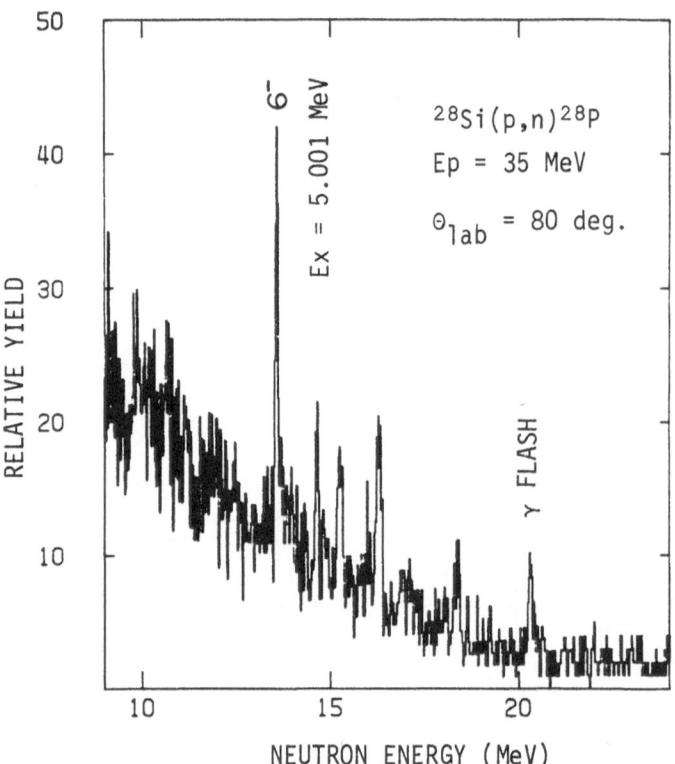

Fig. 3.   Energy spectrum as in Fig. 1 but for ^{28}Si(p,n)^{28}P at Θ(lab)
          = 80° measured with 35-MeV protons. Energy per bin is 25keV.

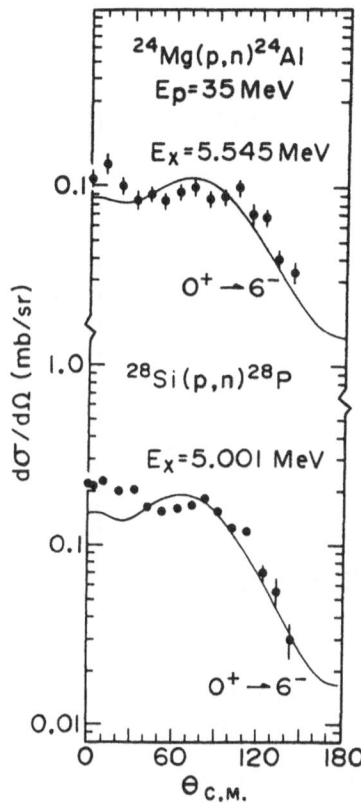

Fig. 4. Same as Fig. 2 but for
the 5.45-MeV state in ^{24}Al and
5.001-MeV state in ^{28}P.

$\sigma(exp)/\sigma(th)$ values are close to those obtained by the (p,p') and
(e,e') experiments.[10]

$0^+ \rightarrow 0^-$ Transition

A sample energy spectrum of neutrons leading to the first
four states in ^{16}F is shown in Fig. 5. The energy resolution
with a neutron flight-path of 30 m was measured to be 60 keV. We
estimate the width of the 0^- ground state is 40 keV. Figure 6 shows
the angular distribution of the cross sections for the ^{16}O(p,n)^{16}F
reaction in a transfer momentum ranging 0.4-0.22 fm^{-1}. Curves in
the figure represent the DWBA predictions by the code DWBA-70 like
the cases of the stretched state. Pure $(\pi p1/2^2, \nu p1/2^2)$ configuration is used for the ground state of ^{16}O, and $(\pi s1/2p1/2^2, \nu p1/2)0^-$
is also assumed for the final state as reported by Donnelly and

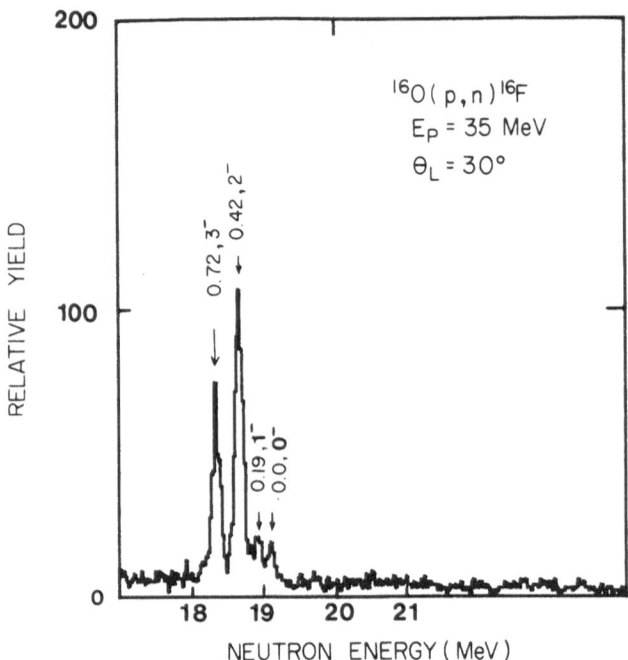

Fig. 5.   Neutron energy spectrum for the reaction $^{16}O(p,n)^{16}F$ at
θ(lab) = 30° measured with 35-MeV protons at a neutron
flight-path of 30 m.   The ordinate is compensated as it
is for Fig. 1.   Energy per bin is 20 keV.

Walker[28] for the 0⁻ T=1 state in ^{16}O.   An effective interaction,
which has been derived[29] from M3Y by switching off the odd-state
parts of the central interaction, was employed for the nucleon-
nucleon force.   A phenomenological interaction of 1-Fermi Yukawa
form (1FY)[30] has also been used to confirm J^{π} assignments for the
low-lying states in ^{16}F.   This interaction gives consistent accounts
for the angular distributions of neutrons leading to the 0⁻ (g.s.)
and 1⁻(0.19 MeV) states.   In order to examine the role of the tensor interaction, three DWBA calculations were carried out with
the more realistic interaction of Ref. 29, i.e. predictions with
tensor force alone, central parts alone and tensor plus central
force all together.   It is concluded that the tensor force is
significant in the analysis of the experimental results for the
0⁺ → 0⁻ transition.

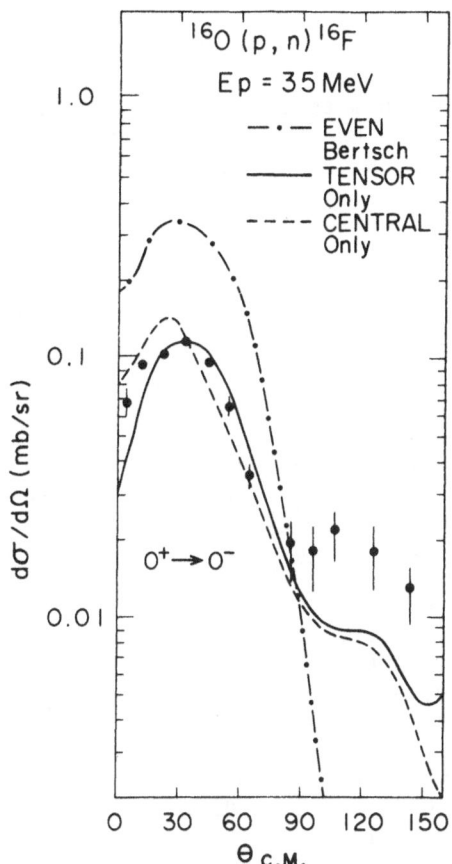

Fig. 6. Differential cross section for 0^- T=1 level in $^{16}O(p,n)^{16}F$. Curves are DWBA predictions obtained by three kinds of the effective interaction. Calculated cross sections are multiplied by a factor 0.71.

SUMMARY

We observed 4^- and 6^- stretched states in the daughter nuclei of light N=Z nuclei. As summarized in Table 1, the energies of the states correspond to analogs of 4^- and 6^- states observed in (e,e') and (p,p'). The observed strengths are much smaller than those calculated using simplified wave functions. The fact that the same strength discrepancy holds for (p,n), (e,e'), and (p,p') suggests that it should not be attributed to the choice of effective inter- action. The $0^+ \to 0^-$ transition is described by the DWBA prediction with the tensor interaction alone.

ACKNOWLEDGEMENTS

     The author is grateful to Professor T. Ishimatsu for his sup-
port in these works. The author wishes to thank Dr. S.-I. Hayakawa
for his helpful discussion. The work has been performed in collaboration with Messers. S. Nishihara, K. Furukawa, K. Miura and
G. C. Kiang, and Drs. T. Murakami, K. Maeda, T. Nakagawa and H.
Ohnuma.

REFERENCES

1.   I. S. Towner and F. C. Khanna, Phys. Rev. Lett. 42, 51 (1979).
2.   J. Speth, V. Klemt, J. Wambach, and G. E. Brown, Nucl. Phys.
     A343, 382 (1980).
3.   M. Rho, Nucl. Phys. A354, 3C (1981).
4.   E. Oset and M. Rho, Phys. Rev. Lett. 42, 47 (1979).
5.   A. Bohr and B. R. Mottelson, Phys. Lett. 100B, 10 (1981).
6.   G. E. Brown and M. Rho, Nucl. Phys. A372, 397 (1981).
7.   F. Petrovich, in The (p,n) Reaction and the Nucleon-Nucleon
     Force, edited by C. D. Goodman et al. (Plenum, New York, 1980),
     p. 115 and references therein.
8.   H. Orihara et al., Phys. Rev. Lett. 47, 301 (1981).
9.   C. Gaarde et al., Nucl. Phys. A369, 258 (1981).
10. R. A. Lindgren, W. J. Gerace, A. D. Bacher, W. G. Love, and
     F. Petrovich, Phys. Rev. Lett. 42, 1524 (1979) and references
     therein.
11. H. Orihara et al., Phys. Rev. Lett. 48, 469 (1982).
12. R. Madey et al., Phys. Rev. C 25, 1715 (1982).
13. H. Ohnuma et al., Phys. Lett. B112, 206 (1982).
14. C. A. Gagliardi, G. T. Garvey, and J. R. Wrobel, Phys. Rev.
     Lett. 48, 914 (1982).
15. E. G. Adelberger et al., Phys. Rev. Lett. 46, 695 (1981).
16. P. Guichon et al., Phys. Rev. C 19, 987 (1979).
17. H. Orihara and T. Murakami, Nucl. Instrum. Methods 188, 15
     (1981).
18. T. W. Donnelly, J. D. Walecka, I. Sick, and E. B. Hughes,
     Phys. Rev. Lett. 21, 1196 (1968).
19. R. Shaeffer and J. Raynal, Saclay Report No. CEA-R 4000.
20. G. Bertsch et al., Nucl. Phys. A284, 399 (1977).
21. F. Petrovich et al., Phys. Lett. 95B, 166 (1980).
22. Petrovich and W. G. Love, Nucl. Phys. A354, 499C (1981).
23. F. C. Barker et al., J. Phys. G7, 657 (1981).
24. G. Marie, G. J. Wagner, P. Doll, K. T. Knöpfle, and H. Breuer,
     Nucl. Phys. A299, 39 (1978).
25. R. S. Henderson et al., Aust. J. Phys. 32, 411 (1979).
26. D. B. Holtkamp et al., Phys. Rev. Lett. 45, 420 (1980).
27. G. S. Adams et al., Phys. Rev. Lett. 38, 1387 (1977).
28. T. W. Donnelly and G. E. Walker, Ann. of Phys. 60, 209 (1970).
29. H. Ohnuma and H. Orihara, Prog. Theor. Phys. 67, 353 (1982).
30. Sam M. Austin, op. cit. Ref. 7, p. 203.

SPIN FLIP 1^+ STRENGTH IN N=28 ISOTONES BY

PROTON INELASTIC SCATTERING AT 65 MeV

M. Fujiwara, S. Imanishi, Y. Fujita, S. Morinobu,
T. Yamazaki, K. Katori*, S.I. Hayakawa[+] and H. Ikegami

Research Center for Nuclear Physics, Osaka University
(Suita Campus), 10-1 Mihogaoka, Ibaraki, Osaka 567
JAPAN
* Laboratory of Nuclear Studies, Faculty of Science, Osaka
University, Toyonaka, Osaka 560, JAPAN
[+] Ashikaga Institute of Technology, Ashikaga 326, JAPAN

INTRODUCTION

Through studies of the Gamow-Teller giant resonance and spin
flip transitions, remarkable progress has been made in understanding the spin-dependence of the effective nucleon-nucleon interaction.
However, the long standing problem of the missing M1 strength in
medium and heavy nuclei still remains. The cleanest spin flip 1^+
(M1) state known to date in medium nuclei is the one at 10.22 MeV
of excitation in ^{48}Ca. This state was first reported by Steffen et
al. with a ^{48}Ca(e,e') experiment[1] and subsequently observed by us
with a ^{48}Ca(p,p') reaction.[2] In both experiments no other strong
1^+ states were found, and this state was considered to be a neutron
particle-hole state of $(f_{7/2}^{-1} f_{5/2})$ configuration. The M1 strength
deduced from the (e,e') data was about one third of the simple shell
model limit.[1] In the case of the (p,p') reaction, a distorted wave
calculation, which used Michigan State three range Yukawa interaction
(M3Y) and the simple $(f_{7/2}^{-1} f_{5/2})$ configuration for the 1^+ state, overestimated the experimental yields by a factor of four.[2] Furthermore,
two years ago Anderson et al.[3] reported observation of the analog of
the 10.22 MeV 1^+ state in ^{48}Ca via ^{48}Ca(p,n)^{48}Sc reaction and pointed
out that this analog at 16.8 MeV of excitation in ^{48}Sc carried only
about 30% of the shell model limit for $T_>$ Gamow-Teller strength.
A recent theoretical study of the 1^+ state in ^{48}Ca by Härting et
al.[4] emphasized the role of virtual Δ-hole excitations which led to
quenching of M1 strength at low excitation energies. However the

effects of core polarization on M1 quenching is expected to be
quite large as was pointed out by McGrory and Wildenthal,[5] and by
Arima.[6] Therefore, acquisition of good experimental strength
distributions of 1^+ states is imperative in order to understand the
M1 quenching in medium weight nuclei. As we see in the program of
this conference, spin flip 1^+ (M1) states in medium and heavy nuclei
are now vigorously studied by many groups using either electromagnetic or hadronic probes. We chose to study the 1^+ states in N=28
isotones of ^{48}Ca, ^{50}Ti, ^{52}Cr and ^{54}Fe by (p,p') reactions at 65 MeV.
In this report we wish to present some preliminary results from our
investigation.

EXPERIMENTAL PROCEDURE

    The experiment was performed using a 65 MeV proton beam from
the cyclotron[7] at the Research Center for Nuclear Physics. All targets used were self-supporting foils with thicknesses in the range
of 0.5~1 mg/cm^2. The inelastically scattered protons were momentum
analyzed with the high resolution spectrograph RAIDEN[8] and detected
with a 1.5 m long two-dimensional position-sensitive proportional
counter system.[9] An overall energy resolution of 15 keV was obtained. The absolute magnitude of the cross sections was determined by
comparing the measured elastic cross sections for ^{48}Ca, ^{50}Ti, ^{52}Cr
and ^{54}Fe targets with optical model calculations that reproduced the
carefully measured elastic cross sections at the same incident
energy.[10] The absolute errors of the cross sections were estimated
to be about 10%.

    Spectra were taken at angles from θ_{lab} = 6° to 70°. The
excitation energies of the states in each nucleus were determined
by calibrating the focal line of the spectrograph using well-
established low-lying levels. The accuracy of the excitation
energies determined in the present experiment was estimated to be
better than 10 keV.

RESULTS AND DISCUSSION

    Figure 1 shows angular distributions for the 1^+ state at 10.22
MeV in ^{48}Ca. The angular distributions for the 9 MeV state also
shown in the figure are almost identical to those for the 10.22 MeV
1^+ state. Thus, we suggest the 9 MeV state to be a 1^+ state. As
we have shown in a recent article,[2] the shape of the angular distributions for a 1^+ state is distinct from those for other states such
as 1^-, 2^+, 2^-, etc. The distorted wave calculation for the 10.22
MeV 1^+ state in ^{48}Ca using the M3Y interaction reproduces the shape
of the experimental angular distributions fairly well. Therefore,
in the present work we adopted the shape of the angular distributions for the 10.22 MeV 1^+ state in ^{48}Ca as the standard shape for
identifying 1^+ states in N=28 isotones. Figure 2 shows proton

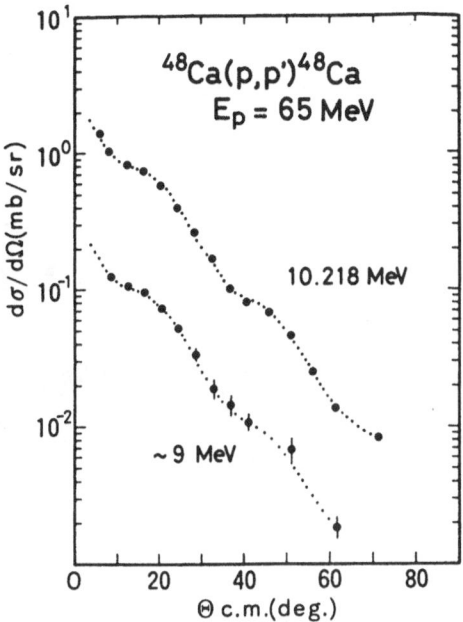

Fig. 1. Experimental angular distributions for 1$^+$ state in ^{48}Ca. Dotted lines are drawn to guide the eye.

spectra from ^{50}Ti, ^{52}Cr, ^{54}Fe (p,p') reactions, respectively. Although the data analysis to decompose partially overlapping peaks seen in the spectra is still in progress, we found that the 8.578, 9.614, 10.163, 10.205 and 10.364 MeV states in ^{50}Ti, the 8.181, 9.143 and 9.221 MeV states in ^{52}Cr, the 7.935 and 9.246 MeV states in ^{54}Fe are of $J^\pi = 1^+$. The other states denoted by their energy values are possible candidates for $J^\pi = 1^+$ states. Typical examples of the angular distribution for the 1$^+$ states are shown in Figs. 3 and 4, where the shapes are compared with that of the 1$^+$ state at 10.22 MeV in ^{48}Ca (denoted by dotted lines in figures).

Among these 1$^+$ states observed in the present experiment, the 1$^+$ assignment for the 9.143 MeV and 9.221 MeV states in ^{52}Cr has already been established by a nuclear resonance fluorescence (NRF) experiment.[11,12] The NRF experiment[12] has shown that the total M1 strength carried by the above two states at 9.143 and 9.221 MeV in ^{52}Cr is about (48±15)% of the experimental value of B(M1) for the transition to the 10.22 MeV 1$^+$ state in ^{48}Ca. In our (p,p') experiment the observed total strength for the two states in ^{52}Cr is about (35±5)% of that for the 10.22 MeV 1$^+$ state in ^{48}Ca. Considering the experimental uncertainties in both the NRF and (p,p') experiments, the agreement between the independent measurements is fairly good.

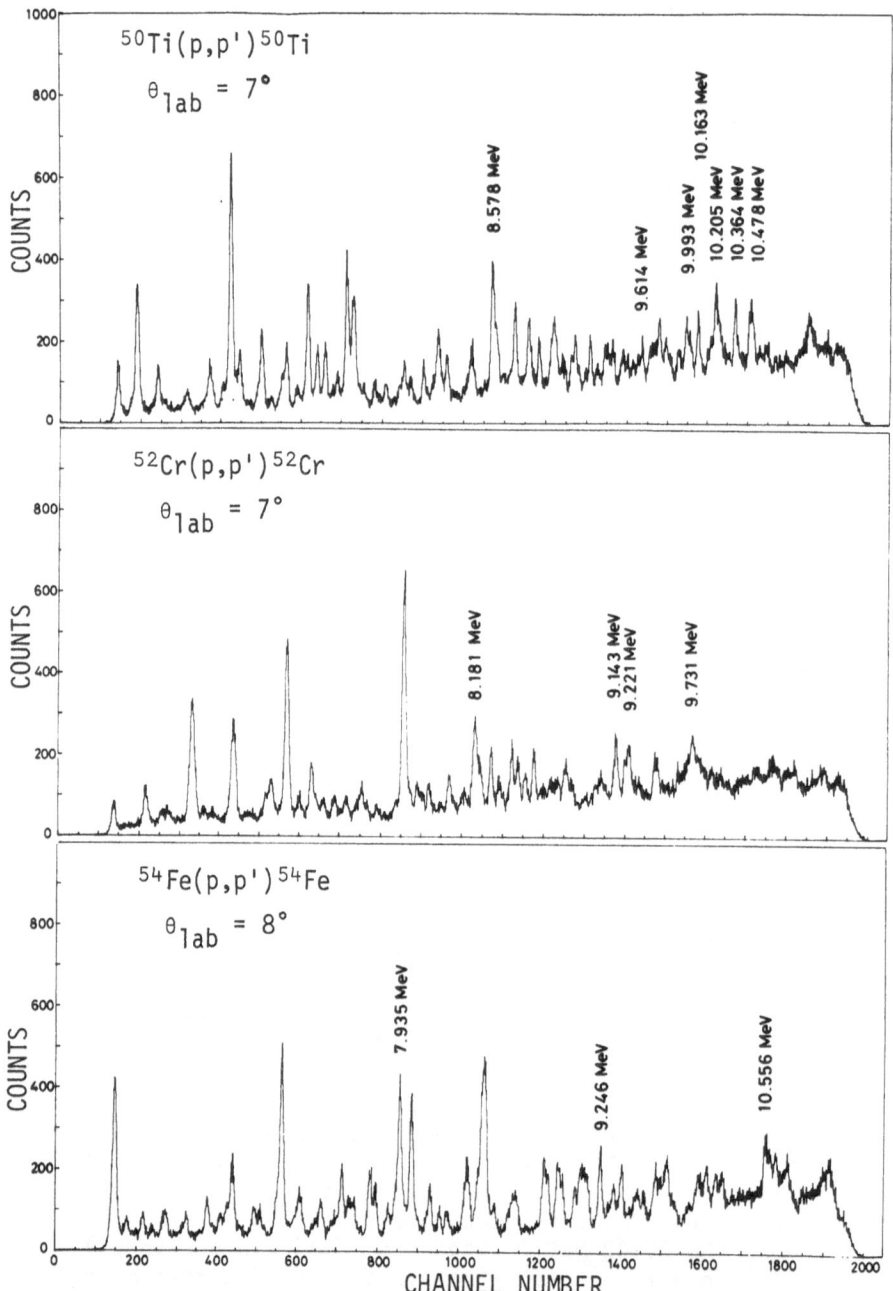

Fig. 2.   Proton spectra for the ^{50}Ti, ^{52}Cr and ^{54}Fe(p,p')
          reactions at 65 MeV. The excitation energies are given
          only for 1^+ states and possible candidates for 1^+ states.

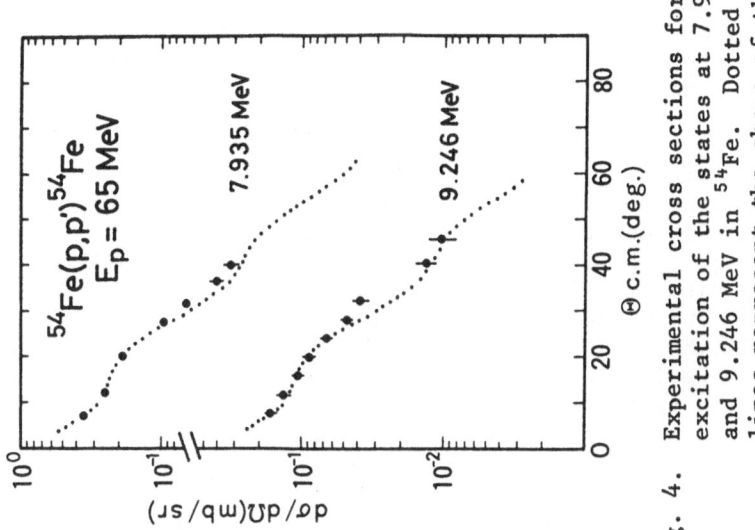

Fig. 4.   Experimental cross sections for excitation of the states at 7.935 and 9.246 MeV in ^{54}Fe. Dotted lines represent the shape of the angular distribution of the cross sections of the 1⁺ state at 10.218 MeV in ^{48}Ca.

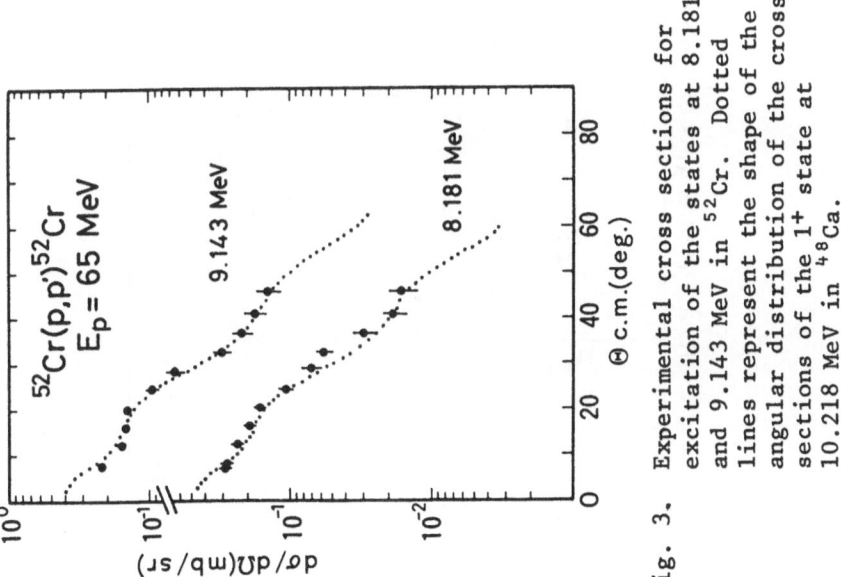

Fig. 3.   Experimental cross sections for excitation of the states at 8.181 and 9.143 MeV in ^{52}Cr. Dotted lines represent the shape of the angular distribution of the cross sections of the 1⁺ state at 10.218 MeV in ^{48}Ca.

Similar agreement in transition strength between the (p,p')
and the (e,e') experiment is also observed in the case of [50]Ti.
The (e,e') experiment at Darmstadt[13] shows that the 1[+] states in
the excitation energy region around E_x=10 MeV carries about 75% of
the M1 strength of the strong 1[+] state in [48]Ca. Our preliminary
analysis for the four 1[+] states at the 9.614, 10.163, 10.205 and
10.364 MeV in [50]Ti sums up to about 50% relative to that of the
strength of the strong 1[+] state in [48]Ca. If the strengths of other
possible 1[+] states at 9.993 and 10.478 MeV were taken into account,
the total strength of about 70% is expected, which is compatible
with the (e,e') results.

In the case of [54]Fe we have rather strongly excited 1[+] states
at 7.935 MeV and 9.246 MeV. The strengths of the states at 7.935
MeV and 9.246 MeV in [54]Fe are about 29% and 13% relative to the
strength of the 10.218 MeV 1[+] state in [48]Ca. It should be noted
that these two states seem to be the parent states of the two
states at E_x~8 MeV and ~9 MeV in [54]Co, which are seen in the zero
degree [54]Fe(p,n)[54]Co spectrum at 160 MeV,[14] and also seen in the
forward (p,n) spectra at 35 MeV.[15]

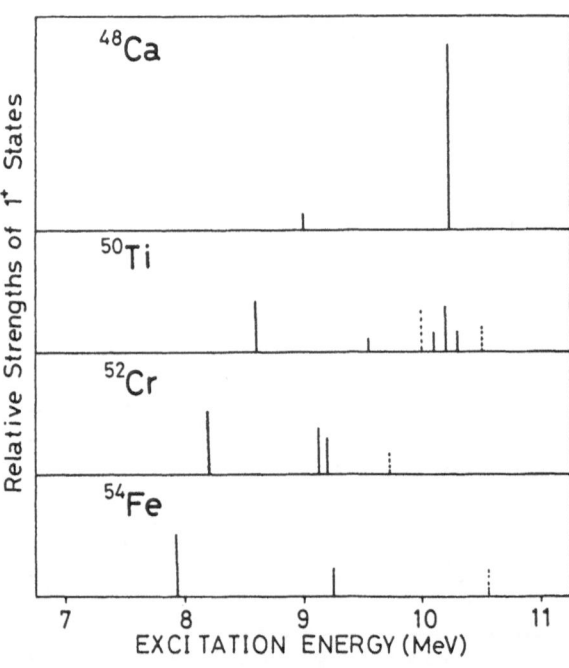

Fig. 5.   Relative strength of the 1[+] states observed in the
          present experiment. The data analysis is still in
          progress in order to find weakly excited 1[+] states.

In Fig. 5, the experimental 1^+ strength distributions so far obtained are summarized for N=28 isotones of ^{48}Ca, ^{50}Ti, ^{52}Cr and ^{54}Fe. Although there is a possibility that we missed other weakly excited 1^+ states, we can make some remarks on the systematics of the 1^+ strength distributions.

The excitation energies of the lowest spin flip 1^+ states (at 9 MeV in ^{48}Ca, 8.578 MeV in ^{50}Ti, 8.181 MeV in ^{52}Cr and 7.935 MeV in ^{54}Fe) change systematically as can be seen in the figure. The strength increases according to the Z number. Considering the actual addition of extra protons in $(\pi f_{7/2})$ shell in the case of ^{50}Ti, ^{52}Cr and ^{54}Fe, these 1^+ states seem to have dominantly proton particle-hole $\pi(f_{7/2}^{-1}, f_{5/2})$ component. In the case of ^{54}Fe the analysis is still incomplete and there is a strong possibility that more 1^+ states may exist at the excitation energies higher than 11 MeV. However, the fragmentation of 1^+ states observed in ^{50}Ti suggests the wave functions of the 1^+ states are complex. Thus, our present study indicates that ^{48}Ca is the best nucleus among the N= 28 isotones for detailed studies of M1 quenching. In ^{48}Ca our most significant finding is the identification of the 1^+ state at 9.0 MeV which has one tenth of the strength of the 10.22 MeV 1^+ state. This finding implies that there is a significant proton two particle-two hole component in the ground state of ^{48}Ca. This core polarization may play an important role in solving the M1 quenching problem in ^{48}Ca.

In summary, we identified several 1^+ states in N-28 isotones of ^{48}Ca, ^{50}Ti, ^{52}Cr, and ^{54}Fe. It was found that the relative strength of the 1^+ states deduced from the (p,p') hadron scattering experiment at 65 MeV corresponded to those deduced from electromagnetic (e,e') and (γ,γ') experiments. Clearly, high resolution (p,p') experiments would be very useful in understanding the magnitude and fragmentation of the M1 strength in nuclei.

ACKNOWLEDGEMENT

We are grateful to the RCNP cyclotron crew for the smooth operation of the cyclotron. This experiment was performed at Research Center for Nuclear Physics, Osaka University under Program No. 13A01.

REFERENCES

1. W. Steffen, H.-D. Gräf, W. Gross, D. Meuer, A. Richter, E. Spamer, O. Titze and W. Knüpfer, Phys. Lett. 95B, 23 (1981).
2. Y. Fujita, M. Fujiwara, S. Morinobu, T. Yamazaki, T. Itahashi S. Imanishi, H. Ikegami and S.I. Hayakawa, Phys. Rev. C25, 678 (1982).
3. B.D. Anderson, J.N. Knudson, P.C. Tandy, J.W. Watson, R. Madey and C.C. Foster, Phys. Rev. Lett. 45, 699 (1980).

4. A. Härting, W. Weise, H. Toki and A. Richter, Phys. Lett. 104B, 261 (1981).

5. J.B. McGrory and B.H. Wildenthal, Phys. Lett. 103B, 173 (1981).

6. A. Arima, these proceedings.

7. M. Kondo et al., Research Center for Nuclear Physics Annual Report, 1976.

8. H. Ikegami, S. Morinobu, I. Katayama, M. Fujiwara and S. Yamabe, Nucl. Instr. and Meth. 175, 335 (1981).

9. Y. Fujita et al., in Proceedings of the Symposium on Nuclear Radiation Detectors, Tokyo, Nucl. Instr. and Meth. (to be published).

10. H. Sakaguchi, M. Nakamura, K. Hatanaka, A. Goto, T. Noro, F. Ohtani, H. Sakamoto and S. Kobayaski, Phys. Lett. 89B, 40 (1979); Noro et al., Nucl. Phys. A366, 189 (1981).

11. N. Kumagai, T. Ishimatsu, E. Tanaka, K. Kageyama and G. Isoyama, Nucl. Phys. A329, 205 (1979).

12. U.E.P. Berg, D. Rück, K. Ackermann, K. Bangert, C. Bläsing, K. Kobras, W. Naatz, R.K.M. Schneider, R. Stock and K. Wienhard, Phys. Lett. 103B, 301 (1981).

13. A. Richter, Nucl. Phys. A374, 177c (1982).

14. C. Goodman, Nucl. Phys. A374, 241c (1982).

15. H. Orihara, T. Murakami, S. Nishihara, T. Nakagawa, K. Maeda, K. Miura and H. Ohnuma, Phys. Rev. Lett. 47, 301 (1981).

DYNAMICAL THEORY OF SPIN EXCITATIONS

J. Speth, S. Krewald, and F. Osterfeld

Institut für Kernphysik
Kernforschungsanlage Jülich
D-5170 Jülich, West Germany

Toru Suzuki

Niels Bohr Institute
Blegdamsvej 17
DK-2100 Copenhagen Ø, Denmark

ABSTRACT

We investigate in a systematic way magnetic resonances in
light, medium and heavy mass nuclei. The numerical calculations
have been performed within the framework of the extended theory of
interacting Fermi systems. Here we incorporate explicitly the π-
and ρ-exchange potential in a generalized spin-dependent particle-
hole interaction. In addition we take into account the effect of
the "dynamical theory of collective states" on the single particle
energies (energy dependence of the effective mass). Special emphasis is given to the question whether the $\Delta(33)$-resonance is responsible for the missing magnetic sum rule strength which experimentally was searched for in electron scattering and hadronic
charge-exchange reactions with highly energetic protons. We point
out that the inclusion of the (Pauli)-exchange terms in the
particle-hole interaction strongly reduces the $\Delta(33)$-hole quenching
effect.

INTRODUCTION

One of the most exciting questions in recent nuclear physics
is whether subnuclear degrees of freedom might play a significant
role in the low energy nuclear excitation spectrum. The most prom-

inent subnuclear degree of freedom is the $\Delta(33)$-resonance which is located approximately 300 MeV above the nucleon mass and which plays a dominant role in intermediate energy physics. A careful analysis of charge-exchange reactions suggests that the $\Delta(33)$-resonance may have to be taken into account explicitly even at nuclear excitation energies of only a few MeV. Using intermediate energy protons (E_p = 160 MeV), the Gamow-Teller resonance was recently discovered in (p,n) charge-exchange reactions at the Indiana Cyclotron[1-4]. The spin-isospin strength, summed over all states, was systematically found to exhaust only approximately 50 % of the Ikeda sum rule strength[5]. Likewise, in inelastic electron scattering and in proton scattering, only a small fraction of the expected Ml-strength has been found so far, particularly in heavy nuclei. Conventional nuclear structure effects, such as many-particle many-hole excitations would conserve and only redistribute the strength. Since 50 % of the strength is missing, however, sub-nuclear degrees of freedom were suggested to play an important role in the quenching of $\sigma\tau$-strength[6-18,31,59]. The $\Delta(33)$-resonance coupled with a nucleon hole (Δ-h), can move part of the isovector $\sigma\tau$-strength into an energy region approximately 300 MeV above the low-lying excitations of the nucleus. The major reason why this mechanism may have a significant effect despite this enormous energy gap is due to the Pauli principle. Since there is no Pauli blocking for the $\Delta(33)$-resonance, virtually all nucleons can participate in building Δ-h states. Therefore the sheer number of possible configurations is able to bridge the energy gap. A polarization effect due to the $\Delta(33)$-resonance was invoked to explain the quenching of the axial vector coupling constant g_A.

In the present contribution we develop a microscopic model which incorporates explicitly the π- and ρ-exchange potential which gives rise to a momentum dependent particle-hole (ph)-interaction. We also include in our theory the effects of the "dynamical theory of collective states[55,57] on the single particle energies". Furthermore, we extend the conventional RPA theory in order to include the Δ-hole configurations in a consistent way. In all cases we compared our theoretical results with the corresponding experimental values.

THEORY OF COLLECTIVE STATES

The Spin-Isospin Dependent Particle-Hole Interaction

There exist only a few many body theories which have been applied to realistic problems. The most successful theory in this respect is Landau's theory of "interacting Fermi systems" and its extension to finite systems by Migdal[19,20]. Starting from the two-particle Green's function one is able to derive e.g. equations which allow one to calculate the excitation energies and the

transition amplitudes of collective states which are in principle
exact. The complications of the many body system, however, are
hidden in a highly renormalized particle-hole (ph) interaction.
Landau and Migdal suggested not to calculate these complicated
expressions but to expand them in terms of Legendre polynomials.

With this procedure one introduces a few free parameters which
have to be extracted from experiments. Basically one is able within
this theory to connect different experimental facts with each other.
As one example we mention isotope shifts and the excitation energy
of the breathing mode. From the investigation of the isotope shifts
in the lead region one deduced the corresponding interaction parameter (f_0 in eq. (2)) which afterwards has been used to calculate
the breathing mode. With this procedure one was able to predict the
excitation energy correctly before it has been found experimentally[21]. In our contribution we shall restrict the discussion on the
excitation energies and transition amplitudes of states in even-
even nuclei. In this connection we have to solve the following
equation

$$\chi^{\mu}_{\nu_1\nu_2} = \frac{n_{\nu_1}-n_{\nu_2}}{\varepsilon_{\nu_1}-\varepsilon_{\nu_2}-\Omega_{\mu}} \sum_{\nu_3\nu_4} F^{ph}_{\nu_1\nu_3,\nu_2\nu_4} \chi^{\mu}_{\nu_3\nu_4} \tag{1}$$

Here, ε_{ν} are the single-particle energies and Ω_{μ} the excitation
energy of the collective state μ. The transition amplitudes χ^{μ} are
directly connected with the transition probability and the scattering cross section, respectively. In the original Landau-Migdal
theory the ph-interaction F^{ph} is expanded in Legendre polynomials.
This expansion is restricted in nearly all cases to the zeroth
order which gives rise to the following expression:

$$F^{ph}_0(\vec{r},\vec{r}') = C_0\{f_0+f'_0\vec{\tau}\cdot\vec{\tau}' + g_0\vec{\sigma}\cdot\vec{\sigma}' + g'_0\vec{\sigma}\cdot\vec{\sigma}'\vec{\tau}\cdot\vec{\tau}'\}\delta(\vec{r}-\vec{r}') \tag{2}$$

C_0 is the inverse of the density of states at the Fermi surface

$$C_0 = \frac{\pi^2\hbar^2}{k_F\cdot m^*} = 302\,\frac{m}{m^*}\,\left[MeV\cdot fm^3\right]\,. \tag{3}$$

This ansatz turned out to be insufficient for unnatural parity
states. We start from the meson exchange picture and choose a
generalized spin-dependent ph-force which includes in addition to
the zero-range part given in eq. (2) also contributions which are
due to the OPEP and also to the one-rho-exchange potential. To
simplify our discussion we restrict ourselves in the following to
the direct part of the spin-isospin interaction (all numerical
calculations have been performed in the way described in ref. 24,

i.e. including the Pauli exchange). This force can be expressed as:

$$F^{ph}_{\sigma\tau} = g'_o - \frac{4\pi f^2_\pi}{m^2_\pi} \frac{\vec{\sigma}\cdot\vec{q}\vec{\sigma}'\cdot\vec{q}}{q^2+m^2_\pi} - \frac{4\pi f^2_\rho}{m^2_\rho} \frac{(\vec{\sigma}x\vec{q})\cdot(\vec{\sigma}'x\vec{q})}{q^2+m^2_\rho} \qquad (4)$$

Since the OPEP and ρ-exchange potential are expected only as a
guide, the coupling parameters are in principle free parameters. It
might be thought that effects from ρ-exchange should be included in
the zero-range parameters, since the ρ-mass is large and the ρ-
interaction, therefore, short-ranged. However, our particle-hole
interaction (eq. (2)) should be generalized to include tensor in-
variants. In practice, these arise almost completely from the OPEP
and ρ-exchange potentials. Introducing these potentials explicitly,
we avoid the complications of tensor terms in eq. (2). Furthermore,
the tensor force from the ρ-exchange potential cuts off that from
the π-exchange at short distances which is of crucial importance
in connection with the precritical phenomena and pion condensation[22,23].

     In Fig. 1 the graphical representation of eq. (4) is given.
Here we have separated the OPEP and ρ-exchange potential into a
central part and tensor part. The zero range term in eq. (2) is a
constant in the Fourier space (g'_o = const). The central part and
tensor part of the π-exchange (thin full line and thin dashed line)

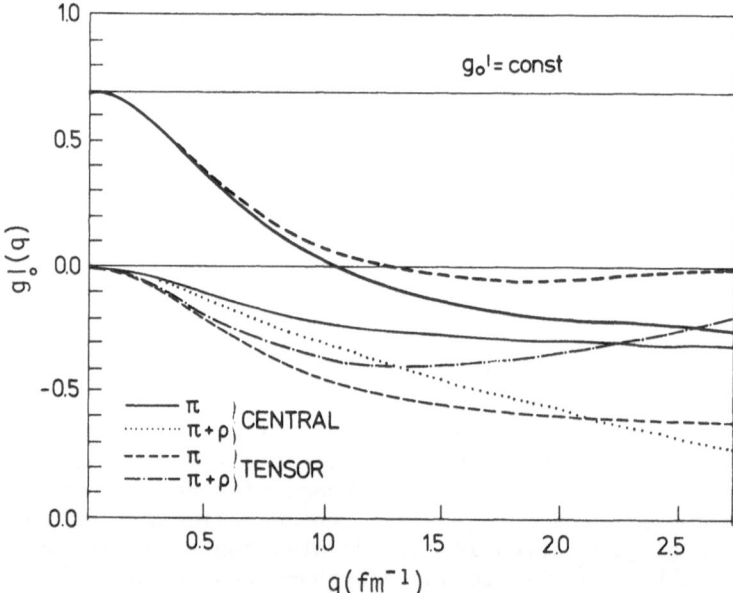

Fig. 1. Graphical representation of the generalized spin-dependent
         interaction in momentum space (eq. (4)).

and of the π- plus ρ-exchange (dotted line and dash-dotted line)
are shown separately. The thick full line is the sum of all contributions. The thick dashed line is again the sum of all contributions, however, with a central part of the ρ-contribution which
is multiplied by a factor of 0.4. The reason for this reduction
factor is due to the very short-ranged ρ-exchange which gets
appreciably modified by the short-range repulsive correlations of
the ω-exchange potential. Following ref. 25, we approximate this
effect by multiplying the ρ-coupling constant by 0.4: $f_\rho^2 = 0.4\ f_\rho^2$.
This factor is not applied to the tensor piece, because the tensor
couples only relative D-states with relative S-states. Therefore
the tensor contribution is only little influenced by the repulsive
short-range correlations.

More recently we have used[31] a more sophisticated version of
the ph-interaction which is of the form:

$$F_{\sigma\tau}^{ph}(q) = \int \frac{d^3k}{(2\pi)^3}\ \Omega(\vec{q}-\vec{k})\left[V_\pi(\vec{k})+V_\rho(\vec{k})\right]\ +\ \delta g_o' \cdot C_o \cdot \vec{\sigma}\cdot\vec{\sigma}'\vec{\tau}\cdot\vec{\tau}' \qquad (5)$$

Here the (bare) one-pion (V_π) and one-rho (V_ρ) exchange potential
is included explicitly. The effect of the other mesons, on the
other hand, is summarized in a two-body correlation function,
$\Omega(\vec{q}-\vec{k})$. In the actual calculation the exact correlation function
$\Omega(\vec{q}-\vec{k})$ is replaced by its dominant Fourier[52] component:

$$\Omega(\vec{q}-\vec{k}) = (2\pi)^3\delta(\vec{q}-\vec{k}) - (\frac{2\pi^2}{q_c^2})\ \delta(|\vec{q}-\vec{k}|-q_c) \qquad (6)$$

with q_o = 3.93 $[fm^{-1}]$, which is of the order of the ω-meson mass.
The momentum dependence of the force of eq. (5) is very similar to
that shown in Fig. 1, which might be considered as a nice justifi-
cation of the previous ansatz (eq. (4)). Since the first term of
eq. (5), in the limit q=0, leads to a Landau parameter $g_o'(q=0)$
= 0.67 as compared to 0.93 obtained in a previous analysis[24], we
added a phenomenological zero range interaction $\delta g_o'$. From the investigation of ref. 34 one knows that the major contribution to
$\delta g_o'$ follows from the second-order effects of the tensor force. In
the following calculations the quantity $\delta g_o'$ is deduced from
experiments.

In ref. 24 the influence of the momentum dependence of the
force on magnetic properties has been investigated. The effect on
a given state can be discussed qualitatively by considering the
corresponding form factors, since the largest (diagonal) contributions to the direct part of the ph-force are simply a double
convolution of the transition density with the interaction. As an

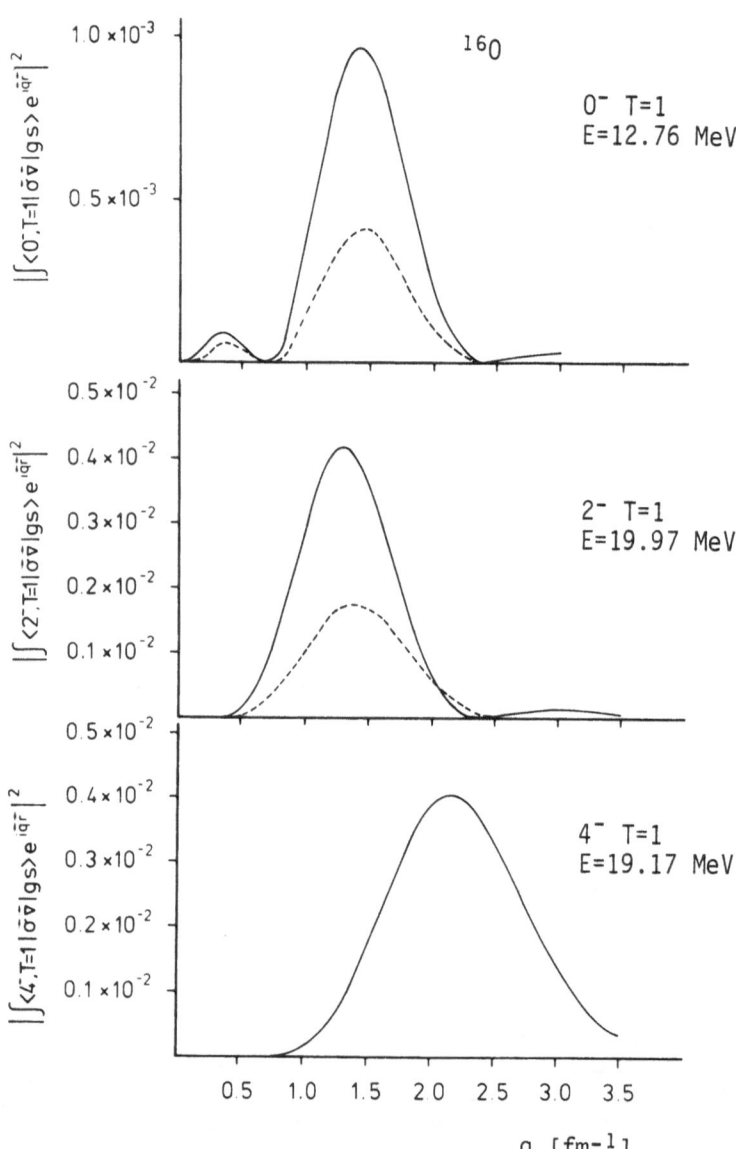

Fig. 2. Fourier transform of the $\vec{\sigma}\cdot\vec{\nabla}$ transition density of
         various unnatural parity states of ^{16}O. The dashed lines
         are the proton contributions of the strongest particle-
         hole configuration.

example we show in Fig. 2 three "pionic" states of ^{16}O. These
states are mainly sensitive to the large momentum transfer behaviour of the ph-interaction. It is obvious from Fig. 1 that in this
momentum region the interaction is weak compared to the zero range
force (g_0' = const in Fig. 1). This fact can also be seen from the
2^- strength distribution for ^{16}O shown in Fig. 3. The distributions shown in Fig. 3a and 3c are the result of RPA calculations
using the force of eq. (4) and a zero range force, respectively.
Here one notices that the major part of the 2^- strength is concentrated around 24 MeV if one uses a zero range force, and it is
shifted to 20 MeV with the force of eq. (4). This theoretical result
is in fair agreement with the experimental situation shown in Fig.
3d. For comparison also the uncorrelated ph-spectrum is shown in
Fig. 3b (F^{ph} = 0). Here one realizes that the uncorrelated strength
is already concentrated around 12 and 18 MeV. The ph-interaction
shifts that strength to higher energies of about 1 and 2 MeV,
respectively (Fig. 3a). Therefore it is clear that the inclusion of
the π^- and ρ-exchange potential reduces the strongly repulsive
zero-range part of that force appreciably.

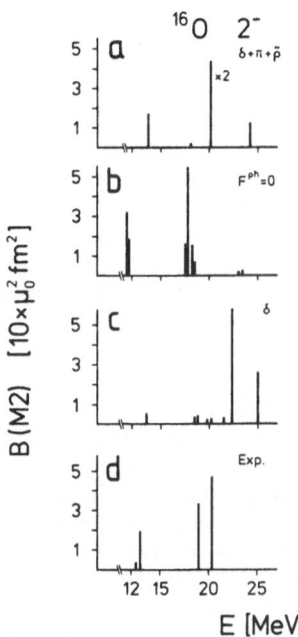

Fig. 3. Strength distribution of the 2^- state in ^{16}O calculated
         with a pure zero range force (c) and zero range plus π-
         and ρ-exchange (a). For comparison the uncorrelated ph-
         result (b) and the experimental data (d) are given.

Dynamical Theory of Collective States

        In order to solve the RPA equations one needs in addition to
the ph-interaction also single particle energies and single particle
wave functions. The calculations of the various groups differ
mainly in the manner in which these input data are determined. Here
one distinguishes between two different methods: shell model RPA
and selfconsistent RPA. There is one basic difference between these
two methods which is basically connected with the single particle
spectrum. It is well known that the single particle spectrum of
medium and heavy mass nuclei calculated within the HF-approach
deviates appreciably from the empirical one. The empirical low-
lying spectrum is more compressed than the theoretical one.
Actually the theoretical result depends on the effective mass used
in a specific HF calculation. Realistic forces derived from a
Brueckner-G-matrix give an effective mass in nuclear matter of
$m^*/m = 0.6-0.7$ whereas the empirical spectrum corresponds in the
average to $m^*/m \gtrsim 1$. It has been shown by Bertsch and Kuo[53] and
Hamamoto and Siemens[54] that the coupling of the single particle
(hole) states to nuclear vibrations gives rise to a compression of
the single particle spectrum which explains essentially the
differences between the HF and the empirical spectrum. The nucleon
effective mass and correspondingly the single particle spectrum
can be regarded as made up of two pieces, as shown in Fig. 4. The
first part (Fig. 4a) is the HF-contribution of the effective
interaction, and the second part (Fig. 4b) is the phonon contribution. In the framework of the selfconsistent RPA only the first
piece is included in the single particle energies. In the shell
model RPA, on the other hand, where one uses an empirical single
particle spectrum (which corresponds to $m^*/m \sim 1$), both contributions are included implicitly. Therefore this version of the RPA
includes (in principle) also processes of the form shown in Fig. 5.
In the original version of the dynamical theory of collective
states, Brown and Speth[57] and Brown, Dehesa and Speth[55] have argued
that the phonon contribution (Fig. 4b) is different for high-lying
and low-lying single particle states. Therefore it is not possible
to include this piece through a constant m^*/m but one has to
introduce an energy dependent effective mass.   This energy
dependence, however, introduces, on the other hand, a serious
problem if such a spectrum is used within the RPA (see Fig. 5).
Here, in general, the phonon contributions are off energy which
means that the corresponding contributions of Figs. 4b and 5 are
different from each other. This difference is small for the low-
lying collective resonances and non-collective states (which are
close to the unperturbed ph-energies). Therefore it is necessary
and meaningful to use here the empirical single particle spectrum.
In the case of giant resonances, however, the compression effect
due to the phonons is much smaller than for low-lying states.
Therefore one has to use a single particle spectrum which is close

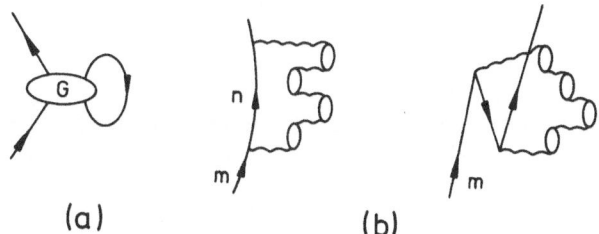

Fig. 4. Interactions leading to the nucleon effective mass:
(a) The Hartree-Fock contribution. (b) The contributions
arising from coupling to collective modes.

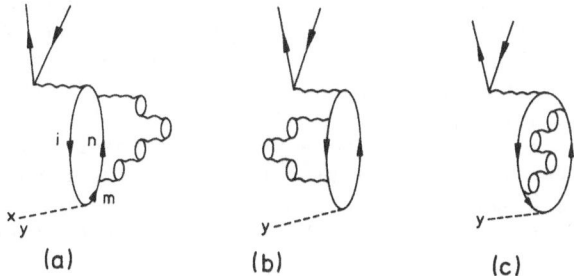

Fig. 5. Coupling of collective modes to a resonance which is
viewed as being started off by an external field denoted
by γ.

to the HF result. In order to simulate this energy dependence the
authors suggested an effective mass which depends on the energy of
the vibration investigated. A more elaborated version of this
approach has been recently given by Brown and Rho (see G.E. Brown's
contribution to this conference). In the original version of the
dynamical theory the authors already pointed out that the energy
dependence introduced through the phonon coupling has to be taken
into account explicitly. The results of such an extended theory
justify the essential features of the dynamical theory (see the
contributions by Bortignon and Wambach to this conference). The
choice of the single particle energies is important for the GT-
resonances in heavy mass nuclei, as we will discuss in the next
section.

MICROSCOPIC DESCRIPTION OF SPIN MODES

Magnetic High-Spin States

A further important test of the high-q behaviour of the
effective interaction is provided by the magnetic high spin states
discovered recently in ^{208}Pb by inelastic electron scattering at
backward angles[26]. These states are of considerable physical
interest because, as a consequence of the high multipolarity, the
cross sections are peaked at a momentum transfer of approximately
$q \sim 2$ $[fm^{-1}]$. Therefore these states are an excellent test of the
high momentum behaviour of our generalized spin- and isospin-
dependent interaction. The number of 1p-1h configurations which
can contribute to these states is severely restricted by the high
multipolarity. Since the experimental excitation energies are close
to the shell model ph-energies, the 12$^-$ state at 6.43 MeV and the
14$^-$ state at 6.74 MeV were tentatively interpreted as pure
$\nu(1j_{15/2}, 1i_{11/2}^{-1})$ ph-excitations while the 12$^-$ state at 7.06 MeV was
assumed to be a pure $\pi(1i_{13/2}, 1h_{11/2}^{-1})$ configuration. This simple
interpretation faces one problem, however, because the experimental
cross section is only 50 % of the single ph-prediction[26]. In the
following we will show that (i) the effects of the OPEP and ρ-
exchange potential give rise to a very weak interaction in this
momentum transfer region, and (ii) that the fragmentation of the
single particle strength due to the phonon coupling is mainly
responsible for the reduction of the cross section (see ref. 27).

It is well known that the coupling to the phonons may modify
the single particle states appreciably; e.g., the $j_{15/2}$ state comes
at an excitation energy of 1.42 MeV relative to the ground state
of ^{209}Pb which is only 1.2 MeV below the $(3^- x \nu 2g_{9/2})_{15/2-}$ configuration. Therefore a considerable mixing of these configurations
has to be expected, which strongly reduces the single particle
strength. This effect is especially large for the so-called spin-
orbit partners which are shifted into the next lower major shell
and which possess therefore the "wrong" parity. All the dominant
configurations of the high spin states are of that special type.

We evaluate the single particle strength by taking into
account explicitly the coupling of phonons in ^{208}Pb to single
particle states, thus obtaining quasi-particle states in the
neighbouring nuclei. These quasi-particle and quasi-hole states
are used to construct a core coupling random-phase wave function
containing the most relevant 2p2h configurations. Details are
given in ref. 27. We mention, however, that we include in the
present case also the interaction between the 2p2h configuration.

In Fig. 6, the inelastic electron scattering cross sections
at $\vartheta = 90^0$ and $\vartheta = 160^0$ are shown for the three magnetic high spin
states. The calculations were performed in Distorted Wave Born

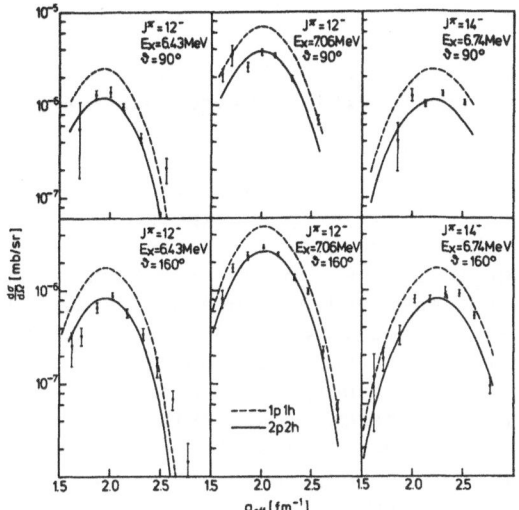

Fig. 6. Inelastic electron scattering cross sections at $\Theta = 90^0$ and $\Theta = 160^0$ of the magnetic high spin states in ^{208}Pb. Experimental cross sections are compared with the RPA calculations (dashed lines) and those including a 2p2h configuration (solid lines). Calculations have been done in DWBA.

Approximation, using the code HEIMAG by J. Heisenberg[29]. The inclusion of 2p2h configurations reduces the cross sections, as expected, so that both shapes and absolute magnitudes of the cross sections are in good agreement with the experimental data in the three cases considered. On a closer inspection, however, one finds that at high momentum transfer, the cross sections of the 12⁻ state at 6.43 MeV and the 14⁻ state at 6.75 MeV are underestimated. In this calculation, the neutron rms-radius has been adjusted to ref. 30.

These calculations clearly prove that we can trust our particle-hole interaction derived from the meson exchange picture up to large momentum transfers of about q = 2.3 fm^{-1}. The structure of this interaction completely rules out any trace of so-called pre-critical phenomena[13,31,32]. No enhancement of magnetic strength is seen, to the contrary, one finds less strength than expected from sum rules. This has been seen now in magnetic states in electron scattering and even more clearly in charge-exchange reactions. Therefore, we now focus our attention to a review of the micro-scopic description of charge-exchange reactions.

Spin-Isospin Modes

Only very recently collective spin-isospin modes have been discovered in (p,n) and (^{3}He,t) charge-exchange experiments[1-4] in medium and heavy mass nuclei. The most prominent of these newly discovered pionic states is the 1^+, $\Delta L=0$, $\Delta S=1$ giant Gamow-Teller resonance (GTR) which was already theoretically predicted many years ago by Ikeda et al.[5]. In (p,n) experiments at very forward angles and with highly energetic protons ($E_p > 100$ MeV) these resonances turn out to be the dominant reaction channel. This fact is connected with the energy-dependence of the isospin-dependent parts of the nucleon-target interaction V as derived by Petrovich and Love[33]: The strength of the τ-dependent part of the interaction is strongly reduced if the energy of the incoming proton is increased so that the excitation of non spin-flip states is very weak for $E_p > 100$ MeV. On the other hand, the $\sigma\tau$-force is nearly independent of the proton energy, therefore beyond $E_p > 100$ MeV mainly spin-flip states are excited in (p,n) reactions. At lower energies ($E_p \lesssim 40$ MeV) the two pieces are nearly equally strong. This surprising effect can be simply understood in terms of one- and two-boson exchange contributions to the interaction[34].

Examples of zero degree (p,n) spectra are shown in Fig. 7. The structure of GTR-states is very similar to the well-known isobaric analog states (IAR). Both resonances can be described in the framework of the RPA as a superposition of proton-particle, neutron-hole states. In the case of the IAR the particle-hole pairs are coupled to 0^+, in the GTR case to 1^+. Both kinds of states are expected to be rather collective in heavy mass nuclei. Since the GTR are connected with a spin- and isospin-flip they allow a selective investigation of the spin-isospin part of the ph-interaction. The same is true also for the 0^-, 1^- and 2^- states ($\Delta\ell = 1$, $\Delta S = 1$ resonances) which dominate the cross sections at slightly larger angles.

In the previous calculations we mainly could test the behaviour of the ph-interaction at larger momentum transfer. The $\Delta\ell = 0$ and $\Delta\ell = 1$ charge-exchange resonances in medium and heavy mass nuclei give us the possibility of testing the momentum dependence of the force at small momentum transfer. As examples we show in the lower part of Fig. 8 the form factors of the GTR and the $\Delta\ell = 1$ resonances in ^{208}Pb. It is obvious from that figure that in the GTR only very small momenta are involved which means that the GTR are sensitive to the strongly repulsive part of the $\sigma\tau$-interaction. In addition the dominant part of the form factor is peaked in a narrow range indicating that the excitation energy of the GTR is insensitive to details of the ph-force.

Results which are very similar to each other were obtained with a separable, constant force[14],[15], zero range force[4],[35], and the momentum dependent force of eq. (4) where also the effects of the dynamical theory were considered[14]. In all cases the major part of the GTR strength is concentrated in one single resonance which is shifted to higher energies compared to the unperturbed ph-energies.

On the other hand, however, the coupling to the $\Delta(1232)$-particle-nucleon hole configurations which modify strongly the transition strength depends sensitively on the form of the interaction used, as will be discussed in the next section.

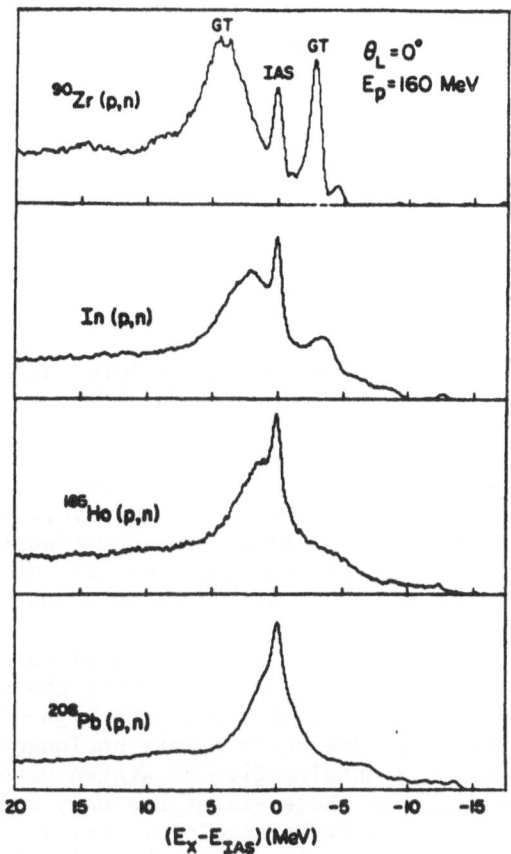

Fig. 7.   Zero degree (p,n) spectra of several nuclei of 160 MeV
          protons. The spectra are plotted on an excitation energy
          scale centered on the isobaric analog state[49].

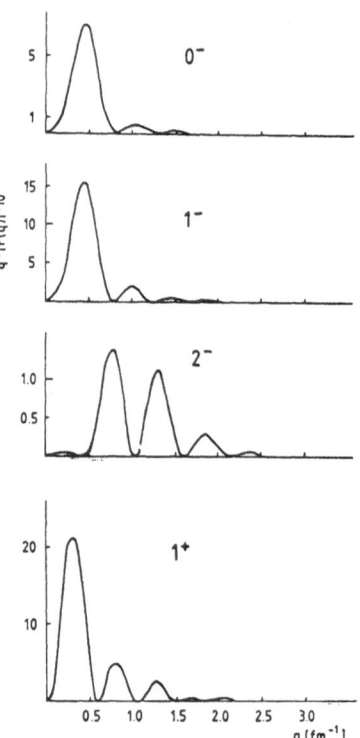

Fig. 8.   Fourier transform of the (p,n) form factors of $\Delta\ell = 0$
(GTR) and $\Delta\ell = 1$ resonances in ^{208}Pb.

     Figure 8 also shows the form factor of the 0^-, 1^- and 2^- reso-
nance. Those of the 0^- and 1^- resonance are similar to the GTR one,
but the maximum is shifted to higher momentum transfer.   Therefore
the repulsion of the ph-force is weaker than in the GTR case[36].
Nevertheless it is still strong enough to build up one single
collective state in which the major fraction of the 0^- and 1^-
strength is concentrated. The 2^- result is qualitatively different.
From the form factor one already realizes that the ph-force in
this specific case should be weak due to the high q-components.
Actually it is that weak that there exists no longer a single
collective state but one obtains five 2^- states with roughly the
same strength. In addition those states are only slightly shifted
from their uncorrelated particle-hole energies. In Fig. 9 a microscopic analysis of (p,n) experiments in ^{208}Pb is shown[36]. In the
experiment using 160 MeV protons the spin-flip $\Delta\ell = 1$ resonances
are strongly excited. Here one indeed sees that the 2^- strength is
much more spread out than the 0^- and 1^- ($\Delta S = 1$) strength and that

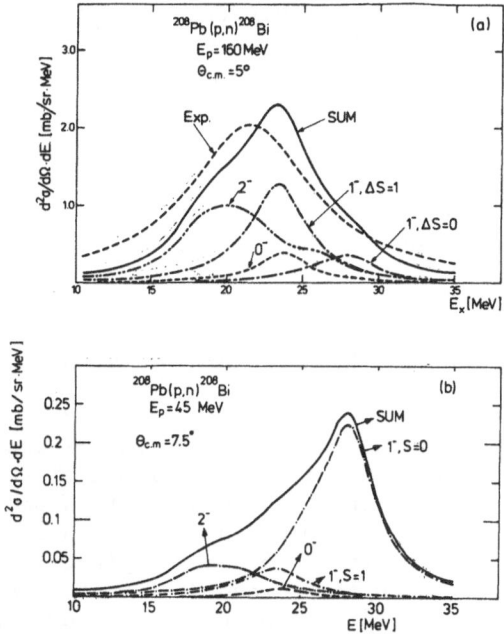

Fig. 9. Charge-exchange spectra for 160 MeV and 45 MeV incident proton energies scattered from ^{208}Pb. In the upper part, the theoretical spectra are compared with the experimental data of ref. 48. The 0^-, the 1^- ($\Delta S = 1$), and the 2^- cross sections have been reduced by a factor of 3 in order to account for quenching effects. The excitation energies are measured with respect to the ground state of ^{208}Bi.

the 2^- strength is several MeV lower in energy than the other two resonances. This explains in a natural way the large experimental width of the $\Delta \ell = 1$ resonance.

The small contribution of the non spin-flip 1^- resonance in the upper part of Fig. 9 is due to the very weak τ-part of the coupling potential at this high proton energy. Actually the 1^- ($\Delta S = 0$) resonance is the most collective one and it is the dominant one in the 45 MeV spectrum (lower part of Fig. 9) while the contributions of the spin-flip states are comparatively small (this is again due to the energy dependence of the τ-part of the coupling potential). One finds, therefore, a shift in the centroid energy of the two spectra of about 2.5 MeV. This is a general feature which is experimentally found in many medium and heavy mass nuclei[37].

For a consistent description of GTR, $\Delta \ell = 1$ and higher magnetic multipole states[24,27] it is crucial to consider the momentum dependence of the $\sigma\tau$-part of the interaction. With a zero range force the excitation energies of the $\Delta \ell = 1$ resonances turned out to be too high compared to experiments, if one takes the strength deduced from the GTR[35].

$\Delta(1236)$-ISOBAR DEGREES OF FREEDOM AND THE STRENGTH OF SPIN-ISOSPIN RESONANCES

The most interesting feature of the spin-isospin modes excited in charge-exchange reactions is the magnitude of the transition strength. In the case of the GTR there exists a well established, model independent sum rule[5] which is simply connected with the number of protons and neutrons: $S_{\beta-} - S_{\beta+} = 3(N-Z)$, where the l.h.s. is the difference between the β^- and β^+ GT-strength. So far only about 50 % of this sum rule strength has been detected experimentally[4]. One example of a microscopic analysis of a (p,n) cross section in ^{208}Pb is shown in Fig. 10 (ref. 14). There is an obvious discrepancy by more than a factor of 2 in the GTR whereas the experimental cross section of the isobaric analog state (IAS) is well reproduced theoretically. It is mainly this "quenching" of the GT-strength which has focused the interest of theorists on these resonances. Missing strength has also been claimed for the M1-strength (refs. 11,12,38). There, however, no model independent sum rule exists. Conventional nuclear structure effects are unable to explain the "missing" GT-strength, e.g. many-particle many-hole excitations give rise only to a redistribution of the strength. Several years ago already Δ-isobar degrees of freedom were suggested to play an important role in the quenching of the axial-vector coupling constant g_A which is directly connected with the GT-strength[6,7,9,10]. (The first study of the "quenching" of g_A was performed by Ericson et al.[17] in terms of the Lorentz-Lorenz effect, without reference to the Δ-resonance.) Investigations of the GTR including the Δ-degrees of freedom have been performed by Bohr and Mottelson[15], Suzuki, Krewald and Speth[31], Brown and Rho[16], and Osterfeld et al.[18]. Similar considerations have also been performed for M1-states by Knüpfer et al.[11] and Härting et al.[12]. In all these approaches the Δ-resonance, coupled with a nucleon hole (Δ-h), can move part of the isovector $\sigma\tau$-strength into an energy region approximately 300 MeV above the low-lying excitations of the nucleus. The major reason why this mechanism may have a significant effect despite this enormous energy gap is due to the Pauli principle. Since there is no Pauli blocking for the Δ-resonance, virtually all nucleons can share in building Δ-h states.

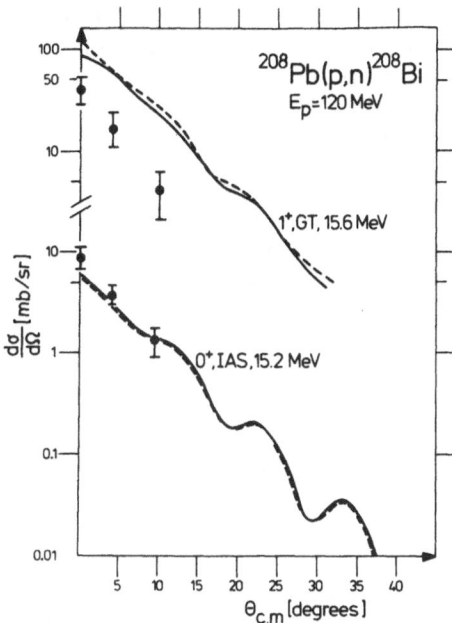

Fig. 10. Theoretical cross sections of the 0^+ (IAS) and 1^+ (GTR) of the reaction $^{208}Pb(p,n)^{208}Bi$. The experimental results are taken from ref. 48.

Therefore the sheer number of possible configurations is able to bridge the energy gap given that the interaction between the nucleon-particle and nucleon-hole pair (F^{ph}_{NN}) and Δ-particle nucleon-hole pair ($F^{ph}_{\Delta N}$) are roughly equally strong. This crucial assumption has been made implicitly in the work by Bohr and Mottelson[15] and Brown and Rho[16] who connected the two different forces (assumed to be constant in k-space) by a scaling factor $F^{ph}_{N\Delta} = f^*_\pi/f_\pi \; F^{ph}_{NN}$, where f_π and f^*_π are the π-nucleon and π-delta coupling constants, respectively. Within this approach the low-lying GT transition probability in ^{208}Pb is reduced by a factor of 2.

There exists, however, a serious problem which is connected with the exchange term of the ph-interaction. By definition, in the Migdal theory and in the schematic models mentioned above F^{ph} includes the direct and the exchange terms of the ph-interaction. Therefore, the scaling assumptions were only justified if the direct and the exchange contribution would have the same structure in the case of $F^{ph}_{N\Delta}$ and F^{ph}_{NN}, respectively. This is, however, not the case. The cancellation between the direct and the exchange term of e.g. the central part of the ρ-exchange (which gives the dominant part to g_o' in eq. (2)) is much larger for Δ-h configurations than for nucleon-hole configurations[31]. Therefore it is essential to develop a microscopic model for the $\sigma\tau$-part of the ph-interaction and to use it in those calculations. For such an interaction one may replace f_π by f_π^* and f_ρ by f_ρ^*, respectively. A first step in this direction has recently been done by Suzuki, Krewald and Speth[31] and Osterfeld et al.[18] who investigated the influence of the Δ-h configurations on spin-isospin resonances in ^{16}O and ^{48}Ca. They explicitly include the Δ-h configurations into a generalized random-phase wave function

$$\psi^J = (\sum_{NN'} \chi^J_{NN'} a_N^+ a_{N'} + \sum_{\Delta N''} \chi^J_{\Delta N''} a_\Delta^+ a_{N''}) |gs> \tag{7}$$

The major improvement with respect to previous calculations which relied on schematic models[15,16], nuclear matter estimates[11] or an effective operator formalism[13] is that now the effect of the Δ-h configurations on the excitation energies, transition probabilities and reaction cross sections can be evaluated simultaneously.

The interaction between nucleons and isobars is obtained by replacing in eq. (5) the spin (isospin) operators σ and τ by the transition spin (isospin) operators S and T and by replacing the coupling constants $f_{\pi NN} \rightarrow f_{\pi N\Delta}$ at the required vertices. For the Δ-N coupling, the Chew-Low value $f_{\pi N\Delta} = 2f_{\pi NN}$ was used[40]. In the case of the phenomenological parameterδ g_o' introduced in eq. (5) we have now to distinguish between three of them,

$$\delta G'_{NN}(q) = \delta g_o' \, C_o \, \vec{\sigma} \cdot \vec{\sigma}' \, \vec{\tau} \cdot \vec{\tau}'$$

$$\delta G'_{\Delta N}(q) = \delta g_o' (f^*/f) C_o \, \vec{S} \cdot \vec{\sigma}' \, \vec{T} \cdot \vec{\tau}' \tag{8}$$

$$\delta G'_{\Delta\Delta}(q) = \delta g_o' (f^*/f)^2 \, C_o \, \vec{S} \cdot \vec{S}' \, \vec{T} \cdot \vec{T}'$$

(with C_o = 301 MeV fm^3), using the same $\delta g_o'$ parameter for the nucleon-nucleon and the nucleon-isobar interaction. From the investigation of ref. 34 we know that the major contribution to $\delta g_o'$ follows from the second-order effects of the tensor force. In the present calculation, the quantity $\delta g_o'$ is fitted to the excitation energies of the two 1^+ states in ^{48}Sc of E_1(exp) = 3.02 MeV and E_2(exp) = 11.1 MeV. This is important since the magnitude of the quenching effect depends on the interaction strength. The transition strength to states excited in charge-exchange reactions is given by the expectation value of the operators:

$$M(GT;\lambda) = \sqrt{\pi}(\sigma Y_\ell)_{\lambda\mu}\,\tau_\pm \qquad \text{for } N\text{-}N^{-1}\,,$$

$$\qquad\qquad = \sqrt{\pi}(f^*/f)(SY_\ell)_{\lambda\mu}\,T_\pm \quad \text{for } \Delta\text{-}N^{-1}\,, \qquad\qquad (9)$$

where $f^*/f = 2$ is assumed as in eq. (8). In the case of unnatural parity states, the standard electromagnetic operator acting in the nucleon space only has to be complemented by

$$M_{\Delta N}(M\lambda) = \mu^*_{IV}\,S\cdot\nabla(r^\lambda Y_\lambda)T_z\,, \qquad\qquad (10)$$

where we have used the experimental value for the transition moment μ^*_{IV} which is 1.3 times the SU(6) value $\sqrt{72/25}\,\mu_{IV}$ predicted by the quark model[39]. In Table 1, the results of our calculations are summarized. In the calculation including the Δ-h degree of freedom, an "additional" Landau parameter of $\delta g'_0 = 0.5$ (eq. (8)) was required to get good agreement with the excitation energies of the 1^+ states in ^{48}Sc as shown in Table 1. If the Δ-h configurations are omitted, this value changes to $\delta g'_0 = 0.4$. The calculations have been performed in a space which includes all ph-configurations up to 6 $\hbar\omega$. For the Δ-h configurations we have included all the isobar orbits from 1s to 1h. The convergence of the calculation has been fairly well accomplished within this model space. The isobar effect is found to reduce the $\sigma\tau$-strength by 29 % and 22 % in the case of the high-lying and low-lying 1^+ state in ^{48}Sc, respectively, and by 36 % for the 1^+ state in ^{48}Ca compared with the corresponding RPA values. It should be noted that a pure $(f_{5/2}\,f\bar{7}/2)_{1+}$ configuration corresponds to a B(M1) value of 12 μ_K^2, whereas the experimental value (summed over the states in the vicinity) is 5.2 μ_K^2. A considerable amount of quenching is given by RPA ground state correlations in the nucleon space alone. Here one obtains B(M1) = 8.2 μ_K^2.

It is interesting to observe that in the case of the 7^+ states in ^{48}Sc practically no quenching due to Δ-h states is obtained. This effect may be explained by the q-dependence of the interaction (see e.g. ref. 34) because the form factor of the 7^+ state is peaked around q $\sim$ 2 (fm^{-1}). The short-range correlations lead to a strong repulsion at small momentum transfer, but at large momentum transfer the effective spin-isospin interaction is weak because of the attractive π and ρ contributions.

Hence the coupling of a state with large multipolarity, such as the 7^+ state in ^{48}Sc, to Δ-h configurations is expected to be small. Therefore the angular momentum dependence is a characteristic experimental signature of the Δ-h model of the quenching effect.

Table 1.   Summary of the calculations for energies and B(Mλ) values.
           We compare the experimental values (column 3) with two
           theoretical results: The numbers in columns 5 and 4 follow
           from RPA calculations in a 6 $\hbar\omega$ space with and without
           the inclusion of Δ-h configurations, respectively.

Nucleus	J^π	E_{exp} (MeV)	B_{exp}	$\delta g_o' = 0.4$		$\delta g_o' = 0.5$		$\dfrac{B_{N+\Delta} - B_N}{B_N} \times 100$
				E_N (MeV)	B_N	$E_{N+\Delta}$ (MeV)	$B_{N+\Delta}$	(%)
^{48}Sc a)	1^+	3.02		3.02	2.42	2.96	1.88	22
^{48}Sc a)	1^+	∿11		11.08	21.09	10.86	14.97	29
^{48}Sc a)	7^+	1.6		1.41	36.62	1.65	33.2	9
^{48}Ca	1^+	10.23	4.0±0.3 b)	10.21	8.20	10.16	5.29	36
			ΣB(M1)∿5.2 c)					

a) ref. 51; b) ref. 45; c) ref. 12.

        We also calculated the form factor for inelastic electron
scattering to the 1^+ state in ^{48}Ca in Born approximation. From the
strongly q-dependent force one would expect naively a strongly q-
dependent quenching effect. It should be largest at q=0, i.e. for
the B(M1) value, and disappear roughly around q∿1 (fm^{-1}). This does
not follow from our calculation. Whereas the B(M1) value is reduced
by 36 % (we compare in the following always the RPA result with and
without Δ-h contributions), we obtain in the form factor (Fig. 11)
a reduction (dashed line compared with dash-dotted line) of 39 %,
22 % and 13 % in the first, second and third maximum, respectively.
It is interesting to see that the form factor in the third maximum
is no longer quenched but increased compared to the pure
$(\nu f_{5/2} \nu f_{7/2}^{-1})_{1^+}$ shell model value. This is due to the admixtures of
the higher ph configurations.

        There are two effects which explain this behaviour qualitative-
ly: (i) The operator for electron scattering is of the form σx∇,
therefore it is more sensitive to the q-dependence of the ρ exchange
potential which is relatively weak because of the large mass of the
ρ-meson. In order to see the strong q-dependence in the στ-channel
one has to use e.g. protons. (ii) The M1 operator does not depend
on the radial coordinate. Therefore it is only sensitive to the
admixture of ph-components which are diagonal in the radial part
of the wave functions (in this sense also spin-orbit partners are
diagonal). The operator in (e,e') does have a radial dependence
(except around the photon point) and therefore the effect of the
off-diagonal matrix elements (e.g. $\Delta \ell = 2$, 2 $\hbar\omega$, 4 $\hbar\omega$ and 6 $\hbar\omega$
admixtures) gets more and more important with increasing momentum
transfer. If we leave out the off-diagonal contributions in the
nucleon and in the Δ-space the reduction of the form factor is 40 %
and 14 % in the first and second maximum, respectively.

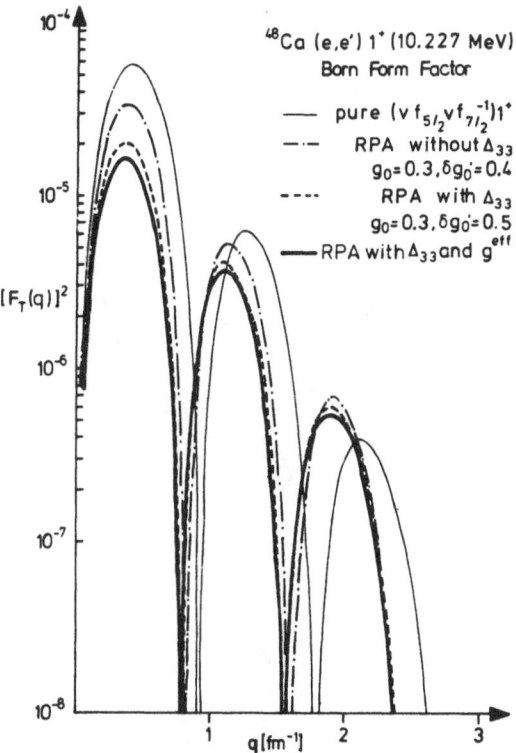

Fig. 11. Theoretical form factors of the inelastic electron scattering to the 1^+, 10.23 MeV state in ^{48}Ca, calculated in Born approximation. The dashed and dash-dotted results correspond to RPA calculations with and without inclusion of Δ-h configurations, respectively. The thin line is the corresponding result using a pure neutron $(f_{5/2} f_{7/2}^{-1})$ configuration. In all three cases the bare magnetic operator was used. In order to simulate the many-particle many-hole effect[47] we also performed a nucleon-hole Δ-h RPA calculation using the effective operator of ref. 20 with $\xi_s = 0.03$, which gives rise to B(M1)$\uparrow$ = 4.27 μ_K^2. The corresponding form factor is shown as full thick line.

Finally, we want to comment about a theoretical uncertainty of all the models which have been used so far within the Δ-h approach. In Fig. 12, some typical matrix elements contributing to the 1^+ states in ^{48}Sc are displayed. The cancellation between the direct and the exchange term of the central ρ-exchange is much larger for Δ-h configurations than for nucleon-hole configurations. For a pure zero-range interaction, the Δ-h force would be completely cancelled. Therefore it is essential to develop a microscopic model for the $\delta g_0'$ term. We expect a large contribution from the two-pion

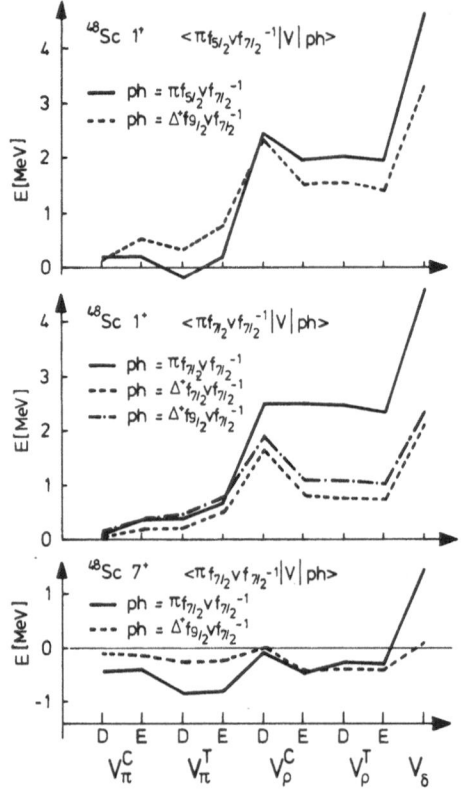

Fig. 12.   Sum of the interaction matrix elements for various
           particle-hole and Δ-h configurations in ^{48}Sc. Starting
           with the direct matrix element of the central part of
           the one-pion exchange interaction each contribution from
           various components of the interaction up to "additional"
           δ-interaction are added up, from the left to the right of
           the figure, to give the total matrix element value which
           is seen in the rightmost point of each line: The
           difference of the energy values of the neighbouring two
           points gives the single contribution denoted in the
           lower part of the figure. D and E denote the direct and
           exchange contributions, respectively. The "additional"
           delta contribution corresponds to the value $\delta g_0' = 0.5$.

exchange[34]. Preliminary calculations indicate that the quenching
will be reduced. In the absence of a full microscopic derivation
of the residual $\delta g_0'$-parameter, it is essential to draw independent
conclusions concerning the validity of the Δ-hole quenching
mechanism. Therefore we investigated the ^{48}Ca(p,n)^{48}Sc cross
sections.

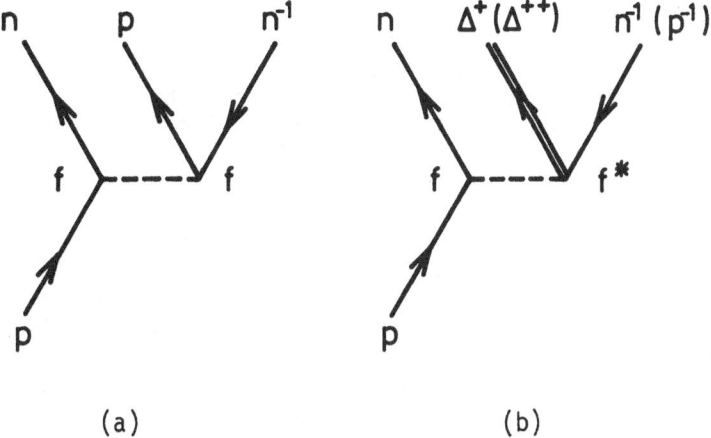

(a)   (b)

Fig. 13.   Graphical representation of reaction processes included
in the DWIA calculations. Only direct graphs are shown.
For the effective projectile-target nucleon interaction
Love's G3Y-force[33] is used (see text). For the isobar
coupling constant the Chew-Low value[40] f* = 2f has been
taken.

    In order to calculate (p,n)-cross sections from the wave
function of eq. (1) we have extended the high speed DWBA-code
FROST-MARS[41] to allow also for the excitation of the (ΔN^{-1})-
components in ψ^J (see Fig. 13). Both direct and knockout exchange
amplitudes are calculated exactly. The exact treatment of exchange
for the (ΔN^{-1}) excitations is very important since direct and
exchange amplitudes interfere destructively for a finite range
projectile-Δ isobar effective interaction. Neglect of exchange
results in an overestimate of the quenching effect!

    In Figs. 14 and 15 we show differential cross sections of
^{48}Ca(p,n)-reactions leading either to the 7^{-}, 1.096 MeV state (Fig.
14) or to the three 1^{+}-states (Fig. 15) in ^{48}Sc. The theoretical
cross sections are antisymmetrized DWIA-calculations using the G3Y-
interaction of Love[33] for the effective projectile-target nucleon
interaction. For the effective projectile-isobar interaction (see
Fig. 13) we simply used the real S=1, T=1 part of the G3Y-inter-
action in which we replaced the spin (σ) and isospin (τ) operators
by the transition operators S and T. Again we assumed for the
nucleon-isobar coupling constant f* = 2f [40]. We have also performed
reaction calculations taking the usual π- and ρ-exchange potential
($f_{\pi N\Delta} = 2f_{\pi NN}$, $f_{\rho N\Delta} = 2f_{\rho NN}$) for the effective projectile-isobar
interaction and found that this force is by a factor of 2 stronger
than that from the G3Y-interaction (see also ref. 33).

The theoretical cross sections in Figs. 14 and 15 are compared
to two sets of data. The open circles correspond to the data of
ref. 42 and the full points are the same data but with the normalization of ref. 43. From Fig. 14 we see that the theoretical cross
sections based on RPA or "RPA with Δ" describe the data of the 7^+-
state, using the normalization of ref. 43, very well. Note, however,
that there might still be an overall normalization factor in the
effective projectile-target nucleon interaction which changes the
magnitude of all theoretical cross sections by the same amount. The
calculations with isobars produce no quenching for the 7^+-state.
This is due to the fact that the finite range residual ph-interaction is weak[24] in the q-transfer region at ~ 2 fm^{-1} where the 7^+
transition density is large. Therefore the residual ph-interaction
has no chance to couple high-lying ΔN^{-1} states from 300 MeV
excitation energy down into the low-lying 7^+-state. Since there is
no quenching for the 7^+-state and since its experimental cross
section is described rather well by the theory one could think that
this state would be a good candidate for calibrating the effective
projectile-target interaction. This argument, however, is ambiguous
because the 7^+-state is dominantly excited by the tensor force
while GT-cross sections are most sensitive to the central forces
(at forward angles). In Fig. 14 we show also a calculation performed

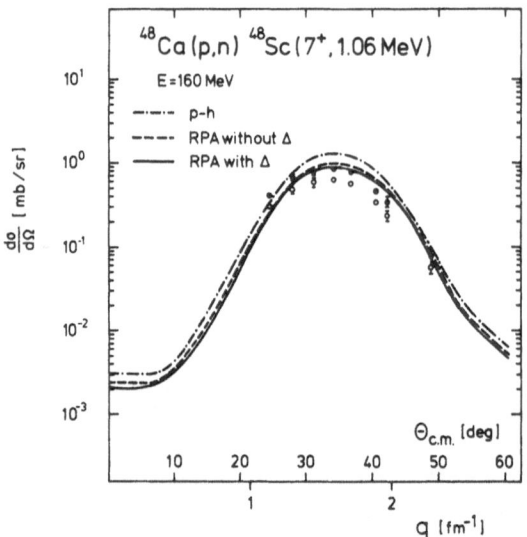

Fig. 14.   Differential cross section for the ^{48}Ca(p,n)^{48}Sc (7^+,
           1.09 MeV) reaction. The open circles are the data of
           ref. 42. The full points are the same data but with the
           normalization of ref. 43. The theoretical angular
           distributions have been calculated as described in the
           text. The optical parameters have been taken from ref. 50.

with a pure $\left[\pi f_{7/2} \, \nu f_{7/2}^{-1}\right]_{7+}$ wave function. The cross section is somewhat larger than that obtained with RPA demonstrating that ground state correlations are still important for this state.

In Fig. 15 we show cross sections for the three 1^+-states at $E_x = 2.52$ MeV, $E_x = 11$ MeV, and $E_x = 16.8$ MeV in ^{48}Sc. Again we compare results obtained either with pure particle-hole, standard RPA, or generalized RPA wave functions (including Δ isobars) with experimental data. By comparison of the results for "p-h" and "RPA" one can see that the residual ph-interaction redistributes the GT-strength in that it shifts a large amount of strength from the low to the high excitation energy region. The theoretical cross section to the low-lying 1^+-state calculated with simple "ph"-wave functions is strongly reduced when RPA wave functions are used, while that for the high-lying 1^+, T=3-state is enhanced. This is due mainly to the repulsive ph-interaction which pushes a large

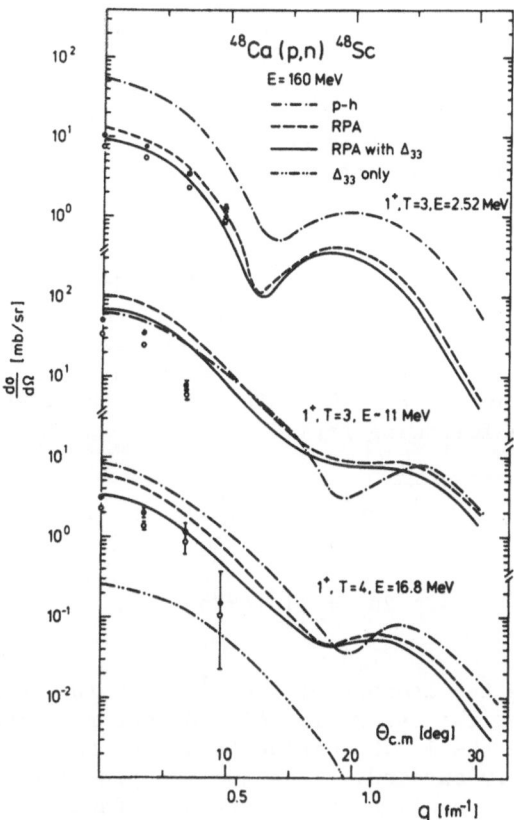

Fig. 15. Angular distributions of the ^{48}Ca(p,n) reactions to the three 1^+, GT-states in ^{48}Sc. The data and calculated angular distributions are explained in Fig. 14.

fraction of the low-lying $\sigma\tau$-strength due to ground state correlations. The introduction of Δ-isobars, which means coupling of these (low-lying) 1^+-states to (ΔN^{-1})-states at ~ 300 MeV excitation energy, has a similar effect in that now part of the GT-strength is robbed from the low-lying states and shifted to the high energy region at ~ 300 MeV. This effect can clearly be seen by comparing the cross sections based on RPA with those obtained with "RPA with Δ". For the 1^+, 2.52 MeV-state the quenching at 0^0 amounts to ~ 28 %, and that of the 1^+, T=3-state at ~ 11 MeV amounts to 30 %. If we compare the calculations including Δ-isobars to the experimental data[42] (with the normalization of ref. 43), then we underestimate the cross section to the 1^+, 2.5 MeV state slightly and overestimate the cross section to the GT-state at ~ 11 MeV by a factor of 1.45. Part of this deviation can be understood from the fact that the RPA wave functions are not coupled to good isospin. So, both wave functions for the 2.52 MeV and the 11 MeV GT-states contain isospin components with T=4, 5 and 6. (The isospin components T=5 and T=6 come from the (ΔN^{-1})-configurations!) These isospin admixtures reduce the 2.52 MeV GT-cross section and enhance the ~ 11 MeV GT-cross section. Since the isospin mixing effect is strongest for the 11 MeV T=3-state, one may, as a rough estimate, multiply the cross sections by 7/8 in order to obtain approximately the T=3-cross section. Doing this we still overestimate the data by a factor of 1.27. Concerning this overestimate one should, however, not forget the background subtraction problems in extracting the experimental data[58]!

In the lower part of Fig. 15 we analyze data of the 1^+, T=4, 16.8 MeV state in ^{48}Sc. The wave function of this state was obtained by simply applying the T_--lowering operator onto the 1^+, T=4-state in ^{48}Ca [44]. (Note: Here we disregard again the coupling to good isospin.) In the case of "RPA with Δ" the wave function has then basically the following form:

$$|1^+,\text{T=4},{}^{48}\text{Sc}> = \sqrt{\frac{1}{2T_o}}|\text{PN}^{-1}> + \sqrt{\frac{3}{2T_o}}|\Delta^{++}\text{P}^{-1}>$$

$$+ \sqrt{\frac{4}{2T_o}}|\Delta^+\text{N}^{-1}> + |2\text{p}-2\text{h}> + |1\text{p}1\Delta-2\text{h}> + \dots , \tag{11}$$

where $T_o = 4$ being the isospin of the 1^+-state in the parent nucleus ^{48}Ca. Under the assumption of a direct reaction mechanism only the 1p-1h and the 1Δ-1h components of the wave function of eq. (11) can be directly excited. Since the isospin factors for the $|\Delta^{++}\text{P}^{-1}>$ and $|\Delta^+\text{N}^{-1}>$ configurations are by a factor of $\sqrt{3}$ or 2, respectively, larger than that for the $|\text{PN}^{-1}>$ configurations and since the $|\text{PN}^{-1}>$ and $|\Delta\text{N}^{-1}>$ excitations interfere destructively we have a particularly strong quenching effect for the $T_>$ GT-states. This effect, however, is partly due to the isospin impurities in the 1^+ wave function of ^{48}Ca. This can actually be seen from the 1^+, T=4 cross section in Fig. 15. The quenching amounts to $\sim$ 50 %. In the lower part of Fig. 15 we show also the contribution to the 1^+, T=4 cross section which is due to the (ΔN^{-1})-components in the wave function. This contribution is only a factor of $\sim$ 10 smaller than that due to the (PN^{-1})-configurations. Note that the q-dependence of the quenching is rather smooth with largest quenching at small q. This effect is quite obvious for the 1^+, T=4 state. In our calculations we find 59 % of the total GT-strength in the three 1^+-states (multiplying the 1^+, 11 MeV cross section with the factor 7/8) while experimentally Gaarde (see C. Gaarde's contribution to this conference) finds $\sim$ 50 %. This agrees also qualitatively with the M1-strength found in (e,e')[45] and (p,p')[46] experiments in ^{48}Ca.

PIONIC STATES IN ^{208}Pb

For many years various experimental groups have searched for 1^+-states in ^{208}Pb. At present only a small fraction of the theoretically expected M1-strength has been found experimentally so far[56]. On the other hand, highly energetic (p,n)-reactions in this nucleus show one of the best examples of a GT-resonance (see Fig. 7). The observed GT-strength, however, exhausts only about 50 % of the Ikeda sum rule. In the following we present theoretical results on the GT-resonance and the 1^+-state in ^{208}Pb which we obtained within the same microscopic model as for ^{48}Ca.

In the following we assume that the $\delta g_0'$-parameters are the same in the Ca- and Pb-region:

$$\delta g_0' (\text{without } \Delta\text{-hole}) = 0.4$$
$$\delta g_0' (\text{with } \Delta\text{-hole}) = 0.5 \tag{12}$$

The Ikeda sum rule gives a GT-strength in ^{208}Pb of

$$B(GT)_{\text{sum rule}} = 132 \tag{13}$$

This number should be compared first with the result of an RPA calculation which does not include Δ-hole configurations. Here we obtain in the GT-region, which includes an energy range of 6 MeV:

$$B(GT)_{\text{GT-region}}^{\Delta=0} = 94 \tag{14}$$

This corresponds to a reduction of the sum rule strength of 22 %. The reduction of the strength in the GT-region is due to the distribution of the strength on the very many ph-states which are present in heavy mass nuclei. The effect is also known as "Landau damping". (Note: The reduction discussed here is not connected with the ground state correlations.) If we include the Δ-hole configurations the GT-strength is <u>further</u> reduced by 27 % with respect to $B(GT)^{\Delta=0} = 94$ and we obtain in the GT-region:

$$B(GT)^{\Delta=0}_{GT-region} = 69 \ . \tag{15}$$

Therefore we obtain theoretically in the GT-region of ^{208}Pb a GT-strength which is only 50 % of the sum rule limit. However, more than half of this reduction is due to the Landau damping.

We also calculated the B(M1)-strength in ^{208}Pb. The corresponding numbers are shown in Table 2. The difference between the Schmidt-value and the RPA($\Delta=0$)-result is due to the ground state correlations. If is interesting to see that the quenching effect due to the Δ-hole configurations is here of the order of 40 % compared to the RPA($\Delta=0$)-value.

Table 2.   B(M1)-strength in ^{208}Pb $\left[\mu_K^2\right]$

Schmidt-value	RPA($\Delta=0$)	RPA($\Delta\neq0$)
50.4	30	16.7

SUMMARY

The most exciting new feature in the field of giant resonances are the spin-isospin modes detected in charge-exchange reactions. This new type of resonances gives important new information on the spin-isospin part of the particle-hole interaction. From the theoretical studies of pion condensation and its precritical phenomena one knows that the particle-hole force in this channel is dominated by the one-pion and one-rho exchange potential. Therefore the spin-isospin modes give us the opportunity to study for the first time the effects of the corresponding exchange potentials inside the nucleus. Moreover, the "quenching" of the Gamow-Teller strength seems to offer a possibility of studying the effects of the Δ-baryon resonances on nuclear structure properties. The amount of the "quenching", however, depends sensitively on the short-range behaviour of the ph-interaction. In particular, the "Pauli" exchange terms in the interaction, which has been considered for the first time in the present approach, give rise to an appreciable reduction. In conclusion: There is no doubt about the importance of the Δ-hole configurations in the $\sigma\cdot\tau$-channel, but certainly they do not give rise to a 50 % quenching of the $\sigma\cdot\tau$-strength.

REFERENCES

1. R.R. Doering, A. Galonsky, D.M. Patterson, and G.F. Bertsch, Phys. Rev. Lett. 35:1961 (1975).
2. D.E. Bainum et al., Phys. Rev. Lett. 44:1751 (1980).
3. C.D. Goodman et al., Phys. Rev. Lett. 44:1755 (1980).
4. C. Gaarde et al., Nucl. Phys. A369:258 (1981).
5. K. Ikeda, S. Fujii, and J.I. Fujita, Phys. Lett. 3:271 (1961).
6. M. Rho, Nucl. Phys. A231:493 (1974).
7. K. Ohta and M. Wakamatsu, Nucl. Phys. A234:445 (1974).
8. J. Delorme, M. Ericson, A. Figureau, and C. Thévenet, Ann. Phys. 102:273 (1976).
9. E. Oset and M. Rho, Phys. Rev. Lett. 42:42 (1979).
10. I.S. Towner and F.C. Khanna, Phys. Rev. Lett. 42:51 (1979).
11. W. Knüpfer, M. Dillig, and A. Richter, Phys. Lett. 95B:349 (1980).
12. A. Härting et al., Phys. Lett. 104B:261 (1981).
13. H. Toki and W. Weise, Phys. Lett. 97B:12 (1980).
14. S. Krewald, F. Osterfeld, J. Speth, and G.E. Brown, Phys. Rev. Lett. 46:103 (1981).
15. A. Bohr and B.R. Mottelson, Phys. Lett. 100B:10 (1981).
16. G.E. Brown and M. Rho, Nucl. Phys. A372:397 (1981).
17. M. Ericson, A. Figureau, and C. Thévenet, Phys. Lett. 45B:19 (1973).
18. F. Osterfeld, S. Krewald, T. Suzuki, and J. Speth, Phys. Rev. Lett. 49:11 (1982).
19. A.B. Migdal, Theory of Finite Fermi Systems and Applications to Atomic Nuclei, Wiley, New York (1967).
20. J. Speth, E. Werner, and W. Wild, Phys. Rep. 33:127 (1977).
21. J. Speth, L. Zamick, and P. Ring, Nucl. Phys. A232:1 (1971).
22. A.B. Migdal, Rev. Mod. Phys. 50:107 (1978) and references therein.
23. G.E. Brown and W. Weise, Phys. Lett. 55B:1 (1975).
24. J. Speth, V. Klemt, J. Wambach, and G.E. Brown, Nucl. Phys. A343:382 (1980).
25. M.R. Anastasio and G.E. Brown, Nucl. Phys. A285:516 (1977).
26. J. Lichtenstadt, J. Heisenberg, C.N. Papanicolas, C.P. Sargent, A.N. Courtemanche, and J.S. McCarthy, Phys. Rev. C20:497 (1979), and Phys. Rev. Lett. 40:1127 (1978).
27. S. Krewald and J. Speth, Phys. Rev. Lett. 45:417 (1980).
28. G.A. Rinker and J. Speth, Nucl. Phys. A306:360 (1978).
29. J. Heisenberg, unpublished.
30. J.W. Negele and D. Vautherin, Phys. Rev. C5:1472 (1972).

31. T. Suzuki, S. Krewald, and J. Speth, Phys. Lett. 107B:9 (1981).

32. F. Osterfeld, T. Suzuki, and J. Speth, Phys. Lett. 99B:75
     (1981), and Phys. Lett. 100B:519 (1981).

33. W.G. Love, in: The (p,n) Reaction and the Nucleon-Nucleon
     Force, C.D. Goodman et al., ed., Plenum, New York (1980) p. 30;
     F. Petrovich and W.G. Love, Nucl. Phys. A354:499c (1981).

34. G.E. Brown, J. Speth, and J. Wambach, Phys. Rev. Lett. 46:1057
     (1981).

35. G.F. Bertsch, C. Cha, H. Toki, Phys. Rev. C24:533 (1981).

36. F. Osterfeld, S. Krewald, H. Dermawan, J. Speth, Phys. Lett.
     105B:257 (1981).

37. W.A. Sterrenburg et al., Phys. Rev. Lett. 45:1839 (1980).

38. N. Anantaraman et al., Phys. Rev. Lett. 46:1318 (1981).

39. M.M. Nagels et al., Nucl. Phys. B147:187 (1979).

40. G.F. Chew and F.E. Low, Phys. Rev. 101:1570 (1956).

41. F. Osterfeld, FROST-MARS-CODE, unpublished.

42. B.D. Anderson et al., Phys. Rev. Lett. 45:599 (1980);
     J.W. Watson et al., Phys. Rev. C23:2373 (1981);
     R. Madey, Kent State University, private communication.

43. C. Gaarde et al., Phys. Lett. B in press.

44. F. Krmpotić and F. Osterfeld, Phys. Lett. 93B:218 (1980).

45. W. Steffen et al., Phys. Lett. 95B:699 (1980).

46. K.E. Rehm et al., Phys. Lett. 114B:15 (1982).

47. J.B. McGrory and B.H. Wildenthal, Phys. Lett. 103B:173 (1981).

48. D.J. Horen et al., Phys. Lett. 95B:27 (1980);
     D.J. Horen et al., Phys. Lett. 99B:383 (1981).

49. C.D. Goodman, Proc. IX Int. Conf. on High Energy Physics and
     Nuclear Structure, Versailles, France, 6-10 July 1981,
     P. Catillon et al., ed., North-Holland, Amsterdam (1982).

50. A. Nadasen et al., Phys. Rev. C23:1023 (1981).

51. C. Gaarde et al., Nucl. Phys. A334:248 (1980).

52. G.E. Brown, S.-O. Bäckman, E. Oset, and W. Weise, Nucl. Phys.
     A286:191 (1977).

53. G.F. Bertsch and T.T.S. Kuo, Nucl. Phys. A112:204 (1968).

54. I. Hamamoto and P. Siemens, Nucl. Phys. A269:199 (1976).

55. G.E. Brown, J.S. Dehesa, and J. Speth, Nucl. Phys. A330:290
     (1979).

56. G.E. Brown and S. Raman, Comments Nucl. Part. Phys. 9:79 (1980).

57. G.E. Brown and J. Speth, Neutron Capture Gamma-Ray Spectroscopy,
     R.E. Chrien and W.R. Kane, ed., Plenum Press, New York (1979).

58. F. Osterfeld, contribution in this volume.

59. E. Grecksch, M. Dillig, and M.G. Huber, Phys. Lett. 72B:11
     (1977).

SPIN EXCITATIONS IN ^{208}Pb

J. Wambach and R. Fiebig

Department of Physics
State University of New York at Stony Brook
Stony Brook, New York 11794

1. ABSTRACT

Various aspects of the response of ^{208}Pb to spin perturbations are considered. Based on a boson exchange picture for the spin channel of the ph interaction the excitation of Gamow-Teller and various transverse magnetic spin states are discussed. We also investigate the role of the tensor force in isoscalar spin states. The dynamics of the coupling of 1p-1h degrees of freedom to multiparticle-multihole channels is analyzed in a 2p-2h doorway state model. Finally we derive energy weighted sum rules for spin operators in the long wavelength limit based on the quark model.

2. THE NUCLEAR RESPONSE TO SPIN- AND SPIN-ISOSPIN PERTURBATIONS

Fluctuations of the spin density $<\vec{\sigma}>$ or the spin-isospin density $<\vec{\sigma}\vec{\tau}>$ in the nucleus induced by an external field are described quantum-mechanically by the propagation of p-h pairs. Since the residual spin-isospin interaction strongly couples nucleons to isobars ΔN^{-1} p-h channels have to be included in a complete description of spin modes in nuclei.[1] In terms of the density fluctuation operators in momentum space

$$\hat{\sigma}_1(\vec{q}) = \sum_{i=1}^{A} \int e^{i\vec{q}\vec{r}} \delta(\vec{r}-\vec{r}_i)\vec{\sigma}\theta \ d\vec{r} \qquad \theta= \begin{cases} 1 & \Delta T=0 \\ \vec{\tau} \\ \vec{T} & \Delta T=1 \end{cases} \qquad (2.1a)$$

$$\hat{\sigma}_2(\vec{q}) = \sum_{i=1}^{A} \int e^{i\vec{q}\vec{r}} \delta(\vec{r}-\vec{r}_i)\vec{S}\theta \ d\vec{r} \qquad\qquad\qquad\qquad\qquad (2.1b)$$

the propagator Π is given by a 2x2 matrix

$$\Pi_{ij}(\vec{q},\vec{q}',\omega) = \sum_{n\neq o} <o|\hat{\sigma}_i(\vec{q})|n> \left(\frac{1}{\omega-\omega_{no}} - \frac{1}{\omega+\omega_{no}}\right) <n|\hat{\sigma}_j(\vec{q}')|o> \quad (2.2)$$

Here $|o>$ and $|n>$ denote eigenstates of the exact Hamiltonian of the system. Because of translational invariance Π is local in an infinite system but nonlocal in a nucleus. For spherical nuclei however the total angular momentum J is a good quantum number. Projecting on the components

$$\Pi_{ij}^J(q,q',\omega) = \int \Pi_{ij}(\underset{\sim}{q},\underset{\sim}{q}',\omega) \; P_J(\hat{q}\cdot\hat{q}') \; d\Omega \qquad (2.3)$$

one obtains for each J an integral equation for Π^J in terms of the unperturbed propagator $\Pi^{o,J}$ and the effective p-h interaction $\mathcal{F}^{ph}$ (Fig. 1)

$$\Pi_{ij}^J(q,q',\omega) = \Pi_{ij}^{o,J}(q,q',\omega)$$

$$+ \sum_{i',j'} \int_0^\infty k^2 dk \; \Pi_{ii'}^{o,J}(q,k,\omega) \; \mathcal{F}_{i'j'}^{ph}(k,\omega)\Pi_{j'j}^J(k,q',\omega) \qquad (2.4)$$

Note that the unperturbed propagator is diagonal in ij

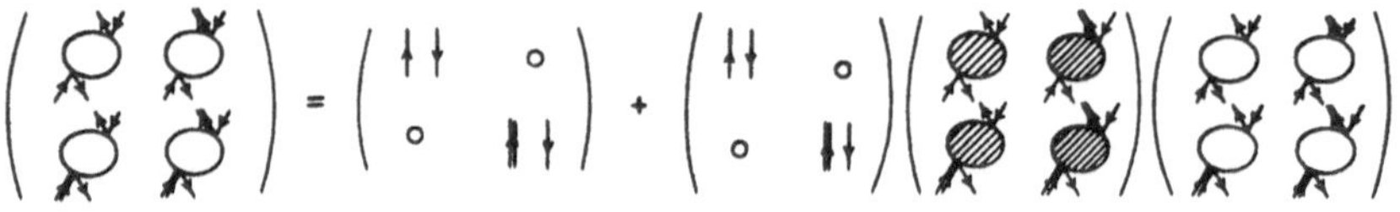

Fig. 1. Graphical representation of the coupled channel equation for the p-h propagator if both nucleon and isobar degrees of freedom are considered.

The generalized p-h interaction $\mathcal{F}^{ph}$ is a complicated object which accounts for the restrictions of the Hilbert space and is hence complex and energy dependent. We split the interaction into a simple static potential term and a remainder which contains the complicated pieces.[2]

$$\mathcal{F}^{ph}(q,\omega) = V(q) + \Delta\mathcal{F}^{ph}(q,\omega) \qquad (2.5)$$

Then eq. (2.4) separates into the usual RPA equation (we omit
momentum variables for brevity)

$$\Pi_{ij}^{RPA,J}(\omega) = \Pi_{ij}^{o,J}(\omega) + \sum_{i',j'} \Pi_{ii}^{o,J}(\omega)\, V_{i'j'}\, \Pi_{j'j}^{RPA,J}(\omega) \qquad (2.6a)$$

and a second equation for the full Π involving Π^{RPA}

$$\Pi_{ij}^{J}(\omega) = \Pi_{ij}^{RPA,J}(\omega) + \sum_{i',j'} \Pi_{ii}^{RPA,J}(\omega)\, \Delta \mathcal{F}_{i'j'}^{ph}(\omega)\, \Pi_{j'j}^{J}(\omega) \qquad (2.6b)$$

which physically describes the decay of the RPA states into more
complicated configurations. We obtain the response of the nucleus
to an external field F $(\vec{q})$, which couples to spin or spin-isospin
density, from the imaginary part of the propagator

$$S_F^J(q,\omega) = \frac{1}{\pi}\, Im \sum_{ij} F_i^{*J}(q)\, \Pi_{ij}^J(q,q,\omega) F_j^J(q) \;. \qquad (2.7)$$

Utilizing the spectral representation for Π (eq.(2.2)) this
reduces to the familiar form

$$S_F(q,\omega) = \sum_n \sum_{ij} F_i^{J*}(q)\sigma_i^{J*}(q)\sigma_j^J(q)F_j^J(q)\delta(\omega - \omega_{no}) \qquad (2.8)$$

The lowest energy moments of S_F^J

$$m_{F,k}^J(q) = \int S_F^J(q,\omega)\, \omega^k d\omega \qquad (2.9)$$

characterize bulk proportions of the response (sum rules). The non-
energy weighted sum rule $m_{F,o}^J(q)$ in the long wavelength limit is of
special interest for Gamow-Teller transitions.[3] The energy weighted
sum rule is given by $m_{F,1}^J(q)$ and the dispersion of strength by

$$\sigma_F^J(q) = (m_{F,2}^J(q) - m_{F,1}^J(q)^2)^{\frac{1}{2}} \qquad (2.10)$$

3. THE RPA RESPONSE

(3.a) The Spin-Isospin Interaction

In the $\sigma\tau$-channel both pion and ρ-meson exchange contribute
strongly to the static potential V extracted from the generalized

p-h interaction $\mathcal{F}^{ph}$. To lowest order the ansatz for V is

$$V(\vec{q}) = V_\pi(\vec{q}) + V_\rho(\vec{q}) \qquad\qquad (3.1)$$

which contains both central and tensor pieces.

$$V_c(\vec{q}) = \frac{-4\pi}{3} \left[\frac{f_\pi^2}{m_\pi^2} \Gamma_\pi^2 \frac{q^2}{q^2+m_\pi^2} + 2 \frac{f_\rho^2}{m_\rho^2} \Gamma_\rho^2 \frac{q^2}{q^2+m_\rho^2} \right] \vec{\sigma}\vec{\sigma}'\vec{\tau}\vec{\tau} \qquad (3.2a)$$

$$V_T(\vec{q}) = -\frac{4\pi}{3} \left[\frac{f_\pi^2}{m_\pi^2} \Gamma_\pi^2 \frac{q^2}{q^2+m_\pi^2} - \frac{f_\rho^2}{m_\rho^2} \Gamma_\rho^2 \frac{q^2}{q^2+m_\rho^2} \right] S_{12}(\hat{q})\vec{\tau}\vec{\tau} \qquad (3.2b)$$

Γ_π and Γ_ρ are formfactors taking into account the hadronic size of the nucleon. As it stands V(q) allows particles at the same point in space to interact, which is of course forbidden by the Pauli principle and short range correlations. For the pionic piece this can be minimally corrected by removing a factor $4\pi f_\pi^2/(3m_\pi)^2$ from V_c (EELL-correction[4]) which takes out the δ-function in r-space. If the ρ-meson is included in the Born term for V_c short range effects can be taken into account more properly by a correlation function obtained from G-matrix or variational calculations for instance.[5] A good fit at low q is given by

$$g(r) = 1 - j_0(q_c r) \qquad\qquad (3.3)$$

where q_c is a cut off momentum of the order of the ω-meson Compton wavelength. (q_c = 3.93 fm^{-1}). Including correlations the lowest order effective potential V is given by a folding of the bare interaction (3.1) with g

$$V(\vec{q}) = \int g(|\vec{k}-\vec{q}|) [V_\pi(\vec{k}) + V_\rho(\vec{k})] \, d\vec{k} \qquad\qquad (3.4)$$

The $\sigma\tau$-strength obtained from 1st-order (Fig. 2) is not sufficient to obtain the observed energy shifts of unnatural parity states in light and medium heavy nuclei.[6] It has been pointed out however[7] that the second order iterated tensor part (eq. (3.2b)) contributes significantly to the effective interaction. For plane wave intermediate states one calculates

$$v^{(2)}(\vec{q},\varepsilon_F) = -16 \, m_N \int_0^\infty dz \, \frac{z^2}{q^2} \frac{u(z)}{-4z^2} \vec{\sigma}\vec{\sigma}'\vec{\tau}\vec{\tau}' \qquad (3.5a)$$

$$u(z) = 2\pi \int_{-1}^{1} d \cos\theta \, (V_T[\tfrac{1}{2}(q^2 + 4z^2 + 4qz \, \cos\theta)^{\frac{1}{2}}])^2 \qquad (3.5b)$$

$V^{(2)}$ gives about 30% of the 1st order at small q (Fig. 2)

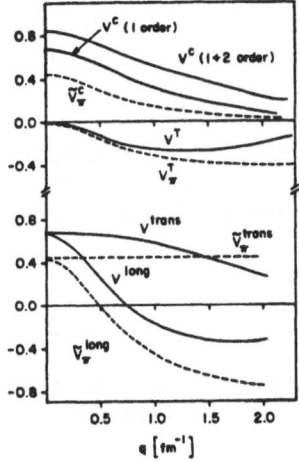

Fig. 2.   Momentum dependence of the effective spin-isospin
          interaction used in the calculations. V_π denotes
          the pure one pion exchange with subtracted δ-function
          in $\vec{r}$-space.

One extracts from the above interaction a Fermi-liquid parameter

$$g_0' = \lim_{\vec{q}\to 0} (V^{(1)}(\vec{q}) + V^{(2)}(\vec{q}))\Big/(\frac{2k_F m^*}{\pi^2}) = 0.9 \qquad (3.6)$$

In the Boson exchange model for $V_{\sigma\tau}$ it is straightforward to construct the transition potentials between NN^{-1} and ΔN^{-1} states
(eq.(2.4) and Fig.1) by substituting the meson-NN coupling vertices
$f_i \Gamma_i$ with meson-ΔN vertices $f_i^* \Gamma_i^*$. To the extent, that the form-
factors Γ_i are the same and the ratios f_π^*/f_π and f_ρ^*/f_ρ are the same,
say f^*/f, we obtain a simple scaling law for the potential V_{ij} in the
NN^{-1} and ΔN^{-1} space ($1 \hat{=} NN^{-1}$, $2 \hat{=} \Delta N^{-1}$)

$$V_{12}(\vec{q})/V_{11}(\vec{q}) = V_{21}(\vec{q})/V_{11}(\vec{q}) = f^*/f \qquad (3.7a)$$

$$V_{22}(\vec{q})/V_{11}(\vec{q}) = (f^*/f)^2 \qquad (3.7b)$$

The constituent quark model predicts a value of $(72/25)^{\frac{1}{2}}$ for the ratio f*/f which is smaller than the empirical value of 2. We adopt the empirical value throughout the calculation.

In connection with various external probes which excite spin modes in the nucleus it is instructive to consider two distinct cases:

(1)   the external field couples longitudinally to the spin density

$$F \propto \vec{q} \cdot <\vec{\sigma}(\vec{q})>$$ (3.8a)

(e.g. a pion field coupling to p-h excitations)

(2)   the probe couples transversely to a spin mode

$$F \propto \vec{q} \times <\vec{\sigma}(\vec{q})>$$ (3.8b)

(e.g. an electromagnetic field coupling to the nuclear magnetization)

We want to demonstrate that the driving forces with which both disturbances propagate are quite different. From the operator identities

$$\vec{\sigma}\vec{\sigma}' = \vec{\sigma}\hat{q}\ \vec{\sigma}'\hat{q} + \vec{\sigma} \times \hat{q}\ \vec{\sigma}' \times \hat{q}$$ (3.9a)

$$S_{12}(q) = 2\vec{\sigma}\hat{q}\vec{\sigma}'\hat{q} - \vec{\sigma} \times \hat{q}\ \vec{\sigma}' \times \hat{q}$$ (3.9b)

one obtains the longitudinal and transverse interaction in terms of V_c and V_T as

$$V_{long}(\vec{q}) = V_c(\vec{q}) + 2V_T(\vec{q})$$ (3.10a)

$$V_{trans}(\vec{q}) = V_c(\vec{q}) - V_T(\vec{q})$$ (3.10b)

which are plotted in the lower part of Fig. 2. While V_{long}, which is the driving force for a pion field in a nuclear medium, shows strong attraction at $q \geq 1$ fm^{-1} V_{trans} remains repulsive even up to large momentum transfer. If the scaling arguments given above apply, the momentum dependence of the transition potentials V_{12} and V_{22} is the same as that of V_{11}.

(3.b)   Gamow-Teller States

We first discuss the long wavelength limit of the spin-isospin response. Since in this limit no momentum is transferred to the nucleus it inhibits the zero sound properties of spin-isospin waves. The long wavelength limit is realized in M1- and GT-excitations. In a nucleus with a big neutron excess like ^{208}Pb one expects a collective GT-state shifted upwards relative to the noninteracting p-h states, since $V_{\sigma T}$ is strong and repulsive at q=o. To obtain collective features in the response one has to solve the RPA equation for the p-h propagator Π (eq.(2.6a)). Including ΔN^{-1} states this amounts to solving the following eigenvalue problem

$$\begin{pmatrix} C_{NN-1,NN-1} & C_{NN-1,\Delta N-1} \\ C_{\Delta N-1,NN-1} & C_{\Delta N-1\,\Delta N-1} \end{pmatrix} \begin{pmatrix} X^n_{NN-1} \\ X^n_{\Delta N-1} \end{pmatrix} = E_n \begin{pmatrix} X^n_{NN-1} \\ X^n_{\Delta N-1} \end{pmatrix} \qquad (3.11a)$$

in which the block matrices $C_{php'h'}$ are of the usual RPA type:

$$C_{php'h'} = \begin{matrix} A_{php'h'} & B_{php'h'} \\ -B^*_{php'h'} & -A^*_{php'h'} \end{matrix} \qquad (3.11b)$$

$$A_{php'h'} = (\varepsilon_p - \varepsilon_h)\,\delta_{pp'}\delta_{hh'} + \langle ph|V_c(\underset{\sim}{q})|p'h'\rangle \qquad (3.11c)$$

$$B_{php'h'} = \langle ph|V_c(\underset{\sim}{q})|h'p'\rangle \qquad (3.11d)$$

Since the p-h propagator $\Pi^{RPA}(q,q',\omega)$ is essentially local in momentum space for ^{208}Pb it is sufficient to replace $V_c(q)$ by $V_c(o)$ in the interaction matrix elements (the error is about 8%). Without further restrictions the coupled eqs. (3.11a) are hard to solve, because of the large p-h dimension in the ΔN^{-1} sector. At q=o however Δ-states can only be created from N-states in the same orbit. This restricts the configuration space considerably such that the matrix equation can be solved. A check for the completeness of the model space at q=o is given by comparing the calculated non-energy-weighted RPA sum rule $\lim\limits_{q \to o} m^1_o(q)$ (eq.(2.9)) to the model independent sum rule[8]

$$\tilde{m}^1_o(o) = 3(N-Z) + \left(\frac{g_A^{\Delta N}}{g_A^{NN}}\right)^2 (Z + \frac{1}{3}N). \qquad (3.12)$$

If the model space is complete one obtains of course

$$R = \frac{\tilde{m}_o{}^1(o) - m_o{}^1(o)^{RPA}}{\tilde{m}_o{}^1(o)} = 0 \qquad\qquad (3.13)$$

From our restricted model space R = 0.06. In Fig. 3 the RPA
GT-strength at q=o is displayed (g_o' = 0.9).

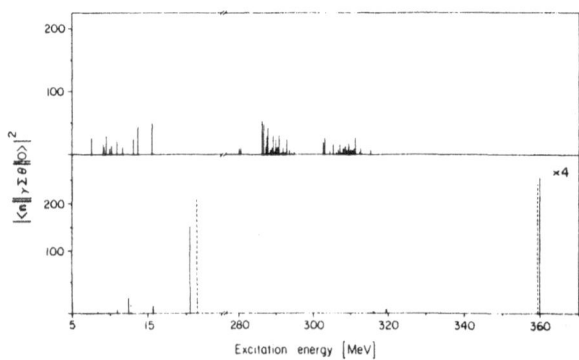

Fig. 3.   GT-transition strength in ^{208}Pb. The upper part shows
          the p-h strength without residual interaction, while
          in the lower part the interaction is turned on.   The
          dashed lines indicate the strength if NN^{-1} and ΔN^{-1}
          states are not allowed to mix.

The calculation has been performed with harmonic oscillator wavefunctions (b=2.4 fm) and single particle energies from a SKIII
Hartree-Fock potential. We obtain 27% quenching of nuclear GT-
strength[2].

(3.c)   Transverse Response at Large Momentum-Transfer

       Elastically and inelastically scattered electrons couple trans-
versely to the spin density. If contributions from the orbital
current of nucleons can be neglected the measured transverse form-
factor to a state of angular momentum J is given by

$$F_T(q) \propto \sum_\mu A(J,\mu)(\mu_p+\mu_N)\int (\vec{Y}_{JJ\,\mu}(\hat{r})\vec{\nabla} \times <\vec{\sigma}>)\,d\vec{r} \qquad (3.14)$$

where A(Jµ) denotes an angular momentum factor. In the calculations
presented below the coupling between NN^{-1} and ΔN^{-1} states has not
been included explicitly as for the GT-states. The effects of isobars on the excitation energies however can be incorporated by

replacing the "bare p-h interaction" V_{11} in the nucleonic sector by an effective potential. In a local approximation for the RPA propagator Π_{ij} one has

$$V_{11}^{eff}(\vec{q},\omega) = V_{11}(\vec{q}) + V_{12}(\vec{q}) \frac{\Pi_{22}^{o}(\vec{q},\omega)}{1-\Pi_{22}^{o}(\vec{q},\omega)V_{22}(\vec{q})} V_{21}(\vec{q}) . \qquad (3.15)$$

A frequency dependence enters via the propagator. Since $\omega \ll m_\Delta - m_N$ for nuclear excitations near the Fermi surface the static approximation

$$V_{11}^{eff}(\vec{q},\omega) \simeq V_{11}^{eff}(\vec{q},o) \qquad (3.16)$$

is justified. To see the ΔN^{-1} polarization effect on V^{eff} we can go to the long wavelength limit in which Π_{22}^{o} has a simple form[9]

$$\Pi_{22}^{o}(o) = - \frac{8}{9} \frac{A}{m_\Delta - m_N} . \qquad (3.17)$$

Using the value for $V_{11}(o)$ given in Sec. 3.a and the scaling property between V_{11}, V_{12} and V_{22} (eqs.(3.7a) and (3.7b)) we find that the matrix elements of V^{eff} are reduced

$$<o|V_{11}^{eff}(o)|NN^{-1}> = 0.77 <o|V_{11}(o)|NN^{-1}> .$$

To obtain the reduction factor one has also to assume that all radial matrix elements of the p-h interaction are equal.[9] This estimate shows that $g_{o,eff}^{1} = 0.7$ has to be used instead of $g_o^1 = 0.9$ working only in the nucleon space. Since $\Pi_{22}^{o}(\vec{q},\omega)$ becomes small at large q because of the finite size of the nucleus and the interaction also decreases one expects $V_{11}^{eff}(\vec{q})$ to approach $V_{11}(\vec{q})$ at large $\vec{q}$. Since V_{trans} is small at large q (Fig. 2) one expects that core polarization and ground state correlations are small. In fact one sees less reduction in the ground state moment distribution of ^{207}Pb for the finite range case (solid line in the left part of Fig. 4) compared to a pure δ-function in $\vec{r}$-space ($g_o^1 = 0.7$, dashed dotted line). To understand why the shell model moment (dashed line) is quenched about equally at $q_{eff} = 1.4$ fm^{-1} and $q_{eff} = 2.4$ fm^{-1} one has to include tensor force effects which are also in our calculation. Similar conclusions as from the ground state moment of ^{207}Pb can be drawn from the 14^--high spin state in ^{208}Pb. Because of the weak transverse p-h interaction RPA ground state correlations are not sufficient to explain the experimentally observed reduction of the shell model strength (right part of Fig. 4).

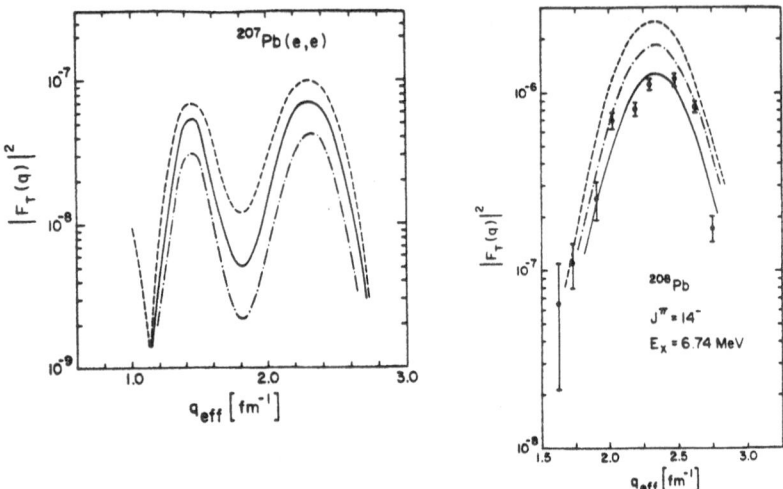

Fig. 4. left: transverse magnetic DWBA form factor for
                the ground state of ^{207}Pb.

        right: transverse (e,e') form factor for the 14-
                state in ^{208}Pb. Dashed line: shell model
                prediction dashed-dotted line RPA-core
                polarization, full line 2p-2h admixtures.
                (Data from ref. 10)

(3.d) p-h Correlations from the Tensor Force

It has been observed first by Bleuler[11] that the ground state
energy of a nuclear system can be lowered by mixing single particle
states of different parity. This instability is caused by the
presence of pions in the nuclear environment mediating the long range
part of the NN-interaction. In the language of Fermi liquid theory
the instability of the normal ground state is signaled by the fact,
that the Fermi-liquid parameters obtained from the Landau expansion
of $\mathcal{F}^{ph}$

$$\mathcal{F}^{ph}(q,\varepsilon_F) = \frac{\pi^2}{2m^*k_F^2} \sum_{\ell} \left\{ f_\ell + f_\ell' \vec{\tau}\vec{\tau}' + [g_\ell + g_e' \vec{\tau}\vec{\tau}'] \vec{\sigma}\vec{\sigma}' \right.$$
$$\left. + \left(\frac{q}{k_F}\right)^2 [h_\ell + h_\ell' \vec{\tau}\vec{\tau}'] \quad S_{12}(\hat{q}) \right\} P_\ell(\cos\theta) \qquad (3.18)$$

do not obey the stability criteria. The lowest order parameters have to satisfy the relation[12]

$$1 + \frac{1}{3} g_1 - \frac{10}{3} h_o + \frac{4}{3} h_1 - \frac{2}{15} h_2 > 0 \; . \tag{3.19}$$

Antisymmetrizing the p-h matrix elements one obtains from the one pion exchange $V_\pi(\vec{q})$ that the LHS of (3.19) is -0.663, i.e. the ground state is unstable. Note that h_o enters with the highest weight which indicates that the phase transition is a long wave-length phenomenon in contrast to pion condensation. Isoscalar magnetic states signal the proximity of nuclei to the instability. The 0^- $\Delta T=0$ state in ^{16}O at 10.95 MeV is shifted downwards by about 2 MeV from the unperturbed p-h energy. The shift is entirely due to the tensor force but it shows that V_T is weak.[12] In the boson exchange picture the weak tensor force is obtained from the ρ-meson which has opposite sign to the pionic part of the tensor interaction and this cuts V_T down at short distances (Fig. 2). Another interest-ing example is the low lying 1^+ $\Delta T=0$ state in ^{208}Pb. In the left part of Fig. 5 the excitation energy of the state is plotted versus the relative ρ-NN coupling strength (f_s denotes the strong ρ-coupling constant). The motion is completely insensitive to the central part of the spin-interaction.

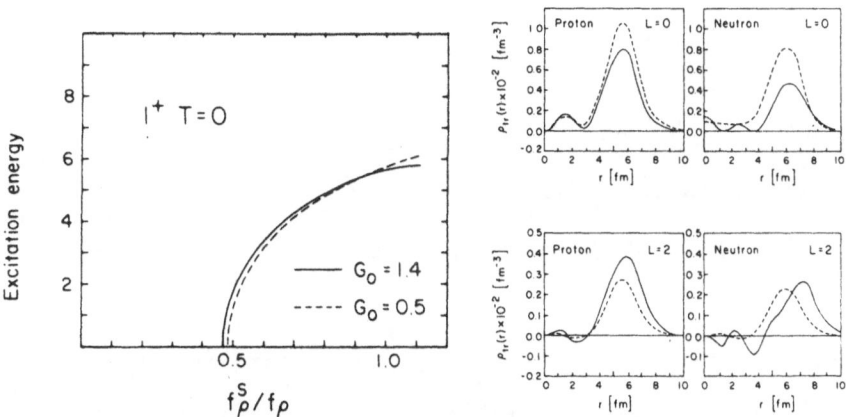

Fig. 5. left: Excitation energy of the low lying 1^+ $\Delta T=0$
                state in ^{208}Pb versus the relative ρ-NN
                coupling strength (f_s is taken from ref. 13)

        right: $[\sigma Y_o]^{1+}$ and $[\sigma Y_2]^{1+}$ transition densities
                for the same states. In the dotted lines the
                tensor interaction is switched off.

Another signature of the tensor force is the ground state transition density given in right part of Fig. 5. While the $[\sigma Y_0]^{1+}$ pieces are strongly quenched the $[\sigma Y_2]^{1+}$ pieces show strong enhancement and a geometrical change in the radial distribution.

4. COUPLING OF P-H EXCITATIONS TO MULTIPARTICLE MULTIHOLE STATES

(4.a) 2p-2h Doorway State Model

Since the 1p-1h excitations of the nucleus are not eigenstates of the full Hamiltonian, they decay into more complicated configurations. To describe this decay one has to consider more complicated pieces of the p-h interaction than just the potential terms discussed so far. In terms of the RPA-solutions to the propagator the full Π is given in a shell model basis representation by a matrix equation

$$\Pi(\omega) = \Pi^{RPA}(\omega) + \Pi^{RPA}(\omega)\,\Delta\mathcal{F}^{ph}(\omega)\,\Pi(\omega) . \qquad (4.1)$$

In terms of the p-h amplitudes of an RPA eigenstate $|n\rangle$

$$\chi_{\alpha\beta}^{n} = \begin{cases} X_{\alpha\beta} & \alpha > \varepsilon_F \quad \beta < \varepsilon_F \\ \\ Y_{\alpha\beta} & \alpha < \varepsilon_F \quad \beta > \varepsilon_F \end{cases} \qquad (4.2)$$

the RPA propagator is given as

$$\Pi_{\alpha\beta\gamma\delta}^{RPA}(\omega) = \sum_{n\neq o}\left(\frac{\chi_{\alpha\beta}^{n}\chi_{\delta\gamma}^{n*}}{\omega-\tilde{\omega}_{no}} - \frac{\chi_{\delta\gamma}^{n}\chi_{\alpha\beta}^{n*}}{\omega-\tilde{\omega}_{no}}\right) . \qquad (4.3)$$

The coupling of an RPA state to np-nh states is described by the self energy matrix element

$$\Sigma_{nn'}(\omega) \equiv \langle n|\Sigma(\omega)|n'\rangle = \sum_{\substack{\alpha,\alpha' \\ \beta,\beta'}} \chi_{\alpha\alpha'}^{n\,*}\,\Delta_{\alpha\alpha'\beta'\beta}^{ph}(\omega)\,\chi_{\beta'\beta}^{h} . \qquad (4.4)$$

In a restricted Q-space of 2p-2h doorway states[14] Σ contains the diagrams displayed in Fig. 6. Next we assume that Σ is diagonal in the RPA-states.[14] This is a good approximation if $|n\rangle$ is a collective state and the average spacing $\Delta\omega_{nn'}$ between several collective states is large compared to the interaction energy $\Sigma_{nn'}$. In the diagonal approximation eq. (4.1) is readily solved to give

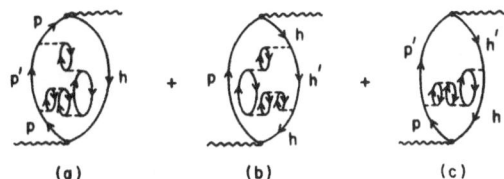

Fig. 6. Goldstone diagrams contained in the 2p-2h doorway
         model for the RPA self energy.

$$\Pi_{\alpha\beta\gamma\delta}(\omega) = \sum_n (\chi^n_{\alpha\beta} \frac{1}{\omega-(\tilde{\omega}_{no}+\Sigma_{nn}(\omega))} \chi^{*n}_{\delta\gamma} - \chi^n_{\delta\gamma} \frac{1}{\omega+(\tilde{\omega}_{no}+\Sigma_{nn}(\omega))} \chi^{n*}_{\alpha\beta}) \,.$$

(4.5)

$\Sigma_{nn}(\omega)$ has poles on the real axis at all energies of 2p-2h intermediate states. At excitation energies of $\sim$ 10 MeV in ^{208}Pb the
pole density is of the order of 10^3/MeV and an average over a small
interval is justified. Mathematically this means that ω can be
moved into the complex plane $\omega \rightarrow \omega - i\Delta/2$. Now the matrix elements
Σ_{nn} become complex. The real part determines the 2p-2h energy shift

$$\Delta E_n(\omega) = \text{Re } \Sigma_{nn}(\omega - i\Delta/2)$$

(4.6a)

and the imaginary part gives the spreading width

$$\Gamma_n\downarrow(\omega) = 2 \text{ Im } \Sigma_{nn}(\omega - i\Delta/2) \,.$$

(4.6b)

It is worth noting that the RPA strength for an operator F

$$S^{RPA}_F(\vec{q},\omega) = \sum_{n\neq o} |\langle n|F(\vec{q})|o\rangle|^2_{RPA} \, \delta(\omega-\tilde{\omega}_{no})$$

(4.7)

is changed in a very simple fashion if we employ the diagonal
approximate for Σ. One obtains for the full strength function

$$S_F(\vec{q},\omega) = \sum_{n\neq o} |\langle n|F(\vec{q})|o\rangle|^2_{RPA} \, P_n(\omega)$$

(4.8)

i.e. the δ-function in the RPA expression for S_F is simply replaced
by the probability distribution[2]

$$P_n(\omega) = \frac{1}{2\pi} \; \frac{\Gamma_n\downarrow(\omega) + \Delta}{(\omega - \tilde{\omega}_{no} + \Delta E_n(\omega))^2 + \frac{1}{4}(\Gamma_n\downarrow(\omega) + \Delta)^2} \; . \qquad (4.9)$$

(4.b) Application of the Model for Pure p-h Excitations

We first consider the mixing of low lying isovector 1^+-states in ^{208}Pb with 2p-2h excitations. From the independent particle model the state has a simple structure, namely it consists of the $\pi h\; 11/2^{-1} \to \pi h\; 9/2$ and $\nu\; i\; 13/2^{-1} \to \nu\; i\; 11/2$ transitions. The residual interaction $V(\vec{q})$ mixes them pushing the isovector state up by about 2.3 MeV (left part of fig. 7).

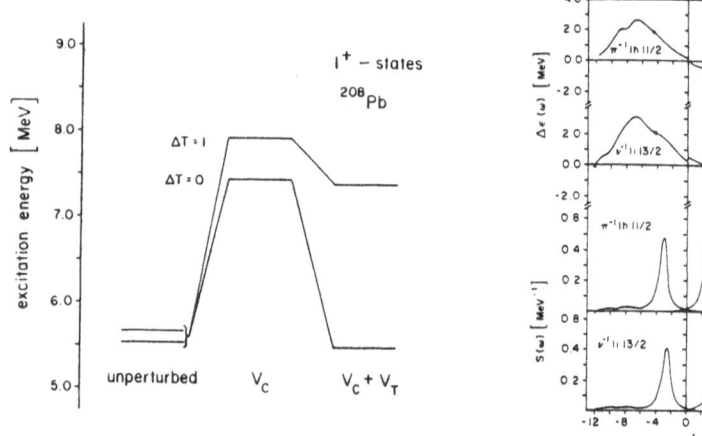

Fig. 7. left: Configuration mixing of the lowlying 1^+-states in ^{208}Pb from $V(\vec{q})$

right: 2p-1h energy shift and strength functions for the single particle states that participate in the motion of the 1^+ $\Delta T=1$ in ^{208}Pb. The dots denote the HF energies.

Since the effect of the tensor interaction is rather weak, the state carries essentially no momentum. Therefore the mixing with ΔN^{-1} states is strong (similar to the GT-states discussed in the last section) resulting in a strong quenching of the pure nucleonic M1-strength . Nevertheless the quenching ($\sim$ 40%) may be not sufficient to explain the present experimental situation. It has been proposed[15] that coupling to 2p-2h states might dissipate

strength to much higher energies such that it escapes observation.
In order to account for the proper off shell propagation of the
p-h excitations one has to start from single particle states in a
HF mean field and couple 2p-2h states explicitly. The results of a
calculation in this spirit is shown in the left part of Fig. 8.
We display the p-h strength function $P_{1+}(\omega)$ (eq. 4.9) as a function
of the excitation energy. Still most of the strength ($\sim$ 80%) is
concentrated below 10 MeV. This result is understood as follows:
the self energy insertions on the particle and hole lines (Fig. 6a, 6b)
introduce an ω-dependent p-h gap energy:

$$\varepsilon_{ph}(\omega) = \varepsilon_p^o + \Sigma_p(\varepsilon_p^o + \Delta\omega) - (\varepsilon_n^o + \Sigma_n(\varepsilon_n^o - \Delta\omega)) \qquad (4.10a)$$

$$\Delta\omega = \tilde{\omega}_{no} - (\varepsilon_p^o - \varepsilon_n^o) \qquad (4.10b)$$

Here $\tilde{\omega}$ denotes the RPA excitation energy of the state $|n>$ and ε_p^o
and ε_n^o are the HF energies (SK III). We display $\varepsilon_{ph}(\omega)$ in the
right part of Fig. 8.

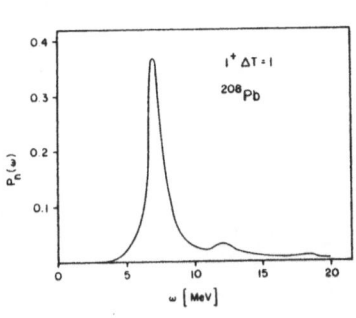 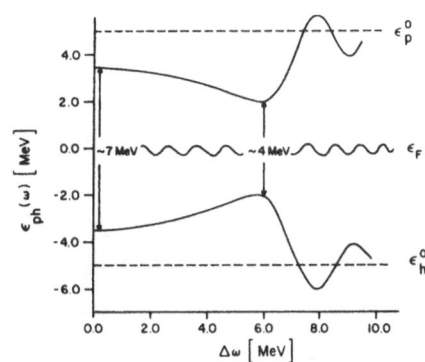

Fig. 8. left: Isovector M1- strength function coupling the
                1p-1h state to 2p-2h excitations of the ^{208}Pb
                core.

        right: Dynamical p-h gap in ^{208}Pb

As the ph-correlation energy $\Delta\omega$ is increased the p-h gap gets
smaller, i.e. the system resists the frequency increase by raising
the effective mass. Only after $\Delta\omega$ has become sufficiently large
($\sim$ 5-6 MeV in ^{208}Pb) this effect vanishes. We conclude that the
correlation energy for the 1^+ ΔT=1 is not sufficient to overcome
this barrier and hence not much strength is dissipated above 10 MeV.

(4.c)   2p-2h Admixture to the GT-Resonance in ^{208}Pb

    Unlike the $1^{\pm}$ ΔT=1 state the nuclear GT-resonance is a collec-
tive state involving many p-h components (Sect. 3.b). This leads
to an energy shift substantially larger (∿ 7 MeV). From the discussion on the dynamical p-h gap in the last section it becomes
clear that such large shifts are sufficient to eliminate essentially
all 2p-2h contributions to the energy of the mode. In fact we find
from the calculation (upper right part of Fig. 9) that the 2p-2h
shift vanishes near 24 MeV (dotted line). The real part of the
induced interaction (diagram (c) in Fig.6) is repulsive below
22 MeV and changes sign above.

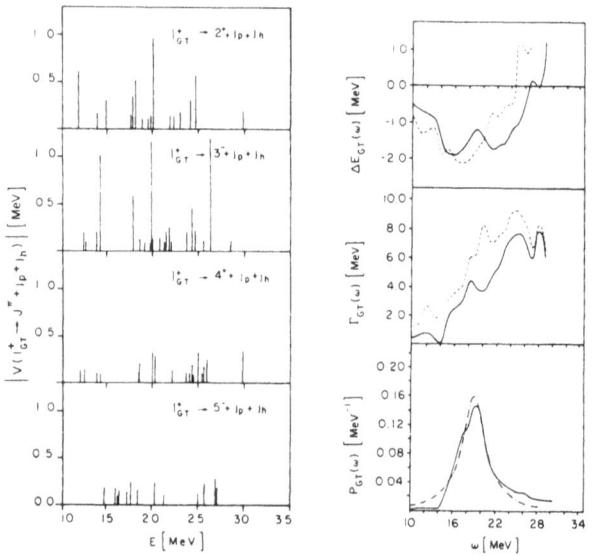

Fig. 9.   left: Energy distribution of the decay matrix
                  elements of the nuclear GT-resonance into
                  a p-h pair and the lowest collective states
                  of ^{208}Pb.

          right: 2p-2h energy shift ΔE_{GT}, spreading width Γ_{GT}
                  and strength function of the GT-resonance.
                  The dashed dotted line indicates the
                  'experimental' strength function[3].

    From independent particle or hole decay into more complex states
one expects a spreading width of ∿ 8 MeV for the GT-state. Observed
are only 4 MeV[3]. The reduction can be understood from the role of
surface modes[2,14] in the damping. In the left part of Fig. 9 we display the GT-decay matrix elements into a surface vibration and an

uncorrelated p-h pair. The decay to the 2^+- and 3^- surface modes is
strongly enhanced. Summing over all possible decay matrix elements
allowed in a 2hω shell model space and averaging over an interval
of 0.5 MeV one obtains Γ↓ as shown in the middle right part of
Fig. 9. The dotted line denotes independent particle and hole
decay while the solid line takes the interference from the induced
interaction into account. The reduction in width is explained by
this process.

5. NUCLEAR SPIN-ISOSPIN SUM RULES BASED ON THE QUARK MODEL

In the discussion of ΔN^{-1} excitations in the nucleus we have
treated nucleons and isobars as distinct elementary particles
characterized by empirical masses, coupling constants, etc. In the
quark picture they are however composite objects of u- and d quarks
in orbital s-states. The only difference is that quark spins and
isospins are flipped in the Δ. The Δ-N mass difference associated
with the transition is given by a residual $\vec{\sigma}\vec{\sigma}$'-interaction. Glashow[16]
proposed a simple Hamiltonian

$$H_o = \sum_i m_q(i) + \frac{B}{2} \sum_{i \neq j} \frac{\vec{\sigma}(i)\vec{\sigma}(j)}{m_q(i)m_q(j)} \tag{5.1}$$

which involves only nonrelativistic quark masses and a coupling constant B. H_o fits the Baryon ground states astonishingly well (Fig.10)

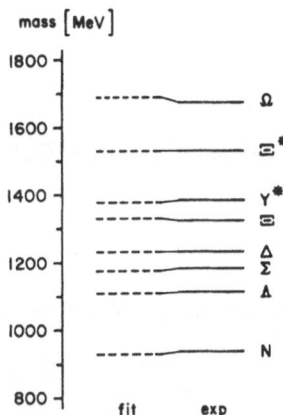

Fig. 10: Baryon ground states in the nonrelativistic quark
          model as obtained by Glashow[16].

Treating nucleons and Δ's in a nucleus as composite objects
we can rewrite nuclear one body transition operators on the quark
level as well. For pure spin-isospin flips without change in the
radial structure of the wavefunctions one has

$$O_{\vec{\sigma}\vec{\tau}} = \sum_{i=1}^{3} \vec{\sigma}_i \vec{\tau}_i \qquad (5.2)$$

where $\vec{\sigma}_i$ and $\vec{\tau}_i$ act on quark i. Since we have a quark Hamiltonian H_o we can evaluate energy weighted sum rules for those operators in nuclei.

$$S(O_{\vec{\sigma}\vec{\tau}}) = \frac{1}{2} <o| [O_{\vec{\sigma}\vec{\tau}}, [O_{\vec{\sigma}\vec{\tau}}, H_o]] |o> \qquad (5.3)$$

(The nuclear ground state $|o>$ consists of A noninteracting bags containing 3 confined quarks.) For the various isospin projections one obtains[17]

$$S(O_{\vec{\sigma}\vec{\tau}_z}) = \frac{2B}{mq^2} \, 8A \qquad (5.4a)$$

$$S(O_{\vec{\sigma}\vec{\tau}_+}) = \frac{2B}{mq^2} \, 24 \, (Z + \frac{1}{3} N) \qquad (5.4b)$$

$$S(O_{\vec{\sigma}\tau_-}) = \frac{2B}{mq^2} \, 24 \, (N + \frac{1}{3} Z) \qquad (5.4c)$$

The connection between $S(\sigma\tau_z)$ and M1-transitions is easily made remembering that the quark magnetic moment is given by

$$\mu_i = \frac{he_i}{2m_i} \sigma_i \qquad (5.5)$$

The quark charge e_i is obtained from the Gell-Mann-Nishijima relation

$$e_i = \frac{1}{2} [\tau_z^i + Y_i]e \qquad (5.6)$$

(For u and d quarks the hypercharge Y is one.) Therefore we obtain the M1-sumrule as

$$S(M1) = S(\vec{\sigma} \tau_z) \, (\frac{m_N}{m_q})^2 \, \mu_N^2 \qquad (5.7)$$

where m_N is the nucleon mass and μ_N the nuclear magnetic moment. Nucleons and isobars interact via the strong 2-body force which we describe in a schematic way[18] by

$$V = \frac{1}{2} g \sum_{i \neq j} \vec{\sigma}_i \vec{\sigma}_j \; \vec{\tau}_i \vec{\tau}_j \tag{5.8}$$

where i and j operate in different baryons. In the presence of V we have to evaluate the double commutator of $O_{\sigma\tau}$ with

$$H = H_o + V \tag{5.9}$$

Instead we may also calculate the RPA expression for the energy weighted sum rule

$$\sum_n (E_n - E_o) |<n|O_{\sigma\tau}|o>|^2_{RPA} = \frac{1}{2} <o|O_{\sigma\tau}[O_{\sigma\tau}, H]|0> \tag{5.10}$$

which follows from Thouless's theorem since H_o plays the role of a Hartree-Fock Hamiltonian[17]. With the schematic interaction the RHS of eq. (5.10) gives the same result as in the uncorrelated case, i.e. the sum rule is conserved.

ACKNOWLEDGEMENT

        This work was supported by USDOE Contract DE-AC02-76ER13001. In addition one of us (R. Fiebig) was supported by Heinrich-Hertz-Stifung.

REFERENCES

1. M. Ericson, Ann. Phys. 63: 562 (1971); E. Oset and M. Rho, Phys. Rev. Lett. 42: 47 (1979); W. Knüpfer, M. Dillig and A. Richter, Phys. Lett. 95B, 349 (1980); H. Toki and W. Weise, Phys. Lett. 97B: 12 (1980); G.F. Bertsch, Nucl. Phys. A354: 157 (1981).
2. H.R. Fiebig and J. Wambach, Nucl. Phys. A386:381 (1982).
3. C. Gaarde, Lectures delivered at the Trieste Nuclear Physics Workshop (October 5-30, 1981) and references therein.
4. M. Ericson and T. Ericson, Ann. Phys. 30: 323 (1966).
5. M.R. Anastasio and G.E. Brown, Nucl. Phys. A285: 516 (1977).
6. T. Suzuki, S. Krewald and J. Speth, Phys. Lett. 107B: 9 (1981).
7. G.E. Brown, J. Speth and F. Wambach, Phys. Rev. Lett. 46: 1057 (1981).
8. S. Krewald and T. Suzuki, private communication.
9. G.E. Brown and M. Rho, Nucl. Phys. A372: 397 (1981).
10. J. Lichtenstadt et al., Phys. Rev. C20: 497 (1979).

11.   K. Bleuler, Seminar at the International School of Physics,
      Varenna (1965), unpublished.
12.   S.O. Bäckman, O. Sjöberg and A.D. Jackson, Nucl. Phys. A321: 10
      (1979); J. Wambach and A.D. Jackson, Nucl. Phys. A348: 221
      (1980).
13.   G. Höhler and E. Pietarinen, Nucl. Phys. B95: 10 (1975).
14.   G.F. Bertsch, P.F. Bertignon, R.A. Broglia and C.H. Dasso,
      Phys. Lett. 80B: 161 (1979); J. Wambach, V. Mishra and
      Li Chu-hsia, Nucl. Phys. A380:285 (1982).
15.   G.E. Brown, J.S. Dehesa and J. Speth, Nucl. Phys. A330: 290
      (1979).
16.   S.L. Glashow, Physika 96A: 27 (1979).
17.   H.R. Fiebig and J. Wambach, Nucl. Phys. A384:371 (1982).
18.   A. Bohr and B.R. Mottelson, Phys. Lett. 100B: 10 (1981).

TRANSVERSE CONTRIBUTIONS TO NATURAL PARITY EXCITATIONS

J. Heisenberg, J. Dawson, and O. Schwentker

Phys. Dept., University of New Hampshire
Durham, N.H. 03824

H.P. Blok

Vrije Universiteit, Amsterdam, The Netherlands

INTRODUCTION

The cross section for inelastic electron scattering in Plane Wave Born Approximation (PWBA) contains one term that in many discussions has been considered unimportant. This term is the subject of this talk.

$$\frac{d\sigma}{d\Omega} = \left(\frac{d\sigma}{d\Omega}\right)_{Mott} \{ \sum_{L>0} |F_L^{\ell}(q)|^2 + (\tfrac{1}{2}+tg^2\tfrac{\theta}{2}) \sum_{L\geq1} (|F_L^{t}(q)|^2 + |F_L^{m}(q)|^2) \} \quad (1)$$

The term we will discuss is the transverse electric form factor $F_L^{t}(q)$. Since this term has the same selection rules as the transition charge operator or the longitudinal form factor this term appears only in natural parity excitations.

DEFINITION OF FORM FACTORS

The natural parity form factors are usually written as Fourier Bessel transforms of the nuclear charge and current densities:

$$F_L^{\ell}(q) = \frac{\hat{J}_f}{\hat{J}_i} \int_0^{\infty} \rho_L(r)\, j_L(qr)\, r^2 dr \qquad (2)$$

$$F_L^{t}(q) = \frac{\hat{J}_f}{\hat{J}_i} \{ \sqrt{\frac{L+1}{2L+1}} \int_0^{\infty} J_{L,L-1}(r) j_{L-1}(qr) r^2 dr + $$
$$\sqrt{\frac{L}{2L+1}} \int_0^{\infty} J_{L,L+1}(r) j_{L+1}(qr) r^2 dr \qquad (3)$$

495

In the event of a natural parity excitation the PWBA cross section shows us that we can measure only two independent quantities while the form factors depend on three nuclear quantities the transition charge ρ and the two transition current densities $J_{L,L-1}$ and $J_{L,L+1}$. These three densities are not independent, since we believe firmly in the continuity equation which ties the divergence of the currents to the transition charge:

$$\hat{L}\,\frac{\omega}{c}\,\rho_L(r) = \sqrt{L}\,\left(\frac{L-1}{r} - \frac{d}{dr}\right)\,J_{L,L-1}(r) - \sqrt{L-1}\,\left(\frac{L+2}{r} + \frac{d}{dr}\right)J_{L,L+1}(r) \quad (4)$$

If we incorporate the condition posed by this continuity equation in these formulae, we can eliminate $J_{L,L-1}$ and write the form factors as Fourier Bessel transforms of only two radial densities, ρ and $J_{L,L+1}$:

$$F_L^{\,t}(q) = \left[\sqrt{\frac{L+1}{L}}\,\frac{\omega}{qc}\,F_L^{\ell}(q) + \frac{\hat{J}_f}{\hat{J}_i}\,\sqrt{\frac{2L+1}{L}}\,\int_0^\infty J_{L,L+1}(r)\,j_{L+1}(qr)\,r^2 dr\right] \quad (5)$$

In this form, we see that the modified Tassie model commonly used for the description of collective levels and which assumes the absence of $J_{L,L+1}$ is equivalent to the assumption that Siegert's theorem holds for all momentum transfers. We also see that $J_{L,L+1}$ can be used as a representation of the transverse electric form factor.

In previous works, this transverse electric contribution has been largely neglected with some exceptions.[1,2] Indeed for very collective states, it requires an experiment at 180^0 to see the effects of that transverse electric form factor. If we look at the size of this current $J_{L,L+1}$ for the individual single ph components:

Charge Density

$$\rho_L(r) = \tilde{S}_{ph}^{L}\,R_p(r)\,R_L(r) \quad (6)$$

$$\tilde{S}_{ph}^{L} = S_{ph}^{L}(-)^{j_p-\frac{1}{2}}\,\frac{\hat{j}_p\hat{j}_h}{\sqrt{4\pi}}\,\begin{pmatrix} j_p & j_h & L \\ \frac{1}{2} & -\frac{1}{2} & 0 \end{pmatrix}$$

Convection Current

$$J_{L,L+1}(r) = \mu_N\,\tilde{S}_{ph}^{L}\,\frac{1}{\hat{L}}\,\{\sqrt{L+1}(R_p(r)R_h'(r) - R_h(r)R_p'(r)) +$$

$$\frac{\ell_p(\ell_p+1) - \ell_h(\ell_h+1)}{\sqrt{L+1}}\,\frac{R_p(r)R_h(r)}{r}\} \quad (7)$$

Magnetization Current

$$J_{L,L+1}(r) = \mu\mu_N \tilde{S}_{ph}^L \frac{1}{\hat{L}} \frac{\chi_p - \chi_h}{\sqrt{L+1}} \left(\frac{d}{dr} - \frac{L}{r}\right) R_p(r) R_h(r) \qquad (8)$$

$$\chi = (\ell-j)(2j+1)$$

we find that they are by no means small. Thus, it is a strong
statement about the nuclear kinematics if we say that the current
$J_{L,L+1}$ becomes vanishingly small. In fact, there are many cases
where this form factor is by no means small, sometimes even dominant. These are the cases we intend to present.

DETERMINATION OF CURRENT DENSITIES FROM EXPERIMENT

Before we show results, we would like to discuss how these current densities are obtained from the experiment. As the PWBA result
shows one can separate the two terms through a comparison of the
cross section measured in forward and those measured in backward
direction. Even though PWBA is not accurate enough for the treatment of inelastic electron scattering, the quite complex DWBA calculations show the same qualitative behavior. It is, therefore, possible to proceed with the determination of these terms in a very similar fashion: First one uses the forward scattering data to determine a first order transition charge. One now considers the backward
scattering data fitting only the transverse density $J_{L,L+1}(r)$ and
keeping the density ρ fixed. With this result, one corrects the forward scattering data for any transverse electric contributions and
performs a new fit to these corrected data to obtain a second order
transition charge. This procedure is iterated and converges quite
rapidly if the data in backward direction are taken at sufficiently
large angles. Figure 1 shows such a result from the first three 5^-
levels in ^{208}Pb.[3] The dashed lines give the cross sections predicted from the transition charge ρ only. The data have been taken
at 160°, and the difference between the data and the dashed curve is
the contribution from $J_{L,L+1}(r)$.

It is obvious from this comparison that the present data are
not very sensitive to the lower Fourier components in J. This is
particularly true for the lowest 5^- level. Thus, the information
obtainable is somewhat limited. With these data one is in a situation where it still makes more sense to use a reasonable model for
the shape of J then to try to determine the density only from the
experiment. As long as the data are reproduced by this model within
the errors, we seem to have a reasonable representation of $J_{L,L+1}$
except that the quoted uncertainties may be underestimated due to
the strong model dependence not considered in the determination of
the uncertainty. These fitted current densities are shown in Figure 2.

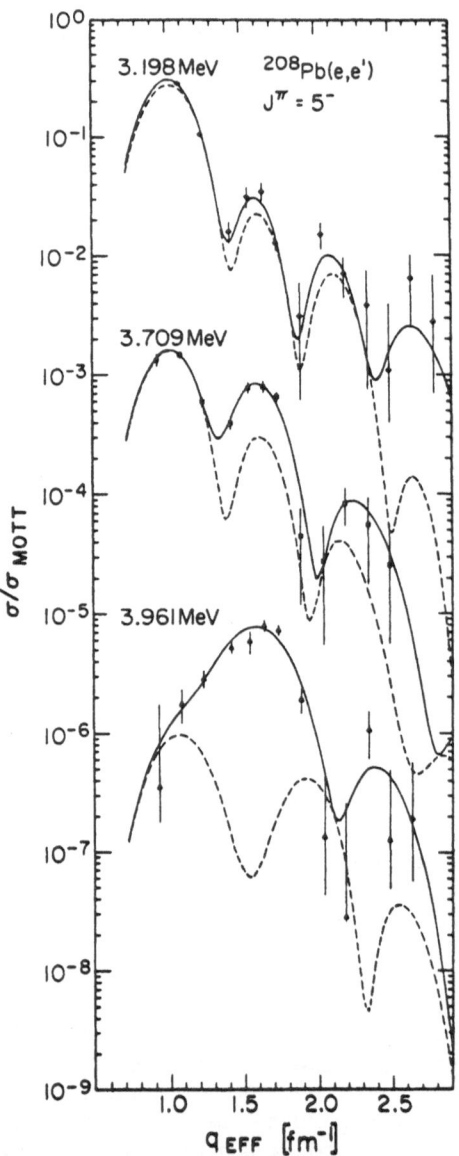

Fig. 1.   Cross sections for the first three 5⁻ levels in ²⁰⁸Pb taken at 160° scattering angle compared to the best fit (solid line) and the prediction with J = 0 (dashed line).

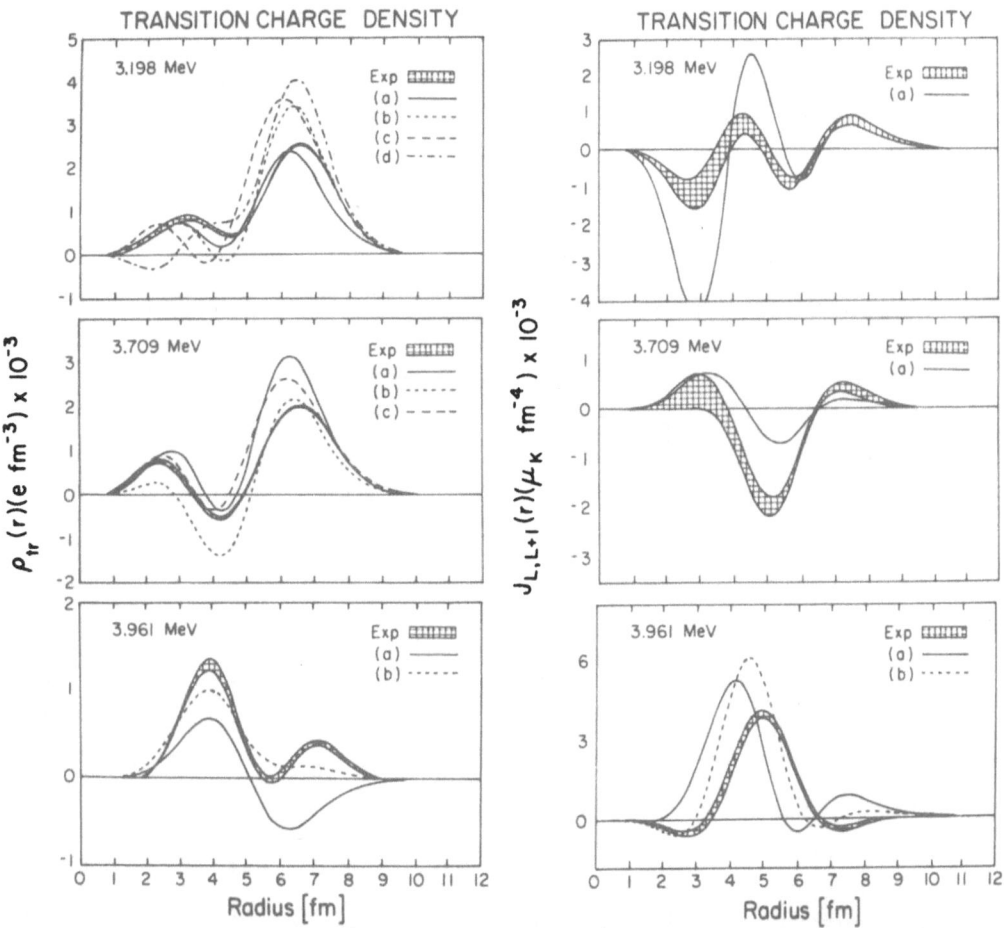

Fig. 2. Transition charge and current densities for the first 5⁻ levels in ^{208}Pb.

GENERAL FEATURES OF CURRENT DENSITIES

Before we go into the discussion of some detailed cases, we should mention some general features. The current is obtained from the general current operator $\bar{j}_n + \bar{\nabla}\times\bar{\mu}_n$. The matrix element of this operator in an unnatural parity transition gives the current $J_{L,L}$ that describes the magnetic transitions while its matrix element in a natural parity transition determines $J_{L,L+1}$. Similar to the magnetic transition one has contributions from the magnetization as well as from the convection current. It has the same properties under time reversal namely: $J_{ab} = -J_{ba}$. The current is also subject to corrections due to exchange currents. Thus, the physics one can learn from this term is essentially identical to what one learns from the magnetic transitions.

Not surprisingly, one finds results for the transverse form factors that are very similar to the results from the magnetic excitations, namely that all observed strengths are considerably quenched. There exists already a large body of experimental results over a wide range of multipolarities that allow us to discuss the systematics.

EXAMPLES

Neutron Excitations in ^{208}Pb

Let us take for instance, the 12^+ level in ^{208}Pb[4] shown in Figure 3:

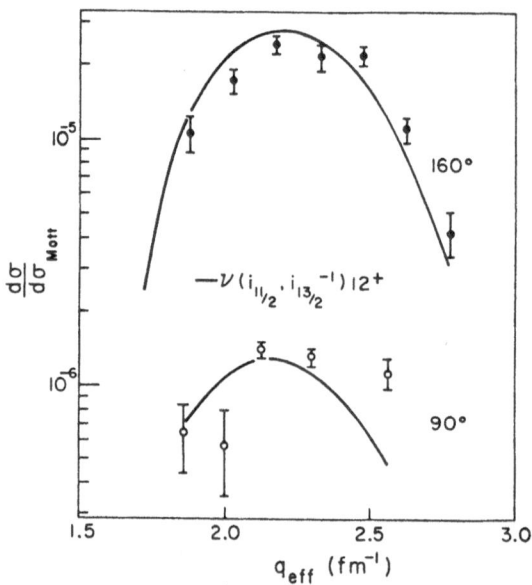

Fig. 3. Reduced (e,e') cross sections from the excited 12^+ state at 6.10 MeV, measured at 90° and 160°. Solid lines are DWBA calculations of the SPH transition $\nu(1i_{11/2}, 1i_{13/2}^{-1})12^+$ scaled down by 0.65 g_{free}.

The observed form factor shows only 40% of the single particle strength, a result that is quite similar to the results from the magnetic excitations of the 14^- and 12^- states in the same nucleus[5], where only 50% of the single particle strength has been observed. Even more quenching is observed in the 7^- level at 4.037 MeV, where only 35% of the single particle strength is seen[3] (Figure 4 and 5).

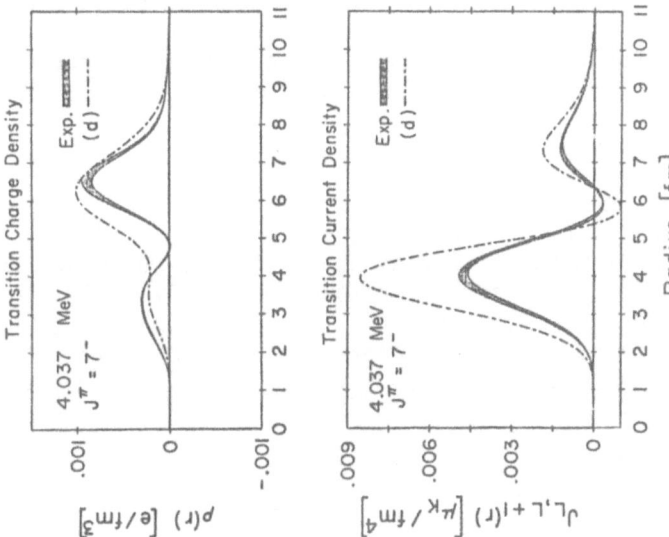

Fig. 5. Transition charge and current densities for the 7⁻ level in ²⁰⁸Pb compared to a RPA prediction.

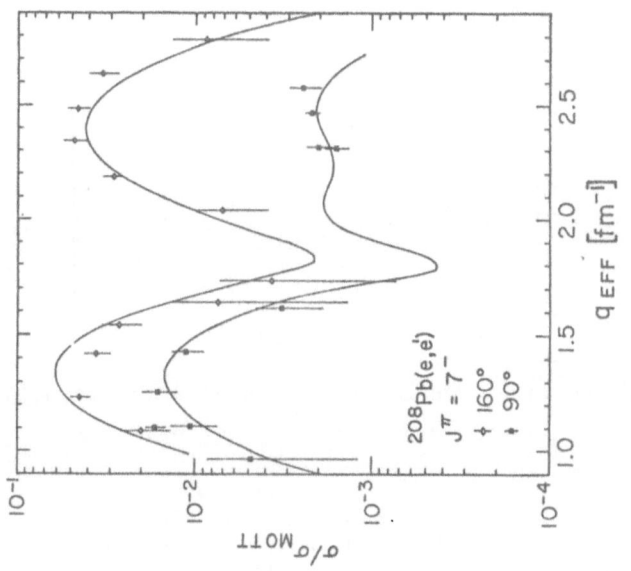

Fig. 4. Reduced cross sections for the 7⁻ level at 4.037 MeV in ²⁰⁸Pb shown with the DWBA fit.

In these two examples, the longitudinal parts are either small
(7⁻) or even negligible (12⁺). The reason is that these states are
built dominantly from neutron configurations. Because of the smallness of the neutron charge form factor, one does not observe much
charge scattering.

Proton Excitations Near Mass 90

This situation changes in a nucleus for which the low lying
states are dominated by proton configurations such as ^{90}Zr. Figure
6 shows the data from the 7⁻ level[6] at 4.373 MeV, which at large
momentum transfers is the strongest level observed. This state is
generated dominantly from the proton structure $\pi(g_{9/2}, f_{5/2}^{-1})$. We
observe in this case, a strong transverse and a large longitudinal
form factor.

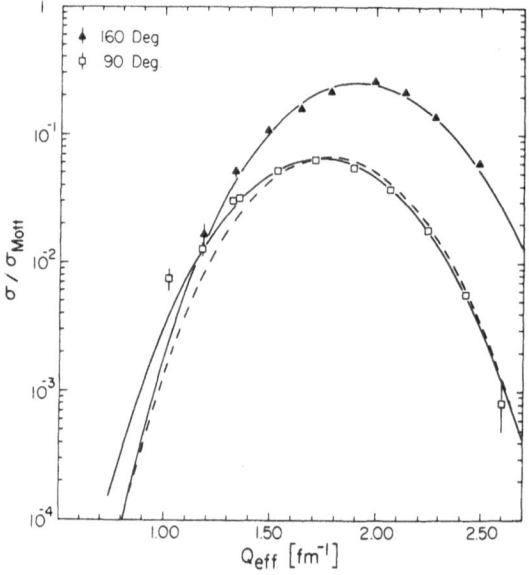

Fig. 6. Reduced cross sections for the 7⁻ level at 4.373 MeV in
^{90}Zr.

A similar result is obtained on the lowest E5 excitations in
^{90}Zr and ^{89}Y[8] illustrated in Fig. 7. Even though the effects
observed with $J_{L,L+1}$ are similar to those observed on magnetic transitions, one has an added advantage in these cases, which comes from
the fact that one observes for the very same transition two separate
properties, namely the charge and the current. This additional
information allows one to rule out certain interpretations.

Krewald and Speth[7] interpreted the reduced strength observed in the high spin states in ^{208}Pb as a consequence of a reduced spectroscopic factor. Such an interpretation in the case of a natural parity state would lead to the same reduction in the transition charge as in the transition current and, in this aspect, can be checked out experimentally.

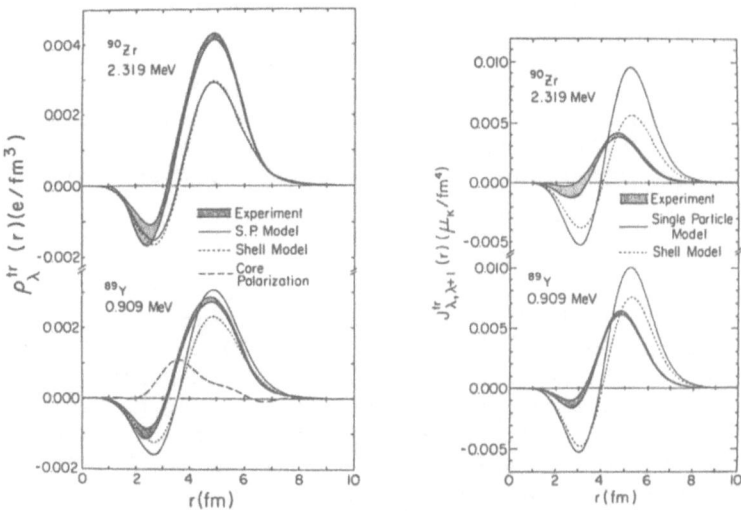

Fig. 7. Transition charge and current densities for the lowest E5 transition in ^{90}Zr and ^{89}Y.

While we might expect that the charge and current densities should scale, we find quite the opposite. In the case of the 7$^-$ level in ^{90}Zr, shown in Figure 6, we observe in the charge scattering only 66% of the single particle strength, which indicates that the spectroscopic factor is smaller than one. In the transverse scattering, we observe only 19% of the single particle strength. Thus, these combined results are inconsistent with just a spectroscopic factor causing the reductions for this case. In addition, the radial shapes observed for the charge and the current scattering are inconsistent with each other if they were produced only by this single particle transition. Scaling the transition current measured for the E5 transition in ^{89}Y up with the same factor needed to match the transition charge, we overpredict the observed current by a factor of 2.4. Thus, we must conclude that we observe core polarization or mesonic effects that affect the charge and current distributions differently. In our opinion, the most likely interpretation in this case is a fairly strong core polarization in the currents that reduces the currents while the charge is much less affected by it. This remains to be verified in detail by nuclear structure calculations, but such effects have been discussed in the literature.[9]

Let us now discuss the E5 transitions observed in ^{89}Y and ^{90}Zr[8] more carefully. From the nuclear structure, these transitions are very similar. Shell model calculations show that both transitions are almost pure $\pi(1g_{9/2}, 2p_{1/2}^{-1})$ transitions with a spectroscopic factor reducing the strength in ^{89}Y considerably. The discussion of Ref. 9 suggests that core polarization effects are smooth in a given mass region. In that spirit, we would expect that the core polarization (or effective charge) should be the same for the E5 transitions in ^{89}Y and ^{90}Zr. Indeed, the shapes for the transition charge and current densities in the two nuclei are very similar, supporting this idea. Core polarization modifies the shape of the transition charge density considerably since it is surface peaked while the single particle density has a node. One can, therefore, take the shift of the node from the single particle position as a measure of the strength of the core polarization. We find the following shifts:

^{89}Y : -0.392 ± 0.045
^{90}Zr : -0.368 ± 0.060

which is identical within the statistical uncertainties. Similarly, we find that the nodes in the currents are also shifted the same amount even though the uncertainties are considerably larger. In addition, from the shell model we expect a strong pairing effect that gives substantial occupation probability for the $g_{9/2}$ orbit. In addition to the amplitude A for the transition $2p_{1/2} \rightarrow 1g_{9/2}$ this gives rise to a large backward going amplitude $\bar{A}$ the transition $1g_{9/2} \rightarrow 2p_{1/2}$ in the excitation of the 5$^-$ level. With these amplitudes, the charge density is given by

$$\rho(r) = (A + \bar{A})\rho_{2p_{1/2} \rightarrow 1g_{9/2}}(r)$$

and the transition current by

$$J(r) = (A - \bar{A})J_{2p_{1/2} \rightarrow 1g_{9/2}}(r).$$

Thus, neglecting the core polarization effect, the measured reduction

$$\frac{A - \bar{A}}{A + \bar{A}} = 0.42 \pm 0.06$$

yields

$$\bar{A}/A = 0.41 \pm 0.06.$$

This ratio can be directly compared to theoretical predictions:

RPA[10]   :   0.088
BCS[11]   :   0.273
Shell Model[12] :   0.220

All of the calculations mentioned predict energy levels that are in rather good agreement with the observed energy levels of ^{90}Zr. Yet they underpredict the actual ground state correlations considerably. This effect is most severe for the RPA. The reason is that for the high spin states such as these 5$^-$ states, the residual interaction is rather small and not the main cause for the ground state 2p2h components. The RPA neglects the coupling between the 2p2h components of different multipolarity in the ground state which enters through the p-p and the h-h interaction in <2p2h|V|2p2h>. For that reason, one expects considerably more quenching in the high spin states than RPA calculations predict. This may indeed be the dominant contribution to the quenching observed in ^{208}Pb.[3,4,5]

A proper combination of core polarization, which is probably contained in the RPA, and the shell model probably can account for the experimental observation. This first example where the effects of ground state correlations can be clearly measured directly, indicates that the importance and the uncertainty in contributions to the quenching of electromagnetic transitions in nuclei due to conventional nuclear structure effects may largely be underestimated. Thus, we feel that the quenched M1 strength seen in ^{48}Ca[13] is at the present time a much less conclusive evidence for the role of the delta in nuclei than presented so far at this meeting.

REFERENCES

1. R.A. Lindgren, et al., Phys. Rev. Lett. <u>41</u>, 1705 (1978).
2. J.B. Flanz, et al., Phys. Rev. Lett. <u>41</u>, 1642 (1978).
3. J. Heisenberg, J. Lichtenstadt, C.N. Papanicolas, and J.S. McCarthy, Phys. Rev. <u>C25</u>, 2292 (1982).
4. J. Lichtenstadt, C.N. Papanicolas, C.P. Sargent, J. Heisenberg, and J.S. McCarthy, Phys. Rev. Lett <u>44</u>, 858 (1980).
5. J. Lichtenstadt, J. Heisenberg, C.N. Papanicolas, C.P. Sargent, A.N. Courtemanche, and J.S. McCarthy, Phys. Rev. <u>C20</u>, 497 (1979).
6. J. Heisenberg, et al., to be published.
7. S. Krewald and J. Speth, Phys. Rev. Lett. <u>45</u>, 417 (1980).
8. O. Schwentker, J. Dawson, S. McCaffrey, J. Robb, J. Heisenberg, J. Lichtenstadt, C.N. Papanicolas, J. Wise, J.S. McCarthy, N. Hintz, and H.P. Blok, Phys. Lett. <u>112B</u>, 40 (1982).
9. F. Petrovich, J.R. Borysowicz, H. McManus, and G.R. Hammerstein, Phys. Rev. <u>C16</u>, 839 (1977).
10. The calculations were done using the programs and the interaction from the Jülich group. See also Ref. 7.
11. K. Akkerman, private communication.
12. J. Dubach and W. Haxton, private communication. The shell model calculation uses the Kuo-Brown matrix elements.
13. W. Steffen, H.D. Graff, W. Gross, D. Meuer, A. Richter, E. Spamer, O. Titze, and W. Knupfer, Phys. Lett. <u>95B</u>, 23 (1980).

CONVECTION CURRENT CONTRIBUTIONS TO SPIN EXCITATIONS

G.E. Walker

Nuclear Theory Center
Physics Department
Indiana University
Bloomington, IN 47405

ABSTRACT

We first briefly review the role of convection currents in
electron scattering and the utility of Rosenbluth plots in separating and studying the familiar longitudinal and transverse form
factors defined for electromagnetically induced nuclear transitions.
Then, noting that an energy-dependent probe-target-nucleon interaction can often be cast, to a good approximation, as an energy independent leading term plus a "convection current" dominated correction term, we study pion-nucleus and nucleon-nucleus transitions
involving these "convection current" corrections. We discuss how,
in both pion and nucleon-nucleus interactions, techniques that separate the (q, E, θ) dependences of the nuclear response (as the Rosenbluth plot does for electron scattering) can be used to advantage.

I. INTRODUCTION

We wish to suggest some ways that convection currents may contribute in nucleon-nucleus inelastic scattering, when these contributions may be important, how they may be separated from other contributions, and what one may learn from their study. In order to
set the stage for this discussion it is useful to first review some
well known results from electron scattering concerning convection
currents and the utility of using a Rosenbluth plot or working near
180° scattering angle to separate longitudinal and transverse form
factor contributions to the electromagnetic nuclear response function. This is done in the next section. In Section III we discuss
similar subjects for pion-nucleus inelastic scattering. More specifically we review how some simple considerations involving the
angle dependence of non-normal parity transitions (compared to the

507

angle dependence associated with non-spin flip normal parity transitions) can provide useful information for identifying the spins
of excited states reached in (π,π'). In this section we also discuss
how the energy-dependence associated with the π-N interaction and
non-localities associated with intermediate isobar propagation and
nucleon recoil at the $\pi N\Delta$ vertex can effectively induce convection
current contributions in pion-nucleus inelastic scattering. Finally
in Section IV we discuss how the two-particle spin-orbit force introduces convection current contributions in nucleon-nucleus reactions.
We show how to use a fixed q, variable energy or scattering angle
plot to identify strong spin-orbit pieces not coming from convection
current contributions (in the region $\sim$ 100 - 300 MeV). We also show
how an energy dependent N-N t-matrix, that would in the usual direct
energy-independent approximation, contribute only to normal parity
transitions, possesses a convection current contribution to non-
normal parity transitions such as those of the Gamow-Teller type.
We discuss this result in some detail, estimating its effect in
special cases, and outline a research program currently underway to
study the associated effects for a wide variety of states and kinematic conditions.

II. ELECTRON SCATTERING

The results contained in this section are generally well known
and so will be summarized quite briefly. For simplicity and definiteness, we consider inelastic electron scattering on a localized
nucleus causing a nuclear transition from the target ground state,
possessing total angular momentum $J_i = 0$, to a nuclear final state
with total angular momentum J_f. The great advantages of electron
scattering are associated with the facts that the electromagnetic
interaction is weak and well known. Assuming a single photon
exchange process and neglecting target recoil, one can write[1] the
expression for the inelastic scattering differential cross section
on a unpolarized target as [q $\equiv$ three momentum transfer, ω $\equiv$ electron energy loss]

$$\frac{d\sigma}{d\Omega} = 4\pi \; \sigma_M \left[\frac{(q^2-\omega^2)^2}{q^4} F_L^2(q) + \left(\frac{q^2-\omega^2}{2q^2} + \tan^2\frac{\theta}{2} \right) F_T^2(q) \right] \quad (1a)$$

$$F_L^2(q) = \sum_{J=0}^{\infty} \left| \langle J_f || M_J(q) || J_i \rangle \right|^2 \quad (1b)$$

$$F_T^2(q) = \sum_{J=1}^{\infty} \left[\left| \langle J_f || T_J^{e\ell}(q) || J_i \rangle \right|^2 + \left| \langle J_f || T_J^{mag}(q) || J_i \rangle \right|^2 \right] \quad (1c)$$

where the nuclear structure information is contained in the quantities defined as the longitudinal $(F_L^2(q))$ and transverse $(F_T^2(q))$ form
factors. The quantity σ_M denotes the Mott scattering cross section
and θ is the laboratory scattering angle. Note the nuclear structure form factors depend only on the three-momentum transfer. Thus
by working at fixed q and ω but varying the incident electron energy
and thus the scattering angle one can separate the dependence of the
cross section on $F_T^2(q)$ and $F_L^2(q)$. For example, at 180° only $F_T^2(q)$
contributes.

Besides working at 180° and thus isolating the transverse form
factor one can also construct a Rosenbluth plot. Suppose we concentrate on a definite nuclear excitation $(\omega^2$ fixed) and work at fixed
q^2 in a situation where $q^2 \gg \omega^2$. Using eq. (1) we find

$$\frac{1}{4\pi\sigma_M} \frac{d\sigma}{d\Omega} = F_L^2(q) + F_T^2(q)\{^1/_2 + \tan^2\theta/_2\} \qquad (2a)$$

$$y \quad = \quad b \quad + \quad m \quad x \quad . \qquad (2b)$$

An (x,y) plot of this straight line yields the longitudinal form
factor as the y intercept and the slope of the straight line yields
the transverse form factor. Deviations from the straight line indicate the single virtual photon exchange approximation is not adequate.
Non-normal parity transitions would yield a y=0 intercept and positive slope, $0^+ \to 0^+$ transitions would yield a horizontal straight
line, while in general, normal parity transitions have a positive y
intercept and non-zero positive slope.

The nucleus interacts with the external electromagnetic four-
vector potential (A,A_o) via[1]

$$H_{int.} = -e \left[\int \hat{j}_N(x) \cdot \vec{A}(x)dx + \int \hat{\mu}_N(x) \cdot (\vec{\nabla}x\vec{A}(x))dx\right.$$

$$\underbrace{\phantom{\int \hat{j}_N(x) \cdot \vec{A}(x)dx}}_{\substack{\text{Nuclear}\\\text{convection}\\\text{density}}} \qquad \underbrace{\phantom{\int \hat{\mu}_N(x) \cdot (\vec{\nabla}x\vec{A}(x))dx}}_{\substack{\text{Nuclear}\\\text{magnetization}\\\text{current}\\\text{density}}}$$

$$\left. - \int \hat{\rho}_N(x)A_o(x)dx \right] \qquad . \qquad\qquad (3)$$

$$\underbrace{\phantom{\int \hat{\rho}_N(x)A_o(x)dx}}_{\substack{\text{Nuclear}\\\text{charge}\\\text{density}}}$$

Often the various nuclear densities listed above are assumed to
arise from a sum of single nucleon operators[1]

$$\hat{\rho}_N(x) = \sum_i \delta(\vec{r}_i - \vec{x})(1 + \tau_z(i))/2 \qquad \text{(proton charge} \qquad (4a)$$
$$\text{density)}$$

$$\hat{\vec{j}}_N(x) = \sum_i [\delta(\vec{r}_i - \vec{x})(1 + \tau_z(i))/2 \frac{1}{iM}\vec{\nabla}(i)]_{sym} \qquad \text{(proton convec-} \qquad (4b)$$
$$\text{tion current density)}$$

$$\hat{\vec{\mu}}_N(x) = \sum_i [\delta(\vec{r}_i - \vec{x})\{\underbrace{(\mu_p + \mu_n)}_{.88}/2 + ([\underbrace{(\mu_p - \mu_n)}_{4.7}/2]\tau_z(i))\}\vec{\sigma}(i). \qquad (4c)$$

$$\text{(magnetization or} \\ \text{spin density)}$$

The nuclear density operators appearing in eqs. (4a-c) appear in the multipole operators M_{JM}, T^{el}_{JM}, and T^{mag}_{JM} as follows[1]:

$$\hat{T}^{el}_{JM}(q) = \frac{1}{q} \int d\vec{x}\{(\vec{\nabla} \times j_J(qx)\vec{Y}^M_{JJ1}) \cdot \hat{\vec{j}}_N(x) + q^2 j_J(qx)\vec{Y}^M_{JJ1} \cdot \hat{\vec{\mu}}_N(x)\}$$

$$(5a)$$

$$\hat{T}^{mag}_{JM}(q) = \int d\vec{x}\{j_J(qx)\vec{Y}^M_{JJ1} \cdot \hat{\vec{j}}_N(x) + (\vec{\nabla} \times j_J(qx)\vec{Y}^M_{JJ1}) \cdot \hat{\vec{\mu}}_N(x)\}$$

$$(5b)$$

$$\hat{M}^{coul}_{JM}(q) = \int d\vec{x} \ j_J(qx) Y_{JM} \ \hat{\rho}_N(x) \quad .$$

$$(5c)$$

Both the convection and magnetization densities appear in each of the transverse operators T^{mag} and T^{el}. Most of the discussion at this conference is associated with the excitation of Gamow-Teller resonances, reached via the isovector spin density operator in eq. (4c); however, we shall be interested in the convection current density operator, eq. (4b), contributions to non-normal parity transitions.

Since we are interested in exploring the role of convection currents in nucleon-nucleus reactions involving non-normal parity transitions it is useful to note where in the nuclear response in the (q,ω) plane, such currents associated with $F^2_T(q)$ play a role in (e,e'). More generally it is useful to remind ourselves of the character of the most strongly excited states in inelastic electron scattering. The longitudinal form factor contains the information regarding the charge density and involves only normal parity transitions. At non-backward angles strong normal parity $\Delta T = 0$ nuclear transitions are generally dominated by the longitudinal charge form factor - high spin states becoming naturally more important as the three momentum transfer increases. As is suggested by eqs. (4b, c), the isovector magnetization density is the most important contributor to prominent excitations reached via the transverse part of the interaction. In particular, prominent "Gamow-Teller" transitions are dominated by this term (although, of course not all $\Delta T = 1$, $J_f = 1^+$ states have small convection current contributions).[2] High spin non-normal parity $\Delta T = 1$ stretched states have no convection current contributions

in the usual models for such states, i.e. the probe must interact
with the intrinsic spin as well as transfer orbital angular momentum
to excite such states. By using the equation of continuity one
relates the most often better-known matrix elements of the charge
density to the corresponding isospin and multipole matrix element of
the convection current in the long-wavelength (small q) limit.[1,3] By
working near 180°, one can study the role of convection currents for
some states thought to have large convection current contributions
such as the T = 0, 3⁻(ΔL = 3, ΔS = 0) 9.6 MeV state of ^{12}C. In a
previous talk Prof. Heisenberg[3,4] reviewed some of the new information becoming available that will considerably expand our information
on convection currents from (e,e'). The general situation at present
is, however, that we have more electromagnetic information on charge
and magnetization densities. Convection current contributions to
non-normal parity (medium to large q) non-stretched states may be
important for pion and proton-nucleus reactions (to be discussed
below). For such transitions, further (e,e') information would be
very useful but exceptionally challenging to obtain because of the
properties of the basic electromagnetic operator.

Before leaving this section we explicitly list an expression,
given in ref. 1, useful for comparing the relative contributions of
the magnetization density and convection current density to the
1⁺ T = 0 and 1(12.71 and 15.11 MeV) states of ^{12}C using a pure
$1p_{1/2}1p_{3/2}^{-1}$ harmonic oscillator particle-hole shell model wavefunction:

$$\frac{d\sigma}{d\Omega} \sim \left(\frac{q}{M}\right)^2 e^{-q^2b^2/2} \left[\underbrace{(\mu_p \pm \mu_n)}_{\substack{\text{magnetization}\\\text{density}}} [1 - \frac{q^2b^2}{8}] + \underbrace{[-^1/_2]}_{\substack{\text{convection}\\\text{current}\\\text{density}}} \right]^2 \qquad (6)$$

where $(\mu_p+\mu_n)$ = 0.88 for ΔT = 0 and $(\mu_p-\mu_n)$ = 4.7 for ΔT = 1 transitions.

Thus for electron scattering the 1⁺ T = 1(G-T) state is strongly
excited and completely dominated by the magnetization density. The
1⁺ T = 0 state is very weakly excited. For the T = 0 state the convection current is predicted to play a role but small effects such as
possible isospin mixing make conclusions difficult from the data.

III. PION-NUCLEUS REACTIONS

We briefly discuss pion-nucleus inelastic scattering because in
this case an energy dependent t-matrix induces strong "convection
current" contributions to selected reactions. Also it has been possible by working at fixed q and varying E or θ to obtain useful additional information even in this case of a strongly interacting probe.
The main feature associated with intermediate energy pion nucleus

elastic and inelastic scattering is the dominance of the (3/2, 3/2)
resonance. This resonance induces a strong energy dependence in the
elastic scattering and in the transition operator responsible for
nuclear transitions. On the other hand for (π,π') since the energy
dependence is approximately the same for elastic and inelastic scattering channels the energy dependent absorption in the elastic channel is largely compensated by the energy dependence of the inelastic
transition operator. This results in significantly less energy dependence in DWIA calculations than might otherwise have been expected.[5] Several authors have studied (π,π') adopting a simple separable form for the π-N t-matrix. Using the DWIA and neglecting terms
of order m_π/M_N allows one to write the (π,π') inelastic cross section
in the form

$$\frac{d\sigma}{d\Omega}(q,E,\theta) = \frac{E(k')k'}{E(k)k}\, e^{q^2 b^2/2A}\; \frac{4\pi}{2J_i+1}\; [F^2_{\rho 0} + \cos\theta\; 2\mathrm{Re}\{F_{\rho 0}F^*_{\rho 1}\}$$

$$+ \cos^2\theta\; F^2_{\rho 1} + \frac{\sin^2\theta}{2}\; F^2_\sigma] \tag{7}$$

where $F^2_{\rho i}$ ($i = 0(1)$ denotes contribution from πN S(P) wave) arises
from the interaction of the pion with the nuclear matter density and
is given by

$$F^2_{\rho i} = \left(\frac{k'}{k}\right)^2 \sum_{J=0}^{\infty}\; |<J_f T_f T_{zf}||M^i_J(q,E)||J_i T_i T_{zi}>|^2 \quad . \tag{8}$$

This term is quite similar to the longitudinal form factor in inelastic electron scattering. Only normal parity isoscalar and isovector
non-spin flip transitions contribute. The magnetization density contribution is contained in the term

$$F_\sigma^{\ 2} = \left(\frac{k'}{k}\right)^2 \sum_{J=1}^{\infty} [\,|<J_f T_f T_{zf}||\hat{S}^N_J(q,E)||J_i T_i T_{zi}>|^2$$

$$+ |<J_f T_f T_{zf}||\hat{S}^U_J(q,E)||J_i T_i T_{zi}>|^2\,] \tag{9}$$

which arises purely from P wave pion-nucleon scattering and, in the
present approximation, contains no convection current term. The
transition operator $S^N(S^U)$ induces only natural (unnatural) parity
isoscalar and isovector spin transitions. Note that the $\hat{M}$ and $\hat{S}$
operators are energy dependent unlike the situation for the charge
and spin density operators in (e,e'). However, it is a good first
approximation to assume that the P_{33} term dominates both the charge
and magnetization operator's energy dependence. Therefore this
energy dependence is roughly the same for all terms. Assuming P_{33}
dominance means that the various spin and isospin dependent operator
magnitudes are related by A(ΔS=0, ΔT=0): B(ΔS=0, ΔT=1): C(ΔS=1,
ΔT=0):D(ΔS=1, ΔT=1) $\rightarrow$ 4:2:2:1. Thus assuming the same degree of collectivity in nuclear wavefunctions, <u>cross sections</u> for spin-flip

isoscalar transitions should be four times as prominent as spin-flip isovector transitions. This is in sharp contrast to the dominance of the spin-flip isovector transition in (e,e') arising because of the large isovector nucleon magnetic moment.

Assuming a common energy dependence and keeping only the P_{33} terms allows eq. (7) to be written schematically as

$$\frac{d\sigma}{d\Omega} \frac{1}{\Gamma(E)} = 4 \ M^2(q) \ \cos^2\theta + S^2(q) \ \sin^2\theta \qquad (10)$$

where $\Gamma(E)$ is a relatively slowly varying function of energy and, in the present approximation, state-independent (see discussion before eq. (7)). In the appropriate angular range ($\theta \leq 90°$) one could divide both sides of eq. (10) by $\cos^2\theta$ and obtain

$$G(E,\theta) = 4M^2(q_o) + S^2(q_o) \ \tan^2\theta \qquad (11)$$

which, for fixed momentum transfer q_o (usually taken near the form factor maximum) has the form of a straight line $y = mx + b$ ($x = \tan^2\theta$), where the slope of the straight line would yield the P wave spin-flip contribution and the contribution at $x = 0$ yields the spin-independent P wave term. For an unnatural parity transition the y intercept at $x = 0$ is zero. Eqs. (10)-(11) have been shown to be useful in identifying the spin-parity of transitions in (π,π'). Some examples, showing the behavior of the data are given in fig. 1.[5] The simple idea works surprisingly well. Naturally one should remain cautious in using eq. (10); however, the success of such ideas in (π,π') motivates us to consider applying these same kinds of ideas for (p,p') in the next section.

Thus far in our discussion convection currents have not entered due to the simplicity of the assumed model. In fact, due to nucleon recoil terms in the basic pion-nucleon coupling as well as an energy dependence in the π-N interaction and medium corrections associated with intermediate Δ propagation and the Δ-nucleus interaction, the effective π-nucleon transition operator is far more complicated (richer) than implied in the preceeding discussion. (See, for example, the contribution by F. Lenz at this conference.[7]) In what follows I will concentrate on the role of nuclear Fermi-motion and the energy dependence of the π-N interaction in inducing an effective convection current contribution in (π,π'). This same effect will be discussed in the next section for (N,N').

We assume the pion-nucleon transition operator is a function of the square of the symmetrized invariant energy $S = [(k_\pi + k_\pi')/2 + (P_N + P_N')/2]^2$. The pion-nucleon transition operator has been evaluated in eq. (7), assuming the nucleon is at rest (fixed scatterer). We can obtain an expression for the pion-nucleon transition operator allowing for nucleon motion by making a Taylor series expansion about $S = S_o$ (nucleon at rest).[5,8] For definiteness we consider the non-spin-flip

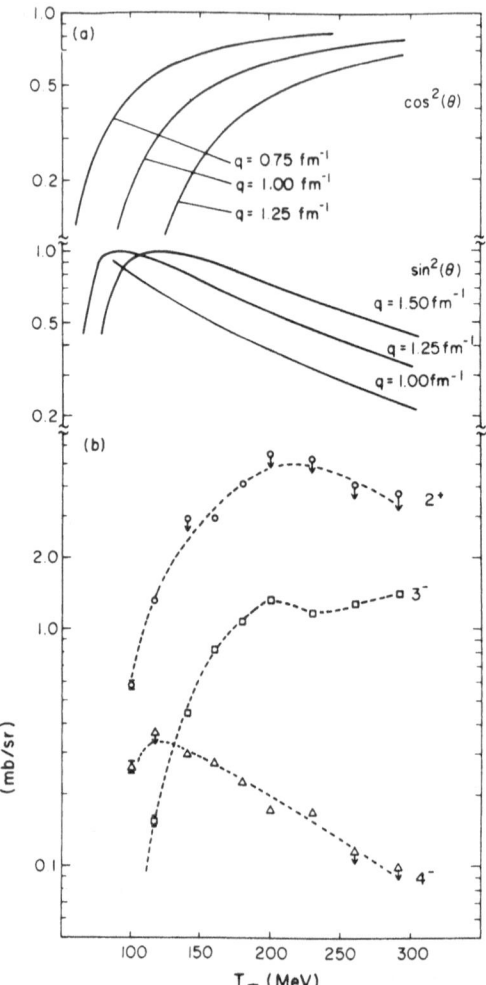

FIG. 1. Differential cross sections at fixed
momentum transfer (q) as a function of pion
laboratory kinetic energy. (a) Plots of $\cos^2(\theta)$
and $\sin^2(\theta)$ normalized to unity and calculated
from Eq. (10) for the indicated values of q.
(b) π^+ inelastic data for the 2^+(4.44 MeV),
3^-(9.64 MeV), and 4^-(19.25 MeV) states of ^{12}C.
The data is from ref. 6. The dashed curves
are to guide the eye.

isoscalar part of the interaction A($\Delta S = 0$, $\Delta T = 0$). We obtain

$$A_\ell(S) = A_\ell(S_0) + (S - S_0) \left.\frac{\partial A_\ell}{\partial S}\right|_{S_0} + \frac{(S - S_0)^2}{2} \left.\frac{\partial^2}{\partial S^2} A_\ell\right|_{S_0} + \cdots$$

$$(12)$$

Numerical estimates indicate that slightly above and below the P_{33} resonance the dominant correction is given by the first derivative term. Using the approximate relationships $S \approx S_0 - (\vec{k}_\pi + \vec{k}_\pi') \cdot (\vec{P}_N + \vec{P}_N')/2$ and $\partial/\partial S \approx (1/2M_N)\partial/\partial T_\pi$ allow the non-spinflip transition operator with Fermi motion corrections to be written as

$$A_1(S)\hat\rho(r) \approx A_1(S_0)\hat\rho(r) - \frac{\partial A_1}{\partial E} \frac{(\vec{k}_\pi + \vec{k}_\pi')}{2} \cdot \hat{j}_N(\vec{r}) \qquad (13)$$

where $\hat{j}_N$ is the usual matter convection current operator

$$\hat{j}_N(\vec{r}) = \sum_i \frac{1}{(iM_N)} [\delta(\vec{r} - \vec{r}_i)\vec{\nabla}(i)]_{sym} \qquad . \qquad (14)$$

As pointed out previously, an important implication for this Fermi motion correction term is that the relatively large non-spin-flip part of the pion-nucleon interaction can now contribute, via the convection current operator, to unnatural parity transitions. This occurs because the total energy in the C.M. system depends on the angle between the direction of nucleon motion and the direction of pion motion in the laboratory frame. The convection current correction, when added to eq. (7) yields

$$\frac{d\sigma'}{d\Omega} = \frac{d\sigma}{d\Omega} + \frac{E(k)k'}{E(k)k} e^{q^2b^2/2A} \frac{2\pi}{(2J_i + 1)} [\cos^2\theta/2 \cos^2\theta F_c^2$$

$$- \sin(2\theta) \cos(\theta/2) \operatorname{Re} \{F_\sigma F_c^*\}] \qquad (15)$$

where F_c^2, the convection current term, is broken up into a natural and unnatural parity transition operator

$$F_c^2 = (k')^2 (\frac{k'}{k}) \sum_{J=1}^\infty [|<J_f T_f T_{zf}||\mathcal{J}_J^N||J_i T_i T_{zi}>|^2 +$$

$$+ |<J_f T_f T_{zf}||\mathcal{J}_J^U||J_i T_i T_{zi}>|^2] \qquad . \qquad (16)$$

The considerable importance of this energy dependent correction is shown in fig. 2 for (π,π') leading to the 12.71 MeV 1^+ T = 0 excited state of ^{12}C. Note that this particular convection current correction does not contribute to $0^+ \rightarrow 0^-$ transitions, to stretched state transitions (because such transitions require $\Delta S = 1$), and that the

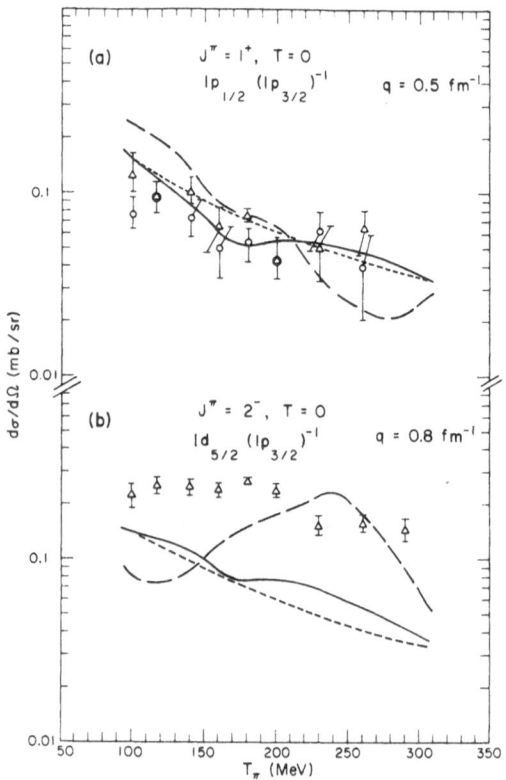

FIG. 2. Differential cross sections at fixed
momentum transfer (q) as a function of pion
laboratory kinetic energy. The solid curves are
calculations without the Fermi-motion terms, the
long-dashed curves are calculations including
the Fermi-motion terms, and the short dashed
curves are the $\sin^2(\theta)$ approximation normalized
to the solid curves at 100 MeV. The data is from
Cottingame (Δ) and Peterson (o) of ref. 6. The
transitions are the (a) $J^\pi = 1^+$, $T = 0$, 12.71 MeV
state and (b) $J^\pi = 2^-$, $T = 0$, 18.36 MeV state in
^{12}C.

convection current and magnetization density interfere for a given
natural parity or unnatural parity transition. The relative size
and angle dependence associated with the convection current and
magnetization density are of course generally quite different than
those obtained in (e,e´).

IV. PROTON-NUCLEUS REACTIONS

Two sources of couplings to the nuclear convection current
that arise in nucleon-nucleus inelastic scattering originate from
the energy dependence of the nucleon-nucleon (N-N) interaction and
the two-particle spin-orbit term in the N-N transition operator.
We first discuss the two-particle spin-orbit term in some detail.
The standard form assumed for the N-N transition operator contains
central, t^C, tensor, t^T, and two-particle spin-orbit, t^{LS}, components.
The two particle spin-orbit term is of the general form

$$t^{LS}(|\vec{r}_1 - \vec{r}_2|)(\vec{r}_1 - \vec{r}_2) \times (\vec{p}_1 - \vec{p}_2) \cdot (\vec{\sigma}_1 + \vec{\sigma}_2) \quad . \qquad (17)$$

Most often that part of the transition operator that is a function
only of $r \equiv |\vec{r}_1 - \vec{r}_2|$ is taken to be a Yukawa or $r^2 \times$ Yukawa shape.
The transition operator is fitted to N-N scattering in the interme-
diate energy region and is to be used where direct and exchange
terms are each calculated explicitly. It is found that most of the
energy dependence seen in N-N scattering can be represented by the
interplay between direct and exchange matrix elements of the transition operator where the actual strength and range parameters can
be either energy independent,[9] or energy dependent.[10] A technique,
to be discussed near the end of this talk, might provide a way for
distinguishing between different assumed parameter energy dependence.
Note that although the two-particle spin orbit interaction provides
a derivative operator that can result in a convection current type
interaction, it also results in interactions with just the
nuclear matter density and/or spin density. It is useful to be
able to identify and in principle, separate these different contributions originating from the two particle spin-orbit term. How
these different terms can arise is especially transparent if one
works with the direct term only and in momentum space where
$\vec{r}_1 - \vec{r}_2 \rightarrow q \equiv \vec{p}_1 - \vec{p}_2'$. For the direct term, for definiteness, let
the subscript 1 (2) refer to the projectile [p] (target [t]) nucleon,
allowing eq. (17) to be rewritten

$$(\vec{p}_p - \vec{p}_{p'}) \times (\vec{p}_p - \vec{p}_t) \cdot [\vec{\sigma}_p + \vec{\sigma}_t] \quad =$$

$$= (\vec{p}_p \times \vec{p}_{p^\prime}) \cdot \vec{\sigma}_p + (\vec{p}_p \times \vec{p}_{p^\prime}) \cdot \vec{\sigma}_t$$

$$\underbrace{\hspace{3cm}}_{\substack{\text{matter} \\ \text{density}}} \qquad \underbrace{\hspace{3cm}}_{\substack{\text{spin} \\ \text{density}}}$$

$$+ (\vec{p}_{p^\prime} - \vec{p}_p) \times \vec{p}_t \cdot \vec{\sigma}_p + (\vec{p}_{p^\prime} - \vec{p}_p) \times \vec{p}_t \cdot \vec{\sigma}_t \qquad (18)$$

$$\underbrace{\hspace{3cm}}_{\substack{\text{current} \\ \text{density}}} \qquad \underbrace{\hspace{3cm}}_{\substack{\text{spin-current} \\ \text{density}}}$$

where the various pieces originate associated-interaction-density contributions as indicated. Except for very forward scattering angles the largest terms are associated with the matter and spin densitycontributions because for those cases the relatively large $|\vec{p}_p \times \vec{p}_p{}^\prime| \sim p^2 \sin\theta$ term is present (here p_p is the medium energy projectile momentum and θ is the projectile scattering angle).

At the first Telluride conference Petrovich[11] wrote the differential cross section for the direct Born amplitude (keeping the major terms) schematically as

$0^+ \to J$(normal parity)

$$\frac{d\sigma}{d\Omega} \alpha \ [\ |\sum_i t_{Ci}(q) \ \rho_M^J(q)\ |^2 + |\sum_i t_{LSi}(q) \ \rho_M^J(q)\ |^2] \qquad (19a)$$

$0^+ \to J$(abnormal parity)

$$\frac{d\sigma}{d\Omega} \alpha \ [\ |\sum_i t_{LSi}(q) \ B \ \rho_S^J(q)\ |^2 \ +$$

$$\sum |\sum_i t_{Ci}(q) \ \rho_S^J(q) + \sum_i t_{Ti}(q) \ \{z \ \rho_S^J(q)\}$$

$$+ \sum_i t_{LSi}(q) \ A\rho_C^J(q)\ |^2] \qquad (19b)$$

where the subscripts M, S, C on the nuclear densities denote matter, spin, and convection current, respectively. Note the convection current piece enters as a major term only in the abnormal parity transitions. Thus $\Delta T = 0$, non-normal parity, non-stretched states may provide the best way to study convection currents via the two-particle spin-orbit interaction. The validity of eq. (19a) is restricted to low-lying, low spin collective normal parity transitions. High spin normal parity transitions, which are not collective in nature, present another possibility for studying convection currents.[12]

It is useful to be able to identify the non-convection current spin-orbit contributions when they are important. One way to do this might be to utilize the $p^2\sin\theta$ dependence at fixed q, for such contributions (much like a Rosenbluth plot is used). Note that in terms of configuration space derivative operators, the exchange term can also be written in the form of eq.(18) where now P_p and P_p' denote the initial and final momentum associated with the energetic incident and detected final particle. In the region 100-300 MeV the important pieces in the elementary interaction decrease or are roughly constant with energy. Thus the $p^2\sin\theta$ piece associated with the non-convection current spin-orbit term, which causes an effective increase with energy of this term (in the many-body system) could be a useful diagnostic tool. That the $p^2\sin\theta$ piece does result in an increase with energy of the fixed q contribution can be seen by noting (assume $p_p' \approx p_p$)

$$q = 2\ p\ \sin\theta/2 \tag{20a}$$

and thus

$$\sin\theta = \,^q/_p(1 - \,^{q^2}/_{4p^2})^{\frac{1}{2}} \tag{20b}$$

so that differential cross sections dominated by such a contribution can be written

$$\frac{d\sigma}{d\Omega}(q) = E_i E_f \frac{k_f}{k_i}|T|^2 = E_i E_f\ p^2 q^2 (1 - \,^{q^2}/_{4p^2})\beta^2(q) \tag{21}$$

or

$$\left.\frac{d\sigma(q)}{d\Omega E_i E_f}\right|_{q=q_o} = (p^2 - a(q))\ b(q) \tag{22}$$

which, for a plot at fixed $q = q_o$ as a function of p^2, would result in a straight line with positive slope $q^2\ \beta^2(q)$ [$\beta^2(q)$ is the square of a nuclear transition form factor] and intercept $-a = -q^4\ \beta^2(q)/4$. An obvious extension of this technique when central and tensor terms are present has been studied by T. Taddeuccu and myself.[13] Using realistic distorted waves and a variety of energy-dependent "realistic" transition operators including both direct and exchange terms we find the idea of an upward slope associated with the dominance of the non-convection current spin-orbit contribution is maintained (for the case of normal parity states where our study has thus far been concentrated) and should be a useful diagnostic tool in the region 100-300 MeV. Olmer, Bacher, and Emery at IUCF are currently looking at (p,p') data in [28]Si to see if this simple behavior is maintained for the appropriate states in that nucleus.

Another interesting place to look for convection current contributions may be associated with Fermi-motion corrections (as discussed earlier for pions). Here because of the different ratio of
projectile to target nucleon mass and the necessity of including
direct and exchange contributions the situation is somewhat altered.
[It should be understood that a more complete theory or model of the
N-N t-matrix is required to determine the dependence on the appropriate variables, both before and after collision and in the off-
shell kinematic region associated with binding in the many-body
environment. What follows is a schematic argument designed to moti-
vate the potential importance of certain considerations if meaningful comparison is to be made between theory and experiment in selected spin-isospin channels.] Denoting the initial (final) four momentum variable of particle i as p_i (p_i') we can write the symmetrized S
variable as $S = [(p_1 + p_2)/2 + (p_1' + p_2')/2]^2$. For $t(S_o)$, the transition operator currently used, we have $\vec{p}_2 = 0$ and, by momentum conservation $\vec{p}_2' = \vec{p}_1 - \vec{p}_1'$. Ignoring binding energy effects in the following we obtain the result

$$S - S_o = <(E_1 + E_1') \frac{\vec{p}_2 \cdot \vec{p}_2'}{2m} > + m\{<T_1> - <T_1'>\}$$

$$- \frac{(\vec{p}_1 + \vec{p}_1') \cdot (\vec{p}_2 + \vec{p}_2')}{2} \tag{23}$$

where, to obtain the first term on the R.H.S. the non-relativistic
approximation for the kinetic energy has been used. This expression
is appropriate for the direct and exchange terms in the DWIA. However we shall apply it for the direct term only in the PWIA where
the separation exhibited by eq. (23) has a particularly simple interpretation in terms of asymptotic variables and corrections to the
usual PWIA can be calculated by hand. For the direct term we have
(in parenthesis below we estimate the size of each term for a 600
MeV/c incident proton, a momentum transfer q of $\sim$ 300 MeV/c and a
projectile energy loss of $\sim$ 10 MeV)

$$S - S_o = (E_1 + E_1') [<T_2> + \frac{\vec{q} \cdot <\vec{p}_2>}{2m}] + m\{<T_1> - <T_1'>\}$$
$$\qquad\qquad (10^5) \qquad\qquad\qquad\qquad (10^4)$$

$$- \frac{(\vec{p}_1 + \vec{p}_1') \cdot (\vec{p}_2 + \vec{p}_2')}{2} \, . \tag{24}$$

$$(3 \times 10^5)$$

We concentrate on the role of the last term in allowing the
large central term in the N-N interaction, through its energy

dependence, to influence non-normal parity transitions. Note in
more detailed calculations the role of the first term should be
investigated also-especially for the large momentum transfers of
special importance for high spin normal parity states. The first
term is about seven times more important here than in the intermediate energy pion case discussed earlier - (compare the discussion
between eqs. (12) and (13) where this term was dropped) because
$E_{proj} = c^2 (p^2 + m^2_{proj})^{1/2}$ and $m_\pi / M_N \sim 1/7$. As mentioned earlier for
the case of pions a contribution from the convection current correc-
tion allows an initially normal parity transition operator to contribute to non-normal parity transitions. We continue to concentrate
here on the 1^+ T = 0, 1 states in ^{12}C. In addition to the relative
sizes of the convection current density contribution and the original
uncorrected transition amplitude, a determining feature in ascertain-
ing the importance of this correction is the multiplicative factor
$(\partial t / \partial S)_S \sim (1/2M_N)(\partial t / \partial T)_{proj}$. Even for the current simple-minded
t-matrix parameterizations there are considerable differences for
this term. For example the N-N t-matrix of Picklesimer and Walker[9]
would yield zero for this term in all channels, whereas the N-N
t-matrix of Franey and Love[10] or that given recently based on the
Paris potential[14] would yield non-zero results, different from each
other and generally different in each spin-isospin channel, even for
the central contributions. Taddeucci and I[13] have estimated $\partial t / \partial S$
for a representative set of parameters using the Paris potential t-
matrix. Utilizing these results for E_p = 160 MeV q = 200 MeV/c and
using the simple p-h model for the 1^+ T = 0, 1 states we find that
the convection contribution from the central spin and isospin independent term t_c, alters the transition form factor associated with
central terms only, by approximately fifty percent! The correction,
at 200 MeV/c, for the 1^+ T = 1 transition, arising from the energy
dependence of the t_τ term is less and is on the order of twenty percent. These corrections most often decrease as the momentum transfer
decreases. The contribution of the first two correction terms in eq.
(24) to transitions of the same parity have been estimated. We find
such corrections are q dependent and on the order of $\sim$ 10% in the
amplitude for t_c and t_c^τ. The Fermi-motion corrections would be
enhanced for transitions having more concentrated convection current
strength. Note again the particular convection current correction
discussed earlier does not contribute to $0^+ \rightarrow 0^-$ transitions or
transitions to stretched states.

   This discussion was only meant to be suggestive. Taddeucci and
Olmer, Emery and Bacher and myself at IUCF are pursuing this research
further. Present investigations are concentrating on estimating the
direct and exchange term corrections, for a range of q and projectile
energies, for all components of the N-N t-matrix including the tensor
and spin-orbit pieces. High and low spin natural and unnatural parity transition are being studied. As mentioned earlier a more detailed N-N t-matrix model would be welcome.

Of course this trivial energy dependence is only one source of "non-locality" that can translate into a convection current contribution. For example, above 200 MeV the importance of a propagating intermediate Δ (interacting with the medium) contribution to the t-matrix may require a more detailed treatment of the medium modifications of the t-matrix resulting in additional corrections in the spirit of Lenz's discussion[7] of the medium corrected π-N t-matrix in (π,π').

REFERENCES

1. T. deForest, Jr. and J.D. Walecka, Adv. Phys. 15, 1 (1966).
2. F. Petrovich, W.G. Love and R.J. McCarthy, Phys. Rev. C21, 1718 (1980).
3. J. Heisenberg, invited contribution to this conference.
4. J. Heisenberg, J. Lichtenstadt, C.N. Papanicolas, and J.S. McCarthy, Phys. Rev. C25, 2292 (1982).
5. E.R. Siciliano and G.E. Walker, Phys. Rev. C23, 2661 (1981).
6. W. Cottingame et al., private communication ; C.L. Morris, Proceedings of the Workshop on Nuclear Structure with Intermediate-Energy Probes, 1980, Los Alamos Scientific Laboratory Report No. LA-8303-C; R.J. Peterson et al., Phys. Rev. C21, 1030 (1980). Error bars are indicated only when they extend beyond the symbol used for the data.
7. F. Lenz, M. Thies and Y. Horikawa, Ann. Phys. 140, 266 (1982) and F. Lenz, invited contribution to this conference.
8. C. Wilkin, Nucl. Phys. A220, 621 (1974).
9. A. Picklesimer and G.E. Walker, Phys. Rev. C17, 237 (1978).
10. W.G. Love and M.A. Franey, Phys. Rev. C24, 1073 (1981).
11. F. Petrovich,in The (p,n) Reaction and the Nucleon-Nucleon Force, edited by C.D. Goodman et al. (Plenum, New York, 1980) p. 115.
12. F. Petrovich and W.G. Love, Nucl. Phys. A354, 499c (1981).
13. T. Taddeucci and G. Walker, unpublished.
14. K. Nakano and H.V. von Geramb, University of Hamburg preprint.

CAN WE LEARN ABOUT THE SPIN FLIP

GIANT DIPOLE RESONANCES WITH PIONS?

Helmut W. Baer

Medium Energy Physics Division
Los Alamos National Laboratory
Los Alamos, NM 87545

ABSTRACT

Data and calculations for the ^{40}Ca($\pi^{\pm}$,π^0) reactions at 164 MeV are shown which indicate that pion scattering possesses a unique signature for separately identifying the 1^- and 2^- spin-isospin components of the giant dipole resonance.

INTRODUCTION

We have heard a great deal about (p,n) charge exchange. Now we come to the part of the program entitled "other reactions," which includes the ($\pi^{\pm}$,π^0) reactions, and later in this session the (π^-,γ) reaction. Since the theme of this conference is spin-excitations in nuclei, I was asked to talk briefly about the possibilities of the ($\pi^{\pm}$,π^0) reactions for study of spin excitations. I must say that at present spin aspects do not constitute the major thrust of our studies.[1] Nevertheless, there was a puzzling feature in the ^{40}Ca($\pi^{\pm}$,π^0) data at 164 MeV which led to an unexpected result with regard to spin excitations with pions. This will be the subject of my talk.

Most of the discussions here have dealt with spin-excitations of unnatural parity states. Of course, one can have spin-transfer $\Delta S = 1$ in the excitation of natural parity states. The surprising phenomenon in pion scattering is for 1^- states. We were led to consider $\Delta L = 1$, $\Delta S = 1$ pion excitations in attempting to understand the observed angle-dependent broadening of the giant dipole resonance (GDR) in the ^{40}Ca($\pi^{\pm}$,π^0) reactions. In March-April 1981 we were performing the first survey experiments

523

on isovector giant resonances with the LAMPF π^0 spectrometer set up at the low-energy pion (LEP) channel. The $^{40}Ca(\pi^{\pm},\pi^0)$ measurements played the role of a calibration experiment. We wanted to see how well the isovector resonances stand out above the continuum and if the cross sections could be understood quantitatively. Since ^{40}Ca has a well-formed GDR at 20 MeV excitation with a width $\Gamma = 4.5$ MeV (FWHM),[2] it was important to see this resonance clearly. Fig. 1 shows the 12^0 and 3.5^0 spectra we had in the counting house during the experiment. There is a good signal to background ratio for the GDR at 12^0 and almost no trace of the GDR at 3.5^0. In the off-line analyses we binned the data into six angular bins as shown for the (π^+,π^0) reaction in Fig. 2. The data displayed in Fig. 2 were taken in two settings of the spectrometer, 0^0 and 20^0, with a total data-taking time of 16 h. With these two settings we covered the angular range 0^0 to 30^0. The 15^0 spectrum shows a nice GDR signal at the expected position for the analog of a 20 MeV state in ^{40}Ca. The observed signal at 15^0 has a width 6.6 ± 0.7 MeV for the (π^+,π^0) spectrum and 6.1 ± 0.5 MeV for the (π^-,π^0) spectrum. These values are larger than the 5.0 ± 0.2 MeV instrumental resolution and are consistent with a GDR width of 4 ± 2 MeV (FWHM). The measured angular distribution is well described by the function[3] $2.9 \, J_1^2(q_\perp R)$ mb/sr where $q_\perp$ is the component of the momentum transfer $\vec{q}$ which is perpendicular to the incident beam direction.

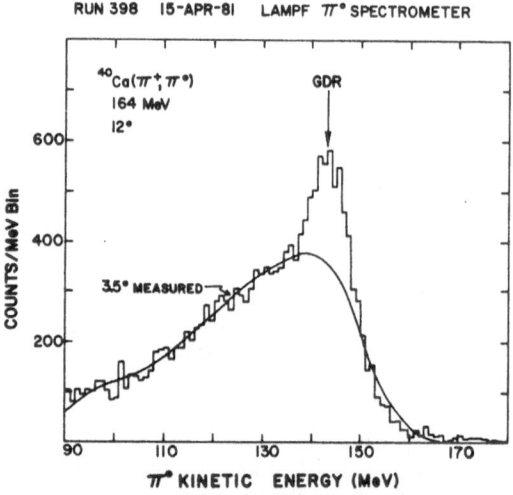

Fig. 1. One of the first spectra measured on the $^{40}Ca(\pi^+,\pi^0)$ reaction which showed that the giant dipole resonance (GDR) is strongly excited in pion single charge exchange. The arrow marks the position for a state corresponding to 20 MeV excitation in ^{40}Ca.

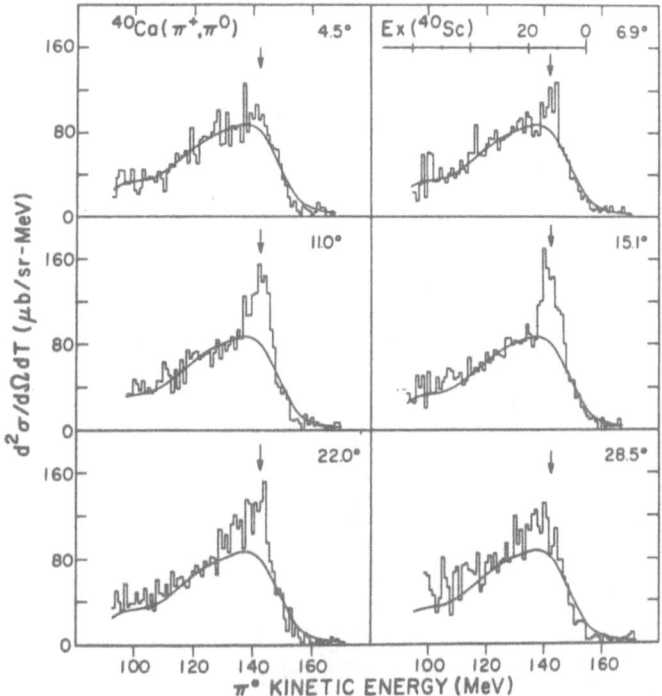

Fig. 2. The measured π^0 spectra for the $^{40}Ca(\pi^+,\pi^0)$ reaction at 164 MeV. The arrow marks the expected position of the analog of the giant dipole resonance at 20-MeV excitation in ^{40}Ca. The solid line in all panels is the smoothed, but not renormalized, 4.5^0 spectrum.

R is the pion interaction radius. A value R = 4.8 fm, déduced from the first minima of elastic π^+ and π^- scattering, gives a good fit of J_1^2 to the GDR angular distribution. The maximum value of 2.9 J_1^2 occurs at 15.4^0 with cross section 0.93 mb/sr. This value is close to what is expected[1,3] in a calculation using the Goldhaber-Teller form of the transition density normalized to exhaust the classical El sum rule. Thus the energy, width, cross section, and angular distribution shape for the signal we see in the $^{40}Ca(\pi^+,\pi^0)$ reaction identifies it as the analog in ^{40}Sc (at 12.34 MeV excitation) of the El photo-resonance observed[1] in ^{40}Ca at 20 MeV.

It is interesting to compare the (p,n) and (π^+,π^0) reactions to see how best to exploit the differences for structure studies. The $^{40}Ca(p,n)$ spectrum at 200 MeV and 4^0 shown in Fig. 3a was presented at an earlier session. Fig. 3b shows the 15^0 spectrum for the $^{40}Ca(\pi^+,\pi^0)$ reaction at 164 MeV. The momentum transfer is

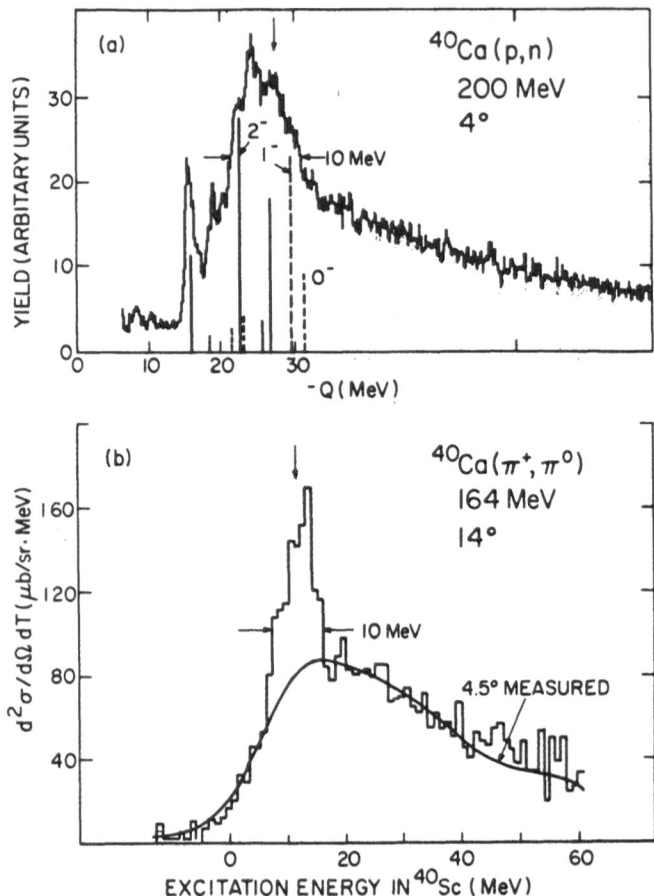

Fig. 3a. ^{40}Ca(p,n) data shown at this conference by K. Gaarde.[4]
3b. The π^0 spectrum for the ^{40}Ca(π^+,π^0) reaction at an angle where the GDR has the maximum cross section. The measured spectrum (smoothed) at 4.5^0 is shown for comparison. The arrow in both (a) and (b) marks the expected positions of the GDR at 20 MeV excitation in ^{40}Ca.

$\approx$65 MeV/c for both spectra. The two spectra look quite similar, and one might be tempted to conclude that the two charge exchange reactions excite the same states when compared at the same q-value. From our present understanding, nothing could be further from the truth. Nearly the entire (p,n) cross section in the GDR region is being interpreted as due to $\Delta L = 1$, $\Delta S = 1$ transitions to 0^-, 1^-, 2^- states. Nearly the entire (π^+,π^0) cross section in the GDR region is being interpreted as due to $\Delta L = 1$, $\Delta S = 0$

transitions to 1^- states. The (π^+,π^0) peak is the parent state $(M_T = T = 1)$ to the photo-resonance $(M_T = 0,\ T = 1)$ of ^{40}Ca, whereas the (p,n) peak is of different origin. Its strength is related to the $(\vec{\sigma} \times \vec{r})_1$ operator which plays a minor role in photoabsorption. From this comparison one can see the complementary roles of pion- and nucleon-charge-exchange scattering in clarifying the full nature of the GDR.

The best resolution that we have obtained in (π^-,π^0) measurements is 2 MeV (FWHM). This still is larger than the few tenths-of-MeV for the (p,n) studies. There is, however, a nice advantage to pion charge exchange measurements. The switch in measurement from (π^+,π^0) to (π^-,π^0) is much simpler than from (p,n) to (n,p). We simply reverse the polarity of the channel magnets. This gives data such as that displayed in Fig. 4. One sees directly the shift in mass between nuclear states due to the addition of 2 units of charge, which for ^{40}Ca is 12.0 MeV. The comparison of the two spectra is useful for distinguishing resonance peaks from artifacts of the continuum. A nuclear eigenstate must shift according to the Coulomb displacement energy, whereas the continuum may differ for the two reactions due to different neutron and proton separation energies, and due to

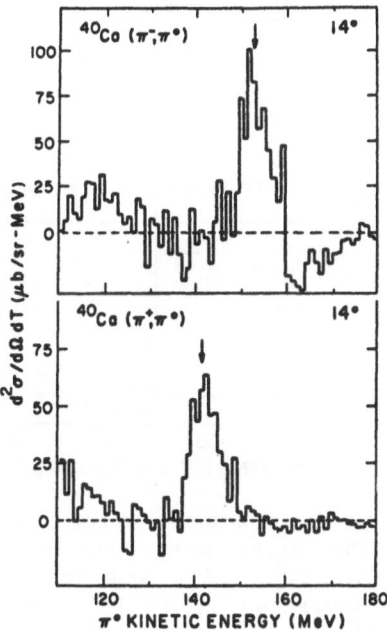

Fig. 4. The π^0 spectra at 14^0 after subtraction of the continuum. The arrows mark the expected position of the GDR. The peaks show the expected displacement of 12 MeV due to the difference in Coulomb energies.

Coulomb effects is suppressing the endpoint of the π^0 spectrum in
the final state of the $\pi^+A \rightarrow (A-1)p\pi^0$ channel. Thus the solution
of the difficult and long-standing problem of separating continuum
and resonance excitation in an experimental spectrum is greatly
aided by comparing the two charge-exchange spectra.

ANGLE-DEPENDENT BROADENING OF THE GDR SIGNAL

There is a puzzling feature in the $^{40}Ca(\pi^\pm,\pi^0)$ data. The GDR
signal has a smaller width in the 14^0 spectra than in the 10^0,
22^0, and 28^0 spectra. The measured width at 14^0 is consistent
with an intrinsic GDR width of 4 ± 2 MeV. At the other angles,

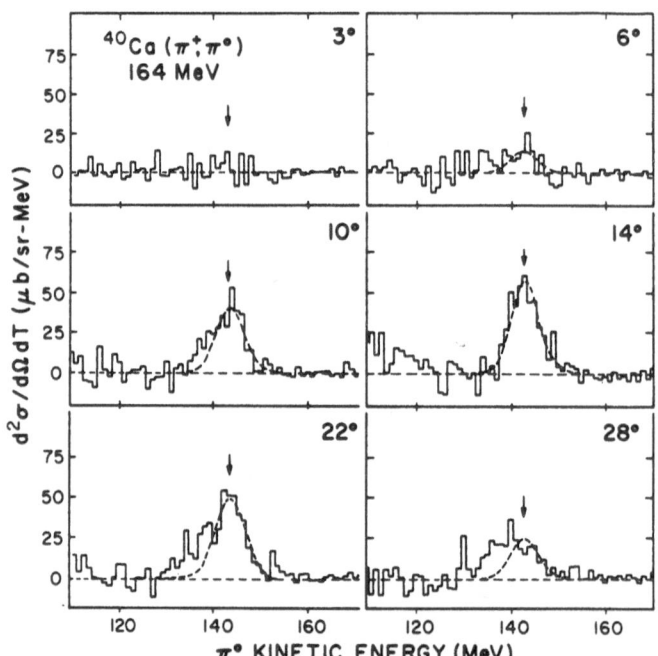

Fig. 5. The $^{40}Ca(\pi^+,\pi^0)$ spectra after subtraction of the
continuum. The arrow marks the expected GDR position.
The spectra at 10^0, 22^0, and 28^0 are suggestive of a
second peak separated by ≈ 5 MeV from the main GDR peak.
The 3^0 and 14^0 spectra do not show this second peak. The
dashed curves are Gaussian functions with parameters held
fixed at the values obtained in the fit at 14^0. (These
data are from an analysis with XCUT = 0.2; the data of
Fig. 2 are with XCUT = 0.1.)

one sees a broadening on the low-energy side of the GDR peak in both the (π^+,π^0) and the (π^-,π^0) data. Fig. 5 shows this effect for the (π^+,π^0) data. In view of this broadening, a further analysis of the (π^+,π^0) data was carried out. The peak structure for each angle was fitted with two Gaussian functions, keeping the main peak parameters fixed at the position and width given by the 14^0 data. The data, together with the Gaussian function for the main peak, are shown in Fig. 5. The excess counts on the low-energy side are clearly evident. The position of this subsidiary peak is 5 ± 2 MeV below the main GDR. The angular distributions of both the main peak and the excess counts are given in Fig. 6. The dashed curve is intended only to guide the eye. Although the uncertainties in the deduced cross sections are large, certain qualitative features are evident: 1) there is a minimum near 14^0; 2) the cross section rises between 14^0 and 28^0; 3) the maximum observed cross section is at 28^0 where it has a value 0.12 ± 0.06 mb/sr which is approximately 15% of the GDR cross section at 14^0. This angular distribution shape was puzzling. The $\Delta S = 0$, $\Delta L = 0$, 1, and 2 transitions are expected to peak at 0^0, 15^0, and 30^0, following closely the functions J_0^2, J_1^2, and J_2^2, respectively. The $0^+ \rightarrow 1^+$ transitions in pion scattering do not peak at 0^0 as they do for the (p,n) reaction. In the absence of more complicated effects than those treated by Siciliano and Walker,[5] transitions to unnatural parity excitations have a negligibly small cross section at 0^0. A $0^+ \rightarrow 1^+$ transition in ^{40}Ca at $T_\pi = 164$ MeV would be expected to have its first

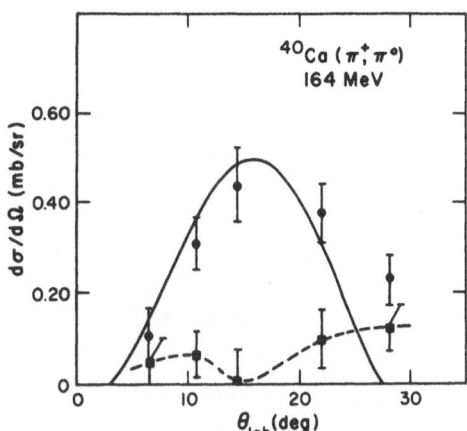

Fig. 6. The measured angular distributions for the main GDR (circles) and for the second state (squares) at ≈ 25 MeV excitation in ^{40}Ca. The solid curve represents the function $\beta[J_1^2(qR) - J_1^2(3^0)]$; the dashed curve is a hand-drawn line to guide the eye.

maximum near 20^0. Thus we were led to examine $\Delta L = 1$, $\Delta S = 1$ transitions.

CONSEQUENCES OF SPIN-TRANSFER IN $\Delta L = 1$ EXCITATIONS WITH PIONS

The shell model calculations of Donnelly and Walker[6] show that there are two 1^- states of quite different character near 20 MeV excitation in ^{40}Ca. Fig. 7 shows the calculated excitation energies and the values of the dipole strength $D = |\int \psi_f^* \vec{r} \psi_i|^2$ and the spin-flip dipole strength $SD = |\int \psi_f^* (\vec{\sigma} \times \vec{r})_1 \psi_i|^2$ for the calculated 1^- states. The dipole strength is largely concentrated in a single state at 18.6 MeV for which $D = 0.88$ and $SD = 0.03$. The spin-flip dipole strength is largest for a state at 22.2 MeV for which $D = 0.08$ and $SD = 0.55$. The separation energy for these two states is 3.6 MeV. The lower state represents the main component of the photonuclear GDR at 20 MeV. The higher 1^- state, if expected at 23.6 MeV, is at about the right energy to be a candidate for our satellite peak. Its wave function, $0.965(d5/2^{-1}f5/2)1^-$ + (small pieces), is dominated by a

^{40}Ca 1^- STATES

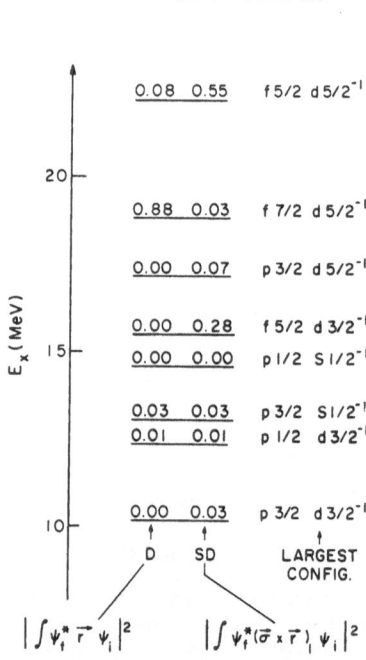

Fig. 7. The spectrum of 1^- states in a 1 $\hbar\omega$ basis calculated by Donnelly and Walker.[6]

configuration of the "spin-flip" type, i.e., $j_h = \ell + 1/2$ and $j_p = \ell - 1/2$. The lower 1^- state has three large components,

$$0.711(d5/2^{-1}f7/2) + 0.503(d3/2^{-1}f5/2) + .362(d5/2^{-1}p3/2)$$

all of which are of the "non-spin-flip" type, i.e., $j_h = \ell + 1/2$ and $j_p = \ell + 1/2$ or $j_h = \ell - 1/2$ and $j_p = \ell - 1/2$. When these two types of configurations were put into a DWIA calculation for pion inelastic scattering, it came as a surprise that the predicted angular distributions were very different. Some representative calculations by Siciliano[5] are shown in Fig. 8. The spin-flip configurations, e.g. $(d5/2^{-1}f5/2)1^-$, have angular distributions which peak at 0^0, and have the first minimum near 20^0. The non-spin-flip configurations, e.g. $(d3/2^{-1}f5/2)1^-$, have the expected J_1^2 angular distribution, with the first maximum near 15^0 and minima at 0^0 and 35^0.

Further DWIA calculations were performed in which the transition amplitudes were decomposed into amplitudes with spin-transfer values $\Delta S = 0$ and $\Delta S = 1$. These contributions add incoherently in the cross section (within the usual DWIA description). Fig. 9 shows the separate contributions for the $(d5/2^{-1}f5/2)1^-$ configuration. From this decomposition we see that the $\Delta S = 1$ amplitude is responsible for the 0^0 maximum. The $\Delta S = 0$ curve peaks at 15^0 and has minima at 0^0 and 35^0. It has the same shape as the $(d5/2^{-1}f7/2)1^-$ angular distribution in Fig. 8 which is dominated by the $\Delta S = 0$ component.

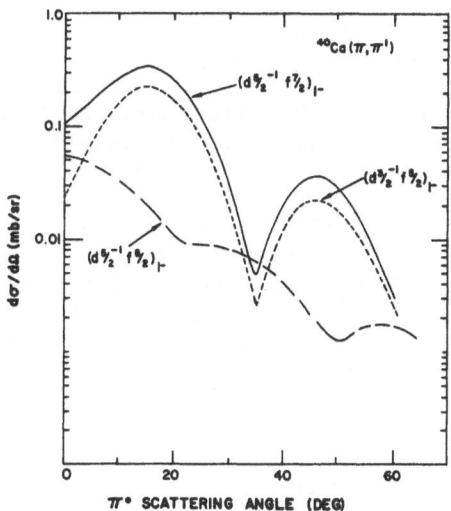

Fig. 8. DWIA calculations performed by E. R. Siciliano[5] for 1^- states excited in pion scattering at 164 MeV. The assumed p-h configurations are indicated.

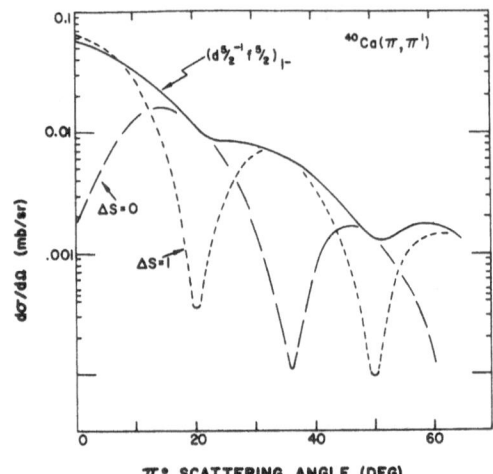

Fig. 9. DWIA calculations[5] showing the separate contributions of
$\Delta S = 0$ and $\Delta S = 1$ amplitudes for a p-h configuration where
$j_h = \ell + 1/2$ and $j_p = \ell - 1/2$.

These features may be understood as follows. The isovector
component of the elementary $\pi - N$ scattering amplitude is well
approximated at the P33 resonance by

$$f(k,k') = \alpha(k)[2\cos\Theta + i\vec{\sigma}\cdot\hat{n}\ \sin\Theta]\vec{t}\cdot\vec{t}\ .$$

The $\Delta S = 0$ transitions are induced by the scalar term ($2\cos\Theta$) and
the $\Delta S = 1$ transitions are induced by the spin-dependent term ($\vec{\sigma}\cdot\hat{n}$
$\sin\Theta$). Thus there is a factor of four in the cross section
favoring $\Delta S = 0$ to $\Delta S = 1$ transitions arising from the elementary
interaction. For any configuration the relative $\Delta S = 0$ to $\Delta S = 1$
amplitudes are determined by the j-j to L-S recoupling
coefficients. When the nuclear p-h excitation is of the
non-spin-flip type, e.g. $(d5/2^{-1}f7/2)1^-$, the ratio of amplitudes
$\Delta S = 0/\Delta S = 1$ arising from the j-j to L-S recoupling is much
larger than one. Thus for these configurations the DWIA
calculations show a nearly pure $\Delta S = 0$ shape. When the
configuration is of the spin-flip type, e.g. $(d5/2^{-1}f5/2)1^-$, the
$\Delta S = 1$ amplitude is much larger. In some cases (all the cases we
investigated) it is sufficiently large to produce an absolute
maximum at 0^0.

For 2^- states the situation is a little different. Angular
momentum and parity conservation force $\Delta S = 1$. However, now there
can be the two values $\Delta L = 1$ or 3. From the DWIA formalism one can
see that the requirement $\Delta S = 1$ forces the 2^- cross section to go

to zero at 0^0 for both L values. The relative amounts of $\Delta L = 1$ to $\Delta L = 3$ affects the angle at which the angular distribution peaks. Representative calculations for $(d5/2^{-1}f7/2)2^-$ and $(d5/2^{-1}f5/2)2^-$ configurations are shown in Fig. 10. They peak at 24^0 and 28^0, respectively. The non-spin-flip type configuration $(d5/2^{-1}f7/2)2^-$ gives a larger cross section by a factor of 5.3. It also has the larger $\Delta L = 1/\Delta L = 3$ amplitude ratio. It is worth noting that the $\Delta S = 1$ cross section differs dramatically for 2^- and 1^- states. The $\Delta S = 1,1^-$ cross section peaks at 0^0 and has a second maximum at 35^0 (Fig. 8). The $\Delta S = 1,2^-$ cross sections are zero at 0^0 and have their first maxima near 25^0 (Fig. 10).

To recapitulate, we see that there are three types of angular distributions involved in the excitation of 1^- and 2^- giant dipole states. The primary maxima of these angular distributions occur at 0^0, 15^0, and 25^0 and are therefore easily distinguishable in an experiment. The 1^- states have two types of angular distribution shapes characterized in the extreme by pure $\Delta S = 0$ or $\Delta S = 1$ transitions. For the p-h configurations involved in ^{40}Ca, the configurations of spin-flip type, e.g., $(d5/2^-f5/2)1^-$, give a large $\Delta S = 1$ amplitude which produces a maximum in the cross section at 0^0. If the configurations are of non-spin-flip type, as they are predominantly in the photonuclear GDR of ^{40}Ca, the $\Delta S = 0$ amplitude dominates and one gets the classical

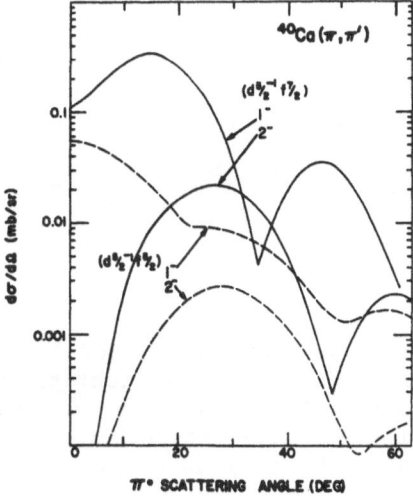

Fig. 10. DWIA calculations[5] for 1^- and 2^- states with pure h-p configurations. These configurations are the major components in the wave functions of Donnelly and Walker[6] for the relevant dipole states.

strong-absorption angular distribution shape $J_1^2(qR)$. This peaks at 15^0.

The relative magnitude of these 3 types of angular distributions for pure p-h states are shown in Fig. 10. The largest cross section is obtained for $\Delta S = 0,1^-$ states. Thus the El GDR is expected to give the largest pion cross section. The 1^- and 2^- states with large $\Delta S = 1$ amplitudes give cross sections which at their maximum values are 10-15% of the GDR cross section.

EIKONAL TREATMENT OF SPIN-FLIP DIPOLE TRANSITIONS

The existence of angle-dependent broadening of the GDR signal in the $^{40}Ca(\pi^+,\pi^0)$ reactions, and the suggestion of possible $\Delta S = 1,1^-$ transitions stimulated A. Gal[3] and M. Johnson[7] to extend the eikonal model treatment to include spin excitations. In the generalized giant dipole resonance[6] there are three states $0^-,1^-,2^-$ reached by $\Delta S = 1$ transitions and one 1^- state reached by a $\Delta S = 0$ transition. Gal obtains the cross section expressions:

$$\frac{d\sigma}{d\Omega}(0^-) = 0$$

$$\frac{d\sigma}{d\Omega}(1^-, \Delta S = 0) = N_1 J_1^2(x)$$

$$\frac{d\sigma}{d\Omega}(1^-, \Delta S = 1) = N_2 |\sigma\alpha J_0(x) + x J_1(x)|^2$$

$$\frac{d\sigma}{d\Omega}(2^-, \Delta S = 1) = N_2 |\sigma\alpha J_2(x) - x J_1(x)|^2$$

where $x = q_\perp R = p'R \sin\Theta$ and $\sigma \approx \sigma_{TOT}(\pi N)$ for reactions at 164 MeV. A sketch of these functions with cross section values as given by Gal is shown in Fig. 11. The only cross section which does not vanish at 0^0 is for the $\Delta S = 1,1^-$ state which has value $N_2(\sigma\alpha)^2$ at 0^0. Gal estimates this to be 0.14 mb/sr. The $\Delta S = 0,1^-$ has a maximum estimated to be 0.94 mb/sr at 15^0. The $\Delta S = 1,1^-$ cross section reaches a minimum as J_0 approaches its first zero and J_1 reaches its maximum. For ^{40}Ca this occurs near 15^0. For larger angles this cross section oscillates but drops off. These eikonal model results are in qualitative agreement with the DWIA calculation (Fig. 10).

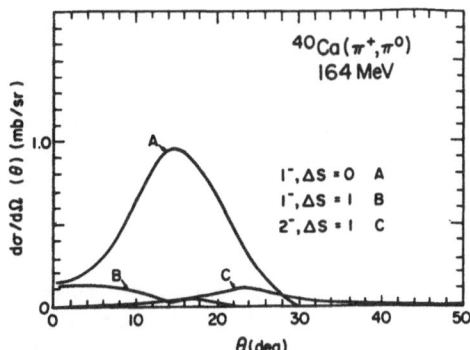

Fig. 11. A qualitative sketch of cross sections predicted by Gal[3]
for giant dipole states in ^{40}Ca. The fourth member of
the spin-isospin GDR multiplet has $\Delta S = 1$, $\Delta L = 1$,
$J^\pi = 0^-$ but its excitation is forbidden in pion
scattering by parity conservation.

Gal points out that the 0^0 peaking of the $\Delta S = 1,1^-$ cross
section depends quadratically on the πN total cross section σ.
Thus, above the (3,3) resonance the $\Delta S = 1,1^-$ cross section
becomes negligible even at 0^0.

The $\Delta S = 1,2^-$ cross section starts at zero at 0^0 and rises
slowly for $\Theta \lesssim 15^0$ (for ^{40}Ca). It reaches a maximum near 24^0,
estimated to be 0.1 mb/sr. The second minimum is near 35^0. This
shape is in qualitative agreement with DWIA calculations
(Fig. 10). From the DWIA calculations one can see that in order
to reach a cross section of 0.1 mb/sr for 2^- states one must have
coherent p-h excitation of the non-spin flip type, e.g.,
$(d5/2^{-1}f7/2)2^-$. These are the same type of configurations
involved in the 1^- GDR state.

COMPARISON WITH EXPERIMENT

Now that the theoretical expectations are quite explicit, we
can go back and ask about the experimental verification and new
measurement possibilities. First we see that if we take Gal's
graph (Fig. 11) for the sum of $\Delta S = 0,1^-$, $\Delta S = 1,1^-$, and $\Delta S = 1,2^-$
transitions, it is in qualitative agreement with the data
(Fig. 6). In the comparison we should plot the theoretical cross
section relative to the 4.5^0 values since this is how the data was
analyzed. The rise in the cross section for the satellite peak
beyond 15^0 would be due to 2^- states. At 0^0 the peak of the
$\Delta S = 1,1^-$ state is obscured by the $\Delta S = 0,1$ GDR. Gal's
calculations show that the two states have comparable cross

sections. We used the 4.5^0 spectrum to give us the shape of the continuum at other angles. The presence of $\Delta S = 1,1^-$ components makes this procedure less accurate. The result is that we cannot for certain identify $\Delta S = 1,1^-$ states in ^{40}Ca. However, the data is consistent with the expectations for $\Delta S = 1,2^-$ states.

It may be possible to enhance the $\Delta S = 1,1^-$ states relative to the $\Delta S = 0,1^-$ states at 0^0 by lowering the beam energy. Excitation functions measured for other nuclei show that the ratio $\sigma(\Delta S = 1)/\sigma(\Delta S = 0)$ measured at constant momentum transfer is a sharp function of pion energy (for a review of this point, see Ref. 8). In the examples studied (2^- and 4^- states) a lower pion energy near 100 MeV is much more favorable for enhancing the spin-flip excitations relative to non-spin flip excitations.

The best experimental evidence for the validity of the theoretical predictions on 1^- state angular distributions comes from the very recent experimental result discussed in Ref. 8. Fig. 12 shows the measured angular distributions for π^+ and π^- inelastic scattering at 162 MeV to a known 1^- state at 4.45 MeV in ^{18}O. The curves are DWIA calculations for $\Delta S = 0$ and $\Delta S = 1$ transitions obtained from a $(p1/2^{-1}d3/2)1^-$ configuration. We see that the $\Delta S = 1$ and $\Delta S = 0$ curves are out of phase, and that the

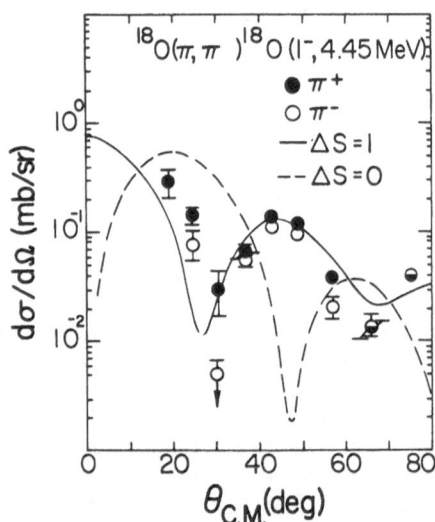

Fig. 12. Preliminary data and calculations of S. Seestrom-Morris et al.[8] on ^{18}O which give evidence that $\Delta S = 0,1^-$ and $\overline{\Delta S} = 1,1^-$ excitations may have very different angular distributions in pion scattering.

data agree quite well with the $\Delta S = 1$ curve. Calculations with other p-h configurations give similar $\Delta S = 1$ and $\Delta S = 0$ curves. The data and calculations taken together indicate a nearly pure $\Delta S = 1$ excitation. Whether this is consistent with realistic shell model calculations remains to be seen. Taken at face value, these preliminary results give a first clue that $\Delta S = 1,1^-$ and $\Delta S = 0,1^-$ transitions in pion scattering might have quite different angular distribution shapes.

SUMMARY

Theoretical studies performed in June-September 1981 predicted there exist large differences in the angular distribution shapes of $\Delta S = 0,1^-$ and $\Delta S = 1,1^-$ transitions. At forward angles these two angular distributions are nearly out of phase. The $\Delta S = 1,1^-$ cross section peaks at 0^0, and the $\Delta S = 0,1^-$ cross section peaks at 15^0. The preliminary data on the $^{18}O(\pi,\pi')^{18}O(1^-,4.45\ \text{MeV})$ angular distribution at 164 MeV and in the region 20^0-60^0 has a shape which looks very much like the calculations for $\Delta S = 1,1^-$ states.

For the study of spin-flip components of the GDR, the differences in angular distribution shapes between $\Delta S = 0,1^-$, $\Delta S = 1,1^-$, and $\Delta S = 1,2^-$ transitions offers a powerful method for separating these components experimentally. However, to exploit this possibility in pion charge exchange scattering requires higher π^0 resolution than 5 MeV(FWHM) and/or an enhancement of the strength of $\Delta S = 1$ transitions relative to $\Delta S = 0$ transitions at other beam energies.

It might be of interest to mention that we at Los Alamos have studied the possibilities for higher resolution. A π^0 resolution of order 0.3 MeV (FWHM) seems quite feasible for a second generation spectrometer based on the present design, with NaI detectors replacing the lead glass Cherenkov detectors.

ACKNOWLEDGEMENTS

The author would like to acknowledge the many discussions of these points with members of the experimental collaboration. In addition, I thank S. Seestrom-Morris for permission to show the preliminary ^{18}O data, and C. Morris for first suggesting that $\Delta S = 1,1^-$ angular distributions may have anomalous shapes. I thank E. R. Siciliano for performing the DWIA calculations shown here, and for numerous informative discussions. Discussions with A. Gal and M. Johnson on the eikonal treatments are also gratefully acknowledged.

REFERENCES

1. The collaborators in the studies of isovector giant resonances
   with $(\pi^{\pm},\pi^0)$ reactions are: R. Bolton, J. D. Bowman,
   M. D. Cooper, F. Cverna, N. S. P. King, M. Leitch,
   H. S. Matis (LANL); J. Alster, A. Doron, A. Erell, and
   M. Moinester (Tel Aviv University); E. Blackmore (TRIUMF).
   The results on ^{40}Ca are available in preprint form and will
   be published shortly. Cross sections given at this
   Conference should be regarded as preliminary values.
2. J. Ahrens, H. Borchert, K. H. Czock, H. P. Eppler, H. Gimm, H.
   Gundrum, M. Kröning, P. Riehn, G. Sita Ram, A. Zieger, and
   B. Ziegler, Nucl. Phys. A251, 479 (1975).
3. A. Gal, Phys. Rev. C25:2680 (1982).
4. C. Gaarde, these proceedings.
5. E. R. Siciliano and G. E. Walker, Phys. Rev. C23, 2661
   (1981); and E. R. Siciliano, private communication.
6. T. W. Donnelly and G. E. Walker, Annals of Physics 60, 209
   (1970).
7. M. Johnson, Los Alamos Workshop on the Study of Giant
   Resonances with Pions, Feb. 1982 (proceedings to be
   published as LANL report).
8. S. Seestrom-Morris, these proceedings.

NOTE added: Since this conference was held, the reports referred
              to in Ref. 1 above have appeared in Phys. Rev. Lett.
              49:1376 (1982) and Nucl. Phys. A396:437c (1983).

FUTURE POSSIBILITIES FOR (n,p) STUDIES AT INTERMEDIATE ENERGIES

F.P. Brady, C.M. Castaneda, and J.L. Romero

Physics Department
University of California
Davis, CA 95616

V.R. Brown and C.H. Poppe

Lawrence Livermore National Laboratory
University of California
Livermore, CA 94550

INTRODUCTION

The study of giant resonances in nuclei via inelastic electron and hadron scattering is currently of great interest.[1,2] These resonances are outstanding manifestations of nuclear vibrational motion in which an appreciable number of nucleons participate. Charge exchange reactions such as (p,n) and (n,p) offer powerful options for studying isovector excitations without interference from isoscalars. In general, if the target has isospin T then an isovector transition leaves the residual nucleus with isospin T-1, T, or T+1. (Refer to fig. 1.) The T-1 excitation requires charge exchange, as in (p,n), so that the 3-component $T_3=T$ (the charge-component) of isospin is also changed to T-1. Only for N=Z (T=0) nuclei are all isovector transitions to states of isospin T+1. In such reactions as (p,p') and (e,e') the isoscalar mode can also be excited and, indeed, these tend to be the strongest transitions at medium energies ($\lesssim$ 100 MeV).

The (p,n) reaction is, in many respects, a unique tool for exciting the T-1 states and many beautiful data from the MSU[3,5] and IUCF[6,8] laboratories have appeared in recent years. At intermediate energies (>100 MeV) the selectivity of (p,n) to spin

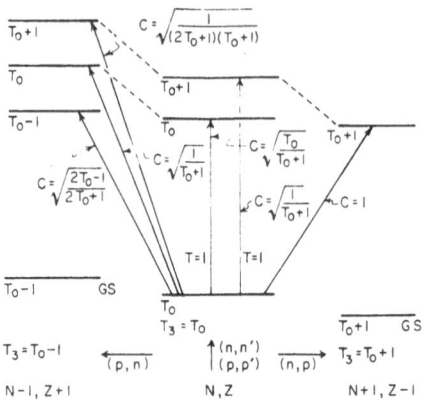

Fig. 1.   Isovector excitations.

mode (s=1 transfer) excitations has been particularly valuable.[6] [10] Currently of great interest is the apparent indication of mesonic and Δ baryonic degrees of freedom in nuclei as manifested[11,12] in (p,n) reactions by the quenching of t=1, s=1, Gamow Teller (GT) strength.

The (n,p) reaction also excites only isovector transitions, and since ΔT_3 = +1 this implies an additional selectivity for (n,p) in N>Z nuclei where only the T+1 analog states will be excited.[13,14]   It can be seen from the Clebsch-Gordan coefficients in fig. 1, that (p,p') favors the population of T states and (p,n) that of T-1 states. For large T, (p,n) transitions to T+1 states are inhibited by a large isospin factor.

For (n,p) the isospin factor (fig. 1) is unity. However, there are important dynamical factors due to nuclear structure and the nuclear interaction which play a key role in determining transition strength. For example, in heavy nuclei where T is large, (n,p) induced $1\hbar\omega$ transitions such as s=0 and s=1 dipole transitions involving neutron particle-proton hole excitations are almost completely blocked.

We have in recent years used the (n,p) reaction to study T+1 components of isovector excitations using $\cong$60 MeV neutrons. Distortion effects have made it difficult to study medium and heavy nuclei at these energies. In this respect there is a great advantage to be gained in going to higher energies. In addition near 60 MeV the spin dependent isovector part of the effective nucleon potential, $V_{\sigma\tau}$, is about equal to the spin independent part V_τ; while at higher energies, $\geq$120 MeV, $V_{\sigma\tau}$ dominates, and so one can isolate the spin-mode excitations.[15,16]

In this paper we attempt to elaborate on the rationale for going to higher energies, to indicate the technical feasibility of (n,p) at higher energies, and to predict its usefulness as a direct reaction tool for studying the effective nucleon-nucleon interaction and nuclear structure.

RATIONALE FOR HIGHER ENERGIES

Distortion Factors

The rationale for using higher energies in (n,p) reactions can be made clear. One of the strongest arguments is based on the effects of distortion which are so much larger at the lower energies. Distortion effects for monopole ($\ell=0$) excitations are particularly dramatic: rather than having the strength concentrated at and near zero degrees, as at higher energies, distortion diminishes $\sigma(0°)$ for lower energies (<100 MeV). The increase in distortion effects with increasing A makes it very difficult to measure T+1 analogs of say GT or dipole strength in medium and heavy nuclei. To begin with, one is looking for, not the huge fragments of T-1 and T isospin seen in (p,n) reactions, but much smaller T+1 pieces. So greater sensitivity is required in order to measure the cross sections for these smaller fragments above the continua.

To illustrate the effect of distortion with increasing A, we show in fig. 2 small-angle energy spectra for (n,p) reaction for ^{6}Li, ^{12}C, ^{28}Si at 60 MeV and for ^{118}Sn at 65 MeV. The arrows show the expected location of GT transitions which are the inverse of β^- decay from the ground state of each residual nucleus to the target ground state. The extrapolated (n,p) 0° cross section decreases from approximately 10 mb/sr for ^{6}Li to 4 mb/sr for ^{12}C to 2 mb/sr for ^{28}Si to an estimated upper limit of $\cong 0.08$ mb/sr for ^{118}Sn.

We used the forward-angle cross sections for these three light targets and the DWIA at small momentum transfer (q $\cong$ 0)[17][19] to extract, $J_{\sigma\tau}$, the volume integral of the isovector, $\ell=0$, spin-flip part of the effective nucleon-nucleon interaction. Distortion factors were calculated from DWBA70. The resulting value of $J_{\sigma\tau}$ is 157 MeV-fm^3. Assuming no distortion, this value can be used along with the β-decay GT matrix element to predict $\sigma(q\cong0)$ $\cong$ 3mb/sr for ^{118}Sn(n,p)^{118}In. With distortion factors estimated from fig. 3a one expects 0.06 mb/sr at 60 MeV and $\cong 1$ mb/sr at 200 MeV.

Figure 3a shows how distortion factors for $\ell=0$ vary with $A^{1/3}$. The curve labelled 120 MeV is the expression from ref. 6 where 120 MeV incident protons were used. The three lowest

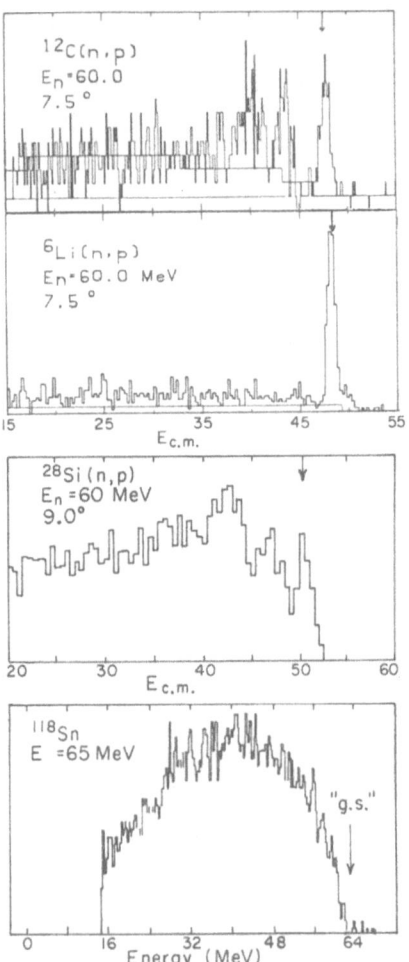

Fig. 2. (n,p) spectra with arrows showing expected GT transitions.

points for ⁶Li, ¹²C, and ²⁸Si on the "60 MeV" curve were taken from ref. 20, in which case 60 MeV neutrons were incident in the (n,p) reactions. The dashed "60 MeV" line passes through $N^D = 1$ at A=1, through the average for values for ⁶Li, ¹²C, and ²⁸Si (mentioned above), and through values for ⁹⁰Zr and ²⁰⁸Pb determined from (p,n) measurements at 45 MeV.[5] The ratios of cross sections at 45 MeV to those at 120 MeV[6 8] were used to determine these N^D values. The N^D values for 200 MeV, also shown in fig. 3a, were determined from the ratio of 200 MeV to 120 MeV[6 8] cross sections.

These values for N^D are probably not good to better than 25% on average. However, they do show, in a rough way, consistent trends, and do indicate why higher nucleon beam energies are very desirable for $\ell=0$, $s=1$ transitions. It might be noted that early work[21] in (n,p) at 156 MeV produced a ^{6}Li(n,p)^{6}He 0° cross section of $\cong$13 mb/sr. For higher multipolarities one expects similar trends for distortion effects. Figure 3b shows trends of N for $\ell=2$ at 60, 120, and 200 MeV based on ^{90}Zr and ^{208}Pb(n,p). These were calculated by taking the ratio of distorted to plane wave predictions at the smallest-angle maximum of the angular distribution.

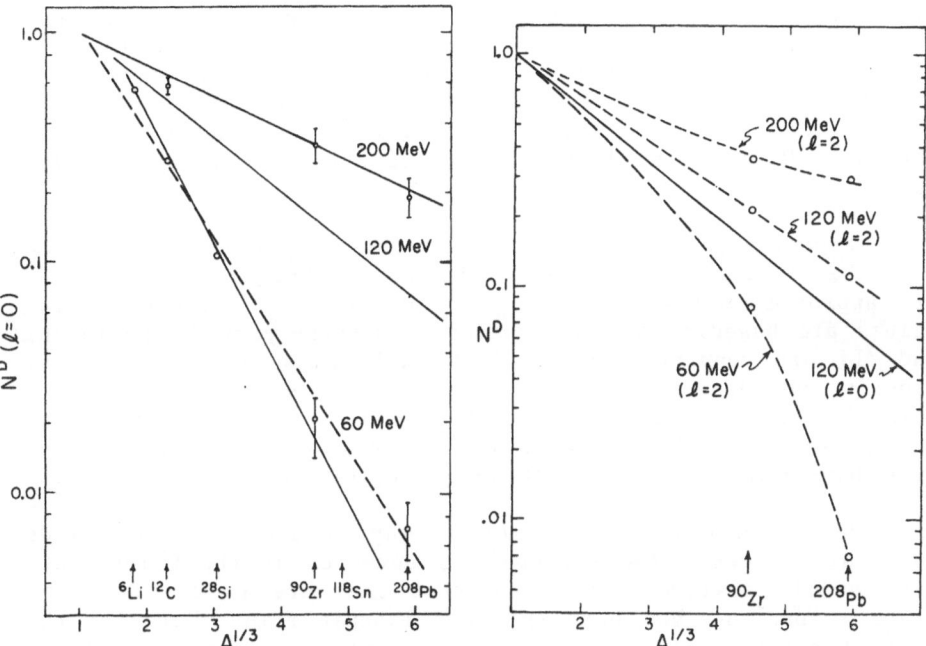

Fig. 3. Distortion factors for $\ell=0$ and $\ell=2$ for (n,p) and (p,n).

Spin flip Sensitivity

A second reason for higher energy neutron beams is the increased selectivity to spin transfer (s=1) excitations. Measurements of (p,n)[15] indicate that $J_{\sigma\tau}/J_\tau$ increases essentially
linearly for proton energies between $\cong 50$ and 200 MeV and that
(other things being equal) spin-flip transitions at 200 MeV will
be an order of magnitude larger in cross section than non-spin flip.

In our 60 MeV (n,p) work we are not able to separate spin-
flip (s=1) from s=0. However, comparisons with photonuclear
cross sections and some (d,^{2}He) measurements seem to indicate
that, at least for the ℓ=1 cases in light nuclei, the spin-flip
modes are not strongly excited (or at least not localized) and
are dominated by the s=0 modes.

In the cases of ^{6}Li and ^{7}Li (and to some degree in ^{9}Be) we
see large fragments of T+1 strength at high excitation energies,
15 and 24.5 MeV in ^{6}He, and at 20 MeV in ^{7}He. The corresponding
target excitations are expected at 19.5 and 29 MeV in ^{6}Li
and at 31 MeV in ^{7}Li. See fig. 4. These excitations are broad,
with estimated Γ widths of 4, 8, and 9 MeV respectively; and show
ℓ=1 angular distributions. They are not seen in the (γ,xn) cross
sections,[22,23] an indication of charged particle decay modes, although there is evidence for the ^{7}Li excitation in the total
photonuclear cross section[24] (the dashed line labeled σ_{tot} in
fig. 4). Data of the latter kind does not seem to be available
for ^{6}Li.

These Li(n,p) cross sections exhaust a large fraction of the
Goldhaber-Teller[25] energy-weighted sum rule. Rough values we obtain
are approximately 45% for ^{6}Li and 40% for ^{7}Li. However, these
values are based on the assumption that these excitations in ^{6}Li
and ^{7}Li are conventional ℓ=1, s=0, GDR components. On general
grounds and because $J_{\sigma\tau} \cong J_\tau$ at these energies, one expects some
spin-flip contribution. Data at higher energy for both (n,p)
and (p,n) where the s=1 mode dominates would be useful, along
with some microscopic particle-hole calculations.

The (d,^{2}He) reaction offers a way of isolating spin mode
(s=1) resonances. Two protons are detected in the final state
with small relative momenta so that they are in the ℓ=0, s=0
state. Thus one has both spin and isospin flip. Data for the
(d,^{2}He) reaction at E_d =55 MeV was obtained[26] at LBL for ^{6}Li and
^{12}C, and more recently for ^{16}O and ^{28}Si by a LBL-UCD collaboration.[27]

In fig. 5 of ^{6}Li(d,^{2}He)^{6}He spectrum (55 MeV) is compared to
one for ^{6}Li(n,p)^{6}He (60 MeV). The momentum transfers are quite
comparable. There is little evidence in (d,^{2}He) for the states
we see in (n,p) at 15 and 24.5 MeV in ^{6}He or for those reported

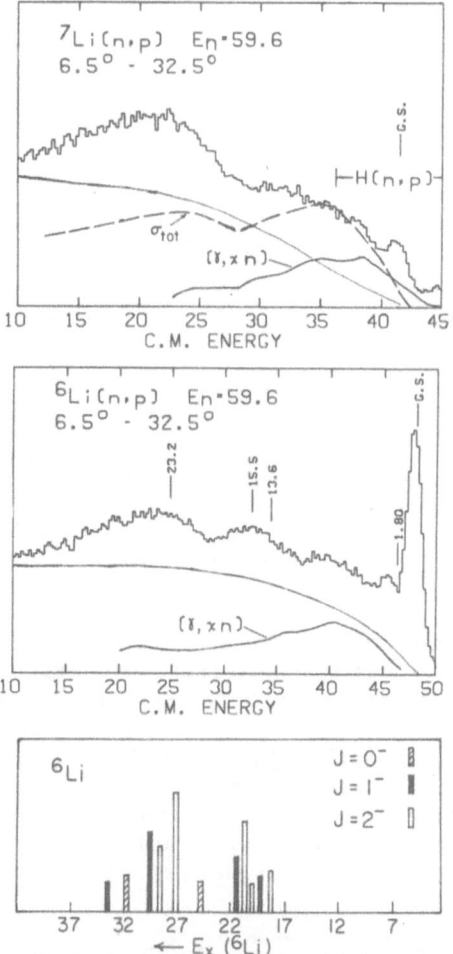

Fig. 4. ^{6}Li and ^{7}Li(n,p) and (bottom) predictions of GDR.

earlier at 13.4, 15.3, and 23.2 MeV.[28]   (In fact, no charge
exchange reactions have previously excited them.) Even allowing
for the energy dependence of $J_{\sigma\tau}/J_{\tau}$, it seems fair to say that
the spin-flip dipole components are not strong in ^{6}Li, and considering the total photonuclear data, σ_{tot}, in fig. 4, this is
probably also true for ^{7}Li.

The case of ^{12}C is also interesting. The ^{12}C(n,p)^{12}B energy
spectrum in fig. 6 shows the strong 1^+ (M1), $(2^-,4^-)$ doublet, and
the broader resonance identified as the GDR. The (d,^{2}He) spectrum
shows the former two strongly but not the latter whose centroid

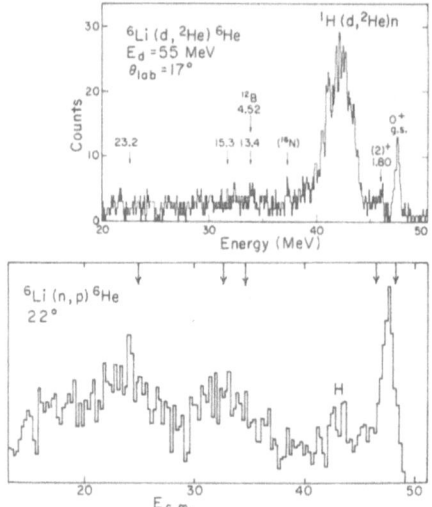

Fig. 5. ^{6}Li(d,^{2}He) and (n,p)^{6}He at comparable momentum transfers.

should be at 7.7 MeV excitation. In this case both the (γ,xn) [shown in fig. 6 so that the excitation energies match] and γ_{tot} excitation functions have an energy dependence very close to that of the GDR in the (n,p) proton energy spectrum integrated over the angles of ℓ=1 strength. So the ℓ=1, s=0 GDR identification appeared reasonable. The 2$^-$ state whose analog is at 19.3 excitation in ^{12}C could be considered to be part of the ℓ=1, s=1 spin-flip dipole. The other fragments are not prominently localized in energy but could be present to some degree as a number of fragments.

The situation for ^{16}O(n,p) (ref. 27) as compared to ^{16}O(d,^{2}He) is similar. Figure 7 compares energy spectra at comparable momentum transfers. The (γ,xn) data[29] which characterizes the (s=0) GDR is shown below. Again s=1 states are seen at lower excitation (below the GDR) in (n,p) and (d,^{2}He) but not much concentrated s=1 strength in the region of the GDR.

It is concluded that the (d,^{2}He) reaction provides some information on spin and isospin flip excitations. However, the nucleon energies are low (for a given E_d), the distortion effects are large, the experiments are difficult, and interpretation is more difficult due to the internal momentum and angular momentum of the deuteron. The (n,p) reaction at higher energies appears to be a better option for s=1 selection.

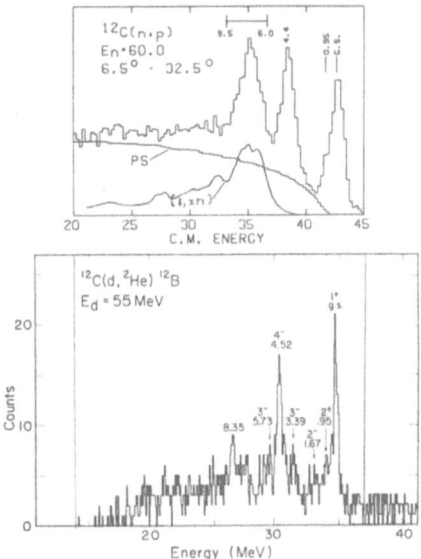

Fig. 6. $^{12}C(n,p)$ and $(d,^2He)^{12}B$ at comparable momentum transfers.

In medium and heavy nuclei the effects of distortion make
measurements at energies below 100 MeV difficult. At 60 MeV in
the Ni isotopes the effects of blocking are evident in the excitation of the GDR.[30] The GDR strength decreases with N-Z and
follows the predictions of Brown and Madsen.[31] See fig. 8(a)
and (b).

EXPERIMENTAL CONSIDERATIONS

The first experimental consideration for (n,p) reactions is
the production of a high-quality, high-intensity neutron beam.
In this respect criteria include neutron intensity, energy resolution, separation of beam peak from low energy neutron tail and
so on. At 60 MeV the best reaction appears to be $^3H(p,n)^3He$.
The 0° cross section is large $\cong 40$ mb/sr and the relative energy
loss of protons in ^3H is small, so that the number of neutrons
per MeV energy loss of the incident protons is large. The neutron
beam energy width, ΔE_n, is determined by the energy loss of the
incident protons and one can reduce ΔE_n by reducing the target
thickness with a corresponding loss in neutron flux. The break
up of ^3He into a deuteron and proton begins more than 5 MeV below
the peak.

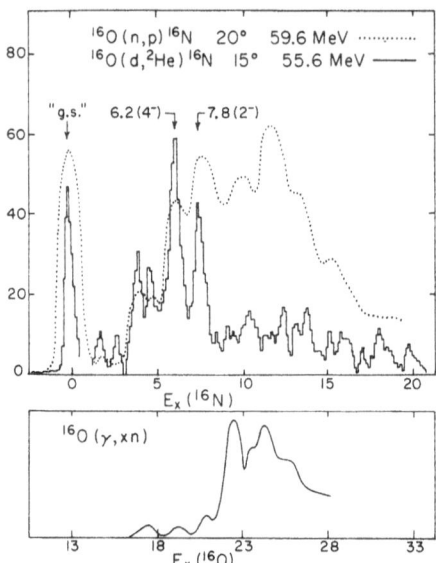

Fig. 7. ^{16}O(n,p) compared to (d,^{2}He) and (γ,xn).

^{7}Li(p,n)^{7}Be has a good 0° cross section. However, we predict that the neutron flux per MeV of energy loss is lower than for ^{3}H by a factor of $\cong$3 at 60 MeV and at 200 MeV. In addition ^{7}Be has an excited state at 0.429 MeV and break-up to ^{3}He+^{4}He begins at 1.586 MeV excitation. Having contributions from the two states limits how small the ΔE of the beam can be.

In the notation of reference 15, the CM (p,n) cross section for Fermi (F) and GT transitions at small q is well-approximated by the factorized form[17,18]

$$\sigma(q) = \sum_{\alpha=F,GT} K_\alpha N_\alpha^D \left| J_\alpha \right|^2 B(\alpha,q) \ .$$

We have used our estimates of N^D, and the $J_{\sigma\tau}/J_\tau$ of ref. 15 in this expression to extrapolate from (p,n) measurements on ^{7}Li and ^{3}H at lower energies to obtain predictions at 200 MeV. Eventually measurements from IUCF will be available. At 60 MeV (200 MeV) transitions to the ^{7}Be first excited state provide about 25% (38%) of the total neutron flux in the beam when it is assumed that both contribute to the peak. Table I gives results for σ(lab) and neutron flux for 60 and 200 MeV. Ten meters is used as a convenient scaling distance and is roughly what is

(a)

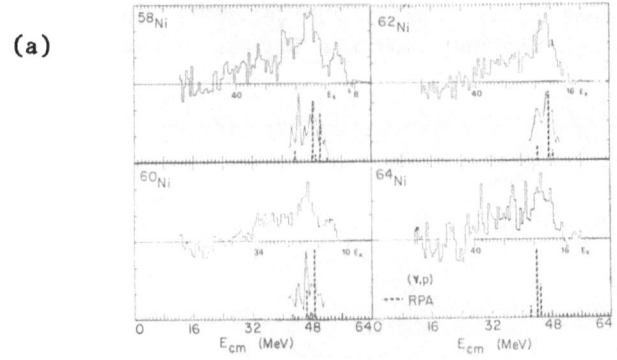

(b)

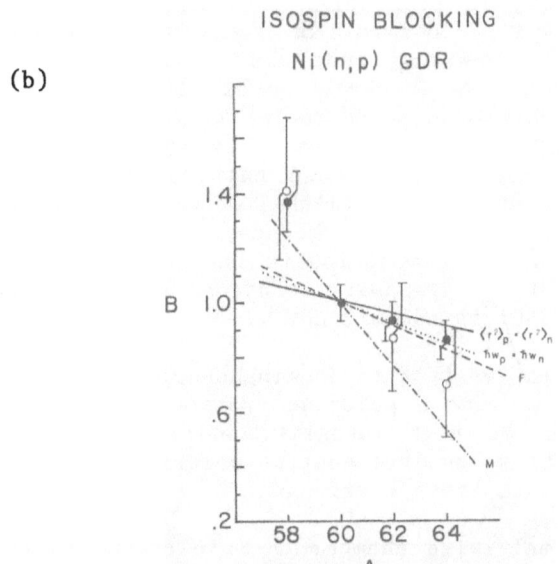

Fig. 8.   (a) Ni(n,p) spectra.   (b) Blocking of the GDR.

needed for a clean TOF cut on the beam peak. At 200 MeV one has the advantage of reduced energy loss in the neutron-production target, and this produces the factor of three larger flux. This is also true for the (n,p) reaction target. Thus at 200 MeV, other things being equal, one should achieve, per µA, almost an order of magnitude larger event rate over that at 60 MeV.

Table I. Neutron beam parameters at 200 (60 MeV) The ^{7}Li values
          of cross section and flux are for the ground plus first
          state.

Target	Q (MeV)	$\sigma_{lab}(0°)$ (mb/sr)	10^3 neutrons/cm^2-μA-MeV at 10 m
^{3}H	-0.76	44 (42)	18 (6)
^{7}Li	-1.66	35 (31)	6 (2)

One concludes that, in principle, the ^{3}H(p,n)^{3}He reaction
provides better flux, smaller possible ΔE_n, and better separation
of the peak from the lower-energy continuum. In practice ^{3}H is
more difficult to handle. At UC Davis we use ^{7}Li for our unpolarized beam, and a high-pressure, LN-cooled T_2 gas target for the
polarized beam. One has to consider the facts that the windows
of the gas target contribute at 0° and that the length of the
target affects the time resolution in the TOF cut used to select
the neutron beam peak. At Harvard, Measday,[32] in a pioneering
effort, used D_2 to produce a monokinetic beam, but flux was a
serious problem. The D_2 beam appears to have a small low energy
tail but the intrinsic ΔE_n is about 2 MeV.

One additional point is that the dominance of GT transitions
at 200 MeV allows one to use a polarized proton beam to make a
polarized neutron beam of high intensity and polarization via
$(\vec{p},\vec{n})$ at 0° or possibly to use the neutron polarization in $(\vec{p},\vec{n})$
as a signal of GT strength above background.

At UCD we use a multiwire chamber ΔE-E telescope to detect
protons and other charged particles (fig. 9). The response of
both the ΔE and E detectors is mapped so that the pulse height
can be corrected for variation with position. Energy resolution
of 1% is achieved, and overall resolution due to beam, target
thicknesses, and detector system is $\cong$1 MeV at 60 MeV with reasonable event rates.

A similar system with intrinsic Ge detectors stacks could be
used at higher energies. Nuclear interactions in the ΔE and E
units, $\cong$20% at higher energies, are a serious problem. A seg-
mented telescope is used to identify such events. At higher
energies it is essential to go to 0° so a magnet and MWC's may be
necessary. To improve count rate one envisages a large area
neutron beam striking an array of targets. Track reconstruction,
using MWC's provides good angular resolution. Energy measurement

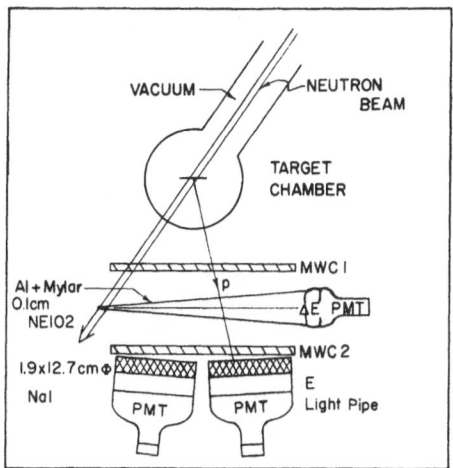

Fig. 9.   MWC system used for (n,p) at UCD.

and particle identification is accomplished with an array of $\Delta E \cdot E$ Ge stacks.

A better alternative at the highest energies might be a magnetic spectrometer which avoids the problems of nuclear inter-actions and multiple scattering in the stacked Ge detectors. Figure 10 shows a possible design. Assuming a position resol-ution of 500 µ in the MWC's, the geometry shown produces a $\Delta p/p \cong 10^{-3}$ which corresponds to $\Delta E_p = 300$ keV at 200 MeV. The angular acceptance is $\cong 22°$ and the corresponding energy bite at one setting is >50 MeV. The cost is unknown. Good statistics for $\cong 10$ targets requires $\cong 50$ hours of beam.

CALCULATIONS IN NEUTRON-EXCESS NUCLEI

In neutron-excess nuclei the $T_>$ strength as seen in (n,p) is expected to diminish with neutron excess. In good isospin schem-atic models[31,33] a distinction is made between matched particle-hole (ph) pairs in which both the neutron and proton ph pairs can be in the same orbit and unmatched ph pairs in which the neutron and proton ph pairs are in different orbits, one orbit of which is in the neutron-excess. Isovector giant multipole resonances have $T_>$ and $T_<$ components, where $T_> = T + 1$ and $T_< = T$, with T the isospin of the ground state. In the schematic model the $T_>$ component is made up purely of matched ph pairs coupled to $\tau=1$. The $T_<$ component is made up of both matched and unmatched ph pairs.   The schematic model gives rough expectations for the

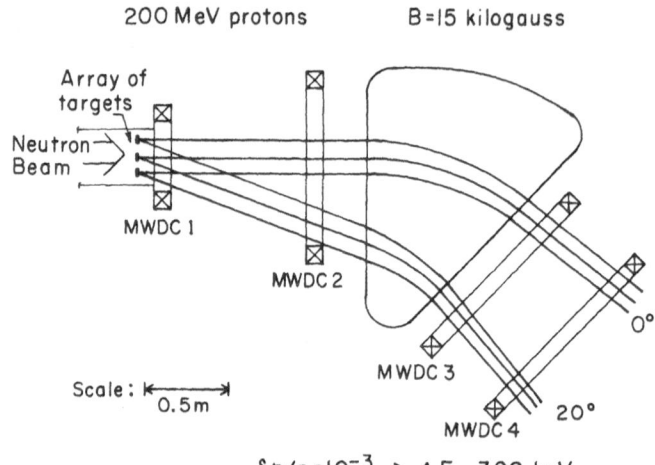

Fig. 10. Magnetic spectrometer system for (n,p) at higher energies.

dipole, quadrupole or other multipole symmetry energy which is defined[31,33] to include the effects of ph contributions. The splitting between $T_>$ and $T_<$ states is a result of the competition between ph contributions which push $T_<$ states higher in energy than $T_>$ states and the Lane isospin term which pushes states of higher T higher in energy. The schematic model also predicts the relative transition strength to $T_>$ and $T_<$ states.

Consider as an example, a schematic-model[31] comparison of the dipole and quadrupole resonances in ^{90}Zr and ^{208}Pb. The isovector energy splitting of the $T_>$ dipole resonance is 3.3 MeV in ^{90}Zr and 4.6 MeV in ^{208}Pb. The $T_>$ strength (transition matrix element squared) compared to the total isovector dipole strength is 0.62 in ^{90}Zr and 0.05 in ^{208}Pb. The isovector quadrupole splitting is 4.7 MeV in ^{90}Zr and 7.6 MeV in ^{208}Pb. The relative $T_>$ quadrupole strength is 0.78 in ^{90}Zr and 0.48 in ^{208}Pb. These results reflect the smaller neutron excess in ^{90}Zr which gives rise to a smaller $T_>$ blocking. The results for ^{208}Pb are intuitively pleasing; the $2\hbar\omega$ quadrupole matched pairs are nearly half blocked as compared to the essentially blocked $1\hbar\omega$ dipole pairs.

The various schematic-model results are useful for interpreting N-Z systematics and for suggesting (n,p) experimental studies. The parent analog of the $T_>$ resonances are seen most favorably in (n,p), where they are shifted down by the Coulomb energy and where the isospin Clebsch Gordan coefficient (see fig. 1) is one. A comparison to inelastic hadron scattering can yield information on the systematics of $T_>$ isovector giant resonances along with the determination of symmetry energies and isospin energy splitting in N>Z nuclei.

Because the $T_>$ dipole or $1\hbar\omega$ resonances are expected to be essentially blocked in ^{208}Pb (n,p), one might hope to have a better look at the various partially blocked $2\hbar\omega$ $T_>$ isovector resonances. In a Davis experiment reported earlier,[14] the experimental results for ^{209}Bi (n,p) were compared to the schematic-model calculations[31] for the $T_>$ quadrupole resonance. The angular distribution was obtained with the Steinwedel-Jensen quadrupole form factor, and the strength was adjusted to correspond to 100% of the isovector $T_>$ E2 strength[31] and the ^{209}Bi (p,p') E2 isoscalar measurement.[1] The experimental results, although somewhat high, were in reasonable agreement with the data. More realistic microscopic RPA calculations have now been performed in collaboration with the Jülich group.

The levels excited in ^{208}Pb (n,p) involve a change of isospin. In the present work we use the same generalized Landau-Migdal interaction as was used earlier[34] for ^{208}Pb (p,n) to the 0^+ isobaric analog resonance (IAR) and the 1^+ Gammow-Teller resonance (GTR), including finite range effects of the one-pion and rho exchanges. The experimental single-particle energies were used as input to the Jülich RPA calculations, which is consistent with earlier[34] calculations for $T_>$ states.

The microscopic treatment of (n,p) to $T_>$ states is a relatively untested area of nuclear physics; consequently, there exist various uncertainties. Our approach has been to compare (n,p) to (p,n) for analogous multipoles such as the 0^+ ground-state IAR seen in (p,n), and (n,p) to the $T_>$ giant isovector monopole. In Born approximation the differential cross section is simply expressed in terms of the momentum transfer dependence of the two-body effective interaction and the appropriate transition densities, also Fourier transformed to q space. For 0^+ charge-exchange transitions the V_τ interaction and the usual particle transition densities completely characterize the Born cross sections. The difference between (n,p) and (p,n) is that the transition densities are different for the two cases. The $0\hbar\omega$ transition density involved in (p,n) is large and related to the Fermi matrix element at q=0. The $2\hbar\omega$ transition density seen in (n,p) peaks at higher q. The diffraction pattern for the Born cross sections is given by the q dependence of the effective interactions, which is the same for (n,p) and (p,n) modified by the q dependence of the transition densities which are quite different for the two cases. Because of this modification, due only to the q dependence of the transition densities, there is clear theoretical evidence that (n,p) reactions can be a valuable tool for studying different regions of the two-nucleon effective

interaction. In particular the exchange contribution plays a new
role in (n,p) as compared to (p,n) and inelastic scattering.

The ingredients for a microscopic calculation of ^{208}Pb (n,p)
at 60 MeV involves the RPA transition density, the microscopic
interaction, including an imaginary part and reliable optical
potentials. At 60 MeV we used the optical potentials of
Patterson et al.[35] These optical potentials have some global
isospin features which are valuable for comparing (p,n) and
(n,p). They were also compared with (n,n') and (p,p') 3 experiments using the appropriate RPA transition density. The agreement was impressive. Our treatment of the imaginary form factor
was to take the real microscopic form factor and modify it by the
ratio of the imaginary to real isovector collective-model form
factor in the Steinwedel-Jensen picture. This prescription
including the appropriate optical potentials was tested for ^{208}Pb
(p,n) to the IAR and the 1^+ GTR. The GTR computation was performed to test our $V_{\sigma\tau}$ spin-flip (n,p) calculation. The results
were excellent, at least as good as previous (p,n) calculations[36]
in comparison with data. The main differences in the cross
sections for (p,n) and (n,p) are provided by the transition
densities which is appropriate in this "good isospin" framework.

Shown in Fig. 11 are the RPA results for ^{208}Pb (n,p) to the
$T_>$ isovector monopole with a Q_{np} = -15 MeV with respect to the
ground state of ^{208}Pb. The Coulomb energy shift, 18.7 MeV, then
puts the major strength at about $200/A^{1/3}$ in ^{208}Pb. Since our
prescription for the imaginary part of the form factor is tested
for (p,n) but not (n,p), we include the cross section with and
without the imaginary form factor for comparison. The data are
preliminary results for the reaction ^{209}Bi (n,p) for Q_{np} = -15 MeV.
The experimental width is of about 4-6 MeV and the cross section
shown are in arbitrary units. The RPA state in ^{208}Pb (n,p) is
at a single energy.

The giant isovector monopole is a volume breathing mode in
which neutrons and protons vibrate in opposition; it is pushed up
with respect to $2\hbar\omega$, the unperturbed ph energy, because the isovector effective interaction is repulsive. The isovector monopole and its systematics are interesting for a variety of reasons.
If the $T_<$ isovector monopole were at $170/A^{1/3}$ as in the hydro-
dynamic model,[37] then the splitting between the $T_<$ and $T_>$ (as
given by the (n,p) data) would be about 5 MeV. The monopole symmetry energy and the splitting can be determined empirically if
along with the (n,p) to the $T_>$, one measures inelastic scattering
to the $T_<$ state.

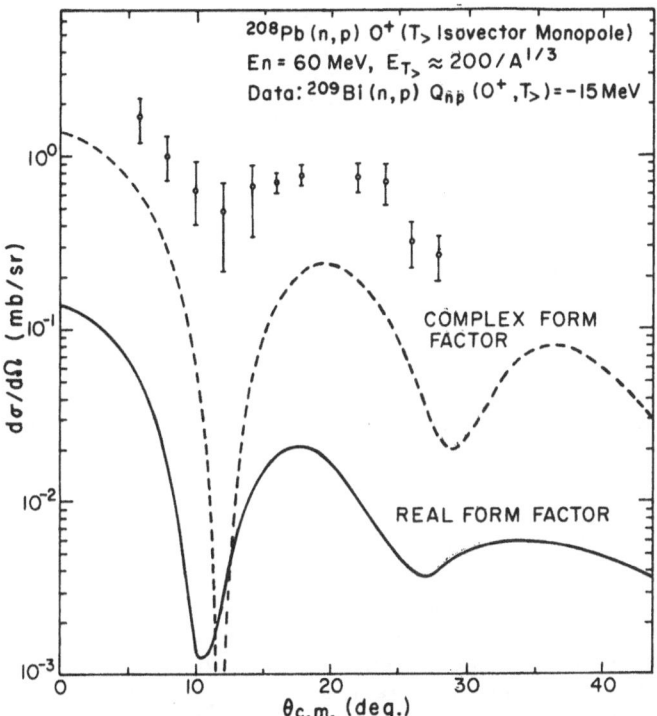

Fig. 11.   Angular distributions of the 0^+, $T_>$ state from the
^{208}Pb(n,p) reaction at 60 MeV incidence neutron energy.
The RPA cross section has been calculated as indicated
in the text. Also shown is the preliminary ^{209}Bi(n,p)
data in arbitrary units. The Q-value of the ^{209}Bi(n,p)
reaction to that state has a centroid of about -15 MeV
with width of 4-6 MeV.

Figure 12 shows the RPA for ^{208}Pb (n,p) to the $T_>$ isovector
quadrupole. In addition to the 2^+ non spin-flip transition we
show the spin-flip transition to a 2^+ state with Q_{np}=-7.98 MeV
with respect to the ground state of ^{208}Pb. The data are from
^{209}Bi(n,p) to the state with Q_{np} = -8 MeV. The experimental
cross section is in arbitrary units. The data are preliminary
and for ^{209}Bi (n,p) not ^{208}Pb (n,p), but the accumulating evidence
points to the fact that there is somewhat more strength seen experimentally than is calculated. Experimental difficulties including continuum subtraction problems make this conclusion
preliminary, but the angular distributions and the experimental
centroids of the peaks are somewhat more certain. From the RPA

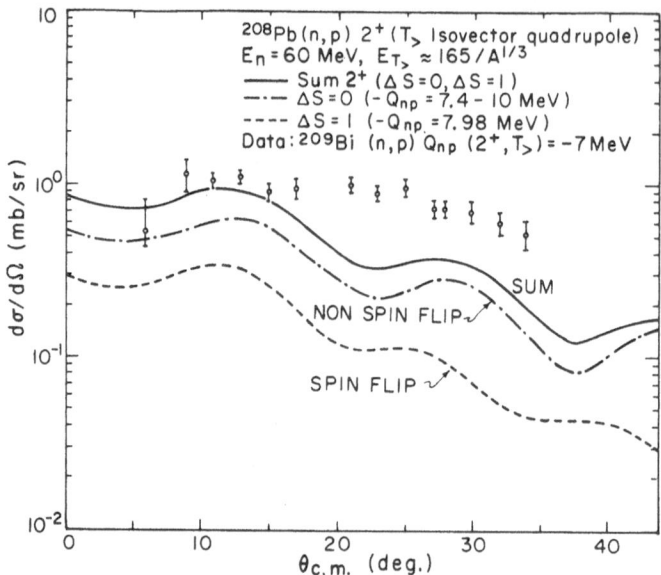

Fig. 12. Angular distributions of the 2^+, $T_>$ state from the ^{208}Pb(n,p) reaction at 60 MeV incident neutron energy. The RPA calculations shown are for the non-spin flip transitions to states at Q_{np}=-7.4, -8., -9.8, and -10.3 MeV. The 2^+ spin flip transitions is at Q_{np}=-8 MeV. Also shown is the preliminary data for ^{209}Bi(n,p) in arbitrary units. The Q-value of the ^{209}Bi(n,p) reaction to the 2^+ state has a centroid of about -7 MeV with a width of 2 MeV.

calculations the $T_>$ isovector quadrupole is around $165/A^{1/3}$ in ^{208}Pb; the centroid of the experimental results is around $160/A$ in ^{209}Bi. If the isovector 2^+ state were at $120/A^{1/3}$ in ^{208}Pb, then the splitting between the $T_>$ and $T_<$ would be about 7.6 MeV in agreement with the schematic model prediction. The measured[1] $T_<$ in ^{208}Pb is closer to $130/A^{1/3}$ giving a splitting of 5.9 MeV.

In this paper we are suggesting that higher energy (n,p) facilities would be very useful for examining $T_>$ spin-flip transitions. The distortion effects go down as does the non spin-flip transition strength. At 60 MeV the 2^+ spin-flip cross section from the single state at 7.98 MeV shown in Fig. 12 is already more than 1/2 the summed strength (4 RPA states) due to the non spin-flip transition. We have done these calculations at 120 MeV and found the spin-flip strength to be a factor of 3 larger in the forward direction than the non spin-flip strength.

Other spin-flip transitions at 120 MeV have also been calculated and the cross sections for the $T_>$ components of the 1^+ GTR, the spin-flip dipole, the 2^- spin-flip and the 3^- spin-flip are all in the millibarn/sr range at forward angles. The details of these results will be reported elsewhere.

In our opinion (n,p) is one of the exciting reactions for the future. It is a useful and sometimes unique tool to learn about nuclear structure and effective interactions. One can use the variation in neutron excess ($T_>$ strength) to learn about the role of isospin in the $1\hbar\omega$ and $2\hbar\omega$ excitations. In addition, (n,p) is a good reaction to learn more about charge exchange with transition densities that do not peak at q=0. The tensor force is more important in $2\hbar\omega$ excitations; the exchange and the odd force components may be studied at higher q. One of the hopes for the future is higher energy (> 100 MeV) (n,p) facilities to study spin-flip modes and reduce distortion factors.

ACKNOWLEDGMENTS

The experimental work described here has been carried out in collaboration with T. Ford, M. Johnson, N.S.P. King, G.A. Needham, T.S. Subramanian, J.L. Ullmann, J.W. Watson, and C. Zanelli. The RPA calculations were done in collaboration with H. Dermawan, J. Speth, S. Krewald, and F. Osterfeld at the IKP, KFA, Jülich, West Germany. This work was performed in part under the U.S.D.O.E. contract W-7405-ENG-48. We also acknowledge theoretical and calculational assistance from F.C. Dietrich, R.H. Howell, V.A. Madsen, F. Petrovich, G.R. Satchler, and W.W. True. We also thank Joseph Cerny, G.R. Jahn, D.P. Stahel, and G.J. Woznick for help at LBL, and the National Science Foundation, Grant No. PHY79-26282, for their support of this work.

REFERENCES

1. F.E. Bertrand, Ann. Rev. Nucl. Sci. 26:468 (1976).
2. G.R. Satchler, Elementary Modes of Excitation in Nuclei LXIV Corso, Soc. Italiana de Fisica, Bologna, Italy (1977).
3. R.R. Doering et al., Phys. Rev. Lett. 35:1691 (1975).
4. A. Galonsky et al., Phys. Lett. 74B:176 (1978).
5. W.A. Sterrenburg et al., Phys. Rev. Lett. 45:1839 (1980).
6. C.D. Goodman et al., Phys. Rev. Lett. 44:1755 (1980).
7. D.E. Bainum et al., Phys. Rev. Lett. 44:1751 (1980).
8. D.H. Horen et al., Phys. Lett. 99B:27 (1980).
9. J. Rapaport et al., Phys. Rev. C24:335 (1981).
10. W.G. Love and M.A. Franey, Phys. Rev. C24:1073 (1981).
11. C.D. Goodman et al., Phys. Lett. 107B:406 (1981).

12. C. Gaarde et al., Nucl. Phys. A369:258 (1981).
13. F.P. Brady and G.A. Needham in "The (p,n) Reaction and the Nucleon-Nucleon Force, edited by C.D. Goodman, S.M. Austin, S.D. Bloom, J. Rapaport, and G.R. Satchler, Plenum Press, N.Y. (1980) p. 121.
14. N.S.P. King and J.L. Ullmann, ibid, p. 372.
15. J. Rapaport et al., B.A.P.S. 24:830 (1979). Taddeucci et al. Phys. Rev. C25:1094 (1982).
16. S.M. Austin et al., Phys. Rev. Lett. 44:972 (1980).
17. F. Petrovich and W.G. Love, Proc. LAMPF Workshop on Pion Single Charge Exchange, Los Alamos, N.M., (1979) Doc. #La-7892C.
18. F. Petrovich, W.G. Love, and R.J. McCarthy, Phys. Rev. C21:1718 (1980).
19. F. Petrovich, ibid, ref. 13, page 121.
20. F.P. Brady, et al., Phys. Rev. Lett. 48:860 (1982).
21. D.F. Measday and J.N. Palmieri, Phys. Rev. 161:1071 (1967).
22. B.L. Berman et al., Phys. Rev. Lett. 15:727 (1965).
23. R.L. Bramblett et al., in Proc. Intl. Conf. on Photonuclear Reactions and Applications, Ed. B.L. Berman, (1973) p. 271.
24. J. Ahrens et al., Nucl. Phys. A251:479 (1975).
25. M. Goldhaber and E. Teller, Phys. Rev. 74:1046 (1948).
26. D.P. Stahel, R. Jahn, G.J. Wozniak, and Joseph Cerny, Phys. Rev. C20:1680 (1980).
27. G.A. Needham, Ph.D. Thesis, U.C. Davis (unpublished).
28. F. Ajzenburg-Selove, Nucl. Phys. A228:1 (1974) and A230:1 (1979).
29. B.L. Berman and S.C. Fultz, Rev. Mod. Phys. 47:713 (1975).
30. J.L. Ullmann et al. (to be published).
31. V.R. Brown and V.A. Madsen, Phys. Rev. C17:1943 (1978).
32. D.F. Measday, Nucl. Inst. and Methods 40:213 (1966).
33. R. Ö. Akyüz and S. Fallieros, Phys. Rev. Lett. 27:1016 (1971).
34. S. Krewald, F. Osterfeld, J. Speth, and G.E. Brown, Phys. Rev. Lett. 46:103 (1981).
35. D.M. Patterson, R.R. Doering, and A. Galonsky, Nuclear Physics A263:261 (1976).
36. F. Osterfeld, S. Krewald, H. Dermawan, and J. Speth, Phys. Lett. 105B:257 (1981).
37. A. Bohr and B. Mottelson, "Nuclear Structure" (Benjamin, Reading, Mass., 1975), Vol. II.

ISOVECTOR SPIN STATES OBERVED IN RADIATIVE PION CAPTURE IN FLIGHT

AND AT REST: RECENT RESULTS FROM SIN

M. Lebrun, C.J. Martoff, U. Straumann, and P. Truöl*

Physik-Institut, Universität Zürich, Zürich, Switzerland

C. Joseph, J.P. Perroud, and M.T. Tran

Institut de Physique Nucléaire, Université de Lausanne
Lausanne, Switzerland

J. Bistirlich and K. Crowe

Lawrence Berkeley Laboratory, University of California
Berkeley, California

J. Deutsch, G. Gregoire, and R. Prieels

Institut de Physique Corpusculaire, Université
Catholique de Louvain, Louvain-La-Neuve, Belgium

W. Dahme

Sektion Physik, Universität München, Garching, Germany

H. Baer

Los Alamos Scientific Laboratory, Los Alamos, New Mexico

INTRODUCTION

Photopion processes on first sight seem like a poor choice,
compared to other charge exchange probes like (n,p) or (p,n) reactions or the traditional electromagnetic probe of nuclear structure

---

*Presented by P. Truöl.

at intermediate momentum transfers, electron scattering. They can
not compete in final state resolution nor in precision. Their usefulness is limited to nuclear targets with mass below A = 40. The
cross-sections for (π^-,γ) reactions in flight, where the momentum
transfer can be varied, are typically 1μb/sr and the requirement to
detect a high-energy photon with good resolution adds further difficulty. If negative pions are captured at rest, data are easier to
obtain and hence exist for a variety of nuclei, but the momentum
transfer is restricted to $q \simeq m_\pi$. Level spins therefore have to be
inferred from other data or from comparison with theory. Despite
all these shortcomings radiative pion absorption has contributed to
the central topic of this conference, the study of spin-isospin
states, since the late sixties when the sensitivity to this particular mode of nuclear excitation was first theoretically predicted and
quickly experimentally demonstrated.[1,2]

Photopion reactions single out isovector transitions with large
spin-density matrix-elements, since the transition operator is basically a Gamow-Teller operator, the dominant part of the elementary
$\pi^-p \leftrightarrow n\gamma$ ($\pi^+n \leftrightarrow p\gamma$) process is proportional to $\vec{\sigma} \cdot \vec{\epsilon}\ \tau^\pm$ where $\vec{\sigma}$
and $\vec{\epsilon}$ are the nucleon spin and photon polarisation vectors and $\tau^\pm$
the isospin ladder operators. Consequently the experimental photon
spectra featured prominent peaks corresponding to analogs of magnetic dipole to octupole transitions in the target nucleus.[3] This is
well known, and will not be discussed here particularly in the light
of the much more detailed information, which has become available on
these states from other charge-exchange processes, in particular the
(n,p)-reaction.[4] The recent interest in magnetic quadrupole transitions, especially those in ^{16}O[5,6], which arises from their role in
scrutinizing the parameters of the effective particle hole interaction, has however stimulated us to extend the previous measurements
on ^{12}C[7], ^{16}O[8] and ^{18}O[8] to other nuclei near the end of the 1p-shell,
namely ^{14}C and ^{15}N. These new measurements, of which the ^{14}C-data
are presented here, when taken together with recent Los Alamos data
on ^{13}C[9], ^{19}F[9] and ^{20}Ne[10] allow now a rather complete interpretation
of these isovector-transitions, which we will discuss below.

Apart from these data, obtained in capture at rest, we report
here the first results from an experiment searching for precursor
phenomena in photopion reactions.[11,12] Such a search is warranted,
since enhancements of up to factor of ten are predicted in the
cross-sections for magnetic transitions near q = 2 to 3 m_π over the
expectations based on the standard Cohen-Kurath model[13]. The case
selected is $^{13}C(\pi^+,\gamma)^{13}N$, where sufficient resolution exists to
allow an exclusive measurement for the ^{13}N-ground state transition
near 120 MeV pion energy. While the latter experiment provides the
first inroad into future applications of radiative pion capture, the
former can be considered a continuation of the traditional use of
the photopion technique.

Before we present out results in detail, we wish to briefly review some aspects of the calculation of capture rates in particular those, which deal with the extraction of model-independent spin-density matrix elements directly from the data. The aim is of course to add credibility to the statement, that photopion reactions are not qualitatively, but quantitatively well understood.

SPIN-DENSITY MATRIX-ELEMENTS

Table 1 lists some M1, mixed M1/E2 and M3 transitions, where (π^-,γ)-branching ratios have been interpreted in a quasi-model independent manner to yield spin-dipole to octopule matrix-elements of the type $\langle J_f||\tau^+[\vec{\sigma}\times\vec{Y}_r]^J||J_i\rangle$. Figure 1 shows the corresponding experimental spectra. The calculation proceeds as follows. The capture rate from a given pionic orbit (n,l) is given in the impulse approximation by

$$\Lambda_\gamma^{n,l} = \frac{1}{(2J_i+1)}\,\frac{1}{(2l+1)}\cdot\frac{kM_a}{m_\pi(M_a+k)}\,\int_\alpha \Sigma\ |\langle J_fM_f|H_{eff}|J_iM_i\rangle|^2 d\hat{k} \quad (1)$$

$$\text{with } H_{eff} = i(1 + \frac{m_\pi}{m_n})\sum_{j=1}^{A} e^{-i\vec{k}\,\vec{r}_j}\, F_{j\lambda}\, \phi_{nl}^m(\hat{r})\tau_j^+,\ \alpha = (M_iM_fm\lambda) \quad (2)$$

and the notation: $\vec{k}$ = photon momentum

q_m = pion momentum

$\phi_{nl}(\vec{r}_j)$ = pion wave function

$\vec{\sigma}_j$ = nucleon spin

m_π, m_A, m_n = pion, target and nucleon mass

The transition amplitude $F_{j\lambda}$ is deduced from the corresponding photopion process on the nucleon, e.g. $\pi^-p \rightarrow n\gamma$

$$F_{j\lambda} = A\vec{\sigma}_j\vec{\epsilon}_\lambda + B(\vec{\sigma}_j\,\vec{\epsilon}_\lambda)(\vec{q}\,\vec{k}) + C(\vec{\sigma}_j\,\vec{k})(\hat{\epsilon}_\lambda\,\vec{q}) - \quad (3)$$

$$iD\hat{\epsilon}_\lambda(\vec{q}\times\vec{k}) + E(\vec{\sigma}_j\vec{q})(\vec{q}\,\hat{\epsilon}_\lambda)$$

Apart from the weak vector term proportional to D (all coefficients are combinations of elementary multipole amplitudes[3]) all other terms contain the nucleon spin. In particular contributions from the first, the well known Kroll-Rudermann term, dominate in general, for 1s capture there is no other term. In this case the

Table 1. Spin-density matrix-elements obtained in the analysis of radiative pion capture data compared to other sources. The last two lines give also matrix-elements of the orbital current, which have been extracted from (π^-,γ) combined with (e,e') and the corresponding theoretical predictions.

Type of Matrix element	Nuclear States	Radiative Pion Capture	Other Source		Ref.
$M1<1^+||[\hat{\vec{\sigma}}\times\hat{Y}]^1_O\tau^+||0^+>=R_{01}$	$^6Li(g.s.)\,^6He(g.s.)$	.627±.034(1s) .685±.120(2p)	.617	β	7
$M1<1^+||[\hat{\vec{\sigma}}\times\hat{Y}]^1_O\tau^+||0^+>=R_{01}$	$^{12}C(g.s.)\,^{12}B(g.s.)$	.285±.22	.282±.002	β	7
$M1<\frac{3}{2}^-||[\hat{\vec{\sigma}}\times\hat{Y}]^1_O\tau^+||\frac{1}{2}^->=R_{01}$	$^{13}C(g.s.)\,^{13}B(g.s.)$	.335±.040	.348±.009	β	9
$M3<0^+||[\hat{\vec{\sigma}}\times\hat{Y}]^3_2\tau^+||3^+>=R_{23}$	$^{10}B(g.s.)\,^{10}Be(g.s.)$	.64±.04	.70±.06	e,e'	7
$M3<0^+||[\hat{\vec{\sigma}}\times\hat{Y}]^3_2\tau^+||3^+>=R_{23}$	$^{10}B(g.s.)\,^{10}Be(3.37)$	1.05±.05	1.09±.06	e,e'	7
$M1<0^+||[\hat{\vec{\sigma}}\times\hat{Y}]^1_O\tau^+||1^+>=R_{01}$	$^{20}Ne(g.s.)\,^{20}F(1.06)$	.12±.04		e,e'	10
$M1<0^+||[\hat{\vec{L}}\times\hat{Y}]^1_O\tau^+||1^+>=L_{01}$	$^{20}Ne(g.s.)\,^{20}F(1.06)$		.62±.21		
Shell model	$^{20}Ne(g.s.)\,^{20}F(1.06)$	.105	.605		14

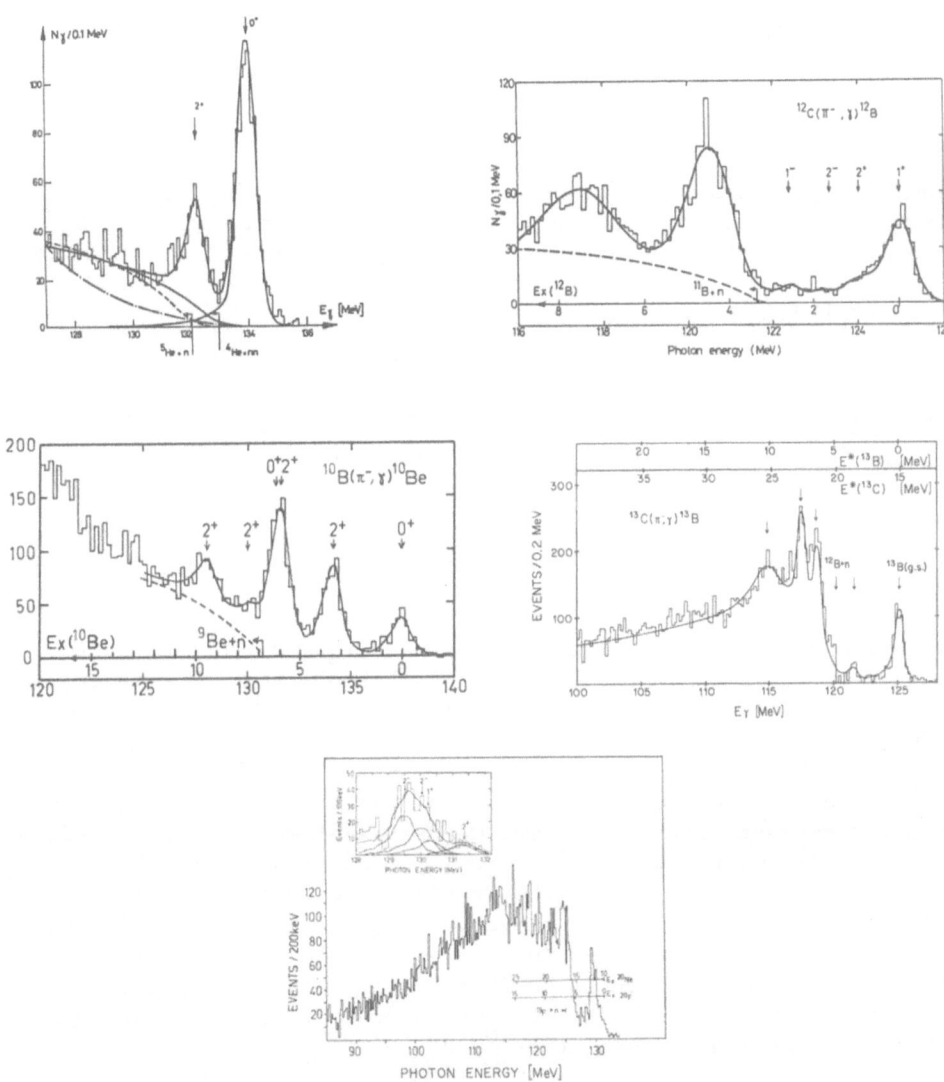

Figure 1. Photon spectra for (π^-,γ) at rest, showing M1 and M3 transitions, for which matrix-elements have been directly determined from the experimental branching ratios: a) ^{6}Li^7 b) ^{12}C^7 c) ^{10}B^7 d) ^{13}C^9 e) ^{20}Ne^{10}.

connection between Gamow-Teller and radiative pion capture strength
can be proven directly through the PCAC hypothesis. A straightforward reduction of the above expression leads to

$$\Lambda_\gamma^{n,1} \sim \sum_{J,L} \left\{ \left| <J_f||O_A(J;L,1)||J_i> \right|^2 + \left| <J_f||O_B(J;L,1)||J_i> \right|^2 \right. \qquad (4)$$

The spherical tensor operators O_A and O_B can be expressed as
sums of operators of the type $R_{LJ} = \tau^+[\vec{\sigma}xY_L]$ for the A,B,C and E
terms and τY_J for the D-term. To relate to the spin-part of the M1-
operators directly we quote here the 1s-part for an L = 1 transition

$$<J_f||O_A(J=L;L,0)||J_i>$$

$$= \frac{1}{\sqrt{4\pi}} \, 2 \, a_\pi^{-\frac{3}{2}} \, \sqrt{C_{1s}} \quad A \left\{ \frac{1}{\sqrt{3}} \, j_2\tau^+[\vec{\sigma}x\hat{Y}_2]^1 - \sqrt{\frac{2}{3}} \, j_0\tau^+[\vec{\sigma}x\hat{Y}_0]^1 \right\} \qquad (5a)$$

$$<J_f||O_B(J=L;L,0)||J_i> = -\frac{1}{\sqrt{4\pi}} \, 2a_\pi^{-\frac{3}{2}} \, \sqrt{C_{1s}} \, Aj_1\tau^+[\vec{\sigma}x\hat{Y}_1]^1 \qquad (5b)$$

with the notation a_π = pion Bohr radius, C_{1s} = distortion factor of
pion wave function at the origin, j_L = Bessel-function with argument
kr. For a 0^+ to 1^+ M1 transition only O_A contributes, and one can
identify the second term with the Gamow-Teller matrix-element from
β-decay, the first term contributes only for higher momentum transfers, e.g. in electron-scattering. In this process in addition the
orbital current contributes. The O_B-matrix-element is of spin-flip
E_1 type, it contributes to a 0^+ to 1^- transition. Since the pion
momentum operator q acts on the 2p-pion wave function, which is proportional to r inside the nuclear volume, one finds basically the
same result for 2p-capture. To give an example, the ^{20}Ne-transition
listed in table 1, which is dominated by 2p-capture, appears with a
branching ratio, after weighting the capture rates with the appropriate absorption probabilities[10], of

$$B_{\pi^-,\gamma}(0^+,1^+) = 6.25\cdot10^{-4}(|R_{01}|^2+0.070|R_{21}|^2+0.013 \, Re(R_{01}R^*_{21})) \qquad (7)$$

Since the interference term and the M1-tensor term are weighted
by small factors, the uncertainty about their magnitude contributes
little to the error on R_{01} extracted from (7) by comparison with the
experimental data. In those cases, where β-decay or (p,n)-data are
available, one may use these data and extract R_{21} from (π^-,γ) branching ratios[7,9].

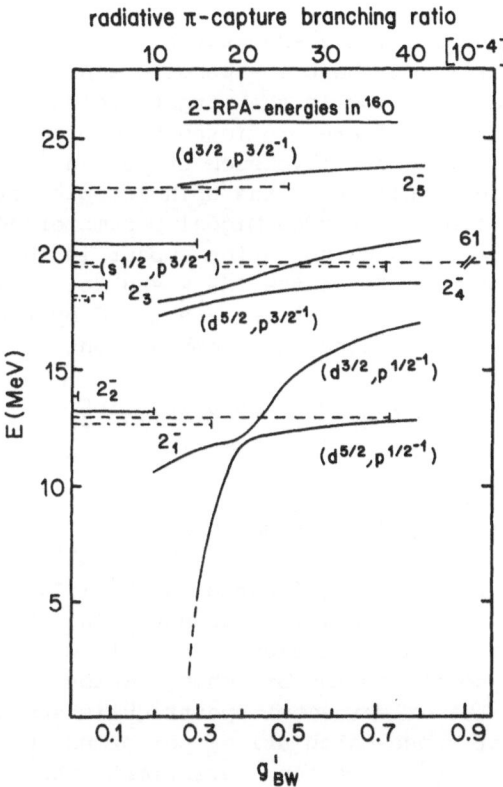

Figure 2. Excitation energies of 2⁻-states in ¹⁶O in function of the Landau-Migdal parameter g'. The top scale indicates the radiative pion capture branching ratio. The broken lines give the theoretical branching ratios (from Table 2) for the calculation of ref. 15. (upper and lower value) the solid lines correspond to the experimental results[8].

ISOVECTOR 2^- STATES IN THE ^{16}O REGION

        To this point we have only considered transitions within the
same shell. Higher multipolarities require the promotion of a particle into the next higher shell. Around ^{16}O the $1\hbar\omega$ excitations
are formed by coupling $1p_{3/2}$ or $1p_{1/2}$ holes to $2s_{1/2}$, $1d_{3/2}$ or $1d_{5/2}$
particles. Angular momentum selection rules limit the possible final
states, if we start from a 0^+ target, to the unnatural parity states
with spin 1^-, 2^- and 3^-. The 2^--M2-states - we label them by the
dominant multipolarity of the corresponding electromagnetic transition to the isotraric analog in the target nucleus - are driven by
the operator $\tau^+[\vec{\sigma} \times Y_1]^2$ and are predicted to dominate the experimental spectrum[15-21]. There are five such states in ^{16}O and their position is shown to be sensitive to the spin-isospin part of the nucleon-nucleon force, i.e. the Landau-Migdal parameter g' as shown in
Figure 2. The largest branching ratios appear in the $|1d_{5/2}1p_{3/2}^{-1}\rangle$
and $|1d_{5/2}1p_{1/2}^{-1}\rangle$ branch. Since the $1d_{5/2}$ and $2s_{1/2}$ single particle
energies are nearly degenerate, they are mixed by the residual in-
teraction and two nearly orthogonal combinations ($\theta \tilde{=} 36°$)

$$|2_3^-\rangle \tilde{=} \cos\theta \, |1d_{5/2}1p_{3/2}^{-1}\rangle - \sin\theta |2s_{1/2}1p_{3/2}^{-1}\rangle \qquad (8a)$$

and

$$|2_4^-\rangle \cong \sin\theta |1d_{5/2}p_{3/2}^{-1}\rangle + \cos\theta |2s_{1/2}1p_{3/2}^{-1}\rangle \qquad (8b)$$

appear. This interference is reflected in the (π^-,γ) branching ratios, the 2_4^--state carries all the strength. The three other states
remain rather pure $1p$-$1h$-excitations. In Table 2 we compare the
actually observed positions and branching ratios to the theoretically
predicted values. The experimental spectrum[8] is shown in Figure 3.
It is apparent that those calculations performed in the Tamm-Dancoff
or the random-phase approximation overestimate the experimental
branching ratios by a factor of three, even if the scattering states
are included in a continuum shell model calculation. The disagreement gets reduced to a factor of two, if ground state correlations
are included, i.e. if ^{16}O is no longer assumed to be a closed shell
nucleus, but $2\hbar\omega$-admixtures are considered. A further but minor
reduction is obtained if selected $3\hbar\omega$ negative parity excitations
are allowed to mix into the final states.

        If one adds further nucleons to ^{16}O one has to consider the cou-
pling of these particles to the ^{16}O base states. It is predicted [22,23]
that this coupling is weak. It leaves the structure unchanged,
but results in a core polarisation or a shift of the energy positions. This is particularly evident for ^{18}O (Figure 5), where a
spectrum nearly identical to ^{16}O is observed. While the ^{19}F case is

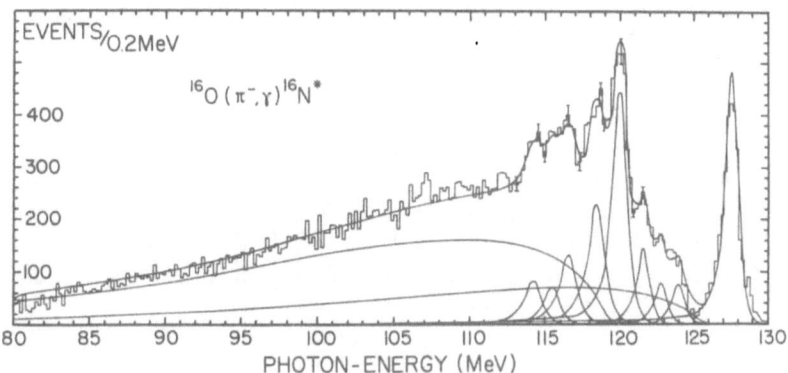

Figure 3. $^{16}O(\pi^-,\gamma)^{16}N*$ photon spectrum for $T_\pi = 0^8$.

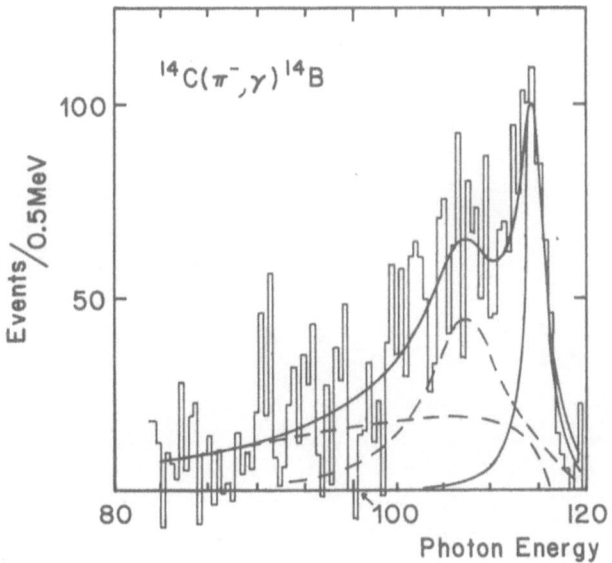

Figure 4. The $^{14}C(\pi^-,\gamma)^{14}B$ photon spectrum

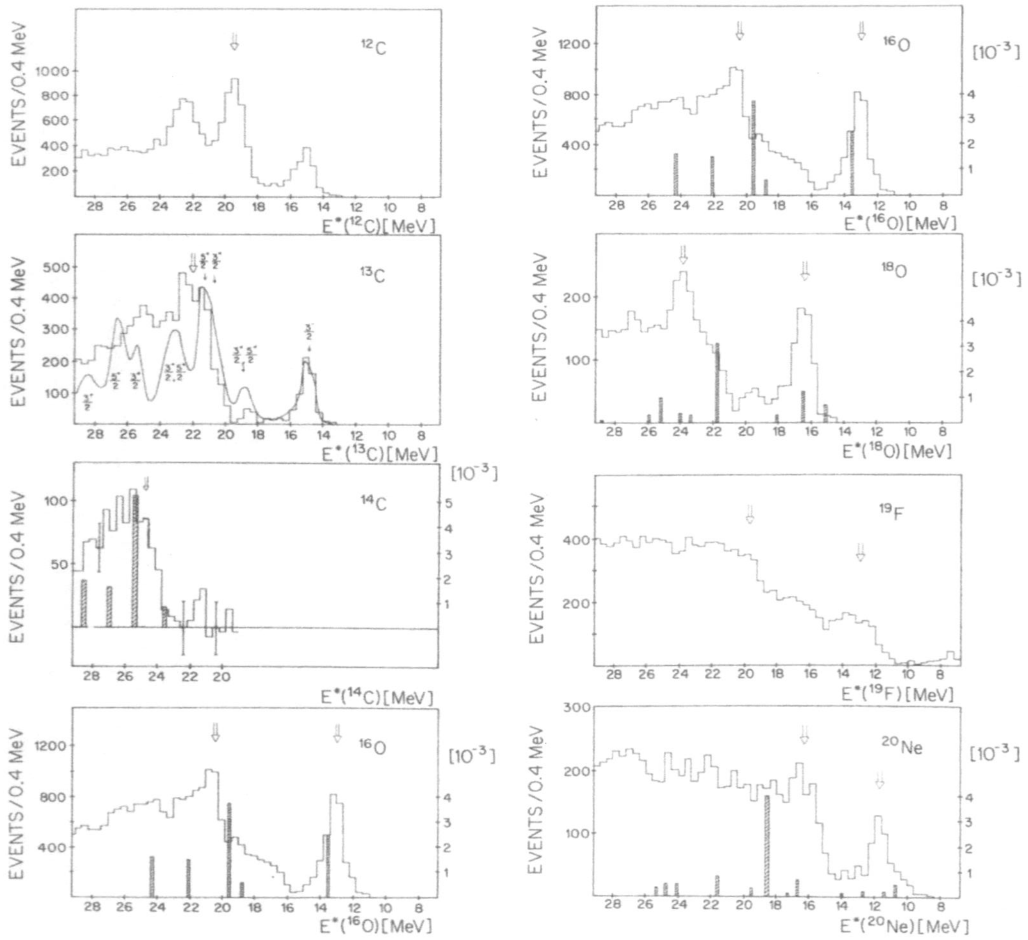

Figure 5.

Spin-quadrupole excita-
tions in A = 12 to A =
20. Displayed are the
photon spectra from
(π^{-},γ) on ^{12}C, ^{13}C,
^{14}C, ^{16}O, ^{18}O, ^{19}F,
^{20}Ne. The photon ener-
gy is converted into
excitation energy rel-
ative to the target
ground-state. The arrows
point to the ($1d5/2$,
$1p\overline{3/2}$) and ($1d5/2,1p1/2$)
configurations. The solid
curve drawn into the ^{13}C-
spectrum represents the
results of Kissener et
al.33, with the yield
normalized to the ^{13}B
ground state 15.1 MeV.
The vertical bars show
the results for the pho-
ton yield of reference 24
for ^{14}C and ref. 22 for
^{16}O, ^{18}O and ^{20}Ne.

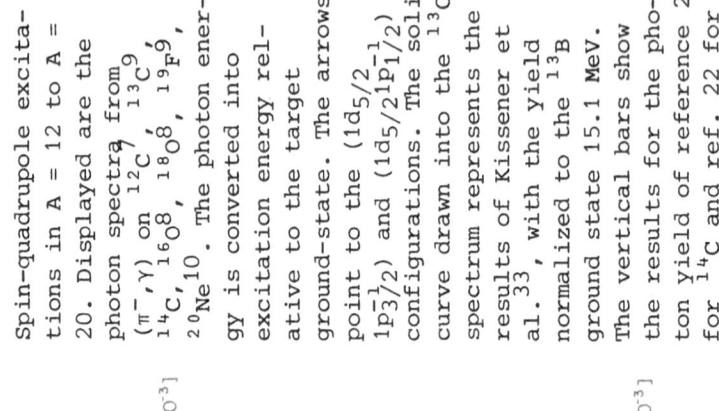

somewhat less clear. In $^{20}\hat{\text{Ne}}$ the M2-strength again is important
and the extra alpha particle only leads to a slight spreading of the
lower structure, built on the 2_1^--level of ^{16}O (Figure 5). Table 2
and Figure 6 demonstrate, that a constant strength of about $30 \cdot 10^{-4}$
or 15 % of all radiative capture strength is concentrated into the
spin-quadrupole branch.

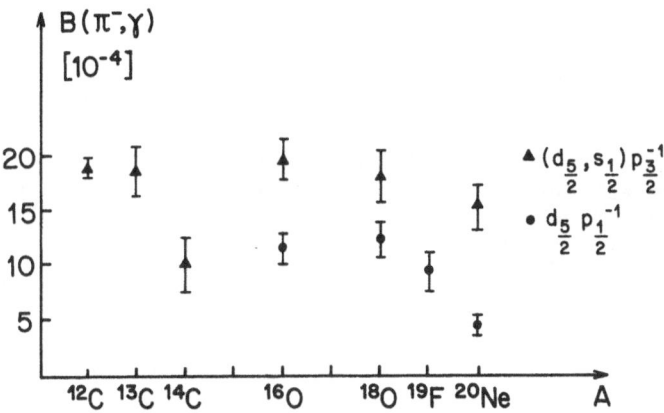

Figure 6. (π^-,γ) branching ratios for M2-type excitations in
             A = 12 to A = 20.

Below ^{16}O the $1p_{1/2}$-hole configurations are no longer available
except for the nitrogen isotopes, which are not discussed here. The
strongest state remains the constructively interferring combination
(8b). This was already observed in the first ^{12}C spectrum[2]. Adding one
neutron again leads to a shift but also to a splitting into a ($3/2^+$,
$5/2^+$)-doublet[9]. A particular interesting case is ^{14}C, where we present data here for the first time. One observes here the T = 2, M2-
states of the A = 14-system. A shell model calculation[24] indicated
a strong 2^- state at 1.8 MeV excitation energy in ^{14}B, which is indeed observed (Figure 4). Its structure is again fairly well represented by the combination given in (8b) above. The experiment is
rather difficult, since only a small target (6 g) of isotopically
enriched (84 % ^{14}C) carbon was available and large subtractions from
the steel container, the ^{12}C content and the beam defining scintillator in front of the target are necessary.

Comparing now the observed M2-strength to the calculated one
(Table 2 and Figure 7), we realize that only 50 % of the expected
branching ratio is observed. The unobserved $|1d_{3/2}1p_{3/2}^{-1}\rangle$ and $|2s_{1/2}$
$1p_{1/2}^{-1}\rangle$ branches cannot account for this, since they have been subtracted from the predictions, too. One is tempted to relate the missing strength to the Δ-mechanisms discussed at this conference in

Table 2. Comparison of experimental and theoretical branching
          ratios for M2-type transitions.

          TDA: Tamm-Dancoff approximation (or standard shell model),
          TDAC: Continuum shell model calculation, RPA: Random phase
                approximation
          EGSA: Extended ground state, Esma: Extanded excited state,
                CESIM: Core excitation model with intermediate
                coupling

          References: a: 8, b: 16, c: 20, d: 22, e: 21, f: 17, g: 15,
                      h: 10, i: this experiment, k: 24, l: 34, m: 7

          The last line in the theoretical columns indicates the
          average ratio of experimental result over theoretal
          prediction.

Nuclei	Dominant Configuration	J^π_n	E^{exp}_Z	E^{exp}_{Z-1} (MeV)	E^{th}_Z	B^{exp} (10^{-4})	B^{th} (10^{-4}) TDA	TDAC	RPA	EGSA	ESMA	CESIM
$^{16}O, ^{16}N$	$\lvert d_{5/2}p^{-1}_{1/2}\rangle$	2^-_1	13.0	0.0	13.0	11.6±1.3[a]	44[b]	25[c]	37[b]	19[b]		25[d]
							35[e]	44[f]	--	--	--	--
							39[g]	--	--	24[g]	18[g]	--
							.30	.37	.32	.54	.65	.47
	$\lvert s_{1/2}p^{-1}_{3/2}\rangle - \lvert d_{5/2}p^{-1}_{3/2}\rangle$	2^-_3	19.1	6.1	18.2	4.4±.6[a]	19[b]	4[c]	18[b]	6[b]		15[d]
							10[e]	2[f]	--	--	--	--
							3[g]	--	--	1	2	--
	$\lvert s_{1/2}p^{-1}_{3/2}\rangle + \lvert d_{5/2}p^{-1}_{3/2}\rangle$	2^-_4	20.4	7.5	19.6	15.1±1.6[a]	50[b]	48[c]	43[b]	57[b]		37[d]
							61[e]	58[f]	--	--	--	--
							62[g]	--	--	39[g]	36[g]	--
							.26	.35	.32	.39	.51	.38
$^{18}O, ^{18}N$	$\lvert d_{5/2}p^{-1}_{1/2}\rangle ; \lvert d_{3/2}p^{-1}_{1/2}\rangle$	2^-_1	16.2	0.0	15.8	12.3±1.6[a]	--	--	--	--	--	20[d]
												.62
	$\lvert d_{5/2}p^{-1}_{3/2}\rangle$	2^-_2	23.1	6.9	21.8	18.1±2.2[a]	--	--	--	--	--	31[d]
												.58
	$\lvert d_{3/2}p^{-1}_{3/2}\rangle$	2^-_3	26.4	10.1	25.5	6.7±.7[a]	--	--	--	--	--	13[d]
												.52
$^{20}Ne, ^{20}F$		2^-	11.6	1.3	10.7	1.7±.6[h]	--	--	--	--	--	5[d]
		2^-	12.1	1.8	11.3	2.6±.3[h]	--	--	--	--	--	2[d]
												.61
		2^-	16.4	6.1	18.6	15.3±2.0[h]	--	--	--	--	--	40[d]
												.38
$^{14}C, ^{14}B$	$\lvert s_{1/2}p^{-1}_{3/2}\rangle - \lvert d_{5/2}p^{-1}_{3/2}\rangle$	2^-_1	23.6	0.0	23.6	---	9[k]	--	--	--	--	--
	$\lvert s_{1/2}p^{-1}_{3/2}\rangle + \lvert d_{5/2}p^{-1}_{3/2}\rangle$	2^-_2	25.4	1.8	25.4	9.9±2.3[i]	53[k]	--	--	--	--	--
							.16					
$^{13}C, ^{13}B$	$\lvert s_{1/2}p^{-1}_{3/2}\rangle - \lvert d_{5/2}p^{-1}_{3/2}\rangle$	$(3/2,5/2)^+_1$	18.6	3.5	18.7	1.0±.5[k]	8[l]	--	--	--	--	--
		$(3/2,5/2)^+_2$	21.6	6.5	20.6	8.3±1.5[k]	11[l]	--	--	--	--	--
	$\lvert s_{1/2}p^{-1}_{3/2}\rangle + \lvert d_{5/2}p^{-1}_{3/2}\rangle$	$(3/2,5/2)^+_2$	22.7	7.6	21.5	9.0±1.6[k]	24[l]	--	--	--	--	--
							.42					
$^{12}C, ^{12}C$	$\lvert s_{1/2}p^{-1}_{3/2}\rangle$	2^-_1	16.6	1.6	16.3	.5±.2[m]	1[b]	0[c]	1[b]	--	--	--
	$\lvert d_{5/2}p^{-1}_{3/2}\rangle$	2^-_2	19.5	4.4	19.3	18.3±.6[m]	50[b]	60[c]	53[b]	--	--	--
							.37	.31	.35			

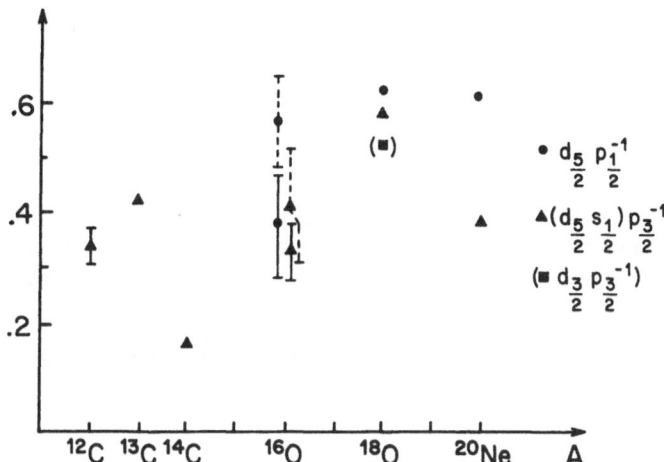

Figure 7. Ratio of observed to calculated strength for M2-type
           excitations.

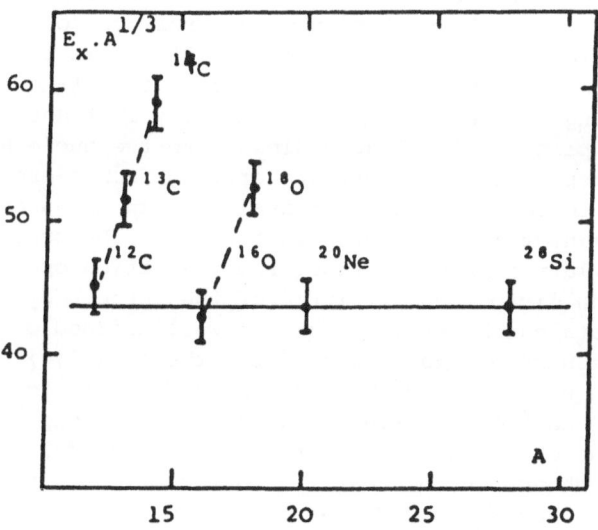

Figure 8. Position of M2-resonance in light nuclei.

connection with the M1-Gamow-Teller states. Discussion along these
lines exist for electron scattering data[5,25-27]. Suzuki et al.[27]
calculate e.g. 25 % quenching of the BM2 values for the 2^- and 8 %
to 11 % for the 2^-_3 and 2^-_4 states of ^{16}O arising from Δ-h admixtures.
Since the study of these high lying excitations in (e,e'), (π,π')
or (p,n) is often troubled by the presence of isoscalar states or
higher spin-multipoles, the (π^-,γ) reaction can at least serve as a
guide, where to search for M2-strength. For ^{18}O use was already made
of this fact[28]. In the s-d-shell further examples are ^{28}Si and ^{32}S.
The ^{28}Si case has been discussed recently by Perroud[29], from whose
report we extract Figure 8, which summarizes the available information on the branching ratio weighted average position of M2-strength
in light nuclei. It follows nicely the Richter-line[26].

ABSENCE OF EVIDENCE FOR PRECRITICAL ENHANCEMENT FROM PHOTOPION
STUDY OF ^{13}C

Delorme[11] has pointed out that photopion reactions should nicely complement other probes which have been proposed or have been
used to search for the possible occurence of critical opalescence
phenomenon at ordinary nuclear densities. This phenomenon which
consists of an enhancement of the virtual pion field in the nuclear
medium takes place around a critical momentum of 2 to 3 m_π and mani-
fests itself in an enhancement of the cross-sections at around this
momentum transfer for magnetic form factors of unnatural parity
states. Most experiments[30] have concentrated on the ^{12}C(15.1 MeV)
M1 transition. The interpretation of the form factors, as e.g. obtained in inelastic electron scattering, if one is trying to isolate
effects related to pion induced spin order, is difficult, since the
photon coupling to the magnetisation density is transverse $(\vec{\sigma}\times\vec{q})$
where as the pion couples longitudinally to the nucleon spin $(\vec{\sigma}\cdot\vec{q}')$.
The photon thus probes best effects arising from ρ-exchange. The
longitudinal component is however available in the nucleonic processes and dominant in the photopion reactions. Delorme argues in
favor of the latter process, since the transition operator is well
known and pion distortion by virtue of the large body of elastic
scattering data can be treated through well defined optical potentials. Figure 9 shows his results[13] for the $^{13}C(\pi^+,\gamma)^{13}N$(g.s.) transition, which was selected, because the separation to the first
excited state in ^{13}N is large enough to allow an experimental separation. With the choice g' = 0.4 for the Landau-Migdal parameter a
spectacular enhancement of nearly a factor of 10 is expected for
pion energies around 100 MeV. The cross section is nearly entirely
given by the longitudinal component. Even with the caution which
with one has to regard a plane wave calculation, an experimental
test seemed feasible.

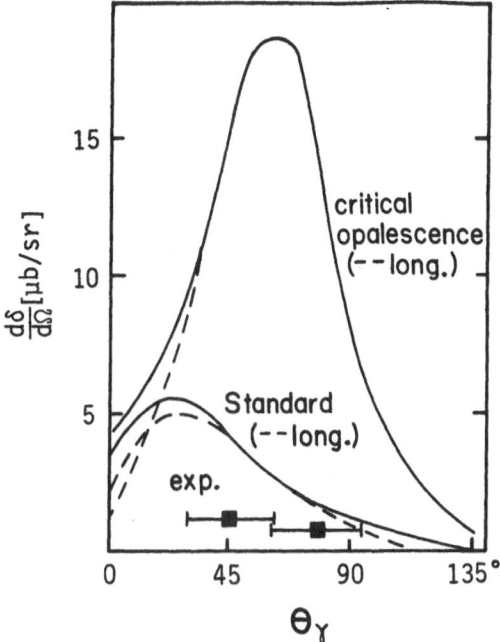

Figure 9.   Predicted $^{13}C(\pi^+,\gamma)^{13}N$(g.s.) cross section. The
curves of ref. 13 are interpolated to T_π = 115 MeV.

    The low cross-section (about 1 to 10 μb/sr) requires high pion
intensities and if possible a large acceptance photon detector.
However for an inflight experiment the final resolution contains,
besides the contribution from the photon detector itself, contributions from the momentum resolution of the beam, the energy loss
inside the target and the angular resolution of the detector, each of
which can only be met at the cost of reduced event rate. Table 3
summarizes the contribution to the resolution. We used the Lausanne-
München-Zürich pairspectrometer at SIN[31] as a photon detector. With
a 6 % radiation length converter we reach a solid angle (including
conversion and detection efficiency) of 0.48 ± 0.07 msr. Three separate calibration experiments were performed. The photon resolution at 130 MeV was measured with the reaction $\pi^-p \to n\gamma$ at T_π = 0.
This was necessary in order to compare the detector response at
maximum magnetic field (1.1 T), where saturation effects are appearing, to that at standard operating field (0.8 T). Secondly the
same reaction was measured at T_π = 115 MeV to compare to the known
cross-section for that reaction[32], respectively its inverse. Good

Table 3. Contribution to the experimental resolution (FWHM).

	$\pi^- p \rightarrow n\gamma$	$^{13}C(\pi^+,\gamma)^{13}N$
angle-energy correlation and target thickness	3.5 MeV	1.3 MeV
spectrometer intrinsic resolution	1.3 MeV	1.3 MeV*
pion beam energy spread	1.5 MeV	1.9 MeV
s u m	4.2 MeV	2.6 MeV

* for standard conditions 3 % radiative length converter and 8 kG field   720 keV

agreement was found. Figures 10 and 11 exhibit the data for this
reaction. The separation of the radiative capture events from the
charge exchange events is clearly visible. The photon angle is re-
constructed from the direction of the electron positron pair (50 mr
FWHM). The rapid variation of the photon energy with angle leads
to a worse resolution then in the ^{13}C case (see Table 3). The last
calibration experiment used the $\pi^+ d \rightarrow pp\gamma$ reaction. This data is
however not yet analysed. Thus we still consider the results given
below for ^{13}C preliminary. Figure 12 shows the experimental spectrum,
which contains all our data. All photon energies were corrected to
a c.m. angle of 90°, the actual data were taken in two angular settings covering 28° to 62° and 60° to 95°. Given a 10 μb cross-section this would have allowed a rather complete angular distribution
in the region, where the 'conventional' and the 'exotic' predictions
differ the most. Unfortunately the angle averaged cross-sections are
1.0 ± 0.2 μb/sr for the lower and 0.60 ± 0.15 μb/sr for the upper
bin, obtained from total 100 events extracted in three weeks of
running time. Though disappointing from an experimental point of
view, the implications for pion opalescence are clear. The conclusions of our experiment do not change, if we compare them to the re-
vised calculations made available to us at the conference. Giraud's[33]
results are shown in Figure 13. His calculation is performed with
the Blomqvist-Laget operator, the pion optical potential of Stricker,
McManus and Carr and RPA transition densities which include π and
ρ-exchange in the particle-hole and Δ-hole channel. The agreement
with the data is quite good for the generally accepted values of
g' around 0.7.

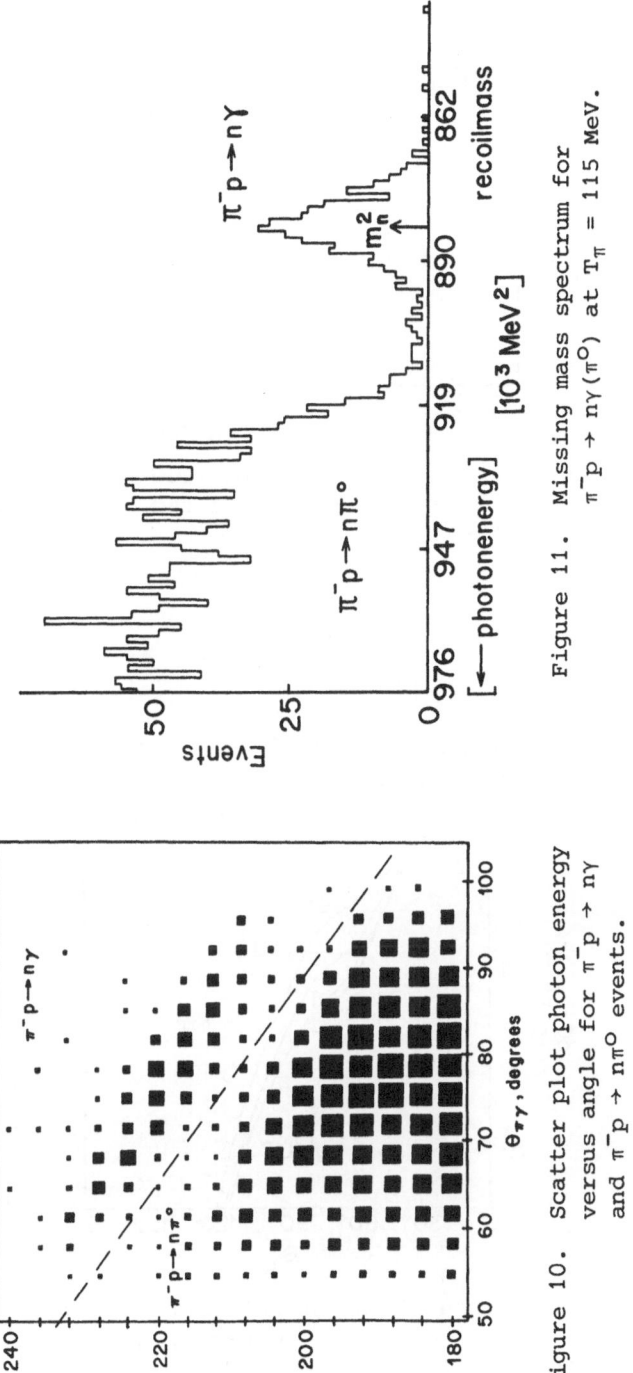

Figure 11. Missing mass spectrum for $\pi^-p \rightarrow n\gamma(\pi^0)$ at $T_\pi = 115$ MeV.

Figure 10. Scatter plot photon energy versus angle for $\pi^-p \rightarrow n\gamma$ and $\pi^-p \rightarrow n\pi^0$ events.

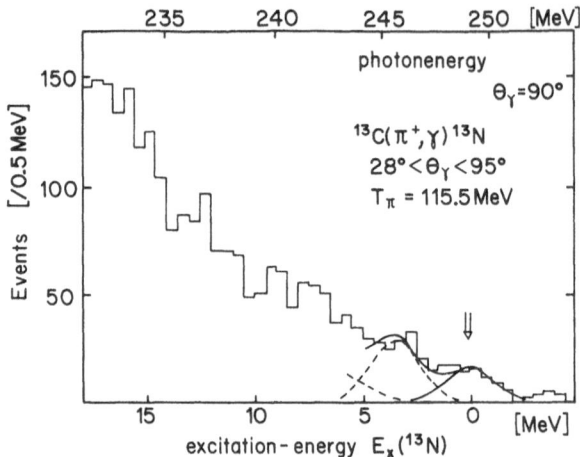

Figure 12.   Angle integrated photon spectrum for $^{13}C(\pi^+,\gamma)^{13}N$.
The energies have been corrected for a center of mass
angle of 90^0.

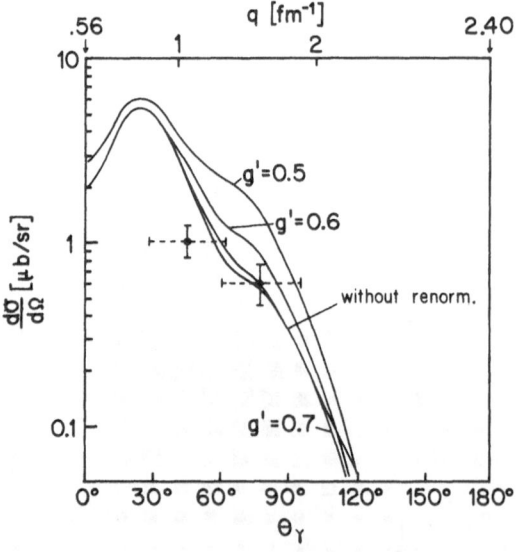

Figure 13.   Comparison of the experimental data for $^{13}C(\pi^+,\gamma)$
$^{13}N(g.s.)$ with the theoretical predictions of
Giraud[33].

REFERENCES

1. D.K. Anderson, J.M. Eisenberg, Phys. Lett. 22:164 (1966);
   J. Delorme, T.E.O. Ericson, Phys. Lett. 21:98 (1966);
   J.D. Murphy et al., Phys. Rev. Lett. 19:714 (1967).
2. J.A. Bistirlich et al., Phys. Rev. Lett. 25:689 (1970).
3. H.W. Baer, K.M. Crowe, P. Truöl, Adv. Nucl. Phys. 9:177 (1977).
4. C.D. Goodman, in: 'The (p,n)-reaction and nucleon-nucleon force'
   Plenum, New York (1980).
5. J. Speth, V. Klemt, J. Wambach, G.E. Brown, Nucl. Phys. A343:
   382 (1980).
6. J. Meyer-ter Vehn, Phys. Rep. 74:323 (1981).
7. A. Perrenoud et al., Université de Lausanne report (1981).
8. G. Strassner et al., Phys. Rev. C20:248 (1979).
9. C.J. Martoff et al., Phys. Rev. C27:1621 (1982).
10. C.J. Martoff et al., Phys. Rev. Lett. 46:891 (1981).
11. J. Delorme, Symposium on 'Perspectives in Electro- and Photo-
    nuclear Physics', Saclay (1980), LYCEN-8081.
12. M. Ericson, in: 'Nuclear and Atomic Physics with Heavy Ions',
    Bucarest (1981), CERN TH 3129.
13. J. Delorme, Workshop 'Intermediate energy nuclear physics with
    monochromatic and polarized photons', Frascati (1980), LYCEN
    8055, J. Phys. G7:L7 (1981).
14. B.H. Wildenthal, W. Chung, Proc. Conf. Nucl. Phys. with Electro-
    magnetic Interactions, Mainz (1979), contr. 1.8.
15. R.A. Eramzhyan et al., Nucl. Phys. A290:294 (1977).
16. N. Ohtsubo, T. Nishiyama, M. Kawaguchi, Nucl. Phys. A224:164
    (1974).
17. R. Wünsch, Nucl. Phys. A345:446 (1980).
18. R.A. Eramzhyan, M. Gmitro, H.R. Kissener, Nucl. Phys. A338:436
    (1980).
19. G.E. Dogotar et al., J. Phys. G5:L221 (1979).
20. N. Ohtsuka, H. Ohtsubo, Nucl. Phys. A306:513 (1978).
21. J.D. Vergados, Phys. Rev. C12:1278 (1975).
22. W. Knüpfer, G. Ender, M. Huber, 8th ICOHEPANS, Vancouver (1979),
    private communication.
23. W. Knüpfer, K. Knauss, M.G. Huber, Phys. Lett. 66B:367 (1977).
24. H.R. Kissener, R.A. Eramzhyan, Nucl. Phys. A326:289 (1979).
25. W. Knüpfer et al., Phys. Lett. 77B:367 (1978).
26. A. Richter, Proc. Conf.'Nucl. Phys. with Electromagnetic Inter-
    actions', Lecture Notes in Phys. 108:19 (1979).
27. T. Suzuki, S. Krewald, J. Speth, Phys. Lett. 107B:9 (1981).
28. E.J. Ansaldo et al., Phys. Lett. 95B:31 (1980).
29. J.P. Perroud, Proc. Seminar on Electromagnetic Interactions of
    Nuclei at Low and Medium Energies, Moscow (1981).
30. See review given by W. Weise at this conference.
31. J.C. Alder et al., Nucl. Instr. and Meth. 160:93 (1979).
32. M.T. Tran et al., Nucl. Phys. A324:301 (1979).
33. N. Giraud, private communication.
34. H.R. Kissener et al., Nucl. Phys. A312:394 (1978).

0^- TO 0^+ BETA DECAY IN A = 16*

C. Gagliardi, G. T. Garvey, J. R. Wrobel

Physics Division
Argonne National Laboratory
Argonne, IL, 60439

and

S. J. Freedman†

Stanford University
Stanford, CA 94305

We have completed a measurement of the beta decay rate of the first excited state of ^{16}N to the ground state of ^{16}O.

$$^{16}N(0^-, \text{120 keV}) \longrightarrow {}^{16}O(0^+, 0.00) + e^- + \tilde{\nu}$$

$J^\pi = 0^-$ to 0^+ transitions have been and continue to be of great interest in nuclear physics. They are extremely restrictive as to the form of the one-body operator that can cause the transition. For example, in first order no vector coupling can be responsible. Weak decays between 0^+ and 0^- levels must arise from the axial vector interaction, and further, only two form factors are involved. Realization of this point over a decade ago[1] focussed attention on measuring <u>both</u> muon capture and beta decay rates between such levels to determine the induced pseudoscalar coupling

---

*This research was supported by the U. S. Department of Energy under Contract W-31-109-Eng-38.
†A. P. Sloan Foundation Fellow.

in nuclei. The transitions are of interest to nuclear structure
studies because they provide a measure of the matrix element
of $(\underset{\sim}{\sigma}_1 \cdot \underset{\sim}{p}_1)$.

Most recently the interest in $0^+ \rightarrow 0^-$ transitions has been
rekindled because of the large role expected for pion exchange
currents. The significant role that pion exchange currents play in
axial vector transitions has been stressed for a long time by M.
Ericson[2] and P. Guichon[3]. A few years ago a compelling argument
indicating where such effects would be most easily observed was made
by Kubodera, Delorme and Rho.[4] They argued that transitions
depending on the time like component of the axial vector current
presented the most favorable case to observe pionic effects. In
this case the nucleonic axial current is reduced by a factor of v/c
while the pion exchange current is not suppressed. Detailed
calculation of this process in A = 16 have recently been carried out
by Towner and Khanna[5] which completely substantiates the points
made in Ref. 4. Although there is a theoretical concensus on this
issue, precious little supporting experimental data exists.[4]

The capture rate

$$\mu^- + {}^{16}O \dashrightarrow {}^{16}N(0^-, 120 \text{ keV}) + \nu_\mu$$

has been well studied[6] and the accepted value is $\Lambda_\mu = 1560 \pm 108$
sec^{-1}. The beta decay of the 0^- level to ${}^{16}O$ has been measured[7]
only once previously and has a large error. The measurement is
difficult owing to the small (order 10^{-6}) branching ratio for beta
decay. It is therefore important to confirm the earlier experiment
and to reduce the error.

Figure 1 shows all the energy levels[8] involved in the decay.
As was recognized by the Louvain group the only characteristic that
can be usefully employed to identify the beta rays from the 0^- level
is their $7.58 \pm .09$ μs lifetime. The ${}^{16}N$ ground state has a half
life of 7.13 sec and must be allowed to decay away in order to avoid
a huge background. A moving target to remove the ${}^{16}N$ ground state
activity in conjunction a pulsed beam are therefore required to do
the experiment. The Argonne National Laboratory Dynamitron was used
to provide a 30 μA beam of $3.4 - 3.8$ MeV D_2^+. The beam was pulsed
using an existing electrostatic system. (The ratio of the target
current for beam off to beam on was $< 6 \times 10^{-8}$). Figure 2 shows a
schematic diagram of the target assembly and reaction chamber. The
target was 99% enriched ${}^{15}NH_4{}^{15}NO_3$ deposited to a thickness of ~ 14
mg/cm^2 into eight 1.6 mm deep grooves in a 3.2 mm thick 50 cm
diameter Al disk. The disk was continuously rotated at 3Hz. Data
was acquired by pulsing the 30 μA D_2^+ beam on for 10 μs, acquiring
data for 60 μs, waiting till the irradiated beam spot moved 1.2 cm
and once again pulsing the beam on. After a complete revolution of

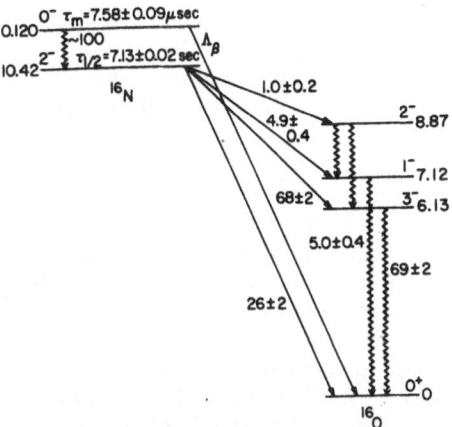

Fig. 1.   Relevant levels and transitions in the A = 16 system
           for ¹⁶N beta decay. The branching ratios are percentages
           and·include the effects of cascades.

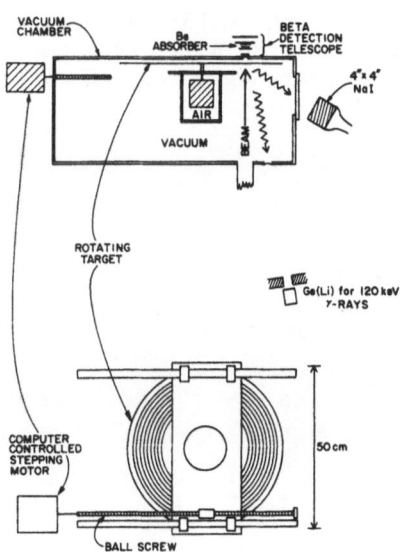

Fig. 2.   Diagram of the ¹⁶N 0⁻ beta decay apparatus.

the wheel a computer controlled stepping motor translated the wheel
so that the next groove was bombarded. The complete cycle time for
all 8 grooves was set at 17 sec so when a particular segment of
target was reused the ^{16}N ground state activity had appreciably
decayed. The beam on duty factor was thus 4.6×10^{-4}. The 120 keV
photons from the dominant (decay) branch of ^{16}N(0^-) were counted in
a Ge(Li) detector located ~200 cm from the target. The beta
particles were detected in the 4 counter plastic scintillator
telescope shown in Fig. 3 in which a valid count required four-fold
fast coincidence. The telescope was collimated in order that it
view only the activity arising from the previous 10 μs beam pulse.
Ground state decays were selected by the use of ^{9}Be adsorber
thickness between the 2nd and 3rd counter. The beam pulsing, data
acquisition, and wheel position were all controlled by a dedicated
LSI-11 micro computer. A 0^- beta rate of approximately 30/hr was
obtained with this system. Figure 4 shows a time spectrum of events
from the beta telescope. The 7.58 μs mean life is easily seen.

    In order to extract a branching ratio, the absolute efficiency
of the beta telescope and gamma ray detector had to be determined.
The efficiency of the Ge(Li) detector was easily measured using
calibrated gamma ray sources. Fixing the efficiency of the beta
detector was more difficult. To do this the absolute efficiency of
a 10 cm × 10 cm NaI detector was determined for the ^{16}O(6.13 MeV)
gamma rays using the ^{19}F(p,α)^{16}O rection on the resonance at E_p =
341 keV. The reaction was observed to be isotropic to 1% and its
yield was simultaneously measured with a Si surface barrier detector
viewing alpha particles to the 6.13 MeV level of ^{16}O.. Leaving the
NaI detector position fixed the 6.13 MeV gamma rays following the
decay of the ^{16}N ground state were counted while the beta telescope

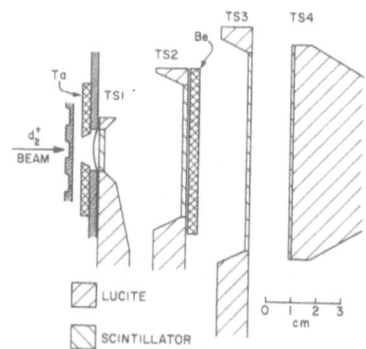

Fig. 3.   Diagram of the beta detector telescope.

recorded the number of beta decays to the ^{16}O ground state. Since the ratio $(2^- \longrightarrow 0^+)/(2^- \longrightarrow 3^-)$ for the ^{16}N ground state decay is presently known to only 10.6% $(.38 \pm .04)$ the accuracy of the present measurement is thereby limited.

With this procedure we determined the ^{16}N 0^- beta decay branching ratio to be

$$\frac{\Gamma^{0^-}_{0^+}(\beta)}{\Gamma^{0^-}(\text{total})} = (3.13 \pm 0.43) \times 10^{-6}$$

The corresponding transition probability is

$$\Lambda^\beta = 0.41 \pm 0.06 \text{ sec}^{-1}$$

This is to be compared to 0.46 ± 0.10 obtained[9] by the Louvain group. They also used the ^{16}N ground state decay to fix their beta detector efficiency. However in their experiment it made a negligible contribution to the final error whereas it is the dominant factor in our more precise measurement. It is gratifying that such good agreement exists between these measurements. Table 1 lists the various sources of error in our result and also indicates the expected improvement in a future experiment. We are also working toward reducing the uncertainty in the $^{16}N(2^-)-\overset{\beta}{\rightarrow}^{16}(0^+)$ branching ratio.

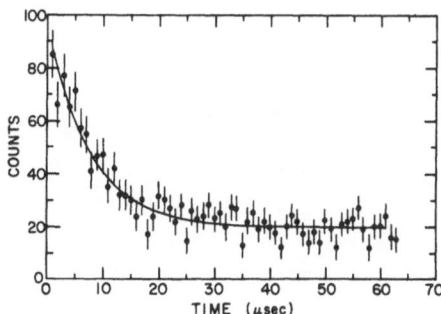

Fig. 4.   Observed rate in the beta telescope as a function of time. The curve is a fit of the form $A_{\exp}(-t/\tau_m) + C$ with $\tau_m = 7.58$ μs. The background is consistent with the rate expected from ^{16}N (2^-) beta decay.

Table 1. List of experimental errors in the present and
planned ^{16}N(0$^-$) beta decay measurements. The upper case S refer to
the error in the calibrated source strength designated by the
subscript. Lower case n's refer to yields, and its subscript
indicates the radiation type while its superscript refers to the
source of the radiation. Unsuperscripted n's are from ^{16}N decays
while s and R refer to sources and the ^{19}F(P,α)^{16}O reaction,
respectively. The ratio $\varepsilon(0^- \to 0^+)/\varepsilon(2^- - 0^+)$ refers to the
uncertainties in the beta spectra shape on the ratio of $0^- \to 0^+$ beta
decays. The T_0 entries refer to uncertainties in the counting
interval of the beta and gamma detectors.

Quantity	December 81 (act. %)	April 82 Run (est. %)
S_γ	1.9	1.5
S_α	1.0	0.5
$n_\beta(0^- \to 0^+)$	6.7	~3.0
$n_\gamma(0^- \to 2^-)$	3.2	<1.0
n_γ^s	2.1	<1.0
$n_\gamma(6.13)$	1.4	<1.0
$n_\beta(2^- \to 0^+)$	0.5 stat. +3.5 var.	<2.0
n_α^R	1.1	1.0
n_γ^R	1.5	
n_α^s	2.3	0.6
$\dfrac{\varepsilon(0^- \to 0^+)}{\varepsilon(2^- \to 0^+)}$	2.0	2.0
T_0 (Total % error)	$\dfrac{1.0}{9.6}$	$\dfrac{0.8}{4.5}$

Using the results from Ref. 5, Fig. 5 compares experiment to
the plot of Λ_μ versus Λ_β calculated with a variety of residual
interactions and different values of the ratio of the induced
pseudoscalar coupling constant (g_p) to the axial vector coupling
constant $(g_A = 1.25)$ with and without meson exchange effects. The
plot shows that meson exchange currents must be included in order to
obtain good agreement with the experimentally observed rates and
values for the ratio. The beta decay is particularly sensitive as
it shows a fourfold increase in rate over the values obtained
without meson exchange effects. Additional work is worthwhile to
verify that the wave functions employed are correct. Improving the
experimental values requires that the branching ratio of the ^{16}N
ground state to the ^{16}O ground state be determined more precisely.

The ^{16}N$(0^-) \to 0^+$ decay provides an excellent example of how
contemporary understanding of nuclear structure extended to include
meson exchange currents allows prediction of sensitive axial vector
decay rates. The A = 16 wave functions should provide reliable
prediction for the isovector matrix elements of $(\underset{\sim}{\sigma}_1 \cdot \underset{\sim}{p}_1)$ involved
in nucleon scattering from ^{16}O. We note that to the extent that the
(p,n) charge exchange reaction arises from pion exchange there is a
formal similarity to the PCAC description of the weak induced
pseudo-scalar coupling.

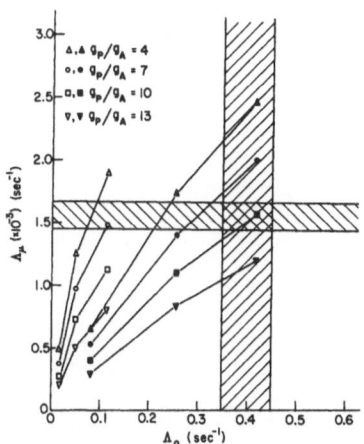

Fig. 5   The measured values of Λ_μ from Ref. 7 and Λ_β from the
           present work. Also shown as points are calculated values
           from Ref. 5. The open points are nucleon only impulse
           approximation while the closed points include the effects
           of pion exchange currents. The calculations employ three
           different residual interactions as well as varying values
           of g_p/g_A.

ACKNOWLEDGEMENT

The authors wish to acknowledge the significant help from Robert Holland and F. Lynch in the execution of this experiment.

REFERENCES

1. A. Maksymowitz, Nuovo Ciemnto $\underline{A48}$, 320 (1967).
2. See for example M. Ericson, Progress in Particle and Nuclear
       Physics, $\underline{1}$, 67 (1978).
3. P. A. M. Guichon, M. Giffen, and C. Samur, Phys. Letts. $\underline{74B}$, 15
       (1978).
4. K. Kubodera, J. Delorme, and M. Rho, Phys. Rev. Lett. $\underline{40}$, 755
       (1978).
5. I. Towner and F. C. Khanna, Nucl. Phys. A $\underline{372}$, 331 (1981).
6. P. Guichon, et al., Phys. Rev. C $\underline{19}$, 987 (1979).
7. L. Palffy, et al., Phys. Rev. Lett. $\underline{34}$, 212 (1975).
8. F. Ajzenberg-Selove, Nucl. Phys. A $\underline{375}$, 1 (1982).
9. Their quoted value of Λ_β=.43 ± .10 did not include a correction
       for 0^- feeding of the 2^- level. Hence Λ_β = .46 ± .10 is a
       more correct value. (J. P. Deutsch private communication).

GAMOW-TELLER STRENGTH DISTRIBUTIONS IN ASTROPHYSICS

G.M. Fuller

Enrico Fermi Institute
University of Chicago
Chicago, Illinois   60637

S.D. Bloom

Lawrence Livermore Laboratory
and the Department of Applied Science
University of California (Davis/Livermore)
Livermore, California   94550

ABSTRACT

The calculation of the weak interaction rates of nuclei in hot
and dense stellar environments is briefly discussed, with emphasis
on the role of Gamow-Teller strength distributions. This discussion
is specialized to the importance of electron capture reactions on
neutron rich nuclei during the collapse of a pre-supernova stellar
core. The results of shell model calculations of Gamow-Teller
strength distributions for electron capture reactions on the ground
and first excited states for each of ^{56}Fe, ^{60}Fe, and ^{64}Fe are presented and the implications for future work in nuclear physics and
the supernova core collapse problem are discussed.

I.   INTRODUCTION

The identification of the collective Gamow-Teller resonance
has changed the way astrophysicists treat the weak interaction
transmutation rates of nuclei in hot and dense stellar environments.
This work begins with a brief exposition of why nuclear weak interaction rates become sensitive functions of temperature and density.
We also discuss the role of collective state systematics in the rate

587

calculations, and the astrophysical sites where weak interaction
rate calculations are important. The discussion then centers on
electron capture reactions on neutron rich nuclei during the collapse of a presupernova stellar core. Such reactions are equivalent to the decay direction probed in a hypothetical (n,p) reaction
on a neutron rich nucleus and, hereafter, these transitions will be
classified $T^< \to T^>$, where the isospin refers to that of the parent
and daughter ground states, respectively. The nuclear physics of
these $T^< \to T^>$ transitions is profoundly different from transitions
characteristic of neutron rich β^--decay or, equivalently, the transition direction probed by (p,n) reactions in neutron rich nuclei
(hereafter classified as $T^> \to T^<$ transitions, where again the isospin refers to that of the parent and daughter ground states, respectively). The results of large scale shell model calculations
of Gamow-Teller strength distributions for astrophysically representative $T^< \to T^>$ electron capture reactions on ^{56}Fe, ^{60}Fe, ^{64}Fe
are presented and discussed.

II. NUCLEAR WEAK INTERACTION RATES IN THE STELLAR INTERIOR

The rate of weak decay from the ith state of the parent to the
j^{th} state of the daughter nucleus is given by

$$\lambda_{ij} = \ln 2 \, \frac{f_{ij}(T,\rho,U_F)}{(ft)_{ij}} \tag{1}$$

where $(ft)_{ij}$ is the comparative half-life, which is related to the
allowed weak-interaction matrix elements by[1]

$$(\log ft)_{GT} = 3.596 - \log |M_{GT}|^2 \tag{2}$$

$$(\log ft)_F = 3.791 - \log |M_F|^2 \tag{3}$$

where $|M_{GT}|^2$ and $|M_F|^2$ are the absolute squares of the Gamow-Teller
and Fermi matrix elements, respectively.

The phase space integral f_{ij} is given by an integral over total
electron energy and expressions for electron or positron emission as
well as continuum positron or continuum electron capture can be
found in Eqs. (3a) and (3b) of Fuller, Fowler, and Newman[2] (hereafter equations from this paper will be denoted with I). In continuum electron capture for example,

$$f_{ij} = (m_e c^2)^{-5} \int_{(Q_n)_{ij}}^{\infty} w^2 (w - (Q_n)_{ij})^2 \, G S_e (1 - S_\nu) dw \tag{4}$$

where G is the slowly varying relativistic Coulomb barrier penetration factor (see Eqs. I-5a through I-5f), S_e and S_ν are the electron and neutrino Fermi-Dirac distribution functions respectively, and $(Q_n)_{ij}$ is the electron capture threshold Q-value. This Q-value is given by the difference of <u>nuclear</u> masses and excitation energies and in units of the electron rest mass $m_e c^2$ is

$$(q_n)_{ij} = \frac{(Q_n)_{ij}}{m_e c^2} = (M_p - M_d + E_i - E_j)/m_e c^2 \qquad (5)$$

where M_p and M_d are the nuclear masses of the parent and daughter, respectively, and E_i and E_j are the appropriate excitation energies of levels i and j above the ground states of the parent and daughter nuclei respectively. Hereafter Q_n, with no indices, will denote the ground state to ground state Q-value, while q_n will denote the same quantity in units of the electron rest mass.

The phase space integrals become sensitive functions of temperature and density through the lepton distribution functions. In most of the conditions where nuclear weak interactions are important, neutrinos stream freely through the stellar material and are not thermalized. In this case $S_\nu = 0$. The electron Fermi-Dirac distribution function is given by Eq. (I-4a) and the electron kinetic Fermi energy U_F is calculated by inverting the integral in Eq. (I-4b). The total electron Fermi energy (including the electron rest mass) is denoted W_F and, neglecting Coulomb plasma interactions, is equivalent to the electron chemical potential μ_e. In the degenerate, zero temperature limit, for $\rho Y_e \gtrsim 10^8$ g cm^{-3} (Y_e is the number of electrons per baryon),

$$\mu_e \approx 5.155 \times 10^{-3} (\rho Y_e)^{1/3} \quad \text{MeV} \qquad (6)$$

where ρY_e is in units of g cm^{-3}. At high density the electron Fermi energy is large, implying Pauli blocking of some β^--transition channels and simultaneously the opening of many otherwise inaccessible electron capture transition channels due to the large continuum electron energy near the top of the Fermi sea.

The total transition rate of the parent nucleus, λ, due to a given weak interaction process is obtained by summing over all daughter states j linked by that process to the initial parent state i, and then summing over each parent state, weighted with the appropriate population index P_i,

$$\lambda = \sum_i P_i \sum_j \lambda_{ij} . \qquad (7)$$

The population index is given by

$$P_i = (2J_i + 1)\exp(-E_i/kT)/G \ , \tag{8}$$

where E_i and J_i are the excitation energy and spin of level i, respectively, and G is the nuclear partition function defined in Eq. (I-9b). At high temperature in the stellar interior many nuclear excited states may participate in a nuclear weak transition, and further temperature sensitivity is introduced in the stellar rate due to the Boltzman factors in Eqs. (7) and (8).

The reader is referred to Fuller, Fowler, and Newman[3,4] for an in-depth discussion of the stellar nuclear weak interaction rate problem. Among other points, these papers discuss the assignment of weak matrix elements to the transitions considered in a stellar decay calculation. In brief, usually only allowed matrix elements are considered; see Ref. 5 for discussion of a more general approach. Many discrete levels, whose excitation energies and spins are taken from experimental tabulations, are included in both the parent and daughter nuclei. In principle, weak transition matrix elements would then be required linking all allowed transitions. Matrix elements for these transitions are taken from experiment where suitable log ft measurements have been made. The set of measured log ft values can be extended by employing isospin symmetry considerations. Unmeasured Fermi matrix elements (and analog state excitation energies) can be calculated with great accuracy. Unmeasured, discrete Gamow-Teller transitions are typically assigned a log ft = 5.0, based on appropriate sd-shell and fp-shell averages. The discrete state transitions serve to pin down the stellar rates at the lower temperatures and densities.

As the recent (p,n) reactions work shows, however, a considerable fraction of the shell model Gamow-Teller strength is located in collective features. These peaked Gamow-Teller distributions will occasionally be referred to as collective resonances in this paper, even though they are sometimes neither resonant nor collective in the usual nuclear physics sense (though $T^> \to T^<$ transitions are usually both in heavy nuclei). The schematic nuclear weak interaction rate problem is shown in Fig. 1. Each nuclear decay-pair consists of a $T^<$ and $T^>$ nucleus as discussed above. The states labeled $E_i^<$ are the discrete states in the $T^<$ nucleus. The groups of states labeled $R_i^>$ represent the centroids of the Gamow-Teller collective states. There is a one-to-one correspondence between each of the discrete states in the $T^<$ nucleus and its collective state in the $T^>$ nucleus. The discrete states $E_i^>$ in the $T^>$ nucleus and their collective states $R_i^<$ in the $T^<$ nucleus are also shown and these behave similarly. The states labelled $A_i^<$ are the isobaric analog states corresponding to the $E_i^>$ discrete states. The energy

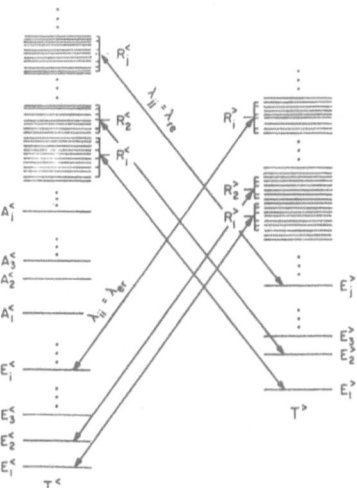

Fig. 1. A schematic representation of a typical exctied state
problem is shown; the notation is described in the text.
Transitions from the discrete states of the $T^<$-nucleus
to Gamow-Teller collective resonances in the $T^>$-nucleus,
and the reverse transitions, are shown with rates λ_{ii} =
λ_{er}. Similarly, transitions from the discrete states of
the $T^>$-nucleus to collective Gamow-Teller resonances in
the $T^<$-nucleus, and the reverse transitions, are shown,
with rates λ_{jj} = λ_{re}. For clarity neither the Fermi transitions λ_{jj}^F between $E_j^> \rightleftarrows A_j^<$ nor the discrete state to discrete state transitions $E_i^< \rightleftarrows E_j^>$ are illustrated.

orderings of the $A_i^<$ and $R_i^<$ states are not necessarily correct, but
merely suggest that in $T^> \rightarrow T^<$ transitions the ground state resonance
$R_1^<$ usually lies above the first IAS, $A_1^<$. The discussion in Ref. 3
gives a simple shell model for the strength and excitation energy
of the Gamow-Teller collective resonances. That paper also describes
how the excited state $\rightleftarrows$ resonance transitions can be taken into account by including only the ground state $\rightleftarrows$ resonance transition and
modifying the way in which the ground state and resonance population
indices (Eq. 8) are calculated.

    The $T^> \rightarrow T^<$ transition direction is that mode corresponding to
conventional β^- decay of a neutron rich nucleus and could be probed
by (p,n) reaction experiments on neutron rich nuclei. In general
these transitions are characterized by a sharp resonance (a $T^<$
feature) located within a few MeV above the first IAS in the $T^<$
daughter. This resonance is usually identified with the Gamow-
Teller spin flip configuration. Depending on the neutron richness
of the $T^>$ parent, some strength may lie in a $T^>$ feature located
above the main resonance by the value of the Lane potential. The

reader is referred to Ref. 3 for a discussion of simple shell model interpretations of these phenomena.

The $T^< \rightarrow T^>$ transition direction corresponds to electron capture on neutron rich nuclei (to make them more neutron rich) and could, in principle, be probed via (n,p) reactions. <u>This transition direction is radically different from $T^> \rightarrow T^<$ transitions</u>. The Gamow-Teller collective strength is located in a relatively broad (but still "narrow" on the scale of astrophysical conditions) feature at <u>low</u> excitation energy in the $T^>$ daughter. Furthermore, there is <u>in</u> general less strength than in the opposite $T^> \rightarrow T^<$ direction; in fact, the total Gamow-Teller strength approaches zero for large enough neutron excess. The reader is again referred to Ref. 3 for a discussion of this transition direction and the paper by Goodman and Bloom[6] in these proceedings.

The different types of weak interaction processes and the two types of transition directions are important in various astrophysical environments. During the hydrostatic phases of stellar evolution the weak interaction rates of intermediate mass nuclei, A = 20 through A = 70, are of interest. These hydrostatic phases include each of the burning stages from hydrogen burning through silicon burning, and encompass conditions from near $kT \sim 1$ keV and $\rho \sim 10$ gcm^{-3}, to $\rho \approx 4 \times 10^9$ gcm^{-3} and $kT \approx 0.7$ MeV at the onset of core collapse.[7,8] Each of electron capture, β^+-decay, β^--decay, and positron capture reactions could conceivably be important here. Both $T^> \rightarrow T^<$ and $T^< \rightarrow T^>$ transitions are important, so that useful experimental information could be obtained from tabulations of discrete-state-transition log ft values (from gamma ray branching and delayed neutron or proton experiments) and from strength distributions measured by, for example, (p,n) reactions for $T^> \rightarrow T^<$ or (n,p) for $T^< \rightarrow T^>$.

Ultimately, nuclear weak interaction rates are important in calculations of nucleosynthesis. The very heavy elements are thought to be produced in neutron addition processes in stars and supernovae: the s-process and r-process.[9] The s-process is thought to take place during the early burning phases of stellar evolution where the conditions are not very extreme, and the main interest is the β^- decay of neutron rich nuclei. The reader is referred to Cosner and Truran[10] for a discussion of the s-process weak interaction rate problem. The astrophysical site of the r-process is not well determined, but is generally though to occur because of the high neutron fluxes following a supernova explosion (see Schramm[11] for a review of r-process problems and Cowan, Cameron and Truran[12] for a non-supernova r-process). In any case, the problem involves the β^--decay of neutron rich nuclei at relatively low temperature and density. Some of these $T^> \rightarrow T^<$ transitions could in principle be probed by (p,n) reactions. However, since the temperatures are thought to be low, only the low lying parent states contribute and it is not hot enough to thermally populate the states corresponding to the $R^>$ resonances in the $T^>$ parent in Fig. 1. The result is that the decays proceed only through the <u>low energy tail</u> of the $R^<$ resonances in the $T^<$ daughter.

The conditions encountered during the later part of silicon burning and during stellar core collapse are considerably more extreme. Temperatures range from about 0.7 MeV at the onset of collapse up to many MeV near core bounce, while densities range from $\rho \approx 3.7 \times 10^9$ gcm^{-3}, $Y_e \approx 0.42$ and $\mu_e \approx 5.9$ MeV (see Ref. 13) to nuclear matter density and beyond. The high electron Fermi energies encountered here imply that electron capture on neutron rich nuclei is the most important weak interaction process and that $T^< \rightarrow T^>$ is the predominant transition direction. In general, the electrons are energetic enough so that the electron capture decay reaction on the $T^<$ parent state <u>proceeds directly through the heart of the $R^>$ resonances</u> in the $T^>$ daughter. The study of these $T^< \rightarrow T^>$ transitions has received scant attention by nuclear physicists, despite their importance for the stellar collapse problem.

III. NEUTRONIZATION IN STELLAR CORE COLLAPSE

Bethe, Brown, Applegate and Lattimer[13] (hereafter, BBAL) first pointed out that due to the large value of the nuclear partition function at higher temperature (kT ~ 1 MeV) nucleons would remain inside nuclei, entropy would remain low during the collapse of the presupernova core, and electron capture on heavy nuclei would be the dominant neutronization mechanism. Since the relativistically degenerate electrons provide most of the pressure, electron capture helps determine the collapse rate. Electron capture on heavy nuclei is also an important source of entropy increase during the collapse of the stellar core.

The pioneering work by BBAL also pointed out that the electron capture rates of iron peak nuclei (around mass 60) should be considerably faster than had previously been calculated. Taking the reaction $^{56}Fe(e^-,\nu_e)^{56}Mn$ as representative, these authors estimated the total strength and excitation energy centroid of the Gamow-Teller strength distribution in ^{56}Mn from simple shell model considerations. The excitation centroid energy was estimated as 3 MeV and the strength corresponded to about log ft $\approx$ 2.58 (see Eq. 11 below). This made the reaction rate roughly 200 times faster than that calculated in the standard tabulation[14] which, like all papers prior to BBAL, did not include the Gamow-Teller collective states for $T^< \rightarrow T^>$ transitions. To date there has been no complete experimental measurement of the low lying Gamow-Teller strength in ^{56}Mn.

A simple zero order shell model can be used to estimate the total strength and centroid excitation energy of the Gamow-Teller collective states, as discussed in Ref. 3. The zero order shell

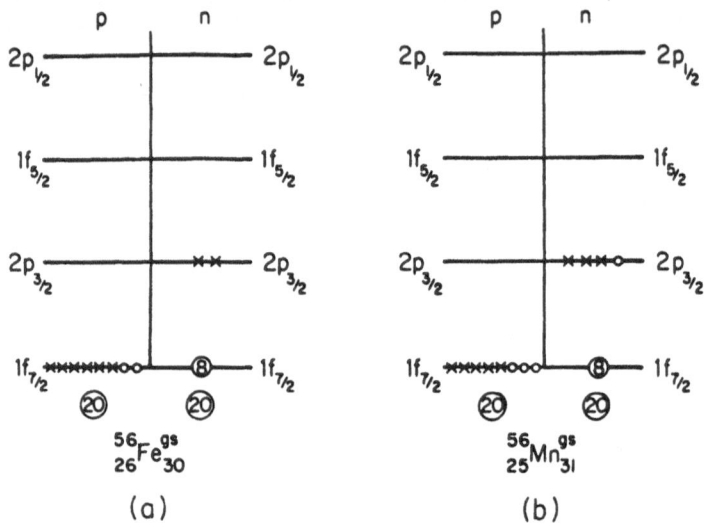

Fig. 2.   (a) The $^{56}_{26}F_{30}$ ground-state configuration zero-order
occupation diagram.   Note the closed proton and neutron
sd and sp-shells (closed at 20 nucleons each) indicated
at the bottom of the diagram. (b) The $^{56}_{25}Mn_{31}$ ground-
state configuration zero-order orbit occupation diagram.

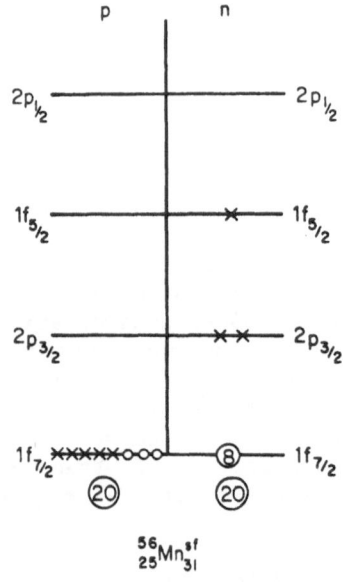

Fig. 3.   The zero-order orbit
occupation diagram for
the spin-flip configura-
tion in ^{56}Mn is shown.
This configuration is
generated from that in
Fig. (2a) by the trans-
formation of a $1f_{7/2}$
proton into a $1f_{5/2}$ neutron hole.

model occupation diagrams for the ground states of ^{56}Fe and ^{56}Mn
are shown in Figs. 2a and 2b, respectively. Both nuclei are taken
as having closed, inert ^{40}Ca cores, with the valence nucleons filling the fp-shell. For example, ^{56}Fe has 6 protons in the $1f_{7/2}$
proton shell, a closed $1f_{7/2}$ neutron shell, and two neutrons in
$2p_{3/2}$. As discussed in Ref. 3, the excitation energy of the centroid of the Gamow-Teller strength distribution corresponds roughly
to the excitation energy of the Gamow-Teller spin-flip configuration. The spin-flip configuration in ^{56}Mn is shown in Fig. 3.
This configuration is generated by having one of the six $1f_{7/2}$ protons in ^{56}Fe transform into one of the five $1f_{5/2}$ neutron holes.
Comparing Fig. 2b and Fig. 3 yields the excitation energy of the
spin-flip configuration,

$$E(^{56}Mn^{sf}) \approx E(1f_{7/2} - 2p_{3/2})_n + 2.0 \quad MeV, \tag{9}$$

where the term in brackets denotes the difference in neutron single
particle orbit energies and the 2.0 MeV is due to the extra particle-
hole repulsion as discussed in Ref. 3. Employing the single particle
energy tabulations of Seeger and Howard[15] gives the result in Ref. 3:

$$E(^{56}Mn^{sf}) \approx 3.777 \quad MeV . \tag{10}$$

The total shell model Gamow-Teller strength can be estimated
from the number of proton particles n_p and neutron holes n_h in each
single particle orbit by employing Eq. (16) of Ref. 3. This approx-
imate expression for the shell model sum rule, $|M_{GT}|^2$, is repeated
here for clarity,

$$|M_{GT}|^2 = \sum_{if} \frac{n_p^i n_h^f}{2j_f + 1} |M_{GT}^{sp}|_{if}^2 , \tag{11}$$

where the sum is over initial single particle orbits i and final or-
bits f, j_f is the spin of final orbit f, and $|M_{GT}^{sp}|_{if}^2$ is the Gamow-
Teller single particle matrix element linking orbits i and f. For
a $1f_{7/2} \rightarrow 1f_{5/2}$ transition in ^{56}Fe the single particle matrix element is

$$|M_{GT}^{sp}|^2(1f_{7/2} \rightarrow 1f_{5/2}) = 12/7 . \tag{12}$$

Thus the total strength for ^{56}Fe (ground state) $\rightarrow$ ^{56}Mn is

$$|M_{GT}|^2 = \frac{6 \cdot 6}{6} \cdot \frac{12}{7} = \frac{72}{7} \approx 10.3 \tag{13}$$

which corresponds to log ft $\approx$ 2.58. These simple estimates will
be useful for comparison with the shell model calculation presented
below.

As the collapse of the stellar core proceeds, electron captures
drive the nuclei in nuclear statistical equilibrium toward increasing neutron richness. Eventually, the mean nucleus will have valence protons filling the proton fp-shell, while the neutrons have
filled the neutron fp-shell and are filling the gd-shell. In this
situation Gamow-Teller transitions are blocked, since the Gamow-
Teller operator has no radial nuclear coordinate dependence and so
cannot mediate an fp $\rightarrow$ gd shell transition which involves a change
of principle quantum number (and, thus, parity). The electron capture transition must then be at least first forbidden and is, therefore, considerably slower than reactions like $^{56}Fe(e^-,\nu_e)^{56}Mn$.
This effect was pointed out by Fuller[5] and is discussed there along
with various unblocking mechanisms: configuration mixing, thermal
unblocking, and forbidden electron capture. In that paper it is
found that the blocking lowers the neutronization rate by a factor
of about 20 from the rate based on taking the strength in ^{56}Fe as
representative throughout the collapse.

The characteristics of the $T^< \rightarrow T^>$ electron capture transitions
are important in determining the outcome of stellar core collapse.
The final electron to baryon ratio (Y_e^f) depends on the amount of
electron capture from silicon burning through neutrino trapping
(near $\rho \sim 10^{12}$ g cm^{-3}). In turn, Y_e^f, and hence the total lepton
to baryon ratio, helps to determine the mass of the homologously
collapsing core which bounces at super nuclear density. As shown
by BBAL[13], Baym, Bethe and Brown[16], and Applegate[17], the mass of
the homologous core is a sensitive function of Y_e^f. In fact, the
larger is Y_e^f, the larger the core mass and the stronger the supernova shock. Present numerical calculations of stellar collapse give
no explosions with small core masses and marginal explosions with
the larger core masses indicated by recent electron capture physics
work.[18] This points up the importance of the $T^< \rightarrow T^>$ Gamow-Teller
strength distribution problem in astrophysics.

IV. SHELL MODEL CALCULATIONS OF GAMOW-TELLER STRENGTH
     DISTRIBUTIONS FOR $T^< \rightarrow T^>$ TRANSITIONS IN SOME IRON ISOTOPES

In this section we discuss the Gamow-Teller strength distribution calculations which were done using the Livermore VMC (vector
method codes). The Lanczos method[19,20] was utilized in the fp-
space with the PMM Hamiltonian.[21] Strength distributions were cal-
culated for ^{56}Fe(ground state) $\rightarrow$ ^{56}Mn, ^{56}Fe(1st excited state) $\rightarrow$
^{56}Mn, ^{60}Fe(ground state) $\rightarrow$ ^{60}Mn, ^{60}Fe(1st excited state) $\rightarrow$ ^{60}Mn,
^{64}Fe(ground state) $\rightarrow$ ^{64}Mn, and ^{64}Fe(1st excited state) $\rightarrow$ ^{64}Mn.

The Lanczos method used here is exactly the same as in Ref. 20. The ground and first excited state-vectors for each of ^{56}Fe, ^{60}Fe, and ^{64}Fe were generated using a shell model with a specified number of particles and holes in the fp-space. For example, the calculation for ^{56}Fe employed a 2-particle/2-hole shell model, with the 2 particles restricted to the $2p_{3/2}$, $2p_{1/2}$, $f_{5/2}$ space. These states are designated the parents, or $|P\rangle$ vectors.

The Gamow-Teller operator is taken as

$$(GT)^+ = \sum_{ij} \langle i \mid \sigma \tau_+ \mid j \rangle \, a_i^\dagger a_j \, , \qquad (14)$$

where $a_i^\dagger$ and a_j are the appropriate shell model creation and destruction operators, σ is the Pauli spin matrix, and τ_+ is the nucleon isospin raising operator (this isospin convention has $\tau_p = -1/2$, $\tau_n = 1/2$ and is opposite to the convention in Ref. 3). This GT operator was then applied to the $|P\rangle$ vectors to produce the collective Gamow-Teller state, or $|CGT\rangle$;

$$(GT)^+ |P\rangle = |CGT\rangle \, . \qquad (15)$$

The total Gamow-Teller strength is then $\langle CGT|CGT\rangle$, which was evaluated in each case.

The strength distribution of $|CGT\rangle$ amongst the eigenstates of the daughter nucleus (^{56,60,64}Mn) was then calculated with the Lanczos method.[19] In the case of the excited state parent vectors, the three different spin-states of $|CGT\rangle$ were sometimes projected out and studied separately.

It should be noted that the Lanczos strength function method does <u>not</u> lead to exact energy eigenstates, which in the cases studied here would have been impractical to calculate anyhow. Furthermore, the detail of exact eigenstates is not important in calculating accurate nuclear weak interaction rates in astrophysics (see Ref. 3). What is important is the proper placement of the centroid of the strength and the general character of its dispersion; i.e., the first few Hamiltonian moments of the eigen-distribution of the Gamow-Teller strength. This information is quickly and easily realized with the Lanczos method which calculates and utilizes these moments in an optimal fashion for the precise purpose of giving the strength distribution of an arbitrary state vector.[19,22]

Typical of the results presented here is the 15 iteration calculation of the Gamow-Teller strength distribution of the ground state of ^{56}Fe. The model for the parent vector $|P\rangle$ (see above) for the ground state was $(1f_{7/2})^{-2}(2p_{3/2}, 2p_{1/2}, 1f_{5/2})^2$. We consider

this the lowest-order realistic model for the problem. It had a
dimension of 1184 basis vectors or Slater determinants in the un-
coupled representation. Since the $|P\rangle$ vector had $J^{\pi}, T = 0^+, 2$ the
application of $(GT)^+$ produced a state of well-defined quantum num-
bers $1^+, 3$ -- the $|CGT\rangle$ vector in ^{56}Mn. The strength distribution
of this state-vector, which had a dimension of 646, was then studied
up to the 15th iteration and 21st iteration with the Lanczos method.
The results are shown in Fig. 4a for the 15 iteration case. The 21
and 15 iteration results were indistinguishable for our purposes,
in not only this case but all the other cases as well.

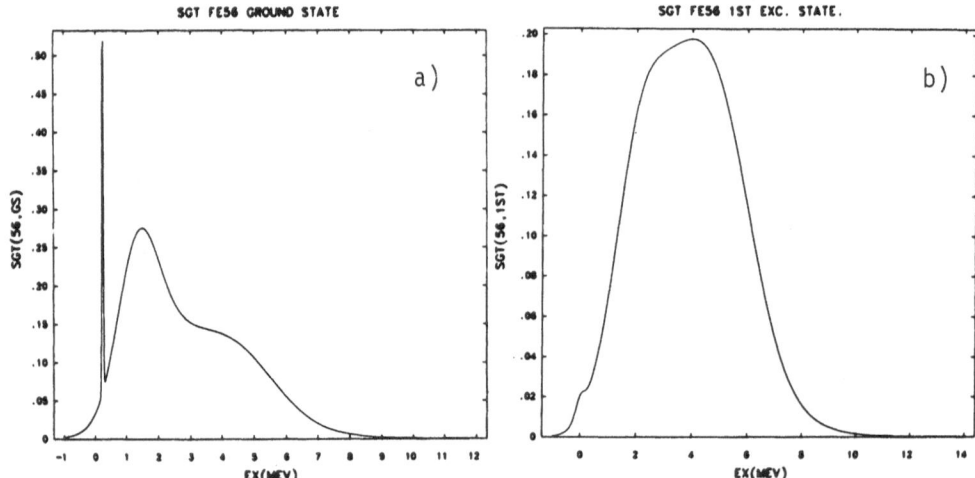

Fig. 4.   (a) Calculated strength function for the electron-capture GT
          transition going from ^{56}Fe(g.s.) to ^{56}Mn. The spike at
          ≈ 0.24 MeV is a single state which carries only 2.6% of
          the total strength. The total strength is $\langle CGT|CGT\rangle =$
          10.003. The centroid of the GT strength distribution is
          at 2.8 MeV. (b) Calculated strength-function for the elec-
          tron capture GT transition going from the first excited
          state of ^{56}Fe to ^{56}Mn. The total strength is $\langle CGT|CGT\rangle =$
          10.067. The centroid of the strength is at 4.3 MeV, shifted
          upward by 1.5 MeV as compared with part (a). The distribu-
          tions in this and the following figures are per MeV and
          are normalized to unity. The approximate eigenstates
          resulting from the Lanczos process are represented by
          Gaussian distributions whose tails may extend to the
          spurious negative energy region. See section IV for other
          details.

The total strength of the Gamow-Teller collective states cor-
responding to the ^{56}Fe ground state is <CGT|CGT> ≈ 10.003 (log ft ≈
2.60). This is to be compared with the simple shell model result[3]
in Eq. (13). The smaller matrix element here reflects the effect
of configuration mixing. The ^{56}Fe first excited state has J^{π},
T = 2^{+},3 and an excitation energy of E_x ≈ 0.847 MeV.[23] The total
strength in the Gamow-Teller collective state corresponding to this
state was found to be <CGT|CGT> ≈ 10.067 (log ft = 2.59). The 15
iteration calculation of the Gamow-Teller strength distribution for
the ^{56}Fe first excited state is shown in Fig. 4b.

Similar calculations have been performed for the ground state
(J^{π} = 0^{+}) and first excited state (E_x ≈ 0.84, J^{π} = 2^{+} in this calculation) of ^{60}Fe. The model of ^{60}Fe used here allows 2 proton
holes in the $1f_{7/2}$ proton orbit, while the $1f_{7/2}$ and $2p_{3/2}$ neutron
orbits are taken as closed, and 2 neutron particles are allowed in
the $2p_{1/2}$ - $1f_{5/2}$ neutron orbit space. The collective states generated from these ^{60}Fe states then correspond to 3p - 3h wave functions. The 15 iteration diagonalization for the ^{60}Fe ground-state
collective wave functions is shown in Fig. 5a. This collective

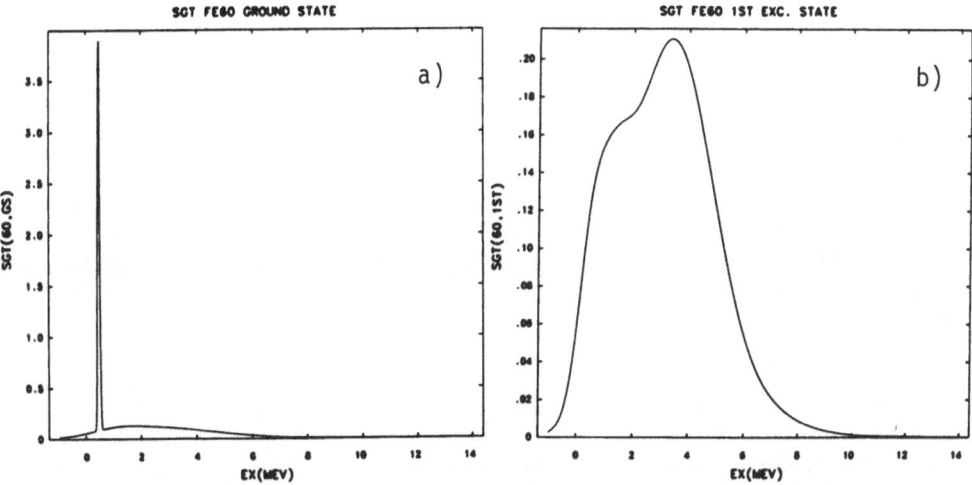

Fig. 5. (a) Calculated strength function for the electron-capture
GT transition going from the ground state of ^{60}Fe to ^{60}Mn.
The total strength is <CGT|CGT> = 9.474. The peak at 0.5
MeV contains 36% of the total strength. (b) Calculated
strength function for the electron capture GT transition
from the first state of ^{60}Fe to ^{60}Mn. The total strength
is <CGT|CGT> = 8.605. See the caption of Fig. 4 for
other details.

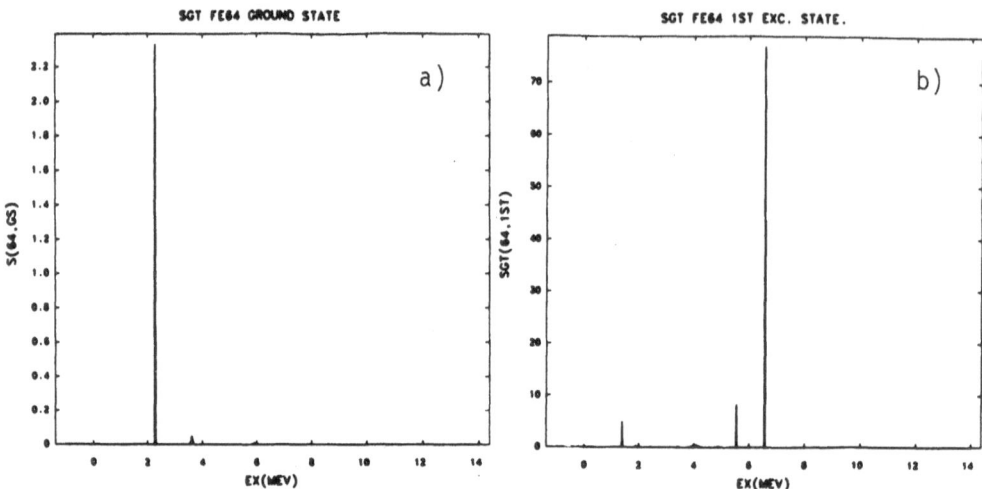

Fig. 6.   (a) Calculated strength function for the electron capture
          GT transition from the ground state of ^{64}Fe to ^{64}Mn.   In
          this case we have suppressed the peak corresponding to the
          lowest-lying $J^\pi = 1^+$ state in ^{64}Mn which absorbs 93% of the
          total strength.   In our model this is the ground state of
          ^{64}Mn.   The total strength is <CGT|CGT> = 1.163.   (b) Cal-
          culated strength function for the electron capture GT tran-
          sition from the first excited state of ^{64}Fe to ^{64}Mn.   The
          total strength is <CGT|CGT> = 1.178.   See the caption of
          Fig. 4 for other details.

state has total Gamow-Teller strength <CGT|CGT> $\approx$ 9.474 (log ft =
2.62).   The 15 iteration diagonalization for the ^{60}Fe first-excited-
state collective wave function is shown in Fig. 5b.   This collective
state has total Gamow-Teller strength <CGT|CGT> $\approx$ 8.605 (log ft =
2.66).

    The Gamow-Teller strength function for ^{64}Fe is especially in-
teresting, since in the zero-order shell model discussed in Ref. 3
this is a blocked nucleus.   The model used for the calculation of
the ^{64}Fe ground-state and first excited state wave functions allows
2 proton holes in the $1f_{7/2}$ proton orbit, while again taking the
$1f_{7/2} - 2p_{3/2}$ neutron orbit space as closed, and now allowing 2
neutron holes in the $2p_{1/2} - 1f_{5/2}$ neutron orbits.   Were the $2p_{1/2}$
neutron orbit to lie above the $1f_{5/2}$ neutron orbit the $1f_{5/2}$ orbit
would be completely filled and, according to the Eq. (13) sum rule,
$|M_{GT}|^2$ = 0.   Were the order of these orbits reversed, the $1f_{5/2}$
orbit would have 2 neutron holes implying $|M_{GT}|^2 = 6 \cdot 2/6 \cdot 12/7 \approx$
3.429 (log ft = 3.06).   The 15 iteration diagonalizations of the
Gamow-Teller collective wave functions for the ^{64}Fe ground-state
and first excited state are presented in Figs. 6a and 6b respective-

1y. In the case of the ^{64}Fe ground-state strength function, the ^{64}Mn ground-state takes up 94% of the Gamow-Teller strength, and is not shown in Fig. 6a.

The strengths calculated in these collective states indicate a situation intermediate to the completely blocked and unblocked cases discussed above. The total Gamow-Teller strengths in the collective wave functions corresponding to the ^{64}Fe ground-state and first excited state are <CGT|CGT> ≈ 1.163 (log ft ≈ 3.53) and <CGT|CGT> ≈ 1.178 (log ft ≈ 3.52), respectively. Note that the ^{64}Fe first-excited state does not show any unblocking relative to the ^{64}Fe ground-state.

The schematic weak decay problem for the ground and first excited states of ^{56}Fe is shown in Fig. 7.

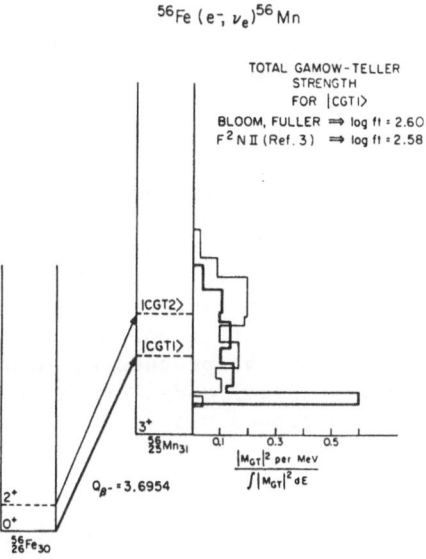

Fig. 7. The envelopes of the 15-iteration calculations of the ^{56}Fe ground and first excited state Gamow-Teller strength functions are shown. The heavy line corresponds to the ^{56}Fe ground state ($J^{\pi} = 0^{+}$), while the light line corresponds to the first excited state ($J^{\pi} = 2^{+}$). The arrows are drawn from the parent states to the approximate centroids of the appropriate Gamow-Teller strength distributions. These arrows denote electron capture transitions possible in a stellar interior. The ground state-to-ground state electron capture Q-value is given as $Q_{\beta} = -Q_{n} = 3.6954$ MeV.

V. CONCLUSIONS AND DISCUSSION

The shell model calculation of Gamow-Teller strength distributions presented here tends to reinforce present ideas on electron capture during stellar collapse. These calculations confirm the large amount of shell model Gamow-Teller strength located at low excitation energy suggested by BBAL[13] for $^{56}Fe(e^-,\nu_e)^{56}Mn$. In addition, the sum rule and excitation energy scaling assumptions for the collective states corresponding to parent excited states used in Ref. 3 are in agreement with the results here for $^{56,60,64}Fe$. It should be pointed out that there is little experimental verification of the total strength and excitation energy of Gamow-Teller collective features in the $T^< \to T^>$ mode. The effect of collective state mixing with the delta isobar has not been considered in these shell model calculations, and presumably the total strength is renormalized by roughly a factor of 2 as observed in the (p,n) reaction work on $T^> \to T^<$ transitions.[6] Such factors in the total strength are in general a small source of error in the stellar electron capture rates compared with uncertainties in the collective state centroid energy, but deserve further consideration. The deleptonization problem in stellar collapse is such that large changes in electron capture rates on heavy nuclei could change significantly the current picture of the supernova explosion. Further work must be done, especially with regard to the effectiveness of neutron shell blocking and calculation of strength distributions for forbidden electron capture.

Gamow-Teller strength distributions are important in various astrophysical problems. In the r-process $T^> \to T^<$ transitions proceed through the very low energy tail of the collective resonance. These transitions can be probed via (p,n) reactions. In contrast, during stellar collapse electron capture on neutron rich nuclei (and therefore $T^< \to T^>$ transitions) are important and the electrons have enough energy so that many decays proceed through the main part of the Gamow-Teller collective shell model strength. In principle, these transitions could be probed via (n,p) or similar reactions.

ACKNOWLEDGEMENTS

We would like to thank W. A. Fowler, R. F. Hausman, and G. J. Matthews for useful suggestions, guidance, and enthusiasm. This work was performed partially under the auspices of the U.S. Department of Energy by the Lawrence Livermore National Laboratory under contract W-7405-ENG-48, by National Science Foundation grant PHY79-23638 at California Institute of Technology and by NASA grant NSG 7212 at the University of Chicago. G. M. Fuller would like to acknowledge support from a Robert R. McCormick Postdoctoral Fellowship at the University of Chicago.

REFERENCES

1. H.S. Wilson, R.W. Kavanagh, and F.M. Mann, Phys. Rev. C23:1696
    (1980).
2. G.M. Fuller, W.A. Fowler, and M.J. Newman, Ap. J. Suppl. 42:447
    (1980).
3. G.M. Fuller, W.A. Fowler, and M.J. Newman, Ap. J. 252:715 (1982).
4. G.M. Fuller, W.A. Fowler, and M.J. Newman, Ap. J. Suppl. 48:279
    (1982).
5. G.M. Fuller, Ap. J. 252:741 (1982).
6. C.D. Goodman and S.D. Bloom, these proceedings.
7. T.A. Weaver, S.E. Woosley, and G.B. Zimmerman, Ap. J., 225:1021
    (1978).
8. W.D. Arnett, Ap. J. 218:815 (1977).
9. E.M. Burbidge, G.R. Burbidge, W.A. Fowler, and F. Hoyle, Rev.
    Mod. Phys. 29:547 (1957).
10. K. Cosner and J.W. Truran, Ap. Space Sci. 78:85 (1981).
11. D.N. Schramm, preprint (1982).
12. J.J. Cowan, A.G.W. Cameron, and J.W. Truran, Ap. J., 252:348
    (1982).
13. H.A. Bethe, G.E. Brown, J. Applegate, and J.M. Lattimer, Nucl.
    Phys. A324:487 (1979).
14. T.J. Mazurek, J.M. Truran, and A.G.M. Cameron, Ap. Space Sci.
    27:261 (1974).
15. P.A. Seeger and W.M. Howard, Nucl. Phys. A238:491 (1975).
16. G.E. Brown, H.A. Bethe and G. Baym, Nucl. Phys. A375:481 (1982).
17. J. Applegate, private communication (1981).
18. W.D. Arnett, Ap. J. Lett. 263:L55 (1983).
19. R.R. Whitehead, A. Watt, and D. Kelvin, Phys. Lett. 89B:313
    (1980).
20. S.D. Bloom, C.D. Goodman, S.H. Grimes, and R.F. Hausman, Phys.
    Lett. 107B:336 (1981).
21. F. Petrovich, H. McManus, V.A. Madsen, and J. Atkinson, Phys.
    Rev. Lett. 22:895 (1969).
22. R.R. Whitehead, "Theory and Applications of Moment Methods in
    Many-Fermion Systems", eds: B.J. Dalton, S.H. Grimes, J.P.
    Vary, and S.A. Williams, Plenum Press (1980)p.235.
23. D.M. Lederer et al., "Table of Isotopes", eds: D.M. Lederer,
    and V.S. Shirley, Wiley and Sons, New York (1978).

GAMOW-TELLER STRENGTH FUNCTIONS AND NEUTRINO PROBLEMS

W. C. Haxton

Theoretical Division
Los Alamos National Laboratory
Los Alamos, New Mexico 87545
and
Department of Physics
Purdue University
West Lafayette, Indiana 47907

ABSTRACT

A quantitative understanding of spin strengths in nuclei is of vital importance in studies of nuclear double beta decay and in solar neutrino spectroscopy. The current status of these problems is outlined.

INTRODUCTION

Fifty years ago Pauli postulated the existence of a new particle, the neutrino or "little neutron," in order to conserve energy in beta decay, $n \rightarrow p + e^- + \bar{\nu}$. The name is suitable, as the neutrino has little or no mass and, like the neutron, spin 1/2 and no charge. Today we know that neutrinos come in three flavors associated respectively with three charged partners: the electron, the muon, and the tauon. Together with these charged partners they participate in a variety of charge-changing weak interactions, including the familiar beta decay.

The discovery a decade ago that a neutrino could scatter off a nucleon without being changed into its charged partner demonstrated the existence of a new class of weak interactions, those mediated by the neutral current. This helped substantiate what is now widely regarded as one of the major theoretical advances in modern physics, the unified description of the weak and electro-

magnetic interactions in the model of Glashow, Weinberg, and Salam
(GWS) [1]. Within this model neutrinos are massless and certain
quantities, such as lepton number, baryon number, and muon number,
are exactly conserved.

Presently great theoretical effort is being expended in an
even more ambitious endeavor, the search for a unified description
of the electroweak and strong interactions [2]. Attempts to construct "grand unified" theories have already met with some success
in that certain models predict the value of the Weinberg angle, a
free parameter in the electroweak theory. The prejudice in grand
unified theories for "naturality," the reluctance to postulate a
priori global conservation laws, also suggests that many of the
conservation laws of the GWS model are in fact only approximate,
reflecting the enormous mass scale governing the strong-electro-
weak unification. The manner in which these conservation laws are
broken should impose important constraints on formulations of
grand unified theories.

How can we obtain these experimental constraints? The situation is quite different from that which prevailed a decade ago,
when the electroweak unification mass ($\sim 10^2$ GeV) appeared directly
accessible, stimulating the development of remarkable accelerator
technology. The expected grand unification mass ($\sim 10^{15}$ GeV) may
condemn us to investigation in the "low energy limit" for some
time. Thus increasingly the future of particle physics will depend on the development of technologies to isolate rare events and
to measure small branching ratios. Present experiments probing
nucleon stability at the level of 10^{32} years may be in the van-
guard of this effort [3].

The opportunities are present for nuclear physics to play an
important role in this quest. The nucleus can serve as a filter
for rare processes, isolating interactions according to spin, isospin, and parity. In Table 1 a few of the ongoing nuclear experiments that may have a profound impact on particle physics are
listed. These and other possibilities for fruitful collaborations
between the nuclear and particle physics communities remind one of
the importance of the β-decay studies that tested the CVC hypothesis and the V-A theory of weak interactions 25 years ago [4,5].
In particular, today I will discuss two problems, ββ-decay and
solar neutrino detection, which promise to constrain possible
descriptions of the neutrino. These are particularly relevant to
the discussions of this conference, as an understanding of spin
excitations in nuclei is a prerequisite to their interpretation.

Table 1

Symmetry/Interaction	Nuclear Test
1. baryon number	decay of bound nucleon
2. lepton number: masses and right-handed couplings of Majorana neutrinos	double beta decay
3. time reversal	nuclear electric dipole moment; nuclear γ-decay
4. flavor mixing; neutrino masses	neutrino oscillations (e.g., solar neutrino detection)
5. separate lepton number	$u \rightarrow e$ conversion in muonic atoms
6. $\Delta S = 0$ weak hadronic current	parity mixing of nuclear levels
7. neutrino mass	tritium β-decay

DIRAC AND MAJORANA NEUTRINOS

One powerful proble of lepton number conservation, of the mass and charge conjugation properties of the electron neutrino, and of possible right-handed admixtures in the weak leptonic current is nuclear $\beta\beta$-decay, $(A,Z) \rightarrow (A,Z+2)$ [6]. This process, a fundamental nuclear decay mode, can be observed in a number of even-even nuclei where, due to the pairing interaction, the competing decay $(A,Z) \rightarrow (A,Z+1)$ is energetically inaccessible (see Fig. 1).

The question historically associated with $\beta\beta$-decay is whether the neutrino should be described by a Dirac or Majorana field. If the neutrino is a Dirac particle, it has a distinct antiparticle; if Majorana, the particle and antiparticle are indistinguishable. The neutrino is unique among the fermions in permitting these alternative descriptions: any fermion having a charge or measurable magnetic moment necessarily has a distinct antiparticle.

If we define the neutrino and antineutrino by

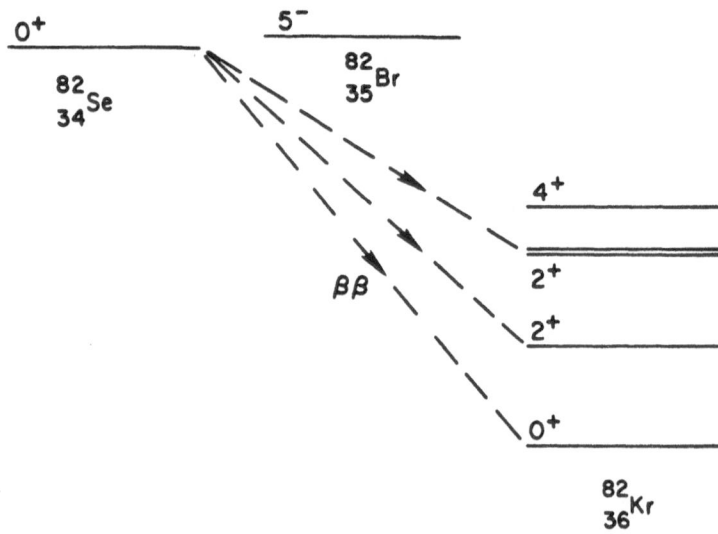

Fig. 1.   Level scheme for the $\beta\beta$-decay of ^{82}Se.

$$n \rightarrow p + e^- + \bar{\nu}_e$$

$$\nu_e + n \rightarrow p + e^- \tag{1}$$

the second order weak interaction

$$2n \rightarrow n + p + e^- + \bar{\nu}_e \rightarrow 2p + 2e^- + 2\bar{\nu}_e \tag{2}$$

will occur. This two-nucleon process will contribute to the decay
$(A,Z) \rightarrow (A,Z+2)$, producing a final state with four leptons, re-
gardless of the charge conjugation properties of the neutrino. If
the neutrino is a Majorana field, a second decay mode is possible

$$2n \rightarrow n + p + e^- + \bar{\nu}_e \equiv n + p + e^- + \nu_e \rightarrow 2p + 2e^- \tag{3}$$

producing a neutrinoless final state. This second process enjoys
a considerable phase space advantage over the reaction in Eq. (2)
and, in the absence of the chirality suppression we will discuss
momentarily, will dominate $\beta\beta$-decay rates for a Majorana electron
neutrino.

Early geochemical measurements of $\beta\beta$-decay showed that the
half lives for the decay $(A,Z) \rightarrow (A,Z+2)$ far exceeded the values

expected for the process in Eq. (3), $\tau_{1/2} \sim 10^{12}\text{-}10^{15}$ years. This was interpreted as a demonstration of the Dirac character of the electron neutrino, and prompted the introduction of lepton number ℓ to distinguish the neutrino from its antiparticle: the electron and neutrino are assigned $\ell = +1$, the positron and antineutrino $\ell = -1$. The assumption that additive lepton number is conserved then allows two-neutrino $\beta\beta$-decay, but forbids neutrinoless $\beta\beta$-decay, for which $\Delta(\Sigma\ell) = 2$.

Yet, with the discovery [5] in 1957 that the weak interaction violates parity conservation maximally (or nearly so), it became apparent that the Majorana/Dirac character of the electron neutrino was still in question. The particles which participate in the reactions of Eq. (1)

$$n + p \rightarrow e^- + \bar{\nu}_e^{(+)}$$

$$\nu_e^{(-)} + n \rightarrow p + e^- \qquad (1a)$$

are the right-handed $\bar{\nu}_e$ and left-handed ν_e. Thus even if the neutrino is a Majorana particle

$$2n \rightarrow n + p + e^- + \bar{\nu}_e^{(+)}$$

$$\equiv n + p + e^- + \nu_e^{(+)} \nrightarrow 2p + 2e^- \qquad (3a)$$

as the neutrino has the wrong helicity for absorption on a neutron. Therefore, if parity violation in the weak interaction is sufficiently close to maximal, the geochemical $\beta\beta$-decay results imply neither a Dirac electron neutrino nor a conserved lepton number.

The great interest today in $\beta\beta$-decay stems from the gauge theory prejudice [7] that a neutrino mass will break the γ_5-invariance of the weak current. Thus neutrinoless $\beta\beta$-decay may occur, though at a rate suppressed by $(m_\nu/m_e)^2$, if the neutrino is a Majorana particle. A careful examination of $\beta\beta$-decay rates then leads to the following conclusions:

(1) Present laboratory limits on neutrinoless $\beta\beta$-decay place an upper bound on the neutrino mass of $\langle m^{Maj}\rangle_\nu \lesssim 10\text{-}50$ eV. This bound may impose a fundamental constraint on the charge conjugation properties of the neutrino if the tritium β-decay mass result 14 eV $\leq m_\nu \leq 52$ eV is correct [8].

(2) There is a hint, in the geochemical total $\beta\beta$-decay rates for ^{128}Te and ^{130}Te, that no-neutrino decay is occurring. The rate, corresponding to $\langle m^{Maj}\rangle_\nu \sim 10$ eV, does not violate any experimental bound on lepton number violation.

(3) Systematic disagreement exists between the geochemical total $\beta\beta$-decay rates and the two-neutrino rates predicted by theory and measured in a single laboratory experiment. The origin of the discrepancy is unclear.

I would now like to summarize the experimental and theoretical work that leads to these results.

NUCLEAR DOUBLE BETA DECAY RATES

Our knowledge of $\beta\beta$-decay rates comes from two classes of experiments, geochemical and laboratory.

Geochemical measurements have determined the total $\beta\beta$-decay half lives for ^{130}Te, ^{128}Te, and ^{82}Se, as shown in Table 2. The noble gases are the rarest of the stable nuclides. Thus, over geologic times these reactions can produce significant elevations in the abundances of the daughter nuclei. The experimental procedure consists of outgasing by stepwise heating of Te- or Se-bearing ore samples, followed by high sensitivity mass spectrometry. The excess of the daughter isotope is determined by comparing the resulting noble gas isotopic distribution to that for the atmosphere. Once the ore age is fixed by geologic arguments or by K-Ar dating, this excess determines the total $\beta\beta$-decay rate.

Laboratory experiments have provided bounds on 2ν and 0ν $\beta\beta$-decay and, in one case, a 2ν half life. Of course, only the electrons are detected. A plot of the sum of the electron kinetic energies T is shown in Fig. 2. For 0ν $\beta\beta$-decay a spike is found at $T = T_o$, the total kinetic energy release; for 2ν decay, the distribution is continuous over the range from $T = 0$ to T_o.

Table 2:   A summary of geochemical $\beta\beta$-decay results. The total kinetic energy carried off by leptons, T_o, is given in units of the electron mass.

Reaction	T_o $(m_e c^2)$	$\tau_{\frac{1}{2}}$ (years)
^{130}Te $\rightarrow$ ^{130}Te	5.0	$(2.0 - 3.1) \cdot 10^{21}$ [9,10]
^{128}Te $\rightarrow$ ^{128}Xe	1.7	$(3.2 - 4.9) \cdot 10^{24}$ [11]
^{82}Se $\rightarrow$ ^{82}Kr	5.9	$2.76 \cdot 10^{20}$ [12]

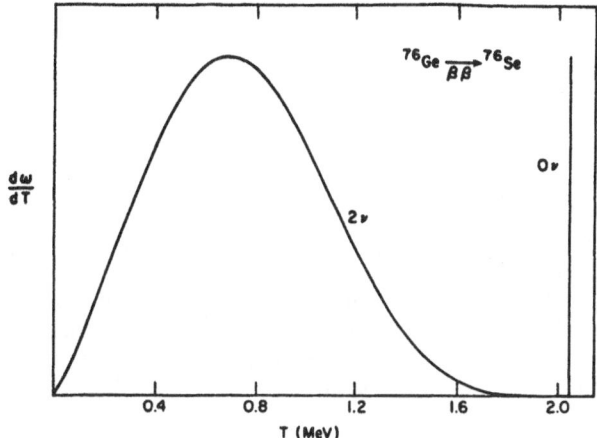

Fig. 2.   Comparison of the differential decay rate dω/dT, where
T is the sum of the kinetic energies carried off by the
electrons, for 0ν and 2ν ββ-decay. The 0ν spectrum is a
line at $T = T_o$.

Table 3:   Laboratory Limits on 2ν and 0ν ββ-decay.

Reaction	To $(m_e c^2)$	$\tau_{\frac{1}{2}}$ (years)	
^{48}Ca → ^{48}Ti	8.4	$\gtrsim 10^{21\cdot 3}$, 0ν	[13]
		$\gtrsim 10^{19\cdot 56}$, 2ν	[13]
^{76}Ge → ^{76}Se	4.0	$\gtrsim 10^{21\cdot 7}$, 0ν	[14]
^{82}Se → ^{82}Kr	5.9	$\gtrsim 10^{21\cdot 5}$, 0ν	[15]
		$10^{19\cdot 0 \pm 0\cdot 2}$, 2ν	[16]

The experimental task of discerning signal from background is
thus simpler in the case of 0ν decay, accounting for the stringent
limits shown in Table 3.

The remaining task is to compare these results with the theoretical predictions for ββ-decay mediated by Dirac and Majorana
neutrinos.   If the neutrino is Dirac

$$\omega = \omega_{2\nu} \tag{4a}$$

and if Majorana

$$\omega = \omega_{2\nu} + \omega_{0\nu}(\eta, \langle m^{Maj}\rangle_\nu) \ . \tag{4b}$$

(I have allowed breaking of the γ_5-invariance of the weak leptonic current by a mass $\langle m^{Maj}\rangle_\nu$, as discussed earlier, and by an explicit right-handed current of strength η.) Clearly the experimental limits on 0ν $\beta\beta$-decay constrain the mass and right-handed coupling of a Majorana electron neutrino provided one can calculate $\omega_{0\nu}$ as a function of η and $\langle m^{Maj}\rangle_\nu$. One can also use total geochemical rates to constrain these parameters.

The lepton number-conserving process of Eq. (2) gives rise to the nuclear decay shown in Fig. 3a. The corresponding decay rate can be evaluated in time-dependent perturbation theory. If the dependence of the energy denominator on the energy of the intermediate nuclear state is approximated by an average value [6]

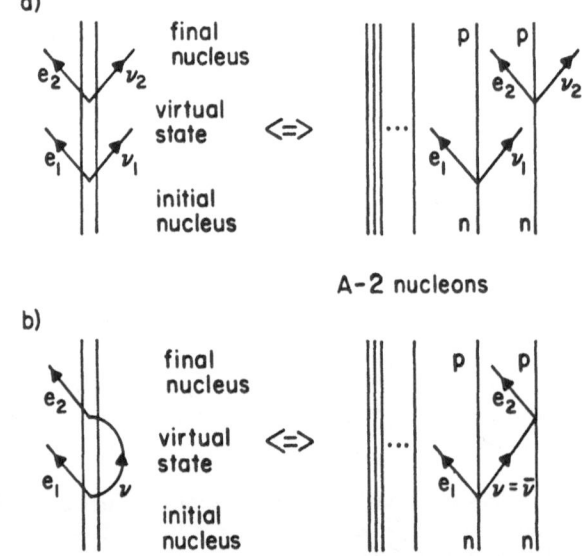

Fig. 3.   Two-nucleon mechanisms for 2ν (a) and 0ν (b) $\beta\beta$-decay.

$$\frac{1}{E_N - E_I + \nu + \varepsilon} \approx \frac{1}{\langle E_N - E_I \rangle + \nu + \varepsilon} \tag{5}$$

the sum over virtual nuclear states can be completed by closure.
[In Eq. (5) E_N and E_I represent the energies of the intermediate
and initial nuclear states, while ν and ε are the energies of the
neutrino and electron emitted in the first β-decay.] Evaluating
each β-decay in the allowed approximation and specializing to
transition between $J^\pi = 0^+$ nuclear states, one finds

$$\omega_{2\nu} = \xi_{2\nu} \frac{16G^4}{\pi^7} [F^{PR}(Z + 2)]^2 \frac{1}{(\langle E_N - E_I \rangle + T_o/2 + m_e)^2} \frac{m_e^{11}}{8!} f(T_o/m_e)$$

$$\times [F_1^4 |M_F|^2 + F_A^4 |M_{GT}|^2 - 2F_1^2 F_A^2 \mathrm{Re}(M_F \cdot M_{GT}^*)] \tag{6a}$$

where

$$F^{PR}(Z) = \frac{2\pi \, \alpha \, Z}{1 - \exp(-2\pi\alpha Z)} \tag{6b}$$

$$f(\varepsilon) = \varepsilon^7 [1 + \frac{\varepsilon}{2} + \frac{\varepsilon^2}{9} + \frac{\varepsilon^3}{90} + \frac{\varepsilon^4}{1980}] \tag{6c}$$

$$M_F = \langle F | \tfrac{1}{2} \sum_{ij} \tau_+(i)\tau_+(j) | I \rangle \tag{6d}$$

$$M_{GT} = \langle F | \tfrac{1}{2} \sum_{ij} \vec{\sigma}(i) \cdot \vec{\sigma}(j) \tau_+(i)\tau_+(j) | I \rangle \tag{6e}$$

The factor $\xi_{2\nu}$ is a phase space correction needed because Eq. (6a)
has been written with the approximate Coulomb correction of Eq.
(6b); it varies from 1.2 to 5, roughly, between $Z = 22$ and $Z = 54$
[17]. The double Fermi matrix element M_F is nonzero only through
isospin impurities in the final state, and quite generally can be
neglected.

The lepton number-violating decay of Fig. 3b can occur via
the two-nucleon process of Eq. (3). The rate can be calculated
for the following choice of the leptonic current

$$j_\mu^{lep}(x) = \bar{\psi}_e(x)\gamma_\mu[(1 - \gamma_5) + \eta(1 + \gamma_5)]\psi_\nu^{Maj}(x) \quad . \tag{7}$$

The γ_5-invariance is broken by an explicit right-handed current
admixture η and by a Majorana neutrino mass $\langle m^{Maj} \rangle_\nu$. The general

result for the decay rate for transitions between $J^\pi = 0^+$ states is somewhat complicated [17,18]. We given only the rate for $\eta = 0$.

$$\omega_{0\nu}^{\eta=0} = \xi_{0\nu} \frac{G^4}{8\pi^5} [F^{PR}(Z + 2)]^2 m_e^7 \left(\frac{<m_\nu^{Maj}>}{m_e}\right)^2$$

$$f_1(T_0/m_e) |F_A^2 M_2'' - F_1^2 M_1''|^2 \qquad (8a)$$

where

$$f_1(\varepsilon) = \varepsilon[1 + 2\varepsilon + \frac{4\varepsilon^2}{3} + \frac{\varepsilon^2}{3} + \frac{\varepsilon^4}{30}] \qquad (8b)$$

$$M_2'' = <F|\tfrac{1}{2} \sum_{ij} \tau_+(i)\tau_+(j)\vec{\sigma}(i)\cdot\vec{\sigma}(j) \frac{g(r_{ij})}{r_{ij}}|I> \qquad (8c)$$

$$M_1'' = <F|\tfrac{1}{2} \sum_{ij} \tau_+(i)\tau_+(j) \frac{g(r_{ij})}{r_{ij}}|I> \qquad (8d)$$

with $g(r_{ij}) \simeq 1$ a slowly-varying function of $r_{ij} = |\vec{r}_i - \vec{r}_j|$. In the general result for $\eta \neq 0$ two additional matrix elements

$$M_3 = <F|\tfrac{1}{2} \sum_{ij} \tau_+(i)\tau_+(j)\hat{r}_{ij}\cdot\vec{\sigma}(i)\hat{r}_{ij}\cdot\vec{\sigma}(j) \frac{g(r_{ij})}{r_{ij}} |I> \qquad (9a)$$

$$M_4 = <F|\tfrac{1}{2} \sum_{ij} \tau_+(i)\tau_+(j)\hat{R}_{ij}\cdot(\hat{r}_{ij} \times (\vec{\sigma}(i) - \vec{\sigma}(j))) \frac{R_{ij}}{r_{ij}^2} g(r_{ij})|I>$$

with $R_{ij} = |\vec{r}_i + \vec{r}_j|$, also appear. Note that the the matrix elements in Eq. (8a) differ from the 2ν operators M_F and M_{GT} only by their radial dependence, $g(r_{ij})/r_{ij}$.

To evaluate the expressions in Eqs. (6) and (8) one must calculate the two-body nuclear matrix elements. Recently the group at Los Alamos (Haxton, G. J. Stephenson, Jr., and D. Strottman) has tackled this structure problem with state-of-the-art shell model techniques [17,19,20]. Although the length of this talk precludes a detailed description of this work, I will provide a brief summary.

The $\beta\beta$-decay transition ^{48}Ca $\rightarrow$ ^{48}Ti appears extremely favorable: the structure is thought to be relatively simple, and the large kinetic energy release T_0 = 4.3 MeV promises considerable phase space enhancement of the rate. However, Lawson showed in a simple Nilsson model that M_{GT} = 0 as the result of a K selection rule [21]. This constraint is relaxed somewhat in more realistic intermediate coupling models. Our shell model diagonalization was performed with the Kuo-Brown full g-matrix [22] in the 2p1f model space. A closed ^{40}Ca core is assumed, and all configurations of eight valence nucleons for which the $1f_{7/2}$ occupation is at least four are allowed.

The treatment of the decays ^{76}Ge $\rightarrow$ ^{76}Se and ^{82}Se $\rightarrow$ ^{82}Kr is more complicated. A direct shell model calculation in the canonical model space, involving the orbitals $1f_{5/2}$, $2p_{3/2}$, $2p_{1/2}$, and $1g_{9/2}$ lying between the magic numbers 28 and 50, is not feasible. We instead employ a weak coupling approximation in which full shell model calculations are performed separately for the valence protons and neutrons [19]. The proton-neutron interaction is then diagonalized in a basis formed from the 50 proton and 50 neutron wave functions lowest in energy, yielding wave functions of the form

$$\psi^{pn}_{J=0} = \sum_{i,j=1}^{50} \xi_{ij} [\psi^p_{J_i \gamma_i} \otimes \psi^n_{J_j \gamma_j}]_{J=0} \quad . \tag{10}$$

We again employ the Kuo g-matrix. It should be noted that certain spin partners, $1f_{7/2}$ and $1g_{7/2}$, are outside the model space. Naively one expects the influence of these subshells to be small. Furthermore, inclusion of these orbitals would introduce spurious center-of-mass wave function components that could have serious effects, as the g-matrix is not translationally invariant.

The calculations for the decays ^{130}Te $\rightarrow$ ^{130}Xe and ^{128}Te $\rightarrow$ ^{128}Xe were also performed in a weak coupling basis. The model space includes the orbitals $1g_{7/2}$, $2d_{5/2}$, $2d_{3/2}$, $3s_{1/2}$, and $1h_{11/2}$ lying between the magic numbers 50 and 82, and the interaction employed is that of Baldridge and Vary [23]. The separate proton and neutron calculations involve sufficiently many configurations that some restrictions must be imposed on the occupation of the less favored orbitals.

The reliability of our limits on η and $<m^{Maj}>_\nu$ will depend on the quality of these wave functions. There is one obvious check suggested by the similarity of the 0ν matrix elements to those governing 2ν $\beta\beta$-decay: do these wave functions properly reproduce the 2ν decay rates?

The results shown in Table 4 are surprising. Theoretical and laboratory rates for 2ν $\beta\beta$-decay in ^{48}Ca and ^{82}Se are in good agreement. However, the upper bounds that can be placed on $|M_{GT}|$ from total geochemical rates are consistently much smaller than values predicted by theory. The laboratory and geochemical rates are also in sharp disagreement for the one case permitting a comparison ^{82}Se $\rightarrow$ ^{82}Kr. The large theoretical matrix elements for ^{76}Ge, ^{82}Se, ^{128}Te, and ^{130}Te come about through a coherent addition of many amplitudes in the two-body density matrix. Very recently Zamick and Auerbach [24] have obtained similar theoretical values for M_{GT} in ^{48}Ca and ^{76}Ge using a Nilsson model with pairing. Importantly, they attribute the coherence found in the shell model to pairing, and demonstrate that large matrix elements result in their treatment for any reasonable choice of the pairing strength.

What is the reason for this disagreement? Perhaps the most troublesome aspect of the theoretical treatment is the replacement of the E^{-1}-weighted sum over intermediate nuclear states by the non-energy-weighted sum. Yet both the coherence described above and tests involving explicit summation over low-lying intermediate states [19] indicate that no significant bias is introduced by the closure approximation. A possibility of great interest in view of the discussions at this conference, the coupling to delta-hole excitations, had little effect on M_{GT} in the calculations of Zamick and Auerbach [24].

Table 4: Calculated and experimental double Gamow-Teller matrix elements M_{GT}.

Nucleus	$|M_{GT}|_{theory}$		$|M_{GT}|_{exp}$	
^{130}Te	1.48	[17]	0.10-0.13*	[9,10]
^{128}Te	1.47	[17]	0.18-0.23*	[11]
^{82}Se	0.94	[19]	1.43	[16]
			0.27*	[12]
^{76}Ge	1.28	[19]		
^{48}Ca	0.22	[20]	$\lesssim$ 0.19	[13]

*Maximum values determined from total geochemical rates.

Alternatively, one can question the geochemical assumptions. Are noble gases retained in the ore over geologic times? The consistency between geologic estimates of the ore age and the results of K-Ar dating demonstrates that this ligher noble gas does remain in the ore [25]. Furthermore, there is reasonable consistency between the geochemical half lives determined from different ores by different geochemists [25]. In summary, we can find no obvious flaw in the geochemistry, in the theoretical treatment, or in the ^{82}Se laboratory experiment of the Moe and Lowenthal. The origin of the inconsistencies in Table IV is unknown.

Observing that the 0ν matrix elements differ from their 2ν counterparts only by the gentle radial dependence $g(r_{ij})/r_{ij}$, Primakoff and Rosen [6] suggested in their early work on $\beta\beta$-decay that a scaling relation might exist between 0ν and 2ν matrix elements, $M''_2/M_{GT} \approx 1/R_0$ with $R_0 = 1.2 A^{1/3}$ the nuclear radius. Our calculations for ^{76}Ge, ^{82}Se, ^{128}Te, and ^{130}Te demonstrate that this scaling holds remarkably well, though with a somewhat different strength, $M''_2/M_{GT} = (0.57 \pm 0.03)/R_0$. One then expects the discrepancies in estimates of 2ν and total $\beta\beta$-decay rates to carry over to 0ν $\beta\beta$-decay. The resulting uncertainties in the bounds on $\langle m^{Maj}\rangle_\nu$ and η for the $\beta\beta$-decay of ^{82}Se are apparent in Fig. 4.

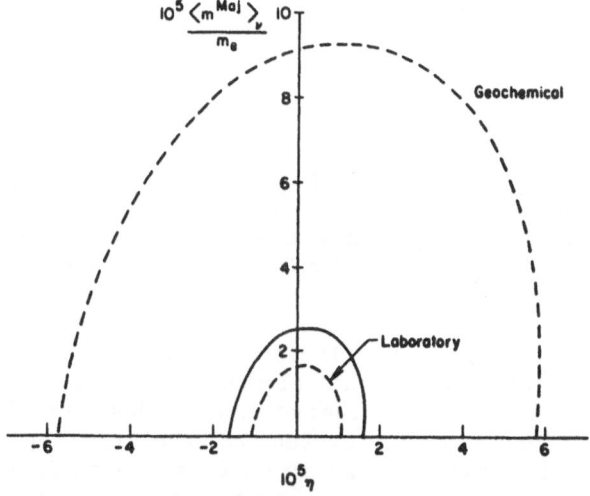

Fig. 4.   Dashed lines show boundaries of allowed regions in η - $\langle m^{Maj}\rangle_\nu$ plane which result if all ^{82}Se $\beta\beta$-decay matrix elements are normalized to reproduce the total geochemical rate [12] and the laboratory 2ν rate of Moe and Lowenthal [16]. Solid line employs theoretical matrix elements.

Pontecorvo suggested that the question of matrix element normalization might be circumvented by comparing $\beta\beta$-decay rates for two different nuclei, ^{128}Te and ^{130}Te. As the structure of these isotopes differs only by a neutron pair, Pontecorvo assumed that the $\beta\beta$-decay nuclear matrix elements would be identical. The Los Alamos calculations (see Table IV) and those of Vergados [26] support this assumption. The ratio of the total $\beta\beta$-decay rates for these isotopes should then be determined by phase space. Because the energy releases for these decays are quite different (see Table 2), one finds

$$\frac{\tau_{\frac{1}{2}}^{2\nu}(128)}{\tau_{\frac{1}{2}}^{2\nu}(130)} = 5100$$

and

$$\frac{\tau_{\frac{1}{2}}^{0\nu}(128)}{\tau_{\frac{1}{2}}^{0\nu}(130)} = 25$$

so that the ratio of half lives tests sensitively the decay mechanism. The experimental result [11]

$$\frac{\tau_{\frac{1}{2}}(128)}{\tau_{\frac{1}{2}}(130)} = 1590$$

suggests that both 2ν and 0ν $\beta\beta$-decay may be contributing. The values for η and $<m^{Maj}>_\nu$ that are consistent with the experimental ratio can be derived given only the relative strength of the 2ν and 0ν matrix elements. The solution is shown in Fig. 5.

I believe that this purported evidence for lepton number violation is really quite weak. It is difficult to accept the theoretical demonstration of matrix element equality in view of the alarming discrepancy between theory and geochemistry in the absolute rates. Furthermore, Kirsten has recently reported [25] a measurement of the half life ratio which is inconsistent with the older Missouri group value [11] given above and consistent with lepton number conservation. While Kirsten's measurement was made with relatively young ore, and thus his statistical error exceeds that of the Missouri group, certainly his result indicates that the experimental situation is unsettled.

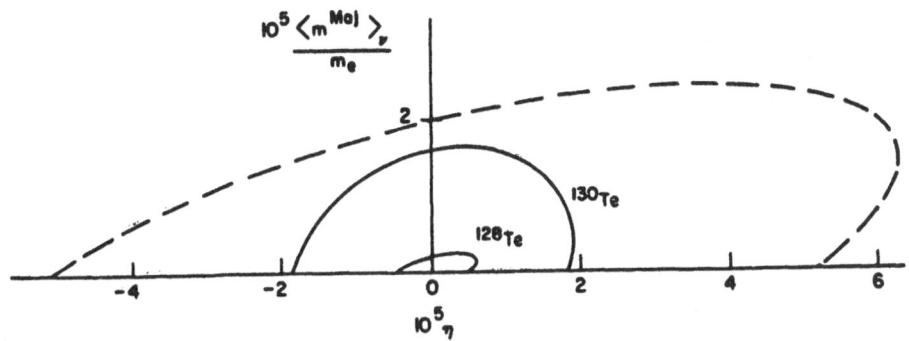

Fig. 5.   Values of η and $\langle m^{Maj}\rangle_\nu$ lying on the dashed line give,
under the assumption of equal matrix elements in ^{130}Te
and ^{128}Te, the ratio of geochemical rates measured by
the Missouri group [11]. The solid lines show the boundaries of the allowed region derived by using theoretical
matrix elements and attributing the entire decay rate to
0ν $\beta\beta$-decay. The inconsistency between solid and dashed
lines reflects the disagreement between theory [17,19]
and the geochemical results [9,10].

In summary, the laboratory limits on 0ν $\beta\beta$-decay yield $\langle m^{Maj}\rangle_\nu$
$\lesssim$ (9-52) eV, with the range reflecting the discrepancy between the
laboratory and geochemical measurements for ^{82}Se. If the magnitudes of matrix elements are taken from the Los Alamos calcula-
tions, $\langle m^{Maj}\rangle_\nu \lesssim 13$ eV; in this case one finds that the limits
on 0ν decay in ^{48}Ca and ^{76}Ge impose somewhat less stringent constraints on $\langle m^{Maj}\rangle_\nu$. The geochemical rates in ^{130}Te and ^{128}Te
give, under the Pontecorvo assumption of equal matrix elements,
$\langle m^{Maj}\rangle_\nu \lesssim 10$ eV. (We choose the inequality because of the conflict between the results of Kirsten and the Missouri group.)

Thus a cautious interpretation of these results indicates
$\langle m^{Maj}\rangle_\nu \lesssim 50$ eV. If, in addition, one chooses to believe the
Pontecorvo assumption, the result of Moe and Lowenthal, or the
theoretical matrix element calculation, then the more stringent
constraint $\langle m^{Maj}\rangle_\nu \lesssim 10$ eV follows. Recently, a measurement of
the endpoint spectrum in the β-decay of the triton yielded $14 \lesssim$
$m_\nu \lesssim 46$ eV [8]. If this is correct, then the more stringent $\beta\bar{\beta}$-
decay limit demonstrates that the electron neutrino cannot be a

Majorana mass eigenstate! This exciting result underscores the
importance of extending current 0ν $\beta\beta$-decay limits one to two
orders of magnitude as, under the most cautious interpretation,
such results could test the charge conjugation properties of the
neutrino.

Finally, I would like to mention a few topics that may convey
some of the flavor of present studies in $\beta\beta$-decay. Rosen [27] and
Doi et al. [18] recently pointed out that $0^+ \rightarrow 2^+$ $\beta\beta$-decay transitions are of particular interest in that the 0ν mechanism can
occur only via the right-handed current. Unfortunately, according
to our Los Alamos work, the matrix elements that govern such
transitions are quite weak, so that $0^+ \rightarrow 0^+$ transitions provide
much more stringent constraints on η. If, however, 0ν decay is
observed, $0^+ \rightarrow 2^+$ transitions may prove a valuable tool for separating mass effects from those of the right-handed current.

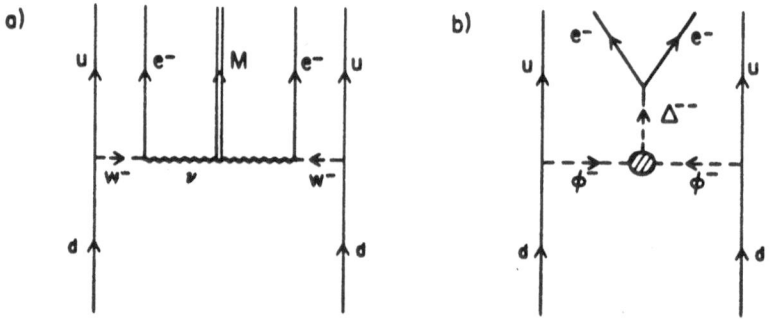

Fig. 6.   Mechanisms for $\beta\beta$-decay involving (a) Majoron production
           [30] and (b) Higgs exchange [20,31,32].

There has also been considerable discussion of mechanisms in-
volving the $\beta\beta$-decay transition $\Delta_{33} \rightarrow n$ within the nucleus [18,28,
29]. It can be shown, in the allowed approximation, that this
amplitude vanishes for $0^+ \rightarrow 0^+$, 2ν and, in the $SU(6)$ limit, 0ν $\beta\beta$-
decay. The possibility of strong $\Delta_{33} \rightarrow n, 0^+ \rightarrow 2^+, 0\nu$ transitions
is presently under study. Other lepton number-violating $\beta\beta$-decay
mechanisms that have been discussed include Majoron production [30]
and Higgs exchange [31] (Fig. 6). Recent work indicates that the

Higgs exchange mechanism is much less important than originally believed [20,32]. Majoron production poses a difficult problem experimentally, as this light scalar would carry off kinetic energy, leaving an electron energy distribution that would be difficult to distinguish from that for 2ν decay.

We have not discussed the definition of $\langle m^{Maj}\rangle_\nu$ except in the case that the neutrino is a mass eigenstate. Wolfenstein [33] has shown that in CP-invariant theories where multiple Majorana neutrinos couple to the electron

$$\langle m^{Maj}\rangle_\nu = \sum_i c_i^2 N_i m_i^{Maj}$$

where N_i is the CP eigenvalue and $\sum_i c_i \leq 1$. Thus it may be a mass difference that is constrained in $\beta\beta$-decay, and this quantity then is not simply related to that mass measured at the tritium $\beta\beta$-decay endpoint. Doi et al. [18] have considered more general mass matrices arising in CP-noninvariant theories.

Finally, in view of the general concerns of this conference, there is an interesting possibility that rigorous upper bounds can be placed on $|M_{GT}|$ by measuring the GT strength distributions in the intermediate nucleus from both parent and daughter. Thus $\beta\beta$-decay nuclei may be attractive candidates for ^{0} (p,n) and (n,p) studies [34]. Also, such GT strength distributions may provide important tests of the nuclear wave functions presently employed in $\beta\beta$-decay studies.

SOLAR NEUTRINOS

I will now briefly discuss the solar neutrino puzzle and the importance of GT strength measurements to future plans for neutrino spectroscopy.

To date only a single solar neutrino experiment, the ^{37}Cl experiment of Ray Davis, Jr., and collaborators, [35] has been mounted. The resulting capture rate, 1.95 ± 0.3 SNU, is in serious disagreement with the predictions of the standard solar and weak interaction models, 8.0 SNU [36] (1 SNU = 10^{-36} captures/^{37}Cl atom/s). If this discrepancy is due to a misunderstanding of the physics of the solar interior, the implications for present theories of stellar evolution could be profound [37]. Alternatively, if the sun does produce the expected neutrino flux, then some mechanism must be altering the character of those neutrinos before they reach earth. This suggestion now seems particularly interesting in view of recent evidence for massive neutrinos [8] and neutrino oscillations [38].

These two classes of solutions to the ^{37}Cl puzzle can be distinguished. Proposed modifications of the standard model to accommodate the ^{37}Cl capture rate result primarily in a reduced flux of high energy (14 Mev endpoint) ^{8}B neutrinos, whose production depends most critically on the central temperature of the sun. Neutrino oscillations or decay would, except under unusual conditions, affect all components on the solar neutrino flux equally. Thus there has been great interest in mounting new experiments to complete the spectroscopy of the neutrino sources shown in Table 5. Today I would like to describe three possibilities for new experiments that I find particularly interesting. I will emphasize the importance of ^{0}O (p,n) Gamow-Teller (GT) strength measurements in eliminating uncertainties in the neutrino capture cross sections estimates for each of these experiments.

Kuzmin [40] suggested a solar neutrino experiment based on the reaction ^{71}Ga$(\nu,e)^{71}$Ge. The calculations of Bahcall [41] and others [42] indicate that the ^{71}Ga capture rate in the standard model is primarily (70%) determined by the flux of neutrinos from the driving reaction of the pp-chain, $p + p \rightarrow {}^{2}H + e^{+} + \nu$. The pp neutrino flux is effectively fixed, provided only that hydrogen burning is the sun's energy source, by the observed solar luminosity. Thus, if a ^{71}Ga experiment yields a capture rate much

Table 5: Reactions (1) - (4) produce solar neutrinos with continuous distributions, while (5) and (6) are line sources. E_{ν}^{max} is the maximum energy of the neutrinos for all reactions except (4), where it has been computed with respect to the center of the broad 2.9 MeV ^{8}Be resonance populated in the β-decay of ^{8}B. Fluxes are taken from the standard solar model calculation of Bahcall et al. [39].

Reaction	E_{ν}^{max} (MeV)	Flux $(10^{10}/cm^2 s)$
(1) $p + p \rightarrow {}^{2}H + e^{+} + \nu$	0.420	6.1
(2) $^{13}N \rightarrow {}^{13}C + e^{+} + \nu$	1.199	4.6×10^{-2}
(3) $^{15}O \rightarrow {}^{15}N + e^{+} + \nu$	1.732	3.7×10^{-2}
(4) $^{8}B \rightarrow {}^{8}Be + e^{+} + \nu$	14.02	5.85×10^{-4}
(5) $^{7}Be + e^{-} \rightarrow {}^{7}Li + \nu$	0.862 (89.6%) 0.384 (10.4%)	4.1×10^{-1}
(6) $p + e^{-} + p \rightarrow {}^{2}H + \nu$	1.442	1.5×10^{-2}

reduced from standard model predictions, our particle physics must be at fault. In this sense the ^{71}Ga experiment will provide a test of neutrino oscillations for small Δm^2 and large mixing angles far beyond that possible with terrestrial neutrino sources.

The primary obstacle to the Brookhaven proposal for a ^{71}Ga experiment is the cost of the requisite quantity of gallium, estimated to exceed $25 million. However, there are in addition nagging uncertainties in the capture cross section. Gamow-Teller transitions to two excited states in ^{71}Ge, the 5/2$^-$ (175 keV) and 3/2$^-$ (500 keV) states, can be excited by ^{7}Be neutrinos. The ^{7}Be neutrino flux, like the ^{8}B neutrino flux, depends sensitively on the sun's central temperature. If the GT strengths for these transitions prove to be unusually strong, one could no longer argue that the ^{71}Ga capture cross was insensitive to solar model assumptions. Bahcall has argued from nuclear systematics that upper limits on the relevant transition strengths are log(ft) = 6.0 (5/2$^-$) and 5.0 (3/2$^-$) [43]. Yet, in view of the importance of this experiment, the need for a definitive measurement of these transition strengths is clear. Presumably $^\circ$0 (p,n) mappings with 175 keV resolution would settle this matter. (Note: In the discussion following this talk Dr. Orihara announced results of (p,n) measurements performed at Tohoku University showing that each of these transitions is quite weak. This substantiates the theoretical work of Bahcall and further strengthens the argument for doing the ^{71}Ga experiment.)

Davis, Sam Hurst, and collaborators [44] have considered designing a solar neutrino experiment based on the reaction ^{81}Br(v,e)^{81}Kr originally discussed by R. D. Scott [45]. The techniques developed by Davis to isolate ^{37}Ar could also be used to collect ^{81}Kr, and noble gas resonance ionization spectrometry might permit ^{81}Kr counting at the necessary sensitivity. The β-decay measurement of Bennett et al. [46] and the calculations of Bahcall [47] and Haxton [48] indicate that the capture rate is predominantly determined by the ^{7}Be neutrino flux. Thus the ^{81}Br and ^{37}Cl experiments would together determine the ^{7}Be/^{8}B neutrino flux ratio, a quantity that also serves to distinguish flaws in solar physics from those in particle physics: the flux ratio is likely to be unaffected by solar oscillations, but is highly sensitive to modifications in the standard solar model affecting the sun's central temperature.

The log(ft) value for the first GT transition, to the 1/2$^-$ (190 keV) state in ^{81}Br, has been measured by Bennett et al. [46]. The strength of a second transition that can be excited by ^{7}Be neutrinos, to the 5/2$^-$ (457 keV) state, is unknown. Thus a $^\circ$0 (p,n) measurement with 270 keV resolution would be important. Unidentified levels exist at 549 keV and 608 keV in ^{81}Kr; if these

states can be excited by the GT operator, an improvement in resolution would be required. The high energy ^{8}B capture rate should
also be appreciable, so a complete mapping of the GT strength distribution below particle breakup in ^{81}Kr will be needed.

An experiment of quite a different kind has been recently
proposed by G. A. Cowan and Haxton: a measurement of the concen-
trations of ^{97}Tc and ^{98}Tc produced by neutrino absorption in a
deeply buried molybdenite ore body over the past several million
years [49]. This experiment would test the long-term stability of
the sun and, in particular, the suggestion that the solar neutrino
puzzle and the recent Pleistocene glacial epoch are both the result of sudden mixing in the solar core four million years ago.

Only the ^{8}B neutrinos can induce the reaction ^{98}Mo$(\nu,e)^{98}$Tc;
as the ^{8}B phase space varies slowly as a function of nuclear excitation energy, no strong restrictions on resolution in GT mappings
exist in this case. In addition to ^{8}B neutrinos, the ^{7}Be neutrinos may contribute importantly to the capture rate for
^{97}Mo$(\nu,e)^{97}$Tc. A resolution of 250 keV should permit an accurate
estimate of this capture rate.

ACKNOWLEDGEMENT

I would like to acknowledge my collaborators in studies of
$\beta\beta$-decay and solar neutrino detection, G. A. Cowan, S. P Rosen,
G. J. Stephenson, Jr., and D. Strottman. In addition to these, I
thank N. Auerbach, S. Austin, F. Avignone, J. N. Bahcall, the Kent
State group, J. Rapaport, and L. Zamick for helpful conversations
and communications. This work was supported in part by the
National Science Foundation (grant no. PHY-8021272) and by the
Department of Energy.

REFERENCES

[1]   See, for example, S. Weinberg, Nobel Prize Address, Science
        210, 1212 (1980).
[2]   See, for example, T. Goldman, talk presented to the APS
        Division of Particles and Fields, Santa Cruz (1981).
[3]   M. Goldhaber, talk presented at the Workshop on Low Energy
        Tests of High Energy Physics, Santa Barbara (1982).
[4]   Y. K. Lee, L. W. Mo, and C. S. Wu, Phys. Rev. Lett. 10, 253
        (1963).
[5]   C. S. Wu, E. Ambler, R. Hayward, D. Hoppes, and R. Hudson,
        Phys. Rev. 105, 1413 (1957).
[6]   For general discussion see H. Primakoff and S. P. Rosen,
        Rep. Prog. Phys. 22, 121 (1950); E. J. Konopinski,
        "Theory of Beta Radioactivity" (Oxford University
        Press, London); talks presented by S. P. Rosen, G. J.
        Stephenson, Jr., and C. S. Wu, Neutrino Mass Mini-
        Conference, Telemark, Wisconsin (1980); S. P. Rosen,
        talk presented at Neutrino '81, Maui, Hawaii (1981);

W. C. Haxton and S. P. Rosen, talks presented at Workshop on Low Energy Tests of High Energy Physics, Santa Barbara (1982); E. Fiorini, Revista del Nuovo Cim. 2, 1 (1971); D. Bryman and C. Picciotto, Rev. Mod. Phys. 50, 11 (1978).

[7]   Talks presented by G. Senjanović, D. Wu, J. Schechter, Y. Tomozana, and L. Wolfenstein, Neutrino Mass Mini-Conference, Telemark, Wisconsin (1980).

[8]   V. P. Lubimov, E. G. Novikov, V. Z. Nozik, E. F. Tretykov, and V. S. Kosik, Yad. Fiz. 32, 301 (1980) and Phys. Lett. 94B, 266 (1980).

[9]   T. Kirsten, W. Gentner, and O. A. Schaeffer, Z. Physik 202, 273 (1967); T. Kirsten, W. Gentner, and O. Müller, Z. Naturforsch 22a, 1703 (1967); T. Kirsten, A. O. Schaeffer, E. Norton, and R. W. Stoenner, Phys. Rev. Lett. 20, 1300 (1968).

[10] E. C. Alexander, Jr., B. Srinivasan, and O. K. Manuel, Earth and Planet. Sci. Lett. 5, 478 (1969); B. Srinivasan, E. C. Alexander, Jr., and O. K. Manuel, Econ. Geol. 67, 592 (1972) and J. Inorg. Nucl. Chem. 34, 2381 (1972).

[11] E. W. Hennecke, O. K. Manuel, and D. D. Sabu, Phys. Rev. C11, 1378 (1972).

[12] B. Srinivasan, E. C. Alexander, Jr., R. D. Beatty, D. E. Sinclair, and O. K. Manuel, Econ. Geol. 68, 252 (1973).

[13] R. Bardin, P. Gollon, J. Ullman, and C. S. Wu, Phys. Lett. 26B, 112 (1967) and Nucl. Phys. A158, 337 (1970).

[14] E. Fiorini, A. Pullia, G. Bertolini, F. Capellani, and G. Restelli, Nuovo Cim. A13, 747 (1973).

[15] B. T. Cleveland, W. R. Leo, C. S. Wu, L. R. Kasday, A. M. Rushton, P. J. Gollon, and J. O. Ullman, Phys. Rev. Lett. 35, 737 (1975).

[16] M. K. Moe and D. Lowenthal, Phys. Rev. C22, 2186 (1980).

[17] W. C. Haxton, G. J. Stephenson, Jr., and D. Strottman, Phys. Rev. D25, 2360 (1982).

[18] M. Doi, T. Kotani, H. Nishiura, K. Okuda, and E. Takasugi, Phys. Lett. 102B, 219 (1981); Prog. Theor. Phys. 66, 1739 (1981); Prog. Theor. Phys. 66, 1765 (1981).

[19] W. C. Haxton, G. J. Stephenson, Jr., and D. Strottman, Phys. Rev. Lett. 47, 153 (1981).

[20] W. C. Haxton, S. P. Rosen, and G. J. Stephenson, Jr., Phys. Rev. D26, 1805 (1982).

[21] R. D. Lawson, Phys. Rev. C24, 1500 (1981).

[22] T. T. S. Kuo, private communication; T. T. S. Kuo and G. E. Brown, Nucl. Phys. A114, 241 (1968).

[23] W. T. Baldridge and J. P. Vary, Phys. Rev. C14, 2246 (1976).

[24] L. Zamick and N. Auerbach, Phys. Rev. C26, 2185 (1982).

[25] T. Kirsten, talk presented at the Workshop on Low Energy Tests of High Energy Physics, Santa Barbara (1982).

[26] J. D. Vergados, Phys. Rev. C12, 865 (1976).

[27] S. P. Rosen, talk presented to Orbis Scientiae 1981, Center

for Theoretical Studies, University of Miami.

[28]   A. Halprin, P. Minkowski, H. Primakoff, and S. P. Rosen,
       Phys. Rev. D13, 2567 (1976).

[29]   C. Picciotto, Can. J. of Phys. 56, 399 (1978).

[30]   H. M. Georgi, S. L. Glashow, and S. Nussinov, Nucl. Phys.
       B193, 297 (1981); G. B. Gelmini and M. Roncadelli,
       Phys. Lett. 99B, 411 (1981).

[31]   R. N. Mohapatra and J. D. Vergados, Phys. Rev. Lett. 47,
       1713 (1981).

[32]   J. Schechter and J. W. F. Valle, Phys. Rev. D25, 2951
       (1982). L. Wolfenstein, private communication.

[33]   L. Wolfenstein, Phys. Lett. 107B, 77 (1981).

[34]   A series of such (p,n) measurements are being made at the
       Indiana University Cyclotron Facility.

[35]   R. Davis, Jr., D. S. Harmer, and K. C. Hoffman, Phys. Rev.
       Lett. 20, 1205 (1968); R. Davis, Jr., Proc. of the
       International Conf. on Neutrino Physics and Astro-
       physics, Moscow, Vol. 2, 99 (1969); R. Davis, Jr., and
       J. M. Evans, Proc. of the 13th International Cosmic
       Ray Conf., Vol. 3, 2001 (1973); R. Davis, Jr., Proc.
       Informal Conf. on the Status and Future of Solar Neu-
       trino Research, ed. G. Friedlander, BNL Report 50879,
       1 (1970).

[36]   J. N. Bahcall, talk presented at Neutrino '81, Maui, Hawaii
       (1981).

[37]   R. T. Rood, in Proc. Informal Conf. on the Status and
       Future of Solar Neutrino Research, ed., G. Friedlander,
       BNL Report 50879, 175 (1978); J. N. Bahcall and R. L.
       Sears, Ann. Rev. Astron. and Astrophys. 10, 25 (1972).

[38]   F. Reines, H. W. Sobel, and E. Pasierb, Phys. Rev. Lett. 45,
       1307 (1980).

[39]   J. N. Bahcall, S. H. Lubow, W. F. Huebner, N. H. Magee, Jr.,
       A. L. Merts, M. F. Argo, P. D. Parker, B. Rozsnyai,
       and R. K. Ulrich, Phys. Rev. Lett. 45, 945 (1980).

[40]   V. A. Kuzmin, Sov. Phys. JETP 22, 1051 (1966).

[41]   J. N. Bahcall, B. Cleveland, R. Davis, Jr., I. Dostrovsky,
       J. C. Evans, Jr., W. Frati, G. Friedlander, K. Lande,
       K. Rowley, and J. Weneser, Phys. Rev. Lett. 40, 1351
       (1978).

[42]   N. Itoh and Y. Kohyama, Astrophys. J. 246, 989 (1981).

[43]   J. N. Bahcall, Rev. Mod. Phys. 50, 681 (1978).

[44]   G. S. Hurst, M. G. Payne, S. Kramer, and C. H. Chen, Phys.
       Today 33, 24; R. Davis, Jr., private communication.

[45]   R. D. Scott, Nature 264, 729 (1976).

[46]   C. L. Bennett, M. M. Lowry, R. A. Naumann, F. Loeser, and
       W. Moore, Phys. Rev. C22, 2245 (1980).

[47]   J. N. Bahcall, Phys. Rev. C24, 2216 (1981).

[48]   W. C. Haxton, Nucl. Phys. A367, 517 (1981).

[49]   G. A. Cowan and W. C. Haxton, Science 216, 51 (1982).

EXPERIMENTAL PERSPECTIVE

William Bertozzi

Department of Physics and
Laboratory for Nuclear Science
Massachusetts Institute of Technology
Cambridge, MA   02139

A great deal of new and important material has been reported at
this conference and many interesting speculations have been presented.
A large number of problems have been exposed and discussed with such
completeness that my task of presenting an experimentalist's perspective for the future has become a difficult one. Partly I feel a
growing compassion for the theorists for whom there is already so
much to understand and it is with great trepidation that I venture
to suggest that the field holds even more promise for the future in
terms of new results of importance. It is with even greater trepidation that I suggest that some of our cherished experimental results
are not as well established as some of us seem to believe. This
feeling of insecurity is, of course, enhanced by the forceful erudi-
tion of the well-known theoretical summary speaker who follows me and
who will surely demonstrate how all the new physics can be derived
most interestingly from a few simple and general ideas and that once
again all is well with the world. Perhaps experimentalists should
learn to leave well enough alone but, unfortunately, we carry in our
genes some kind of force that compels us to turn knobs and make ad-
justments just to be sure that all is well with our measurements and
with our ideas. With all due respect, such is the nature of some of
my remarks.

The most spectacular result of this field has been the discovery
of the Gamow-Teller resonance and its predominance in the (p,n) reaction at energies above 100 MeV. Our attention has been concentrated
on the strength of this resonance and a comparison to the sum rule,
$S^- - S^+ = 3(N-Z)$. We have learned at this meeting that the process of
integrating the experimental (p,n) strength in the region of the
resonance develops problems with respect to defining the background

associated with the continuum process and that angular distributions
are necessary to extend the experimental sums to zero momentum transfer and to isolate the $\ell = 0$ components. The presentations all seem
to concur with the conclusion that integrating with care over approximately an 8 MeV band about the resonances yields systematically 50-
60% of the theoretical sum rule. Some theorists have interpreted
this result as a clear exhibition of Δ-hole configurations depleting
the strength normally associated with the nucleonic degrees of freedom. The resultant quenching of the nucleon magnetization has been
associated with the suppression of M1 strengths and this quenching
also has been offered as an explanation of the suppression observed
in the excitation of stretched high-spin Mλ states. The connection
to the quark structure of nucleons has caught our imagination.
Appropriately, the excitement in the community is very high.

Nevertheless, I would like to come back to the question that
some of us have asked throughout the meeting about the GT-sum. Experimentally, we must ask how much GT strength is located far above
the few strong and sharp resonances that are now included in integra-
tions over intervals of about 8 MeV. If we integrate over the next
100 MeV of excitation can we rule out another 25% of the strength,
or, for that matter, most of what is missing? All that we need is a
continuum intensity of a few percent of that in the peaks to provide
most of the missing strength. These are continuum states that must
surely have "tails" of strength at higher excitation. Experience
with the giant electric-dipole resonance would lead me to be very
cautious. I also remind you of (e,e'p) measurements at Saclay that
only find about 60% of the single particle strength when up to 60 MeV
of excitation is examined. No one doubts that Δ-hole mixing has some
influence on GT strengths. The importance of the missing GT strength
result lies in its large size and we must not take for granted that
nature would like us to have as clear and as simple a signature of
Δ-hole mixing as the one that we have observed. Experimentally, I am
not convinced that we know how much of the sum is missing and how
much is contained in small amplitudes at excitation energies high in
the continuum. This is a difficult experimental challenge that we
cannot overlook, for the stakes are high as we begin to couple nucleonic structure to collective phenomena.

On the theoretical side we must inquire into dynamic effects
other than Δ-hole mixing that can cause a suppression of the GT
strength. We know that nuclei are not simply mean-field ground
states and that multi-particle-hole configurations play important
roles in much of what we see in nuclear reactions. How do these
features affect the GT response function and the distribution of
strength? We have heard some suggestions from theorists that Δ-hole
effects can be responsible for only a fraction of the observed problem and that these more complex aspects of nuclear structure play
an important role.

The curtain has not closed on this issue of missing GT strength and it is one of the central themes of our field for the foreseeable future requiring experimental care and theoretical insight.

Turning to the other examples of missing strength, I remark on the situation with respect to the excitation of stretched high spin states. Examples are the 4^- states in ^{16}O and 8^- states in ^{58}Ni. Throughout the periodic table, only about 50% of the strength has been observed in a few states. It has been mentioned that the Δ-hole quenching might be responsible for this phenomenon as well, and that we have an explicit general quenching of the single nucleon magnetization of large proportions. In addition to the problems already mentioned, we must caution against such a view on experimental grounds. I remind you that the highest multipole elastic magnetic scattering from ^{17}O, ^{89}Nb and at least two other nuclei has the full single particle strength. In view of these results, it is much more likely that we are observing rather severe nuclear structure effects in these stretched $M\lambda$ excitations that reduce the strength or distribute the strength among many small continuum states at higher excitation energies that are not detectable. At the high momentum transfers that these high multipoles manifest themselves, quenching of nucleon magnetism does not seem to be a large universal mechanism.

Missing strength is very common for processes dominated by a single nucleon in our simple mean field picture. It is generally a puzzling and an interesting feature and possibly an important one. Deciding that it is largely due to modifications of nucleonic structure rather than complications due to nuclear structure is a large step we should take only with extreme care. In this connection it is relevant to remind ourselves of the situation with respect to the central charge density of ^{208}Pb, which appears to require about a 40% reduction of the occupation of the 3s proton orbit. I suspect this is due to the very approximate and defective description that a single Slater determinant in Hartree-Fock theory provides for nuclear ground states. Yet, we have no quantitative theory that can provide this size of suppression in a natural computation. There is no question that we should look for more accurate models of nuclei in the conventional nucleonic many-body sense. As a first priority, experimentalists must perform those measurements that are able to provide systematic guidance for theorists in this most difficult problem.

Along the same lines, it is very important to understand the effect of the nuclear medium on the nucleon-nucleon interaction. We have observed at intermediate energies the strong influence of the nuclear medium on the central, spin-independent isoscalar interaction as demonstrated by the interweaving of inelastic electron scattering and inelastic proton scattering. The other modes of the nucleon-nucleon interaction involving spin flip do not appear to exhibit a strong influence. At least it is not clearly observable given the state of many-body scattering theory. Another variation on this

theme involves the introduction of a minimal amount of relativity
with some rather striking successes at higher energies without medium
modifications.

    One important message of this work is that something new is
emerging in our field that promises to enlarge our view of nuclear
structure. For many years we have accepted the mean-field idea for
the ground state as the best independent-pair approximation available.
For excited states we have dealt with various truncated shell models.
Data from all our probes are compared with the predictions of these
models accepting also for each probe the uncertainties of the reaction mechanism. No connection of substance exists generally between
the shell models used to describe excited states and the mean-field
theory used to describe the ground states. Variation of shell model
parameters really does not tell us much about the more fundamental
many-body problem, let alone how to improve our approximate treatments. At long last, we are imbedding the scattering problem in the
fundamental many-body theory of nuclear matter and we are observing
some striking partial success. This more intimate connection should
be pushed for all its worth and more. Let us not return to that dark
age of optical model phenomenology.

    As an aside but relevant to this more general issue, I remark on
the importance of the interplay that is developing with the use of
various probes to their best advantage. Electrons measure one-body
densities of charge and electromagnetism. Comparisons of cross sections for electrons and hadrons leading to the same excitations clearly give us hints about the reaction process for the hadrons and demonstrate the essential one-step process for proton inelastic scattering
at intermediate energies. The electromagnetic densities liberate us
from inaccurate theories and we can focus more accurately on the
nature of the nucleon-nucleon interaction in the medium. In addition
to allowing a more intimate connection to fundamental many-body theory
as discussed earlier, we enrich coherently our understanding of details. Perhaps we can reach eventually that stage where the panoply
of one-body densities that the nucleus possess can be probed by the
sagacious use of all the probes available. Numbered are the days when
an experimenter can be content to study a nucleus with one probe alone.

    The interplay and complementarity of electromagnetic and hadronic
probes can be summarized in many ways and perhaps the most useful is
to demonstrate the many one-body densities of matter and spin [$\rho_{JL}(r)$,
$\mu_{JL}(r)$] that can be studied. This is beyond the scope of my talk and
I will only comment on the local and non-local character of the densities that are studied by these probes in single arm experiments
$\rho(r,r')$. Electron scattering is limited to the diagonal or local densities. They are well determined experimentally and form the testing
ground for the hadron studies which extend the measurements to non-
local properties averaged by the range of the interaction.

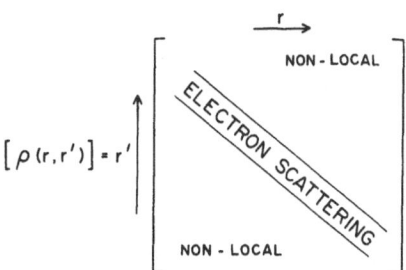

One-body Density Matrix

The (p,n) process deserves some special attention in terms other than the GT-sum rule. There can be no controversy about the fact that the $\sigma\tau$ operator seems to function with great clarity in the (p,n) process at intermediate energies as a one-body probe. I would suggest that we should have a mini gold-rush in this connection and use it to study all those nuclear structure features that are available. To the theorists I would ask a question about exchange currents. How do they show up? Are they as clear as we have observed them in the (e,e') experiments? Another point relates to electromagnetic processes wherein both current and σ terms are mixed. Perhaps the (p,n) reaction can be used to allow the separation of the convective current parts from the spin-flip parts. The (p,n) reaction should be liberated from q = 0 as it is a marvelous tool at intermediate energies for studies of nuclear structure.

Turning to the issue of pion condensates and g', I believe that it is fair to say that g' continues to grow well past the limit we expect for condensates to appear. Although we may be disappointed or pleased, depending on one's persuasion, it is also fair to say that precursor phenomena are not likely to exhibit themselves with clarity. Certainly, we have no positive evidence for these collective extranucleon phenomena in nuclei. The search has involved the (e,e') study of the Ml excitations in ^{12}C and other electromagnetic experiments and also some (p,p') and the (p,n) reactions. One lesson is clear. The hadronic probes are uniquely able to study the $\vec{\sigma}\cdot\vec{q}$ response function which is very sensitive to these phenomena, compared to the $\vec{\sigma}\times\vec{q}$ response function. In this connection I remind you of the curious result of deeply inelastic (e,e') from nuclei, in particular the results from Saclay on $^{12}C(ee')$ ranging over the quasi-elastic peak into the Δ-region. The longitudinal or charge part of the (e,e') reaction calibrates the one-body character of the process and shows us that the transverse electromagnetic process ($\vec{\sigma}\times\vec{q}$) has an extra contribution that is as much as 40% in excess of the one-body reaction mechanism in the quasi-elastic region. Perhaps the added strength is due to mesic exchange currents or modifications of nucleon magnetic structure, but we do not understand this result. It is clear that a thorough study of the quasi-elastic $\vec{\sigma}\cdot\vec{q}$ and $\vec{\sigma}\times\vec{q}$ nu-

cleonic response functions may prove to be very interesting for the
discovery of extranucleon degrees of freedom in the nuclear medium.
Unifying an analysis of the hadronic processes with the electromagnetic reaction should be very valuable.

     We have many mysteries; some that we have dealt with are general
in character, some are specific to a particular reaction. These
latter problems when elaborated experimentally and theoretically
often become the starting ventures exploring a new phenomenon. I
mention quickly some that strike my fancy: the interesting energy-
dependent behavior of the polarization of protons scattered from
complex nuclei, deeply inelastic, between the quasi-elastic peak and
the Δ-resonance; the $\Delta S = 1$ behavior of pion scattering leading to
the excitation of the low-lying 1^- states in ^{18}O which the electron
scatterers know with great precision to be a longitudinal excitation;
and the one-body character of a large component of pion double charge
exchange. These are a few more indications of the complexities of
the many-body story that we have begun to uncover in more detail.
The richness of our field is abundantly clear. I believe it is fair
to say that a new enthusiasm for nuclear physics has emerged as we
have come to make closer contact with fundamental theory. In no
small part this situation has been enhanced by our ability to separate spin excitations at intermediate energies and the clear signature that has emerged for the collective GT-excitations. Our field
has been strengthened by these recent discoveries and developments
and once again emerges as exploratory in character. As an experimentalist I can give only one comment of advice in such a situation
— assume very little and measure everything.

UNRESOLVED THEORETICAL PROBLEMS

Erich Vogt
TRIUMF
4004 Wesbrook Mall
Vancouver, B.C., Canada V6T 2A3

Perhaps as much as any recent conference on nuclear physics this Telluride conference has brought together a number of very important ideas in nuclear physics and has pointed to future directions of the whole field. In this summary I want to give some personal impressions of the ideas about which we heard and also to try to put them into a perspective within which we can look at unresolved theoretical problems – questions which could have importance in the future. We have had the kind of vigorous discussions which are appropriate to a frontier town like Telluride. There were some moments of glory and some heinous crimes. I want to provide, as appropriate, some quick accolades and some summary frontier justice. The first such quick and harsh justice should probably be dispensed to the previous summary speaker who has, in his usual style, stolen most of the prize nuggets unearthed at this conference.

It has been the history of nuclear physics that quantum jumps forward have occurred when the right experiments and the right ideas came forward at the right time. Usually the right time came not as the result of any vigorous theoretical prediction but rather when some experimental evidence – for example, some enormous resonance behaviour – hit us on the head at a time when a climate of ideas existed in terms of which the resonance behaviour could be understood. Before looking at the Gamow-Teller resonance and the climate of ideas in which it has flourished let's take a brief look at earlier similar revolutions of our field.

For the nuclear shell model individual physicists made predictions from the mid 1930's onward. However, it wasn't until about 1950 that the resonance and the climate of ideas quite suddenly came together. The resonances, of course, were the nuclear size resonance

first clearly seen in the Wisconsin neutron cross sections. The
ideas pertained to spin-orbit coupling but even more to the beginning of the phenomenological understanding of the shell model for
systems of nucleons and its basis in the appropriate many-body
theory.

Next, in the late 1950's, came the understanding of nuclear
collective motion in terms of individual nucleons in single-particle
orbits with a residual interaction. This breakthrough was heralded
as much as anything by the data for giant dipole resonances observed
in photonuclear reactions. The climate of ideas arose from the experience of handling interacting nucleons and nucleon holes in
single-particle configurations. The ideas for coherent single-particle motion in the giant dipole state came from David Brink and
Gerry Brown and have led to two decades of very great progress on
collective motion of nuclear many-body systems.

Cluster degrees of freedom came into focus when, in 1960, heavy
ion accelerators first achieved high resolution and the strong
molecular resonances of the ^{12}C and ^{16}O ions were discovered. The
climate of ideas again arose from the shell model. Sorting out true
cluster correlations from these correlations imposed on the nuclear
system by Pauli principle and by attractive residual interactions
among the nucleons of a single shell has taken a long time.

In the mid 1960's isobaric analogic resonances were a somewhat
startling experimental discovery which led to a much deeper understanding of nuclear symmetry laws and how little they were violated
by Coulomb forces.

In each of these cases there were some prior theoretical speculations. In each there were beautiful experiments perceived with a
large element of surprise in spite of the earlier speculations. In
each there was a climate of ideas within which the significance of
the results was quite quickly comprehended. In each there were,
suddenly, many questions and unresolved problems - one knew where
one was and what had to be done. In each case there was an important conference when the experiments and the ideas came together.
All of these conditions appear to be present now about the experiments pertaining to the Gamow-Teller resonance and the perspective
of ideas within which it exists. It is my opinion that this conference will be judged to have been the important one for this
conjunction.

I want to summarize the nuclear physics ideas which form the
perspective within which the spin excitations in nuclei should be
considered and to give some indications of the directions in which
we must head, as elucidated in the many interesting review talks and
discussions which we have heard.

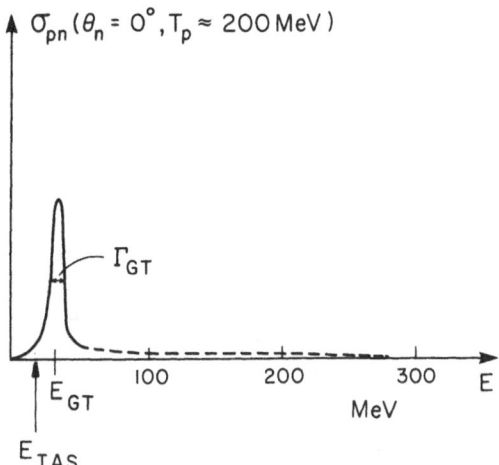

Fig. 1. A qualitative description of the collective GT resonance
         observed in σ_{pn} at $0°$. The primary interest here is
         neither in the level energy, E_{GT}, nor in its width, Γ_{GT},
         but rather in the wings of the resonance.

Let me begin by recognizing some of the individuals who predicted some of the phenomena which we have discussed but whose
ideas came before the time was ripe - before the Telluride epoch.
In the early 1960's Anderson, Wong and McClure anticipated the
existence of the collective Gamow-Teller resonance and correctly
predicted its energy. Also Fujita and his colleagues in Japan discussed the collective Gamow-Teller resonance.

A little later Migdal foresaw the basic ideas of pion condensates although he did not have the insights provided by QCD. Much
credit should be given to Magda Ericson and her collaborators
(echoed by the Paris group - fortunately for Rho and his collaborators echoing is not a crime in the San Juan mountains of
Telluride) for predicting in the 1970's the quenching ideas related
to the role of delta-hole interactions. Some early experimental
work at low energy discovered the Gamow-Teller resonance. However,
the full experimental exploration required higher energies and came
about, four years ago, in the beautiful Indiana work of Goodman and
his collaborators.

Let's use the Indiana experiment to epitomize the physics
ideas confronted here at Telluride. I have sketched in Fig. 1 the
idealized (p,n) double differential cross section (at $0°$ and as a
function of excitation energy in the residual nucleus), for an incident proton energy of about 200 MeV. We know that for such proton

energies the cross section at 0° is dominated by the magnetic isovector spin-flip process. The Gamow-Teller collective giant resonance is shown (any fine structure is assumed to be averaged over)
and we enquire about its properties.

Generally the most important properties of any resonance are
its energy and its width. With regard to these the Gamow-Teller
resonance is garden variety. We have heard at this conference about
some very fine and sophisticated calculations of the energy and
width, but frankly, in 1982, there is no reason to be enormously
excited about these. The resonance position is not far from the
place immediately above the isobaric analogue resonance where
Anderson, Wong and McClure predicted it to be in 1962. There are
some modest contributions to the energy shift arising from current
ideas.

Let me spend a little more time on the width of the GTR which
is, as I said, also garden variety and not nearly as startling as
those for the resonances on which our previous revolutions in
nuclear physics were based.

The width of any giant resonance is the sum of the natural
width of the state and its spreading or damping width. In the situations to which I shall refer - including the GTR resonance - it is
the spreading width, $\Gamma^{\downarrow}$, which dominates. The spreading width was
discussed at this conference by Wambach and by Bortignon, and it
pertains to the way in which a residual interaction distributes the
simple wave function of a model Hamiltonian among the eigenstates of
the full Hamiltonian (or among the eigenstates of a doorway Hamiltonian). The amount of the spreading is, phenomenologically, equal
to twice the imaginary part of the optical potential, $\Gamma^{\downarrow} = 2\langle W\rangle$, of
the model state or, at a deeper level,

$$\Gamma^{\downarrow} \approx \sqrt{M_2} = 2\sqrt{\langle v_{res}^2\rangle}$$

where M_2 is the second moment.

If, for example, one considers the possible single-particle
motion of nucleons and if one takes the residual interaction, v_{res},
to be the sum of all the two-body interactions with neighbouring
nucleons, it was already known before 1940 that one obtained an
answer of about 40 MeV for $\Gamma^{\downarrow}$. This came simply from the strength
(depth $\simeq$ 40 MeV, range $\simeq$ 1 fm) of the nucleon-nucleon interaction
needed to bind the deuteron. Because this spreading width of 40 MeV
was roughly equal to the spacing of single-particle levels in a
potential well of nuclear size, it seemed impossible to have a viable
shell model. This was, perhaps, the strongest reason why no reasonable physicist entertained seriously the thought of a shell model

until overwhelming experimental evidence established it. Why were
the theorists wrong? Why did the spreading width – as measured by
the low energy neutron and proton optical potential ($W \approx 4$ MeV,
$\Gamma^\uparrow \approx 8$ MeV) – turn out to be so much smaller? The answer lay in
the correlations imposed on the system of nucleons by the Pauli
principle and by short-range repulsion.

The narrow width of the giant dipole resonance (about 4 MeV in
carbon) was not as much of a surprise since one already understood
the comparable widths of nuclear single-particle levels.

The situation for molecular levels of heavy ion systems was,
however, totally astonishing. In $^{12}C+^{12}C$ these states – probably
alpha cluster doorway states – have widths of about 100 keV. Since
doorway states are spread by the same strong nuclear interaction as
single particle and collective giant resonance, there was no reason
why they should be much narrower – let alone two orders of magnitude
narrower. Two decades later we have no real understanding of the
long lifetime of molecular levels, even though a whole industry has
flourished to elucidate their spectroscopy. Spreading width problems are difficult and have been notoriously neglected, but surely
some very deep information about nuclear structure resides in understanding the physics of the various factors which conspire to
produce such narrow doorway states.

The isobaric analogue resonances presented a similar surprise.
The Coulomb force in heavy nuclei is, on the average, only three
times smaller than the nuclear force so the widths were expected to
be about three times smaller than those of nucleon size resonances –
or about 3 MeV. No one anticipated the 50 keV widths found experimentally. The spreading of such states by Coulomb forces into
states of lower isotopic spin is reasonably understood in terms of
the long range of the Coulomb interaction.

For a Gamow-Teller resonance we would have expected spreading
widths similar to those for other states (nucleon size-resonances,
giant dipole resonance, etc.) of about 5 MeV or so and that is what
is found. So we have no enormous excitement caused by the width but
some good physics. At first sight it might seem surprising that
such collective particle-hole states should be narrower than the
square root of the sum of the squares of the spreading widths of the
particle and hole states on which they are based. Recently George
Bertsch and his collaborators have studied the spreading of simple
low energy nuclear excitations into doorway states made up of surface vibrations. Bortignon reported on how the moderate narrowing
of the GTR was accomplished by this mechanism. Similar results
were achieved by Wambach et al. with beautiful RPA calculations including coupling to Δ-hole states. However, this is not where the
current revolution lies.

The drama resides in the area under the resonance and in the shape of its wings. It is interesting that the question of line shape has been totally unimportant in all of the earlier nuclear revolutions to which I referred, and even the question of area under the resonance – generally quantified in terms of sum rules – has been of only moderate interest. Here, because of the remarkable spin and isospin selectivity of the (p,n) reaction at intermediate energies one has been able to probe the resonance shape over a reasonably wide energy interval – and one has been able to quantify the area under the resonance accurately by normalization to established β-decays. Both the quenching of the GTR area and the shape of its wings involve exciting physics. It is that excitement which has brought us together. Let's look at the ideas which are the source of this excitement.

About half the strength has gone from the peak. Where it goes tells us something immediately about the nucleon. If the missing strength corresponds to the dissipation of the simple excitation into other shell model configurations – such as 2p-2h or 4p-4h – then it should be found in the "near wings" of the resonance. If it involves short distances and very high energy states – such as the Δ-hole excitations lying 300 MeV higher – then one might expect the missing strength to go into the "far wings". Osterfeld's analysis as reported at this conference shows that microscopic particle-hole calculations may give some strong energy dependence to the non-collective background in the vicinity of the Gamow-Teller resonance so that some care must be taken in interpreting the "near wings". Even allowing for this there appears to be a real quenching effect – a substantial missing portion of the collective resonance.

Extracting tangible physics from analysis of line shapes is not easy. We would not have been drawn to the main questions of our Telluride conference if the recent development of nuclear and particle physics had not led us there. Quantum chromodynamics (QCD) should matter for atomic nuclei. There are compelling reasons why QCD effects might show up in the GT resonance. We want it to be there.

It is the compelling conjunction of particle physics (QCD) with nuclear physics which constitutes the climate of ideas within which the GT resonance is discussed and which may possibly herald another revolutionary change in our thinking about nuclear physics. Almost everything that transpired at the conference can be put in terms of a series of questions, few of which have any solid answers at present. It is the fact that so many questions have suddenly presented themselves with such urgency that should make us optimistic for the next decade of nuclear physics.

We are looking for ways in which QCD effects can augment our standard single-particle picture of atomic nuclei which has proved

so fertile for three decades. In the single-particle picture,
Fig. 2(a), protons and neutrons in single-particle orbits interact
through meson exchange (π, ρ ...) with the presence of meson-exchange
currents (MEC). Although so much has been accomplished phenomenologically with this model we are only now learning how to properly
handle systems of mixed fermions, to treat their mean fields, residual interactions, etc. while Pauli is probably smiling benignly at
our efforts to handle the effects of his exclusion principle on
neutrons and protons with path integrals and Grassman algebras.

In the QCD nucleus, Fig. 2(b), each nucleon is a colour-neutral
bag of quarks embedded in a medium. Goldstone bosons (pions) exist
in the medium but not inside the bags. Efforts are made to conserve
chiral symmetry at the bag surface which is the interface between
the bag and the medium. The very fact that we talk about a bag — a
sharp interface — highlights the fact that at present we know so
little about the physics of confinement.

Although QCD tells us about quark interactions at small distance in a nucleus we must confront the little-known physics associated with confinement. The average internucleon spacings (~2 fm)
are only little larger than the bag radius (~0.8 fm — Rho and Brown
may, in the future, retreat from their smaller bag radius or face
frontier justice). There are, then, immediately a host of questions,
in addition to the size of the bag radius. Do the bags coalesce or

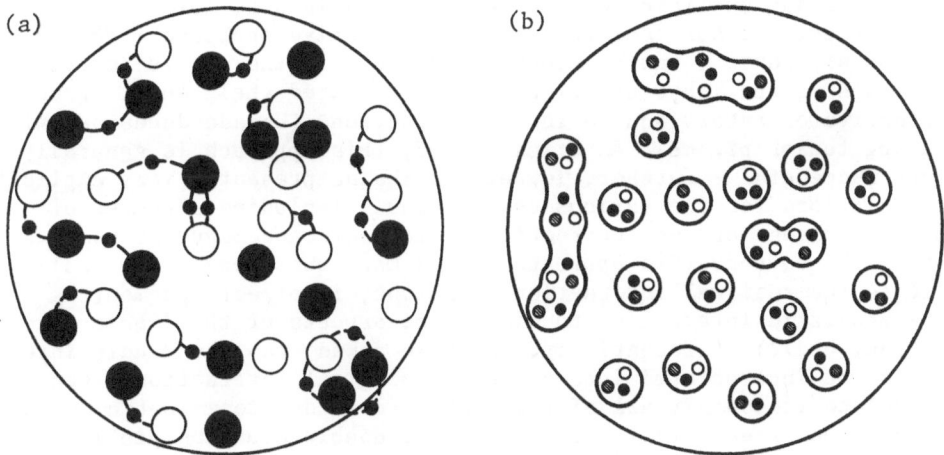

Fig. 2. (a) The familiar single-particle picture of the nucleus in
which protons (large solid circles) and neutrons (open
circles) interact through meson (small closed circles) exchange; (b) the QCD nucleus in very qualitative terms;
nucleons are considered as colour-neutral bags of quarks
with some of the quark bags overlapping.

percolate? What are the consequences of percolation? Is the
medium in between the bags polarized? If so, how are the pions
affected? What are the effects of bag deformation? Can one bring
"real" deltas into the nucleus? Can one introduce ρ-mesons into the
interactions? [Introducing ρ-mesons into effective interactions now
appears to require more larceny than is acceptable in a frontier
town. Possibly because of the fear of summary justice Gerry Brown
did not appear in person in Telluride to give his talk on effective
interactions which, instead, was given by another member of the
Brown gang who was not seen subsequently.] At what densities or
temperatures do phase transitions occur from normal nuclear matter
to quark matter?

        Although all of these questions were raised at the conference
the real interest was, of course, the search for any nuclear properties which might possibly have some bearing on QCD effects. Rho
gave an analysis of nuclear response functions some of which might
act as filters in being especially sensitive to such effects. In
his opening address Teller presented the simple picture why the
magnetic isovector spin-flip transition should involve Δ-nucleon
excitations. The stage was set then for considering the Gamow-
Teller strength function and other detailed nuclear properties
which might provide clues to the elusive quark degrees of freedom.

        The present evidence for the effect of Δ-hole interactions or
other manifestations of quark degrees of freedom is sufficiently
oblique that it is clearly important to be sure how far the standard
model without explicit quark effects can go in describing nuclear
properties. At our conference the conservative approach - the role
of honest broker - was presented by Arima. In the single-particle
approach of the Tokyo school one uses truncated shell model spaces,
effective operators and residual interactions (Hamada-Johnston) with
strong tensor pieces. As Arima showed, this approach is generally
very successful. Further, Towner and Khanna presented very explicit
calculations for the various corrections - including a number of
meson-exchange current corrections - to isovector magnetic moments.
Still the conservative approach raises many questions. How realistic is the residual interaction? (Is one, in effect, augmenting
the realistic interaction to mimic some aspects of the Δ-hole interaction, etc.?) Alternatively, as Weise asked, can the Δ-hole interaction be incorporated into a phenomenological interaction? Can one
calculate reasonably all of the many corrections to magnetic moments,
transition rates, etc.? Does one avoid double-counting problems
(for renormalized M1 and GT operators)? These questions about the
standard model are not necessarily new but they have now gained a
new urgency.

        Focusing, as our conference did, on the Δ-hole interaction as
an example of nuclear QCD effects there are a number of questions
which remain open. Is the Δ-hole interaction justified by QCD? Is

the coupling constant, g', the same for the Δ as for the nucleon?
Is the Δ-hole interaction reduced by exchange? Is there a good
experimental signature for the Δ-hole interaction which distinguish
its effects from those of the N-hole and the tensor force? (Here
possible candidates are the quenching in low and high spin stages,
magnetic moment corrections, etc.)

    Turning to the quenching itself - the physical phenomenon
which was the central interest of our Telluride conference - there
remain a host of open problems. How well do we really understand
the Δ-hole interaction? Considering the background subtraction
problems, what is the magnitude of the quenching of the Gamow-
Teller resonance? (The answer for the amount of strength, relative
to the sum rule limit, found under the resonance and its near wings
appeared to increase as the conference progressed from about 33% to
well over 50%.) What does QCD (Adler-Weisburger) say about the sum
rule? What is the energy dependence and total angular momentum
dependence of quenching? Has any GT strength been found (or should
any such strength be expected to manifest itself) at 300 MeV, for
example in the very new (^{3}He,t) work?

    In turn, the GT-quenching and Δ-hole interaction problems give
rise to questions about real Δ's in the nucleus which were addressed
by Lenz and others. How good is the Δ-dominance picture for the
pion-nucleus optical potential? How does the propagating Δ interact
with the nucleus? What is its mean free path? Does the ^{12}C(π,π')^{12}C*
(15.11 MeV) excitation function tell us something about real Δ's in
the final state? (Probably not!) What is the Δ-N interaction?
What reactions might inform us about real Δ's in the nucleus? (One
candidate might be recoilless Δ-production in (p,d) reactions.)

    Perhaps one of the most important results of Telluride is the
impetus it has given to the general study of nuclear response functions with a variety of probes. The response function has a large
number of different components varying in their multipolarity and
isospin. Unscrambling the cross section data with various probes
to obtain the individual components or strength functions is not a
trivial task. Walker, Love, Heisenberg and others showed how it
might be done and at what energies it is possible and desirable.
Clearly one of the most important tasks for nuclear physics in the
next decade is to unravel the individual strength functions from
the inelastic scattering and reactions of electrons, nucleons and
pions at intermediate energies.

    In the analysis of nuclear strength functions a number of
important questions remain. What is the difference between the
longitudinal and transverse particle-hole interaction and their
effect on the corresponding nuclear response function? How good a
testing ground are the stretched states, especially considering the
importance of initial and final state correlations?

Although the whole GT business began with the study of (p,n) reactions - at zero degrees and rather low energy - there is clearly now a need to explore (p,p'). Is this a very rich field? How good is DWIA at intermediate energies? [In particular, what is the validity of the one-step assumption and of a local t-matrix at energies as low as 150 MeV?] Can one sort out the longitudinal modes at large q? Is there a window of visibility for spin-excitations?

Similar questions face inelastic pion field which is just emerging with important data from the EPICS spectrometer of LAMPF. Here one can begin with the questions raised above. Is the π-A (or the Δ-A) interaction known well enough? Can one look at T=1 final states? Do inelastic excitation functions identify ΔS=1? Is the angular distribution sensitive to spin transfer? How does one sort out convection current contributions?

It is undoubtedly the long list of questions which is the lasting legacy of Telluride. The GT quenching effect is, at best, an oblique and somewhat illusory manifestation of QCD effects in nuclei, but the phenomenon itself survived the searching analysis of the conference. Accepting the evidence then, the open problems point to the direction in which nuclear physics will now turn.

PARTICIPANTS

Ken Amos
School of Physics
University of Melbourne
Parkville 3052
Victoria, AUSTRALIA

Bryon D. Anderson
Department of Physics
Kent State University
Kent, Ohio 44242

A. Arima
Department of Physics
University of Tokyo
Tokyo, JAPAN

Sam M. Austin
Department of Physics
Michigan State University
East Lansing, Michigan 48824

Andrew Bacher
Physics Department
Indiana University
Bloomington, Indiana 47401

Helmut Baer
MP-4, MS 846
Los Alamos National Laboratory
Los Alamos, New Mexico 87545

Peter D. Barnes
Department of Physics
Carnegie-Mellon University
Pittsburgh, Pennsylvania 15213

Georg P.A. Berg
IKP-KFA Julich
Postfach 1913, D-5170 Julich
WEST GERMANY

Ulrich E.P. Berg
Institut fur Kernphysik
Leihdesterner Weg 217
D6300 Giessen
WEST GERMANY

Peter M. Bernays
Chemical Abstracts Service
Box 3012
Columbus, Ohio 43210

William Bertozzi
Physics Department, Rm. 26-411
Massachusetts Institute of
  Technology
Cambridge, Massachusetts 02139

Marek Bleszynski
Physics Department
UCLA
Los Angeles, California 90024

H.P. Blok
Natuurkundig Laboratorium
Vrije Universitet
P.O. Box 7161
Amsterdam 107MC
THE NETHERLANDS

Stewart D. Bloom
L 405
Lawrence Livermore National
  Laboratory
P.O. Box 808
Livermore, California 94550

P. Blunden
Physics Department
Queen's University
Kingston, Ontario K7L 3N6
CANADA

643

Pier Francesco Bortignon
Istituto Difisica
University Padova
Via Marzolo 8
Padova, ITALY   35100

F. Paul Brady
Department of Physics
University of California
Davis, California 95616

Virginia R. Brown
L 71 202
Lawrence Livermore National
  Laboratory
P.O. Box 808
Livermore, California 94550

Thomas A. Carey
MS 456
Los Alamos National Laboratory
P.O. Box 1663
Los Alamos, New Mexico 87545

James A. Carr
Physics Department
Florida State University
Tallahassee, Florida 32306

Bunny C. Clark
Department of Physics
Ohio State University
174 W. 18th St.
Columbus, Ohio 43210

Joseph R. Comfort
Physics Department
Arizona State University
Tempe, Arizona 85281

Sidney A. Coon
Physics Department
University of Arizona
Tucson, Arizona 85721

Gerard M. Crawley
Cyclotron Lab
Michigan State University
East Lansing, Michigan 48824

John F. Dawson
Rm. 105 DeMeritt Hall
University of New Hampshire
Durham, New Hampshire 03824

Dietrich Dehnhard
School of Physics and Astronomy
University of Minnesota
116 Church Street, SE
Minneapolis, Minnesota 55455

Raymond Devito
Indiana University Cyclotron
  Facility
Milo B. Sampson Lane
Bloomington, Indiana 47405

Frank S. Dietrich
L 405
Lawrence Livermore National
  Laboratory
P.O. Box 808
Livermore, California 94550

Hiroyasu Ejiri
Department of Physics
Osaka University
Toyonaka, Osaka 560
JAPAN

Guy T. Emery
Physics Department
Indiana University
Bloomington, Indiana 47405

M. Ericson
Inst. de Phys. Nuc. de Lyon
43, BD., Du 11 Novembre 1918
F 69622 Villeurbanne, Cedex
FRANCE

Lawrence W. Fagg
Physics Department
Catholic University of America
Washington, DC 20064

Roger W. Finlay
Physics Department
Ohio University
Athens, Ohio 45701

H.T. Fortune
Department of Physics
University of Pennsylvania
Philadelphia, Pennsylvania 19104

Charles C. Foster
Indiana University Cyclotron
  Facility
Milo B. Sampson Lane
Bloomington, Indiana 47405

Mamoru Fujiwara
Research Center for Nuclear Physics
Osaka University, Suita
Osaka 565, JAPAN

George M. Fuller
Enrico Fermi Institute
University of Chicago
5640 South Ellis Avenue
Chicago, Illinois 60637

Herbert A. Funsten
Physics Department
College of William and Mary
Williamsburg, Virginia 23185

Carl Gaarde
Niels Bohr Institute
University of Copenhagen
DK 2100
Copenhagen Ø
DENMARK

A. Galonsky
Cyclotron Lab
Michigan State University
East Lansing, Michigan 48824

Umesh Garg
Cyclotron Institute
Texas A & M University
College Station, Texas 77843

G. Garvey
Argonne National Laboratory
9700 South Cass Avenue
Argonne, Illinois 60439

George Ciangaru
Department of Physics and
  Astronomy
University of Maryland
College Park, Maryland 20742

Charles Glashausser
Serin Physics Laboratory
Rutgers University
Box 849
Piscataway, New Jersey 08854

Norman K. Glendenning
Lawrence Berkeley Laboratory
Berkeley, California 94720

Chuck Glover
Indiana University Cyclotron
  Facility
Milo B. Sampson Lane
Bloomington, Indiana 47405

Charles D. Goodman
Indiana University Cyclotron
  Facility
Milo B. Sampson Lane
Bloomington, Indiana 47405

Charles N. Goulding
E G & G Los Alamos National
  Laboratory
P.O. Box 809
Los Alamos, New Mexico 87544

Mark Bentley Greenfield
Physics Department
Florida A & M University
Tallahassee, Florida 32307

Stanley S. Hanna
Department of Physics
Stanford University
Stanford, California 94305

W. Haxton
Department of Physics
Purdue University
West Lafayette, Indiana 47907

Shun-Ichiro Hayakawa
Ashikaga Institute of Technology
Ashikaga 326
JAPAN

Jochen H. Heisenberg
Physics Department
University of New Hampshire
Durham, New Hampshire 03824

Ross Hicks
Department of Physics
University of Massachusetts
Amherst, Massachusetts 01003

Norton Hintz
School of Physics
University of Minnesota
Minneapolis, Minnesota 55455

G.W. Hoffman
Department of Physics, RLM 5.208
University of Texas
Austin, Texas 78712

Richard H. Howell
L 405
Lawrence Livermore National
  Laboratory
P.O. Box 808
Livermore, California 94550

Michael Hynes
Los Alamos National Laboratory
P.O. Box 1663
Los Alamos, New Mexico 87545

Munetake Ichimura
Institute of Physics
University of Tokyo
Komaba, Tokyo 153
JAPAN

George J. Igo
UCLA
405 Hilgard Avenue
Los Angeles, California 90024

M. Johnson
Los Alamos National Laboratory
P.O. Box 1663
Los Alamos, New Mexico 87545

James Kelly
Physics Department, Rm. 26-445
Massachusetts Institute of
  Technology
Cambridge, Massachusetts 02139

Nicholas S.P. King
MS 406
Los Alamos National Laboratory
P.O. Box 1663
Los Alamos, New Mexico 87545

Michio Kohno
Department of Physics
McMaster University
Hamilton, Ontario
CANADA L8S 4M1

Dieter Kurath
Physics Department
Argonne National Laboratory
Argonne, Illinois 60439

Jens S. Larsen
Niels Bohr Institute
University of Copenhagen
2100 Copenhagen Ø
DENMARK

F. Lenz
Swiss Institute for Nuclear
  Research
SIN, CH 5234
Villigen, SWITZERLAND

David A. Lind
Nuclear Physics Lab
University of Colorado
Boulder, Colorado 80309

Richard Lindgren
Department of Physics
University of Massachusetts
Amherst, Massachusetts 01003

Paul W. Lisowski
MS 442
Los Alamos National Laboratory
P.O. Box 1663
Los Alamos, New Mexico 87545

Tim Londergan
Department of Physics
Indiana University
Bloomington, Indiana   47405

W. G. Love
Department of Physics
University of Georgia
Athens, Georgia   30602

Richard Madey
Department of Physics
Kent State University
Kent, Ohio   44242

Victor A. Madsen
Physics Department
Oregon State University
Corvallis, Oregon   97331

T. G. Masterson
Department of Physics
University of Colorado
Boulder, Colorado   80309

John B. McClelland
MP 10
Los Alamos National Laboratory
P.O. Box 1663
Los Alamos, New Mexico   87545

John A. McGill
MS 841
Los Alamos National Laboratory
P.O. Box 1663
Los Alamos, New Mexico   87545

H. McManus
Department of Physics
Michigan State University
East Lansing, Michigan   48824

James A. McNeil
Department of Physics
University of Pennsylvania
Philadelphia, Pennsylvania   19139

D. John Millener
Brookhaven National Laboratory
510A
Upton, New York   11973

Gary Mitchell
Physics Department
North Carolina State University
Raleigh, North Carolina   27650

Christopher L. Morris
MS 841
Los Alamos National Laboratory
P.O. Box 1663
Los Alamos, New Mexico   87545

Joel M. Moss
MS 456
Los Alamos National Laboratory
P.O. Box 1663
Los Alamos, New Mexico   87545

Steven A. Moszkowski
Department of Physics
UCLA
Los Angeles, California   90024

K. Nagatani
Cyclotron Institute
Texas A & M University
College Station, Texas   77843

J. Rayford Nix
MS 452
Los Alamos National Laboratory
P.O. Box 1663
Los Alamos, New Mexico   87545

Catherine Olmer
Physics Department
Indiana University
Bloomington, Indiana   47405

Hikonojo Orihara
Cyclotron and Radioisotope
   Center
Tohoku University, Sendai 980
JAPAN

Franz Osterfeld
IKP (Theorie)
KFA Julich
D-5170 Julich, WEST GERMANY

Gerald A. Peterson
Nuclear Physics Group
University of Massachusetts
Amherst, Massachusetts   01003

Fred Petrovich
Department of Physics
Florida State University
Tallahassee, Florida   32306

Alan Picklesimer
Physics Department
Case Western Reserve University
Cleveland, Ohio   44106

Carl H. Poppe
L 405
Lawrence Livermore National
  Laboratory
P.O. Box 808
Livermore, California   94550

Jack Rapaport
Physics Department
Ohio University
Athens, Ohio   45701

Mannque Rho
Service de Physique Theorique
CEN Saclay
91191 Gif-Sur-Yvette
FRANCE

N. Russell Roberson
Department of Physics
Duke University
Durham, North Carolina   27706

George Rosensteel
Department of Mathematics
Arizona State University
Tempe, Arizona   85287

Hideyuki Sakai
Indiana University Cyclotron
  Facility
Milo B. Sampson Lane
Bloomington, Indiana   47405

Peter Schwandt
Department of Physics
Indiana University
Bloomington, Indiana   47405

Susan J. Seestrom-Morris
MS 456
Los Alamos National Laboratory
P.O. Box 1663
Los Alamos, New Mexico   87545

Kamal K. Seth
Physics Department
Northwestern University
Evanston, Illinois   60201

Edward R. Siciliano
Department of Physics
University of Colorado
Boulder, Colorado   80309

Josef Speth
IKP (Theorie)
KFA Julich
D5170 Julich, WEST GERMANY

Evan R. Sugarbaker
Ohio State University
OSU Van de Graaff
1302 Kinnear Road
Columbus, Ohio   43212

Terry N. Taddeucci
Indiana University Cyclotron
  Facility
Milo B. Sampson Lane
Bloomington, Indiana   47405

Peter C. Tandy
Department of Physics
Kent State University
Kent, Ohio   44242

Edward Teller
Lawrence Livermore Laboratory
P.O. Box 808
Livermore, California 94550

Peter Truol
Physik Institut
University of Zurich
Schoenberggassee 9
8001 Zurich
SWITZERLAND

Hiroshi Toki
Cyclotron Lab
Michigan State University
East Lansing, Michigan 48824

I. S. Towner
Chalk River Nuclear Lab
AECL Research Co.
Chalk River, Ontario
CANADA K0J 1J0

W. T. H. Van Oers
TRIUMF
4004 Wesbrook Mall
Vancouver, British Columbia
CANADA V6T 2A3

E. W. Vogt
TRIUMF
4004 Wesbrook Mall
Vancouver, British Columbia
CANADA V6T 2A3

George E. Walker
Physics Department
Indiana University
Bloomington, Indiana 47405

Jochen Wambach
Department of Physics
State University of New York
Stony Brook, New York 11794

Wolfram Weise
Institute of Theoretical Physics
University of Regensburg
D-8400 Regensburg
WEST GERMANY

Henry R. Weller
Physics Department
Duke University
Durham, North Carolina 27706

Robert T. Westervelt
Department of Physics
Stanford University
Stanford, California 94305

Chris Zafiratos
Nuclear Physics Laboratory
University of Colorado
Boulder, Colorado 80309

Kevin Jones
Serin Physics Laboratory
Rutgers University
Piscataway, New Jersey 08854

INDEX